Carmen alexander

Principles of Digital and Analog Communications

Second Edition

Principles of Digital and Analog Communications

Jerry D. Gibson

Department of Electrical Engineering
Texas A&M University

Macmillan Publishing Company
NEW YORK

Maxwell Macmillan Canada
TORONTO

Maxwell Macmillan International
NEW YORK OXFORD SINGAPORE SYDNEY

Editor: John Griffin
Production Supervisor: Elaine W. Wetterau
Production Manager: Paul Smolenski
Text Designer: Susan Bierlein
Cover Designer: Thomas Mack
Illustrations: Precision Graphics

This book was set in Times Roman by Syntax International,
and printed and bound by Book Press, Inc.
The cover was printed by Book Press, Inc.

Printed in the United States of America

Macmillan Publishing Company is part
of the Maxwell Communication Group of Companies.

Macmillan Publishing Company
866 Third Avenue, New York, New York 10022

Maxwell Macmillan Canada, Inc.
1200 Eglinton Avenue East
Suite 200
Don Mills, Ontario M3C 3N1

Library of Congress Cataloging in Publication Data
Gibson, Jerry D.
Principles of digital and analog communications
Jerry D. Gibson. [2nd ed.]
p. cm.
Includes bibliographical references and index.
ISBN 0-02-341860-5
1. Telecommunication systems. 2. Digital communications.
I. Title.
TK5101.G53 1993
621.382--dc20 91-39287
CIP

Printing: 2 3 4 5 6 7 8 Year: 3 4 5 6 7 8 9 0 1

To Erin and Chuck

Preface

This book is an outgrowth of over fifteen years of teaching analog and digital communications classes to undergraduate and graduate students at two universities. It is intended to serve as a textbook for juniors and seniors majoring in electrical engineering and the presentation reflects this undergraduate orientation by containing a minimum of advanced mathematics. This is not to say that there is little mathematics; indeed, mathematics is an important tool for communication systems design and analysis, and most undergraduates find that the study of communications requires greater mathematical maturity than many other undergraduate courses.

Like the first edition, this second edition covers both analog and digital communications, but the emphasis is clearly on digital communications. The treatment of analog communications is complete, but to keep the development concise, many straightforward proofs and results are relegated to the problems. Two new digital communications chapters have been added to the second edition, one on spread spectrum and one on computer communication networks. Such chapters are becoming relatively standard in undergraduate texts. To make room for these chapters, the material on speech and image compression in the first edition has been removed.

Another major change between editions is that the material on random variables and stochastic processes has been moved to an appendix (but not shortened) for the second edition. This improves the flow of the principal communications topics and gives the instructor greater flexibility as to when this material is covered. Other improvements include a new section on QAM, a section on cyclic codes, an appendix on PN sequences, and a section on cyclostationary processes, as well as revised treatments of pulse shaping, equalization, modems, ASK, FSK, and PSK.

The book is suitable as a text for a variety of course offerings depending upon the required prerequisites. Several possibilities are indicated in the table. For example, the book can serve as a text for a two-semester sequence on analog and digital communications with only an introductory circuit analysis prerequisite, by which I mean a relatively weak background in transforms and linear systems and no probability, or as a one-semester course on digital and analog communications for students with a solid background in the contents of Chapters 2–4 and Appendix A. How much is covered out of Chapters 11–15 in any course depends upon the instructor's interests and the preparation and maturity of the student. The appendixes contain supplementary material and are easily integrated into the various course offerings. I typically include Section B.4 and Appendix F on television in analog communications classes and Appendix E on switching systems in digital communications classes.

One-Semester Course	Prerequisites	Chapters
Analog Communications	Introductory Circuit Analysis	2–6, sparse coverage of 7 and Appendix A
Analog Communications	Fourier Transforms, Linear Systems	2–4 (briefly), 5, 6, 7, Appendix A
Digital Communications	Chapters 2–6	8–10, Appendix A, some from 12–15 (Instructor's Choice)
Digital Communications	Chapters 2–6, Appendix A	8–10, 11 (briefly), 13, 14
Analog and Digital Communications	Chapters 2–4, Appendix A	5–7, 8–10, some from 12–15 (Instructor's Choice)

This book, like any book of this type, draws upon all of the textbooks, monographs, and research papers that have gone before it. The communications literature is rich, and the debt I owe goes far beyond the cited references. I also gladly acknowledge the students at the University of Nebraska–Lincoln and Texas A&M University, who used and commented upon the many drafts of my class notes, and the reviewers selected by the editors at Macmillan whose criticisms and suggestions made this a better book.

Additionally, a number of friends and colleagues that used the first edition have provided corrections and suggestions that have been incorporated. In particular, Peter Mathys, University of Colorado–Boulder, supplied numerous critical comments on the first half of the book that greatly improved the presentation. I would also like to acknowledge Khalid Sayood at the University of Nebraska–Lincoln, Tom Fischer at Washington State University, and Hossein Mousavinezhad at Western Michigan University for their insights and corrections. Larry Milstein of the University of California–San Diego and Mike Pursley of the University of Illinois read and provided comments on the new spread spectrum chapter, and Nader Mehravari of AT&T Bell Laboratories and Pierce Cantrell of Texas A&M University commented on the new networking chapter. Although all of their suggestions were not implemented, and I am sure that they would write much different chapters of their own, the book benefited enormously from their insights.

The writing of a book is a long, painstaking process, and many secretaries typed and retyped various versions of the manuscript. For the first edition, which is the foundation for this edition, I gratefully acknowledge the exceptional efforts of Ms. Karen Balke in producing the book and the efforts of Ms. Kay Yocham in proofreading, generating the index, and preparing the Solutions Manual. Ms. Niki Harris has taken the lead in all phases of producing this second edition. Her enthusiasm and attention to detail have been instrumental in bringing this project to fruition. I also gratefully acknowledge Stan McClellan for his help in generating several of the figures.

Finally, I thank my family for their unwavering support and encouragement throughout my career.

Jerry D. Gibson
College Station, Texas

Contents

CHAPTER 1

Introduction to Communications

1.1 Introduction

The Telecommunications Age is upon us. In our private lives, we listen to the radio, watch television, and talk on the telephone regularly. To enhance our television viewing, we may have cable TV or a satellite dish in the yard. Further, we may have a CB radio or mobile telephone in the car, and we may connect our personal computer with other computers over telephone lines using a modem. In our jobs, we may rely on communication networks that interconnect computers throughout the world to provide inventory, financial, and other planning data to conduct our business, and in addition to the telephone, we may employ other communication services, such as teleconferencing, to reduce travel while maintaining vital personal contacts. The list of communication services available to us is seemingly endless and growing almost daily, and the demand for expanded communication services continues to be high.

Since we are talking about electrical communication, all of this is particularly relevant to electrical engineers. In this book we introduce electrical engineers to the basic principles of existing communication systems. The purpose of this brief introductory chapter is to orient the reader by providing some historical and philosophical underpinnings and by outlining what can be expected in the remainder of the book.

1.2 What Is a Communication System?

It is common today in both the public media and the technical literature to refer to the various communication technologies, services, and systems as *telecommunication* technologies, services, and systems. This is probably because the prefix *tele* implies "at a distance," and its inclusion allows us to distinguish between electrical communication and face-to-face oral communication. However, since there seems to be no real reason to prefer either *communications* or *telecommunications*, we use them interchangeably in this book, with a preference toward the simpler *communications*. We focus on the very general communication system block diagram shown in Fig. 1.2.1. The *source* block represents

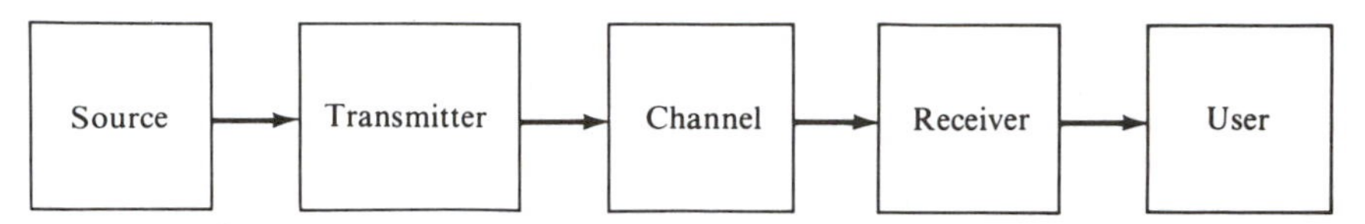

FIGURE 1.2.1 General communication system block diagram.

messages to be transmitted, such as a speech signal, a color television signal, a black-and-white document, a stream of binary ones and zeros, or an electrocardiogram signal. We assume that the source output is an electrical signal of some kind, so the transducer that changes these quantities into an electrical waveform, such as a microphone or cameras, is included within the source block. The *transmitter* operates on the source output and prepares it for propagation over the channel. The *channel* may consist of a pair of wires, coaxial cable, optical fibers, a microwave link, a mobile radio channel, electromechanical or electronic switches, or deep space, among many other possibilities. The *receiver* attempts to retrieve the source output from the channel output and presents the resulting signal to the *user*, which may be the human eye or ear, a remote-controlled device, or a computer. The input to the user is assumed to be an electrical signal, and therefore the user block may contain a transducer, such as a speaker or display unit, so that the source can be returned to its "original" physical form.

Communications engineers usually have much control over the transmitter and receiver but little or no control over the source, channel, and user. The essence of communications is the proper selection of the transmitter and receiver in order to achieve reliable high-fidelity transmission of the source output to the user, and the design and analysis of suitable transmitters and receivers is the central topic of this book. The size, weight, complexity, cost, and performance of the transmitters and receivers studied vary drastically, as do the functions performed by these blocks.

In studying communication systems, we work with mathematical or statistical models of the source, channel, and user. Indeed, a considerable portion of communication system design is concerned with developing adequate but analytically tractable models of these entities. If the source, channel, or user model is poor, even the most careful transmitter and receiver designs may be doomed to failure. Furthermore, even if the models are extremely accurate, it is possible that the models are so complicated mathematically that the analysis required to find good transmitter and receiver designs may be impossible. Thus there is a continual trade-off between model accuracy and analytical tractability.

To carry our development a little further, we can expand the block diagram in Fig. 1.2.1 as shown in Fig. 1.2.2. Each block can represent many operations and every communication system does not have all of the blocks. However, the source, channel, and user blocks are always present. An example of a communication system that has only these three blocks is a telephone conversation between two people in the same neighborhood.

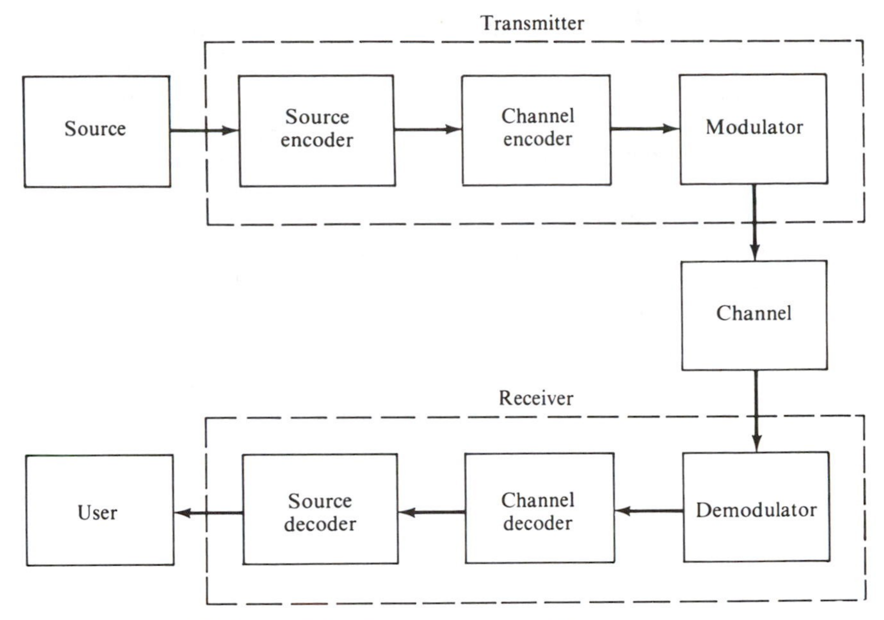

FIGURE 1.2.2 Expanded system block diagram.

Although communication systems can be classified in a wide variety of ways, one common classification is as an *analog* system or a *digital* system. An *analog signal* can take an infinite number of possible amplitudes in a given range, while a *discrete signal* can take only a finite number of amplitudes in a fixed range. A communication system is usually classified as analog or digital, depending on whether the signal transmitted over the channel is analog or discrete, respectively. This distinction becomes imprecise for those communication systems that transmit discrete amplitude signals using analog modulation methods.

Most of the communication systems familiar to the layperson contain the modulator and demodulator blocks but do not include the source encoder, channel encoder, channel decoder, and source decoder blocks. The *modulator* changes the signal into a form suitable for transmission over the channel, while the *demodulator* reverses the process. A common example of such a system is commercial AM radio, in which the voice or music signal is moved to a higher frequency by the modulator so that it will propagate through the air and not interfere with signals at other frequencies.

Many sources of interest, such as speech and images, are analog signals. However, for numerous applications there may be advantages to changing these signals into digital form prior to transmission. The *source encoder* block represents such an operation. A source encoder can also be used to encode an existing digital signal more efficiently. The *source decoder* block returns the source signal to its original form as nearly as is possible. An example of a system that uses source encoding is the medium- to long-distance transmission of telephone speech via pulse code modulation (PCM).

The *channel encoder* block operates on an input digital signal in such a way as to reduce the probability that the digital signal will be decoded erroneously at the receiver. The *channel decoder* attempts to reproduce the input to the channel encoder as reliably as possible. The channel encoder/decoder blocks are present in many commercial and military communication systems that transmit digital signals, deep-space communication systems, computer systems, and even consumer products such as the compact disc.

We are interested in both the analysis and design of the source encoder/decoder, channel encoder/decoder, and modulator/demodulator blocks in Fig. 1.2.2. Further, we are interested in the analysis and modeling of physical sources, channels, and users, since such information is necessary to evaluate and design efficient communication systems. Thus the field "communications" involves many disciplines, from signal processing, detection theory, estimation theory, information theory, and coding theory to switching, transmission, and terminal design. Indeed, it is this great diversity of disciplines that makes working in the communications field both challenging and exciting.

To answer the question posed in the title of this section, we shall call a *communication system* that which transfers the source output to the user over an unspecified distance with an unspecified fidelity. Obviously, this is a very imprecise statement, but we must cover a great many physical situations by this "definition" and we can be more specific only when these physical situations are known. Also, an exact definition is not really important here as long as the concept is evident to the reader.

1.3 System Performance and Fundamental Limits

The key parameters of a communication system are its performance, bandwidth, and complexity (or cost). A thorough analysis of a communication system examines all of these parameters simultaneously to evaluate the utility of a system for a given application. The performance indicator for an analog communication system is usually the output signal-to-noise ratio (SNR). For a digital communication system, performance is generally measured in terms of probability of bit error (P_e) or, equivalently, bit error rate (BER), although P_e can be translated into SNR when it is desirable to do so. When the source signal is speech or images, more subjective measures of performance, such as listening tests or visual examinations by human subjects, can play an important role.

Figure 1.3.1 is a plot of typical performance curves for analog communication systems. Shown in the figure are the output SNR versus the normalized channel SNR for frequency modulation (FM), coherently demodulated amplitude modulation for both double-sideband (AMDSB-SC) and single-sideband (AMSSB-SC) transmission, and conventional amplitude modulation (AMDSB-TC) with envelope detection, as is common in AM radio. For a given channel SNR, one system has better performance than another system if the former has a higher output SNR. Thus, from Fig. 1.3.1, for a channel SNR of 20 dB, FM offers better performance than double-sideband or single-sideband AM and con-

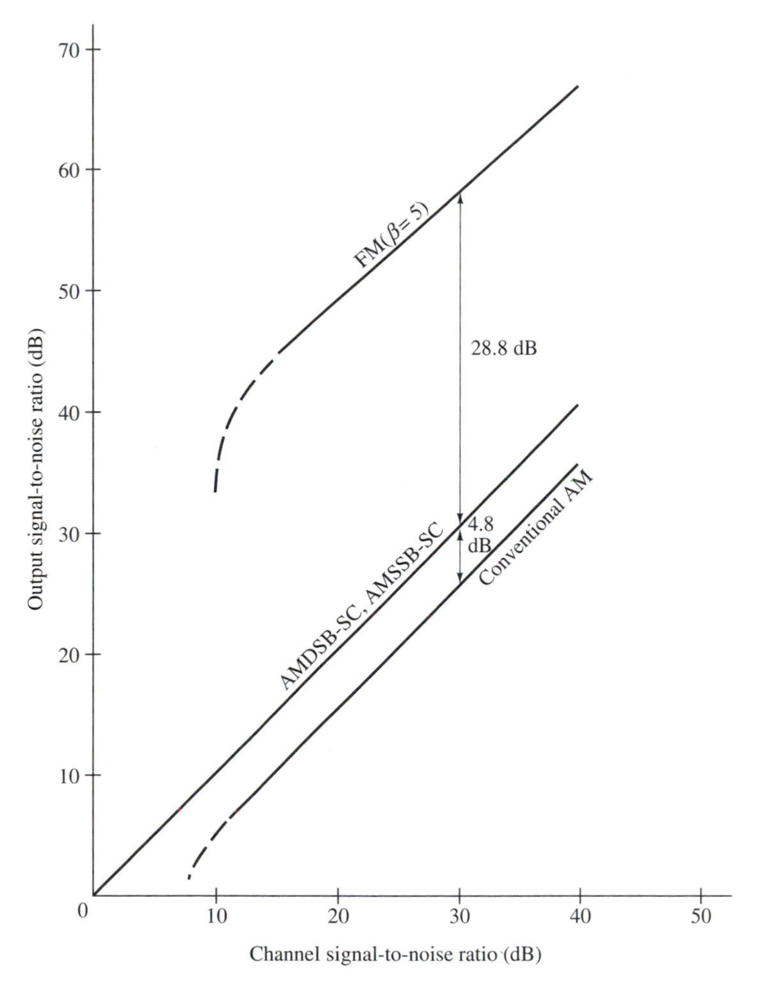

FIGURE 1.3.1 Performance comparison of analog communication systems. From S. Haykin, *Communication Systems*, 2nd ed., New York: John Wiley & Sons, Inc., © 1983, p. 353. Reprinted by permission of John Wiley & Sons, Inc.

ventional AM. FM obtains this performance advantage at the expense of an increased transmission bandwidth requirement.

From the figure, one may ask why conventional AM was chosen for AM radio if it performs as poorly (relatively) as it does. The answer is that conventional AM can use a very simple demodulator, called an *envelope detector*, that allows AM radio receivers to be constructed inexpensively in comparison to the other methods. This brief discussion highlights the types of trade-offs involved among performance, bandwidth, and complexity for common analog communication systems.

Figure 1.3.2 shows performance curves for several digital communication systems. In particular, this figure shows plots of bit error probability versus signal-to-noise ratio ($E_b/\mathcal{N}_0$) for coherently detected frequency shift keying (FSK),

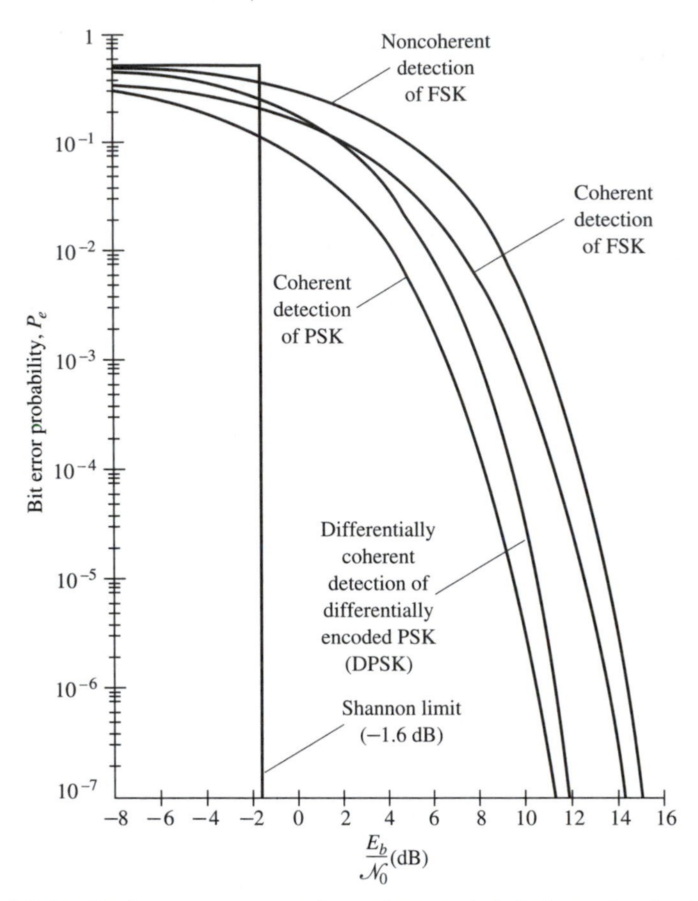

FIGURE 1.3.2 Performance comparison of several digital communication systems. From Bernard Sklar, *Digital Communications: Fundamentals and Applications*, 1988, p. 160. Reprinted by permission of Prentice Hall, Englewood Cliffs, New Jersey.

noncoherently detected FSK, coherently detected phase shift keying (PSK), and differential PSK (DPSK). The signal-to-noise ratio here is transmitted energy per bit (E_b) divided by the noise power per unit bandwidth ($\mathcal{N}_0$). For a given $E_b/\mathcal{N}_0$, the system with the lowest bit error probability has the best performance. Coherent PSK is often used in many applications, and the relative performance shown in Fig. 1.3.2 is one of the principal reasons for its popularity. Also, noncoherent FSK is widely used in many low-speed modems (300 bits/sec), even though it is evident from the figure that coherent FSK out-performs noncoherent FSK. This is because noncoherent detection is simpler than coherent detection and thus results in less expensive modems at this low rate.

A common goal of communication theory is to establish what fundamental limits there are to the performance of a communication system. When such fundamental limits are known, they can be used as guidelines to determine if more time and money should be expended on a particular communications problem or whether the existing system operates so close to the theoretical optimum that further effort would be poorly rewarded. For analog systems, the question is usually: What is the minimum bandwidth required to transmit the given message for a specified SNR? For digital communications, it is desired to determine the maximum bit rate possible for a specified P_e in a fixed bandwidth. Of course, the preceding requirements could be restated somewhat depending on the particular application. For example, in digital communications it may be desirable to find the minimum P_e for a fixed bandwidth and specified bit rate. It is important to note that in this discussion on fundamental limits, system complexity is not mentioned. Such is the case in general. Complexity requirements must be ascertained by finding (designing) systems that achieve or approach the fundamental limits.

For digital communication systems, such fundamental limits have been discovered. In 1924, Nyquist proved that the number of noninterfering pulses that can be transmitted over a channel of bandwidth W hertz in a T-second time interval is kTW, where $k \leq 2$, with the exact value of k dependent on the transmitted pulse shape and the chosen definition of bandwidth. If each pulse represents 1 bit, the resulting bit rate is kW bits/sec. Hartley then determined in 1928 that if each transmitted pulse can take on N distinguishable amplitudes, the total number of distinct signals in T seconds for a W-hertz bandwidth channel is $M = N^{kTW}$. The number of bits transmitted in a T-second interval is thus $\log_2 M = kTW \log_2 N$ bits, which yields a bit rate of $kW \log_2 N$ bits/sec. Note that if $N = 2$ levels, we get Nyquist's result.

In 1948, Shannon took a giant step in computing fundamental performance limits on communication systems when he established the discipline of *information theory*. The two key concepts provided by Shannon were a definition of information in terms of the entropy of a random process and the idea of allowing delay in communication systems to improve performance. Using these concepts, Shannon showed that an arbitrarily small P_e could be attained over a given communication channel as long as data were transmitted at a rate below a fundamental limit called *channel capacity* and denoted by C bits/sec. The possibility of achieving a near-zero P_e is acquired by allowing very large, possibly infinite, delays at the receiver. In this theory, increasing delay is somewhat synonymous with increasing complexity. This theory characterizes how much performance improvement we can expect by including the channel encoder/decoder blocks in our digital communication system.

The most widely known result of Shannon's is the formula for the capacity of the additive white Gaussian noise (AWGN) channel with a bandwidth of W hertz given by

$$C = W \log_2 \left(1 + \frac{S}{N}\right), \tag{1.3.1}$$

where S is the received signal power, N the noise power, and C the channel capacity in bits/sec. The implications of Eq. (1.3.1) are that it is theoretically possible to communicate over any AWGN channel of bandwidth W hertz at a rate $R \leq C$ bits/sec with an arbitrarily small bit error probability. On the other hand, if we try to transmit at a rate $R > C$ bits/sec, reliable communication is not possible. Therefore, we see that for the given channel model, Eq. (1.3.1) sets a fundamental limit on communication system performance.

Generally, when investigated further, Shannon's results show that reliable communication over a noisy channel with a specified signal-to-noise ratio is possible at a sufficiently low information rate. It can be shown using Eq. (1.3.1) that in the limit as the normalized rate $C/W \to 0$, the smallest possible $E_b/\mathcal{N}_0$ for reliable transmission is -1.6 dB. This hard limit on performance is shown in Fig. 1.3.2 as a "brick wall," where the error probability jumps from 0 to $\frac{1}{2}$. A plot of Eq. (1.3.1) for appropriate definitions of S and N is given in Fig. 1.3.3. Communication systems can be designed that operate in the region below this curve, but no amount of ingenuity and complexity can produce a communication system that operates above this curve.

Shannon also developed, in 1948 and 1959, a discipline called *rate distortion theory*, which permits fundamental limits on the data rate required to represent a source with a specified distortion to be calculated. Again, long delays (this time at the transmitter) may be required to achieve the fundamental limits provided by rate distortion theory. Rate distortion theory establishes the limits on performance afforded by the source encoder/decoder blocks in Fig. 1.2.2. The results of Nyquist, Hartley, and Shannon play important roles in the design and analysis of digital communication systems, and we will see evidence of these fundamental limits throughout our study.

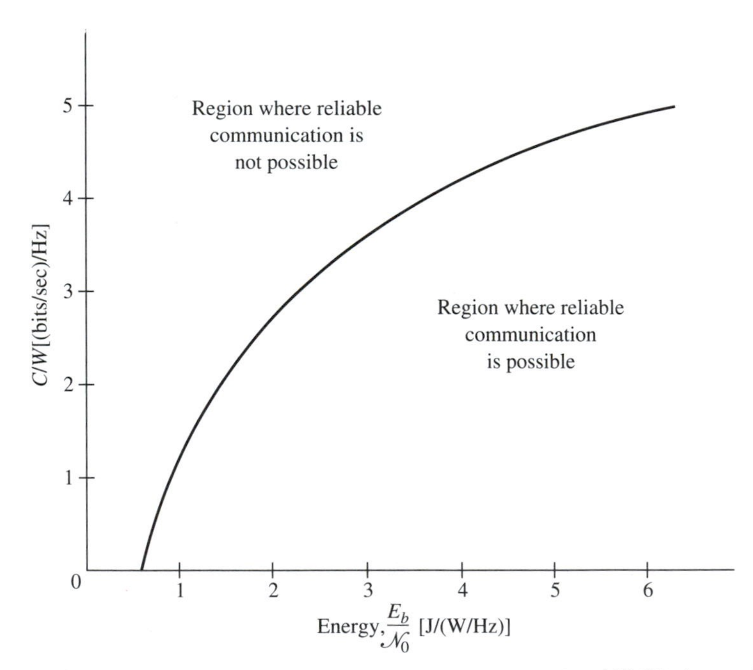

FIGURE 1.3.3 Fundamental limit on performance over the AWGN channel. From R. E. Blahut, *Digital Transmission of Information*, © 1990 by Addison-Wesley Publishing Company, Inc. Reprinted with permission of the publisher.

1.4 Theory and Practice

Communication systems have been changing and developing rapidly in recent years, with the overall goal of providing new or improved services at a reasonable or a lower price. On the technical side, changes in communication systems have come from advances in both circuit and device technology and theoretical understanding. Clearly, the development of devices such as the vacuum tube, the transistor, the integrated circuit, and microprocessors had a profound impact on communication systems, since relatively soon after each was developed, it appeared in a communication system. Also, theoretical discoveries such as Nyquist's work, Kotel'nikov's signal space (advanced in the United States by Wozencraft and Jacobs [1965]), Shannon theory, Wiener's work on optimal filtering, the Kalman filter, Lucky's equalization work, and Lender's duobinary techniques, to name a few, have all had an impact on communication systems.

In the past, developments in device and circuit technology seem to have led the way, with communication theory being left to play an explanatory role [Wozencraft and Jacobs, 1965]. However, because device and circuit technology have advanced so far, very complicated communication systems can be built, leaving circuit designers and engineers with the dominating question of *what* should be built rather than *how* it should be built. With previous complexity restrictions on both implementations and ideas either relaxed or removed, the way has been cleared for theoretical results to affect communication system and network design. This is not to imply that theoretical results lead automatically to complex systems. On the contrary, theoretical results may produce insights that lead to elegant, simple solutions of problems. What has happened in the past, however, is that complexity limitations, real or imagined, have often prevented difficult problems from being posed carefully and their optimum solutions from being sought.

Communication systems were originally viewed as having two parameters, *power* and *bandwidth*. Shannon [1948, 1949] demonstrated that a third parameter is *delay*, which, in essence, translates into additional *signal processing*. Unfortunately, in the past, the complexity of this additional signal processing precluded its widespread use in communication systems. Now, however, complexity is much less of an issue, and communication systems can take advantage of this third parameter, signal processing, and as a consequence, theoretical results that lead to new signal processing algorithms play a prominent role in the development of today's communication systems.

There is another very important aspect of theoretical work that has not yet been mentioned. Many communication systems today are parts of large networks, such as the telephone network, and include many diverse components and subsystems such as satellite, microwave, and fiber optic links and large switches. It is not possible to take these networks into the laboratory for experimentation. In fact, just taking any part of a network out of service for even a short period of time may be expensive (loss of revenue) or even impossible. Therefore, advanced theoretical and simulation analyses of these networks must be conducted in order to plan and develop new services before they are introduced into the communication system. Thus, in many situations, theoretical studies are an integral part of developing practical systems.

To conclude this section, then, theory and practice are much less separable today than they were, say, 40 years ago, and this will become evident even in this introductory communications textbook.

1.5 Topical Coverage of the Book

The topical coverage of the book begins in Chapter 2 with a development of Fourier and other orthogonal series. Orthogonal series in general, and Fourier series in particular, have played an important role in signal and system analyses. Fourier transforms and their properties are presented in Chapter 3, where the critical ability to move between the time and frequency domains is developed. Techniques for the analysis of linear systems in the time and frequency domains are studied in Chapter 4, with particular emphasis on the system transfer function. Finally, in Chapter 5, the ideas of the three preceding chapters are employed to study amplitude modulation systems, including double-sideband, single-sideband, and vestigial sideband modulation. Chapter 6 then considers frequency and phase modulation, with both Chapters 5 and 6 indicating how these various modulation methods can be represented in the time and frequency domains. The effects of channel noise on analog modulation system performance are derived in Chapter 7 based on input versus output signal-to-noise ratios, and some simple system comparisons are discussed. In Chapter 8 analog modulation methods are applied to the transmission of digital messages. Emphasis is placed on the transmission of digital sequences through a bandlimited channel with deterministic distortion, and hence on the topics of pulse shaping, amplitude and delay distortion, equalization, and scrambling. Modems are also introduced here. In Chapter 9 we examine pulse code modulation (PCM) in some detail, providing information on quantizer design, quantization level and line coding, PCM system performance, and the digital transmission hierarchy in the telephone network. In Chapter 10 we consider the effects of noise in digital communication systems, where the concepts of signal space, optimum receivers, and signal design are introduced, and the performance of the data transmission and digital communication systems presented in Chapter 8 is derived. Relevant performance comparisons are given. Fundamental limits on digital communication system performance are addressed in Chapter 11, where rate distortion theory and information theory are introduced. Simple error control coding techniques are developed in Chapter 12, which is followed by a treatment of the relatively new concepts of joint modulation and coding and trellis-coded modulation in Chapter 13. The basic principles of spread spectrum communications are presented and studied in Chapter 14, and an overview of computer communication networks and methods for their analysis is given in Chapter 15. Appendix A contains a careful development of probability, random variables, and stochastic processes at a level suitable for this book. Several additional appendixes complete the book.

Much of the material is relatively standard for this type of book; however, there are topics in Chapters 8 to 15 that go somewhat beyond existing textbooks. This newer material tends to emphasize digital communications and to indicate those fields that are becoming crucial to the design and analysis of communication systems. As usual, working the problems is an integral part of understanding the material presented in each chapter.

Since this is an introductory text, it is obvious that many if not all of the chapters could easily be expanded so that each filled an entire book. The reader will probably notice this at times, and may wish for more derivations, more design details, or more performance comparisons. If this occurs, the reader should first examine the problems. However, if this fails, the reader is encouraged to consult the references and to visit the nearest technical library. The field of communications is a dynamic and exciting one, and it is hoped that this fact will become evident to the reader as he or she proceeds through the book.

CHAPTER 2

Orthogonal Functions and Fourier Series

2.1 Introduction

Fundamental to the practice of electrical engineering and the study of communication systems is the ability to represent symbolically the signals and waveforms that we work with daily. The representation of these signals and waveforms in the time domain is the subject of this chapter. On first thought, writing an expression for some time function seems trivial; that is, everyone can write equations for straight lines, parabolas, sinusoids, and exponentials. Indeed, this knowledge is important, and it is precisely this ability to write expressions for and to manipulate familiar functions that we will draw on heavily in the sequel. Here, however, we are interested in the extension of this ability to include more complex, real-world waveforms, such as those that might be observed using an oscilloscope at the input or output of a circuit or system.

It is intuitive that these more complex signals cannot be fully described by a single member of our set of familiar mathematical functions. Further, often a situation occurs where a signal may not have a unique representation; that is, there may be several different combinations of functions that accurately represent the signal. In this case, however, it usually turns out that one or more of the representations has some advantage over the others, such as fewer terms, ease of manipulation, or being physically more meaningful. Of course, it is not always clear at the outset whether a certain combination of functions has any or all of the advantages above. Of even greater concern is which functions to select and how we are to combine them to obtain a given wave shape without resorting to the cut-and-try method. It is clear that some guidelines are needed here.

We begin our development of these guidelines in Section 2.2 by establishing some properties that are desirable for the basic building block functions to have and by introducing particular functions that have these properties. How some of these functions are combined to obtain a given waveform and two specific forms of these representations—trigonometric and complex exponential Fourier series—are developed in Sections 2.3 and 2.4, respectively. In Section 2.5 several properties of waveforms that can greatly simplify the evaluation of Fourier series coefficients are presented and illustrated. Since in many problems an approximation to a waveform is adequate, in Section 2.6 we demonstrate that a truncated Fourier series can be used to approximate a function in the least squares sense. Starting with the idea of a least squares approximation, additional details concerning general Fourier series are also given in Section 2.6, which provide a much firmer mathematical basis for the Fourier series developments in the preceding sections. In Section 2.7 we introduce the important concept of spectral or frequency content of a signal that is used throughout the book and is basic to most facets of electrical engineering.

2.2 Signal Representation and Orthogonal Functions

The ultimate goal of this chapter is to provide the reader with the ability to represent "nonstandard" wave shapes symbolically. One powerful way to approach this problem is to specify a set of basic functions that are then combined in some way to produce expressions for less familiar waveforms. What properties are desirable when selecting this set of basic functions? The answer is available from our knowledge of two- and three-dimensional vector spaces. The fundamental building blocks of these geometrical spaces are unit vectors in the x, y, and z directions, which by definition are *orthonormal*, that is, orthogonal with magnitudes normalized to 1. By analogy, then, it would seem useful to require that our basic building block functions be orthogonal, or even better, orthonormal.

What is the definition of orthogonality in terms of functions? Two real functions $f(t)$ and $g(t)$ are said to be *orthogonal* over the interval (t_0, t_1) if the integral (called the inner product)

$$\int_{t_0}^{t_1} f(t)g(t)\, dt = 0 \tag{2.2.1}$$

for $f(t) \neq g(t)$. Let us test the set of functions $\{1, t, t^2, t^3, \ldots, t^n, \ldots\}$ to see if they are orthogonal. With $f(t) = 1$ and $g(t) = t$, we have

$$\int_{t_0}^{t_1} (1)t\, dt = \frac{t^2}{2}\bigg|_{t_0}^{t_1} = \frac{1}{2}[t_1^2 - t_0^2].$$

These functions will be orthogonal over symmetrical limits, that is, with $t_0 = -t_1$. Continuing the investigation with $f(t) = 1$ and $g(t) = t^2$, we find that

$$\int_{t_0}^{t_1} (1)t^2\, dt = \frac{t^3}{3}\bigg|_{t_0}^{t_1} = \frac{1}{3}[t_1^3 - t_0^3].$$

The functions 1 and t^2 are not orthogonal over a symmetrical interval, and indeed, they do not seem to be orthogonal over any interval, excluding the trivial case when $t_0 = t_1$. This is the general result. The set of functions consisting of powers of t is not orthogonal and hence does not seem to be a good set for representing general wave shapes.

Before continuing the search for orthogonal functions, we extend the definition of orthogonality in Eq. (2.2.1) to complex signals by noting that two complex signals $f(t)$ and $g(t)$ are said to be orthogonal if

$$\int_{t_0}^{t_1} f(t)g^*(t)\,dt = \int_{t_0}^{t_1} f^*(t)g(t)\,dt = 0, \tag{2.2.2}$$

where the superscript * indicates the complex conjugate operation. Furthermore, two possibly complex time functions $f(t)$ and $g(t)$ are said to be ortho*normal* if Eq. (2.2.2) holds and they satisfy the additional relations

$$\int_{t_0}^{t_1} f(t)f^*(t)\,dt = 1 \qquad \text{and} \qquad \int_{t_0}^{t_1} g(t)g^*(t)\,dt = 1. \tag{2.2.3}$$

Notice that Eqs. (2.2.3) are valid for $f(t)$ and $g(t)$ real or complex, since for $f(t)$ and $g(t)$ real we have $f^*(t) = f(t)$ and $g^*(t) = g(t)$.

Now that definitions of orthogonality for both real and complex functions are available, the task remains to find some familiar functions that possess this property. A very important group of functions that are orthogonal is the set of functions $\cos n\omega_0 t$ and $\sin m\omega_0 t$ over the interval $t_0 \le t \le t_0 + 2\pi/\omega_0$ for n and m nonzero integers with $n \ne m$. The demonstration of the orthogonality of these functions is given in the following example.

EXAMPLE 2.2.1

We would like to investigate the orthogonality of the set of functions $\{\cos n\omega_0 t, \sin m\omega_0 t\}$ over the interval $t_0 \le t \le t_0 + 2\pi/\omega_0$ with n and m nonzero integers and $n \ne m$. Since these functions are real, either Eq. (2.2.1) or (2.2.2) is applicable here. To include all possible combinations of functions, it is necessary that we demonstrate orthogonality for three separate cases. We must show that for $n \ne m$:

(1) $\displaystyle\int_{t_0}^{t_0+2\pi/\omega_0} \cos n\omega_0 t \cos m\omega_0 t\,dt = 0$

(2) $\displaystyle\int_{t_0}^{t_0+2\pi/\omega_0} \sin n\omega_0 t \sin m\omega_0 t\,dt = 0$

(3) $\displaystyle\int_{t_0}^{t_0+2\pi/\omega_0} \cos n\omega_0 t \sin m\omega_0 t\,dt = 0.$

For the first case we have

(1) $\displaystyle\int_{t_0}^{t_0+2\pi/\omega_0} \cos n\omega_0 t \cos m\omega_0 t\,dt$

$$= \frac{1}{2}\int_{t_0}^{t_0+2\pi/\omega_0} \cos (n+m)\omega_0 t\,dt + \frac{1}{2}\int_{t_0}^{t_0+2\pi/\omega_0} \cos (n-m)\omega_0 t\,dt, \tag{2.2.4}$$

since $n \ne m$, $n + m$ and $n - m$ are nonzero integers, and the functions $\cos (n+m)\omega_0 t$ and $\cos (n-m)\omega_0 t$ have exactly $n + m$ and $n - m$ complete periods, respectively, in the interval $[t_0, t_0 + 2\pi/\omega_0]$. The integrals in Eq. (2.2.4) thus encompass a whole number of periods and hence both will be zero, since the integral of a cosine over any whole number of periods is zero.

Cases (2) and (3) follow by an identical argument, and hence they are left as an exercise. The reader should note that the case (3) result is also true for $m = n$ (see Problem 2.4).

A very useful set of complex orthogonal functions can be surmised from the set of orthogonal sine and cosine functions just discussed. That is, since $e^{j\theta} = \cos\theta + j\sin\theta$, we are led to conjecture from the immediately preceding results that the functions $e^{jn\omega_0 t}$, $e^{jm\omega_0 t}$ for m and n integers and $m \neq n$ are orthogonal over the interval $[t_0, t_0 + 2\pi/\omega_0]$. The orthogonality of these functions is demonstrated in the next example.

EXAMPLE 2.2.2

To show that $e^{jn\omega_0 t}$ and $e^{jm\omega_0 t}$ are orthogonal over $t_0 \leq t \leq t_0 + 2\pi/\omega_0$ for m and n integers and $m \neq n$, we employ Eq. (2.2.2) with $t_1 = t_0 + 2\pi/\omega_0$. Substituting $f(t) = e^{jn\omega_0 t}$ and $g^*(t) = e^{-jm\omega_0 t}$ into the integral on the left-hand side of Eq. (2.2.2) produces

$$\int_{t_0}^{t_0 + 2\pi/\omega_0} e^{jn\omega_0 t} e^{-jm\omega_0 t}\, dt = \int_{t_0}^{t_0 + 2\pi/\omega_0} e^{j(n-m)\omega_0 t}\, dt$$

$$= \frac{e^{j(n-m)\omega_0 t_0}}{j(n-m)\omega_0} \{e^{j(n-m)(2\pi)} - 1\}.$$

Since $n - m$ is an integer,

$$e^{j(n-m)(2\pi)} = \cos 2\pi(n-m) + j \sin 2\pi(n-m) = 1,$$

so

$$\int_{t_0}^{t_0 + 2\pi/\omega_0} e^{jn\omega_0 t} e^{-jm\omega_0 t}\, dt = 0,$$

and hence the set of functions $\{e^{jn\omega_0 t}\}$ for all integral values of n are orthogonal over the interval $[t_0, t_0 + 2\pi/\omega_0]$.

The two examples in this section have demonstrated the orthogonality of two sets of functions that are very important in the analysis and design of communication systems. Some common sets of functions that are not critical to our development but which are orthogonal in the sense that they are orthogonal with respect to a weighting function are Bessel functions, Legendre polynomials, Jacobi polynomials, Laguerre polynomials, and Hermite polynomials. The concept of orthogonality with respect to a weighting function is not required in the sequel and therefore is not considered further here. For additional details, see Jackson [1941], Kaplan [1959], and Problem 2.5.

Sets of functions that possess the important property of orthogonality were defined and investigated briefly in this section. The task remains for us to demonstrate how the functions in each set can be combined to represent signals and waveforms that occur in communication systems. This development is the subject of the following sections.

2.3 Trigonometric Fourier Series

In the present section we investigate the representation of a waveform in terms of sine and cosine functions. Initially limiting consideration to periodic functions, that is, functions with $g(t) = g(t + T)$ where T = period, we express the periodic function $f(t)$ with period $T = 2\pi/\omega_0$ in terms of an infinite trigonometric series given by

$$f(t) = a_0 + \sum_{n=1}^{\infty} \{a_n \cos n\omega_0 t + b_n \sin n\omega_0 t\}, \tag{2.3.1}$$

where a_0, a_n, and b_n are constants. Notice two things about Eq. (2.3.1). First, the reason we are working only with periodic functions is that each of the terms on the right-hand side of Eq. (2.3.1) is periodic. Second, an infinite number of terms are included in the representation. Why this is necessary is discussed later in the section and in more detail in Section 2.6.

Before the trigonometric series in Eq. (2.3.1) can be called a Fourier series, it remains to specify the constant coefficients a_0, a_n, and b_n, $n = 1, 2, \ldots$. These coefficients can be determined as follows. To find a_0, multiply both sides of Eq. (2.3.1) by dt and integrate over one (arbitrary) period $t_0 \le t \le t_0 + T$, to obtain

$$a_0 = \frac{1}{T} \int_{t_0}^{t_0+T} f(t)\, dt. \tag{2.3.2}$$

To derive an expression for a_n, we multiply both sides of Eq. (2.3.1) by $\cos k\omega_0 t\, dt$ and integrate over a period, which produces (letting $k \to n$)

$$a_n = \frac{2}{T} \int_{t_0}^{t_0+T} f(t) \cos n\omega_0 t\, dt \tag{2.3.3}$$

for $n = 1, 2, 3, \ldots$. Similarly, an expression for the b_n coefficients is obtained by multiplying Eq. (2.3.1) by $\sin k\omega_0 t\, dt$ and integrating over one period, which yields

$$b_n = \frac{2}{T} \int_{t_0}^{t_0+T} f(t) \sin n\omega_0 t\, dt. \tag{2.3.4}$$

Equation (2.3.1) with the coefficients defined by Eqs. (2.3.2), (2.3.3), and (2.3.4) is called a *trigonometric Fourier series.* Any periodic function can be expanded in a Fourier series simply by determining the period T and using these equations. As an illustration of the procedure, consider the following example.

EXAMPLE 2.3.1

We would like to write a Fourier series representation of the waveform shown in Fig. 2.3.1. To do this, we first find the period of the waveform, which is $T = \tau$. Next we must pick the period that we wish to integrate over, which is equivalent to selecting t_0 in the expressions for the coefficients. Since

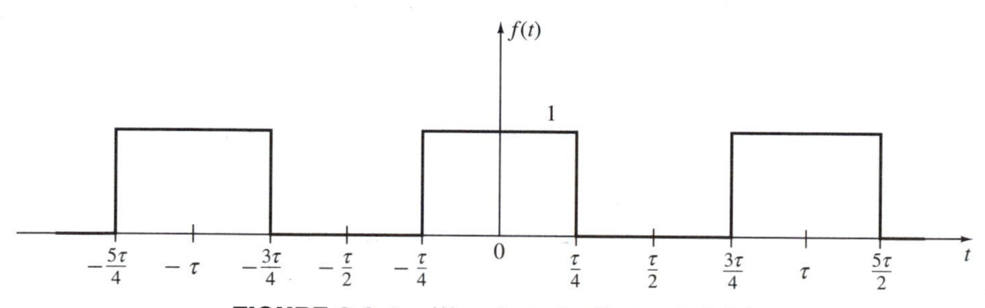

FIGURE 2.3.1 Waveform for Example 2.3.1.

this choice is arbitrary, it is wise to select that period which most simplifies the evaluation of the integrals. For this particular case, we choose $t_0 = -\tau/2$.

We are now ready to evaluate the coefficients. Using Eq. (2.3.2) with $T = \tau$ and $t_0 = -\tau/2$, we have

$$a_0 = \frac{1}{\tau}\int_{-\tau/2}^{\tau/2} f(t)\,dt = \frac{1}{\tau}\int_{-\tau/4}^{\tau/4} (1)\,dt = \frac{1}{2}. \tag{2.3.5}$$

Calculating a_n using Eq. (2.3.3), we see that

$$\begin{aligned} a_n &= \frac{2}{\tau}\int_{-\tau/2}^{\tau/2} f(t)\cos n\omega_0 t\,dt \\ &= \frac{2}{\tau}\cdot\frac{1}{n\omega_0}\sin n\omega_0 t\Big|_{-\tau/4}^{\tau/4} = \frac{1}{n\pi}\left[\sin\frac{n\pi}{2} - \sin\frac{-n\pi}{2}\right]. \end{aligned} \tag{2.3.6}$$

Noting that the sine function is odd, that is, $\sin(-x) = -\sin x$, Eq. (2.3.6) can be written compactly as

$$a_n = \begin{cases} \dfrac{2}{n\pi}(-1)^{(n-1)/2}, & \text{for } n \text{ odd} \\ 0, & \text{for } n \text{ even} \end{cases} \tag{2.3.7}$$

since the sine of integral multiples of π radians is zero and the sine of integral multiples of $\pi/2$ is ± 1.

For the b_n coefficients, we find from Eq. (2.3.4) that

$$b_n = \frac{2}{\tau}\int_{-\tau/2}^{\tau/2} f(t)\sin n\omega_0 t\,dt = \frac{2}{\tau}\int_{-\tau/4}^{\tau/4} (1)\sin n\omega_0 t\,dt = 0. \tag{2.3.8}$$

By substituting Eqs. (2.3.5), (2.3.7), and (2.3.8) into Eq. (2.3.1), the desired Fourier series representation is

$$f(t) = \frac{1}{2} + \sum_{\substack{n=1 \\ n \text{ odd}}}^{\infty} \frac{2}{n\pi}(-1)^{(n-1)/2}\cos n\omega_0 t, \tag{2.3.9}$$

where $\omega_0 = 2\pi/\tau$.

It is important to notice that the Fourier series expansion of a periodic function is valid for all time, $-\infty < t < \infty$, even though the integration when computing the coefficients is carried out over only one period. This is because since the function is periodic, if we accurately represent the function over one period, we have an accurate representation for all other periods, and hence for all time.

To obtain a Fourier series representation of a nonperiodic function over a given finite interval, the approach is to let the time interval of interest be the period, T, and proceed exactly as before. That is, the coefficients are evaluated as if the function were periodic. The resulting Fourier series is an exact representation of the function within the time interval, which was assumed to be one period. The Fourier series may be totally inaccurate outside this time interval; however, this is of no consequence to us. The principal difference between writing a Fourier series for a periodic or a nonperiodic waveform is that in the periodic case, the series is an accurate expression for all time, whereas in the nonperiodic case, the series is valid only over the time interval assumed to be one period.

2.4 Exponential (Complex) Fourier Series

The exponential or complex form of a Fourier series is extremely important to our study of communication systems, although just how important this form of Fourier series is will not become clear until Section 2.7 and Chapter 3. The complex form can be obtained from the trigonometric Fourier series in Eq. (2.3.1) by some simple manipulations. What is required is to note that sine and cosine functions can be written in terms of complex exponentials as

$$\sin x = \frac{1}{2j}\left[e^{jx} - e^{-jx}\right]$$

and

$$\cos x = \frac{1}{2}\left[e^{jx} + e^{-jx}\right].$$

Substituting these expressions into Eq. (2.3.1) produces

$$\begin{aligned} f(t) &= a_0 + \sum_{n=1}^{\infty}\left\{a_n\left[\frac{1}{2}(e^{jn\omega_0 t} + e^{-jn\omega_0 t})\right] + b_n\left[\frac{1}{2j}(e^{jn\omega_0 t} - e^{-jn\omega_0 t})\right]\right\} \\ &= a_0 + \sum_{n=1}^{\infty}\left\{\left(\frac{a_n - jb_n}{2}\right)e^{jn\omega_0 t} + \left(\frac{a_n + jb_n}{2}\right)e^{-jn\omega_0 t}\right\} \end{aligned} \tag{2.4.1}$$

after grouping like exponential terms. If we define the complex Fourier coefficients

$$c_0 = a_0, \qquad c_n = \frac{a_n - jb_n}{2}, \qquad c_{-n} = \frac{a_n + jb_n}{2} \tag{2.4.2}$$

for $n = 1, 2, \ldots$, Eq. (2.3.1) can be written as

$$f(t) = \sum_{n=-\infty}^{\infty} c_n e^{jn\omega_0 t}, \tag{2.4.3}$$

which is the exponential or complex form of a Fourier series. Notice in Eq. (2.4.3) that the summation over n extends from $-\infty$ to $+\infty$.

We still need expressions for the complex coefficients, c_n, in terms of $f(t)$, the function to be expanded. The required relationship can be determined from the definitions in Eq. (2.4.2) or by starting with Eq. (2.4.3) and using the orthogonality property of complex exponentials proved in Example 2.2.2. The final result is that

$$c_n = \frac{1}{T} \int_{t_0}^{t_0+T} f(t) e^{-jn\omega_0 t}\, dt, \tag{2.4.4}$$

where $n = \ldots, -2, -1, 0, 1, 2, \ldots$.

EXAMPLE 2.4.1

To illustrate the calculation of the complex Fourier coefficients, let us obtain the exponential Fourier series representation of the waveform for Example 2.3.1 shown in Fig. 2.3.1. Since $T = \tau$ and $\omega_0 = 2\pi/\tau$, again let $t_0 = -\tau/2$, so that Eq. (2.4.4) becomes

$$c_n = \frac{1}{\tau} \int_{-\tau/4}^{\tau/4} e^{-jn\omega_0 t}\, dt = \frac{1}{n\pi} \sin \frac{n\pi}{2}. \tag{2.4.5}$$

The complex Fourier coefficients are thus given by

$$c_n = \begin{cases} 0, & \text{for } n \text{ even} \\ \dfrac{1}{n\pi}(-1)^{(n-1)/2}, & \text{for } n \text{ odd} \\ \frac{1}{2}, & \text{for } n = 0 \end{cases} \tag{2.4.6}$$

using L'Hospital's rule (see Problem 2.12) to evaluate c_n for $n = 0$. Actually, it is always preferable to evaluate Eq. (2.4.4) for the $n \neq 0$ and the $n = 0$ cases separately to minimize complications. The exponential Fourier series is given by Eq. (2.4.3), with the coefficients displayed in Eq. (2.4.6).

As in the case of trigonometric Fourier series, a complex Fourier series can be written for a nonperiodic waveform provided that the time interval of interest is assumed to be one period. Just as before, the resulting Fourier series may not be accurate outside the specified time interval of interest. The trigonometric Fourier series in Section 2.3 and the complex Fourier series discussed in this section are not actually two different series but are simply two different forms of the same series. There are situations where one form may be preferred over the other, and hence these alternative forms provide us with additional flexibility in signal analysis work if we are familiar with both versions. No other forms of Fourier series are treated in detail in the body of the text, since the trigonometric and complex Fourier series will prove completely satisfactory for our purposes. In the following sections we investigate Fourier series in more depth and study ways to simplify the evaluation of Fourier series coefficients.

2.5 Fourier Coefficient Evaluation Using Special Properties

The calculation of the Fourier series coefficients by straightforward evaluation of the integrals as was done in Sections 2.3 and 2.4 can sometimes be quite tedious. Fortunately, in a few special cases, we can obtain relief from part of this computational burden, depending on the properties of the waveform, say $f(t)$, that we are trying to represent.

Probably the most used properties when evaluating trigonometric Fourier series are even and odd symmetry. A waveform has *even symmetry* or is *even* if $f(t) = f(-t)$, and a waveform is *odd* or has *odd symmetry* if $f(t) = -f(-t)$. The trigonometric Fourier series representation of an even waveform contains no sine terms, that is, the $b_n = 0$ for all n, and the representation of an odd function has no constant term or cosine terms, so the $a_n = 0$ for $n = 0, 1, 2, \ldots$. It is straightforward to demonstrate these results by using the even and odd symmetry properties in the expressions for a_n and b_n in Eqs. (2.3.3) and (2.3.4).

EXAMPLE 2.5.1

The waveform in Fig. 2.3.1 for Example 2.3.1 is even, since $f(t) = f(-t)$ for all values of t. As a result, we should have found in Example 2.3.1 that $b_n = 0$ for all n. This is exactly as we determined, as can be seen from Eq. (2.3.8).

EXAMPLE 2.5.2

Suppose that we wish to obtain a trigonometric Fourier series representation for the function shown in Fig. 2.5.1. If possible, we would like to simplify the required calculations by using either the even or the odd symmetry property. Checking to see whether the function is even or odd, we find that $f(t) \neq f(-t)$ but $f(t) = -f(-t)$, and hence the waveform is odd. From our earlier results, then, we conclude that $a_n = 0$ for $n = 0, 1, 2, \ldots$. The calculation of the b_n coefficients is left as an exercise for the reader.

Although it is not necessary for any of the results obtained thus far in the book, we have implicitly assumed throughout that all the waveforms and functions that we are representing by a Fourier series are purely real. This assumption is entirely acceptable since all the waveforms that we can observe in communication systems are real. There are times, however, when it is mathematically convenient to work with complex signals. When we use complex functions in any of our mathematical developments, this fact will be stated explicitly or it will be clear from the context.

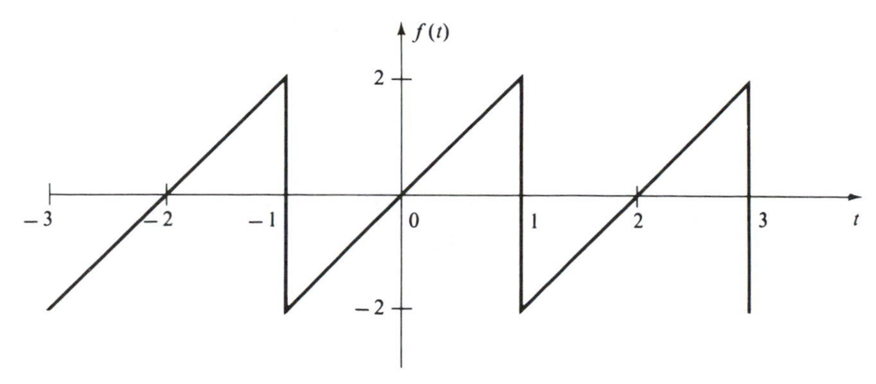

FIGURE 2.5.1 Waveform for Example 2.5.2.

Limiting consideration to purely real functions, we can also use even and odd symmetry to simplify or check the calculation of complex Fourier coefficients for waveforms that have either of these properties. Specifically, for $f(t)$ real and even, the complex Fourier coefficients given by Eq. (2.4.4) are all real, while for $f(t)$ real and odd, the c_n are all imaginary. These statements can be proven in a number of ways. The most transparent approach (assuming no knowledge of the a_n and b_n properties) is to use Euler's identity to rewrite Eq. (2.4.4) and then apply the definitions of even and odd functions. Of course, if it is known that $a_n = 0$ for $f(t)$ odd and $b_n = 0$ for $f(t)$ even, the c_n properties can be established by inspection of Eq. (2.4.2).

EXAMPLE 2.5.3

In Example 2.5.1 the waveform in Fig. 2.3.1 was shown to be even. Since this function is also real, we conclude that the complex Fourier coefficients should be purely real. Examining the results of Example 2.4.1, we see that the c_n are indeed purely real, just as expected.

EXAMPLE 2.5.4

The waveform in Fig. 2.5.1 was concluded to be an odd function in Example 2.5.2. The waveform is also real, and hence the complex Fourier coefficients should be purely imaginary. Noting that the period of the waveform is $T = 2$ and hence $\omega_0 = \pi$, we have from Eq. (2.4.4) with $t_0 = 1$,

$$c_n = \frac{1}{2}\int_{-1}^{1} 2te^{-jn\omega_0 t}\,dt = \frac{j2}{n\pi}\cos n\pi, \tag{2.5.1}$$

which is clearly purely imaginary, as predicted.

Another interesting and useful property is that of *rotation symmetry*, which means that $f(t) = -f(t \pm T/2)$, where T is the period of $f(t)$. A function that has rotation symmetry is also said to be *odd harmonic*, since its complex Fourier coefficients are nonzero only for n odd.

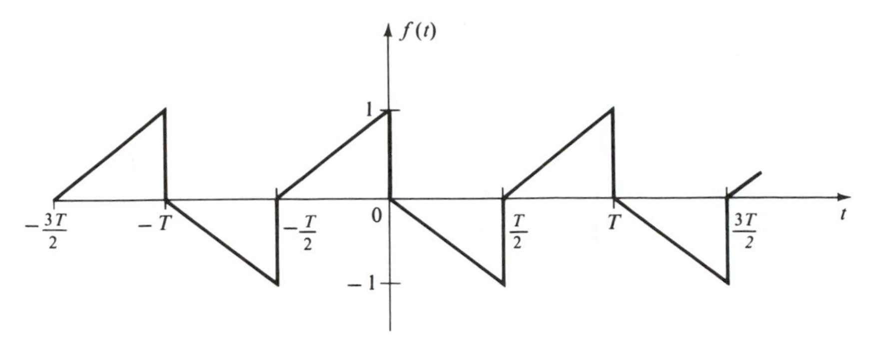

FIGURE 2.5.2 Waveform for Example 2.5.5.

EXAMPLE 2.5.5

The signal in Fig. 2.5.2 has the property that $f(t) = -f(t \pm T/2)$, and thus its complex Fourier coefficients should be zero for $n = 0, 2, 4, \ldots$. By direct evaluation from Eq. (2.4.4) for $n \neq 0$,

$$c_n = \frac{1}{T} \int_{-T/2}^{T/2} f(t) e^{-jn\omega_0 t}\, dt$$

$$= \frac{1}{T} \left\{ \frac{1}{jn\omega_0} - \frac{4}{Tn^2\omega_0^2} \right\} [\cos n\pi - 1]. \tag{2.5.2}$$

When n is even, $\cos n\pi = +1$ and the $c_n = 0$; further, when n is odd, $\cos n\pi = -1$ and the c_n are nonzero. Finally, for $n = 0$ in Eq. (2.4.4), $c_0 = 0$. This is the desired result.

As has been demonstrated by examples, the recognition that a signal waveform has one of the special properties mentioned can save substantial amounts of time and effort in the evaluation of Fourier series coefficients. The properties can also provide excellent checks on the accuracy of coefficient calculations. Because of these facts, the reader should become as familiar as possible with the special properties and their application.

2.6 Least Squares Approximations and Generalized Fourier Series

An important question to be answered concerning Fourier series, especially for some practical applications, is: What kind of approximation do we have if we truncate a Fourier series and retain only a finite number of terms? To answer this question, let the truncated trigonometric Fourier series of $f(t)$ be denoted by

$$f_N(t) = a_0 + \sum_{n=1}^{N} \{a_n \cos n\omega_0 t + b_n \sin n\omega_0 t\}. \tag{2.6.1}$$

It is possible to show that the coefficients of the truncated Fourier series in Eq. (2.6.1) are precisely those coefficients that minimize the integral squared error (ISE):

$$\text{ISE} = \int_{t_0}^{t_0+T} [f(t) - h_N(t)]^2 \, dt, \tag{2.6.2}$$

where

$$h_N(t) = p_0 + \sum_{n=1}^{N} [p_n \cos n\omega_0 t + q_n \sin n\omega_0 t]. \tag{2.6.3}$$

In other words, the partial sum $f_N(t)$, which is generated by truncating the Fourier series expansion of $f(t)$, is *the* one of all possible trigonometric sums $h_N(t)$ of order N or less that minimizes the ISE in Eq. (2.6.2). The sum $f_N(t)$ is usually said to approximate $f(t)$ in the least squares sense.

The proof of this result can be approached in several ways. One way is just direct substitution; another way is to use Eq. (2.6.3) for $h_N(t)$ in Eq. (2.6.2), take partial derivatives with respect to each of the coefficients, equate each derivative to zero, and solve for p_n and q_n. Note that this approach yields only necessary conditions on the coefficients to minimize Eq. (2.6.2) (see Problem 2.21).

For some applications an approximation to a waveform may prove adequate, and in these cases it is necessary to determine the accuracy of the approximation. A natural indicator of approximation accuracy is the ISE in Eq. (2.6.2) with $h_N(t)$ replaced by the approximation being used. In our case, $h_N(t) = f_N(t)$, which minimizes Eq. (2.6.2). A simplified form of the minimum value of the integral squared error in terms of $f(t)$ and the Fourier coefficients can be shown by straightforward manipulations to be

$$\int_{t_0}^{t_0+T} [f(t) - f_N(t)]^2 \, dt = \int_{t_0}^{t_0+T} f^2(t) \, dt - \left\{ Ta_0^2 + \frac{T}{2} \sum_{n=1}^{N} [a_n^2 + b_n^2] \right\}. \tag{2.6.4}$$

Hence for a given signal $f(t)$, Eq. (2.6.4) can be used to calculate the error in the approximation $f_N(t)$, and the number of terms N can be increased until the approximation error is acceptable. Notice that since we have obtained Eq. (2.6.4) by letting $h_N(t) = f_N(t)$ in Eq. (2.6.2), Eq. (2.6.4) is the minimum value of the ISE possible using trigonometric sums.

The following example illustrates the approximation of a given waveform to a prespecified degree of accuracy.

EXAMPLE 2.6.1

We would like to obtain a truncated trigonometric series approximation to $f(t)$ in Fig. 2.6.1 such that the ISE in the approximation is 2% or less of the integral squared value of $f(t)$. Note that since $f(t)$ is periodic, it exists for all time, and hence the integral squared value of $f(t)$ is actually infinite. However, if we consider only one period, as was done in deriving Eq. (2.6.4), no difficulties arise.

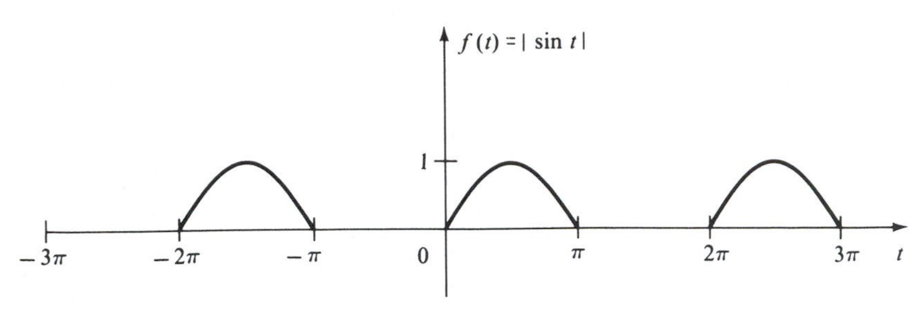

FIGURE 2.6.1 Waveform for Example 2.6.1.

Referring to Fig. 2.6.1, we see that with $t_0 = 0$,

$$\int_{t_0}^{t_0+T} f^2(t)\,dt = \int_0^{2\pi} f^2(t)\,dt = \int_0^{\pi} \sin^2 t\,dt = \frac{\pi}{2} = 1.571. \tag{2.6.5}$$

The trigonometric Fourier series coefficients for $f(t)$ are

$$a_0 = \frac{1}{\pi}, \qquad a_n = \frac{(-1)^{n+1} - 1}{\pi(n^2 - 1)}, \qquad b_1 = \frac{1}{2}, \qquad b_n = 0 \quad \text{for } n \neq 1.$$

First letting $N = 1$ in Eq. (2.6.1), we have $a_0 = 1/\pi$, $a_1 = 0$, $b_1 = \frac{1}{2}$, so Eq. (2.6.4) becomes, upon using Eq. (2.6.5),

$$\int_0^{2\pi} [f(t) - f_1(t)]^2\,dt = 1.571 - \left\{2\pi\left(\frac{1}{\pi}\right)^2 + \pi\left(\frac{1}{2}\right)^2\right\} = 0.149. \tag{2.6.6}$$

Dividing by Eq. (2.6.5), we find that the ISE is 9.5% of the integral squared value of $f(t)$, which is above our goal.

Upon letting $N = 2$, the required Fourier series coefficients are $a_0 = 1/\pi$, $a_1 = 0$, $a_2 = -2/3\pi$, $b_1 = \frac{1}{2}$, and $b_n = 0$ for $n \neq 1$. Substituting into Eq. (2.6.4), we find that

$$\int_0^{2\pi} [f(t) - f_2(t)]^2\,dt = 1.571 - \left\{2\pi\left(\frac{1}{\pi}\right)^2 + \pi\cdot\left[\frac{4}{9\pi^2} + \frac{1}{4}\right]\right\} = 0.0075, \tag{2.6.7}$$

which is only 0.5% of the integral squared value of $f(t)$. The desired approximation is thus

$$f_2(t) = \frac{1}{\pi} + \frac{1}{2}\sin t - \frac{2}{3\pi}\cos 2t. \tag{2.6.8}$$

Thus far in this section, the discussions have been limited to trigonometric Fourier series. Of course, similar results can be obtained for the exponential Fourier series defined in Section 2.4. However, rather than consider several specific cases, let us define what is called a generalized Fourier series given by

$$f(t) = \sum_{n=-\infty}^{\infty} \gamma_n \phi_n(t), \tag{2.6.9}$$

where $f(t)$ is some waveform defined over an interval $t_0 \le t \le t_0 + T$, the γ_n are the coefficients yet to be determined, and the $\{\phi_n(t)\}$, $n = \ldots, -2, -1, 0, 1, 2, \ldots$, are a set of possibly complex orthogonal functions over $t_0 \le t \le t_0 + T$. Since the $\phi_n(t)$ are orthogonal, they satisfy the relation

$$\int_{t_0}^{t_0+T} \phi_n(t)\phi_m^*(t)\, dt = \begin{cases} 0, & \text{for } m \neq n \\ K_n, & \text{for } m = n. \end{cases} \tag{2.6.10}$$

Expressions for the coefficients γ_n can be obtained by multiplying both sides of Eq. (2.6.9) by $\phi_m^*(t)$ and integrating from t_0 to $t_0 + T$, so that

$$\gamma_n = \frac{1}{K_n} \int_{t_0}^{t_0+T} f(t)\phi_n^*(t)\, dt \tag{2.6.11}$$

for all n. The derivation of Eq. (2.6.11) is left as an exercise for the reader (see Problem 2.25).

In deriving Eq. (2.6.11) and the equivalent expressions for the trigonometric and complex Fourier series coefficients, it is necessary to integrate the postulated series, Eq. (2.6.9) here, term by term. We have yet to address the validity of this approach. Although we will not prove the result, the only requirement is that the original series formulation, Eqs. (2.3.1), (2.4.3), and (2.6.9), be uniformly convergent for all t in the interval being considered. Hence assuming uniform convergence of the various forms of Fourier series, our derivations of the Fourier coefficients are justified. (See Kaplan [1959] for a discussion of uniform convergence.)

It is possible to show that the truncated generalized Fourier series given by

$$g_N(t) = \sum_{n=-N}^{N} \gamma_n \phi_n(t) \tag{2.6.12}$$

minimizes the integral squared error

$$\begin{aligned} \varepsilon^2 &= \int_{t_0}^{t_0+T} |f(t) - h_N(t)|^2\, dt \\ &= \int_{t_0}^{t_0+T} [f(t) - h_N(t)][f(t) - h_N(t)]^*\, dt \end{aligned} \tag{2.6.13}$$

for all $h_N(t)$ of the form $h_N(t) = \sum_{n=-N}^{N} p_n \phi_n(t)$. Furthermore, the minimum value of the ISE can be shown to be

$$\varepsilon_{\min}^2 = \int_{t_0}^{t_0+T} |f(t)|^2\, dt - \sum_{n=-N}^{N} K_n |\gamma_n|^2. \tag{2.6.14}$$

It is interesting to note that using Eq. (2.6.9),

$$\begin{aligned} \int_{t_0}^{t_0+T} |f(t)|^2\, dt &= \int_{t_0}^{t_0+T} f(t) \left\{ \sum_{n=-\infty}^{\infty} \gamma_n \phi_n(t) \right\}^* dt \\ &= \sum_{n=-\infty}^{\infty} \gamma_n^* \int_{t_0}^{t_0+T} f(t)\phi_n^*(t)\, dt \\ &= \sum_{n=-\infty}^{\infty} K_n \gamma_n \gamma_n^* = \sum_{n=-\infty}^{\infty} K_n |\gamma_n|^2, \end{aligned} \tag{2.6.15}$$

which is known as *Parseval's theorem.* As a consequence of Eq. (2.6.15), we see that

$$\lim_{N \to \infty} \varepsilon_{\min}^2 = 0, \tag{2.6.16}$$

which simply states that the generalized Fourier series in Eq. (2.6.9) represents $f(t)$ exactly over the specified interval in the sense of providing a minimum integral squared error. Equation (2.6.15) also indicates that the integral squared value of a signal $f(t)$ can be expressed in terms of the generalized Fourier series coefficients.

Finally, we would like to consider the question: How many terms are necessary to represent a waveform exactly? We have, of course, already answered this question by letting our Fourier series representations have an infinite number of terms. What we would like to do now is to indicate why this was done. Central to the discussion are the concepts of *completeness* and *uniqueness.* Definitions of these terms are available in advanced calculus books [Kaplan, 1959]. Completeness is concerned with the fact that the $ISE \to 0$ as $N \to \infty$ [see Eq. (2.6.16)], while uniqueness is related to the requirement of having enough functions to represent a given waveform. If a set of orthogonal functions is complete, the set of functions also has the uniqueness property. However, uniqueness alone does not guarantee that a set of functions is complete. What we were doing then, when we included an infinite number of terms in the Fourier series representations in this and earlier sections, was to ensure that the set of orthogonal functions we were using was complete. The trigonometric functions including the constant term in Section 2.3 and the complex exponential functions in Section 2.4 are all complete, and therefore, Fourier series representations in terms of these functions are unique.

2.7 Spectral Content of Periodic Signals

The trigonometric and complex exponential Fourier series representations in Sections 2.3 and 2.4 are methods of separating a periodic time function into its various components. In this section we explicitly emphasize the frequency content interpretation of the trigonometric and complex Fourier series coefficients. Although both the trigonometric and complex forms contain identical information, the complex exponential form is used as a basis for the development here. The primary reasons for this selection are that the complex coefficients require the evaluation of only a single integral, and more important, the exponential form leads us rather directly to the definition of the Fourier transform, as will be seen in Chapter 3.

The set of complex Fourier coefficients c_n completely describe the frequency content of a periodic signal $f(t)$, and as a group constitute what is usually called the *line spectrum* or simply spectrum of $f(t)$. The motivation for the former terminology will be clear soon. Each coefficient specifies the complex amplitude of a certain frequency component. For example, the coefficient c_0 is the amplitude of the zero-frequency value, usually called the *average* or *direct-current* value.

The coefficient c_1 indicates the amplitude of the component with the same period as $f(t)$, and hence is called the *fundamental frequency*. The other c_n represent the complex amplitudes of the n harmonics of $f(t)$.

To aid in visualizing the frequency content of a periodic signal, it is possible to sketch several different graphical representations of the complex Fourier coefficients. Since the c_n are in general complex, two graphs are necessary to display all of the information completely. One possible pair of graphs would be to plot the real and imaginary parts of c_n; however, it is more common to plot the magnitude and phase of the c_n given by Eqs. (2.7.1) and (2.7.2), respectively,

$$|c_n| = \{[\mathrm{Re}(c_n)]^2 + [\mathrm{Im}(c_n)]^2\}^{1/2} \tag{2.7.1}$$

and

$$\underline{/c_n} = \tan^{-1} \frac{\mathrm{Im}(c_n)}{\mathrm{Re}(c_n)}. \tag{2.7.2}$$

Equations (2.7.1) and (2.7.2) are usually called the *amplitude spectrum* and *phase spectrum* of $f(t)$, respectively. The magnitude and phase plots are important, since they can be related to the amplitude and phase of a sinusoid (see Problem 2.11). Of course, both pairs of graphs contain the same information. Important properties of the amplitude and phase spectrum are that the amplitude is even,

$$|c_n| = |c_{-n}| \tag{2.7.3}$$

and the phase is odd,

$$\underline{/c_n} = -\underline{/c_{-n}}. \tag{2.7.4}$$

These results provide very tangible benefits, since it is only necessary to determine the magnitude and phase for positive n and then use Eqs. (2.7.3) and (2.7.4) to determine the magnitude and phase for negative n.

EXAMPLE 2.7.1

We would like to sketch the amplitude and phase spectra of the square wave in Fig. 2.7.1. Of course, to determine the amplitude and phase spectra, it is first necessary to compute the complex Fourier series coefficients. Using Eq. (2.4.4) directly with $t_0 = 0$ and $T = \tau$, we have for $n \neq 0$,

$$c_n = \begin{cases} 0, & \text{for } n \text{ even} \\ \dfrac{-j2A}{n\pi}, & \text{for } n \text{ odd} \end{cases} \tag{2.7.5}$$

and $c_0 = 0$.

Let us now use some of the special properties developed in Section 2.5 and this section to check these results. From Fig. 2.7.1 we can see immediately that $f(t)$ is real and odd and hence the c_n should be purely imaginary. This checks with Eq. (2.7.5). Observe, too, that $f(t)$ satisfies the rotation symmetry property, that is, $f(t) = -f(t + T/2)$, and hence the c_n should be nonzero only for n odd, which is in agreement with our results. We also know that for $f(t)$ real, $c_{-n} = c_n^*$, which is seen to be true from Eq. (2.7.5) by replacing n with $-n$.

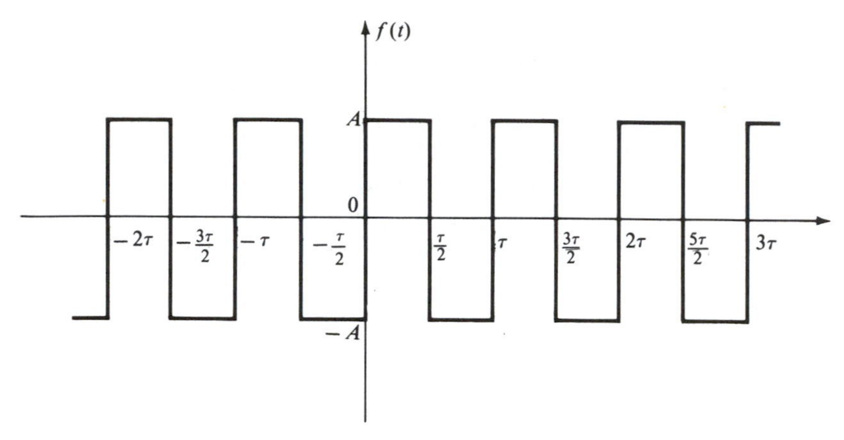

FIGURE 2.7.1 Waveform for Example 2.7.1.

We now calculate the magnitude and phase of the c_n. It is only necessary to find the spectra for $n > 0$, since Eqs. (2.7.3) and (2.7.4) will then give us the results for $n < 0$. Using Eq. (2.7.1), we obtain the amplitude spectrum

$$|c_n| = \{[\mathrm{Re}(c_n)]^2 + [\mathrm{Im}(c_n)]^2\}^{1/2} = \left\{(0)^2 + \left(\frac{-2A}{n\pi}\right)^2\right\}^{1/2} = \frac{2A}{n\pi} = |c_{-n}| \tag{2.7.6}$$

for n odd and employing Eq. (2.7.3). For n even,

$$|c_n| = |c_{-n}| = 0. \tag{2.7.7}$$

The phase spectrum for $n > 0$ is given by Eq. (2.7.2) as

$$c_n = \tan^{-1}\frac{\mathrm{Im}(c_n)}{\mathrm{Re}(c_n)} = \tan^{-1}\left(\frac{-2A/n\pi}{0}\right). \tag{2.7.8}$$

Our first impulse is to conclude from Eq. (2.7.8) that $c_n = \tan^{-1}(-\infty)$. However, a sticky problem suddenly arises since $\tan^{-1}(-\infty)$ is a multivalued function. (See any set of mathematical tables for a plot of the tangent function: e.g., Selby and Girling [1965].)

The problem can be resolved in an unambiguous manner by sketching the complex plane and writing the values of the tangent function at 90° intervals as shown in Fig. 2.7.2. Upon reconsidering Eq. (2.7.8), the negative imaginary part puts us either in quadrant III or IV. In these two quadrants, the tangent takes on the value of $-\infty$ at only one angle, namely, $-\pi/2$ radians. Hence for $n > 0$ and n odd,

$$\angle c_n = -\frac{\pi}{2}. \tag{2.7.9}$$

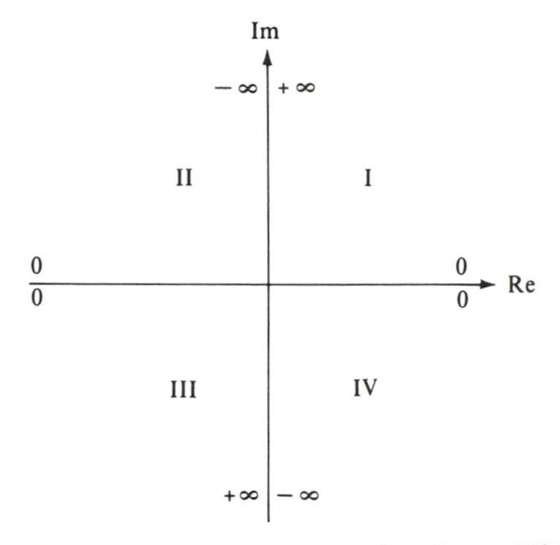

FIGURE 2.7.2 Values of the tangent function at 90° intervals.

We are now ready to plot the amplitude and phase spectra of $f(t)$. What parameter should be used for the abscissa? We could use n, but we can also use $n\omega_0$, since c_n is the complex amplitude of the nth harmonic. Since we desire a frequency spectrum interpretation here, $n\omega_0$ will be used as the abscissa, where $\omega_0 = 2\pi/\tau$. Figures 2.7.3 and 2.7.4 show sketches of the amplitude and phase spectra of $f(t)$, respectively, obtained from Eqs. (2.7.6), (2.7.7),

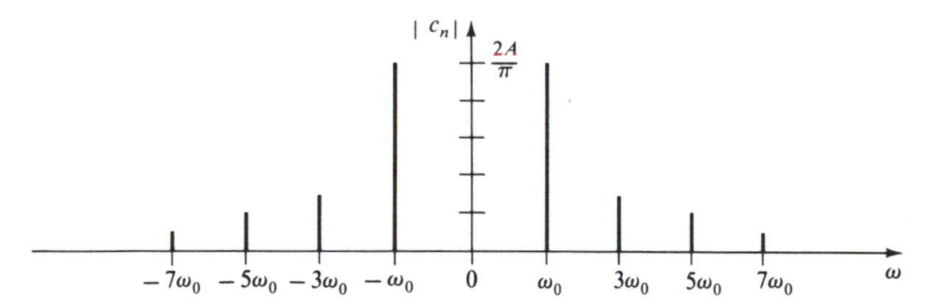

FIGURE 2.7.3 Amplitude spectrum for Example 2.7.1.

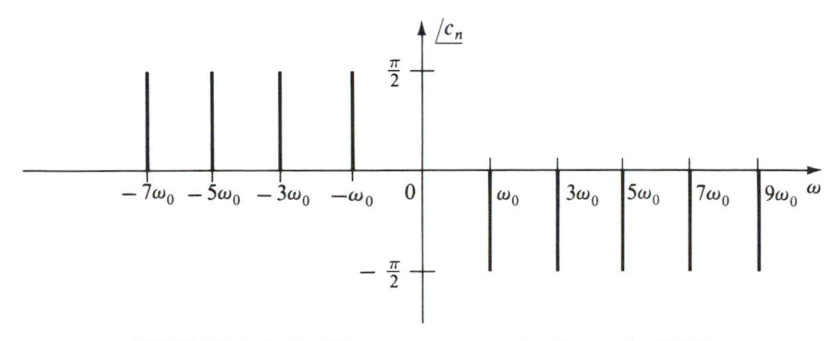

FIGURE 2.7.4 Phase spectrum for Example 2.7.1.

and (2.7.9). The most noticeable thing about Figs. 2.7.3 and 2.7.4 is that both the amplitude and phase spectra are discrete; that is, they are defined only for discrete values of frequency that are integral multiples of ω_0. This is not surprising, though, since any periodic signal has discrete amplitude and phase spectra because only integral multiples of the fundamental frequency are required to synthesize the waveforms via a Fourier series. The terminology *line spectrum* is drawn from the distinctive appearance illustrated by Figs. 2.7.3 and 2.7.4. As mentioned earlier, the entire set of complex Fourier coefficients is usually called the spectrum of $f(t)$.

Since the last example was somewhat drawn out, let us work an additional example in a very concise fashion.

EXAMPLE 2.7.2

We desire to sketch the amplitude and phase spectra of $f(t)$ in Fig. 2.3.1. In Example 2.4.1 the complex Fourier series coefficients for this waveform were found to be

$$c_n = \begin{cases} 0, & \text{for } n \text{ even} \\ \dfrac{1}{n\pi}(-1)^{(n-1)/2}, & \text{for } n \text{ odd} \\ \frac{1}{2}, & \text{for } n = 0. \end{cases}$$

Since the amplitude spectrum is even and the phase spectrum is odd, it is only necessary to consider $n \geq 0$.

The c_n are purely real for this waveform, so from Eq. (2.7.1) for n odd,

$$c_n = \left\{\left[\frac{1}{n\pi}(-1)^{(n-1)/2}\right]^2\right\}^{1/2} = \frac{1}{n\pi}, \tag{2.7.10}$$

since $(-1)^{n-1} = +1$ for n odd. The values for $n = 0$ and n even can be obtained by inspection, so in summary

$$|c_n| = \begin{cases} 0, & \text{for } n \text{ even} \\ \dfrac{1}{n\pi}, & \text{for } n \text{ odd} \\ \frac{1}{2}, & \text{for } n = 0. \end{cases} \tag{2.7.11}$$

For the phase spectrum with n odd, we have from Eq. (2.7.2),

$$\underline{/c_n} = \tan^{-1}\frac{0}{(1/n\pi)(-1)^{(n-1)/2}}. \tag{2.7.12}$$

Again, we must proceed carefully in determining the appropriate angles from Eq. (2.7.12). For $n = 1, 5, 9, 13, \ldots$, the denominator of the arctangent argu-

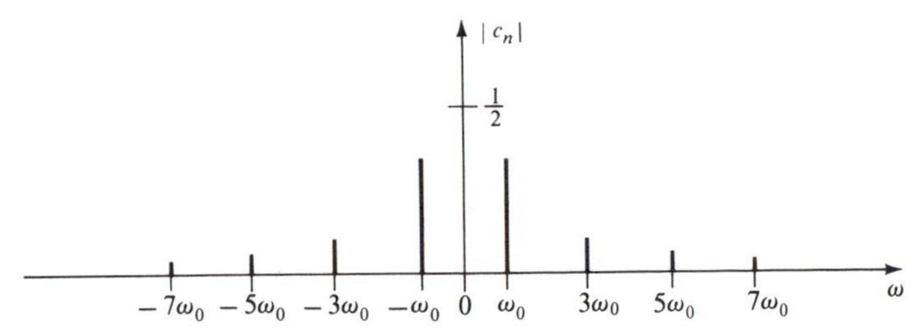

FIGURE 2.7.5 Amplitude spectrum for Example 2.7.2.

ment is positive, and hence from Fig. 2.7.2, the required angle lies in quadrant I or IV of the complex plane. Since the numerator is zero, we have immediately that $c_n = 0$ for these values of n. For $n = 3, 7, 11, 15, \ldots$, the denominator is negative, which locates the angle in the left half of the complex plane. Since the numerator is zero, $\angle c_n = \pm\pi$ radians for these values of n. By inspection for $n = 0$, $\angle c_n = 0$, and therefore to summarize,

$$\angle c_n = \begin{cases} 0^\circ, & \text{for } n \text{ even} \\ 0^\circ, & \text{for } n = 0 \\ 0^\circ, & \text{for } n = 1, 5, 9, 13, \ldots \\ \pm 180^\circ, & \text{for } n = 3, 7, 11, 15, \ldots. \end{cases} \tag{2.7.13}$$

Notice that we associate an angle of zero degrees with those coefficients with a magnitude of zero.

Sketches of the amplitude and phase spectra of $f(t)$ given by Eqs. (2.7.11) and (2.7.13) are shown in Figs. 2.7.5 and 2.7.6. The phase is plotted as alternating between $\pm\pi$ since this seems to be a matter of convention. We could just as well have sketched the phase as $+\pi$ radians or $-\pi$ radians without alternating and still conveyed the same information.

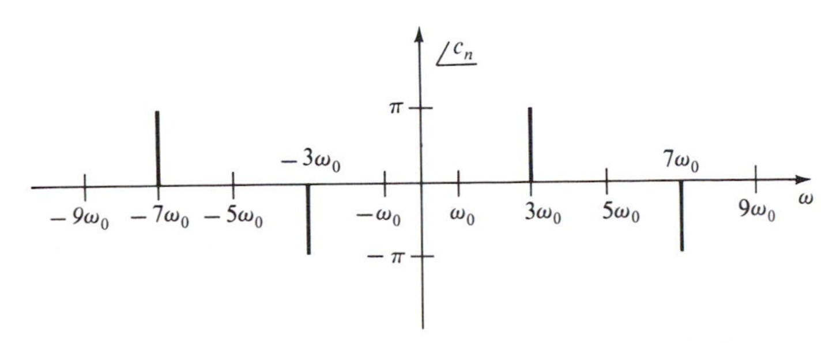

FIGURE 2.7.6 Phase spectrum for Example 2.7.2.

In both of the preceding examples, the real and imaginary parts of the complex coefficients could have been sketched instead of the amplitude and phase spectra without any loss of information. However, as noted previously, the amplitude and phase are more meaningful for our purposes, since they can be interpreted as the amplitude and phase of a sinusoid.

SUMMARY

In this chapter we have expended substantial time and effort in motivating, defining, and developing the idea of a Fourier series representation of a signal or waveform. In Section 2.2 we provided the initial motivation and introduced sets of functions that possess the property of orthogonality. The two most important forms of Fourier series for our purposes, the trigonometric and the complex exponential forms, were presented in Sections 2.3 and 2.4, respectively, and their use illustrated. In Sections 2.5 and 2.6 we concentrated on simplifying the Fourier series calculations and supplying some previously ignored mathematical details. The important concept of a spectrum of a periodic signal was introduced in Section 2.7 and examples were given to clarify calculations. As we shall see, Section 2.7 serves as a critical stepping-stone to the definition of the Fourier transform in Chapter 3.

PROBLEMS

2.1 For the two vectors $\mathbf{B}_0 = 3\hat{\mathbf{a}}_x + 4\hat{\mathbf{a}}_y$ and $\mathbf{B}_1 = -\hat{\mathbf{a}}_x + 2\hat{\mathbf{a}}_y$, determine the component of $\mathbf{B}_0$ in the $\mathbf{B}_1$ direction and the component of $\mathbf{B}_1$ in the $\mathbf{B}_0$ direction.

2.2 Determine whether each of the following sets of vectors is orthogonal. Are they orthonormal?
(a) $\mathbf{B}_0 = -\hat{\mathbf{a}}_x + \hat{\mathbf{a}}_y$ and $\mathbf{B}_1 = \hat{\mathbf{a}}_x - \hat{\mathbf{a}}_y$.
(b) $\mathbf{B}_0 = -\hat{\mathbf{a}}_x + \hat{\mathbf{a}}_y$ and $\mathbf{B}_1 = -\hat{\mathbf{a}}_x - \hat{\mathbf{a}}_y$.
(c) $\mathbf{B}_0 = (\hat{\mathbf{a}}_x + \hat{\mathbf{a}}_y)/\sqrt{2}$ and $\mathbf{B}_1 = (\hat{\mathbf{a}}_x - \hat{\mathbf{a}}_y)/\sqrt{2}$.

2.3 We desire to approximate the vector

$$\mathbf{A}_1 = 4\hat{\mathbf{a}}_x + \hat{\mathbf{a}}_y + 2\hat{\mathbf{a}}_z$$

by a linear combination of the vectors $\mathbf{y}_1$ and $\mathbf{y}_2$,

$$\mathbf{A}_1' = d_1\mathbf{y}_1 + d_2\mathbf{y}_2,$$

where $\mathbf{y}_1 = -\hat{\mathbf{a}}_x + \hat{\mathbf{a}}_y$ and $\mathbf{y}_2 = -\hat{\mathbf{a}}_x - \hat{\mathbf{a}}_y$. Find the coefficients d_1 and d_2 such that the approximation minimizes the least squares loss function given by $\varepsilon^2 = |\mathbf{A}_1 - \mathbf{A}_1'|^2$.

2.4 Complete Example 2.2.1 by considering cases (2) and (3).

2.5 A set of functions $\{f_n(x)\}$ is said to be orthogonal over the interval (a, b) with respect to the weighting function $\rho(x)$ if the functions $\rho^{1/2}(x)f_n(x)$ and $\rho^{1/2}f_m(x)$, $m \neq n$, are orthogonal, and thus

$$\int_a^b \rho(x)f_n(x)f_m(x)\,dx = 0.$$

The set of polynomials $\{H_n(x)\}$ defined by the equations

$$H_n(x) = (-1)^n e^{x^2/2} \frac{d^n}{dx^n} e^{-x^2/2}, \qquad \text{for } n = 0, 1, 2, \ldots$$

are called *Hermite polynomials* and they are orthogonal over the interval $-\infty < x < \infty$ with respect to the weighting function $e^{-x^2/2}$. Specifically, this says that the two functions $e^{-x^2/4}H_m(x)$ and $e^{-x^2/4}H_n(x)$, $n \neq m$, are orthogonal, so that

$$\int_{-\infty}^{\infty} e^{-x^2/2}H_m(x)H_n(x)\,dx = 0.$$

Show that $H_0(x)$ and $H_1(x)$ satisfy the relation above.

2.6 A set of polynomials that are orthogonal over the interval $-1 \leq t \leq 1$ without the use of a weighting function, called *Legendre polynomials*, is defined by the relations

$$P_n(t) = \frac{1}{2^n n!} \frac{d^n}{dt^n} (t^2 - 1)^n, \qquad \text{for } n = 0, 1, 2, \ldots$$

and thus

$$P_0(t) = 1, \qquad P_1(t) = t, \qquad P_2(t) = (\tfrac{1}{2})(3t^2 - 1),$$

and so on. Legendre polynomials are very closely related to the set of polynomials $\{1, t, t^2, \ldots, t^n, \ldots\}$ that we found to be nonorthogonal in Section 2.2. In fact, by using a technique called the Gram–Schmidt orthogonalization process [Jackson, 1941; Kaplan, 1959], the normalized Legendre polynomials, which are thus orthonormal, can be generated.

Calculate for the Legendre polynomials with $n = 0, 1, 2$, and so on, as necessary the value of

$$\int_{-1}^{1} P_n^2(t)\,dt$$

and hence use induction to prove that

$$\int_{-1}^{1} P_n^2(t)\,dt = \frac{2}{2n + 1}.$$

Notice that this shows that the Legendre polynomials are not orthonormal. Based on these results, can you construct a set of polynomials that are orthonormal?

2.7 Show that the trigonometric functions in Example 2.2.1 are not orthonormal over $t_0 \leq t \leq t_0 + 2\pi/\omega_0$. From these results deduce a set of orthonormal trigonometric functions. Repeat both of these steps for the exponential functions in Example 2.2.2.

2.8 Notice that any polynomial in t can be expressed as a linear combination of Legendre polynomials. This fact follows straightforwardly from their definition in Problem 2.6, since we then have

$$1 = P_0(t), \qquad t = P_1(t), \qquad t^2 = \tfrac{2}{3}P_2(t) + \tfrac{1}{3}P_0(t), \qquad t^3 = \tfrac{2}{5}P_3(t) + \tfrac{3}{5}P_1(t),$$

and so on. The reader should verify these statements. Using this result, then, we observe that the Legendre polynomial $P_n(t)$ is orthogonal to any polynomial of degree $n - 1$ or less over $-1 \leq t \leq 1$. That is, for $g(t)$ a polynomial in t of degree $n - 1$ or less,

$$\int_{-1}^{1} P_n(t)g(t)\,dt = 0.$$

Demonstrate the validity of this claim for $n = 4$ and $g(t) = 5t^3 - 3t^2 + 2t + 7$.

2.9 Determine the trigonometric Fourier series expansion for the periodic function shown in Fig. P2.9 by direct calculation.

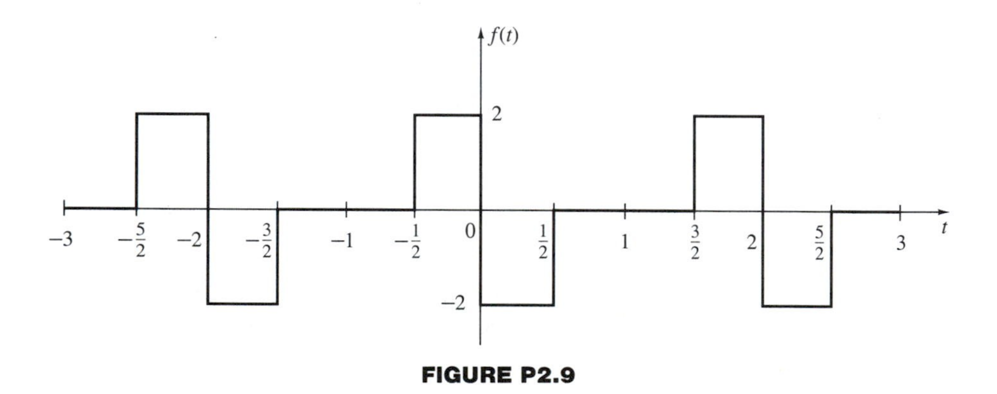

FIGURE P2.9

2.10 Obtain the trigonometric Fourier series representation of the waveform in Fig. P2.10.

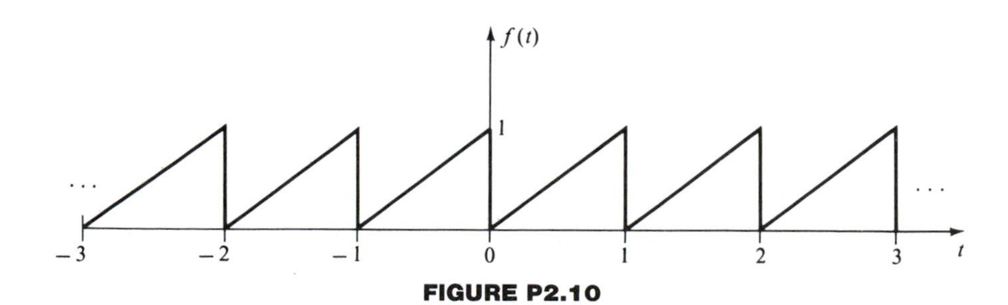

FIGURE P2.10

2.11 Derive the magnitude–angle form of a trigonometric Fourier series given by

$$f(t) = a_0 + \sum_{n=1}^{\infty} d_n \cos\left[\frac{2\pi nt}{T} + \theta_n\right]$$

with

$$d_n = 2|c_n| = \sqrt{a_n^2 + b_n^2}$$

and

$$\theta_n = \tan^{-1} \frac{\mathrm{Im}\{c_n\}}{\mathrm{Re}\{c_n\}},$$

where the c_n are the complex Fourier series coefficients and the a_n and b_n are the trigonometric Fourier series coefficients. Start with Eq. (2.4.3) and use the fact that

$$c_n = |c_n| e^{j\angle c_n} \qquad \text{when } n > 0$$

and

$$c_n = |c_n| e^{-j\angle c_n} \qquad \text{when } n < 0$$

for $f(t)$ real.

2.12 The trigonometric Fourier series coefficients for a half-wave rectified version of $f(t) = \sin t$ are

$$a_n = \frac{1}{2\pi}\left[\frac{\cos(n-1)t}{n-1} - \frac{\cos(n+1)t}{n+1}\right]\Bigg|_0^{\pi}$$

and

$$b_n = \frac{1}{2\pi}\left[\frac{\sin(n-1)t}{n-1} - \frac{\sin(n+1)t}{n+1}\right]\Bigg|_0^{\pi}.$$

These expressions are indeterminate when $n = 1$.

(a) Use Eqs. (2.3.3) and (2.3.4) with $n = 1$ to show that $a_1 = 0$ and $b_1 = \frac{1}{2}$, respectively.

(b) In evaluating indeterminate forms, a set of theorems from calculus called L'Hospital's rules sometimes proves useful. Briefly, these rules state that if $\lim_{x \to x_0} f(x)/g(x)$ is indeterminate of the form $0/0$ or ∞/∞ and $f(x)$ and $g(x)$ are differentiable in the interval of interest with $g'(x) \neq 0$, then

$$\lim_{x \to x_0} \frac{f(x)}{g(x)} = \lim_{x \to x_0} \frac{f'(x)}{g'(x)}.$$

Use this rule to find a_1 and b_1, respectively. (See an undergraduate calculus book such as Thomas [1968] for more details on L'Hospital's rules.)

2.13 Find the complex Fourier series representation of a nonperiodic function identical to $f(t)$ in Fig. 2.3.1 over the interval $-\tau/2 \leq t \leq \tau/2$.

2.14 Evaluate the complex Fourier series coefficients for the waveform in Fig. P2.9.

2.15 Determine the complex Fourier series representation for the waveform in Fig. P2.10.

2.16 Calculate the complex Fourier series coefficients for the periodic signal in Fig. P2.16.

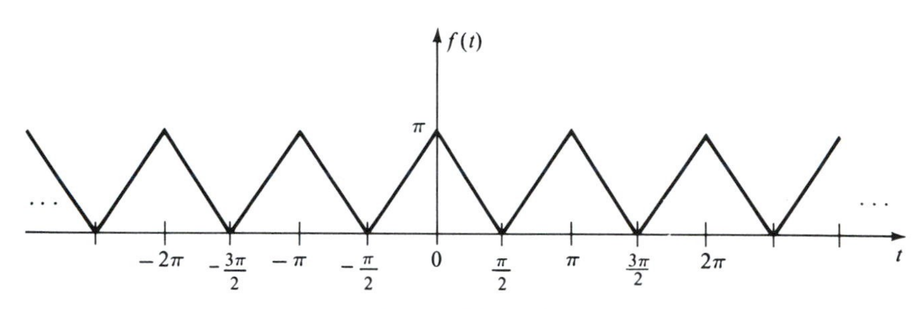

FIGURE P2.16

2.17 Determine whether each of the waveforms in Fig. P2.17 is even, odd, odd harmonic, or none of these. Substantiate your conclusions.

2.18 For the waveforms specified below, determine if any of the symmetry properties discussed in Section 2.5 are satisfied. State any conclusions that can be reached concerning the trigonometric and complex Fourier series in each case.
(a) Fig. P2.9.
(b) Fig. P2.10.
(c) Fig. P2.16.

2.19 Prove that if $f(t) = -f(t - T/2)$, the complex Fourier series coefficients are zero for $n = 0, 2, 4, \ldots$.

2.20 Prove that the integral of an odd function over symmetrical limits is zero; that is, show that if $f(t) = -f(-t)$, then

$$\int_{-a}^{a} f(t)\, dt = 0.$$

2.21 Obtain necessary conditions on the coefficients p_0, p_n, and q_n, $n = 1, 2, 3, \ldots$, in Eq. (2.6.3) to minimize Eq. (2.6.2). Do these coefficients have any special significance?

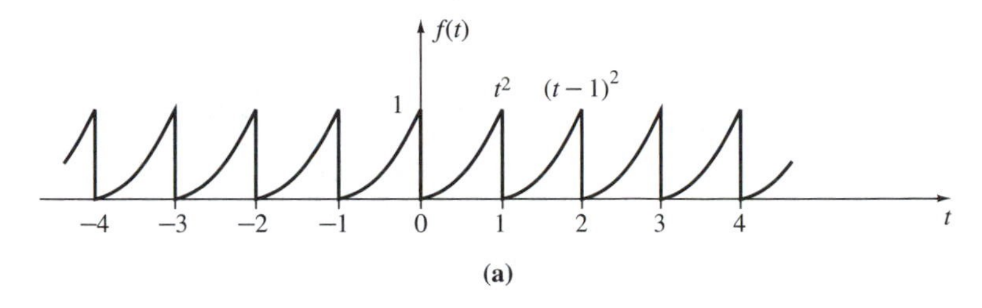

(a)

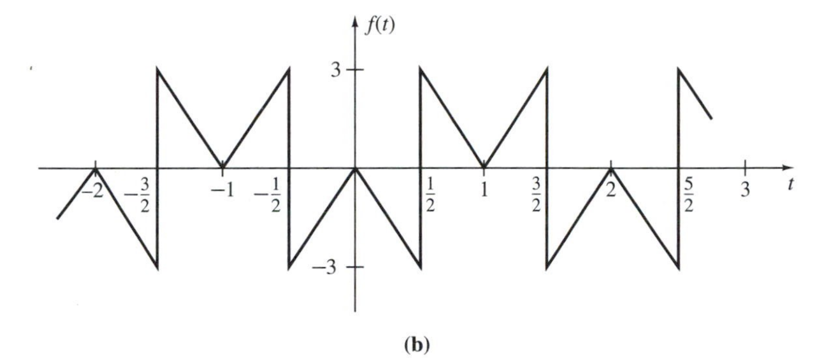

(b)

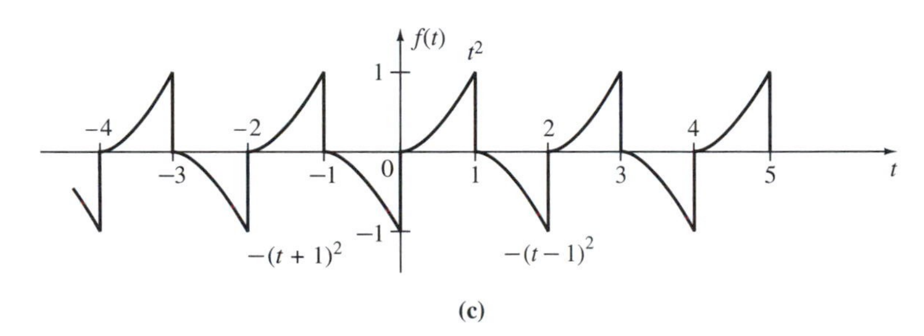

(c)

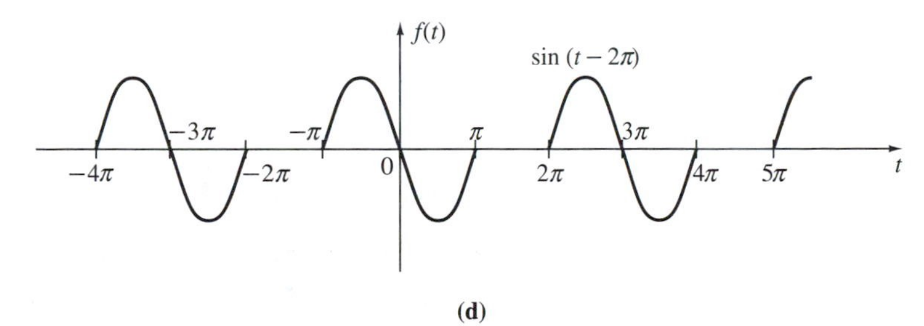

(d)

FIGURE P2.17

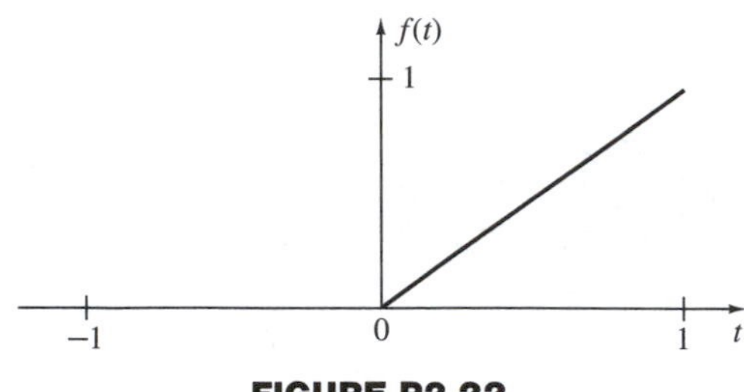

FIGURE P2.22

2.22 Obtain a trigonometric series approximation to $f(t)$ in Fig. P2.22 such that the total energy remaining in the error is 5% or less of the total energy in $f(t)$. Do not include any more terms than is necessary.

2.23 Plot the first three partial sums for the Fourier series of Fig. P2.22,

$$f(t) = \frac{1}{2} + \sum_{n=1}^{\infty} \left(\frac{-1}{n\pi}\right) \sin 2n\pi t,$$

on a large sheet of graph paper superimposed on the waveform in Fig. P2.22. What is the value of the integral squared error in this approximation?

2.24 What is the trigonometric series approximation to $f(t)$ in Fig. P2.9 such that the energy in the error is 15% or less of the total energy in $f(t)$? Include the fewest terms possible. Plot the resulting approximation over one period and compare to $f(t)$.

2.25 Derive Eq. (2.6.11).

2.26 Define $r_N(t) = h_N(t) - f_N(t)$, and note that $f(t) - h_N(t) = f(t) - f_N(t) - r_N(t)$. Use this last expression in Eq. (2.6.2) to show that $f_N(t)$ is the one out of all possible trigonometric sums of order N that minimizes the ISE.

2.27 Derive the expression for the minimum value of the integral squared error in Eq. (2.6.14).

2.28 Approximate the waveform in Problem 2.13 by a series of Legendre polynomials of the form $f(t) = \sum_{n=0}^{2} \delta_n P_n(t)$. What percentage of the energy in $f(t)$ remains in the approximation error if we let $\tau = 1$?

2.29 Show that $\text{Re}\{c_n\}$ is even and $\text{Im}\{c_n\}$ is odd by finding the real and imaginary parts of $c_n = |c_n| e^{j\angle c_n}$ and then using Eqs. (2.7.3) and (2.7.4).

2.30 Repeat Problem 2.29 by finding the real and imaginary parts of Eq. (2.4.4).

2.31 Find expressions for and sketch the amplitude and phase spectra of $f(t)$ in Fig. P2.9.

2.32 Repeat Problem 2.31 for $f(t)$ in Fig. P2.10.

2.33 Repeat Problem 2.31 for $f(t)$ in Fig. P2.16.

CHAPTER 3

Fourier Transforms

3.1 Introduction

As we saw in Chapter 2, trigonometric and complex exponential Fourier series are extremely useful techniques for obtaining representations of arbitrary periodic time waveforms. These methods also can be applied to nonperiodic signals if there is a specific time interval of interest outside which the accuracy of the representation is unimportant. No problems with this approach are evident until one compares the spectrum of a periodic signal and a nonperiodic signal expressed in this manner. We then quickly discover that since the two waveforms have the same Fourier series coefficients, they have the same spectrum! This is particularly disturbing since it is necessary that the spectrum of a function be unique, that is, not exactly the same as any other function, if the concept of a spectrum is to be of any utility to us. The reason for the occurrence of this problem is that the representation for the nonperiodic waveform was obtained somewhat artificially by assuming that the waveform was actually periodic for all time. It is our purpose in this chapter to develop a unique representation for nonperiodic signals that is valid for all time, $-\infty < t < \infty$, and therefore has a unique spectrum.

In Section 3.2 the Fourier transform pair, the Fourier transform and its inverse, is obtained from the complex exponential form of the Fourier series via a limiting argument. A sufficient condition for the existence of the Fourier transform is established in Section 3.3, and a special set of mathematical functions called generalized functions, which are very useful in communication systems analysis, are developed in Section 3.4. The Fourier transforms of certain signals involving generalized functions are defined in Section 3.5. Properties that can aid in the evaluation of Fourier transforms are stated and proven in Section 3.6. We complete the discussion of Fourier transforms in Section 3.7 by developing the concept of a spectrum for nonperiodic signals and by illustrating how the Fourier transform information can be presented in graphical form.

3.2 Fourier Transform Pair

There are numerous possible approaches for arriving at the required expressions for the Fourier transform pair. The approach used here is probably the simplest, most direct, and most transparent of any of these possibilities. We begin by considering slightly modified versions of the complex Fourier series and its coefficients given by

$$f_T(t) = \frac{1}{2\pi} \sum_{n=-\infty}^{\infty} c_n' e^{j\omega_n t} \, \Delta\omega_n \tag{3.2.1}$$

and

$$c_n' = \int_{-T/2}^{T/2} f_T(t) e^{-j\omega_n t} \, dt, \tag{3.2.2}$$

respectively. In Eqs. (3.2.1) and (3.2.2), $\omega_n = n\omega_0 = 2n\pi/T$, $\Delta\omega_n = \Delta(n\omega_0) = n\omega_0 - (n-1)\omega_0 = \omega_0$ is an incremental change in the frequency variable ω_n, and $c_n' = Tc_n$. With these definitions we see that Eqs. (3.2.1) and (3.2.2) are exactly equivalent to Eqs. (2.4.3) and (2.4.4) with $t_0 = -T/2$.

Holding the shape of $f_T(t)$ in the interval $-T/2 \leq t \leq T/2$ fixed, if we take the limit of $f_T(t)$ as $T \to \infty$, we obtain

$$f(t) = \lim_{T\to\infty} f_T(t) = \frac{1}{2\pi} \int_{-\infty}^{\infty} c_n' e^{j\omega t} \, d\omega \tag{3.2.3}$$

and

$$c_n' = \int_{-\infty}^{\infty} f(t) e^{-j\omega t} \, dt, \tag{3.2.4}$$

where we have let $\omega_n \to \omega$ and $\Delta\omega \to d\omega$. Noting that the complex Fourier series coefficients defined the discrete spectrum in Chapter 2, we define $F(\omega) = c_n'$, so Eqs. (3.2.3) and (3.2.4) become

$$f(t) = \frac{1}{2\pi} \int_{-\infty}^{\infty} F(\omega) e^{j\omega t} \, d\omega \tag{3.2.5}$$

and

$$F(\omega) = \int_{-\infty}^{\infty} f(t) e^{-j\omega t} \, dt, \tag{3.2.6}$$

respectively.

Equation (3.2.6) is called the *Fourier transform* of the time function $f(t)$ and is sometimes denoted by $F(\omega) = \mathscr{F}\{f(t)\}$, while Eq. (3.2.5) is called the *inverse Fourier transform* of $F(\omega)$, sometimes written as $f(t) = \mathscr{F}^{-1}\{F(\omega)\}$. The two expressions together, Eqs. (3.2.5) and (3.2.6), are called the *Fourier transform pair*, and this relationship is sometimes indicated by $f(t) \leftrightarrow F(\omega)$.

A portion of the importance of the Fourier transform and its inverse can be attributed to the fact that the transform is *unique*. That is, every time function for which Eq. (3.2.6) is defined has a unique Fourier transform, and conversely, given its Fourier transform, we can exactly recover the original time function. This uniqueness property is critical, since without it, the transform would be useless. Of course, the Fourier transform possesses additional characteristics that give it advantages over other unique transforms. These characteristics are developed in the remainder of this chapter and Chapter 4.

The following two examples illustrate the calculation of the Fourier transform for two common time functions.

EXAMPLE 3.2.1

Consider the signal $f(t)$ specified by

$$f(t) = \begin{cases} Ve^{-t/\tau}, & \text{for } t \geq 0 \\ 0, & \text{for } t < 0. \end{cases} \tag{3.2.7}$$

We desire to find the Fourier transform of $f(t)$ by using Eq. (3.2.6). By direct substitution, we have

$$F(\omega) = \int_{-\infty}^{\infty} f(t)e^{-j\omega t}\,dt = \int_{0}^{\infty} Ve^{-t/\tau}e^{-j\omega t}\,dt$$
$$= \frac{V\tau}{1 + j\tau\omega}. \tag{3.2.8}$$

A natural thing to do to complete the cycle would be to obtain $f(t)$ from $F(\omega)$ in Eq. (3.2.8) using Eq. (3.2.5). Plugging $F(\omega)$ into Eq. (3.2.5), we find that

$$f(t) = \frac{1}{2\pi}\int_{-\infty}^{\infty} \frac{V\tau}{1 + j\omega\tau}\, e^{j\omega t}\,d\omega, \tag{3.2.9}$$

which is a nontrivial integral. Of course, the integral in Eq. (3.2.9) can be evaluated using integral tables; however, this is not too instructive. Instead, we emphasize that since the Fourier transform is unique, once we find $F(\omega)$ given $f(t)$ in Eq. (3.2.7), we know that Eq. (3.2.9) must equal Eq. (3.2.7) exactly.[1] Because of this fact and since extensive tables of the Fourier transform are available, we are not often required to evaluate difficult inverse transform integrals similar to Eq. (3.2.9).

EXAMPLE 3.2.2

We desire to find the Fourier transform of the time function defined by

$$f(t) = \begin{cases} V, & \text{for } -\frac{\tau}{2} \leq t \leq \frac{\tau}{2} \\ 0, & \text{otherwise.} \end{cases} \tag{3.2.10}$$

[1] Depending on how $f(0)$ is defined, this equality may not be exact at $t = 0$; however, such an occurrence has little physical significance (see Section 3.3).

From Eq. (3.2.6), we find immediately that

$$F(\omega) = \int_{-\tau/2}^{\tau/2} Ve^{-j\omega t}\,dt = V\tau\,\frac{\sin(\omega\tau/2)}{\omega\tau/2}. \tag{3.2.11}$$

The reason for manipulating the result into this last form is explained in Section 3.7. Again, the inverse Fourier transform can be obtained by the use of Eqs. (3.2.5) and (3.2.11) and integral tables, or by inspection.

One final point should be made that $F(\omega)$ in each of Eqs. (3.2.8) and (3.2.11) is a continuous function of ω, as opposed to the discrete nature of the c_n in Chapter 2. As noted before, the expression of the Fourier transform in graphical form is considered in Section 3.7, after the Fourier transforms of many other common time functions are derived.

3.3 Existence of the Fourier Transform

Although we have obtained the general equations for the Fourier transform and its inverse in Section 3.2, we have not yet considered under what conditions the Fourier transform integral in Eq. (3.2.6) exists. This may seem like a strange statement to the reader. That is, what exactly do we mean when we say that an integral does or does not exist? We say that an integral *exists* or *converges* if

$$\int_a^b g(x)\,dx < \infty \tag{3.3.1}$$

and an integral *does not exist* or *diverges* if

$$\int_a^b g(x)\,dx = \infty. \tag{3.3.2}$$

Thus we see that in order for the Fourier transform to exist, we must have

$$\int_{-\infty}^{\infty} f(t)e^{-j\omega t}\,dt < \infty. \tag{3.3.3}$$

Since the integral in Eq. (3.3.3) is not always a simple one, we would like to have a condition that we can check without completely evaluating the Fourier transform of a function.

By using the Weierstrass M-test for integrals [Kaplan, 1959], it is possible to prove that the Fourier transform exists, or equivalently, Eq. (3.3.3) is true, if

$$\int_{-\infty}^{\infty} |f(t)|\,dt < \infty \tag{3.3.4}$$

[Chen, 1963] (see Problem 3.5). In words, a time function $f(t)$ has a Fourier transform if $f(t)$ is absolutely integrable, that is, if Eq. (3.3.4) holds. As we shall see, Eq. (3.3.4) is a fairly simple condition to test for and does not require that we calculate the Fourier transform completely. Actually, a couple of additional conditions are required to ensure convergence of the Fourier integral. These conditions are (1) that $f(t)$ must have a finite number of maxima and minima in any finite interval, and (2) that $f(t)$ must have a finite number of discontinuities in any finite interval. Under these last two conditions, the inverse Fourier transform converges to the average of $f(t)$ evaluated at the right- and left-hand limits at jump discontinuities. All functions that we consider will satisfy these last two requirements on maxima and minima and discontinuities.

Equation (3.3.4) and the last two conditions are sometimes called the *Dirichlet conditions*, and they are sufficient for the Fourier transform to exist, but they are not necessary. For instance, every time function that satisfies the Dirichlet conditions has a Fourier transform; however, there are some functions that have a Fourier transform but are not absolutely integrable. We shall encounter examples of such functions shortly.

EXAMPLE 3.3.1

Using Eq. (3.3.4) to test $f(t)$ from Example 3.2.1, we find that for $V > 0$,

$$\int_{-\infty}^{\infty} |f(t)|\, dt = \int_{0}^{\infty} Ve^{-t/\tau}\, dt = -V\tau e^{-t/\tau}\Big|_{0}^{\infty} = V\tau < \infty$$

for V and τ finite. Since by inspection we see that $f(t)$ in Eq. (3.2.7) has a finite number of maxima and minima and a finite number of discontinuities in any finite interval, we conclude that the Fourier transform exists for $f(t)$.

Notice that for the discontinuity at $t = 0$, the inverse Fourier transform converges to $[f(0+) + f(0-)]/2 = V/2$, which does not agree with Eq. (3.2.7) at $t = 0$. We can alleviate this discrepancy by letting $f(t = 0) = V/2$ in Eq. (3.2.7); however, since the value of $f(t)$ at precisely $t = 0$ is of little importance physically, we will not belabor the point.

EXAMPLE 3.3.2

For $f(t)$ in Example 3.2.2 and $0 < V < \infty$,

$$\int_{-\infty}^{\infty} |f(t)|\, dt = \int_{-\tau/2}^{\tau/2} V\, dt = V\tau < \infty$$

for τ finite, which satisfies Eq. (3.3.4). Again, we see by inspection that the additional conditions are satisfied, and thus we conclude that $F(\omega) = \mathscr{F}\{f(t)\}$ exists for $f(t)$ as given.

The only two functions that we have tested thus far satisfy Eq. (3.3.4). Unfortunately, however, there are some very common and hence important signals that are not absolutely integrable. Two of these functions are considered in the following examples.

EXAMPLE 3.3.3

The unit step function is defined by

$$u(t) = \begin{cases} 1, & \text{for } t \geq 0 \\ 0, & \text{otherwise,} \end{cases} \tag{3.3.5}$$

and is an extremely common and vital signal for testing system response. Directly, we discover that

$$\int_{-\infty}^{\infty} |f(t)|\,dt = \int_{0}^{\infty} (1)\,dt = \infty,$$

and therefore Eq. (3.3.4) is violated. The unit step function may still have a Fourier transform, but we cannot arrive at this conclusion from the Dirichlet conditions.

EXAMPLE 3.3.4

The cosine function plays an important role in electrical engineering as a test signal and is used many times in communication systems as a carrier signal. However, it is easy to show that this function is not absolutely integrable; for any ω,

$$\int_{-\infty}^{\infty} |f(t)|\,dt = \int_{-\infty}^{\infty} |\cos \omega t|\,dt = (\infty)2 \int_{-T/4}^{T/4} \cos \omega t\,dt,$$

where $T = 2\pi/\omega$ and since there are an infinite number of periods from $-\infty$ to $+\infty$. Continuing, we find that

$$\int_{-\infty}^{\infty} |f(t)|\,dt = (\infty)\frac{2}{\omega} \sin \omega t \Big|_{-T/4}^{T/4} = (\infty)\frac{4}{\omega} = \infty$$

and thus Eq. (3.3.4) is not satisfied. As in Example 3.3.3, it is therefore not possible to conclude whether or not a Fourier transform exists for this function.

Although Examples 3.3.3 and 3.3.4 indicate that we may have some difficulty in defining Fourier transforms for a step function and periodic waveforms, it is possible to obtain Fourier transforms for these important signals using the theory of generalized functions. We develop the necessary concepts in Section 3.4.

3.4 Generalized Functions

As mentioned previously, the unit step function and the sine and cosine functions are of fundamental importance to the study of electrical engineering. These functions are of no less import in the design and analysis of communication systems, and hence it is imperative that we be able to obtain Fourier transforms for these functions. Fourier transforms can be written for these and other functions that are not absolutely integrable by allowing the transforms to contain impulses or delta functions. The *impulse* or *delta function*, denoted by $\delta(t)$, is usually defined by the statements that

$$\delta(t) = \begin{cases} 0, & \text{for } t \neq 0, \quad (3.4.1) \\ \infty, & \text{for } t = 0, \quad (3.4.2) \end{cases}$$

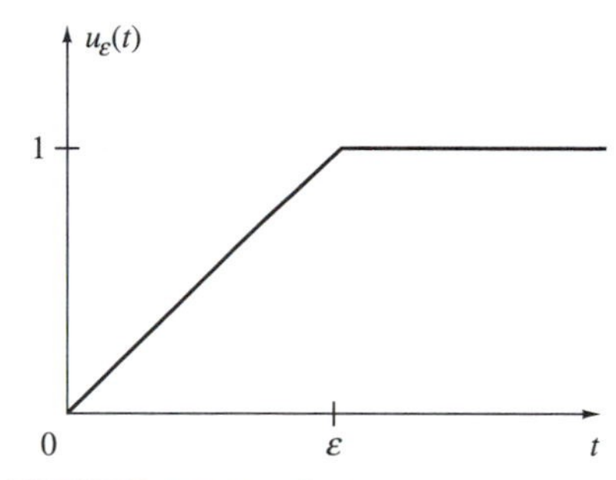

FIGURE 3.4.1 Finite ramp function.

and the function is infinite at the origin in the very special way that

$$\int_{-\infty}^{\infty} \delta(t)\, dt = 1. \tag{3.4.3}$$

Equations (3.4.1)–(3.4.3) are not consistent with the mathematics of ordinary functions, since any function that is zero everywhere except at one point cannot have a nonzero integral. A rigorous mathematical development of the delta function in the ordinary sense is thus not possible. However, a completely rigorous mathematical justification of the delta function is possible by using the theory of distributions developed by Schwartz [1950] or the theory of generalized functions introduced by Temple [1953]. Because of the restricted scope of this book, a rigorous mathematical development will not be pursued here.

The delta function is usually considered to be the derivative of the familiar unit step function defined by Eq. (3.3.5). Of course, this relationship can be proven rigorously. However, rather than do so, we motivate the relationship between the impulse and unit step function by considering the unit step to be the limit of a finite ramp function with slope $1/\varepsilon$ as $\varepsilon \to 0$. The finite ramp is shown in Fig. 3.4.1. Thus we see that

$$u(t) = \lim_{\varepsilon \to 0} u_\varepsilon(t). \tag{3.4.4}$$

If we take the derivative of $u_\varepsilon(t)$ with respect to t (at all points where the derivative exists), we obtain the function illustrated in Fig. 3.4.2. Notice that this

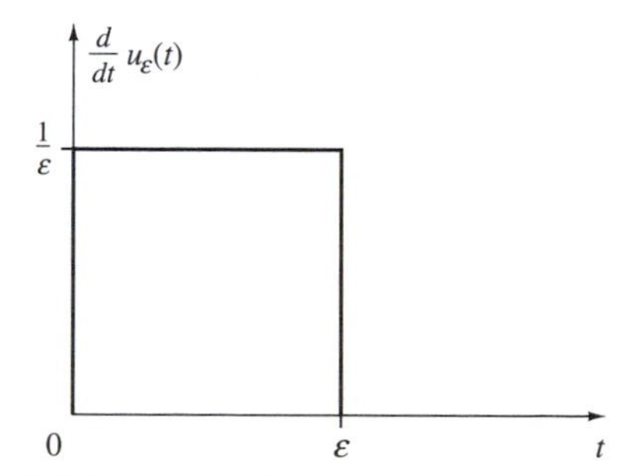

FIGURE 3.4.2 Derivative of the finite ramp function.

function is a pulse of magnitude $1/\varepsilon$ and duration ε with an area of 1. As ε becomes smaller and smaller, the magnitude of the pulse becomes larger and the width decreases but the area remains constant at unity. Hence in the limit as $\varepsilon \to 0$ and $u_\varepsilon(t) \to u(t)$, the derivative of $u_\varepsilon(t)$ approaches an impulse, so we have obtained heuristically that

$$\delta(t) = \lim_{\varepsilon \to 0} \frac{d}{dt} u_\varepsilon(t). \tag{3.4.5}$$

There are numerous interesting properties of the delta function that can be derived. Here we simply state without proof some of the more important ones for our purposes.

Sifting Property

$$\int_{-\infty}^{\infty} \delta(t - t_0)\phi(t)\, dt = \phi(t_0) \tag{3.4.6}$$

for $\phi(t)$ continuous at $t = t_0$.

The sifting property is probably the most used and most important property of the delta function. Notice that Eq. (3.4.3) is just a special case of Eq. (3.4.6), and therefore Eq. (3.4.6) is sometimes used to define a delta function.

The limits on the integral in Eq. (3.4.6) need not extend from $-\infty$ to $+\infty$ as long as the point $t = t_0$ lies within the limits of integration; that is, we can write

$$\int_{t_1}^{t_2} \delta(t - t_0)\phi(t)\, dt = \phi(t_0) \tag{3.4.7}$$

as long as $t_1 < t_0 < t_2$. If $t = t_0$ is outside the integration limits, the integral is zero.

Derivative Property

$$\int_{-\infty}^{\infty} \delta^{(n)}(t - t_0)\phi(t)\, dt = (-1)^n \phi^{(n)}(t)\Big|_{t=t_0} \tag{3.4.8}$$

for $\phi(t)$ continuous as $t = t_0$ and where the notation "superscript (n)" denotes the nth derivative of the function. The first derivative of the delta function $(n = 1)$ occurs somewhat frequently, and hence is given a special name, the *doublet*.

Scaling Property

$$\int_{-\infty}^{\infty} \delta(at)\phi(t)\, dt = \frac{1}{|a|}\phi(0). \tag{3.4.9}$$

Observe that with Eq. (3.4.6) this implies that

$$\int_{-\infty}^{\infty} \delta(at)\phi(t)\, dt = \frac{1}{|a|}\int_{-\infty}^{\infty} \delta(t)\phi(t)\, dt. \tag{3.4.10}$$

The following examples illustrate the application of the important properties developed in this section.

EXAMPLE 3.4.1

We desire to evaluate the following integrals involving impulses:

(a) $\int_{-\infty}^{\infty} \delta(t)[t+1]\,dt$

(b) $\int_{-\infty}^{\infty} \delta(t)e^{-3t}\,dt$

(c) $\int_{-\infty}^{\infty} \delta(t-2)[t^2-3t+2]\,dt$

Solutions

(a) Using Eq. (3.4.6) with $\phi(t) = t + 1$, we find that

$$\int_{-\infty}^{\infty} \delta(t)[t+1]\,dt = [t+1]\Big|_{t=0} = 1.$$

(b) Again from Eq. (3.4.6) with $\phi(t) = e^{-3t}$, we obtain

$$\int_{-\infty}^{\infty} \delta(t)e^{-3t}\,dt = e^{-3t}\Big|_{t=0} = 1.$$

(c) Using the sifting property with $t_0 = 2$ and $\phi(t) = t^2 - 3t + 2$ yields

$$\int_{-\infty}^{\infty} \delta(t-2)[t^2-3t+2]\,dt = 0.$$

EXAMPLE 3.4.2

Evaluate the following integrals involving impulses:

(a) $\int_{-2}^{1} \delta(t)\,dt$

(b) $\int_{5}^{10} \delta(t-7)e^{-t}\,dt$

(c) $\int_{-3}^{-1} \delta(t)[t^2+2]\,dt$

(d) $\int_{-3}^{0} \delta(t+1)t^3\,dt$

Solutions

(a) Directly from Eq. (3.4.6),

$$\int_{-2}^{1} \delta(t)\,dt = 1.$$

(b) Using Eq. (3.4.7) with $t_1 = 5$, $t_0 = 7$, and $t_2 = 10$, we obtain

$$\int_{5}^{10} \delta(t-7)e^{-t}\,dt = e^{-7}.$$

(c) Here we have $t_0 = 0$, which lies outside the limits of integration; hence the integral is zero.

(d) We have $t_0 = -1$, so from Eq. (3.4.7),

$$\int_{-3}^{0} \delta(t+1)t^3\, dt = t^3\Big|_{-1} = -1.$$

EXAMPLE 3.4.3

We desire to evaluate the following integrals using special properties of the impulse function.

(a) $\int_{-\infty}^{\infty} \delta^{(1)}(t)\, dt$

(b) $\int_{-\infty}^{\infty} \delta(2t)e^{-t/2}\, dt$

(c) $\int_{-\infty}^{\infty} \delta^{(2)}(t-1)[t^3+t+1]\, dt$

Solutions

(a) From the derivative property in Eq. (3.4.8) with $n = 1$, $t_0 = 0$, and $\phi(t) = 1$,

$$\int_{-\infty}^{\infty} \delta^{(1)}(t)\, dt = (-1)\frac{d}{dt}(1) = 0.$$

Notice that this is the integral of the doublet.

(b) Using the scaling property, Eq. (3.4.9), gives

$$\int_{-\infty}^{\infty} \delta(2t)e^{-t/2}\, dt = \frac{1}{|2|}\, e^{-t/2}\Big|_{t=0} = \frac{1}{2}.$$

(c) A straightforward application of Eq. (3.4.8) produces

$$\int_{-\infty}^{\infty} \delta^{(2)}(t-1)[t^3+t+1]\, dt = (-1)^2\left\{\frac{d^2}{dt^2}[t^3+t+1]\right\}\Big|_{t=1} = 6.$$

In this section we have introduced two special functions, the unit impulse and the unit step function, and have developed the properties of the impulse. Later we shall make liberal use of these functions and their properties.

3.5 Fourier Transforms and Impulse Functions

In this section we return to the problem of finding the Fourier transforms of functions that are not absolutely integrable. We consider the various singularity functions involved as limits of ordinary functions.[2] Although such an ap-

[2] A singularity function is defined as a function that does not possess ordinary derivatives of all orders.

proach is nonrigorous, it allows the correct results to be obtained more easily and transparently.

Let us first calculate the Fourier transform of a delta function. By straightforward use of Eqs. (3.2.6) and (3.4.6), we find that

$$\mathscr{F}\{\delta(t)\} = \int_{-\infty}^{\infty} \delta(t)e^{-j\omega t}\,dt = e^{-j\omega t}\Big|_{t=0} = 1. \tag{3.5.1}$$

Since Eq. (3.5.1) holds for all values of radian frequency ω, the delta function thus contains the same "amount" of all frequencies. Using the fact that the Fourier transform is unique, we have immediately from Eq. (3.5.1) that

$$\mathscr{F}^{-1}\{1\} = \frac{1}{2\pi}\int_{-\infty}^{\infty} (1)e^{j\omega t}\,d\omega = \delta(t). \tag{3.5.2}$$

Hence we have the Fourier transform pair of the delta function,

$$\delta(t) \leftrightarrow 1. \tag{3.5.3}$$

Let us turn now to the calculation of the Fourier transform of a constant function of amplitude A that exists for all time, $-\infty < t < \infty$. Proceeding in a straightforward fashion, we have

$$\mathscr{F}\{A\} = \int_{-\infty}^{\infty} Ae^{-j\omega t}\,dt = \frac{-A}{j\omega}\,e^{-j\omega t}\Big|_{-\infty}^{\infty} = \frac{A}{j\omega}\,[e^{+j\infty} - e^{-j\infty}]. \tag{3.5.4}$$

Using Euler's identity on the exponentials in Eq. (3.5.4), we see that this integral does not converge; in fact, it oscillates continually for all values of ω. Note that we could have predicted this from Eq. (3.3.4), since a constant that exists for all time is not absolutely integrable.

To compute this Fourier transform, it is necessary to employ the unit impulse. From Eq. (3.5.2) we observe that

$$\int_{-\infty}^{\infty} e^{j\omega t}\,d\omega = 2\pi\delta(t). \tag{3.5.5}$$

Again writing the integral expression for the Fourier transform of A, we have

$$\mathscr{F}\{A\} = \int_{-\infty}^{\infty} Ae^{-j\omega t}\,dt = A\int_{-\infty}^{\infty} e^{-j\omega t}\,dt. \tag{3.5.6}$$

Making a change of variable in this last integral by letting $t = -x$ and using Eq. (3.5.5) yields

$$\mathscr{F}\{A\} = 2\pi A\delta(\omega). \tag{3.5.7}$$

Note that in Eq. (3.5.5), ω is simply a dummy variable of integration.

Another important function in the study of communication systems is the signum or sign function given by

$$\operatorname{sgn}(t) = \begin{cases} 1, & \text{for } t > 0 \\ 0, & \text{for } t = 0 \\ -1, & \text{for } t < 0. \end{cases} \tag{3.5.8}$$

This function is also not absolutely integrable, and therefore an alternative method other than the direct integration of Eq. (3.2.6) is necessary to determine its Fourier transform. The approach taken here is to let

$$\operatorname{sgn}(t) = \lim_{a \to 0} [e^{-a|t|} \operatorname{sgn}(t)], \tag{3.5.9}$$

so that from Eq. (3.2.6) we obtain

$$\mathscr{F}\{\operatorname{sgn}(t)\} = \mathscr{F}\left\{\lim_{a \to 0} [e^{-a|t|} \operatorname{sgn}(t)]\right\} = \lim_{a \to 0} \int_{-\infty}^{\infty} e^{-a|t|} \operatorname{sgn}(t) e^{-j\omega t}\, dt \tag{3.5.10}$$

upon interchanging the integration and limiting operations. Notice that we have not justified mathematically the removal of the limit from under the integral sign. Although such a justification is required to ensure the validity of the derivation, we will not do so here, since the correct result is obtained in a direct and intuitive manner.

Returning to Eq. (3.5.10), we see that $e^{-a|t|} \operatorname{sgn}(t)$ is absolutely integrable, and hence we proceed to evaluate the transform directly to find

$$\mathscr{F}\{\operatorname{sgn}(t)\} = \lim_{a \to 0} \left\{\frac{-1}{a - j\omega} + \frac{1}{a + j\omega}\right\} = \frac{2}{j\omega}. \tag{3.5.11}$$

The Fourier transform pair for the signum function is thus

$$\operatorname{sgn}(t) \leftrightarrow \frac{2}{j\omega}. \tag{3.5.12}$$

If we observe that the unit step function can be written as

$$u(t) = \frac{1}{2} + \frac{1}{2} \operatorname{sgn}(t), \tag{3.5.13}$$

it is now possible to determine the Fourier transform of $u(t)$ using Eqs. (3.5.7) and (3.5.12) and the linearity of the Fourier transform [see Eq. (3.6.6)] as

$$\mathscr{F}\{u(t)\} = \mathscr{F}\left\{\frac{1}{2}\right\} + \frac{1}{2}\mathscr{F}\{\operatorname{sgn}(t)\} = \pi\delta(\omega) + \frac{1}{j\omega}. \tag{3.5.14}$$

Continuing the development of Fourier transforms for time functions that are not absolutely integrable, we consider the function $e^{j\omega_0 t}$ for $-\infty < t < \infty$. Although direct integration using Eq. (3.2.6) will not yield a convergent result, the desired Fourier transform can be calculated using an approach analogous to that employed in obtaining the transform of a constant. Substituting into Eq. (3.2.6) yields

$$\mathscr{F}\{e^{j\omega_0 t}\} = \int_{-\infty}^{\infty} e^{j\omega_0 t} e^{-j\omega t}\, dt = \int_{-\infty}^{\infty} e^{-j(\omega - \omega_0)t}\, dt. \tag{3.5.15}$$

By making the change of variable $y = -t$ and evaluating this integral by using Eq. (3.5.5), the Fourier transform of the complex exponential is found to be

$$\mathscr{F}\{e^{j\omega_0 t}\} = 2\pi\delta(\omega - \omega_0). \tag{3.5.16}$$

It remains for us to determine the Fourier transform of a sine or cosine function that exists for all time. Although back in Example 3.3.4 it seemed that this might be a sticky problem, we can now determine the transform almost trivially. Since $\cos \omega_0 t$ can be expressed as

$$\cos \omega_0 t = \tfrac{1}{2}[e^{j\omega_0 t} + e^{-j\omega_0 t}], \tag{3.5.17}$$

we find immediately from Eq. (3.5.16) that

$$\mathscr{F}\{\cos \omega_0 t\} = \pi[\delta(\omega - \omega_0) + \delta(\omega + \omega_0)]. \tag{3.5.18}$$

In a similar fashion we can show that the Fourier transform of $\sin \omega_0 t$ for $-\infty < t < \infty$ is

$$\mathscr{F}\{\sin \omega_0 t\} = j\pi[\delta(\omega + \omega_0) - \delta(\omega - \omega_0)]. \tag{3.5.19}$$

The reader may have noticed by this point that even though the Fourier transform originally was defined in Section 3.2 only for nonperiodic functions, we have been able to write transforms for certain periodic functions, namely the complex exponential, sine, and cosine functions, by using delta functions. The question, then, naturally arises as to whether it is possible to obtain the Fourier transform of any general periodic time function. The answer to this question is in the affirmative. The procedure is first to write the complex exponential Fourier series for the function of interest, and then take the Fourier transform of this series on a term-by-term basis. That is, for some general periodic function $f(t)$,

$$\begin{aligned}\mathscr{F}\{f(t)\} = \mathscr{F}\left\{\sum_{n=-\infty}^{\infty} c_n e^{jn\omega_0 t}\right\} &= \sum_{n=-\infty}^{\infty} c_n \mathscr{F}\{e^{jn\omega_0 t}\} \\ &= 2\pi \sum_{n=-\infty}^{\infty} c_n \delta(\omega - n\omega_0),\end{aligned} \tag{3.5.20}$$

where the summation and integration can be interchanged since the series is uniformly convergent and the transform is a linear operation, and we have again used Eq. (3.5.5).

EXAMPLE 3.5.1

As an illustration of the application of Eq. (3.5.20), let us find the Fourier transform of the periodic function shown in Fig. 2.3.1. From Example 2.3.1 we know that $\omega_0 = 2\pi/\tau$ and since we have previously shown that the c_n for this waveform are given by Eq. (2.4.6), we have that

$$\mathscr{F}\{f(t)\} = 2\pi \sum_{n=-\infty}^{\infty} c_n \delta\left(\omega - \frac{2n\pi}{\tau}\right) \tag{3.5.21}$$

with c_n as specified above.

An extremely important periodic signal for the analysis of communication systems is the train of unit impulses given by

$$\delta_T(t) = \sum_{n=-\infty}^{\infty} \delta(t - nT), \tag{3.5.22}$$

where T is the period. To find the Fourier transform of this function from Eq. (3.5.20), we first must find the complex Fourier series coefficients. Directly, using Eq. (2.4.4) we obtain

$$c_n = \frac{1}{T}\int_{-T/2}^{T/2} \delta_T(t)e^{-jn\omega_0 t}\,dt = \frac{1}{T}\int_{-T/2}^{T/2} \delta(t)e^{-jn\omega_0 t}\,dt = \frac{1}{T}. \tag{3.5.23}$$

Since $\omega_0 = 2\pi/T$, we have for the desired transform,

$$\mathscr{F}\{\delta_T(t)\} = \frac{2\pi}{T}\sum_{n=-\infty}^{\infty} \delta(\omega - n\omega_0). \tag{3.5.24}$$

Therefore, a train of unit impulses in the time domain with period T has as its Fourier transform a train of impulses with magnitude $\omega_0 = 2\pi/T$ spaced every ω_0 rad/sec. We will use this signal and its Fourier transform extensively in later analyses.

Transforms of several important time signals that are not absolutely integrable were obtained in this section by allowing the transforms to contain delta functions. Although the development of the transforms was nonrigorous, the results are nonetheless quite accurate, and we shall employ the Fourier transforms derived here time and again throughout our work.

3.6 Fourier Transform Properties

In calculating Fourier transforms and inverse transforms, several special situations and mathematical operations occur frequently enough that it is well worth our time to develop general approaches to handling them. In the following the most useful of the approaches are stated as properties of the Fourier transform pair, and a proof or method of proof is given for each property. When appropriate, an example of the application of the properties is presented immediately following each proof. The reader should note that when stating these properties, $g(\cdot)$ denotes a function of either time or frequency that has the shape g. Similarly, $G(\cdot)$ indicates a function that has the shape G, independent of whether it is a function of time or frequency. Hence $g(t)$ and $g(\omega)$ have the same shape, even though the first is a time signal and the latter is a function of frequency.

Symmetry Property

If

$$g(t) \leftrightarrow G(\omega),$$

then

$$G(t) \leftrightarrow 2\pi g(-\omega). \tag{3.6.1}$$

Proof: Use Eq. (3.2.5) and a change of variables.

EXAMPLE 3.6.1

In Example 3.2.2 we found that

$$g(t) = \begin{cases} V, & \text{for } -\frac{\tau}{2} \le t \le \frac{\tau}{2} \\ 0, & \text{otherwise} \end{cases} \tag{3.6.2}$$

has the Fourier transform

$$G(\omega) = \mathscr{F}\{g(t)\} = V\tau \frac{\sin(\omega\tau/2)}{\omega\tau/2}. \tag{3.6.3}$$

Suppose now that we wish to find the Fourier transform of

$$G(t) = Vv \frac{\sin(vt/2)}{vt/2}. \tag{3.6.4}$$

Direct calculation of $\mathscr{F}\{G(t)\}$ requires that we evaluate the integral

$$\mathscr{F}\{G(t)\} = \int_{-\infty}^{\infty} Vv \frac{\sin(vt/2)}{vt/2} e^{-j\omega t}\, dt.$$

However, the symmetry property tells us from Eq. (3.6.1) that if $g(t)$ is known, then $\mathscr{F}\{G(t)\} = 2\pi g(-\omega)$. Using Eq. (3.6.2), we thus find that

$$\mathscr{F}\{G(t)\} = \begin{cases} 2\pi V, & \text{for } -\frac{v}{2} \le \omega \le \frac{v}{2} \\ 0, & \text{otherwise.} \end{cases} \tag{3.6.5}$$

Linearity Property

If

$$g_1(t) \leftrightarrow G_1(\omega)$$

and

$$g_2(t) \leftrightarrow G_2(\omega),$$

then for arbitrary constants a and b,

$$ag_1(t) + bg_2(t) \leftrightarrow aG_1(\omega) + bG_2(\omega). \tag{3.6.6}$$

Proof: Straightforward.

Scaling Property

If

$$g(t) \leftrightarrow G(\omega),$$

then for a real constant b,

$$g(bt) \leftrightarrow \frac{1}{|b|} G\left(\frac{\omega}{b}\right). \tag{3.6.7}$$

Proof: Use change of variables.

EXAMPLE 3.6.2

Let $g_1(t) = g(t)$ given by Eq. (3.6.2), which has the Fourier transform $G_1(\omega) = G(\omega)$ as specified by Eq. (3.6.3). Let us find the Fourier transforms of the time functions

$$(i) \quad g_2(t) = g_1(2t) = \begin{cases} V, & \text{for } -\dfrac{\tau}{4} \le t \le \dfrac{\tau}{4} \\ 0, & \text{otherwise} \end{cases}$$

and

$$(ii) \quad g_3(t) = g_1\left(\frac{t}{2}\right) = \begin{cases} V, & \text{for } -\tau \le t \le \tau \\ 0, & \text{otherwise.} \end{cases}$$

1. Notice that $g_2(t)$ is compressed in time when compared to $g_1(t)$. The Fourier transform of $g_2(t)$ can be found from $G_1(\omega)$ and Eq. (3.6.7) by letting $b = 2$, so

$$\mathscr{F}\{g_2(t)\} = \frac{1}{2} G_1\left(\frac{\omega}{2}\right) = \frac{V\tau}{2} \frac{\sin(\omega\tau/4)}{\omega\tau/4}. \tag{3.6.8}$$

2. Here $g_3(t)$ is expanded in time compared to $g_1(t)$. Again using Eq. (3.6.7), this time with $b = \frac{1}{2}$, we have

$$\mathscr{F}\{g_3(t)\} = 2V\tau \frac{\sin(\omega\tau)}{\omega\tau}. \tag{3.6.9}$$

The scaling property is a very intuitive result, since a function that is compressed in time ($b > 1$) has a Fourier transform that extends to higher frequencies. Similarly, a function that is expanded in time varies more slowly with time and hence is compressed in terms of frequency content. A comparison of Eqs. (3.6.8) and (3.6.9) with $G_1(\omega)$ illustrates these last statements.

Time Shifting Property

If

$$g(t) \leftrightarrow G(\omega),$$

then

$$g(t - t_0) \leftrightarrow G(\omega)e^{-j\omega t_0}. \tag{3.6.10}$$

Proof: Straightforward change of variables.

Frequency Shifting Property

If

$$g(t) \leftrightarrow G(\omega),$$

then

$$g(t)e^{j\omega_0 t} \leftrightarrow G(\omega - \omega_0). \tag{3.6.11}$$

The notation $G(\omega - \omega_0)$ indicates the Fourier transform $G(\omega)$ centered about $+\omega_0$. Hence multiplication by $e^{j\omega_0 t}$ in the time domain produces a frequency shift of ω_0 radians per second in the frequency domain.

Proof: Direct substitution into Eq. (3.2.6).

The frequency shifting property is sometimes called the *modulation theorem.* The following example illustrates why this nomenclature is appropriate.

EXAMPLE 3.6.3

To send information over long distances, it is necessary to translate the message signal to a higher frequency band. One way of accomplishing this translation is to form the product of the sum $s_0 + m(t)$, where s_0 is a constant and $m(t)$ is the message signal, and a carrier signal $\cos \omega_c t$. By using the frequency shifting property, it is possible to demonstrate that multiplication by $\cos \omega_c t$ does in fact achieve the desired frequency translation.

First working in the time domain, we have upon substituting the exponential form of $\cos \omega_c t$,

$$\begin{aligned}[s_0 + m(t)] \cos \omega_c t &= s_0 \cos \omega_c t + m(t) \cos \omega_c t \\ &= \frac{s_0}{2}\left[e^{j\omega_c t} + e^{-j\omega_c t}\right] + \frac{m(t)}{2}\left[e^{j\omega_c t} + e^{-j\omega_c t}\right]. \end{aligned} \tag{3.6.12}$$

From Eq. (3.5.7) we know that

$$\mathscr{F}\left\{\frac{s_0}{2}\right\} = \pi s_0 \delta(\omega) \tag{3.6.13}$$

and we denote by $M(\omega)$ the Fourier transform of the general message signal $m(t)$. Invoking the frequency shifting property, we find from Eq. (3.6.12) that

$$\begin{aligned}\mathscr{F}\{[s_0 + m(t)] \cos \omega_c t\} &= \pi s_0[\delta(\omega - \omega_c) + \delta(\omega + \omega_c)] \\ &\quad + \tfrac{1}{2}[M(\omega - \omega_c) + M(\omega + \omega_c)]. \end{aligned} \tag{3.6.14}$$

Since $M(\omega - \omega_c)$ and $M(\omega + \omega_c)$ denote the Fourier transform $M(\omega)$ centered about $+\omega_c$ and $-\omega_c$, respectively, we see that the multiplication has had the desired result. The delta functions at $\pm\omega_c$ are also of great importance and their meaning is discussed in detail later.

Time Differentiation Property

If

$$g(t) \leftrightarrow G(\omega),$$

then

$$\frac{d}{dt} g(t) = j\omega G(\omega). \tag{3.6.15}$$

Proof: Direct differentiation of the inverse Fourier transform.

Time Integration Property

If

$$g(t) \leftrightarrow G(\omega),$$

then

$$\int_{-\infty}^{t} g(\lambda)\, d\lambda \leftrightarrow \frac{1}{j\omega} G(\omega) + \pi G(0)\delta(\omega). \tag{3.6.16}$$

Proof: The proof is not presented here, since it requires the use of the time convolution theorem presented in Section 4.3.

It is sometimes possible to use the special properties of the delta function and the time integration property to simplify greatly the evaluation of Fourier transforms. That is, since integrals involving impulses are quite easy to evaluate, if a function can be reduced by repeated differentiation to a sum of delta functions, the transform of this sum of impulses can be found and the time integration property employed to determine the transform of the original function. This procedure is examined in Problem 3.19.

In this section a host of important properties of the Fourier transform were stated and proven, and their application illustrated. These properties are used almost continually in the study of communication systems, and the student will soon have them committed to memory.

3.7 Graphical Presentation of Fourier Transforms

The Fourier transform of a time signal specifies the spectral content of the signal or the "amount" of each frequency that the signal contains. Thus far in this chapter we have concerned ourselves only with obtaining analytical expressions for Fourier transforms and have not discussed at all the graphical presentation of Fourier transform information. Since such graphical presentations can increase the understanding of Fourier transforms and many times constitute the most effective way of describing a signal's Fourier transform, we turn our attention to these graphical procedures in this section.

As mentioned in Example 3.2.1, the Fourier transform is, in general, a complex function, and hence two different graphs are necessary to present all of the information completely. Just as in the case of the complex Fourier series coefficients, we have a choice of which pair of graphs to plot. The real and imaginary parts are exactly that, the real and imaginary parts of the Fourier transform, and are denoted by $\text{Re}\{F(\omega)\}$ and $\text{Im}\{F(\omega)\}$, where $F(\omega)$ is given by Eq. (3.2.6). The *amplitude spectrum* is defined by the relation

$$|F(\omega)| = [(\text{Re}\{F(\omega)\})^2 + (\text{Im}\{F(\omega)\})^2]^{1/2} \tag{3.7.1}$$

and the *phase spectrum* is given by

$$\underline{/F(\omega)} = \tan^{-1} \frac{\text{Im}\{F(\omega)\}}{\text{Re}\{F(\omega)\}}. \tag{3.7.2}$$

The amplitude and phase spectra are usually chosen over the real and imaginary parts for graphical presentation, since Eqs. (3.7.1) and (3.7.2) have a natural interpretation as the amplitude and phase, respectively, of an elementary sinusoid with radian frequency ω. Before proceeding, let us illustrate the calculation and graphical presentation of the amplitude and phase spectra.

EXAMPLE 3.7.1

We wish to sketch the amplitude and phase spectra of $f(t)$ in Eq. (3.2.7). The Fourier transform for this signal was found in Example 3.2.1 to be

$$F(\omega) = \frac{V\tau}{1 + j\omega\tau} = \frac{V\tau(1 - j\omega\tau)}{1 + \omega^2\tau^2}. \tag{3.7.3}$$

Directly from Eq. (3.7.3), we have

$$\text{Re}\{F(\omega)\} = \frac{V\tau}{1 + \omega^2\tau^2} \tag{3.7.4}$$

and

$$\text{Im}\{F(\omega)\} = \frac{-V\omega\tau^2}{1 + \omega^2\tau^2}. \tag{3.7.5}$$

Notice that in writing $\text{Im}\{F(\omega)\}$, the j is not included.

Substituting these last two results into Eq. (3.7.1), we obtain the amplitude spectrum of $F(\omega)$,

$$|F(\omega)| = \frac{V\tau}{[1 + \omega^2\tau^2]^{1/2}}. \tag{3.7.6}$$

From Eq. (3.7.2) we find for the phase spectrum,

$$\underline{/F(\omega)} = \tan^{-1} \frac{-V\omega\tau^2/(1 + \omega^2\tau^2)}{V\tau/(1 + \omega^2\tau^2)} = \tan^{-1}(-\omega\tau) = -\tan^{-1} \omega\tau. \tag{3.7.7}$$

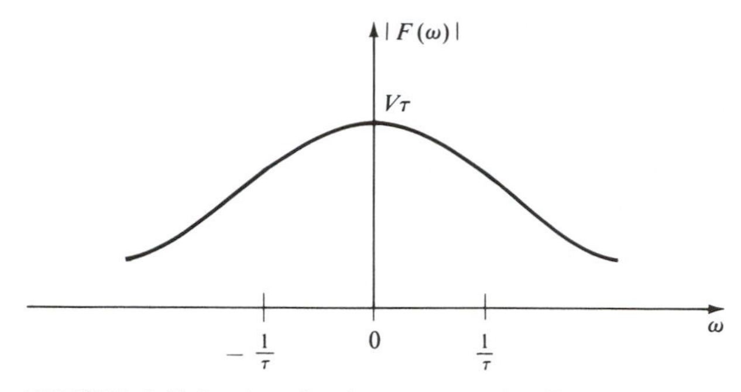

FIGURE 3.7.1 Amplitude spectrum for Example 3.7.1.

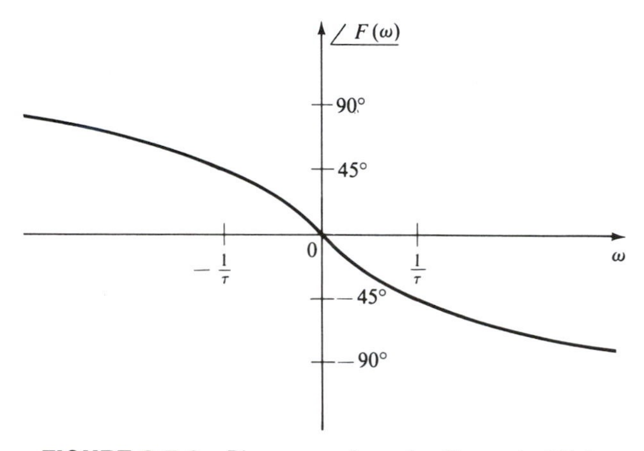

FIGURE 3.7.2 Phase spectrum for Example 3.7.1.

Sketches of Eqs. (3.7.6) and (3.7.7) are given in Figs. 3.7.1 and 3.7.2, respectively. These sketches contain all of the information in the analytical expression for $F(\omega)$, and by inspection of Figs. 3.7.1 and 3.7.2, we can determine the amplitude and phase associated with any frequency found in $f(t)$.

Just as was done in Section 2.5 for the Fourier series coefficients, it is also possible to prove various properties of $\text{Re}\{F(\omega)\}$ and $\text{Im}\{F(\omega)\}$ and the amplitude and phase spectra, depending on the form of $f(t)$. These properties are summarized in Table 3.7.1 and can simplify the sketching of the graphical presentation and can serve as a check on our analytical calculations. For instance, for purely real signals, which we are primarily interested in, we can show that the real part of $F(\omega)$ is an even function of ω and the imaginary part is odd. These facts can be quickly demonstrated by starting with Eq. (3.2.6), using Euler's identity, and working with the real and imaginary parts.

Considering now the amplitude spectrum of $F(\omega)$, we see by inspection of Eq. (3.7.1) that for $f(t)$ real, the amplitude spectrum is even since the square of

TABLE 3.7.1 Properties of the Fourier Transform for Various Forms of $f(t)$

If $f(t)$ is a:	Then $F(\omega)$ is a:
Real and even function of t	Real and even function of ω
Real and odd	Imaginary and odd
Imaginary and even	Imaginary and even
Complex and even	Complex and even
Complex and odd	Complex and odd

an even or odd function is even. Note that we always assign the positive sign to the radical. Letting ω become $-\omega$ in Eq. (3.7.2) yields

$$\underline{/F(-\omega)} = \tan^{-1}\frac{\text{Im}\{F(-\omega)\}}{\text{Re}\{F(-\omega)\}} = \tan^{-1}\frac{-\text{Im}\{F(\omega)\}}{\text{Re}\{F(\omega)\}}$$

$$= -\tan^{-1}\frac{\text{Im}\{F(\omega)\}}{\text{Re}\{F(\omega)\}} = -\underline{/F(\omega)} \tag{3.7.8}$$

and hence the phase spectrum is odd for $f(t)$ real.

These properties can be useful in preparing sketches of the spectra, since once we have the sketch for $\omega > 0$, we also know the shape of the spectra for $\omega < 0$. Notice that $f(t)$ is real in Example 3.7.1 and that the real part of its Fourier transform and its amplitude spectrum are even and that the imaginary part of its Fourier transform and its phase spectrum are odd, just as we would predict. It is not difficult to substantiate each of the statements in Table 3.7.1.

Let us now sketch the amplitude and phase spectra of some of the transforms calculated in earlier sections.

EXAMPLE 3.7.2

Let us sketch the real and imaginary parts of the Fourier transform and the amplitude and phase spectra for $f(t)$ given by Eq. (3.2.10) and shown in Fig. 3.7.3. The Fourier transform was found in Example 3.2.2 to be

$$F(\omega) = V\tau\frac{\sin\omega\tau/2}{\omega\tau/2}. \tag{3.7.9}$$

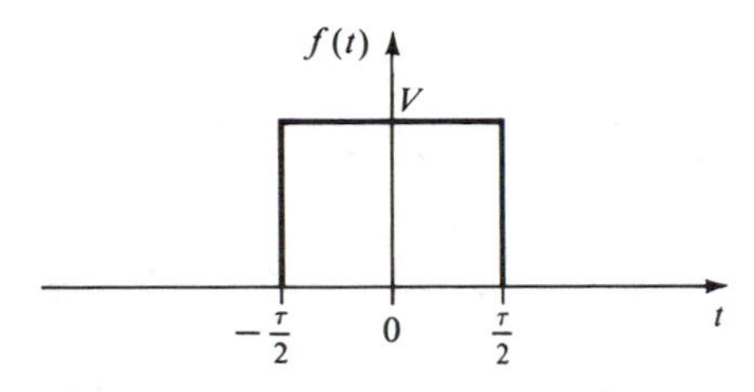

FIGURE 3.7.3 Time waveform for Example 3.7.2.

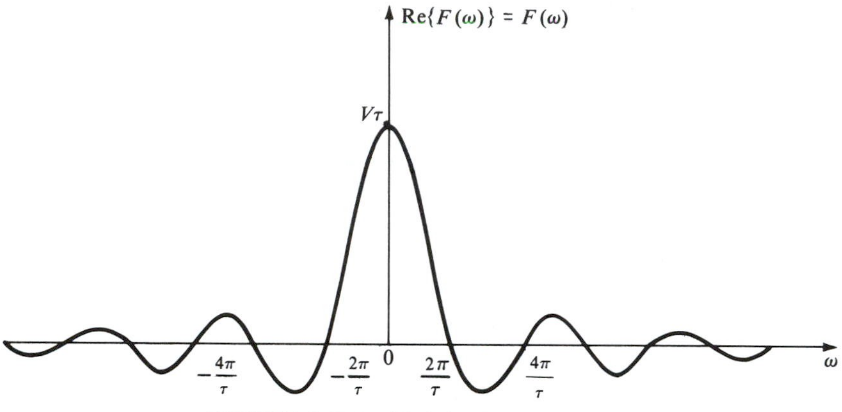

FIGURE 3.7.4 $F(\omega)$ for Example 3.7.2.

This Fourier transform is purely real and hence the real part is just $F(\omega)$ and the imaginary part is zero. The real part of $F(\omega)$ is sketched in Fig. 3.7.4. A function of the form of Eq. (3.7.9), which looks like the waveform in Fig. 3.7.4, is very common and is sometimes called the sampling function. This is the reason for the extra manipulations in Example 3.2.2 to obtain this form.

Only slightly more effort is required to obtain the amplitude and phase spectra. Since $F(\omega)$ is purely real, the amplitude spectrum is

$$|F(\omega)| = V\tau \left| \frac{\sin \omega\tau/2}{\omega\tau/2} \right| \tag{3.7.10}$$

for $V > 0$, and it is sketched in Fig. 3.7.5. Although the phase spectrum can almost be obtained by inspection, it is instructive to perform the required manipulations carefully. Using Eqs. (3.7.2) and (3.7.9), we have

$$\underline{/F(\omega)} = \tan^{-1} \frac{0}{\operatorname{Re}\{F(\omega)\}}. \tag{3.7.11}$$

From Eq. (3.7.11) it is tempting to conclude that the phase is zero for all ω; however, recall from similar situations in Chapter 2 that we must ascertain the sign of the denominator in order to determine the angle correctly.

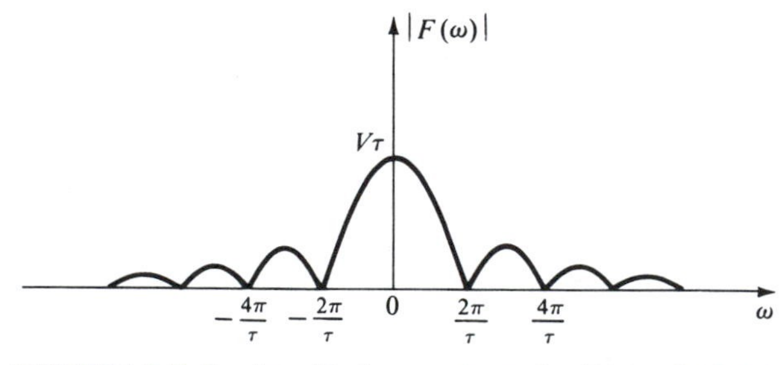

FIGURE 3.7.5 Amplitude spectrum for Example 3.7.2.

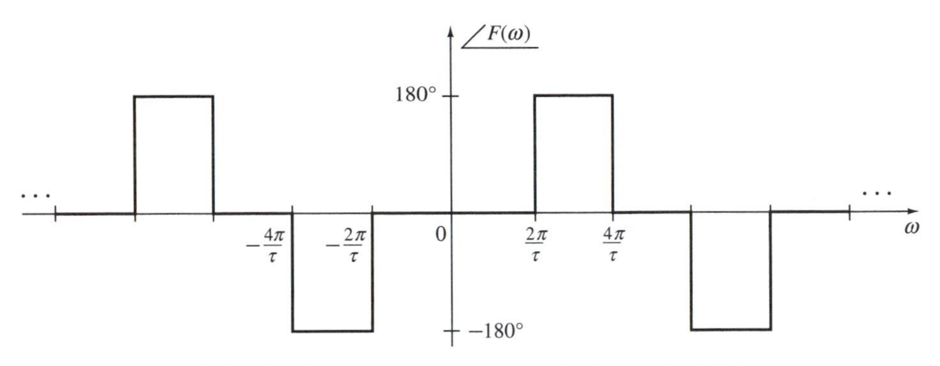

FIGURE 3.7.6 Phase spectrum for Example 3.7.2.

Since $\text{Re}\{F(\omega)\} = F(\omega)$, we see that the desired phase is

$$\underline{/F(\omega)} = \begin{cases} \tan^{-1} \dfrac{0}{+1} = 0°, & \text{for } F(\omega) \geq 0 \\ \tan^{-1} \dfrac{0}{-1} = \pm 180°, & \text{for } F(\omega) < 0 \end{cases} \tag{3.7.12}$$

since only the sign of the denominator is important. The phase spectrum is sketched in Fig. 3.7.6. The angle for $F(\omega) < 0$ is alternated between $+180°$ and $-180°$ by convention.

EXAMPLE 3.7.3

Let us display graphically the Fourier transforms of $\cos \omega_0 t$ and $\sin \omega_0 t$. Working with $\cos \omega_0 t$ first, we see from Eq. (3.5.18) that its Fourier transform is always real and positive, and thus $\text{Re}\{F(\omega)\} = |F(\omega)|$ and $\text{Im}\{F(\omega)\} = \underline{/F(\omega)} = 0$. Hence only the one plot shown in Fig. 3.7.7 is necessary to convey all of the information.

The Fourier transform for $\sin \omega_0 t$ is given by Eq. (3.5.19) and is purely imaginary. Therefore, only one plot is necessary again to display the Fourier transform information graphically, but this time the plot is of $\text{Im}\{F(\omega)\}$, which is sketched in Fig. 3.7.8.

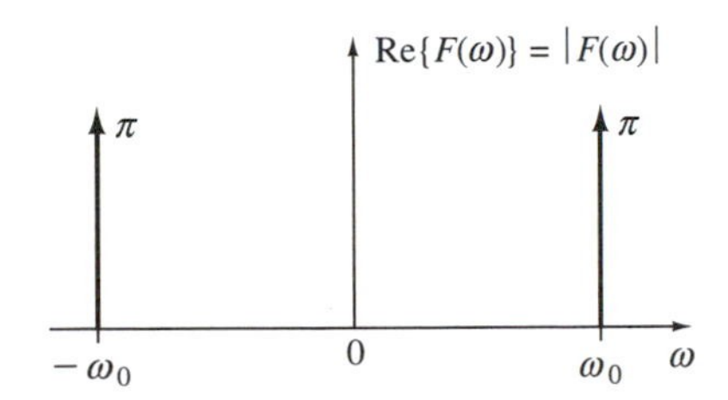

FIGURE 3.7.7 Amplitude spectrum of $\cos \omega_0 t$.

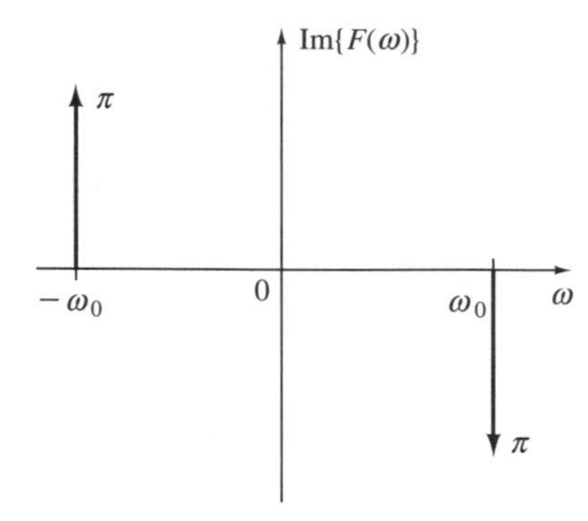

FIGURE 3.7.8 Imaginary part of $\mathscr{F}\{\sin \omega_0 t\}$.

The graphical display of Fourier transform information allows important comparisons to be made almost at a glance. The fundamental importance of the amplitude and phase spectra will become clearer as we progress through Chapter 4.

SUMMARY

The Fourier transforms of a wide variety of time signals were calculated in this chapter and properties of the Fourier transform were presented to enable us to obtain those transforms not specifically given here. The definition of the Fourier transform pair and an intuitive development of the Fourier transform were given in Section 3.2. In Section 3.3 it was shown that a sufficient condition for the Fourier transform to exist is that the time function be absolutely integrable. After developing some special functions in Section 3.4, we were able to determine the Fourier transform of many important signals that are not absolutely integrable in Section 3.5 by allowing the transforms to contain impulses. In Section 3.6 we focused on several important properties that can greatly simplify the evaluation of some Fourier transforms. The techniques for presenting Fourier transforms graphically were given in Section 3.7 and the utility of these graphical methods was discussed and demonstrated. We employ the transforms and transform methods contained in this chapter continually throughout the remainder of the book.

PROBLEMS

3.1 Find the Fourier transform of the function

$$f(t) = \begin{cases} V, & \text{for } 0 \le t \le \tau \\ 0, & \text{otherwise} \end{cases}$$

for $0 < V < \infty$ by direct application of Eq. (3.2.6).

3.2 Determine the Fourier transform of the double-sided exponential function $f(t) = e^{-|t|/\tau}$ for $-\infty < t < \infty$ by direct application of Eq. (3.2.6).

3.3 Find the Fourier transform of the function $f(t)$ shown in Fig. P3.3.

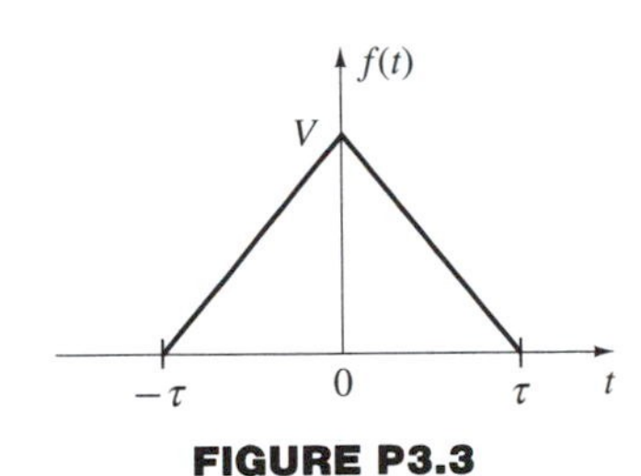

FIGURE P3.3

3.4 Find the Fourier transform of the function

$$f(t) = \begin{cases} \dfrac{2V}{\tau}\, t, & \text{for } -\dfrac{\tau}{2} \le t \le \dfrac{\tau}{2} \\ 0, & \text{otherwise.} \end{cases}$$

3.5 The Weierstrass M-test for integrals states that for $M(t)$ continuous in $a \le t \le \infty$ and $f(t, \omega)$ continuous in $a < t < \infty$ and $\omega_1 \le \omega \le \omega_2$ if

$$|f(t, \omega)| \le M(t)$$

for $\omega_1 \le \omega \le \omega_2$ and

$$\int_a^\infty M(t)\, dt$$

converges, then

$$\int_a^\infty f(t, \omega)\, dt$$

is uniformly and absolutely convergent for $\omega_1 \le \omega \le \omega_2$ [Kaplan, 1959]. Starting with the Fourier transform integral in Eq. (3.2.6), use the M-test to show that Eq. (3.3.4) is sufficient for Eq. (3.3.3) to hold.

Hint: Break the Fourier transform integral into a sum of two integrals and apply the M-test to each one.

3.6 Show that each of the following functions satisfies the Dirichlet conditions and hence has a Fourier transform: $f(t)$ in
(a) Problem 3.1.
(b) Problem 3.2.
(c) Problem 3.3.
(d) Problem 3.4.

3.7 Demonstrate that the function $f(t) = \sin \omega t$ for $\omega = 2\pi/T$ is not absolutely integrable.

3.8 In an heuristic fashion the waveform $\delta_\varepsilon(t)$ given in Fig. P3.8a can be shown to approach a delta function in the limit as $\varepsilon \rightarrow 0$. Show that the derivative of $\delta_\varepsilon(t)$ approaches the function illustrated in Fig. P3.8b, which is the doublet, as $\varepsilon \rightarrow 0$.

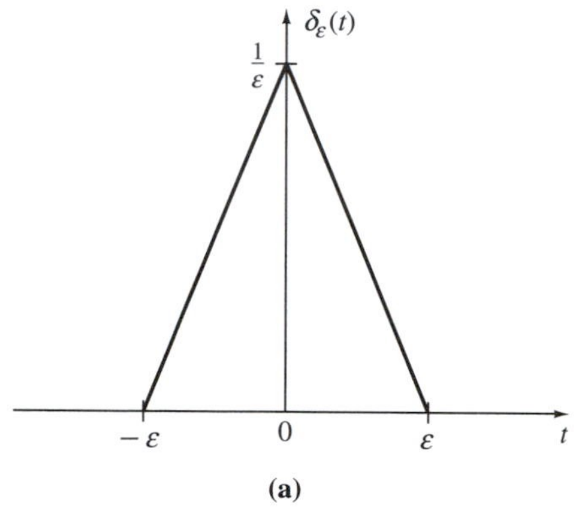

(a)

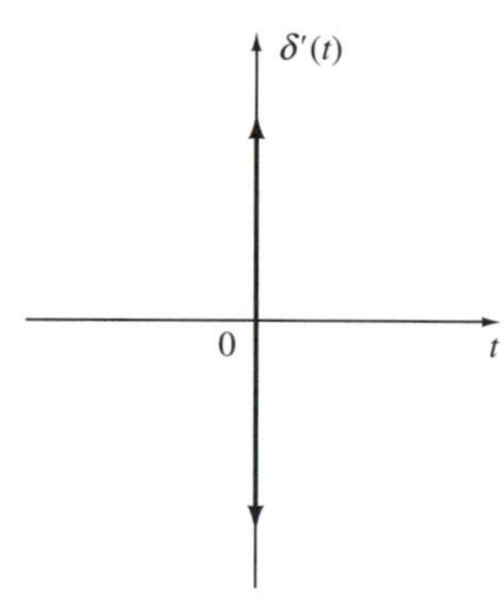

(b)

FIGURE P3.8

3.9 Evaluate the following integrals involving impulses.

(a) $\int_{-\infty}^{\infty} \delta(t)t^3\, dt$ **(b)** $\int_{0}^{\infty} \delta(t+1)e^{-t}\, dt$

(c) $\int_{-2}^{5} \delta(t-4)[t+1]^2\, dt$ **(d)** $\int_{-\infty}^{\infty} \delta(6t)e^{-3t}\, dt$

(e) $\int_{-\infty}^{\infty} \delta^{(1)}(t-2)[t^3-3t^2+1]\, dt$ **(f)** $\int_{-8}^{-1} \delta(t)\, dt$

(g) $\int_{-\infty}^{6} \delta(t-1)e^{-t/2}\, dt$

3.10 Derive the Fourier transform of a constant A that exists for all time, given by Eq. (3.5.7), using an approach identical to that employed in obtaining $\mathscr{F}\{\text{sgn}(t)\}$ in Eqs. (3.5.10)–(3.5.12).

Hint: Use L'Hospital's rule at $\omega = 0$.

3.11 Determine the Fourier transform of the periodic function in Fig. 2.5.1.

3.12 Find the Fourier transform of the periodic square wave in Fig. 2.7.1.

3.13 Use the results of Example 3.2.1 and the scaling property of the Fourier transform to find $\mathscr{F}\{e^{-t}u(t)\}$.

3.14 Determine the Fourier transform of $\delta(t - t_0)$ by using the time shifting property.

3.15 If the message signal $m(t)$ has the transform $M(\omega) = \mathscr{F}\{m(t)\}$, employ the frequency shifting property to find $\mathscr{F}\{m(t) \sin \omega_c t\}$.

3.16 Calculate the Fourier transform of $d(\sin \omega_0 t)/dt$ using Eq. (3.5.21) and the time differentiation property. Verify your result by the direct evaluation of $\mathscr{F}\{\omega_0 \cos \omega_0 t\}$.

3.17 Rework Problem 3.1 using the result of Example 3.2.2 and the time shifting property.

3.18 Determine the Fourier transform of the signal shown in Fig. P3.18 using the linearity and time shifting properties of the Fourier transform.

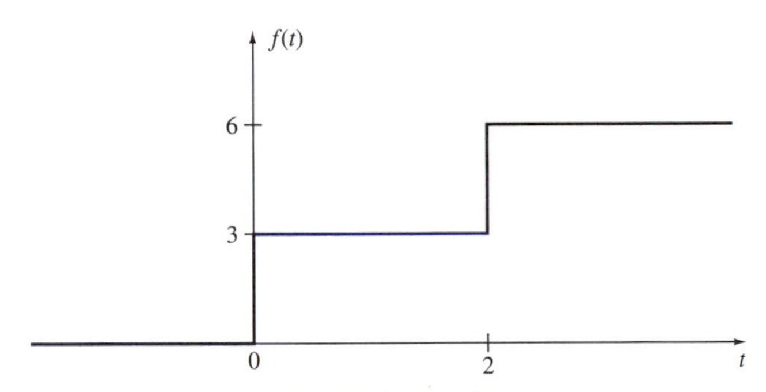

FIGURE P3.18

3.19 Compute the Fourier transform of the function

$$f(t) = \begin{cases} t + \dfrac{t^2}{2T}, & -T \le t < 0 \\ t - \dfrac{t^2}{2T}, & 0 \le t \le T \end{cases}$$

by using the linearity and time integration properties of the Fourier transform.

3.20 Show that the Fourier transform of the Gaussian pulse

$$f(t) = Ve^{-t^2/2\tau^2}, \qquad -\infty < t < \infty$$

is

$$F(\omega) = \tau V \sqrt{2\pi}\, e^{-\omega^2\tau^2/2}, \qquad -\infty < f < \infty,$$

which is also a Gaussian pulse.

3.21 Use the frequency shifting property and Eq. (3.2.8) to find the Fourier transform of

$$f(t) = e^{-t/\tau} \sin \omega_0 t \, u(t).$$

3.22 Determine the Fourier transform of the signal $f(t) = e^{-t/\tau} \cos \omega_0 t \, u(t)$. Let $\tau \to \infty$ and compare the resulting transform to that obtained by direct evaluation of $\mathscr{F}\{\cos \omega_0 t \, u(t)\}$.

3.23 Show that an arbitrary function can always be expressed as the sum of an even function $f_e(t)$ and an odd function $f_0(t)$. What are the even and odd components of $u(t)$?

3.24 Using the results of Problem 3.23 that any function can be written as $f(t) = f_e(t) + f_0(t)$, show that for $f(t)$ real (a) $\text{Re}\{F(\omega)\} = \mathscr{F}\{f_e(t)\}$, and (b) $j\,\text{Im}\{F(\omega)\} = \mathscr{F}\{f_0(t)\}$.

3.25 Show that $\mathscr{F}\{f^*(t)\} = F^*(-\omega)$.

Hint: Let $f(t) = f_r(t) + jf_i(t)$, where $f_r(t)$ is the real part of $f(t)$ and $f_i(t)$ is the imaginary part (both purely real).

3.26 Use the result of Problem 3.25 to prove that for $f(t)$ a general complex function:

(a) $\mathscr{F}\{f_r(t)\} = \dfrac{1}{2}[F(\omega) + F^*(-\omega)]$.

(b) $\mathscr{F}\{f_i(t)\} = \dfrac{1}{2j}[F(\omega) - F^*(-\omega)]$.

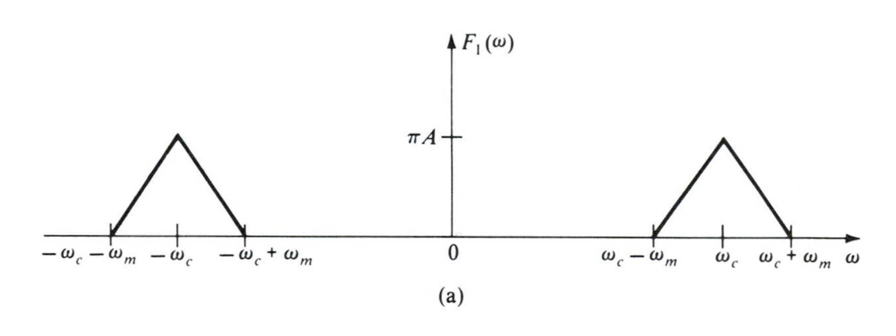

(a)

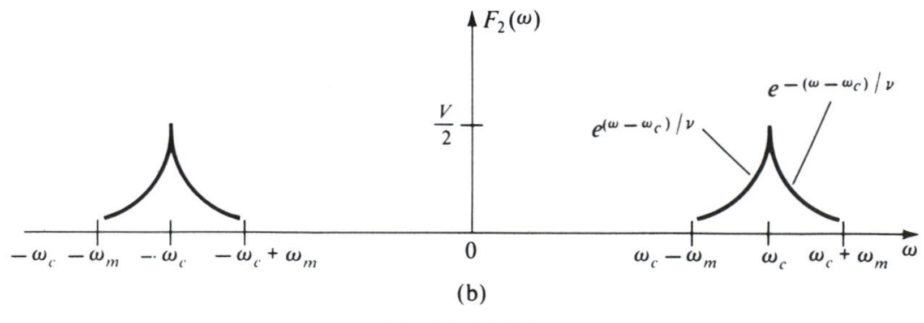

(b)

FIGURE P3.28

3.27 Find the Fourier transform of

$$f(t) = \begin{cases} V \cos \omega_0 t, & \text{for } -\dfrac{\tau}{2} \le t \le \dfrac{\tau}{2} \\ 0, & \text{otherwise} \end{cases}$$

by using the frequency shifting property and by direct evaluation.

3.28 Determine the time functions whose Fourier transforms are shown in Fig. P3.28.

3.29 Sketch the amplitude and phase spectra for $f(t)$ in Problem 3.1.

3.30 Repeat Problem 3.29 for $f(t)$ in Problem 3.4.

Linear Systems, Convolution, and Filtering

4.1 Introduction

The representation of signals in both the time and frequency domains is extremely important, and this is exactly why two chapters have been devoted to this subject. Of equal importance, however, is the ability to analyze and specify circuits and systems that operate on, or process, signals to achieve a desired result. It is our purpose in this chapter to develop several techniques for system representation and analysis that are necessary for the study of the communication systems to follow.

We focus our investigation on the input/output behavior of systems and how this behavior can be expressed in the time and frequency domains. The discussion begins in Section 4.2, where we define what is meant by a linear system. A time-domain analysis method for calculating the output of a linear system for a given input, called convolution, is presented in Section 4.3, followed in Section 4.4 by a development of a graphical approach for performing the required convolution. The topic of filters and their use in signal processing is covered in Sections 4.5 and 4.6, while the relationship between time response and system bandwidth is specified in Section 4.7. In Section 4.8 we discuss the important procedure of analog-to-digital conversion and the mathematical basis for such a procedure, the time-domain sampling theorem. Finally, the vital concepts of power and energy are defined in Section 4.9 for later use in system evaluation and comparison.

4.2 Linear Systems

All the discussions in this chapter are limited to linear time-invariant systems. Although this class of systems does not encompass all the systems that we will come in contact with while studying communication systems, there are several sound reasons for limiting consideration to such systems. First, many physical systems are linear and time invariant or can be approximated accurately by linear time-invariant systems over some region of interest. Second, linear time-invariant systems are more easily analyzed than are other types of systems, and the analyses remain valid for very general conditions on the systems. Third, a thorough understanding of linear time-invariant systems is necessary before proceeding to study more advanced time-varying or nonlinear circuits and systems.

The adjectives *linear* and *time invariant* specify two separate and distinct properties of a system. We now define clearly what is meant by the terms *linear* and *time invariant*. There are numerous ways to determine whether a system is linear or nonlinear. One approach is to investigate the differential equation that represents the system. If all derivatives of the input and output are raised only to the first power and there are no products of the input and output or their derivatives, the system is said to be linear. An alternative definition requires only that we be able to measure or compute the output response for different inputs rather than the describing differential equation.

Definition Given that $y_1(t)$ and $y_2(t)$ are the system output responses to inputs $r_1(t)$ and $r_2(t)$, respectively, a system is said to be *linear* if the input signal $ar_1(t) + br_2(t)$ produces the system output response $ay_1(t) + by_2(t)$, where a and b are constants.

The reader may recognize this definition as the statement of a possibly more familiar concept called *superposition.*

The decision as to whether a system is time invariant or time varying can also be made by inspecting the system differential equation. If any of the coefficients in the differential equation are a function of time, the system is time varying. If the coefficients are all constants, the system is time invariant. Time invariance can also be established by observing the system output for a given input applied at different time instances, as stated by the following definition.

Definition Given that the output response of a system is $y_1(t)$ for an input $r(t)$ applied at $t = t_0$, a system is said to be *time invariant* if the output is $y_2(t) = y_1(t - t_1)$ for an input signal $r(t - t_1)$ applied at time $t = t_0 + t_1$.

In words, this definition says that a system is time invariant if the shape of the output response is the same regardless of when, in time, the input is applied.

Unless it is possible to measure the system response for various inputs in a laboratory, it is necessary that we work with a mathematical model of the system. There are four mathematical operations that are the basic building blocks of our mathematical models. These operations are scalar multiplication, differentiation, integration, and time delay, and the outputs for an input $r(t)$ applied at $t = t_0$ for systems that perform these operations are indicated as follows:

Scalar Multiplication

$$y(t) = \alpha r(t), \tag{4.2.1}$$

where α is an arbitrary constant.

Differentiation

$$y(t) = \frac{d}{dt} r(t) \tag{4.2.2}$$

Integration

$$y(t) = \int_{t_0}^{t} r(\tau)\, d\tau \tag{4.2.3}$$

Time Delay

$$y(t) = r(t - t_1) \tag{4.2.4}$$

Each of these operations describes a linear time-invariant system. The proofs of this statement are left as exercises (see Problems 4.1 and 4.2).

Let us investigate for a moment the implications of linearity or superposition. If a system is linear, the output response due to a sum of several inputs is the sum of the responses due to each individual input. Linear time-invariant systems also possess the significant property that the system output can contain only those frequencies that are present in the input. That is, no new frequencies are generated. The combination of the superposition and frequency preservation properties will prove to be of fundamental importance in determining the response of a linear system.

4.3 Linear Systems Response: Convolution

We now direct our attention toward the calculation of the output response of a linear time-invariant system to a given input waveform or excitation. We would like to keep the development as general as possible, so that the final result will be valid over a wide range of excitation signals and systems. Toward this end, consider the general waveform shown in Fig. 4.3.1, which is defined for $-\infty < t < \infty$. This signal is applied to the input of some general linear time-invariant system, and we wish to obtain an expression for the system output $y(t)$ in terms of $r(t)$ and some as yet unspecified characteristic of the system.

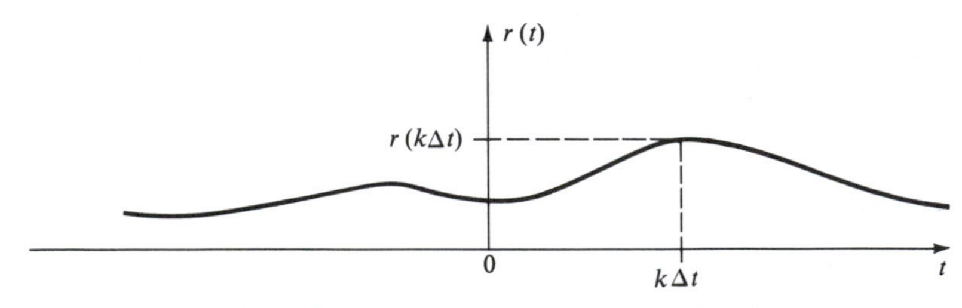

FIGURE 4.3.1 General excitation signal.

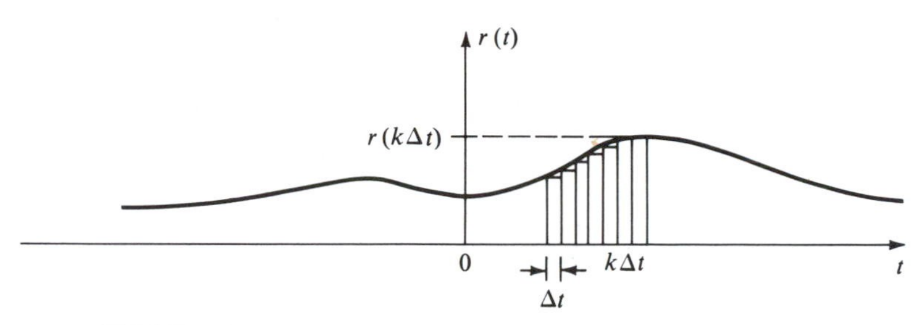

FIGURE 4.3.2 Approximation of $r(t)$ by a sum of pulses.

To do this, the signal $r(t)$ is written as an infinite sum of pulses of width Δt, as illustrated by Fig. 4.3.2. Limiting consideration to only the input pulse occurring at time $k\Delta t$ and assuming that the system response to such a pulse has the shape denoted by $h(\cdot)$, the output response to this one input pulse is

$$y(t) = r(k\Delta t)h(t - k\Delta t)\,\Delta t. \tag{4.3.1}$$

Equation (4.3.1) is the output due only to the pulse at $t = k\Delta t$ and consists of the magnitude of the input pulse times the response of the system to the pulse delayed by $k\Delta t$ multiplied by the duration of the input. Since the system is assumed to be linear, it obeys superposition and the total output can be obtained by summing the responses to each of the component pulses of $r(t)$. Hence we have

$$y(t) = \sum_{k=-\infty}^{\infty} r(k\Delta t)h(t - k\Delta t)\,\Delta t. \tag{4.3.2}$$

However, for the sequence of pulses of width Δt to represent $r(t)$ exactly, it is necessary that $\Delta t \to 0$. The final expression for the system output response is thus

$$y(t) = \lim_{\Delta t \to 0} \sum_{k=-\infty}^{\infty} r(k\Delta t)h(t - k\Delta t)\,\Delta t, \tag{4.3.3}$$

which formally yields

$$y(t) = \int_{-\infty}^{\infty} r(\tau)h(t - \tau)\,d\tau. \tag{4.3.4}$$

Equation (4.3.4) is an extremely important result and is usually called the *convolution integral.* Since Eq. (4.3.4) was derived by letting $\Delta t \to 0$, $h(\cdot)$ is thus the system response to an impulse input, and $h(t)$ is called the *impulse response* of the system being considered. Therefore, if the input signal and the system impulse response are known, the output can be calculated by the convolution operation indicated in Eq. (4.3.4). The symbol $*$ is commonly used as a shorthand notation for the convolution operation, and hence Eq. (4.3.4) can also be written as $y(t) = r(t) * h(t)$.

Before illustrating the use of the convolution integral, let us consider further the impulse response, $h(t)$. A more quantitative indication of why $h(t)$ is called the impulse response can be produced by returning to Eq. (4.3.4) and letting

$r(\tau) = \delta(\tau)$, so that we are calculating the system output response to an impulse excitation. By applying the sifting property of the impulse, the output is found as

$$y(t) = \int_{-\infty}^{\infty} \delta(\tau)h(t - \tau)\, dt = h(t). \tag{4.3.5}$$

Equation (4.3.5) makes the issue crystal clear; that is, $h(t)$ is called the impulse response because it is the system response to a unit impulse excitation.

One additional question concerning the convolution integral should be addressed. That is, Eq. (4.3.4) seems to be a very general expression for the output of a linear, time-invariant system, yet it depends on the system response to a particular input signal, the unit impulse. Does the impulse function have some property that uniquely qualifies it for the job, or could Eq. (4.3.4) just as well have been written in terms of the unit step response or response to some other excitation function? The answer is that the unit impulse does possess a special property that is evident from inspecting its Fourier transform in Eq. (3.5.1). The unit impulse contains all frequencies of equal magnitudes! Therefore, when we know the impulse response of a system, we have information concerning the system response to all frequencies. This fact emphasizes the importance of $h(t)$ in system response studies and further justifies its presence in Eq. (4.3.4).

EXAMPLE 4.3.1

The (voltage) impulse response of the simple RC network shown in Fig. 4.3.3 is given by

$$h(t) = \begin{cases} \dfrac{1}{RC} e^{-t/RC}, & \text{for } t > 0 \\ 0, & \text{for t} < 0. \end{cases} \tag{4.3.6}$$

We desire to determine the output response of this circuit to the input pulse

$$r(t) = \begin{cases} 1, & \text{for } 0 < t \leq 2 \\ 0, & \text{otherwise.} \end{cases} \tag{4.3.7}$$

Using unit step functions to express $r(t)$ and $h(t)$, we have by direct substitution into Eq. (4.3.4) that

$$\begin{aligned} y(t) &= \int_{-\infty}^{\infty} [u(\tau) - u(\tau - 2)] \frac{1}{RC} e^{-(t-\tau)/RC} u(t - \tau)\, d\tau \\ &= \frac{1}{RC} \int_{-\infty}^{\infty} [u(\tau)u(t - \tau) - u(\tau - 2)u(t - \tau)]e^{-(t-\tau)/RC}\, d\tau \\ &= \frac{1}{RC} \int_{0}^{\infty} u(t - \tau)e^{-(t-\tau)/RC}\, d\tau - \frac{1}{RC} \int_{2}^{\infty} u(t - \tau)e^{-(t-\tau)/RC}\, d\tau. \end{aligned} \tag{4.3.8}$$

Notice that there are three different intervals for t which must be considered. For $t < 0$, $y(t) = 0$, since the factor $u(t - \tau)$ is zero for these values of

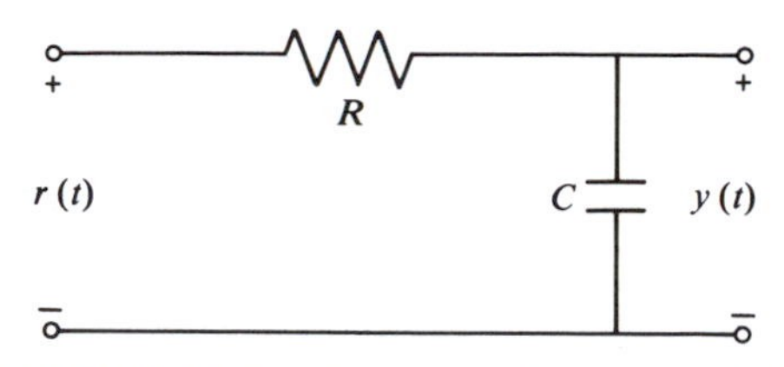

FIGURE 4.3.3 Circuit for Example 4.3.1.

t. For $0 \le t < 2$, Eq. (4.3.8) yields

$$y(t) = \frac{1}{RC}\int_0^t e^{-(t-\tau)/RC}\,d\tau = [1 - e^{-t/RC}]. \tag{4.3.9}$$

When $t \ge 2$, the output is given by

$$y(t) = \frac{1}{RC}\int_0^2 e^{-(t-\tau)/RC}\,d\tau = e^{-t/RC}[e^{2/RC} - 1]. \tag{4.3.10}$$

Summarizing the results, we find that

$$y(t) = \begin{cases} 0, & \text{for } t < 0 \\ [1 - e^{-t/RC}], & \text{for } 0 \le t < 2 \\ e^{-t/RC}[e^{2/RC} - 1], & \text{for } t \ge 2. \end{cases} \tag{4.3.11}$$

Just as for the mathematical operation of multiplication, it is possible to show that the convolution operation is commutative, distributive, and associative. These properties are stated in the following paragraphs.

Commutative Law

$$r(t) * h(t) = h(t) * r(t) \tag{4.3.12}$$

Proof: Use a change of variables.

Distributive Law

$$r(t) * [h_1(t) + h_2(t)] = r(t) * h_1(t) + r(t) * h_2(t) \tag{4.3.13}$$

Proof: Straightforward.

Associative Law

$$r(t) * [h_1(t) * h_2(t)] = [r(t) * h_1(t)] * h_2(t) \tag{4.3.14}$$

Proof: Deferred until later in this section.

Although we have a straightforward method of calculating the system response given the impulse response and the input signal, the evaluation of the convolution integral sometimes may be very challenging. In many cases, difficulties of this sort can be avoided by employing what is called the *time convolution theorem.*

Time Convolution Theorem

If

$$r(t) \leftrightarrow R(\omega)$$

and

$$h(t) \leftrightarrow H(\omega),$$

then

$$\mathscr{F}\{r(t) * h(t)\} = R(\omega)H(\omega). \tag{4.3.15}$$

Proof: From the definition of the Fourier transform,

$$\begin{aligned}\mathscr{F}\{r(t) * h(t)\} &= \mathscr{F}\left\{\int_{-\infty}^{\infty} r(\tau)h(t-\tau)\,d\tau\right\} \\ &= \int_{-\infty}^{\infty}\left\{\int_{-\infty}^{\infty} r(\tau)h(t-\tau)\,d\tau\right\}e^{-j\omega t}\,dt.\end{aligned} \tag{4.3.16}$$

If we interchange the order of integration and make the change of variables $\lambda = t - \tau$, so that $d\lambda = dt$ in the inner integral,

$$\begin{aligned}\mathscr{F}\{r(t) * h(t)\} &= \int_{-\infty}^{\infty} r(\tau)\left\{\int_{-\infty}^{\infty} h(\lambda)e^{-j\omega(\lambda+\tau)}\,d\lambda\right\}d\tau \\ &= \int_{-\infty}^{\infty} r(\tau)\left\{\int_{-\infty}^{\infty} h(\lambda)e^{-j\omega\lambda}\,d\lambda\right\}e^{-j\omega\tau}\,d\tau \\ &= H(\omega)\int_{-\infty}^{\infty} r(\tau)e^{-j\omega\tau}\,d\tau = R(\omega)H(\omega)\end{aligned}$$

upon employing the definition of the Fourier transform twice.

The time convolution theorem states that the Fourier transform of the convolution of two time functions is the product of their Fourier transforms. Therefore, if we know the impulse response of a system and we want to calculate the output response to a given input signal, one approach is to compute the Fourier transforms of the input and $h(t)$, form the product of these transforms, and then find the inverse transform. The power of the time convolution theorem is that frequently this last step is not necessary, since the frequency content of the response may be all the information required in many situations.

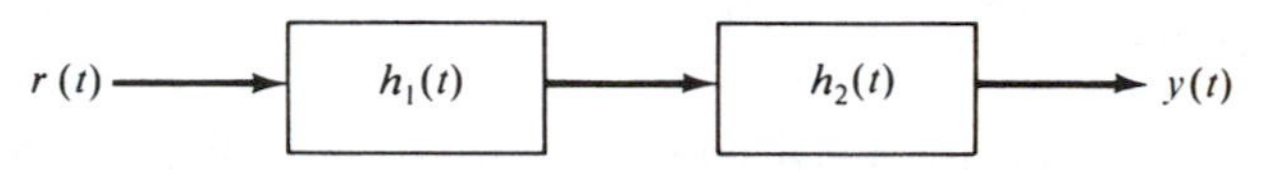

FIGURE 4.3.4 Two systems in cascade.

The time convolution theorem underscores the utility of the Fourier transform in systems analysis. For instance, suppose that it is desired to find the output response of two systems with impulse responses $h_1(t)$ and $h_2(t)$ that are connected in cascade as illustrated by Fig. 4.3.4. If the input signal is $r(t)$, we could, of course, use the associative law and perform two successive convolution operations. However, an alternative approach is simply to multiply the respective Fourier transforms of $r(t)$, $h_1(t)$, and $h_2(t)$ to obtain $Y(\omega) = \mathscr{F}\{y(t)\}$, and then, if necessary, take the inverse transform to get $y(t)$. We utilize such an approach repeatedly in the later chapters.

Just as it is possible to define the convolution operation in the time domain, it is also meaningful to consider the convolution of two signals in the frequency domain, as indicated by the following theorem.

Frequency Convolution Theorem

If

$$f(t) \leftrightarrow F(\omega)$$

and

$$g(t) \leftrightarrow G(\omega),$$

then

$$\mathscr{F}\{f(t)g(t)\} = \frac{1}{2\pi} F(\omega) * G(\omega). \tag{4.3.17}$$

Proof: Left as an exercise (see Problem 4.9).

Since most communication systems require the multiplication of two time signals, Eq. (4.3.17) is extremely useful for the analysis of these systems.

Let us now return to the proof of the associative law of convolution given by Eq. (4.3.14). Letting $\mathscr{F}\{r(t)\} = R(\omega)$, $\mathscr{F}\{h_1(t)\} = H_1(\omega)$, and $\mathscr{F}\{h_2(t)\} = H_2(\omega)$ and using the time convolution theorem to take the Fourier transform of the left-hand side of Eq. (4.3.14) yields

$$\begin{aligned} \mathscr{F}\{r(t) * [h_1(t) * h_2(t)]\} &= R(\omega)[H_1(\omega)H_2(\omega)] \\ &= [R(\omega)H_1(\omega)]H_2(\omega), \end{aligned} \tag{4.3.18}$$

where the last step is possible, since multiplication is associative. Taking the inverse transform of both sides of Eq. (4.3.18) establishes Eq. (4.3.14).

4.4 Graphical Convolution

As might be expected, graphical convolution consists of performing exactly those operations indicated by the convolution integral graphically. Consider the convolution of the two time functions $f_1(t)$ and $f_2(t)$ given by

$$f_1(t) * f_2(t) = \int_{-\infty}^{\infty} f_1(\tau) f_2(t - \tau)\, d\tau. \tag{4.4.1}$$

The first operation involved in Eq. (4.4.1) is that of replacing t in $f_2(t)$ by $-\tau$, which constitutes a reflection about the $\tau = 0$ axis. The result of this first operation is then shifted by an amount of t seconds. Next, after letting $t = \tau$ in $f_1(t)$, the product of $f_1(\tau)$ and $f_2(t - \tau)$ is formed to obtain $f_1(\tau) \cdot f_2(t - \tau)$. Finally, this product is integrated over all values of τ, which is equivalent to computing the *area* under the *product*. These operations are repeated for all possible values of t to produce the total convolution waveform represented by Eq. (4.4.1). This last point must be emphasized. The analytical convolution in Eq. (4.4.1) yields a function that is defined for all values of t. Therefore, when performing the convolution graphically, it is necessary to calculate the area under the product for $-\infty < t < \infty$. To illustrate the concept of graphical convolution, consider the following brief example.

EXAMPLE 4.4.1

We desire to convolve $f_1(t)$ graphically in Fig. 4.4.1 with $f_2(t)$ in Fig. 4.4.2. Replacing t in $f_2(t)$ with $-\tau$ and letting $t = \tau$ in $f_1(t)$, we are ready to begin the calculation. Shifting $f_2(-\tau)$ by an amount t and letting t increase from negative infinity, we find that the product $f_1(\tau) f_2(t - \tau) = 0$ for $-\infty < t < 0.5$.

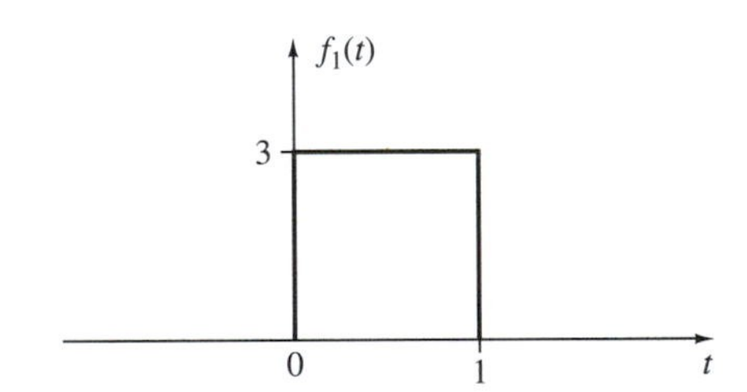

FIGURE 4.4.1 $f_1(t)$ for Example 4.4.1.

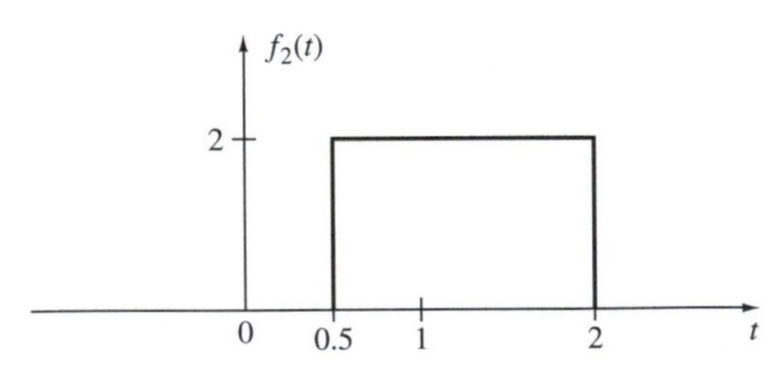

FIGURE 4.4.2 $f_2(t)$ for Example 4.4.1.

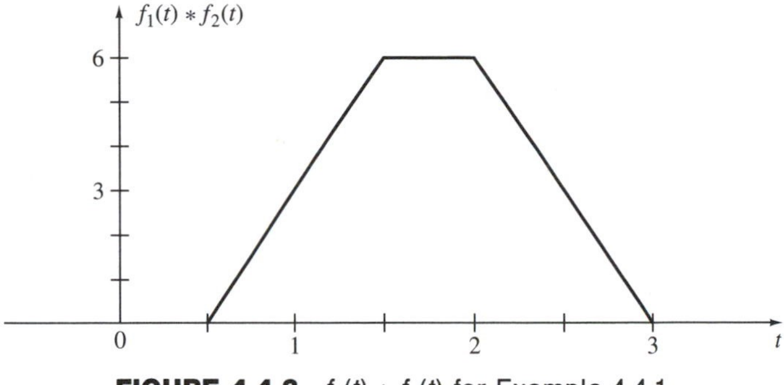

FIGURE 4.4.3 $f_1(t) * f_2(t)$ for Example 4.4.1.

Further, over the interval $0.5 \leq t \leq 3$, $f_1(\tau)f_2(t-\tau) \neq 0$ while for $3 < t < \infty$, the product is again zero. The remaining details are left to the reader. The final result for all t is shown in Fig. 4.4.3.

Although there are situations when it is advantageous to perform the convolution graphically and there are occasions when the direct analytical approach is preferred, it often turns out that a combination of both methods is most efficient. The graphical approach can be used first to determine the values of t over which the convolution is nonzero and to get an idea of the shape of the resulting waveform. The convolution can then be computed precisely using the limits deduced from the graphical analysis to evaluate the convolution integral.

4.5 Ideal Filters

The word *filter* is used by electrical engineers to denote a circuit or system that exhibits some sort of frequency selective behavior. Of course, every circuit or system fits this description to some extent; however, if for the specific application being considered all frequencies within the interval of interest are passed undistorted, the circuit or system is not acting as a filter. Before proceeding, however, it is necessary to specify exactly what we mean when we say that a signal is passed undistorted. For distortionless transmission through a circuit or system, we require that the exact input signal *shape* be reproduced at the output. It is not important whether or not the exact amplitude of the signal is preserved, and within reasonable limits, we do not care if the signal is delayed in time. Only the shape of the input must be passed unchanged. Therefore, for distortionless transmission of an input signal $r(t)$, the output is

$$y(t) = Ar(t - t_d). \tag{4.5.1}$$

Notice that Eq. (4.5.1) is a combination of the scalar multiplication and time-delay operations in Section 4.2, and can be shown to describe a linear time-invariant system.

Taking the Fourier transform of Eq. (4.5.1) with $\mathscr{F}\{y(t)\} = Y(\omega)$ and $\mathscr{F}\{r(t)\} = R(\omega)$ yields

$$Y(\omega) = AR(\omega)e^{-j\omega t_d}. \tag{4.5.2}$$

We know from the time convolution theorem that we also have

$$Y(\omega) = H(\omega)R(\omega), \tag{4.5.3}$$

where for $h(t)$ the impulse response of the system, $H(\omega) = \mathscr{F}\{h(t)\}$ and is usually called the *system transfer function.* By comparing Eqs. (4.5.2) and (4.5.3), we see that for distortionless transmission the system must have

$$H(\omega) = Ae^{-j\omega t_d} \tag{4.5.4}$$

over the frequency range of interest.

Filters are usually characterized as low pass, high pass, bandpass, or band stop, all of which refer to the shape of the amplitude spectrum of the filter's impulse response (transfer function). Drawing on the previous results concerning distortionless transmission, we find that an ideal low-pass filter (LPF) is defined by the frequency characteristic

$$H_{\text{LPF}}(\omega) = \begin{cases} Ae^{-j\omega t_d}, & \text{for } |\omega| \le \omega_s \\ 0, & \text{otherwise.} \end{cases} \tag{4.5.5}$$

The amplitude and phase of $H_{\text{LPF}}(\omega)$ are sketched in Figs. 4.5.1 and 4.5.2. Thus an ideal low-pass filter passes without distortion all input signal components with radian frequencies below ω_s, which is called the cutoff frequency. All signal components above the cutoff frequency are rejected.

The impulse response of the ideal LPF can be found by taking the inverse transform of Eq. (4.5.5), which yields

$$h_{\text{LPF}}(t) = \frac{A \sin \omega_s(t - t_d)}{\pi(t - t_d)} \tag{4.5.6}$$

by using the Fourier transform tables in Appendix K and the time shifting property. The impulse response has the form shown in Fig. 4.5.3 for $t_d \gg 1/2f_s$. One example of an application of the ideal LPF is given in the following example.

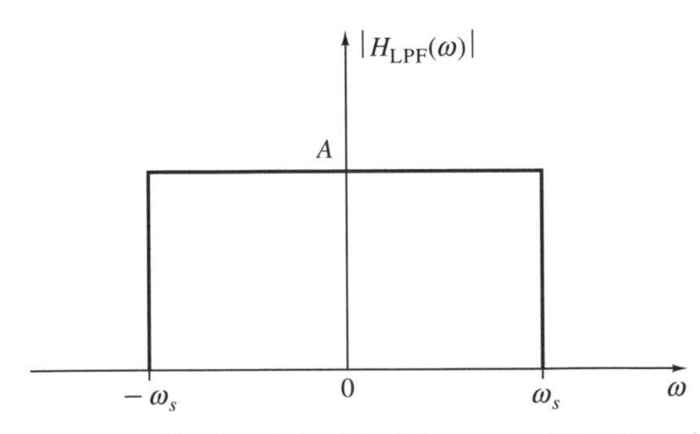

FIGURE 4.5.1 Amplitude of the ideal low-pass filter transfer function.

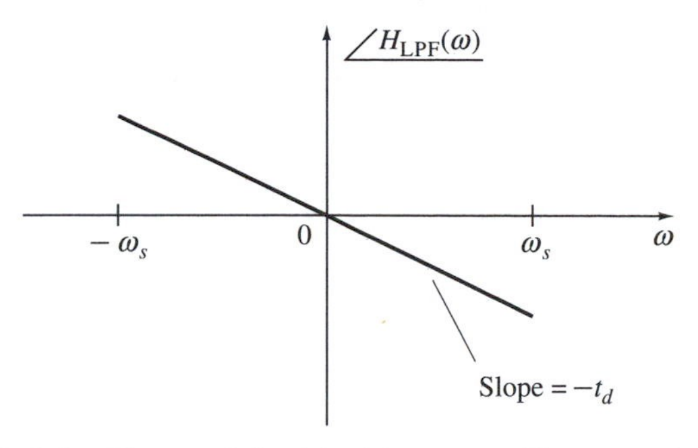

FIGURE 4.5.2 Phase of the ideal low-pass filter transfer function.

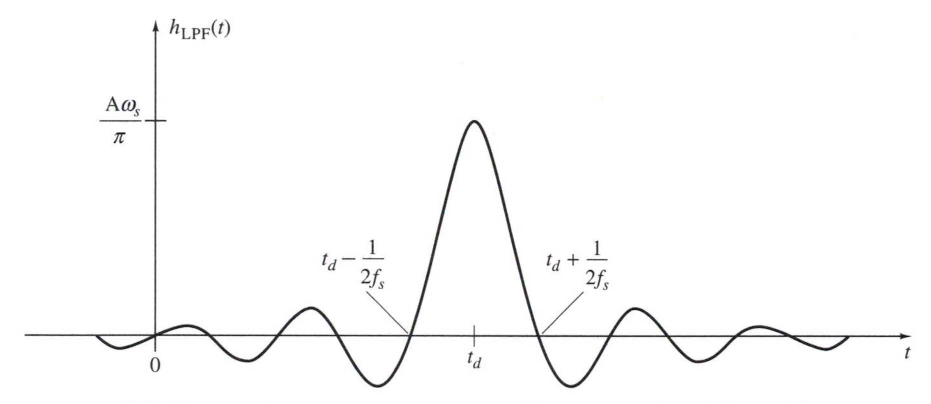

FIGURE 4.5.3 Impulse response of the ideal LPF with $t_d \gg 1/2f_s$.

EXAMPLE 4.5.1

A signal $f(t)$ is applied to the input of an ideal LPF with $A = 2$, $t_d = 0$, and a cutoff frequency of $\omega_s = 40$ rad/sec. The magnitude of the transfer function of the filter is shown in Fig. 4.5.4. We would like to determine the output of this filter if the input signal is given by

$$r(t) = 3 + \sin 3t + \sin 12t - \cos 30t + 5 \cos 47t + \sin 85t + 2 \sin 102t + \cos 220t + \sin 377t. \tag{4.5.7}$$

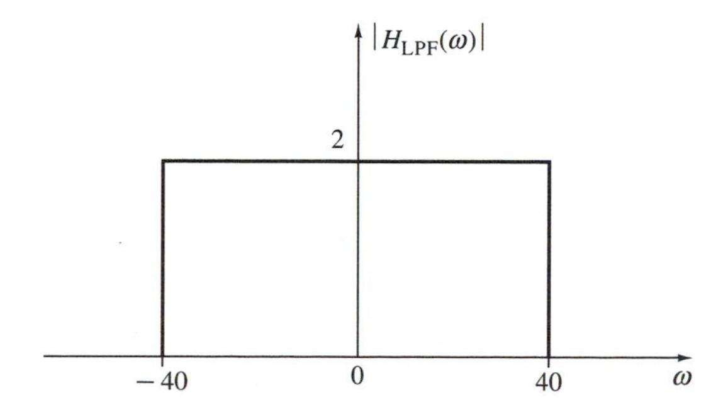

FIGURE 4.5.4 Magnitude of the LPF transfer function for Example 4.5.1.

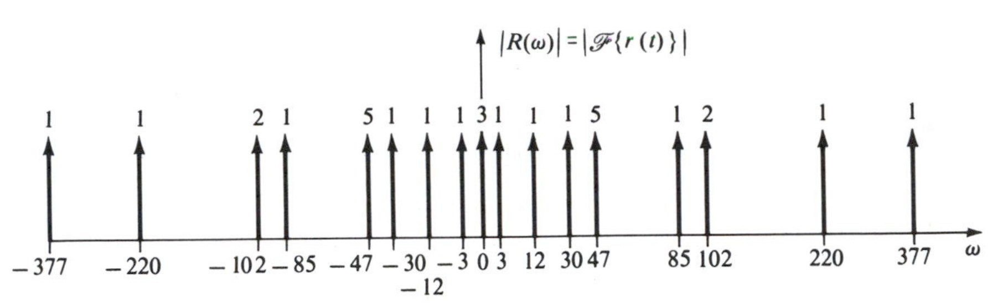

FIGURE 4.5.5 Amplitude spectrum of the input for Example 4.5.1.

Taking the Fourier transform of $r(t)$, we find that the amplitude spectrum of $r(t)$ is as illustrated in Fig. 4.5.5.

From the time convolution theorem, we know that if $y(t)$ is the output waveform, then

$$\begin{aligned}\mathscr{F}\{y(t)\} &= Y(\omega) = H_{\text{LPF}}(\omega)R(\omega) \\ &= |H_{\text{LPF}}(\omega)|\,|R(\omega)|e^{j[\underline{/H_{\text{LPF}}(\omega)} + \underline{/R(\omega)}]}.\end{aligned} \tag{4.5.8}$$

If we are only interested in the magnitudes of the various components in the output, we have that

$$|Y(\omega)| = |H_{\text{LPF}}(\omega)| \cdot |R(\omega)|. \tag{4.5.9}$$

By performing the multiplication in Eq. (4.5.9) graphically, $|Y(\omega)|$ is found to be as shown in Fig. 4.5.6. Hence the output contains a dc term and components at ± 3, ± 12, and ± 30 rad/sec. All terms with frequencies above 40 rad/sec have been eliminated.

It is not possible to determine whether the impulses at ± 3, ± 12, and ± 30 rad/sec represent a sine or cosine function and whether the terms are positive or negative from Fig. 4.5.6 without also computing their phases. However, if we return to Eq. (4.5.8) and realize that since $t_d = 0$, $\underline{/H_{\text{LPF}}(\omega)} = 0$, we see that the phase of each component is unchanged by the filter. The filter output can thus be found from Eq. (4.5.7) to be

$$y(t) = 6 + 2 \sin 3t + 2 \sin 12t - 2 \cos 30t. \tag{4.5.10}$$

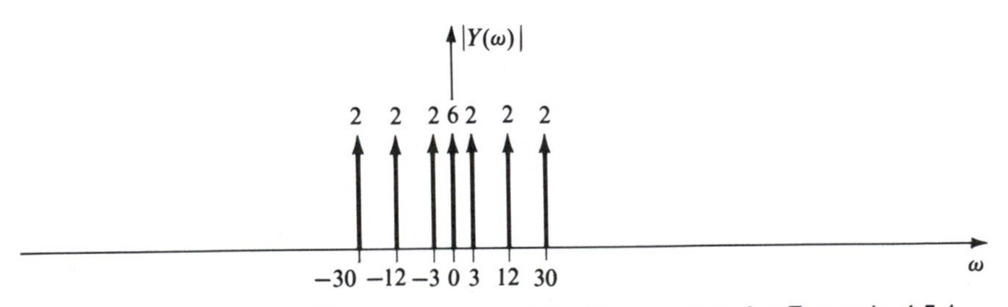

FIGURE 4.5.6 Amplitude spectrum of the filter output for Example 4.5.1.

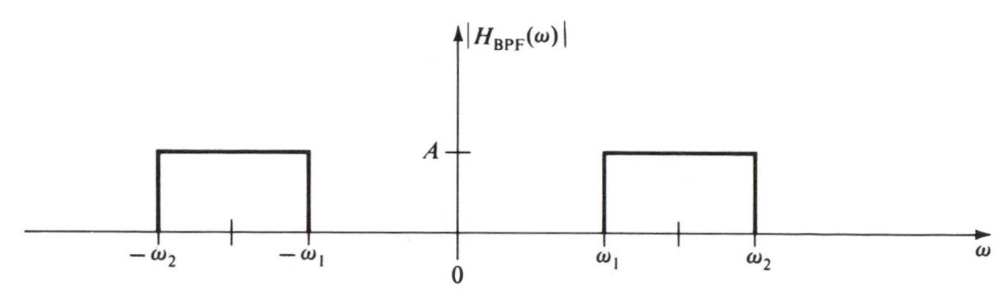

FIGURE 4.5.7 Amplitude of the ideal bandpass filter transfer function.

There are many situations in communication systems where it is necessary to pass signals that have frequency components between two nonzero frequencies, say ω_1 and ω_2, and reject all signals outside this range. A system that accomplishes this operation is called a bandpass filter. The transfer function of an ideal bandpass filter (BPF) is given by

$$H_{BPF}(\omega) = \begin{cases} Ae^{-j\omega t_d}, & \text{for } \omega_1 \leq |\omega| \leq \omega_2 \\ 0, & \text{otherwise.} \end{cases} \tag{4.5.11}$$

The magnitude and phase of Eq. (4.5.11) are shown in Figs. 4.5.7 and 4.5.8, respectively. From these figures and Eq. (4.5.11), we see that the ideal BPF passes undistorted any input signal components in the range $\omega_1 \leq |\omega| \leq \omega_2$ and rejects all other components.

The impulse response of the ideal BPF can be computed quite simply by noting that $H_{BPF}(\omega)$ is just

$$H_1(\omega) = \begin{cases} Ae^{-j\omega t_d}, & \text{for } |\omega| \leq \dfrac{\omega_2 - \omega_1}{2} \\ 0, & \text{otherwise,} \end{cases} \tag{4.5.12}$$

shifted in frequency by the amounts $\pm[(\omega_1 + \omega_2)/2]$. Hence

$$H_{BPF}(\omega) = H_1\left[\omega + \frac{\omega_1 + \omega_2}{2}\right] + H_1\left[\omega - \frac{\omega_1 + \omega_2}{2}\right]. \tag{4.5.13}$$

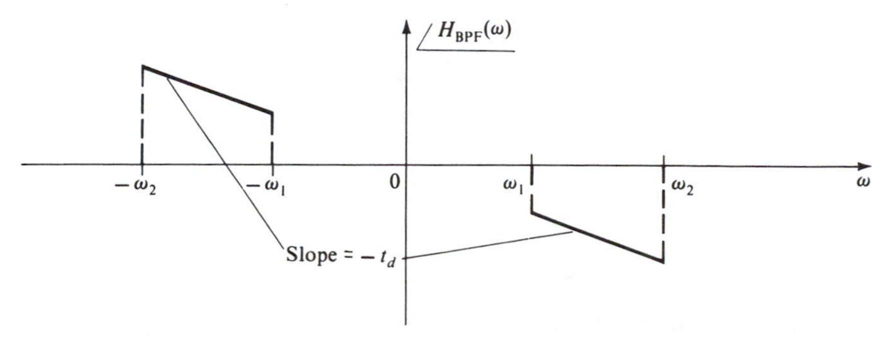

FIGURE 4.5.8 Phase of the ideal bandpass filter transfer function.

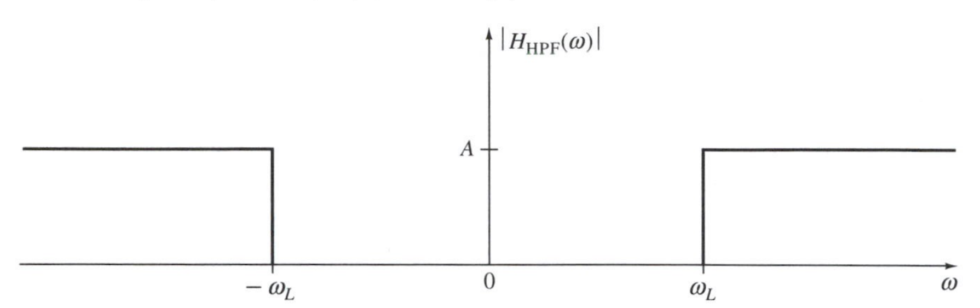

FIGURE 4.5.9 Amplitude characteristic of an ideal high-pass filter.

If we use the frequency shifting property and the table of Fourier transforms in Appendix K at the end of the book, the impulse response of the ideal BPF can be found from Eq. (4.5.13) to be

$$h_{\text{BPF}}(t) = \frac{2A}{\pi} \cdot \frac{\sin\left[\left(\frac{\omega_2 - \omega_1}{2}\right)(t - t_d)\right]}{(t - t_d)} \cdot \cos\left[\frac{\omega_1 + \omega_2}{2}(t - t_d)\right]. \quad (4.5.14)$$

The magnitude and phase characteristics of an ideal high-pass filter (HPF) are shown in Figs. 4.5.9 and 4.5.10, respectively, and the entire transfer function is given by

$$H_{\text{HPF}}(\omega) = \begin{cases} Ae^{-j\omega t_d}, & \text{for } |\omega| > \omega_L \\ 0, & \text{otherwise.} \end{cases} \quad (4.5.15)$$

The ideal HPF rejects all input signal components at frequencies less than ω_L and passes all terms above ω_L with a multiplicative gain of A and a linear phase shift.

The last ideal filter we consider is the ideal band-stop filter (BSF), which has the amplitude and phase responses sketched in Figs. 4.5.11 and 4.5.12, respectively. As is evident from Fig. 4.5.11, this filter is designed to reject only

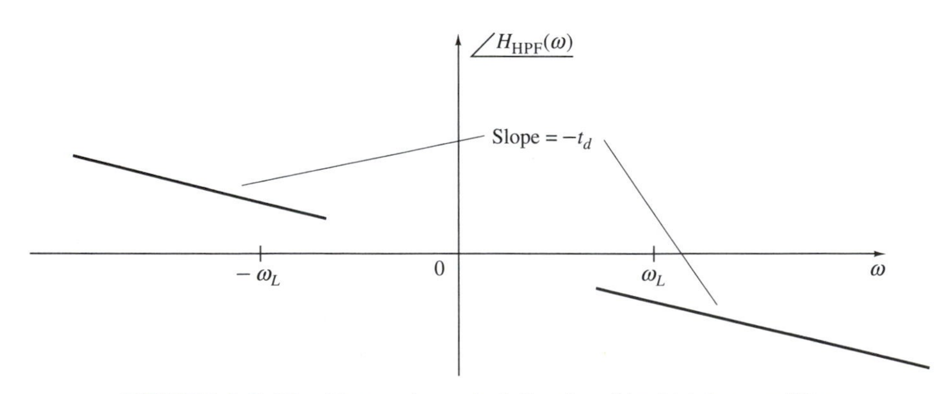

FIGURE 4.5.10 Phase characteristic of an ideal high-pass filter.

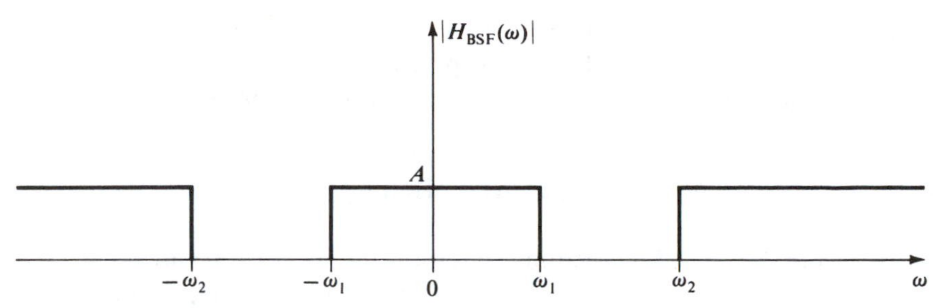

FIGURE 4.5.11 Amplitude response of an ideal band stop filter.

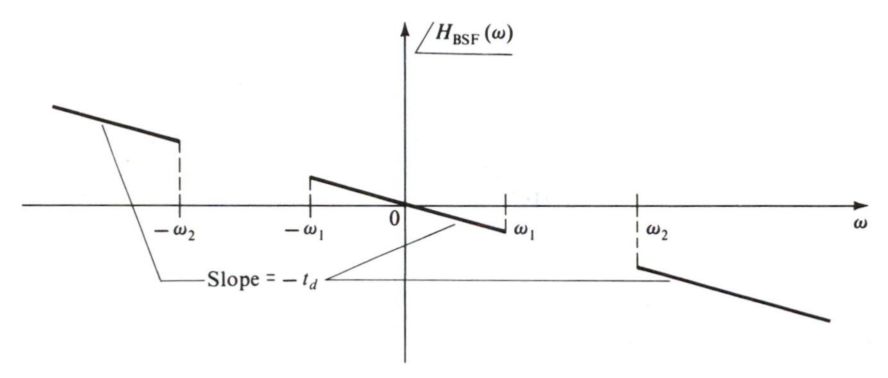

FIGURE 4.5.12 Phase response of an ideal band stop filter.

those signals in a specified band of frequencies between ω_1 and ω_2 and pass the remaining components undistorted. An analytical expression for the ideal BSF transfer function can be written as

$$H_{BSF}(\omega) = Ae^{-j\omega t_d} - H_{BPF}(\omega), \tag{4.5.16}$$

where $H_{BPF}(\omega)$ is given by Eq. (4.5.13). Although we shall not have an occasion to use it, the impulse response of the ideal BSF can be found quite easily from Eq. (4.5.16).

It now seems that we have at our disposal virtually every kind of filter necessary for the design and analysis of communication systems, and hence we should be ready to leave the subject of filtering and proceed to the next topic. There is one very important detail that has been ignored. This detail is that the ideal filters discussed in the present section cannot actually be built in the lab. The ideal filters are not physically realizable. The reason for this is best illustrated by considering the ideal LPF that has the amplitude and phase responses in Figs. 4.5.1 and 4.5.2. If a unit impulse is applied to this LPF at $t = 0$, the output will be as given by Eq. (4.5.6), which is sketched in Fig. 4.5.3. Notice that the filter has a nonzero output *before* the impulse excitation signal is applied.

Such a response is a physical impossibility. In the following section we investigate this situation in more detail and specify some physically realizable filters that adequately approximate the ideal filters in the present section.

4.6 Physically Realizable Filters

At the end of the preceding section, we reasoned that since the ideal LPF has a nonzero output prior to the application of an excitation signal, the filter cannot be constructed with physical components. This important condition, that is, the property that the system output cannot anticipate the input, is called the *causality condition.* A circuit or system is said to be *causal* or nonanticipatory if its impulse response $h(t)$ satisfies

$$h(t) = 0 \qquad \text{for } t < 0. \tag{4.6.1}$$

By examining the impulse responses of the ideal LPF and BPF in Eqs. (4.5.6) and (4.5.14), respectively, both of these filters can be seen to violate Eq. (4.6.1). Additionally, since the ideal HPF and BSF transfer functions can be written in terms of the ideal LPF and BPF transfer functions, the impulse responses of these filters do not satisfy Eq. (4.6.1). All of these ideal filters are thus noncausal and hence not physically realizable.

Equation (4.6.1) expresses the causality requirement in the time domain. In terms of frequency-domain concepts, a system is said to be causal or *realizable* if

$$\int_{-\infty}^{\infty} \frac{\left|\log |H(\omega)|\right|}{1+\omega^2}\, d\omega < \infty, \tag{4.6.2}$$

where $H(\omega)$ is the system transfer function. Before applying Eq. (4.6.2), it is also necessary to establish that

$$\int_{-\infty}^{\infty} |H(\omega)|^2\, d\omega < \infty. \tag{4.6.3}$$

The condition specified in Eq. (4.6.2) is called the *Paley–Wiener criterion.* Notice that the ideal LPF and BPF satisfy Eq. (4.6.3) and therefore the Paley–Wiener criterion can be used to determine the realizability of these filters. Both filters violate Eq. (4.6.2), however, since $|H_{\text{LPF}}(\omega)|$ and $|H_{\text{BPF}}(\omega)|$ are identically zero over a range of ω values. The ideal HPF and BSF transfer functions fail Eq. (4.6.3) and thus the Paley–Wiener criterion is not applicable to these filters.

Communication systems require the extensive use of all types of filters, and since it is not reasonable for an engineer to design a system using unrealizable elements, practical filters that achieve our goals must be found. Limiting consideration to the ideal LPF transfer function in Figs. 4.5.1 and 4.5.2, we see that there are three very stringent requirements that this filter satisfies:

1. Constant gain in the passband.
2. Linear phase response across the passband.
3. Perfect attenuation (total rejection) outside the passband.

We already know that it is not possible to realize a filter that exactly achieves all of these characteristics; and in fact, it is not possible to design a physically realizable filter that accurately approximates all three requirements. In practice, what has been done is to design three different types of filters, each of which provides a good approximation to one of the ideal LPF properties.

A type of practical filter called a *Butterworth filter* approximates the requirement of a constant gain throughout the passband. The amplitude characteristic of Butterworth low-pass filters can be expressed as

$$|H(\omega)| = \frac{1}{[1 + (\omega/\omega_c)^{2n}]^{1/2}}, \tag{4.6.4}$$

where ω_c is the 3-dB cutoff frequency and $n = 1, 2, 3, \ldots$ is the number of poles in the system transfer function, usually called the *order* of the filter. A sketch of Eq. (4.6.4) for $n = 1$, 3, and 5 and $\omega > 0$, along with the amplitude response of an ideal LPF, are given in Fig. 4.6.1. Notice that for $n = 1$ and ω_c appropriately defined in Eq. (4.6.4), we obtain the magnitude of the transfer function for the *RC* network in Fig. 4.3.3 [see Eq. (3.7.6)].

Although Butterworth filters provide what is called a *maximally flat* amplitude response, their attenuation outside the desired passband may not be sufficient for many applications. A class of filters called *Chebyshev filters* provide greater attenuation for $\omega > \omega_c$ than Butterworth filters, and hence may prove useful in such situations. The amplitude response of an nth-order Chebyshev

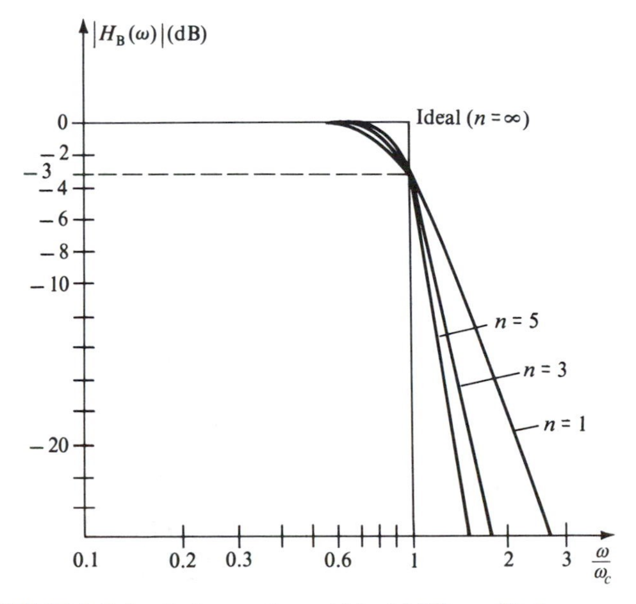

FIGURE 4.6.1 Butterworth and ideal LPF amplitude response.

filter is given by

$$|H(\omega)| = \frac{1}{[1 + \varepsilon^2 C_n^2(\omega/\omega_c)]^{1/2}}, \tag{4.6.5}$$

where $C_n(\cdot)$ denotes the nth-order Chebyshev polynomial from which the filter takes its name. Since Chebyshev polynomials are cosine polynomials, the amplitude response in Eq. (4.6.5) is not flat across the passband like the Butterworth filter, but contains ripples. The magnitude of the ripple depends on ε, since Eq. (4.6.5) oscillates between 1 and $[1 + \varepsilon^2]^{-1/2}$ for $0 \leq \omega/\omega_c \leq 1$.

Neither the Butterworth nor the Chebyshev filters exhibit a linear phase response across the passband, although the Butterworth approximation is not too bad. However, if true distortionless transmission of the signal phase is required, a class of filters called *Bessel filters* can be used. These filters, based on the Bessel polynomials, achieve a linear phase response at the expense of the other two ideal LPF requirements of constant gain across the passband and infinite attenuation outside the passband. The transfer function of a Bessel LPF is not written here, but it is pointed out that the Bessel LPF filters are obtained by truncating series expansions of the ideal LPF transfer function in Eq. (4.5.5).

Further discussion of filters and their design is not included here. For a more detailed development, the reader is referred to the many excellent books on the topic (see, e.g., Kuo [1962]).

4.7 Time-Domain Response and System Bandwidth

The duality between the time and frequency domains is evident if one peruses the Fourier transform properties in Section 3.6. As a specific example, consider the time shifting and frequency shifting properties. This duality is important, since it sometimes provides intuition as to how a problem might be solved when otherwise there would be none. More important than the duality relationship, however, is the simple fact that for most signals and systems there exists both a time-domain and a frequency-domain representation. This fact, coupled with the time convolution theorem, allows us to perform analyses that would not be possible limited to either the time or frequency domain alone.

There are situations, unfortunately, where it may be difficult to transition from one domain to the other, and in these instances it is necessary to use information from one domain to make inferences concerning the other domain. One particularly common and crucial example of such a situation is the determination of system bandwidth from the system time response to some input. What is meant by the *bandwidth* of a system? For the present development we need only state that bandwidth is an indication of the range of real (positive) frequencies passed by the system, deferring a more quantitative definition until later in this section. Returning to the original discussion, there are many times in the laboratory when it is desired to obtain an approximate idea of a system's bandwidth. Even though it may be possible to observe on an oscilloscope the system impulse response (or a close approximation), it may be difficult to express $h(t)$ analytically so that $\mathscr{F}\{h(t)\} = H(\omega)$ can be calculated. Of course, another method of determining the system bandwidth is to plot a frequency response of the system. This approach is time consuming, and in some cases, may be impossible due to system restrictions. In these circumstances it may be possible to estimate the bandwidth from the system response to a step function input.

To be more explicit, let us calculate the response of the ideal LPF in Eqs. (4.5.5) and (4.5.6) to a unit step input. There are several ways to approach this problem, one of which is to employ the time convolution theorem and then find the inverse transform (Problem 4.23). Another method is to use the property that the output of a system due to the integral of some signal is the integral of the output due to the nonintegrated signal (Problem 4.24). This allows the step response to be found from the system impulse response (Problem 4.25). The approach taken here is that of direct computation using the convolution integral.

The step response of the ideal LPF in Eq. (4.5.6) is given by

$$\begin{aligned} y(t) = u(t) * h(t) &= \int_{-\infty}^{\infty} \frac{A \sin \omega_s(\tau - t_d)}{\pi(\tau - t_d)} u(t - \tau)\, d\tau \\ &= \int_{-\infty}^{t} \frac{A \sin \omega_s(\tau - t_d)}{\pi(\tau - t_d)}\, d\tau \\ &= \int_{-\infty}^{t-t_d} \frac{A \sin \omega_s x}{\pi x}\, dx \end{aligned} \tag{4.7.1}$$

upon making the change of variable $x = \tau - t_d$. The last integral in Eq. (4.7.1) cannot be evaluated in closed form, but the integral

$$\text{Si}(t) = \int_0^t \frac{\sin x}{x}\, dx \tag{4.7.2}$$

is tabulated extensively [Jahnke and Emde, 1945], and hence we manipulate Eq. (4.7.1) to get this last form. This produces

$$\begin{aligned} y(t) &= \int_{-\infty}^{t-t_d} \frac{A}{\pi} \frac{\sin \omega_s x}{\omega_s x} \omega_s\, dx = \int_{-\infty}^{\omega_s(t-t_d)} \frac{A}{\pi} \frac{\sin v}{v}\, dv \\ &= \frac{A}{\pi}\left\{\text{Si}[\omega_s(t - t_d)] + \frac{\pi}{2}\right\}, \end{aligned} \tag{4.7.3}$$

since $\text{Si}(-\infty) = -\pi/2$. The function $\text{Si}(t)$ in Eq. (4.7.2) has the shape shown in Fig. 4.7.1, and therefore the output $y(t)$ in Eq. (4.7.3) has the form sketched in Fig. 4.7.2.

Notice that in Fig. 4.7.2, the elapsed time from when the minimum of $y(t)$ occurs to t_d is $1/2f_s$ seconds, and the elapsed time from t_d until the maximum of $y(t)$ occurs is $1/2f_s$ seconds. These values can be deduced from a sketch of the $\sin \omega_s t/\omega_s t$ function since the minimum and maximum of the integral of this function occur at $t = -1/2f_s$ and $t = +1/2f_s$, respectively. If we define the *rise time* (t_r) as the time it takes for the system output to reach its maximum value

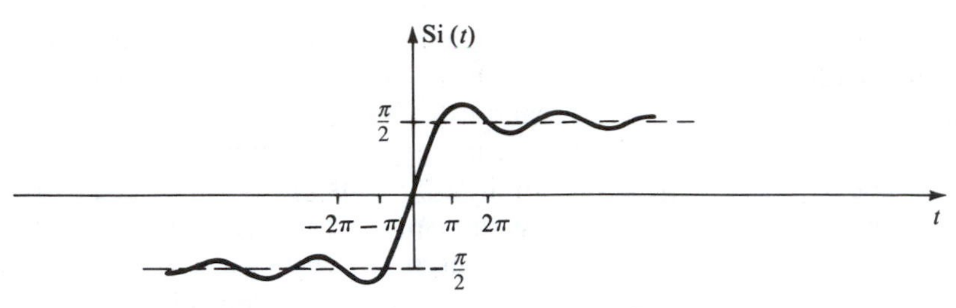

FIGURE 4.7.1 Graph of $Si(t)$ in Eq. (4.7.2).

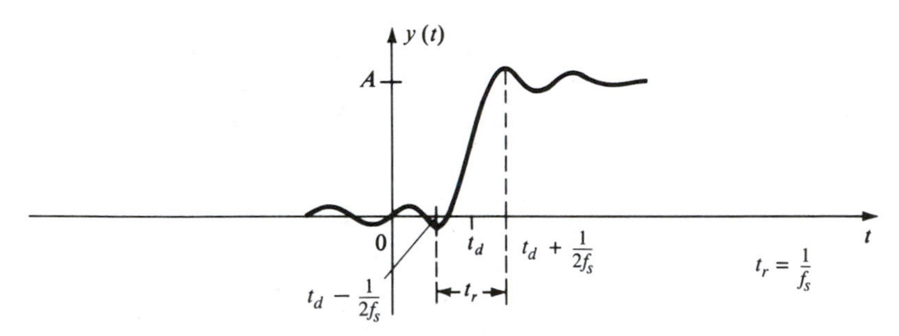

FIGURE 4.7.2 Unit step response of an ideal low-pass filter.

from its minimum value, the rise time of the ideal LPF is $1/f_s$ seconds. Although this definition of rise time is somewhat arbitrary, any other definition simply changes the rise time by some multiplicative constant. The important point is that the system rise time to a unit step input is inversely proportional to the system bandwidth. This last statement remains true regardless of the definition of rise time. This relation between rise time and system bandwidth is very important, since the output rise time to a unit step input is simple to compute in the lab even when a mathematical calculation or a frequency response plot is not feasible.

It is now necessary that we be more precise as to what is meant by the *bandwidth* of a system. If bandwidth (BW) is defined as the difference between the highest and lowest frequencies passed by a system, there is no difficulty in specifying the bandwidth of any of the ideal filters, since a frequency component is either passed undistorted or rejected. However, when attention is shifted to one of the practical filters, such as a Butterworth filter, the amplitude response is not perfectly flat within the desired band nor does it provide infinite attenuation outside this band of frequencies. Hence some quantitative definition is required. The almost universally accepted definition of bandwidth is the difference between the highest and lowest positive frequencies such that $|H(\omega)|$ is 0.707 of the magnitude at the filter's middle frequency (called *midband*). These upper and lower frequencies are generally called the *cutoff frequencies* of the filter. The following example illustrates the important concepts in this rather long-winded discussion.

EXAMPLE 4.7.1

For the RC network in Fig. 4.3.3, let us first estimate the bandwidth from the output rise time to a unit step input, and then compare this result to the bandwidth computed from the system amplitude response. The unit step response of the network is easily shown to be

$$y(t) = [1 - e^{-t/RC}]\,u(t) \tag{4.7.4}$$

and is sketched in Fig. 4.7.3. If we apply our earlier definition of rise time, which is the elapsed time from the minimum to the maximum of $y(t)$, we find that $t_r \cong 10RC$ (say) $- 0 \cong 10RC$, since $e^{-10} \cong 0$. Therefore, we have as an estimate of the bandwidth in hertz:

$$\text{BW} \cong \frac{1}{t_r} = \frac{1}{10RC} \qquad \text{hertz.} \tag{4.7.5}$$

The amplitude response of the RC network is available from Eq. (3.7.7) by letting $\tau = RC$ and $V \to 1/RC$ as

$$|H(\omega)| = \frac{1}{[1 + \omega^2R^2C^2]^{1/2}}. \tag{4.7.6}$$

For a low-pass filter, midband is usually defined as $\omega = 0$, and thus we must determine the highest and lowest frequencies such that

$$|H(\omega)| = 0.707. \tag{4.7.7}$$

Since the lowest frequency passed by the filter is $\omega = 0$ and $|H(\omega = 0)| = 1 > 0.707$, $\omega = 0$ is the low-frequency value we are searching for. To find the higher frequency, we equate Eqs. (4.7.6) and (4.7.7) and solve for ω to produce $\omega = 1/RC$ rad/sec and $f = 1/2\pi RC$ hertz. The upper frequency is thus

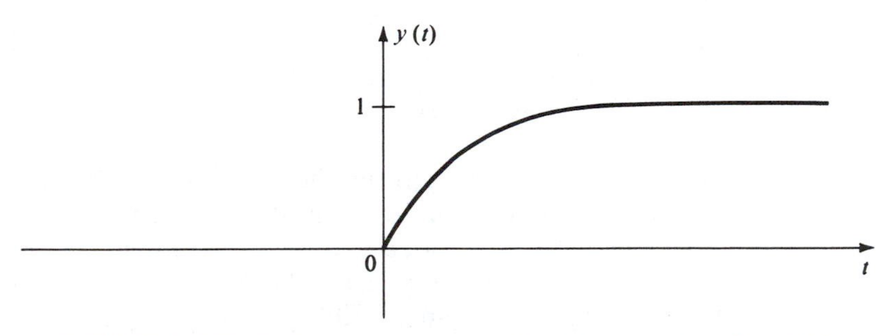

FIGURE 4.7.3 Unit step response of the *RC* network in Fig. 4.3.3.

$1/2\pi RC$ hertz and the lower frequency is 0 Hz, and therefore,

$$\text{BW} = \frac{1}{2\pi RC} \quad \text{hertz.} \tag{4.7.8}$$

Comparing Eqs. (4.7.5) and (4.7.8), we see that the reciprocal of the rise time to a unit step response is indeed a reasonable estimate of the system bandwidth as we have defined it. Other definitions of rise time and bandwidth will change the accuracy of the estimate, but the basic reciprocal relationship between t_r and BW will remain valid.

Notice in Example 4.7.1 that the system bandwidth has a pronounced effect on the shape of the output waveform. That is, if the bandwidth of the system is narrow, then from Eq. (4.7.8), RC is large, and the rise time is very slow. As a result, the output given by Eq. (4.7.4) and sketched in Fig. 4.7.3 is very different from the input signal shape, $u(t)$. On the other hand, if the system has a wide bandwidth, RC is small and hence t_r is small. The output in this case greatly resembles the unit step input, and as $RC \to 0$ (BW $\to \infty$), $y(t) \to u(t)$. This same phenomenon is evident in Example 4.3.1.

4.8 The Sampling Theorem

Of all the theorems in communication theory, the one that is probably most often applied today is the time-domain sampling theorem. Electrical engineers, geologists, computer engineers, statisticians, and many other workers employ this theorem on a daily basis. The reason for the theorem's importance is that the digital computer is a powerful and commonly used tool for signal processing, and since we live in a world that is basically analog or continuous-time, some guidelines are necessary to allow analog signals to be digitized without loss of information. These guidelines are specified by the time-domain sampling theorem.

Time-Domain Sampling Theorem

If the Fourier transform of a time function is identically zero for $|\omega| > 2\pi B$ rad/sec, the time function is uniquely determined by its samples taken at uniform time intervals less than $1/2B$ seconds apart.

The sampling theorem thus states that any time function $f(t)$ with $\mathscr{F}\{f(t)\} = 0$ for $|\omega| > 2\pi B$ [such an $f(t)$ is usually said to be *bandlimited*] can be recovered exactly from its samples taken at a rate of $2B$ samples/sec or faster. A sampling rate of exactly $2B$ samples/sec is generally called the *Nyquist rate.* Although several proofs of the sampling theorem are possible, only one is presented here. This proof is probably the most illustrative of those available and is actually as much an example as it is a proof.

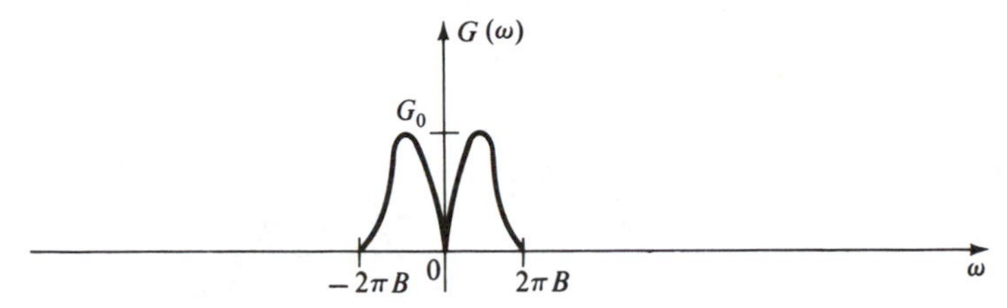

FIGURE 4.8.1 $G(\omega)$ for proof of the sampling theorem.

Proof: Given a bandlimited time signal, we desire to show that the signal can be uniquely recovered from its samples taken $1/2B$ seconds apart or less. Let $g(t)$ be a time signal bandlimited to B hertz, so that

$$\mathscr{F}\{g(t)\} = G(\omega) = 0 \qquad \text{for } \omega > 2\text{n}B, \tag{4.8.1}$$

where $G(\omega)$ is arbitrarily selected to be as shown in Fig. 4.8.1. Sampling $g(t)$ every T seconds by multiplying by the infinite train of impulses $\delta_T(t)$ in Eq. (3.5.22) yields

$$g_s(t) = g(t)\delta_T(t) = g(t) \sum_{n=-\infty}^{\infty} \delta(t - nT) = \sum_{n=-\infty}^{\infty} g(nT)\delta(t - nT) \tag{4.8.2}$$

with $T < 1/2B$.

Taking the Fourier transform of $g_s(t)$, we find that

$$\mathscr{F}\{g_s(t)\} = G_s(\omega) = \mathscr{F}\{g(t)\delta_T(t)\} = \frac{1}{2\pi} \mathscr{F}\{g(t)\} * \mathscr{F}\{\delta_T(t)\} \tag{4.8.3}$$

by the frequency convolution theorem in Eq. (4.3.17). Since $\mathscr{F}\{g(t)\} = G(\omega)$ and $\mathscr{F}\{\delta_T(t)\}$ is given by Eq. (3.5.24), we obtain for $\omega_0 = 2\pi/T$,

$$\begin{aligned} G_s(\omega) &= \frac{1}{2\pi} G(\omega) * \omega_0 \sum_{n=-\infty}^{\infty} \delta(\omega - n\omega_0) \\ &= \frac{\omega_0}{2\pi} \int_{-\infty}^{\infty} \left[\sum_{n=-\infty}^{\infty} G(\omega - v)\delta(v - n\omega_0) \right] dv \\ &= \frac{1}{T} \sum_{n=-\infty}^{\infty} \left[\int_{-\infty}^{\infty} G(\omega - v)\delta(v - n\omega_0)\, dv \right] \\ &= \frac{1}{T} \sum_{n=-\infty}^{\infty} G(\omega - n\omega_0) \end{aligned} \tag{4.8.4}$$

by the sifting property of the impulse. The Fourier transform of $g_s(t)$ is therefore just $G(\omega)$ scaled by $1/T$ and centered about $\pm n\omega_0$. A sketch of $G_s(\omega)$ for $G(\omega)$ in Fig. 4.8.1 is sketched in Fig. 4.8.2 with $\omega_B = 2\pi B$.

We now have both time- and frequency-domain representations of the sampled time function $g(t)$, which are given by Eqs. (4.8.2) and (4.8.4), respectively. The question is whether or not $g(t)$ can be recovered undistorted from $g_s(t)$. The

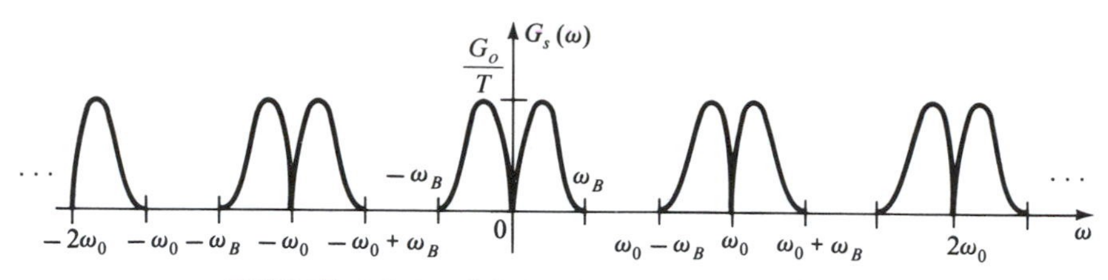

FIGURE 4.8.2 $G_s(\omega)$ in Eq. (4.8.4) for $T < 1/2B$.

answer is "yes" if $g_s(t)$ is appropriately low-pass filtered. That is, if $g_s(t)$ is applied to the input of a ideal LPF with a cutoff frequency ω_s such that $\omega_B < \omega_s < \omega_0 - \omega_B$, only the center lobe of $G_s(\omega)$ in Fig. 4.8.2, which is simply $(1/T)G(\omega)$, will not be rejected. Since $\mathscr{F}^{-1}\{(1/T)G(\omega)\} = (1/T)g(t)$, we have indeed recovered $g(t)$ from $g_s(t)$ with only an amplitude change, and hence the sampling theorem is proved.

To add the finishing touches to the proof, let us consider the importance of the requirement that the signal be sampled at least every $1/2B$ seconds. First, did we make explicit use of this specification on the sampling rate in proving the theorem? Yes, we did, since we assumed that the period of $\delta_T(t)$ satisfies $T < 1/2B$. This in turn implies that

$$\omega_0 > 2\omega_B. \tag{4.8.5}$$

Equation (4.8.5) thus says that the sampling frequency ω_0 must be greater than twice the highest frequency in $g(t)$, the signal being sampled. Why this restriction on sampling frequency is necessary becomes evident upon considering Fig. 4.8.2.

In this figure, the sketch indicates that $\omega_0 - \omega_B > \omega_B$ or $\omega_0 > 2\omega_B$, as we expect. What happens, however, if $\omega_0 \leq 2\omega_B$ (or equivalently, $T \geq 1/2B$) so that $\omega_0 - \omega_B \leq \omega_B$? When this situation occurs, $G_s(\omega)$ is as shown in Fig. 4.8.3. Notice the regions of overlap. Because of this overlap, it is no longer possible to recover $g(t)$ from $g_s(t)$ by low-pass filtering. In fact, it is not possible to obtain

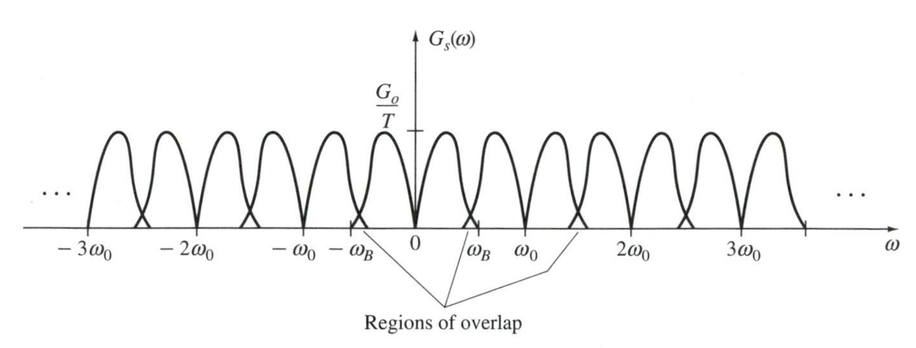

FIGURE 4.8.3 $G_s(\omega)$ in Eq. (4.8.4) for $T \geq 1/2B$.

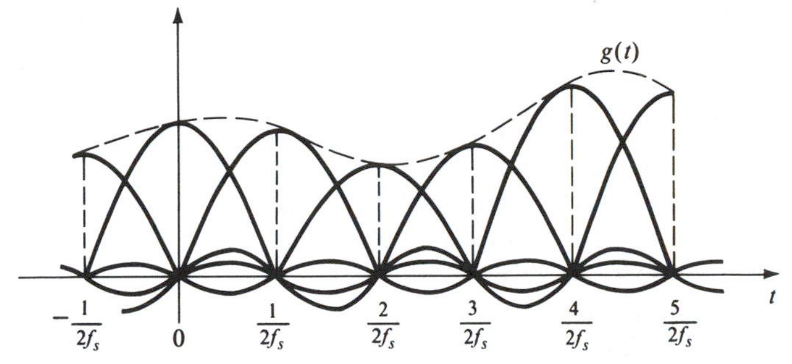

FIGURE 4.8.4 $g(t)$ Reconstructed from its samples.

$g(t)$ from $g_s(t)$ by any means, since the frequency content in these regions of overlap adds, and therefore the signal is distorted. The distortion that occurs when a signal is sampled too slowly is called *aliasing*.

To complete the process begun in the proof, it is necessary to demonstrate how $g(t)$ is reconstructed from its samples by passing $g_s(t)$ through an ideal LPF. The exact calculation is left as an exercise. The final result is given by

$$g(t) = \sum_{n=-\infty}^{\infty} g(nT) \frac{\sin(\omega_s t - n\pi)}{\omega_s t - n\pi}, \tag{4.8.6}$$

which is displayed graphically in Fig. 4.8.4.

EXAMPLE 4.8.1

With the increased availability of low-cost integrated circuits in recent years, digital voice terminals and digital transmission systems have become commonplace in the U.S. telephone network. The reason for this is simply that for some applications, the digital techniques provide improved performance at a lower cost than previous analog methods.

Since human speech is a continuous or analog signal, it is necessary that the speech be sampled before transmission through the digital systems. How often the speech signal must be sampled is dictated by the voice signal bandwidth and the sampling theorem. Since the frequency content of a speech signal can vary substantially depending on the speaker, the signal is lowpass filtered to some bandwidth less than 4 kHz prior to sampling. This bandlimited signal is then sampled 8000 times per second to satisfy the sampling theorem. Further discussion of digital voice transmission is included in Chapter 9.

The application of the sampling theorem described in Example 4.8.1 brings forth two important practical considerations. First, the sampling theorem requires that the time function be strictly bandlimited; that is, the Fourier transform of the function must be identically zero above some value of ω. Physical signals, however, are never absolutely bandlimited, since if a time signal is nonzero only over some finite interval, it can be shown that the signal contains components of all frequencies [see, e.g., Eqs. (3.2.10) and (3.2.11)]. Nonetheless, for all practical purposes it is reasonable to assume that most physical signals are bandlimited, since the majority of the signal "energy" (see Section 4.9) is contained in a finite frequency interval.

A second practical consideration is that even though a signal is low-pass filtered to a 3-dB bandwidth of, say, f_0 hertz, the signal is usually sampled at a rate greater than $2f_0$ samples/sec. For instance, as mentioned in Example 4.8.1, in the telephone network speech signals are low-pass filtered to less than 4 kHz (actually, bandpass filtered to 200 to 3400 Hz) but are sampled 8000 times a second, which is greater than the Nyquist rate of 6800 samples/sec. This is called *oversampling*, and it is, of course, permissible, since the sampling theorem is satisfied. However, why would one place an additional burden on the equipment by sampling faster than the Nyquist rate?

Part of the answer lies in the fact, illustrated by Fig. 4.8.2, that the higher the sampling rate, the more separation between adjacent lobes in the sampled signal's Fourier transform. The amount of the separation becomes critical when we realize two things: (1) The original analog signal is never perfectly bandlimited, and (2) we must use a physically realizable filter, not an ideal one, to recover the original signal from its sampled version. Since a physical filter does not have perfect attenuation outside the passband, if the lobes are too close, part of the (unwanted) frequency content centered about $\pm\omega_0$ will also be recovered. This results in a distorted signal.

Another idealization employed in the proof of the sampling theorem is that of sampling with perfect impulses. Since in practice we cannot generate a true impulse or delta function, the sampling must be accomplished using finite width pulses. The question is: What effect, if any, will this have on recovering the message signal from the samples? The answer to this question is that as long as the sampler pulse width is much smaller than the time between samples, the distortion is negligible. The derivation of this result is considered in Problem 4.37.

As one might surmise, there is a dual to the time-domain sampling theorem designated the frequency-domain sampling theorem.

Frequency-Domain Sampling Theorem

A time-limited signal that is identically zero for $|t| > T$ is uniquely determined by samples of its Fourier transform taken at uniform intervals less than π/T rad/sec apart.

Proof: Left as an exercise.

It should be noted that the sampling theorem discretizes the signal in time but not in amplitude. Thus, to complete the analog-to-digital conversion process, we must also discretize the amplitude of the samples. This process is fully developed in Chapter 9.

4.9 Power and Energy

As the reader is well aware, power and energy are two very important concepts in the study of electrical circuits. These ideas also prove useful for signal and system analysis. In signal analysis, the *total energy* of a possibly complex time signal $f(t)$ in the time interval $t_1 < t < t_2$ is defined as

$$E = \int_{t_1}^{t_2} f(t)f^*(t)\,dt = \int_{t_1}^{t_2} |f(t)|^2\,dt. \tag{4.9.1}$$

Although Eq. (4.9.1) is not dimensionally correct for energy in the electrical sense if $f(t)$ is a voltage, this definition is universal in signal and system analysis, since it allows a broader interpretation of the concept. If $t_1 = -\infty$ and $t_2 = +\infty$, Eq. (4.9.1) becomes the total energy in $f(t)$ for all time, given by

$$E = \int_{-\infty}^{\infty} |f(t)|^2\,dt. \tag{4.9.2}$$

There are some signals for which Eq. (4.9.2) becomes infinite, and for these signals it is possible to define a related concept called the average power. The *average power* of a signal $f(t)$ in the time interval $t_1 < t < t_2$ is defined as

$$P = \frac{1}{t_2 - t_1}\int_{t_1}^{t_2} f(t)f^*(t)\,dt. \tag{4.9.3}$$

Equation (4.9.3) must be modified slightly to represent the average power over all time, which is given by

$$P = \lim_{\tau\to\infty} \frac{1}{2\tau}\int_{-\tau}^{\tau} f(t)f^*(t)\,dt. \tag{4.9.4}$$

The definitions of average power in Eqs. (4.9.3) and (4.9.4) can be motivated by recalling that "power measures the *rate* at which energy is transformed" [Smith, 1971]. Hence, if Eq. (4.9.1) represents the energy of $f(t)$ over the time interval $t_1 < t < t_2$, then Eq. (4.9.3) expresses the average rate of change in energy over the time interval. Of course, identical comments are appropriate for Eqs. (4.9.2) and (4.9.4). From Eqs. (4.9.2) and (4.9.4), it can be seen that a signal with finite energy has $P_\infty = 0$, while a signal with $E_\infty = \infty$ has finite average power over all time.

The following two examples illustrate the calculation of the energy and average power of a signal.

EXAMPLE 4.9.1

It is desired to compute the energy over all time of the signal $f(t) = Ae^{-\alpha t}\, u(t)$, where $\alpha > 0$. This signal is purely real, and hence, substituting into Eq. (4.9.2) produces

$$E_\infty = \int_{-\infty}^{\infty} A^2 e^{-2\alpha t}\, u(t)\, dt = \frac{A^2}{2\alpha} \tag{4.9.5}$$

for $-\infty < A < \infty$. By inspection of Eq. (4.9.4), we see that $P_\infty = 0$.

EXAMPLE 4.9.2

We wish to compute the energy and average power over all time of the periodic signal $f(t) = A \sin \omega_c t$. From Eq. (4.9.2),

$$E_\infty = A^2 \int_{-\infty}^{\infty} \sin^2 \omega_c t\, dt = A^2 \int_{-\infty}^{\infty} [\tfrac{1}{2} - \tfrac{1}{2}\cos 2\omega_c t]\, dt = \infty. \tag{4.9.6}$$

The average power is, from Eq. (4.9.4),

$$\begin{aligned} P_\infty &= \lim_{\tau \to \infty} \frac{A^2}{2\tau} \int_{-\tau}^{\tau} \sin^2 \omega_c t\, dt = \lim_{\tau \to \infty} \frac{A^2}{2\tau}\left[\frac{t}{2} - \frac{1}{4\omega_c}\sin 2\omega_c t\right]\Bigg|_{-\tau}^{\tau} \\ &= \lim_{\tau \to \infty}\left[\frac{A^2}{2} - \frac{A^2}{4\omega_c \tau}\sin 2\omega_c \tau\right] = \frac{A^2}{2}. \end{aligned} \tag{4.9.7}$$

As is suggested by this example, all periodic signals have $E_\infty = \infty$, but they may have P_∞ finite.

If energy and average power could be computed only by using Eqs. (4.9.1)–(4.9.4), their utility would be limited. This is because many situations occur where the frequency content of the signal is known, but not the time-domain waveform. To calculate the energy or average power would thus require that the inverse Fourier transform be taken first. Fortunately, frequency-domain expressions for the energy and average power over the infinite time interval can be obtained.

From the inverse Fourier transform relationship in Eq. (3.2.5), we have that

$$f^*(t) = \frac{1}{2\pi} \int_{-\infty}^{\infty} F^*(\omega) e^{-j\omega t}\, d\omega, \tag{4.9.8}$$

which when substituted into Eq. (4.9.2) yields

$$\begin{aligned} E_\infty &= \int_{-\infty}^{\infty} f(t) \left\{ \frac{1}{2\pi} \int_{-\infty}^{\infty} F^*(\omega) e^{-j\omega t}\, d\omega \right\} dt \\ &= \frac{1}{2\pi} \int_{-\infty}^{\infty} F^*(\omega) \left\{ \int_{-\infty}^{\infty} f(t) e^{-j\omega t}\, dt \right\} d\omega \\ &= \frac{1}{2\pi} \int_{-\infty}^{\infty} F^*(\omega) F(\omega)\, d\omega = \frac{1}{2\pi} \int_{-\infty}^{\infty} |F(\omega)|^2\, d\omega. \end{aligned} \tag{4.9.9}$$

Equation (4.9.9) is usually called *Parseval's theorem*, and it provides a method for calculating the energy in a signal directly from the signal's Fourier transform.

The quantity $|F(\omega)|^2$ under the integral in Eq. (4.9.9) is usually called the *energy density spectrum* of $f(t)$, since when it is multiplied by $1/2\pi$ and integrated over all ω, the total energy in $f(t)$ is obtained. The importance of the energy density spectrum and Eq. (4.9.9) is underscored by the following result. From Section 4.3 we know that the Fourier transform of the output of a linear system with transfer function $H(\omega)$ is given by $Y(\omega) = H(\omega)R(\omega)$, where $R(\omega)$ is the Fourier transform of the input. The energy density spectrum of the output is thus

$$\begin{aligned} |Y(\omega)|^2 &= Y(\omega)Y^*(\omega) = [H(\omega)R(\omega)][H^*(\omega)R^*(\omega)] \\ &= |H(\omega)|^2|R(\omega)|^2. \end{aligned} \tag{4.9.10}$$

Therefore, given the system transfer function and the energy density spectrum of the input, the output energy density spectrum can be found from Eq. (4.9.10).

EXAMPLE 4.9.3

Let us use Eq. (4.9.9) to calculate the energy in the time signal $f(t) = Ae^{-\alpha t}\,u(t)$ from Example 4.9.1. The Fourier transform of $f(t)$ is, from Example 3.2.1,

$$\mathscr{F}\{f(t)\} = \frac{A}{\alpha + j\omega} = F(\omega),$$

so that

$$|F(\omega)|^2 = \frac{A^2}{\alpha^2 + \omega^2}. \tag{4.9.11}$$

Substituting Eq. (4.9.11) into Eq. (4.9.9) yields

$$E_\infty = \frac{1}{2\pi}\int_{-\infty}^{\infty} \frac{A^2}{\alpha^2 + \omega^2}\,d\omega = \frac{A^2}{2\pi\alpha}\tan^{-1}\frac{\omega}{\alpha}\bigg|_{-\infty}^{\infty} = \frac{A^2}{2\alpha}, \tag{4.9.12}$$

which agrees with the result of Example 4.9.1.

Since for some signals E_∞ is infinite, it is advantageous to obtain frequency-domain expressions for the average power of these signals, which are analogous to the energy density results in Eq. (4.9.9). Given a signal $f(t)$ for which E_∞ is infinite, we define a new time function

$$f_\tau(t) = \begin{cases} f(t), & |t| < \tau \\ 0, & \text{otherwise}, \end{cases} \tag{4.9.13}$$

which has the Fourier transform $F_\tau(\omega) = \mathscr{F}\{f_\tau(t)\}$. The average power of $f(t)$ is then given by

$$\begin{aligned} P_\infty &= \lim_{\tau\to\infty}\frac{1}{2\tau}\int_{-\tau}^{\tau}|f(t)|^2\,dt = \lim_{\tau\to\infty}\frac{1}{4\pi\tau}\int_{-\infty}^{\infty}|F_\tau(\omega)|^2\,d\omega \\ &= \frac{1}{2\pi}\int_{-\infty}^{\infty}\left\{\lim_{\tau\to\infty}\frac{1}{2\tau}|F_\tau(\omega)|^2\right\}d\omega \end{aligned} \tag{4.9.14}$$

upon interchanging the integration and limiting operations. If the limit exists, the quantity in the braces in Eq. (4.9.14) is called the *power density spectrum*, and is usually denoted by $S_f(\omega)$.

The result for the power density spectrum analogous to Eq. (4.9.10) can be found by noting that the average output power over all time is

$$P_\infty = \lim_{\tau \to \infty} \frac{1}{2\tau} \int_{-\tau}^{\tau} |y(t)|^2 \, dt$$

$$= \frac{1}{2\pi} \int_{-\infty}^{\infty} |H(\omega)|^2 \left[\lim_{\tau \to \infty} \frac{|F_\tau(\omega)|^2}{2\tau} \right] d\omega. \tag{4.9.15}$$

The integrand in Eq. (4.9.15) is the output power density spectrum and the quantity in brackets is the input power density spectrum. Therefore, the output power density spectrum of a system with transfer function $H(\omega)$ can be found from the input power density spectrum using the equation

$$S_y(\omega) = |H(\omega)|^2 S_f(\omega). \tag{4.9.16}$$

It is pointed out that the expressions for the power density spectra in Eqs. (4.9.14) and (4.9.15) are valid only for deterministic signals, since interchanging the integration and limiting operations is not possible for random signals.

To illustrate the calculation of the average power of signals using Eq. (4.9.14), consider the following example.

EXAMPLE 4.9.4

Let us employ Eq. (4.9.14) to calculate the average power over all time of $f(t) = A \sin \omega_c t$. The Fourier transform of the truncated signal is

$$F_\tau(\omega) = \int_{-\tau}^{\tau} A \sin \omega_c t e^{-j\omega t} \, dt = \frac{A}{j2} \int_{-\tau}^{\tau} [e^{-j(\omega - \omega_c)t} - e^{-j(\omega + \omega_c)t}] \, dt$$

$$= \frac{A\tau}{j} \left[\frac{\sin (\omega - \omega_c)\tau}{(\omega - \omega_c)\tau} - \frac{\sin (\omega + \omega_c)\tau}{(\omega + \omega_c)\tau} \right]. \tag{4.9.17}$$

Substituting $F_\tau(\omega)$ into Eq. (4.9.14) yields

$$P_\infty = \frac{1}{2\pi} \int_{-\infty}^{\infty} \left\{ \lim_{\tau \to \infty} \frac{1}{2\tau} \cdot A^2 \tau^2 \left[\frac{\sin (\omega - \omega_c)\tau}{(\omega - \omega_c)\tau} - \frac{\sin (\omega + \omega_c)\tau}{(\omega + \omega_c)\tau} \right]^2 \right\} d\omega$$

$$= \frac{A^2}{2} \int_0^{\infty} \left\{ \lim_{\tau \to \infty} \frac{\tau}{\pi} \left[\frac{\sin (\omega - \omega_c)\tau}{(\omega - \omega_c)\tau} \right]^2 + \lim_{\tau \to \infty} \frac{\tau}{\pi} \left[\frac{\sin (\omega + \omega_c)\tau}{(\omega + \omega_c)\tau} \right]^2 \right\} d\omega. \tag{4.9.18}$$

The cross-term falls out of Eq. (4.9.18) when the limit is taken, since the two $\sin x/x$ components become nonoverlapping.

One possible limiting expression for the unit impulse is

$$\delta(\omega) = \lim_{\tau \to \infty} \frac{\tau}{\pi} \left[\frac{\sin \omega\tau}{\omega\tau} \right]^2, \tag{4.9.19}$$

and hence Eq. (4.9.18) can be written as

$$P_\infty = \frac{A^2}{2} \int_0^\infty [\delta(\omega - \omega_c) + \delta(\omega + \omega_c)]\, d\omega = \frac{A^2}{2}, \tag{4.9.20}$$

since for $\omega_c \neq 0$, only the first impulse falls within the limits of integration. Notice that Eq. (4.9.20) is the same value as that obtained in Example 4.9.2, which it must be.

Illustrations of the use of Eqs. (4.9.10) and (4.9.16) are not presented here, since the calculations required are much the same as demonstrated in Examples 4.9.3 and 4.9.4. The ability to compute the energy or average power will prove very important in later communication system comparative studies.

SUMMARY

This chapter constitutes the final preparatory step before proceeding to the analysis of communication systems in the absence of noise. In Sections 4.2 through 4.4, the concept of a linear time-invariant system was developed and time- and frequency-domain methods for determining the system output for a specified input were derived and illustrated. The input/output behavior of ideal and physically realizable filters was investigated in Sections 4.5 and 4.6, followed in Section 4.7 by a discussion of system bandwidth and its relationship to the system time response. The important topics of analog-to-digital and digital-to-analog conversions were introduced in Section 4.8 with a development of the time-domain sampling theorem. Expressions for the calculation of the energy and average power of signals from both time- and frequency-domain representations were presented and illustrated in Section 4.9. Each of the ideas and concepts discussed in this chapter is employed repeatedly in the analyses that follow.

PROBLEMS

4.1 Prove that the differentiation operation in Eq. (4.2.2) describes a linear time-invariant system.

Hint: For time-invariance, use the definition of the derivative.

4.2 Show that the time-delay operation in Eq. (4.2.4) represents a linear time-invariant system.

4.3 Find and sketch the output response of the RC network in Fig. 4.3.3 to the input signal $r(t) = \delta(t) + 2\delta(t-3) + \delta(t-4)$ if $R = 1\ \mathrm{M}\Omega$ and $C = 1\ \mu\mathrm{F}$.

4.4 For $RC = 1$, sketch the system response given by Eq. (4.3.11).

4.5 Use Eq. (4.3.4) to find the response of the RC network in Fig. 4.3.3 to the input signal $r(t) = \sin \omega_0 t$ for $-\infty < t < \infty$.

4.6 Evaluate the following convolution integrals.
(a) $e^{-t}\,u(t) * e^{-2t}\,u(t)$.
(b) $e^{-t}\,u(t) * t\,u(t)$.

4.7 Use Eq. (4.3.4) to compute $f_1(t) * f_2(t)$, where $f_1(t)$ and $f_2(t)$ are shown in Fig. P4.7. Sketch your result.

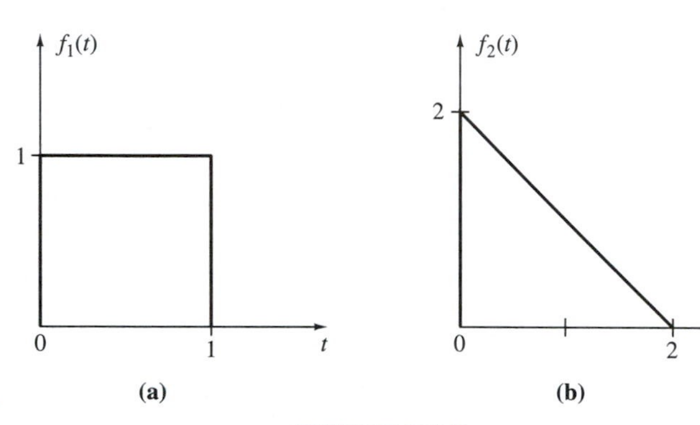

FIGURE P4.7

4.8 Rework Problem 4.5 using the time convolution theorem. Note how easy it is to find the frequency content of the output.

4.9 Prove the frequency convolution theorem by showing that the inverse Fourier transform of the right side of Eq. (4.3.17) is $f(t)g(t)$.

4.10 Rework Problem 4.7 using graphical convolution.

4.11 Graphically convolve $f_1(t)$ with $f_2(t)$, where $f_1(t)$ and $f_2(t)$ are given by

$$f_1(t) = 3[u(t+2) - u(t-2)]$$

and

$$f_2(t) = t[u(t) - u(t-2)] + (4-t)[u(t-2) - u(t-4)].$$

4.12 Graphically convolve $f_1(t)$ with $f_2(t)$, where $f_1(t)$ and $f_2(t)$ are shown in Fig. P4.12(a) and (b), respectively.

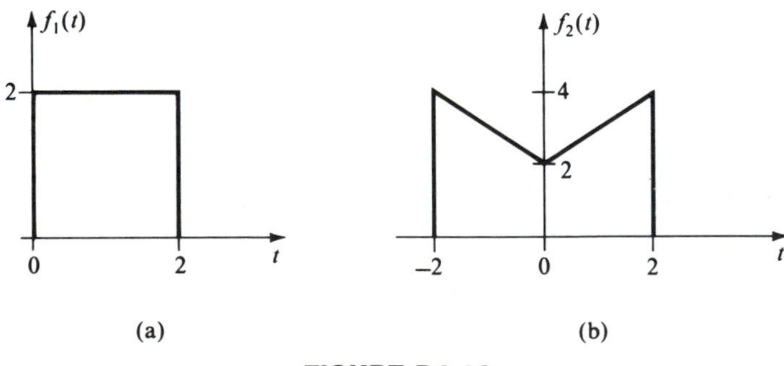

FIGURE P4.12

4.13 Find the output of an ideal LPF with $\omega_s = 377$ rad/sec, $A = 2.2$, and $t_d = 0$, for an input signal $r(t) = 10 + 2 \cos 72t - 14 \sin 300t + \sin 428t + \cos 1022t$.

4.14 Determine the output of an ideal BPF with $A = 1, t_d = 0, \omega_1 = 200$ rad/sec, and $\omega_2 = 500$ rad/sec, to the input signal $r(t)$ in Problem 4.13.

4.15 Repeat Problem 4.14 if $t_d = 10$ msec.

4.16 For an ideal HPF with $A = 3$, $\omega_L = 1000$ rad/sec, and $t_d = 0$, find the output for an input $r(t) = \sin 211t - \cos 306t - 8 \sin 830t + 7 \sin 1011t - \cos 3200t$.

4.17 Find the output of an ideal BSF with $A = 1$, $t_d = 0$, $\omega_1 = 500$ rad/sec, and $\omega_2 = 1000$ rad/sec, to the input $r(t)$ in Problem 4.13.

4.18 Find the output of the RC network in Fig. 4.3.3 for the input signal $r(t)$ in Problem 4.13 if $1/RC = 377$ rad/sec.

4.19 For a realizable BPF with the transfer function

$$H(\omega) = H_1(\omega + 250) + H_1(\omega - 250),$$

where

$$H_1(\omega) = \frac{1}{1 + j\omega RC}$$

and $RC = 1/377$, find the output time response for the input $r(t)$ in Problem 4.13.

4.20 The transfer function of the RC network in Fig. P4.20 is $H(\omega) = j\omega RC/(1 + j\omega RC)$. Does this network function as a LPF, BPF, BSF, or HPF? Which one? Find the output of this circuit to the input $r(t)$ in Problem 4.16 if $RC = 1/1000$.

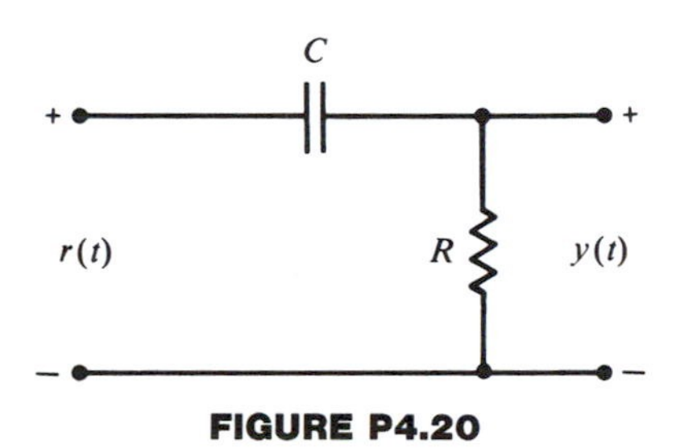

FIGURE P4.20

4.21 The pulse shown in Fig. P4.21 is applied to the input of an ideal LPF with a step response given by Eq. (4.7.3). Write an expression for the filter output, $y(t)$.

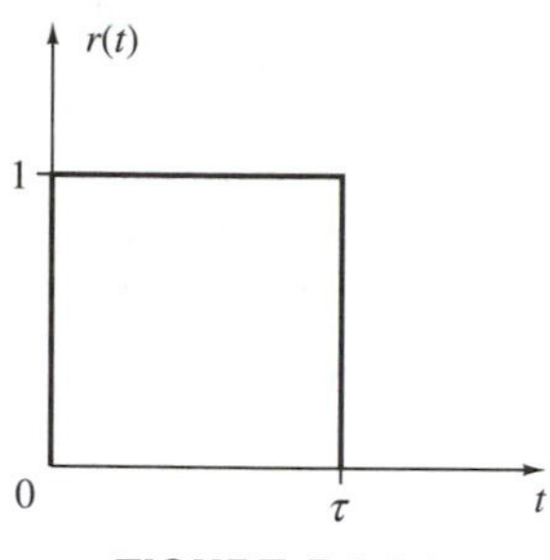

FIGURE P4.21

4.22 By referring to Fig. 4.7.2 and a suitable table of Si(t) [Jahnke and Emde, 1945], sketch the system output in Problem 4.21 when $\tau = 100\ \mu$sec for $\omega_s \gg 10$ kHz, $\omega_s = 10$ kHz, and $\omega_s \ll 10$ kHz. Note the relationship between bandwidth and output pulse shape.

4.23 Calculate the step response of the ideal LPF using the time convolution theorem, Eq. (4.5.5) and the Fourier transform of $u(t)$.

4.24 For a linear time-invariant system, it is possible to show that if

$$r_1(t) * h(t) = y_1(t),$$

then

$$\left[\frac{d}{dt} r_1(t)\right] * h(t) = \frac{d}{dt} y_1(t).$$

Using this result, demonstrate that if

$$r_2(t) * h(t) = y_2(t),$$

then

$$\left[\int_{-\infty}^{t} r_2(\tau)\, d\tau\right] * h(t) = \int_{-\infty}^{t} y_2(\tau)\, d\tau.$$

4.25 Use the result of Problem 4.24 to find the unit step response of the ideal LPF from $h_{\text{LPF}}(t)$ in Eq. (4.5.6).

4.26 **(a)** Consider the time function $g_1(t)$ that has the Fourier transform $\mathscr{F}\{g_1(t)\} = G_1(\omega)$ shown in Fig. P4.26(a). Can $g_1(t)$ be recovered from appropriately placed samples of $g_1(t)$? If so, what is the required sampling rate?
(b) Repeat part (a) for $G_2(\omega)$ in Fig. P4.26(b).

4.27 The time function $g(t)$ with a Fourier transform $G(\omega)$ shown in Fig. P4.27 is sampled using shaped-top pulses by the infinite pulse train in Fig. 2.3.1. If $\tau = 0.1$ msec, write an expression for the Fourier transform of the sampled signal, $G_s(\omega)$. Can $g(t)$ be recovered undistorted from $g_s(t)$?

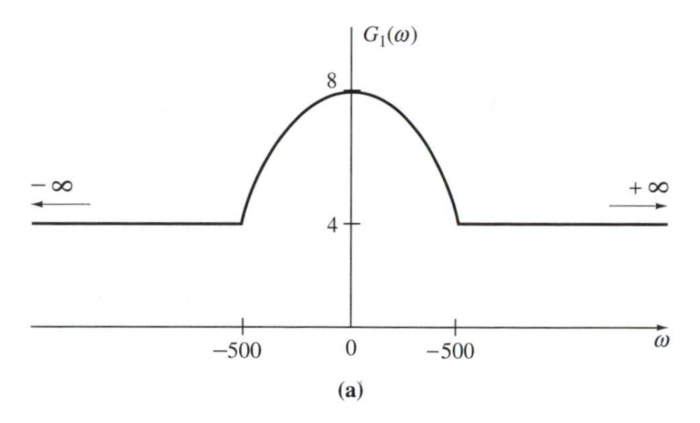

(a)

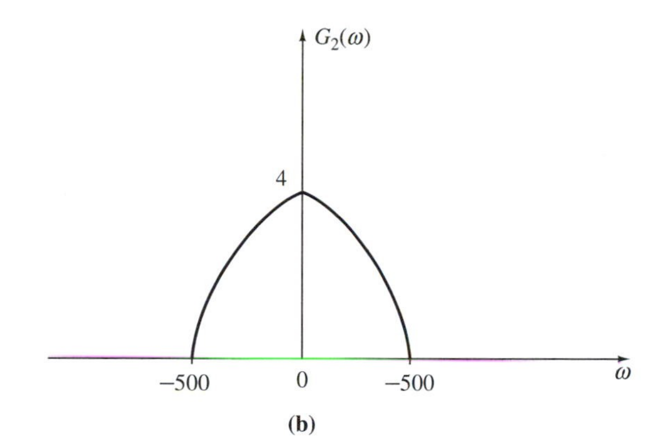

(b)

FIGURE P4.26

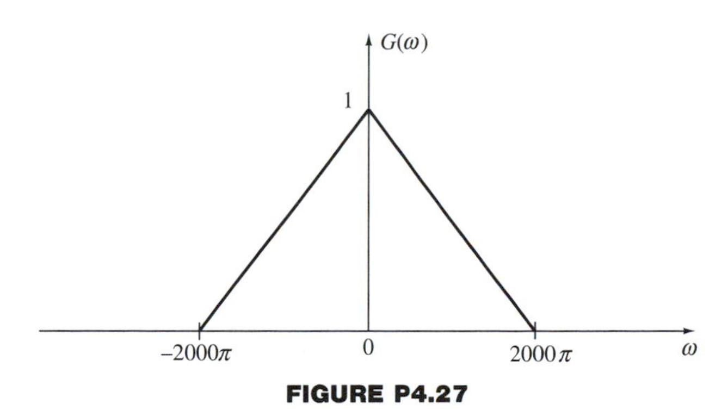

FIGURE P4.27

4.28 A voice signal, $m(t)$, which is bandlimited to 3200 Hz, multiplies the function $\cos \omega_c t$, where $\omega_c = 2\pi(10{,}000)$ rad/sec. Assuming ideal sampling, specify a sampling rate such that $m(t) \cos \omega_c t$ can be uniquely recovered using an ideal BPF. What are the ideal BPF cutoff frequencies?

4.29 A time signal $g(t)$ is ideally sampled by the infinite train of impulses

$$\delta_T\left(t - \frac{T}{2}\right) = \sum_{n=-\infty}^{\infty} \delta\left(t - nT - \frac{T}{2}\right).$$

Assuming that T satisfies the sampling theorem, can the original time function be reconstructed from these samples undistorted? Justify your answer.

4.30 Use Eq. (4.9.2) to find the energy in $y(t)$ in Eq. (4.3.11). Find an expression for the energy density spectrum of $y(t)$ and set up the integral for the total energy in $y(t)$ in terms of its energy density.

4.31 A signal $f(t)$ that is bandlimited to ω_m rad/sec has a Fourier transform denoted by $F(\omega)$. Find the energy density spectrum of $f(t) \cos \omega_c t$, where $\omega_c \gg \omega_m$.

4.32 Find the power density spectrum of $f(t) \cos \omega_c t$ if $f(t)$ is bandlimited to ω_m rad/sec and has the power density spectrum $S_f(\omega) = \lim_{\tau \to \infty} (1/2\tau)|F_\tau(\omega)|^2$. Let $\omega_c \gg \omega_m$.

4.33 The signal $A \cos \omega_c t$ is applied to the input of a system with impulse response $h(t) = e^{-t/\beta}\, u(t)$. Find the power density spectrum of the output.

4.34 If the signal $f(t) = Ae^{-\alpha t}\, u(t)$ is applied to the input of a differentiator, find the energy density spectrum of the output.

4.35 Find the energy in the signal $f(t) = \sqrt{t}\, e^{-\alpha t}\, u(t)$.

4.36 Finish the proof of Eq. (3.6.16) and show what happens if $\int_{-\infty}^{\infty} g(\lambda)\, d\lambda = 0$.

4.37 We would like to show that we can sample with flat-topped, finite width pulses with negligible distortion. The signal to be sampled is $g(t)$, and the flat-top pulses are generated by passing $g(t)$ through a sample-and-hold circuit. We assume that the sample-and-hold circuit behaves like a system with impulse response

$$h(t) = \begin{cases} 1, & \text{for } -\dfrac{\tau}{2} \leq t \leq \dfrac{\tau}{2} \\ 0, & \text{otherwise} \end{cases}$$

excited by the signal

$$r(t) = g(t)\delta_T(t) = \sum_{n=-\infty}^{\infty} g(nT)\delta(t - nT).$$

If the output of the sample-and-hold circuit is $g_s(t)$, show that

$$G_s(\omega) = \mathscr{F}\{g_s(t)\} = \omega_0 \tau \,\frac{\sin(\omega\tau/2)}{\omega\tau/2} \sum_{n=-\infty}^{\infty} G(\omega - n\omega_0).$$

Sketch this result for some $G(\omega)$.

CHAPTER 5

Amplitude Modulation

5.1 Introduction

A common lay definition of the verb *to modulate* is "to tune to a certain key or pitch." In amplitude modulation, this is precisely what is being done; namely, the information to be transmitted, hereafter called the *message signal*, is moved to another location in the frequency spectrum. As we know from Chapter 3, this relocation is easily accomplished by multiplying the message signal by a pure sinusoid; however, one may wonder why we wish to modulate the message signal at all. A little thought produces several excellent reasons.

First, the original frequency content of many signals that we wish to transmit overlap. For example, speech signals fall in the range 0 to 4000 Hz, music contains frequencies in the range 0 to 20,000 Hz, and television video signals originally occupy the band 0 to 5 MHz. If several of these signals were transmitted simultaneously over the same medium (cable, wire pairs, etc.) without modulation, they would interfere with each other and unintelligible signals would be received. Modulation allows us to send these signals simultaneously without interference by moving each signal to a different frequency band during transmission. A second reason for relocating message signals in frequency is that there are numerous sources of noiselike signals at low frequencies, such as car ignitions, electric lights, and electric motors. A third reason relates to antenna size for free-space transmission. For the efficient radiation of electromagnetic waves, the antenna size should be proportional to or larger than the transmitted wavelength. If the frequency to be transmitted is 5000 Hz, the wavelength is 60,000 meters! Of course, an antenna anywhere near this size is impractical. There are additional reasons for using various types of modulation, and they will become evident as the development progresses.

The discussion of amplitude modulation (AM) begins with a treatment of AM double-sideband suppressed carrier (AMDSB-SC) systems in Section 5.2. The development includes sketches of a typical time-domain AMDSB-SC waveform as well as its frequency content, a discussion of coherent demodulation, and a derivation of the effects of frequency and phase errors in coherent demodulation. Conventional AM (i.e., AMDSB-transmitted carrier) is presented in Section 5.3, which includes envelope detection, sketches of the time-domain waveform and its Fourier transform, and a discussion of percent modulation and efficiency. In Sections 5.4 and 5.5 we develop single-sideband and vestigial sideband AM, respectively. Time-domain waveform expressions and their Fourier transforms, demodulation of the received signals, frequency and phase errors in coherent demodulation, and the selection of filter shape are addressed. Superheterodyne systems are discussed in Section 5.6, which includes such topics as the choice of intermediate-frequency (IF) filter center frequency and bandwidth, image stations and radio-frequency (RF) filter bandwidth, and tuning range. AM radio examples are given. Quadrature amplitude modulation is defined in Section 5.7, and in Section 5.8 we introduce the concept of frequency-division multiplexing and how it is used in communication systems. Binary and multilevel amplitude shift keying (ASK) are considered in Section 5.9, with sketches of time-domain waveforms and a discussion of demodulation methods.

5.2 Double-Sideband Suppressed Carrier

From Chapter 3 we know that one way of relocating a signal in the frequency domain is to multiply its time-domain waveform by a pure sinusoid. More specifically, let the message signal be denoted by $m(t)$ and let the multiplying sinusoid or *carrier signal* be $A_c \cos \omega_c t$. Forming the product of these two signals yields

$$s_{\mathrm{DSB}}(t) = A_c m(t) \cos \omega_c t. \tag{5.2.1}$$

From Eq. (3.6.11) we know that the frequency content of $s_{\mathrm{DSB}}(t)$ is given by

$$S_{\mathrm{DSB}}(\omega) \triangleq \mathscr{F}\{s_{\mathrm{DSB}}(t)\} = \frac{A_c}{2}[M(\omega + \omega_c) + M(\omega - \omega_c)], \tag{5.2.2}$$

where $\mathscr{F}\{m(t)\} = M(\omega)$. The relocation of $m(t)$ in the frequency domain is easily seen by sketching Eq. (5.2.2). If $|M(\omega)|$ is as shown in Fig. 5.2.1(a), then $|S_{\mathrm{DSB}}(\omega)|$ has the form illustrated in Fig. 5.2.1(b). For simplicity, the phases are assumed to be zero and are not shown. Nonzero phases do not alter the development.

The waveform in Eq. (5.2.1) is called an AM double-sideband, suppressed carrier (AMDSB-SC) wave, or more commonly, just a double-sideband wave. The justification for this name comes directly from Fig. 5.2.1(b). The frequency content in the range $\omega_c \le |\omega| \le \omega_c + \omega_m$ is called the upper sideband (USB), and the frequency content in the range $\omega_c - \omega_m \le |\omega| < \omega_c$ is called the lower sideband (LSB). Since physical signals are defined only for positive frequencies, $m(t)$ is totally described by the shape of the frequency content in Fig. 5.2.1(a) in the range $0 \le \omega \le \omega_m$. Concentrating only on positive frequencies, note that the USB in Fig. 5.2.1(b) has the same shape as $M(\omega)$ and that the LSB has the same shape except that it is rotated about $\omega = \omega_c$. Because this shape is preserved, each sideband alone has enough information[1] to reconstruct $m(t)$. Since we have

[1] The term *information* as used here is very vague and ill-defined. A precise definition of information is presented in Chapter 11.

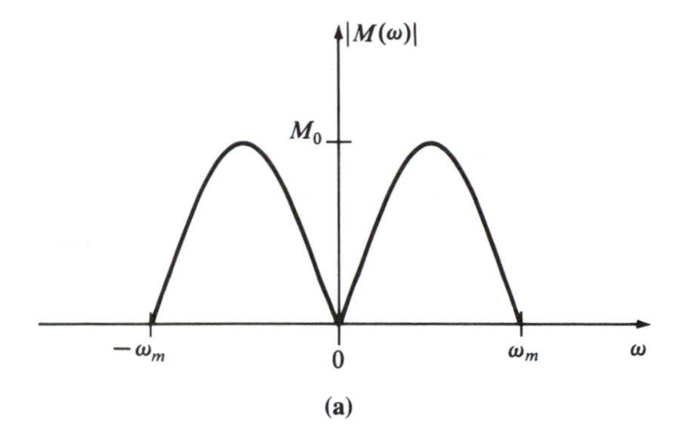

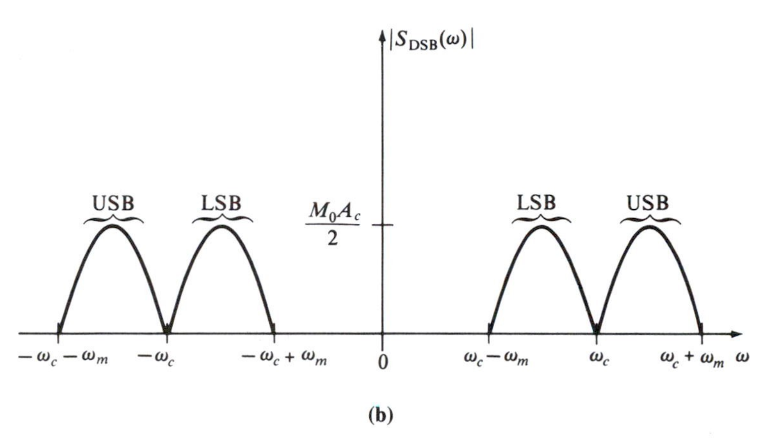

FIGURE 5.2.1 Frequency content of the message and double sideband signals: (a) Message signal; (b) double sideband signal.

two sidebands, we use the term *double sideband*. The terminology *suppressed carrier* comes from the fact that although cos $\omega_c t$ appears explicitly in Eq. (5.2.1), there is no frequency content in $S_{\text{DSB}}(\omega)$ at $\omega = \omega_c$ that was not originally at $\omega = 0$ in $M(\omega)$. In the following section, where a different type of AM is discussed, a carrier term will be clearly evident in the modulated waveform frequency content.

In sketching $S_{\text{DSB}}(\omega)$, it has been implicitly assumed that $\omega_c > \omega_m$. If this is not the case, then $M(\omega - \omega_c)$ and $M(\omega + \omega_c)$ will overlap, thus causing distortion. For most free-space transmission applications, $\omega_c \gg \omega_m(\omega_c > 10\omega_m)$.

Although the multiplication in Eq. (5.2.1) has in fact relocated $M(\omega)$ in the frequency domain, the multiplication has also doubled the required bandwidth. Again considering only positive frequencies, the bandwidth occupied by $M(\omega)$ is ω_m, but the frequency band occupied by $S_{\text{DSB}}(\omega)$ is $2\omega_m$. Since bandwidth is at a premium in many applications, this fact is important for design trade-offs with other modulation methods.

A sketch of $s_{\text{DSB}}(t)$ in Eq. (5.2.1) is shown in Fig. 5.2.2(b) for the assumed time-domain message signal sketched in Fig. 5.2.2(a). Notice that the dashed

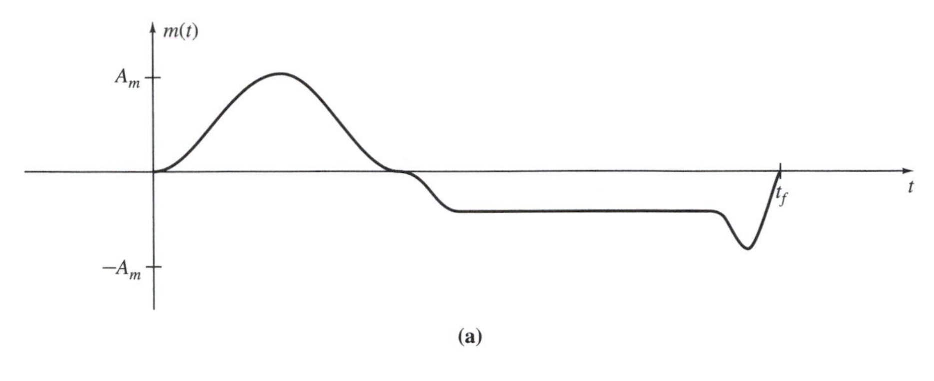

(a)

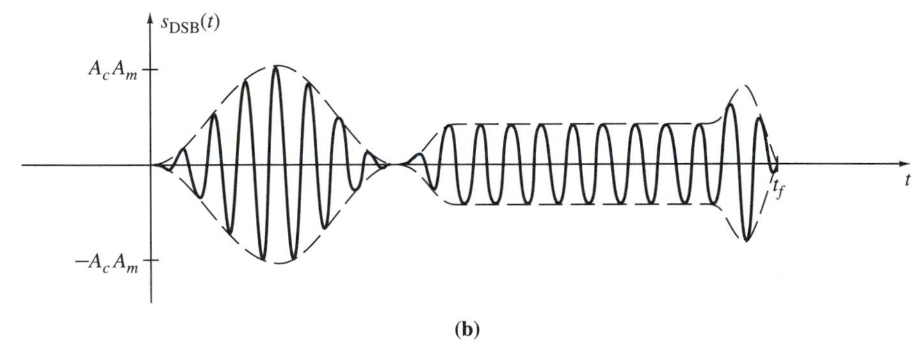

(b)

FIGURE 5.2.2 Time-domain waveforms for Eq. (5.2.1): (a) Message signal; (b) product of $m(t)$ and $A_c \cos \omega_c t$.

outline in Fig. 5.2.2(b) is simply $m(t)$ and $m(t)$ flipped about the time axis. This dashed line is, of course, not physically observable, but the outline is shown in the figure, since it is commonly called the *envelope* of the waveform. Notice that in sketching Fig. 5.2.2(b), we have again used the fact that $\omega_c > \omega_m$. This is evident since the inner waveform has frequency ω_c and oscillates several times while the amplitude of $m(t)$ is changing only slightly.

The signal $s_{DSB}(t)$ is the modulated waveform and would be the transmitted signal in a communication system that uses AMDSB-SC modulation. However, we have yet to demonstrate that $m(t)$ can be recovered from $s_{DSB}(t)$, and certainly this is a requirement for any useful system. To illustrate the recovery of $m(t)$, consider Fig. 5.2.3. Figure 5.2.3(a) represents the transmitter, while (b) is a block diagram of the receiver. Following Fig. 5.2.3(b), we form

$$x(t) = s_{DSB}(t) \cos \omega_c t = A_c m(t)[\tfrac{1}{2} + \tfrac{1}{2} \cos 2\omega_c t]. \qquad (5.2.3)$$

From Eq. (5.2.3) we see that in the frequency domain we have a scaled version of $M(\omega)$ at $\omega = 0$ and at $\omega = \pm 2\omega_c$. Thus if the low-pass filter in Fig. 5.2.3(b) is distortionless for $0 \leq |\omega| \leq \omega_m$ and has a cutoff frequency (ω_{co}) such that $\omega_m < \omega_{co} < 2\omega_c - \omega_m$, then $m(t)$ can be recovered exactly (within a scale factor).

The carrier signal at the receiver, $\cos \omega_c t$, is given the special name *local oscillator* (LO). The accuracy of the LO is extremely important if $m(t)$ is to be

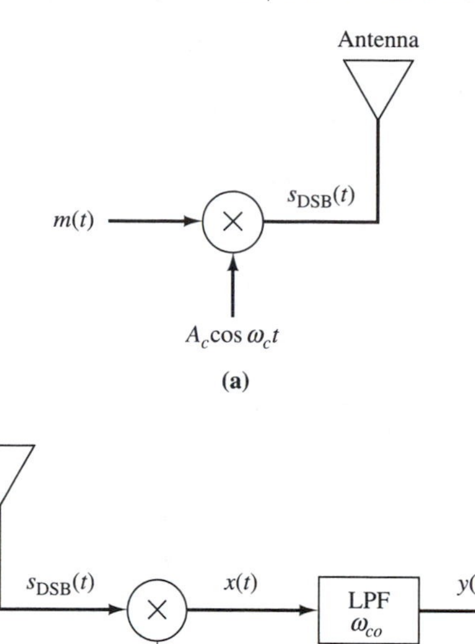

FIGURE 5.2.3 A MDSB-SC System block diagram: (a) Transmitter; (b) receiver.

recovered undistorted. For instance, in Fig. 5.2.3 we assumed that the carrier used at the transmitter and the LO used at the receiver had exactly the same frequency and phase. We therefore say that the carrier and the LO are synchronized in frequency and phase, and as a result, this type of demodulation is sometimes called *synchronous* or *coherent* demodulation. If the LO is not synchronized with the carrier, distortion will result. To illustrate this claim, let the transmitted signal be $s_{DSB}(t)$ in Eq. (5.2.1) but let the LO signal be $\cos[(\omega_c + \Delta\omega_c)t + \theta]$, where we have included a frequency error $\Delta\omega_c$ and a phase error θ. Then

$$\begin{aligned} s_{DSB}(t)\cos[(\omega_c + \Delta\omega_c)t + \theta] &= A_c m(t) \cos \omega_c t \cos[(\omega_c + \Delta\omega_c)t + \theta] \\ &= \frac{A_c}{2} m(t)\{\cos[\Delta\omega_c t + \theta] \\ &\quad + \cos[(2\omega_c + \Delta\omega_c)t + \theta]\}. \end{aligned} \tag{5.2.4}$$

If this signal is applied to the low-pass filter in Fig. 5.2.3(b), the distorted output will be (assuming that $\Delta\omega_c$ is not too large)

$$y_d(t) = \frac{A_c}{2} m(t) \cos[\Delta\omega_c t + \theta]. \tag{5.2.5}$$

The frequency error causes $m(t)$ to be modulated, which can cause a wide variety of distortions depending on the relationship between ω_m and $\Delta\omega_c$. If we assume that $\Delta\omega_c = 0$, we see that the phase error causes $m(t)$ to be attenuated. As θ gets larger and approaches 90°, $y_d(t)$ approaches zero! Certainly, inaccurate LO frequency and phase are to be avoided if possible.

5.3 Conventional AM

The most familiar application of amplitude modulation is to AM radio. In this application, we have many receivers (radios) and relatively few transmitters (radio stations). As a result, it is highly desirable that the receivers be as simple and hence as cheap as possible. If we use AMDSB-SC, we must design the receiver such that the LO frequency and phase track that of the transmitter. Since this can be expensive, another type of amplitude modulation is preferred.

For this type of AM, the transmitted signal is given by

$$s_{\text{AM}}(t) = [A + aA_c m(t)] \cos \omega_c t = A[1 + am_n(t)] \cos \omega_c t, \tag{5.3.1}$$

where $m_n(t) = m(t)/|m(t)|_{\max} = m(t)/A_m$ is a normalized version of $m(t)$, $A = A_c A_m$, and $0 \leq a \leq 1$ is called the *modulation index*. We require that $A > aA_c|m(t)|_{\max}$. The signal $s_{\text{AM}}(t)$ in Eq. (5.3.1) is called AMDSB-TC (transmitted carrier), *conventional AM*, or simply AM. The reasons for the designation AMDSB-TC are most evident from the Fourier transform of $s_{\text{AM}}(t)$; however, it is clear by inspection of Eq. (5.3.1) that in addition to the product of $m(t)$ and $\cos \omega_c t$ there is another term which consists of only $\cos \omega_c t$ (the carrier). The signal in Eq. (5.3.1) is called *conventional AM*, or just AM, because of its widespread use in commercial AM radio.

If we assume that $\mathscr{F}\{m(t)\} = M(\omega)$ is as shown in Fig. 5.2.1(a), the Fourier transform of $s_{\text{AM}}(t)$ is shown in Fig. 5.3.1 and is given by

$$S_{\text{AM}}(\omega) = \pi A[\delta(\omega + \omega_c) + \delta(\omega - \omega_c)] + \frac{aA_c}{2}[M(\omega + \omega_c) + M(\omega - \omega_c)]. \tag{5.3.2}$$

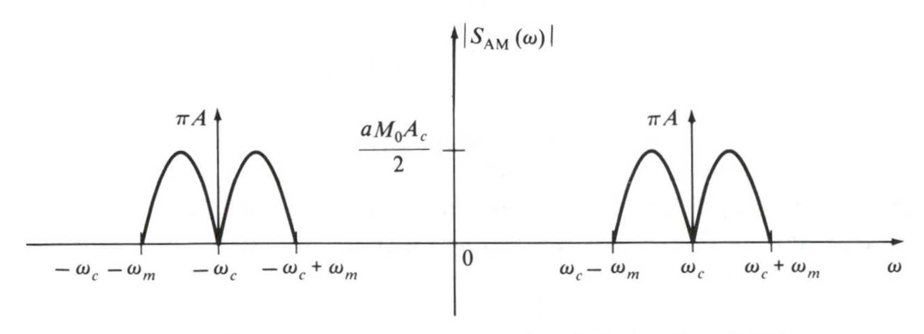

FIGURE 5.3.1 Frequency content of conventional AM.

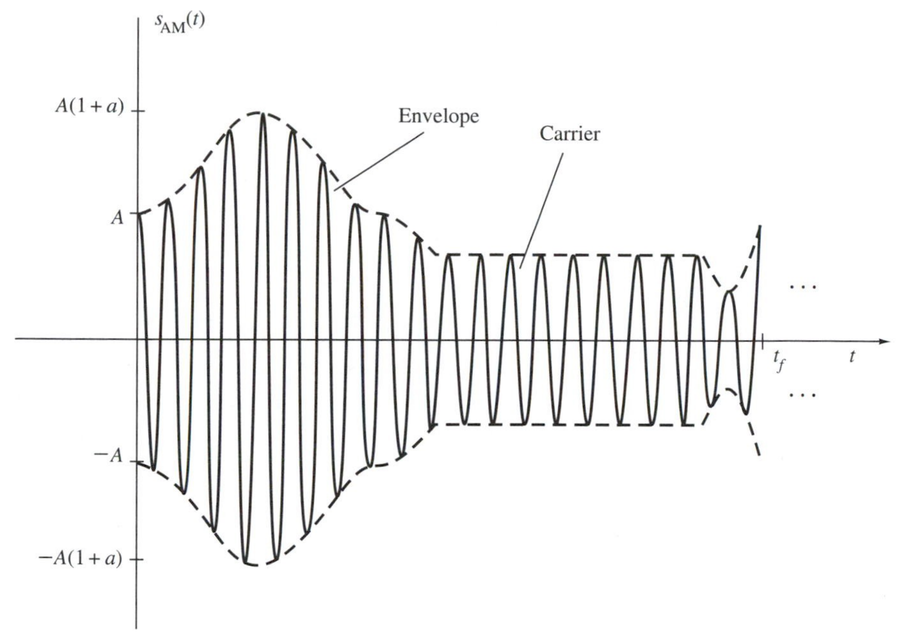

FIGURE 5.3.2 Time-domain sketch of $s_{AM}(t)$.

It is clear both from Fig. 5.3.1 and from Eq. (5.3.2) that there is now a separate carrier component not present in $S_{DSB}(\omega)$ [compare Fig. 5.2.1(b) and Eq. (5.2.2)]. Of course, the two sidebands remain.

Although we have established that there is, in fact, a separate carrier term in $s_{AM}(t)$, the question is: Why is it there? To address this question, let us assume that $m(t)$ is as shown in Fig. 5.2.2(a) and sketch $s_{AM}(t)$ in the time domain. This waveform is illustrated by Fig. 5.3.2. The key point of interest in this figure is that the positive outer envelope of $s_{AM}(t)$ has the exact shape of $m(t)$. As a result, if we could construct a receiver that simply follows this envelope, $m(t)$ can be recovered. That this can be achieved will be discussed shortly. Continuing our discussion of Fig. 5.3.2, we note that we have made use of the requirement that $A > aA_c|m(t)|_{\max} = aA_cA_m$. The purpose of this requirement is to guarantee that the positive envelope never goes negative, and hence that the positive envelope has the exact shape of $m(t)$. This could just as well have been guaranteed by making A larger than the modulation index times the amplitude of the most negative swing of $A_cm(t)$. However, since most waveforms of interest, for example, speech and music, have approximately equal positive and negative voltage amplitude ranges, our requirement on A is logical.

We have now established that the purpose of the extra carrier term in $s_{AM}(t)$ is to preserve the positive envelope of the transmitted signal in the shape of $m(t)$. The way we take advantage of this fact is to use an *envelope detector* at the receiver. A circuit diagram of an envelope detector is shown in Fig. 5.3.3. The principle behind this circuit is extremely simple. During a positive cycle of the AM wave, the diode is foward biased and the capacitor charges up to the peak

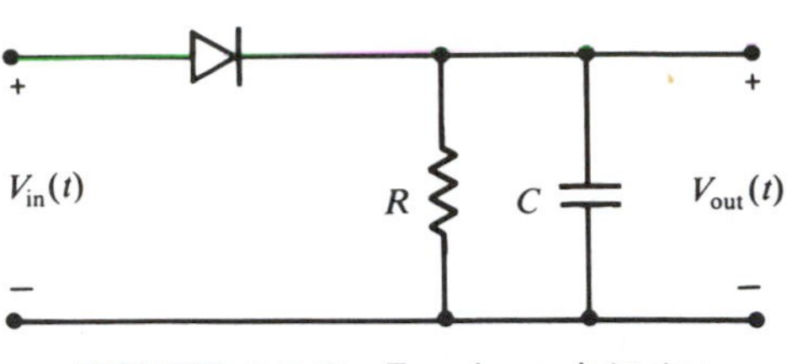

FIGURE 5.3.3 Envelope detector.

value, which is the envelope. During a negative cycle of $s_{AM}(t)$, the diode is back biased and the capacitor discharges through the resistor. The principal possible source of distortion is the time constant of the discharge during negative cycles. If the capacitor discharges too rapidly, the output voltage will fall far below the envelope before the next positive cycle. This situation will cause excessive ripple at the detector output. If $\omega_c \gg \omega_m$, however, the ripples will be slight. Another possible source of distortion is called "failure to follow," and it occurs when the capacitor discharges too slowly to follow a rapid decrease in the envelope. Of course, by a judicious choice of the RC time constant these problems can be avoided.

The output of the envelope detector is thus $A + aA_c m(t)$, and since it is not difficult to remove the dc component A, we can easily obtain a scaled version of $m(t)$. Note that the envelope detector requires no accurate knowledge of the carrier frequency or phase, and hence the requirement for an accurate LO signal as in AMDSB-SC has been eliminated.

It is common for conventional AM systems to be discussed in terms of "percent modulation." Such terminology relates to the modulation index, a, in Eq. (5.3.1). If $a = 1$, it is called 100% modulation, and if $a = 0.25$, it is called 25% modulation. The reason for this nomenclature is that it gives an indication of the relative maximum amplitudes of the carrier term alone and the message-carrying term in Eq. (5.3.1). If $a = 1$, then the two terms have equal amplitudes, since $A = A_c A_m$. If $a = 0.25$, the message carrying term has only one-fourth the amplitude of the free carrier term. It is intuitive that for maximum system efficiency, as much energy as possible should be allocated to the information-bearing term. For 100% modulation, this is exactly what is occurring.

5.4 Single Sideband

While both AMDSB-SC and conventional AM are successful methods for relocating a message signal in frequency, it is clear that both techniques do so by doubling the bandwidth over that required by the baseband message signal. Since the spectral shape of the message is preserved in both the upper and lower sidebands, only one of these sidebands is necessary to represent exactly the message at the receiver. Therefore, to conserve bandwidth, we now investigate modulation methods that require half the frequency allocation of the DSB techniques discussed in Sections 5.2 and 5.3.

Modulation methods that transmit only the upper or lower sideband of the DSB signal are called *single sideband* (SSB). There are two common methods for generating SSB signals. One method consists of generating a DSB-SC signal and then filtering out the unwanted sideband. This approach requires an excellent filter and the filtering is usually performed at a lower frequency and then translated to a higher frequency. The filtering approach is most attractive for message signals that have little low-frequency content. Since frequencies below 100 to 200 Hz are not critical to the intelligibility of voice signals, speech can be transmitted successfully using this filtering technique for SSB generation.

The second method for generating SSB signals is the phase shift method. This approach is best introduced by an example.

EXAMPLE 5.4.1

Let the message signal be a single tone, $m(t) = \cos \omega_m t$. The Fourier transform of $m(t)$ is illustrated in Fig. 5.4.1(a). Multiplying $m(t)$ by the carrier wave, we get $s_{\text{DSB}}(t) = m(t) \cos \omega_c t$, which has the spectrum shown in Fig. 5.4.1(b). If we pass $s_{\text{DSB}}(t)$ through a filter that retains only the lower sideband, the remaining spectral content is as shown in Fig. 5.4.1(c). In the time domain the LSB signal is $s_{\text{LSB}}(t) = \cos(\omega_c - \omega_m)t = \cos \omega_m t \cos \omega_c t + \sin \omega_m t \sin \omega_c t$. Thus we can produce $s_{\text{LSB}}(t)$ by generating $\cos \omega_m t \cos \omega_c t$ and $\sin \omega_c t \sin \omega_m t$ and adding them together. The first term is just the product of the message and carrier signals, while the latter can be written as $\sin \omega_m t \sin \omega_c t = \cos(\omega_m t - \pi/2) \cdot \cos(\omega_c t - \pi/2)$, and hence can be obtained by separately shifting the phases of the message and carrier waves by $-90°$ and then forming their product.

Although Example 5.4.1 exhibits only a single special case, the result holds in general for all message signals, since any signal can be expressed as a sum of sinusoids over a given finite interval. The general result is that for a message signal $m(t)$, the SSB signal is

$$s_{\text{SSB}}(t) = m(t) \cos \omega_c t \pm m_h(t) \sin \omega_c t, \tag{5.4.1}$$

where $m_h(t)$ is obtained by shifting the phase of each frequency component of $m(t)$ by $-90°$. In Eq. (5.4.1), the positive sign is associated with the LSB, while the USB is gotten with the negative sign. The component $m_h(t)$ is called the *Hilbert transform* of $m(t)$ and is given by

$$\mathscr{H}\{m(t)\} = m_h(t) = \frac{1}{\pi} \int_{-\infty}^{\infty} \frac{m(\tau)}{t - \tau} \, d\tau. \tag{5.4.2}$$

Since $m_h(t)$ is obtained by shifting the phase of each frequency component of $m(t)$ by $-90°$, we know that

$$\mathscr{F}\{m_h(t)\} = M_h(\omega) = \begin{cases} -jM(\omega), & \omega \geq 0 \\ jM(\omega), & \omega < 0. \end{cases} \tag{5.4.3}$$

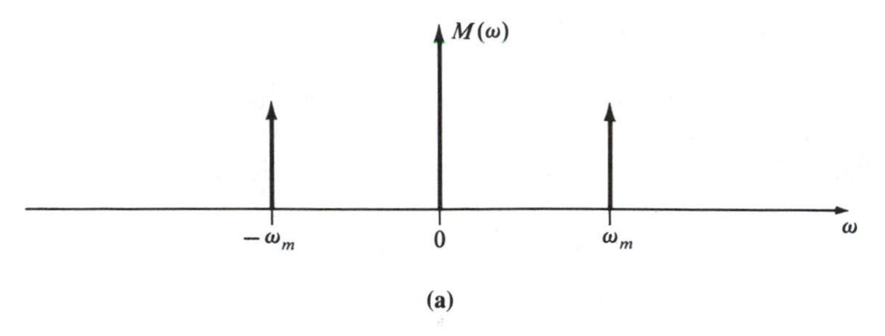

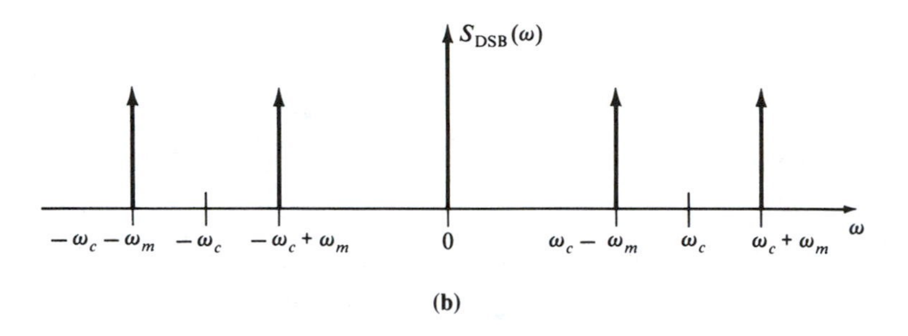

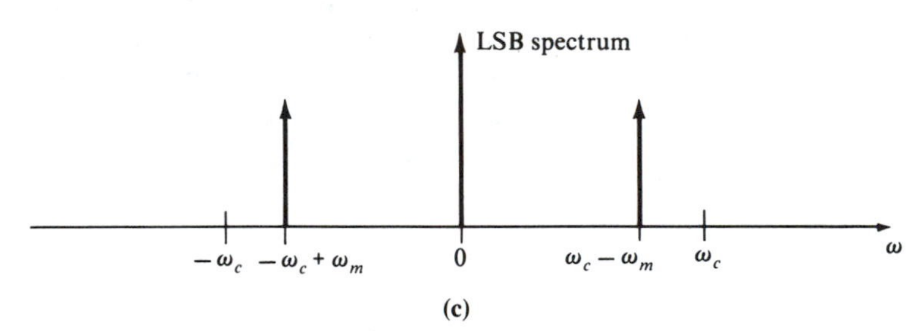

FIGURE 5.4.1 Spectra for Example 5.4.1: (a) Message signal; (b) double sideband signal; (c) lower sideband.

Using Eq. (5.4.3), we can generalize Example 5.4.1 and demonstrate that Eq. (5.4.1) represents a single-sideband signal for a general $m(t)$.[2] Consider the Fourier transform of a general message signal $M(\omega) = \mathscr{F}\{m(t)\}$, which is sketched in Fig. 5.4.2(a). Defining

$$M_p(\omega) = M(\omega), \qquad \omega \geq 0, \tag{5.4.4}$$

and

$$M_n(\omega) = M(\omega), \qquad \omega < 0, \tag{5.4.5}$$

[2] This development is due to J. L. LoCicero [private correspondence, 1984]. A similar treatment appears in Lathi [1968].

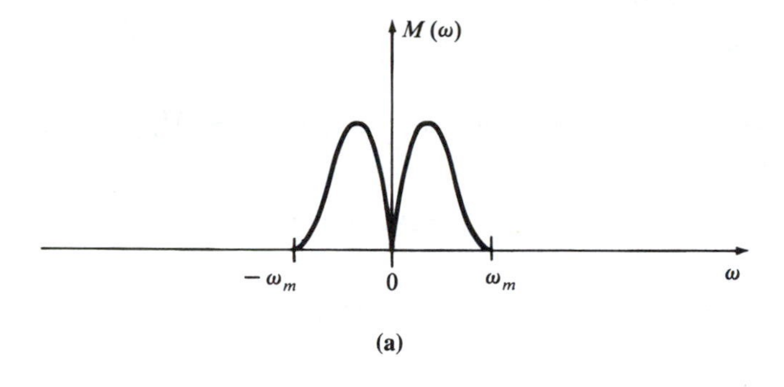

(a)

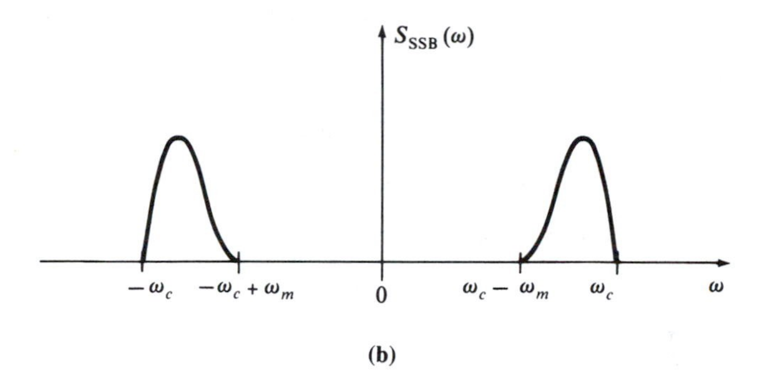

(b)

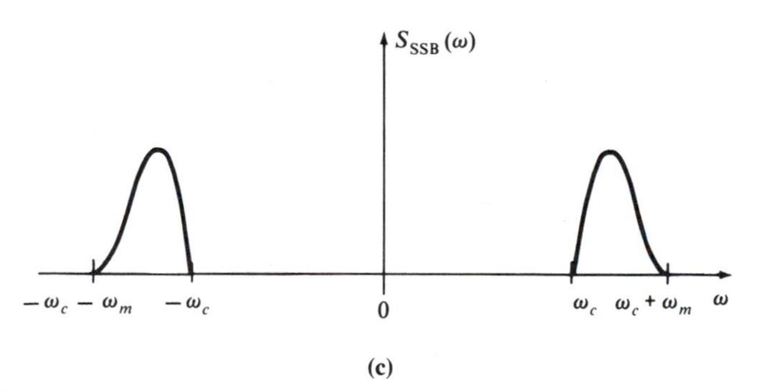

(c)

FIGURE 5.4.2 Single sideband spectra for a general message: (a) Message spectrum; (b) spectrum of a lower sideband SSB signal; (c) spectrum of an upper sideband SSB signal.

we can write

$$M(\omega) = M_p(\omega) + M_n(\omega), \tag{5.4.6}$$

and from Eq. (5.4.3),

$$M_h(\omega) = -jM_p(\omega) + jM_n(\omega). \tag{5.4.7}$$

Taking the Fourier transform of Eq. (5.4.1), we have

$$
\begin{aligned}
S_{SSB}(\omega) &= \mathscr{F}\{s_{SSB}(t)\} \\
&= \frac{1}{2}\{M_p(\omega+\omega_c) + M_n(\omega+\omega_c) + M_p(\omega-\omega_c) + M_n(\omega-\omega_c)\} \\
&\quad \pm \frac{1}{2j}\{-j[-M_p(\omega+\omega_c) + M_n(\omega+\omega_c) + M_p(\omega-\omega_c) - M_n(\omega-\omega_c)]\} \\
&= \frac{1}{2}\{M_p(\omega+\omega_c) + M_n(\omega+\omega_c) + M_p(\omega-\omega_c) + M_n(\omega-\omega_c)\} \\
&\quad \pm \frac{1}{2}\{M_p(\omega+\omega_c) - M_n(\omega+\omega_c) - M_p(\omega-\omega_c) + M_n(\omega-\omega_c)\}.
\end{aligned}
\tag{5.4.8}
$$

For the positive sign on the second set of braces in Eq. (5.4.8), we obtain

$$S_{SSB}(\omega) = M_p(\omega+\omega_c) + M_n(\omega-\omega_c), \tag{5.4.9}$$

which is sketched in Fig. 5.4.2(b). Similarly, we have for the negative sign,

$$S_{SSB}(\omega) = M_p(\omega-\omega_c) + M_n(\omega+\omega_c), \tag{5.4.10}$$

which is sketched in Fig. 5.4.2(c). Clearly, Eq. (5.4.9) represents a LSB signal, while Eq. (5.4.10) is an USB signal, and thus Eq. (5.4.1) is valid for any $m(t)$, as claimed.

To recover $m(t)$ from the SSB signal in Eq. (5.4.1), we can use coherent demodulation, which consists of multiplying by the carrier waveform to get

$$
\begin{aligned}
S_{SSB}(t)\cos\omega_c t &= m(t)\cos^2\omega_c t \pm m_h(t)\sin\omega_c t\cos\omega_c t \\
&= \frac{m(t)}{2} + \frac{1}{2}[m(t)\cos 2\omega_c t \pm m_h(t)\sin 2\omega_c t].
\end{aligned}
\tag{5.4.11}
$$

The $2\omega_c$ components can be removed from Eq. (5.4.11) by simple low-pass filtering, and thus $m(t)$ is recovered. Of course, as in AMDSB-SC demodulation, phase or frequency inaccuracies in the LO will result in distortion of the recovered message.

The SSB signal discussed thus far is a suppressed carrier signal in that no separate carrier term is available in the transmitted signal. Having illustrated the utility of envelope detection in Section 5.3, we find that it is of interest to investigate the possibility of using envelope detection in conjunction with SSB transmission. Such a combination would conserve bandwidth and still be simple to demodulate. To begin, consider the (LSB) SSB-TC (transmitted carrier) signal

$$s_{SSBTC}(t) = A\cos\omega_c t + m(t)\cos\omega_c t + m_h(t)\sin\omega_c t, \tag{5.4.12}$$

which can be rewritten in the form

$$s_{SSBTC}(t) = \{[A + m(t)]^2 + m_h^2(t)\}^{1/2}\cos[\omega_c t + \theta], \tag{5.4.13}$$

where

$$\theta = -\tan^{-1}\left\{\frac{m_h(t)}{A + m(t)}\right\}. \tag{5.4.14}$$

If the SSB-TC is applied to an envelope detector, the envelope detector output will be

$$\begin{aligned}\rho(t) &= \{A^2 + m^2(t) + 2Am(t) + m_h^2(t)\}^{1/2} \\ &= A\left\{1 + \frac{m^2(t)}{A^2} + \frac{2m(t)}{A} + \frac{m_h^2(t)}{A^2}\right\}^{1/2}.\end{aligned} \tag{5.4.15}$$

If $A \gg |m(t)|$ and $A \gg |m_h(t)|$, the envelope is approximately

$$\rho(t) \cong A\left\{1 + \frac{2m(t)}{A}\right\}^{1/2}. \tag{5.4.16}$$

Using a series expansion for the square root and neglecting terms of second order and higher, we obtain

$$\rho(t) \cong A + m(t). \tag{5.4.17}$$

Removing the dc term (A), we are left with $m(t)$, as desired.

Unfortunately, to arrive at Eq. (5.4.17) we had to assume that $A \gg m(t)$ and $A \gg m_h(t)$. Although this can usually be achieved in practice (but not always), it is extremely inefficient to allocate so much power to the carrier term. As a result, envelope detection is not nearly as important in SSB transmission as it is for conventional AM.

5.5 Vestigial Sideband

A single-sideband signal can be generated by first producing an AMDSB-SC signal and then filtering out one of the sidebands, or the phase shift method demonstrated in Example 5.4.1 can be used. The former method is most successful on messages that have little low-frequency content and hence do not require exceptionally sharp filter cutoff characteristics. The phase shift method can be employed for messages with frequency content down to dc, but the design of a 90° phase shift network over a wide range of frequencies is not an easy task. Therefore, in those instances where SSB generation is difficult, we must search for another modulation method, which is implementable but still bandwidth efficient.

Vestigial sideband (VSB) AM transmission has both of these characteristics. VSB relaxes the stringent sharp cutoff requirements of SSB by retaining a vestige or trace of the unwanted sideband in the transmitted signal. As long as only a portion of the "extra" sideband is sent, VSB occupies less bandwidth than DSB; however, VSB always requires greater bandwidth than SSB. Vestigial sideband signals are generated by approximately filtering a DSB-SC waveform.

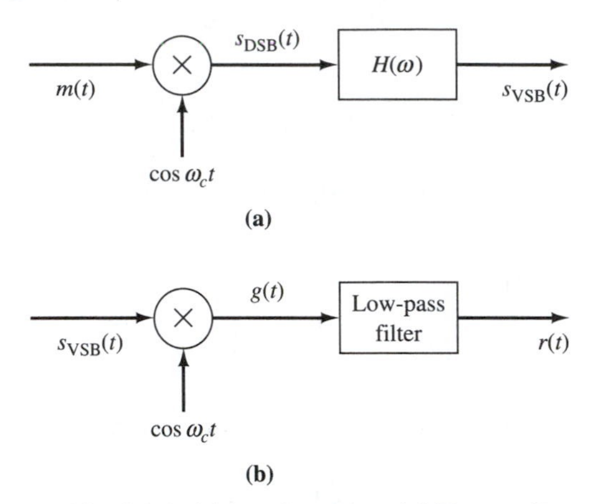

FIGURE 5.5.1 Vestigial sideband system: (a) Transmitter; (b) receiver.

Of course, the filter characteristics for obtaining a VSB signal from a DSB signal cannot be chosen arbitrarily.

The constraints on the filter transfer functions for VSB can be established by considering the communication system block diagram shown in Fig. 5.5.1. In this diagram, $H(\omega)$ produces a VSB signal from the just-generated DSB signal, and we desire to determine any requirements on $H(\omega)$ such that $m(t)$ can be recovered by synchronous or coherent demodulation, as indicated in Fig. 5.5.1(b). By inspection of Fig. 5.5.1(a), we have that $\mathscr{F}\{s_{\mathrm{VSB}}(t)\} = S_{\mathrm{VSB}}(\omega)$ is given by

$$S_{\mathrm{VSB}}(\omega) = \tfrac{1}{2}[M(\omega + \omega_c) + M(\omega - \omega_c)]H(\omega), \tag{5.5.1}$$

where $\mathscr{F}\{m(t)\} = M(\omega)$. At the receiver we have $g(t) = s_{\mathrm{VSB}}(t) \cos \omega_c t$, so

$$\begin{aligned} G(\omega) = \tfrac{1}{4}\{&[M(\omega + 2\omega_c) + M(\omega)]H(\omega + \omega_c) \\ &+ [M(\omega) + M(\omega - 2\omega_c)]H(\omega - \omega_c)\}. \end{aligned} \tag{5.5.2}$$

The receiver low-pass filter removes the components at $\pm 2\omega_c$, so that the Fourier transform of $r(t)$ is

$$R(\omega) = \tfrac{1}{4}M(\omega)[H(\omega + \omega_c) + H(\omega - \omega_c)]. \tag{5.5.3}$$

For distortionless reception of $m(t)$, we require that

$$R(\omega) = CM(\omega)e^{-j\omega t_d}, \qquad |\omega| \le \omega_m, \tag{5.5.4}$$

where C is a constant, ω_m is the highest frequency present in $m(t)$, and t_d is a time delay. Directly from Eq. (5.5.4), we find that

$$H(\omega + \omega_c) + H(\omega - \omega_c) = Ce^{-j\omega t_d} \tag{5.5.5}$$

for $|\omega| \le \omega_m$.

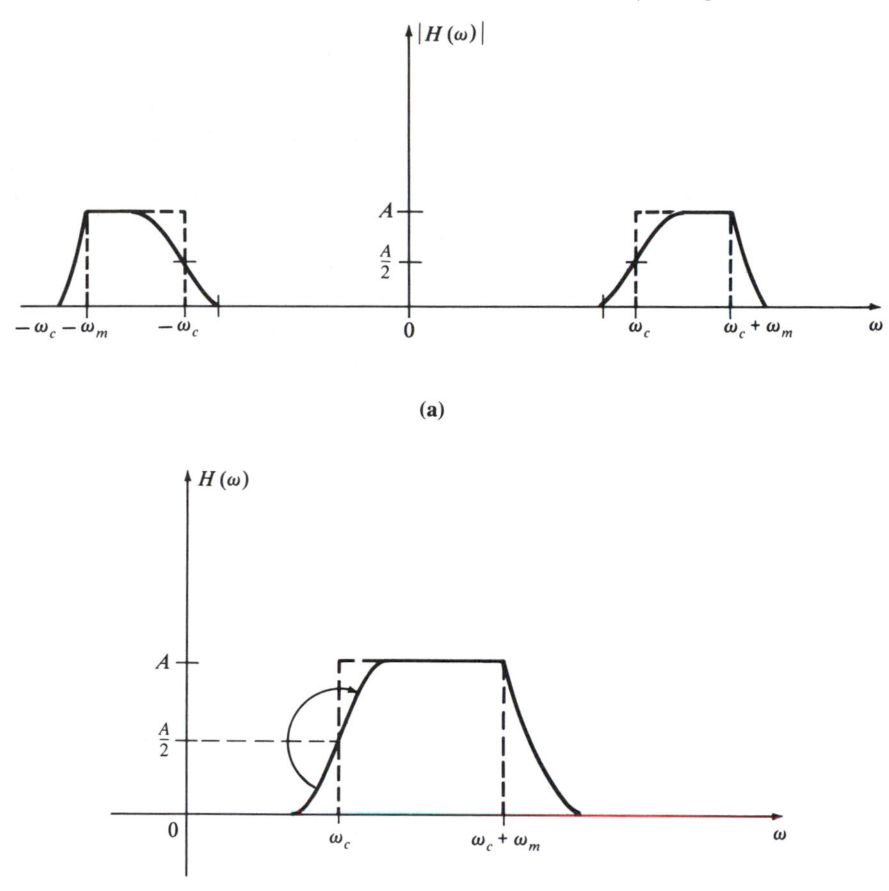

FIGURE 5.5.2 VSB filter magnitude requirement: (a) Magnitude of $H(\omega)$; (b) complementary symmetry.

If we ignore the phase of $H(\omega)$ [assume that $\underline{/H(\omega)} = 0$], the requirement in Eq. (5.5.5) necessitates that $|H(\omega)|$ have complementary symmetry about ω_c, as demonstrated graphically in Fig. 5.5.2. If the lower-sideband response is pivoted about the point $|H(\omega_c)|$ and an ideal filter response is obtained in the range $\omega_c \leq \omega \leq \omega_c + \omega_m$, the filter has the complementary symmetry requirement. The linear phase characteristic is less easy to attain; however, in many applications, the phase requirement may prove unimportant. Although we have carried out the development for an upper-sideband bandpass filter, the results are of course valid for the lower-sideband case.

Vestigial sideband transmission can also be used with envelope detection if the necessary carrier term is injected. As in SSB, though, envelope detection of VSB is an inefficient use of available transmitted power, since most of the power must be in the carrier signal.

5.6 Superheterodyne Systems

Broadcast AM radio uses conventional AM modulation (AMDSB-TC). Each pair of adjacent radio stations is spaced 10 kHz apart in the frequency spectrum. To select the radio station of our choice, we need to be able to tune our receiver in frequency, and we also need a receiver that will reject all stations except the one to which we wish to listen. As a minimum, this requires a tunable, highly selective bandpass filter. Unfortunately, such filters are not particularly easy to build, and since it is desirable to keep radio receivers as inexpensive as possible, an alternative approach is used.

Most AM radio receivers are of the superheterodyne type shown in Fig. 5.6.1. The main components of the superheterodyne receiver are (1) the radio frequency (RF) amplifier, which is tuned to the desired radio frequency; (2) the intermediate-frequency (IF) amplifier, which is fixed and provides most of the gain and selectivity; (3) the tunable local oscillator (LO); (4) the envelope detector; and (5) the audio-frequency (AF) amplifier, which matches the power level out of the envelope detector to that required by the speaker. The way this receiver works is that the desired radio signal is translated down to the IF center frequency by the tunable LO. The IF amplifier then rejects all other signals except the one radio station in its passband. Since the IF filter is fixed, the problem of building a tunable, highly selective filter is avoided. Further, since the carrier frequency supplied to the envelope detector is always the same, the envelope detector performance does not vary for different radio stations.

Commercial AM radio occupies the frequency band from 535 to 1605 kHz with radio stations located in the range 540 to 1600 kHz, spaced 10 kHz apart. Since double-sideband AM transmission is used, the baseband message bandwidth is 5 kHz. The bandwidths of the amplifiers in Fig. 5.6.1 are $10\text{ kHz} < \text{BW}_{\text{RF}} < 910\text{ kHz}$, $\text{BW}_{\text{IF}} = 10\text{ kHz}$, and $\text{BW}_{\text{AF}} = 5\text{ kHz}$, where BW_{RF} refers to the RF amplifier bandwidth; and so on. The choices of the IF and AF amplifier

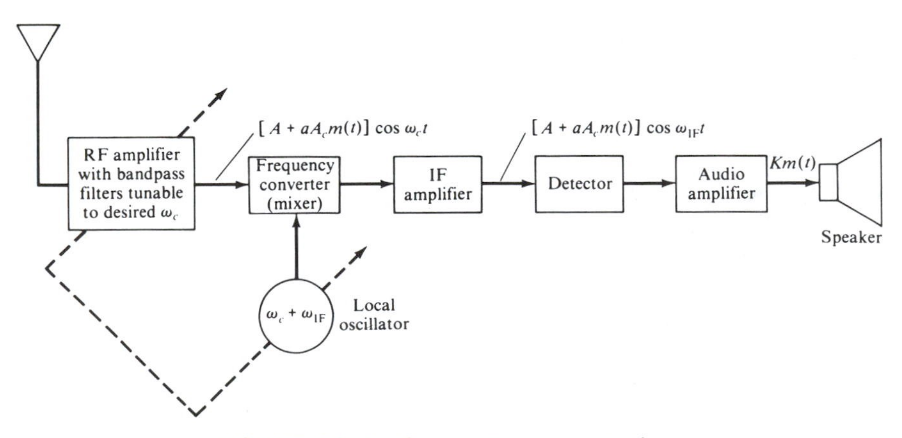

FIGURE 5.6.1 Superheterodyne receiver.

bandwidths are obvious, and the bandwidth of the RF amplifier will be justified shortly.

The center frequency of the IF amplifier is 455 kHz, and each carrier frequency from 540 to 1600 kHz must be shifted to this frequency. Thus the LO could be tuned over the range 995 to 2055 kHz or over the range 85 to 1145 kHz. The former tuning range was chosen for commercial application, since it constitutes only a 2:1 tuning ratio as opposed to the 13:1 ratio required by the latter range. As the LO is tuned, the center frequency of the RF amplifier is also varied. The purpose of the tunable RF amplifier is to reject the *image station.*

The concept of an image station is best illustrated by an example.

EXAMPLE 5.6.1

We wish to select the radio station at 550 kHz. To do this, we tune the LO to $550 + 455 = 1005$ kHz. However, without an RF amplifier, the radio station at 1460 kHz will also be translated to the IF center frequency, since $1460 - 1005 = 455$. The station at 1460 kHz is called an *image station.* With the tunable RF amplifier, all stations outside the band $550 \pm \frac{1}{2}(\mathrm{BW_{RF}})$ are eliminated before they get to the tunable LO. Thus the choice of the RF amplifier bandwidth is not critical as long as the image station is rejected, and the RF amplifier need not be highly selective, since all stations except the image station will be eliminated by the fixed IF filter.

Generalizing the discussion in the preceding example, we see that the center frequency of the RF amplifier is f_c, and the image station will be located at $f_c + 910$ kHz. Therefore, $\frac{1}{2}\mathrm{BW_{RF}}$ only needs to be less than 910 kHz. Usually, the RF amplifier passband is stated to be $f_c \pm 455$ kHz, so $\frac{1}{2}\mathrm{BW_{RF}} = 455$ kHz. In practice, this value is not critical, since the mixer bandwidth is commonly less than this value. The minimum allowable bandwidth of the RF amplifier is determined by the bandwidth occupied by a single station, which is 10 kHz. We thus get the specification that $10 \text{ kHz} < \mathrm{BW_{RF}} < 910 \text{ kHz}$. Of course, since this filter must be tunable, it is not designed to have steep cutoff characteristics.

5.7 Quadrature Amplitude Modulation

In many applications, bandwidth is an important commodity. A widely used, bandwidth-efficient modulation technique called *quadrature amplitude modulation* (QAM) allows two message signals to be transmitted in the same frequency band without mutual interference. Given two distinct low-pass message signals $m_1(t)$ and $m_2(t)$ of bandwidth f_m hertz, message $m_1(t)$ amplitude modulates a carrier waveform $\cos \omega_c(t)$ and $m_2(t)$ modulates a carrier waveform $\sin \omega_c(t)$, and these two AMDSB-SC signals are summed to yield the QAM signal

$$s_{\mathrm{QAM}}(t) = A_c m_1(t) \cos \omega_c t + A_c m_2(t) \sin \omega_c t. \tag{5.7.1}$$

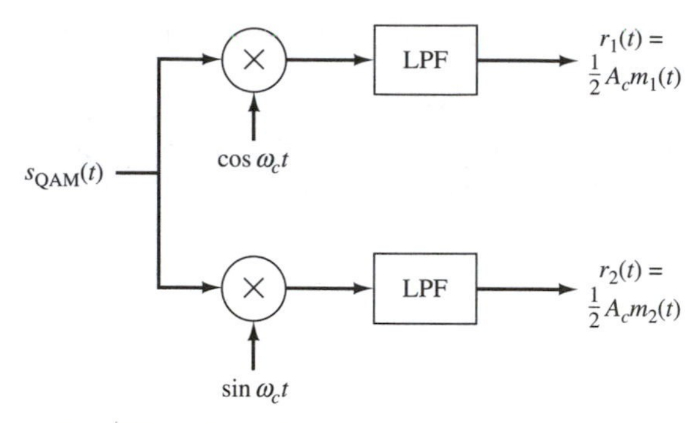

FIGURE 5.7.1 QAM receiver block diagram.

This waveform is called quadrature amplitude modulation, since the two message signals amplitude modulate separate carriers that are in phase quadrature with each other. Clearly, $s_{QAM}(t)$ occupies no greater bandwidth than an AMDSB-SC signal for either message alone, so the information-carrying content of the available bandwidth has effectively been doubled.

A block diagram of a QAM receiver is shown in Fig. 5.7.1, where $s_{QAM}(t)$ is fed to two parallel branches. The upper branch multiplies by $\cos \omega_c t$ and the lower branch by $\sin \omega_c t$. After low-pass filtering these products in each branch, the output of the upper branch is $(A_c/2)m_1(t)$ and the output of the lower branch is $(A_c/2)m_2(t)$. Absolutely accurate frequency and phase tracking is essential to prevent distortion. For example, if the two local oscillators have phase errors of, say, θ_1 and θ_2 in the upper and lower branches, respectively, the demodulated outputs of the low-pass filters are

$$r_1(t) = \frac{A_c}{2} m_1(t) \cos \theta_1 + \frac{A_c}{2} m_2(t) \sin \theta_1 \tag{5.7.2}$$

and

$$r_2(t) = \frac{A_c}{2} m_1(t) \sin \theta_2 + \frac{A_c}{2} m_2(t) \cos \theta_2. \tag{5.7.3}$$

Thus phase errors in the local oscillators cause *crosstalk* between the two quadrature components.

Noncoherent demodulation of $s_{QAM}(t)$ is not possible since the envelope of $s_{QAM}(t)$ is $A_c\sqrt{m_1^2(t) + m_2^2(t)}$. Can some sort of transmitted carrier QAM be used, as in Eq. (5.4.12) for SSB, to facilitate envelope detection?

5.8 Frequency-Division Multiplexing

It often occurs that the particular transmission channel available to us, whether it is a wire pair, coaxial cable, or free space, will support a much wider bandwidth than that required by a single baseband message waveform. In these

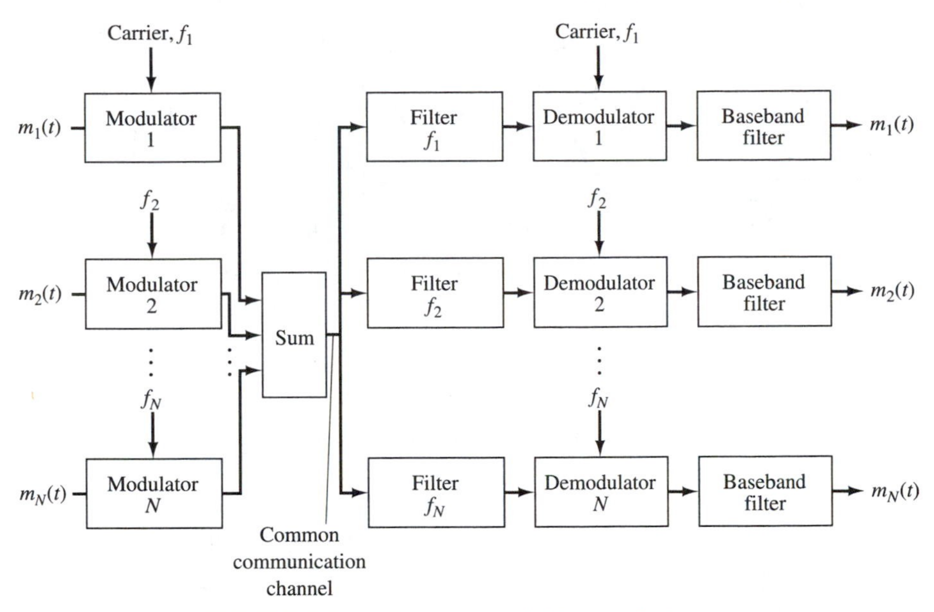

FIGURE 5.8.1 Frequency-division multiplexing.

instances it is possible to employ *frequency-division multiplexing* (FDM) to allow simultaneous transmission of numerous unrelated message signals over the single communications channel. The FDM concept is illustrated in general by Fig. 5.8.1.

In a FDM system, each baseband message signal is transmitted using the same type of modulation (e.g., AMDSB-SC, conventional AM, SSB, or VSB), and the carrier frequencies, ω_1 through ω_k, must be chosen such that none of the transmitted spectra overlap. Although the receiver could take a variety of forms, for the structure shown in Fig. 5.8.1, the bandpass filters are chosen to reject all messages except the one with the indicated carrier frequency. Demodulation then proceeds in a straightforward fashion.

The following example illustrates the details of FDM transmission.

EXAMPLE 5.8.1

Four message signals, $m_1(t)$, $m_2(t)$, $m_3(t)$, and $m_4(t)$, with the Fourier transforms sketched in Fig. 5.8.2(a) are to be transmitted in the frequency range $2\pi(100{,}000) \leq \omega \leq 2\pi(120{,}000)$ over a single communications channel. To do this using FDM, we must select a modulation technique. Clearly, DSB transmission is not possible, since it would require a bandwidth of 32 kHz and only 20 kHz has been allocated. Although it is possible to use VSB here with properly chosen vestige bandwidths, we choose SSB modulation instead.

By using SSB transmission, the message signals will occupy only 16 kHz of the total 20 kHz available. We thus have some flexibility in choosing carrier frequencies. A simple choice would be to select carrier frequencies such that

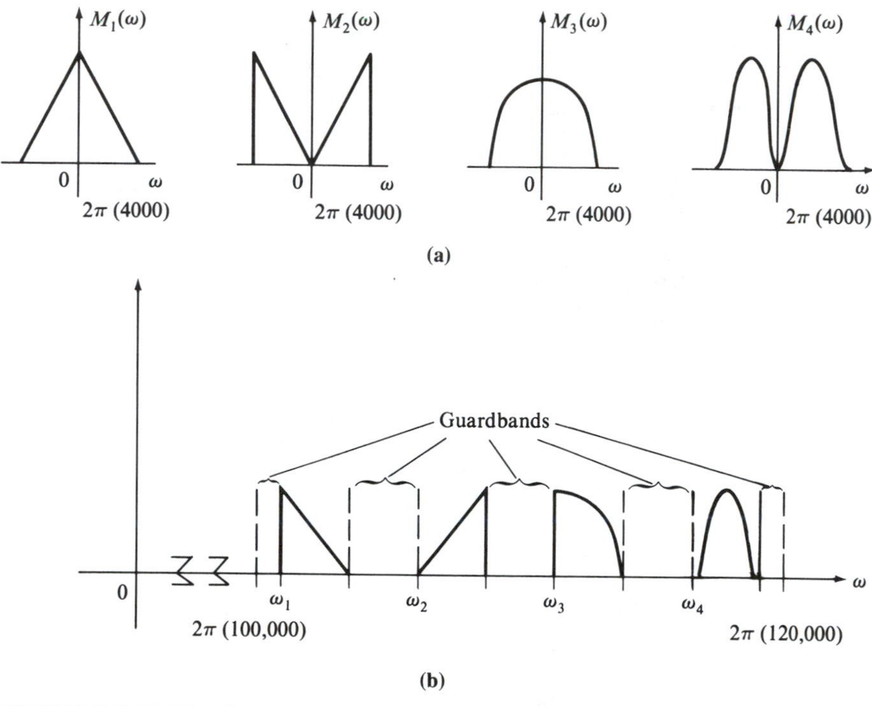

FIGURE 5.8.2 Fourier transforms for Example 5.8.1: (a) Fourier transforms of baseband message signals; (b) frequency division multiplexing with guardbands.

the transmitted messages are located contiguously at one end or the other of the allocated band. Instead, we will use the excess bandwidth to provide *guardbands.* Guardbands are vacant bands of frequencies between messages that "guard" against adjacent channel interference by providing extra spectral separation.

With this concept in mind and assuming USB SSB transmission, we assign the carrier frequencies $\omega_1 = 2\pi(100{,}500)$, $\omega_2 = 2\pi(105{,}500)$, $\omega_3 = 2\pi(110{,}500)$, and $\omega_4 = 2\pi(115{,}500)$. The multiplexed transmitted signal has the frequency content shown in Fig. 5.8.2(b). We have thus allowed guardbands of 1000 Hz between adjacent messages, a 500-Hz guardband below message 1, and a 500-Hz guardband above message 4. Clearly, these choices are not unique and many other selections are possible.

5.9 Amplitude Shift Keying/On–Off Keying

Digital signals constitute an ever-growing portion of the message or information signals we are required to transmit over our communication systems. Many existing communication channels are inherently analog in nature, and some

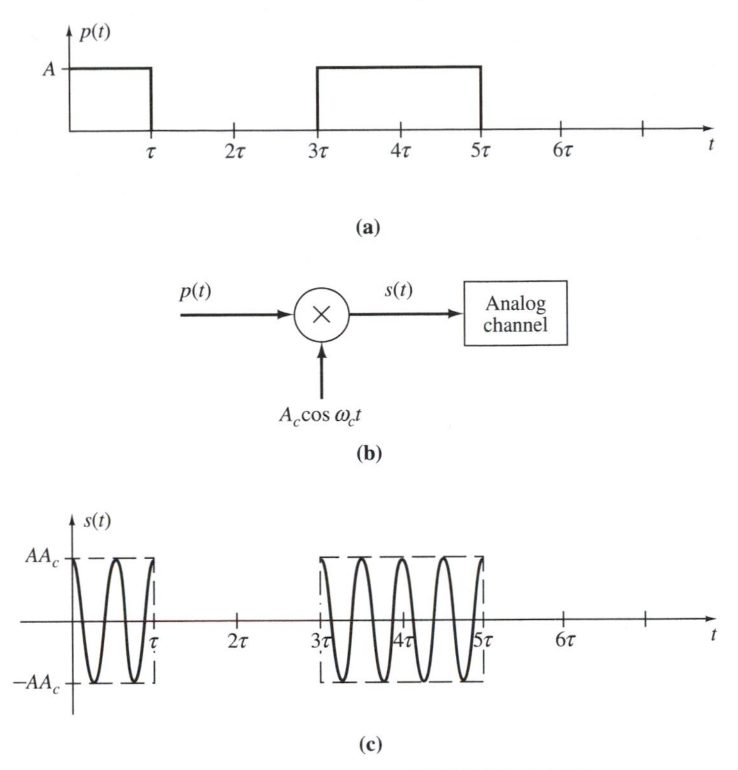

FIGURE 5.9.1 Binary transmission using AMDSB-SC: (a) Binary message sequence; (b) AMDSB-SC system; (c) modulated carrier (OOK).

important communication channels, such as those found in the telephone network, have been designed primarily to carry analog signals. Of course, the telephone network and many other systems are evolving toward an all-digital or mostly digital configuration.

To transmit digital information over an analog channel, we use analog modulation methods. Although all types of AM modulation are employed in data transmission systems, we begin our discussion with AMDSB-SC. Suppose that we wish to send the binary sequence shown in Fig. 5.9.1(a) using the AMDSB-SC system represented by Fig. 5.9.1(b). The input to the analog channel, $s(t)$, for this message signal is sketched in Fig. 5.9.1(c). Thus, for each rectangular pulse in the original data stream, we get a pulsed version of the carrier with a rectangular envelope.

The frequency content of $s(t)$ is not as easy to ascertain. Certainly, if we have a single rectangular pulse multiplied by $A_c \cos \omega_c t$, we know that we have a $\sin x/x$ shape centered about $\pm\omega_c$. However, $p(t)$ is not just a single binary pulse, and there are some intervening spaces. Furthermore, we would like to analyze our system's frequency content for all possible data signals that might be sent, not just the single sequence currently being investigated. In general, whether we have a pulse or no pulse in any given interval is a random variable, and we must treat it as such in our analysis. We learn how to do this in Appendix A, and we present the resulting spectral density in this section shortly.

To get some idea of the frequency content of a data signal transmitted using AMDSB-SC modulation, consider the periodic binary sequence sketched in Fig. 5.9.2(a). The Fourier transform of this sequence is a series of spectral lines spaced $2\pi/2\tau = \pi/\tau$ rad/sec apart. When $g(t)$ is passed through the AMDSB-SC system shown in Fig. 5.9.1(b), the magnitude of the Fourier transform of $s(t)$ is as sketched in Fig. 5.9.2(b). The envelope of the spectral lines depends on the pulse shape, while the spacing of the spectral lines depends on the period of the pulse train. When other pulse shapes are used, the time-domain signal in Fig. 5.9.1(c) and the frequency content in Fig. 5.9.2(b) both change accordingly.

The frequency content of a random sequence of pulses, with q = probability of a pulse of amplitude A and $1 - q$ = probability of no pulse, can be found using Eq. (A.10.13) from Appendix A. In the notation of Appendix A, we have

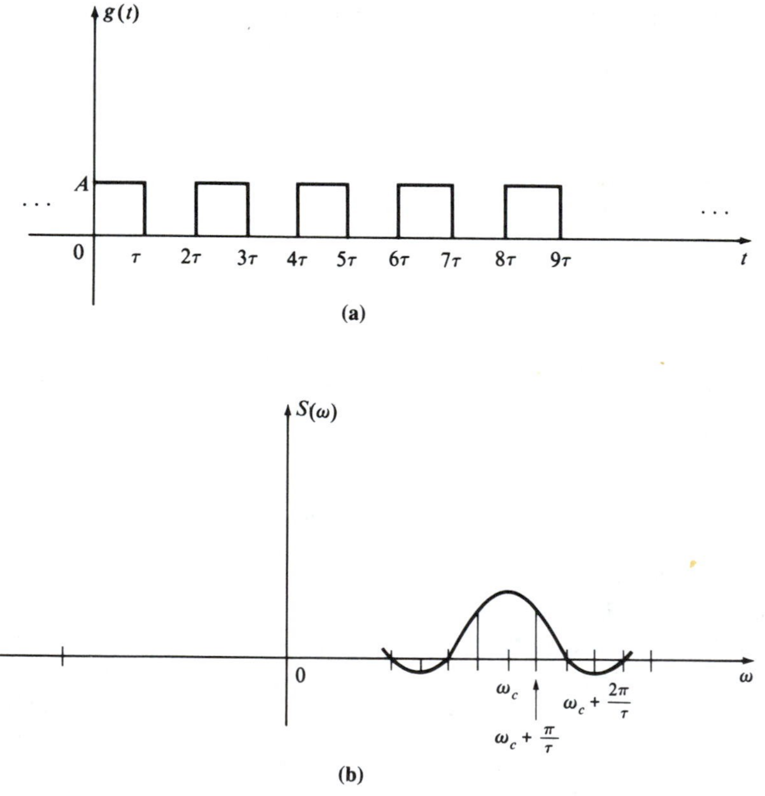

FIGURE 5.9.2 Transmission of a periodic binary sequence using AMDSB-SC: (a) Periodic binary sequence; (b) frequency content of transmitted AMDSB-SC signal with message $g(t)$.

here that $\mu_a = qA$ and $E[a_n^2] = qA^2$, so $\sigma_a^2 = qA^2(1-q)$. Further,

$$T_s = \tau \qquad \text{and} \qquad P(\omega) = \tau e^{-j\omega\tau/2} \frac{\sin(\omega\tau/2)}{\omega\tau/2}.$$

Thus, using Eq. (A.10.13), we have the power spectral density of the pulse sequence,

$$\begin{aligned} S(\omega) &= qA^2(1-q)\tau \left| \frac{\sin(\omega\tau/2)}{(\omega\tau/2)} \right|^2 + 2\pi q^2 A^2 \sum_{n=-\infty}^{\infty} \left| \frac{\sin n\pi}{n\pi} \right|^2 \delta\left(\omega - \frac{2n\pi}{\tau}\right) \\ &= qA^2(1-q)\tau \left| \frac{\sin(\omega\tau/2)}{(\omega\tau/2)} \right|^2 + 2\pi q^2 A^2 \delta(\omega). \end{aligned} \tag{5.9.1}$$

A sketch of this power spectral density is shown in Fig. 5.9.3. If the random pulse sequence is used to modulate a carrier, $\cos \omega_c t$, then copies of this power spectral density will be located about $\pm\omega_c$.

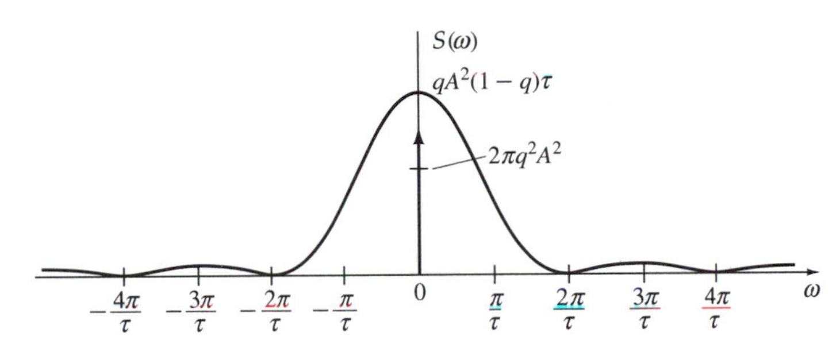

FIGURE 5.9.3 Power spectral density of a random pulse sequence.

To compare with the deterministic result in Fig. 5.9.2, we need to take the power density spectrum (Section 4.9) of the alternating pulse sequence in Fig. 5.9.2(a), and then we see that the first term in Eq. (5.9.1) compares with the envelope of the spectral lines for $|\mathscr{F}\{g(t)\}|^2 = |G(\omega)|^2$. The alternating deterministic pulse sequence $g(t)$ thus allowed us to get a crude idea of the required bandwidth, but the more realistic result is that in Eq. (5.9.1) and Fig. (5.9.3).

The sketch of the time-domain waveform in Fig. 5.9.1(c) clearly indicates the motivation behind the nomenclature *on–off keying* (OOK), since the transmitted signal is either on or off. When more than two-level pulses are used, called m-ary transmission for m-level signals, the transmitted signal is usually called *amplitude shift keying* (ASK). The "keying" terminology is a remnant of telegraph transmission days.

It is instructive to sketch an example of ASK transmission of data. Figure 5.9.4(a) shows a sequence of pulses each with four allowable output levels. If we use AMDSB-SC to transmit this pulse sequence, the modulated carrier signal

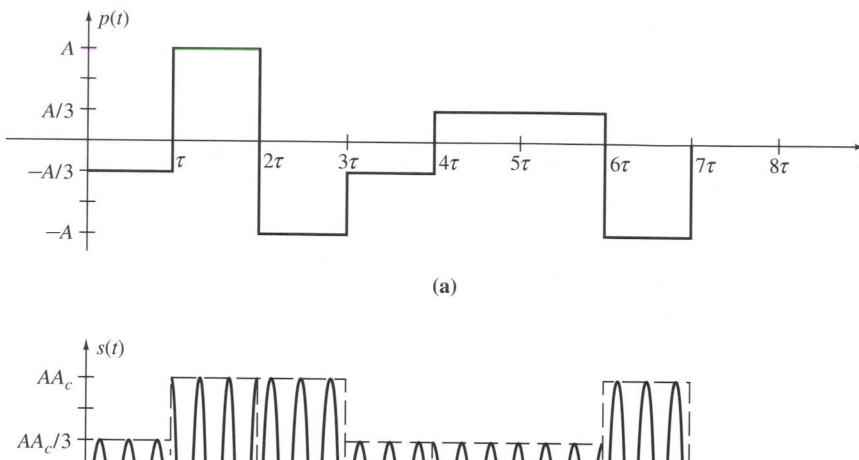

(a)

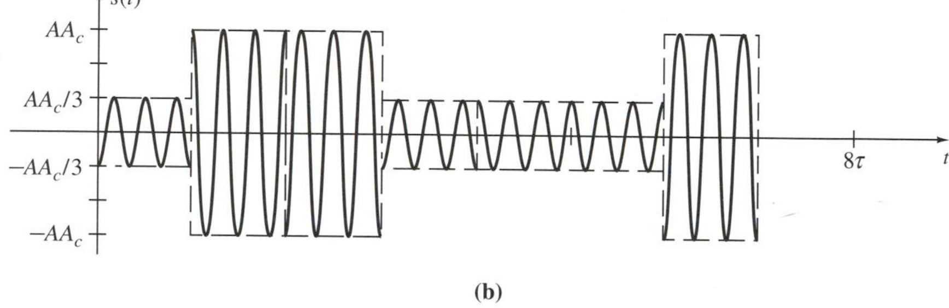

(b)

FIGURE 5.9.4 Four-level transmission using AMDSB-SC: (a) Sequence of four-level pulses; (b) modulated carrier (ASK).

appears as shown in Fig. 5.9.4(b). It is interesting to note that simply by observing $s(t)$ in Fig. 5.9.4(b) alone, we cannot tell whether $s(t)$ represents a four-level ASK sequence using AMDSB-SC modulation or a binary sequence using conventional AM modulation with a modulation index $a = 2/3$. Of course, when we design a communication system, the transmitter and receiver are jointly chosen and we do not have to guess what type of modulation was used.

The message sequence polarity in Fig. 5.9.4(a) is preserved in the phase of the carrier in Fig. 5.9.4(b). Other AM methods are frequently used for data transmission, and these applications are developed in Chapter 8.

SUMMARY

In this chapter we have presented the fundamental principles of amplitude modulation, including AMDSB-SC, conventional AM, SSB, VSB, superheterodyne systems, and OOK/ASK. The use of these modulation methods for data transmission is developed in Chapter 8, and the effects of noise on AM modulation systems are discussed in Chapter 7.

PROBLEMS

5.1 A message signal, $m(t)$, is transmitted using AMDSB-SC modulation; thus the transmitted waveform is given by $s_{\mathrm{DSB}}(t) = m(t) \cos \omega_c t$. During transmission, the frequency and phase of the carrier signal are distorted, so that the received signal is $r_{\mathrm{DSB}}(t) = m(t) \cos [(\omega_c + \Delta\omega)t + \phi]$. If the LO signal is $A_c \cos \omega_c t$:

(a) Find an expression for the low-pass filter output of the receiver.

(b) If $\Delta\omega = 0$, find an expression for the total energy in the low-pass filter output, and plot the energy as a function of ϕ for $\int_{-\infty}^{\infty} m^2(t)\, dt = 1$.

(c) Let $\phi = 0$, and describe the effect of the erroneous frequency reference. Sketch a typical Fourier transform of the low-pass filter output. Assume a shape for $M(\omega) = \mathscr{F}\{m(t)\}$.

5.2 Given the message signal $m(t) = 5 \sin 100\pi t$ and the carrier waveform $2 \cos 2000\pi t$, sketch $s_{\mathrm{DSB}}(t)$ and its Fourier transform.

5.3 For the message and carrier signals in Problem 5.2, sketch $s_{\mathrm{AM}}(t)$ and its Fourier transform if $a = 0.5$. See Eq. (5.3.1).

5.4 Show that for the conventional AM waveform in Eq. (5.3.1), the fraction of the total average power in the sidebands is

$$\eta = \frac{a^2 A_c^2 \langle m^2(t) \rangle}{A^2 + a^2 A_c^2 \langle m^2(t) \rangle},$$

where $\langle m^2(t) \rangle$ is the average power in $m(t)$. η is called the *efficiency*.

5.5 Let $m(t) = \cos \omega_m t$ and plot η in Problem 5.4 as a function of the modulation index a. Compare these results to the fraction of the total average power in the sidebands for AMDSB-SC.

5.6 A conventional AM waveform can be generated by the switching modulator circuit shown in Fig. P5.6. Assuming that $|m(t)|_{\max} \ll A_c$, we have under

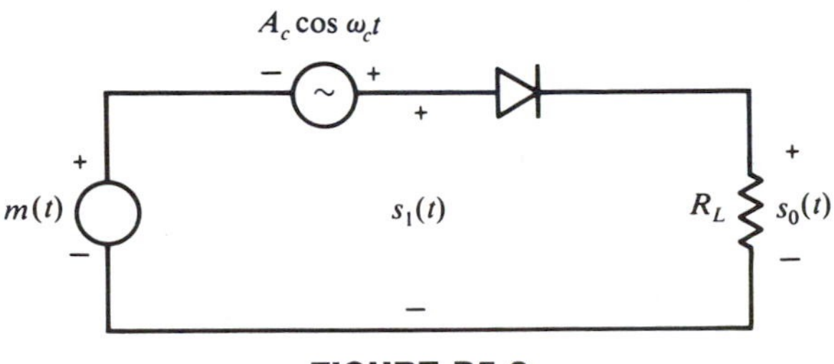

FIGURE P5.6

ideal conditions that

$$s_0(t) = \begin{cases} s_1(t), & \cos \omega_c t > 0 \\ 0, & \cos \omega_c t < 0. \end{cases}$$

Show how a conventional AM waveform can be obtained from $s_0(t)$ by appropriate filtering.

5.7 Consider the chopper modulator circuit shown in Fig. P5.7. An analysis of this circuit reveals that

$$s_1(t) = \begin{cases} m(t), & \cos \omega_c t > 0 \\ 0, & \cos \omega_c t < 0. \end{cases}$$

Show that $s_0(t)$ can represent an AMDSB-SC signal. Specify the filter type, center frequency, cutoff frequencies, and bandwidth.

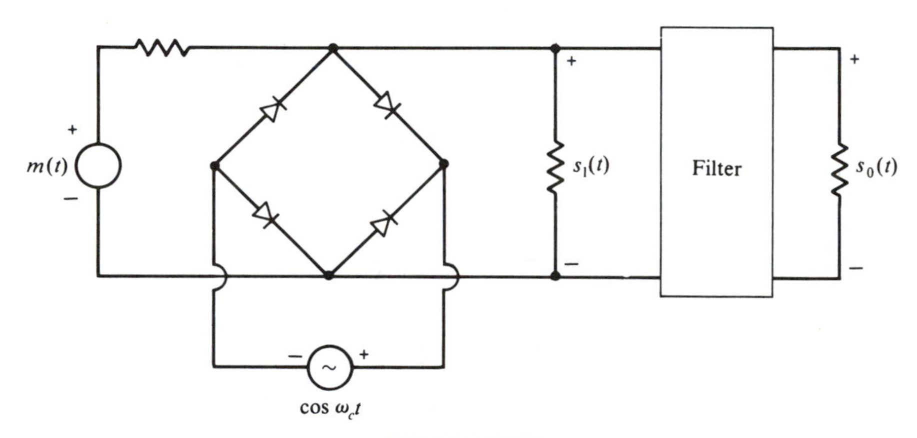

FIGURE P5.7

5.8 Consider the block diagram shown in Fig. P5.8. If $m(t)$ is the message signal and $y(t) = x^2(t)$, show how an AMDSB-SC wave can be obtained from $y(t)$. What happens if $y(t) = x^3(t)$?

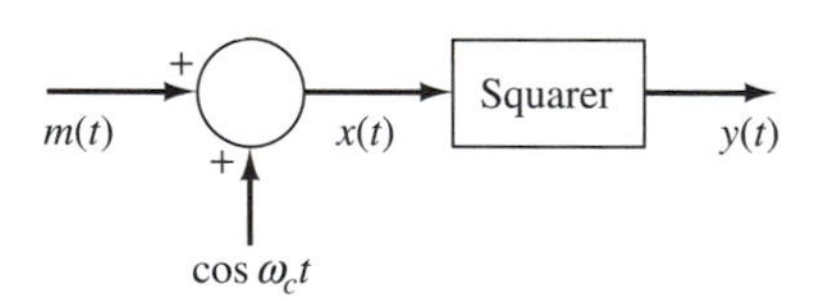

FIGURE P5.8

5.9 A nonlinear channel with input $x(t)$ and output $y(t)$ has the input/output relationship

$$y(t) = c_1 x(t) + c_2 x^2(t) + c_3 x^3(t).$$

If $x(t) = s_{DSB}(t) = m(t)A_c \cos \omega_c t$, can the AMDSB-SC wave be obtained undistorted at the receiver?

5.10 The signal

$$f(t) = \rho(t) \cos [\omega_c t + \theta(t)]$$

can be expanded using trigonometric identities to obtain

$$\begin{aligned} f(t) &= \rho(t) \cos \theta(t) \cos \omega_c t - \rho(t) \sin \theta(t) \sin \omega_c t \\ &= f_i(t) \cos \omega_c t - f_q(t) \sin \omega_c t. \end{aligned}$$

$\rho(t)$ and $\theta(t)$ are called the envelope and phase of $f(t)$, respectively, while $f_i(t)$ is called the in-phase component and $f_q(t)$ is called the quadrature component. Show that

$$\rho(t) = \sqrt{f_i^2(t) + f_q^2(t)}$$

and

$$\theta(t) = \tan^{-1} \frac{f_q(t)}{f_i(t)}.$$

5.11 Sketch the output of an envelope detector for each of the input signals listed.

(a) $f_1(t) = \sin t \cos 20{,}000t$.
(b) $f_2(t) = [1 + \sin t] \cos 20{,}000t$.
(c) $f_3(t) = \cos [20{,}000t - \cos t]$.

5.12 Given the periodic, normalized message waveform shown in Fig. P5.12:

(a) Find the total average power in $m_n(t)$.
(b) For a conventional AM waveform, what is the value of the modulation index that maximizes the efficiency η? (See Problem 5.4.)
(c) Calculate the maximum value of η for the message signal $m_n(t)$.

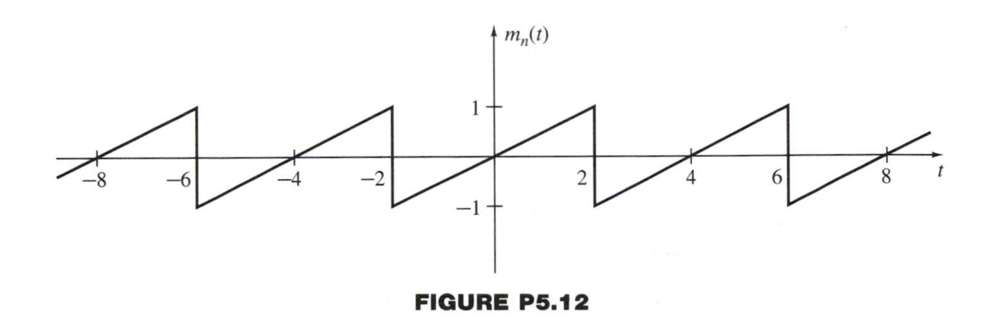

FIGURE P5.12

5.13 Consider the message, modulator, and channel shown in Fig. P5.13(a). Sketch the Fourier transform at each labeled point in the demodulators in Fig. P5.13(b) and (c).

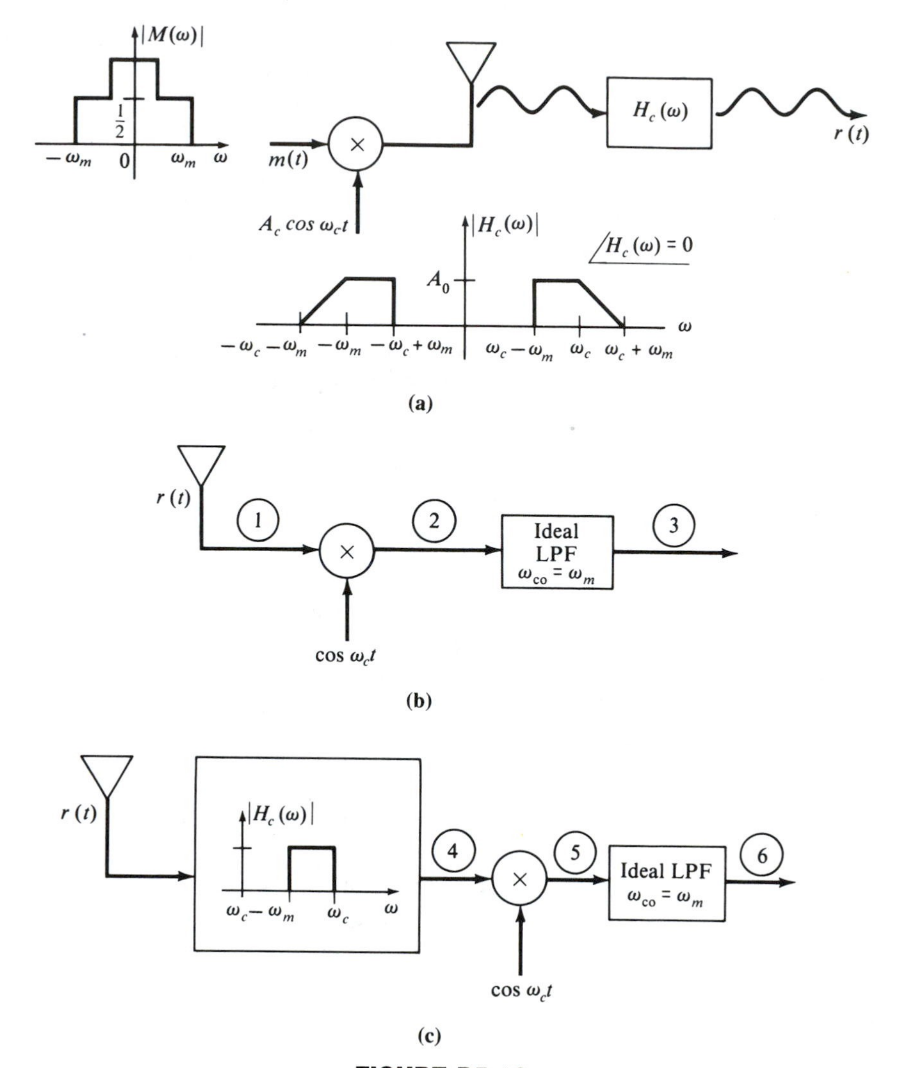

FIGURE P5.13

5.14 For $m(t)$ with the Fourier transform shown in Fig. P5.14(a), show that the block diagram in Fig. P5.14(b) functions as a data scrambler.

5.15 Let $g_h(t)$ be the Hilbert transform of the function $g(t)$. Prove the following properties.

(a) If $g(t)$ is even, $g_h(t)$ is odd.

(b) If $g(t)$ is odd, $g_h(t)$ is even.

(c) $\int_{-\infty}^{\infty} g^2(t)\,dt = \int_{-\infty}^{\infty} g_h^2(t)\,dt$.

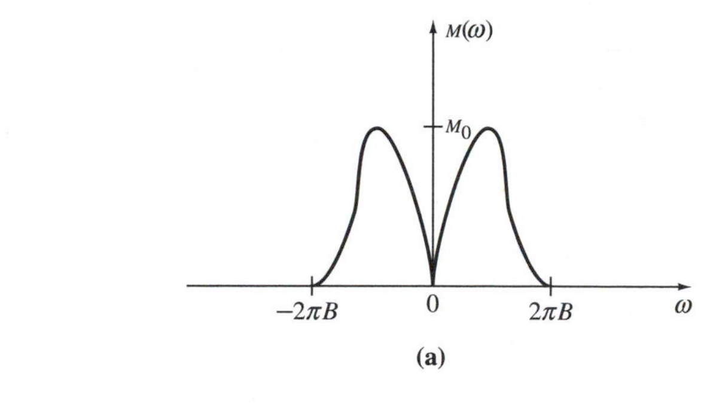

(a)

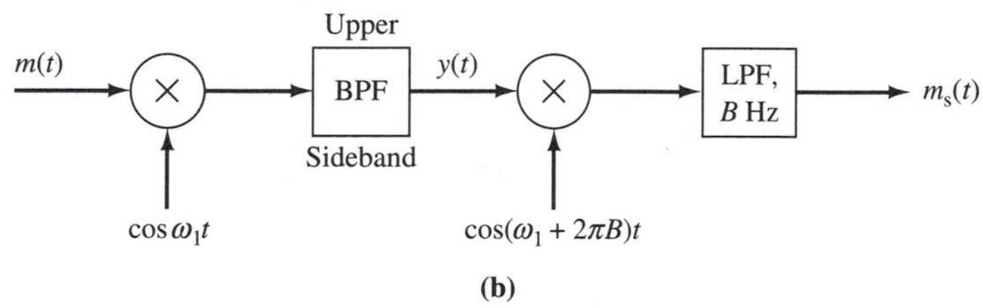

(b)

FIGURE P5.14

5.16 Given the time function

$$m(t) = \begin{cases} 1, & |t| \leq 1 \\ 0, & \text{otherwise,} \end{cases}$$

find $m_h(t) = \mathscr{H}\{m(t)\}$. Sketch $m(t)$ and $m_h(t)$.

5.17 Given that SSB-TC modulation is used to transmit $m(t)$ in Problem 5.16, sketch the output of an envelope detector. Contrast this result with coherent demodulation of a SSB-SC wave carrying the same $m(t)$.

5.18 State the principal advantage of each of the following modulation methods.

(a) Conventional AM.
(b) AMDSB-SC.
(c) AMSSB-SC.

5.19 (This problem was provided by J. L. LoCicero.) Given the block diagram of a SSB phase shift modulator, sometimes called a *Hartley modulator*, in Fig. P5.19, show that this system can generate an upper- or lower-sideband SSB wave for a general message signal $m(t)$.

Hint: Use Eqs. (5.4.3)–(5.4.5).

5.20 (This problem was provided by J. L. LoCicero.) For a general $m(t)$, show that the Weaver modulator in Fig. P5.20 can be used to generate an upper- or lower-sideband SSB wave.

Hint: Use Eqs. (5.4.3)–(5.4.5).

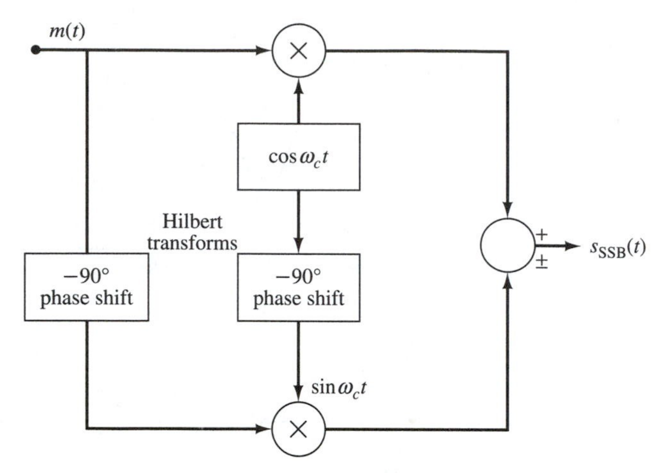

FIGURE P5.19

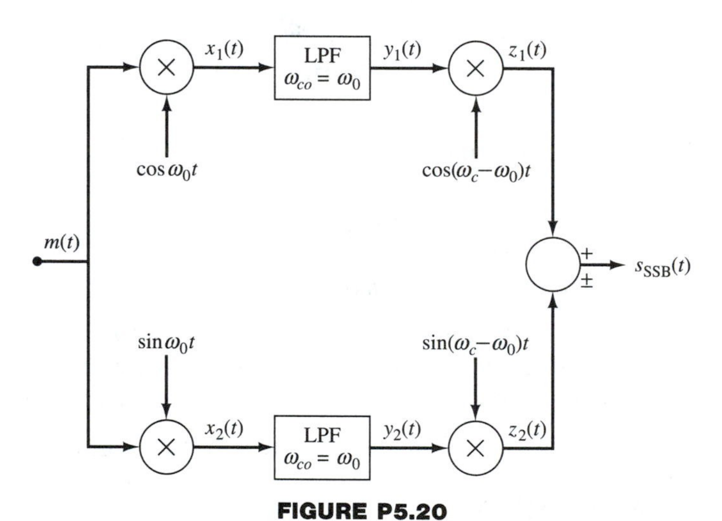

FIGURE P5.20

5.21 Noting that Eq. (5.4.3) can be written as $M_h(\omega) = -jM(\omega)\ \mathrm{sgn}\ (\omega)$, derive Eq. (5.4.2).

5.22 A periodic signal is given by

$$f(t) = \sum_{n=-\infty}^{\infty} (-1)^n p(t - nT),$$

where $p(t)$ is a rectangular pulse of amplitude A and length T seconds.

(a) Write an expression for the Hilbert transform of $f(t)$, denoted $f_h(t)$.

(b) Sketch $f(t)$ and $f_h(t)$.

(c) If $f(t)$ is transmitted by SSB-SC and the demodulator local oscillator has a phase error of $\theta°$, what is the demodulator output?

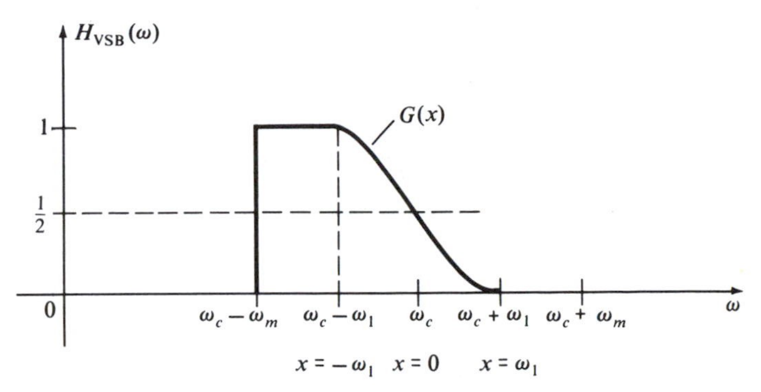

FIGURE P5.23

5.23 The complementary symmetry requirement of VSB can be specified in a variety of ways. Referring to Fig. P5.23, one way we can specify this complementary symmetry is by considering only positive ω, and defining a new variable, say x, with its origin at $\omega = \omega_c$. Letting $G(x)$ be the filter shape about $x = 0$, we have for the complementary symmetry requirement that $|G(-\alpha)| = 1 - |G(\alpha)|$. Verify that the function

$$G(x) = \begin{cases} \frac{1}{2}\left[1 - \sin\frac{\pi x}{2\omega_1}\right], & |x| \le \omega_1 \\ 0, & x > \omega_1 \\ 1, & -\omega_m \le x < -\omega_1, \end{cases}$$

satisfies the VSB complementary symmetry requirements.

5.24 It is also possible to represent the VSB complementary symmetry requirement as a sum of an ideal filter response and a filter with odd symmetry about ω_c, as shown in Fig. P5.24. Show that we can write a time-domain expression for a VSB wave of the form

$$s_{VSB}(t) = m(t) \cos \omega_c t \pm [m_h(t) + m_\beta(t)] \sin \omega_c t.$$

Specify $m_\beta(t)$.

5.25 Using the expression for $s_{VSB}(t)$ given in Problem 5.24, demonstrate that coherent demodulation can regain $m(t)$. Exhibit what can happen if the local oscillator has an inaccurate frequency or phase reference.

5.26 Starting with the expression for $s_{VSB}(t)$ in Problem 5.24, show that envelope detection of VSB can be effective if a suitable carrier term is injected or transmitted. State the required assumptions.

5.27 For broadcast AM radio, show that image stations can be eliminated without RF filtering by an appropriate choice of the IF center frequency.

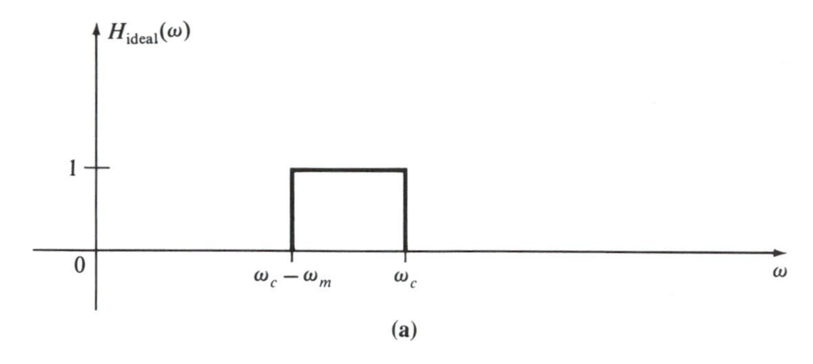

(a)

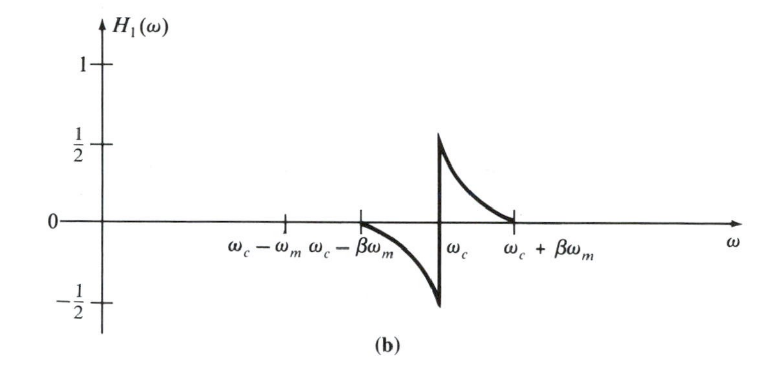

(b)

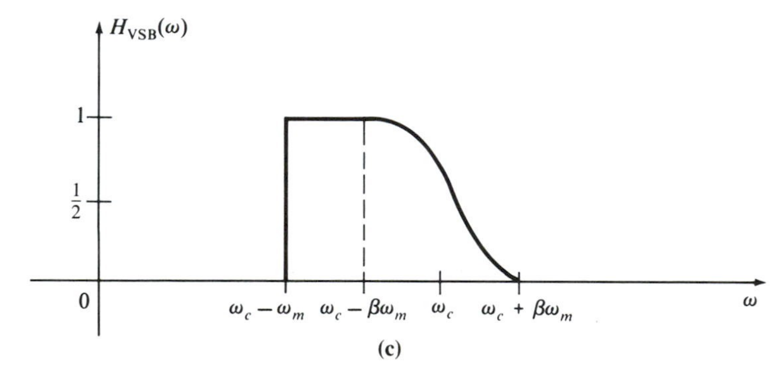

(c)

FIGURE P5.24

5.28 We wish to receive the AM radio station at 980 kHz. What is the local oscillator frequency? What is the frequency of the image station? What is the center frequency of the tunable RF amplifier?

5.29 The sequence of pulses shown in Fig. P5.29 is to be transmitted using AMDSB-SC. If the equation for the pulse shape when centered at $t = 0$ is $p(t) = V \cos(\pi t/\tau)$ for $|t| \leq \tau/2$, and 0 otherwise, sketch the modulated carrier. Assume that $\omega_c \gg 2\pi/\tau$.

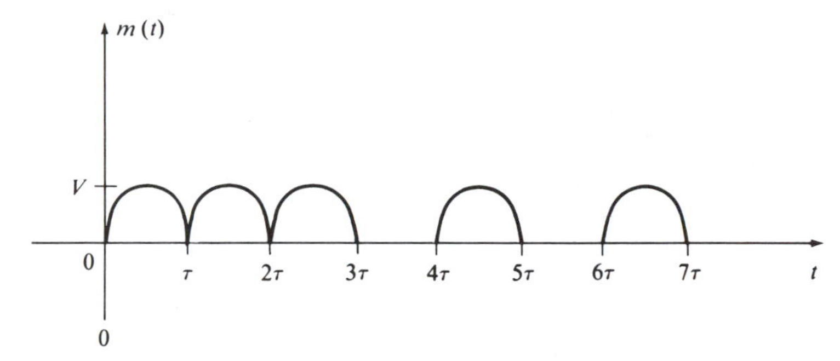

FIGURE P5.29

5.30 If the message to be transmitted using AMDSB-SC is

$$f(t) = \sum_{n=-\infty}^{\infty} p(t - 2n\tau),$$

where $p(t)$ is given in Problem 5.29, sketch the frequency content of the transmitted waveform. Assume that $\omega_c \gg 2\pi/\tau$.

5.31 A three-level pulse sequence shown in Fig. P5.31 is to be transmitted using AMDSB-SC. Sketch the transmitted waveform. Assume that $\omega_c \gg 2\pi/\tau$.

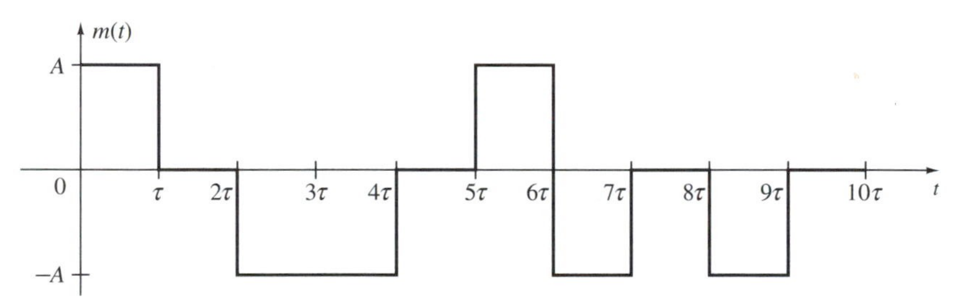

FIGURE P5.31

5.32 Five message signals, $m_i(t)$, $i = 1, 2, 3, 4, 5$, each with a low-pass bandwidth of 5 kHz, are to be transmitted using frequency-division multiplexing in the band $2\pi(100{,}000) \leq \omega \leq 2\pi(130{,}000)$. Can AMDSB methods be used? Why or why not? If lower-sideband AMVSB transmission is used, specify the carrier frequency and the band of frequencies occupied for each message assuming that no bandwidth is allocated to guardbands.

5.33 Spread spectrum communication systems reduce the effects of narrow bandwidth interference by spreading the bandwidth of the transmitted signal out over a range of frequencies much greater than the bandwidth occupied by an AMDSB-SC signal alone. Consider the AMDSB-SC signal in Eq. (5.2.1), where $m(t)$ is the message with $|M(\omega)|$ as in Fig. 5.2.1(a).

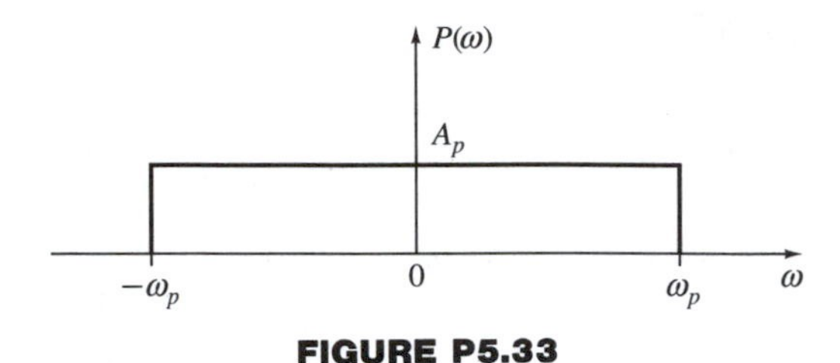

FIGURE P5.33

Let $p(t)$ be a sequence of ± 1 pulses with $\mathscr{F}\{p(t)\} = P(\omega)$ as in Fig. P5.33, where $\omega_p \gg \omega_c \gg \omega_m$. The spreading of the bandwidth is accomplished by multiplying by $p(t)$ to obtain $s_{\text{DSB}}(t)p(t)$.

(a) Find the band of frequencies occupied by $s_{\text{DSB}}(t)p(t)$.

(b) If there is no distortion, the transmitted signal $s_{\text{DSB}}(t)p(t)$ appears at the receiver input. Show that multiplying by $p(t)$ again produces $s_{\text{DSB}}(t)$, from which we can obtain the message $m(t)$.

(c) Assume that there is additive distortion $n(t)$ with $\mathscr{F}\{n(t)\} = N(\omega)$ such that

$$N(\omega) = \begin{cases} N_0, & \omega_c - \omega_m \leq |\omega| < \omega_c + \omega_m \\ 0, & \text{elsewhere.} \end{cases}$$

Show that if the received waveform is $r(t) = s_{\text{DSB}}(t)p(t) + n(t)$, then multiplying by $p(t)$ at the receiver reduces the noise power in the bandwidth of interest by a factor proportional to $1/\omega_p$.

Phase and Frequency Modulation

6.1 Introduction

Amplitude modulation is often called *linear modulation*, since superposition applies for AMDSB-SC systems. For instance, in AMDSB-SC if a message signal $m_1(t)$ generates a modulated waveform $s_1(t)$, and message $m_2(t)$ generates the waveform $s_2(t)$, the sum of the messages, $m_1(t) + m_2(t)$, will produce the modulated waveform $s_1(t) + s_2(t)$. If we ignore the constant added to the message signal, AMDSB-TC, or conventional AM, also has this property. In this chapter we turn our attention to angle modulation, where instead of varying the carrier amplitude in proportion to the message, we vary the carrier phase or frequency in relation to the message signal. Since the phase and frequency are in the argument of a sine or cosine function, angle modulation does not obey superposition, and hence it is sometimes called *nonlinear modulation.* This nonlinearity will become clear as the development progresses.

6.2 Phase and Frequency Modulation

In *phase modulation* (PM) we vary the phase of the carrier signal in a linear relationship with the message signal, $m(t)$. That is, given the carrier signal

$$s_{PM}(t) = A_c \cos [\omega_c t + \theta(t)], \tag{6.2.1}$$

we choose

$$\theta(t) = c_p m(t), \tag{6.2.2}$$

where c_p is a constant. Certainly, Eqs. (6.2.1) and (6.2.2) do not represent a linear modulation method, since we can rewrite Eq. (6.2.1) as

$$s_{PM}(t) = A_c \cos [c_p m(t)] \cos \omega_c t - A_c \sin [c_p m(t)] \sin \omega_c t \tag{6.2.3}$$

by using a trigonometric identity. The carrier frequency terms are thus modulated by nonlinear transformations of $c_p m(t)$. Equation (6.2.3) is not the best way to visualize what the time-domain waveform of a phase-modulated signal looks like, however. In fact, in order to sketch the time-domain waveform in all but the simplest cases, we need to introduce the concept of *frequency modulation* (FM).

When first considering frequency modulation, we are led immediately to the problem of what we mean by the frequency of a waveform. Intuitively, the frequency of a sine or cosine waveform makes sense only if the frequency is fixed (constant) and the sine or cosine wave exists for all time. This problem arises primarily because it only makes sense to talk about the sine or cosine of an angle (not a frequency), and generally we consider the angle to vary linearly with time. For example, in

$$s(t) = A_c \cos \phi(t) = A_c \cos [\omega_c t + \psi], \tag{6.2.4}$$

where ψ is a constant, the angle $\phi(t)$ varies in linear proportion to the fixed radian frequency ω_c. Thus, in this specific case, we can interpret the radian frequency as the derivative of the angle; that is, $\omega_c = d\phi(t)/dt$, and we get an answer that coincides with our intuition.

If $\phi(t)$ does not vary linearly with time, the justification for defining the frequency as the derivative of the angle is not so clear. Since we are going to be talking about signals whose frequency is changing with time, we must have a definition of frequency that allows us to speak of the frequency of a signal at any time instant. For this reason we define the *instantaneous radian frequency*, denoted by ω_i, as

$$\omega_i \triangleq \frac{d}{dt} \phi(t), \tag{6.2.5}$$

where $\phi(t)$ is the angle of the carrier waveform. If $\phi(t) = \omega_c t + \psi$, this definition agrees with our intuition. Otherwise, however, the result may not be so intuitive. For Eqs. (6.2.1) and (6.2.2), we have

$$\omega_i = \frac{d}{dt} [\omega_c t + c_p m(t)] = \omega_c + c_p \frac{d}{dt} m(t). \tag{6.2.6}$$

This result implies that for PM, the instantaneous frequency varies about the fixed value ω_c in linear proportion to the derivative of the message signal.

A logical definition of a frequency-modulated waveform would be to have the instantaneous frequency vary linearly with respect to $m(t)$ itself; hence

$$\omega_i = \omega_c + c_f m(t), \tag{6.2.7}$$

where c_f is a constant, represents the instantaneous frequency of an FM signal. To write a time-domain expression for the FM waveform, we note Eq. (6.2.5) and integrate both sides of Eq. (6.2.7) to find

$$\phi(t) = \int \omega_i \, dt = \omega_c t + c_f \int m(t) \, dt + \psi, \tag{6.2.8}$$

where ψ represents a constant of integration. Note that it is customary to write the instantaneous frequency as ω_i rather than $\omega_i(t)$, even though generally it is a function of time. Using Eq. (6.2.8), we can write the FM signal as

$$s_{\text{FM}}(t) = A_c \cos\left[\omega_c t + c_f \int m(t)\, dt\right], \tag{6.2.9}$$

where ψ is chosen to be zero for simplicity.

By comparing Eqs. (6.2.1) and (6.2.2) with Eq. (6.2.9), it is evident that FM and PM are different modulation methods, but it is equally evident that they are both angle modulation methods and hence closely related. The following examples illustrate the instantaneous frequency concept and the relationship between FM and PM.

EXAMPLE 6.2.1

Consider the message signal sketched in Fig. 6.2.1(a). This message can be transmitted by both FM and PM. For PM let $c_p = \pi/4$ rad/V and assume for this example that $T = 2T_c$, where $\omega_c = 2\pi/T_c$. Under these assumptions, the PM waveform can be sketched as shown in Fig. 6.2.1(b).

To transmit $m(t)$ using FM, let $c_f = \pi/2T$ rad/sec per volt, so that

$$\omega_i = \omega_c + \frac{\pi}{2T} m(t) \tag{6.2.10}$$

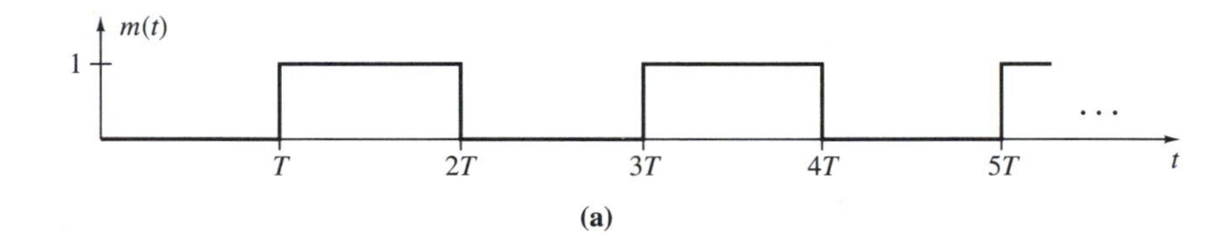

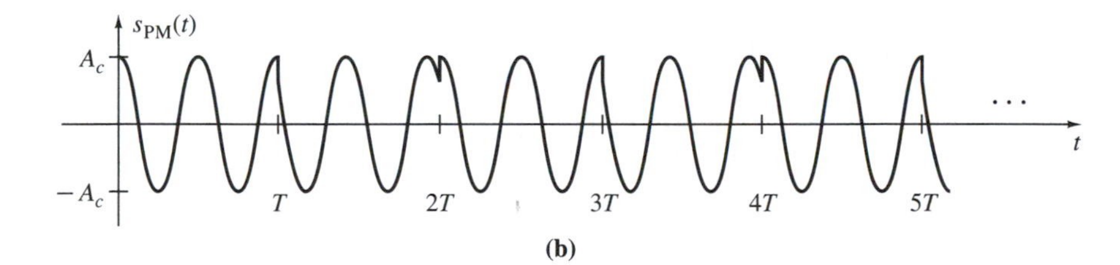

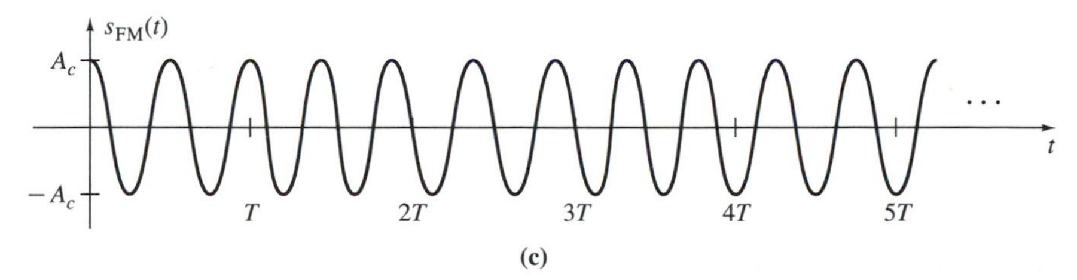

FIGURE 6.2.1 Waveforms for Example 6.2.1: (a) Message signal for Example 6.2.1; (b) PM waveform for $m(t)$ in part (a); (c) FM waveform for $m(t)$ in part (a).

or

$$\phi(t) = \omega_c t + \frac{\pi}{2T} \int m(t)\, dt. \tag{6.2.11}$$

We can sketch $s_{FM}(t)$ by referring to either Eq. (6.2.10) or (6.2.11). From Eq. (6.2.10) we see that the instantaneous frequency is changing between ω_c and $\omega_c + \pi/2T = 2\pi/T_c + \pi/4T_c = \frac{9}{8}\omega_c$. Thus $s_{FM}(t)$ can be sketched as shown in Fig. 6.2.1(c). Note that the same result is evident from Eq. (6.2.11), which implies that the phase linearly increases when $m(t)$ is nonzero up to 90° in the interval $T \leq t \leq 2T$, and in every interval where $m(t)$ is 1.

EXAMPLE 6.2.2

Given the waveform

$$s(t) = 10 \cos [90 \times 10^6 t + 200 \cos 2000t],$$

(a) If $s(t)$ is a PM signal, what is $m(t)$? From Eqs. (6.2.1) and (6.2.2) we see immediately that $c_p m(t) = 200 \cos 2000t$, so that

$$m(t) = \left(\frac{200}{c_p}\right) \cos 2000t.$$

(b) If $s(t)$ is an FM signal, what is $m(t)$? Following Eq. (6.2.5), we obtain

$$\omega_i = 90 \times 10^6 - 4 \times 10^5 \sin 2000t,$$

which by comparison with Eq. (6.2.7) reveals that $c_f m(t) = -4 \times 10^5 \sin 2000t$ or $m(t) = (-4 \times 10^5/c_f) \sin 2000t$.

Using the phase and instantaneous frequency, we can thus determine and sketch (if we so desire) the corresponding FM and PM time-domain waveforms for a given message, but we have not yet specified the frequency-domain representation of an angle-modulated signal. This is the topic of the following section.

6.3 Bandwidth Requirements

To begin the investigation of FM bandwidth requirements, we restrict the message signal to be a pure tone given by

$$m(t) = A \cos \omega_m t. \tag{6.3.1}$$

Although virtually all messages of interest are much more complicated than $m(t)$ in Eq. (6.3.1), we shall discover that the bandwidth requirements of an FM wave carrying this $m(t)$ are representative of the bandwidth requirements of many other physically important signals. By assuming a message signal of this form, we can thus illustrate several critical concepts without undue mathematical difficulty.

If we use $m(t)$ in Eq. (6.3.1) to frequency modulate a carrier signal, the instantaneous frequency of the carrier can be expressed as

$$\omega_i = \omega_c + \Delta\omega \cos \omega_m t, \tag{6.3.2}$$

where $\Delta\omega$ is a constant such that $\Delta\omega \ll \omega_c$. The instantaneous frequency of the FM wave thus varies about the fixed carrier frequency ω_c at a rate of ω_m rad/sec with a maximum frequency change of $\Delta\omega$ rad/sec. The quantity $\Delta\omega$ is called the *maximum frequency deviation* or just the *frequency deviation* of the FM waveform. Using Eq. (6.2.8), we find that

$$\phi(t) = \int \omega_i \, dt = \omega_c t + \frac{\Delta\omega}{\omega_m} \sin \omega_m t + \psi, \tag{6.3.3}$$

where we normally let $\psi = 0$. A time-domain equation for the FM signal is therefore

$$s_{\mathrm{FM}}(t) = A_c \cos\left[\omega_c t + \frac{\Delta\omega}{\omega_m} \sin \omega_m t\right] = A_c \cos\left[\omega_c t + \beta \sin \omega_m t\right], \tag{6.3.4}$$

where we have defined

$$\beta = \frac{\Delta\omega}{\omega_m}, \tag{6.3.5}$$

which is called the *modulation index*. Since β depends on $\Delta\omega$ and ω_m, and $\Delta\omega$ indicates the frequency range over which the FM signal will vary and ω_m indicates how rapidly this variation takes place, β is clearly related to the required bandwidth of the FM wave.

Two different classes of FM waves are of interest, depending on the value of β. For $\beta \ll \pi/2$ (usually, $\beta < 0.2$ or 0.5), called *narrowband FM*, we use trigonometric identities to write

$$s_{\mathrm{FM}}(t) = A_c \cos \omega_c t \cos\left[\beta \sin \omega_m t\right] - A_c \sin \omega_c t \sin\left[\beta \sin \omega_m t\right], \tag{6.3.6}$$

which for β small we can approximate as

$$s_{\mathrm{FM}}(t) \cong A_c \cos \omega_c t - A_c \beta \sin \omega_m t \sin \omega_c t, \tag{6.3.7}$$

since $\cos\left[\beta \sin \omega_m t\right] \cong 1$ and $\sin\left[\beta \sin \omega_m t\right] \cong \beta \sin \omega_m t$. Using a trigonometric identity on the second term in Eq. (6.3.7) reveals that the bandwidth of a narrowband FM wave is approximately $2\omega_m$, the same as AMDSB methods.

When $\beta > \pi/2$ the frequency-modulated signal is called *wideband FM*, which is the type of FM used most often in analog communication systems. To determine the bandwidth requirements of wideband FM, we build on Eq. (6.3.6) by expanding the terms $\cos\left[\beta \sin \omega_m t\right]$ and $\sin\left[\beta \sin \omega_m t\right]$ in a Fourier series. Since $\mathrm{Re}\{e^{j\beta \sin \omega_m t}\} = \cos\left[\beta \sin \omega_m t\right]$ and $\mathrm{Im}\{e^{j\beta \sin \omega_m t}\} = \sin\left[\beta \sin \omega_m t\right]$, we can find both expressions by writing a Fourier series for

$$g(t) = e^{j\beta \sin \omega_m t}, \qquad -\frac{T}{2} \le t \le \frac{T}{2}, \tag{6.3.8}$$

where $T = 2\pi/\omega_m$.

Straightforwardly, the complex Fourier series coefficients are given by

$$c_n = \frac{1}{T}\int_{-T/2}^{T/2} e^{j(\beta \sin \omega_m t - n\omega_m t)}\, dt = \frac{1}{2\pi}\int_{-\pi}^{\pi} e^{j(\beta \sin x - nx)}\, dx \tag{6.3.9}$$

by making the change of variable $x = \omega_m t$. The integral in Eq. (6.3.9) can be evaluated only by means of an infinite series, but fortunately, because it often appears in many physical problems, it is tabulated extensively. The integral in Eq. (6.3.9) is called the *Bessel function of the first kind* and is represented by

$$J_n(\beta) = \frac{1}{2\pi}\int_{-\pi}^{\pi} e^{j(\beta \sin x - nx)}\, dx, \tag{6.3.10}$$

which is a function of both n and β. The functions in Eq. (6.3.10) have the properties that

$$J_n(\beta) = J_{-n}(\beta) \qquad \text{for } n \text{ even} \tag{6.3.11a}$$

and

$$J_n(\beta) = -J_{-n}(\beta) \qquad \text{for } n \text{ odd.} \tag{6.3.11b}$$

Plots of the Bessel functions of the first kind are shown in Fig. 6.3.1, and a table is given in Appendix I.

Writing out the Fourier series term by term and employing Eqs. (6.3.11) to combine the positive and negative terms with equal magnitudes of n yields

$$\begin{aligned} e^{j\beta \sin \omega_m t} = J_0(\beta) &+ 2[J_2(\beta)\cos 2\omega_m t + J_4(\beta)\cos 4\omega_m t + \cdots] \\ &+ j2[J_1(\beta)\sin \omega_m t + J_3(\beta)\sin 3\omega_m t + \cdots]. \end{aligned} \tag{6.3.12}$$

Using Euler's identity on the left side of Eq. (6.3.12) and equating real and imaginary parts produces the desired expressions,

$$\begin{aligned} \cos[\beta \sin \omega_m t] &= J_0(\beta) + 2J_2(\beta)\cos 2\omega_m t + 2J_4(\beta)\cos 4\omega_m t + \cdots \\ &= J_0(\beta) + 2\sum_{\substack{n=2 \\ n \text{ even}}}^{\infty} J_n(\beta)\cos n\omega_m t \end{aligned} \tag{6.3.13}$$

and

$$\begin{aligned} \sin[\beta \sin \omega_m t] &= 2J_1(\beta)\sin \omega_m t + 2J_3(\beta)\sin 3\omega_m t + \cdots \\ &= 2\sum_{\substack{n=1 \\ n \text{ odd}}}^{\infty} J_n(\beta)\sin n\omega_m t. \end{aligned} \tag{6.3.14}$$

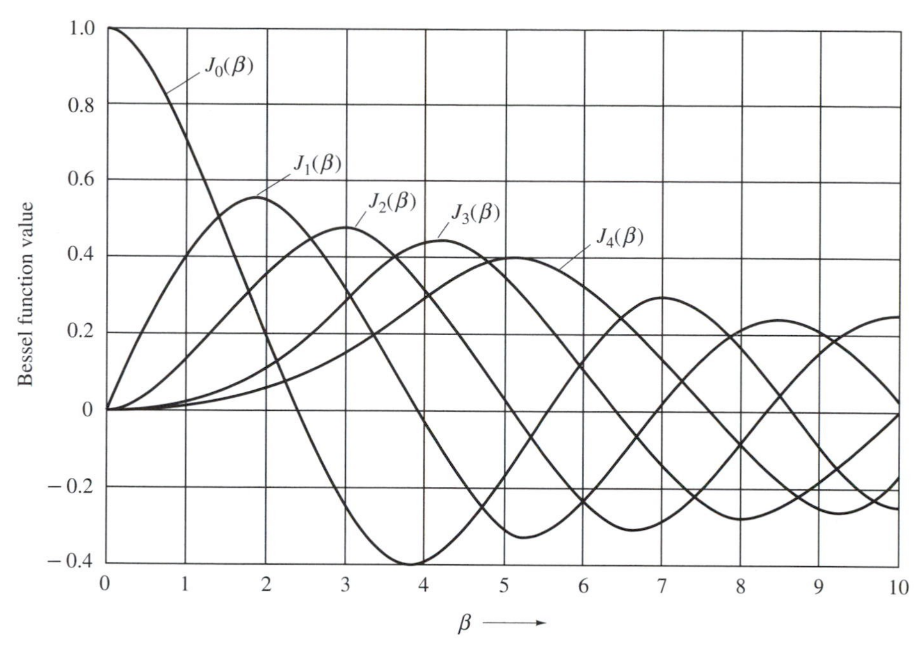

FIGURE 6.3.1 Plots of Bessel functions.

Substituting Eqs. (6.3.13) and (6.3.14) into Eq. (6.3.6) and using trigonometric identities gives

$$\begin{aligned} s_{\mathrm{FM}}(t) &= A_c\{J_0(\beta)\cos\omega_c t - J_1(\beta)[\cos(\omega_c - \omega_m)t - \cos(\omega_c + \omega_m)t] \\ &\quad + J_2(\beta)[\cos(\omega_c - 2\omega_m)t + \cos(\omega_c + 2\omega_m)t] \\ &\quad - J_3(\beta)[\cos(\omega_c - 3\omega_m)t - \cos(\omega_c + 3\omega_m)t] + \cdots\} \\ &= A_c\left\{J_0(\beta)\cos\omega_c t + \sum_{n=1}^{\infty}(-1)^n J_n(\beta)[\cos(\omega_c - n\omega_m)t\right. \\ &\quad \left. + (-1)^n\cos(\omega_c + n\omega_m)t]\right\}. \end{aligned} \tag{6.3.15}$$

Since for a specific application β is fixed, it is evident from Eq. (6.3.15) that the FM wave consists of a carrier plus an infinite number of sidebands spaced at radian frequencies $\pm\omega_m$, $\pm 2\omega_m, \ldots$, about the carrier. In comparison to AMDSB methods with the same message signal, which requires a bandwidth of only $2\omega_m$, we see that the nonlinear angle modulation has generated numerous additional sidebands.

Equation (6.3.15) is valid for all β, and although there are in general an infinite number of sidebands, the number of *significant* sidebands depends on the value of β. More specifically, for $n > 1$ and $\beta \ll n$, $J_n(\beta) \cong 0$, but for $\beta \gg 1$, the number of significant sidebands is approximately $n = \beta$. Hence, for β very large, the approximate bandwidth requirement for (wideband) FM is

$$\text{BW} \cong 2n\omega_m = 2\beta\omega_m = 2\Delta\omega, \tag{6.3.16}$$

while for β small we have narrowband FM with

$$\text{BW} \cong 2\omega_m. \tag{6.3.17}$$

A general rule of thumb for the bandwidth of FM for any β is

$$\text{BW} = 2(\Delta\omega + \omega_m) = 2\omega_m(1 + \beta), \tag{6.3.18}$$

which is called *Carson's rule*. All of the bandwidths can be expressed in hertz simply by replacing $\Delta\omega$ with Δf and ω_m with f_m.

EXAMPLE 6.3.1

In broadcast FM, the FCC restricts Δf to be 75 kHz and $(f_m)_{\max}$ = maximum frequency in $m(t)$ to be less than or equal to 15 kHz. The smallest value of the modulation index for Δf = 75 kHz is

$$\beta = \frac{\Delta f}{(f_m)_{\max}} = \frac{75 \text{ kHz}}{15 \text{ kHz}} = 5. \tag{6.3.19}$$

For message signals with lower values of f_m, β will be larger. Using $\beta = 5$ and Carson's rule, we find that

$$\text{BW} = 2f_m(1 + \beta) = 180 \text{ kHz}. \tag{6.3.20}$$

Since the total bandwidth allocated to each FM station is 200 kHz, we are safely within the specified band. Note that for smaller values of f_m, we shall have a large β, or wideband FM, so then $\text{BW} \cong 2\,\Delta f = 150$ kHz.

EXAMPLE 6.3.2

Given the angle-modulated waveform

$$s(t) = 10\cos[90 \times 10^6 t + 200 \cos 2000t],$$

what is its required bandwidth? Whether $s(t)$ represents a PM or FM wave, the bandwidth can be found from Carson's rule. The instantaneous frequency is given by

$$\omega_i = 90 \times 10^6 - 4 \times 10^5 \sin 2000t,$$

so $\omega_m = 2000$ and $\Delta\omega = 4 \times 10^5$. Therefore,

$$\text{BW} = 2[4 \times 10^5 + 2000] = 8.04 \times 10^5 \text{ rad/sec}.$$

Note that $\beta = 4 \times 10^5/2 \times 10^3 = 200$, which is wideband FM, hence

$$\text{BW} \cong 2\Delta\omega = 8 \times 10^5 \text{ rad/sec}.$$

Although all of the results in this section concerning the bandwidth of an FM signal have been derived under the assumption of a single-tone modulating signal, we are fortunate in that the expressions can be extended to yield bandwidth guidelines for more general messages. In particular, if we wish to use Eqs. (6.3.16), (6.3.17), and (6.3.18) to determine the bandwidth of a general message signal $m(t)$ with a highest frequency present in $m(t)$ of $2\pi W_m$, then we need only let $\omega_m = 2\pi W_m$ and $\Delta\omega = c_f[\max|m(t)|]$, for which we find

$$\beta = \frac{\Delta\omega}{\omega_m} = \frac{c_f[\max|m(t)|]}{2\pi W_m}. \tag{6.3.21}$$

Of course, Eqs. (6.3.16) to (6.3.18) are only approximations to the bandwidth of any FM signal, and other rules of thumb are found in the literature.

6.4 Modulation and Demodulation Methods

Just as in developing bandwidth requirements, modulation methods for FM and PM can be separated into narrowband and wideband categories. Since we considered only single-tone messages in Section 6.3, we need to illustrate the narrowband concepts for general messages and both PM and FM. To begin, we can write a general angle-modulated signal as

$$s(t) = A_c \cos\left[\omega_c t + \phi(t)\right], \tag{6.4.1}$$

which can then be expanded using a trigonometric identity as

$$s(t) = A_c \cos\phi(t)\cos\omega_c t - A_c \sin\phi(t)\sin\omega_c t. \tag{6.4.2}$$

If $|\phi(t)|_{\max}$ is small, then Eq. (6.4.2) is approximately

$$s(t) = A_c \cos\omega_c t - A_c\phi(t)\sin\omega_c t. \tag{6.4.3}$$

For PM, we have that $\phi(t) = c_p m(t)$, so the condition for narrowband PM is $c_p|m(t)|_{\max} \ll \pi/2$, and the narrowband PM wave is

$$s_{\mathrm{PM}}(t) = A_c \cos\omega_c t - A_c c_p m(t)\sin\omega_c t. \tag{6.4.4}$$

For FM, $\phi(t) = c_f \int m(t)\,dt$, and the FM waveform is

$$s_{\mathrm{FM}}(t) = A_c \cos\omega_c t - A_c c_f\left[\int m(t)\,dt\right]\sin\omega_c t, \tag{6.4.5}$$

where $c_f|\int m(t)\,dt|_{\max} \ll \pi/2$. From Eqs. (6.4.4) and (6.4.5), we can implement narrowband PM or FM as illustrated by the block diagram in Fig. 6.4.1.

The generation of wideband FM can be separated into two types: indirect FM and direct FM. *Indirect FM* starts with a narrowband frequency-modulated signal and uses frequency multiplication to generate a wideband FM signal. This procedure is illustrated by Fig. 6.4.2. If the narrowband FM signal is

$$x(t) = A_c \cos\left[\omega_c t + \phi(t)\right],$$

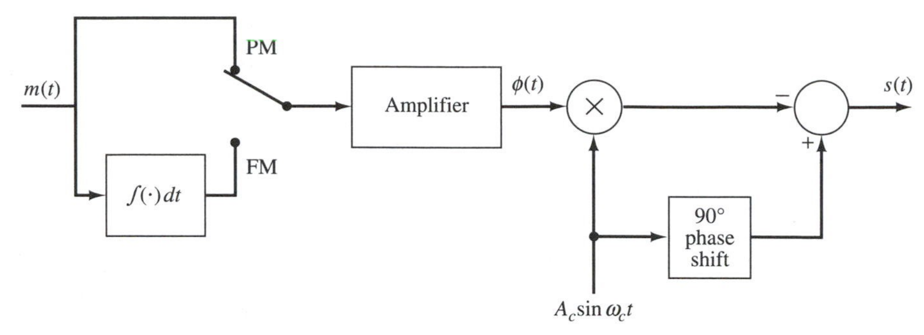

FIGURE 6.4.1 Generation of narrowband PM or FM.

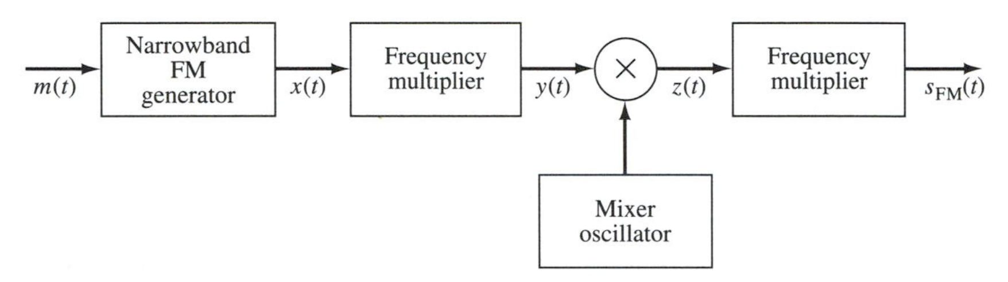

FIGURE 6.4.2 Indirect generation of wideband FM.

the signal after frequency multiplication is

$$y(t) = A_c \cos\left[n\omega_c t + n\phi(t)\right],$$

so that the instantaneous frequency of $y(t)$ is

$$\omega_i = n\omega_c + n\frac{d}{dt}\phi(t) = n\omega_c + nc_f m(t).$$

As a result, the new frequency deviation is n times the narrowband signal's frequency deviation, and thus a wideband FM wave is generated. The frequency multiplication also increases the carrier frequency by a factor of n. If this is not desirable, the carrier frequency of the frequency multiplier output can be adjusted by the mixer oscillator as shown in Fig. 6.4.2. Of course, the mixing operation only relocates the wideband FM waveform in the frequency spectrum; it does not affect the frequency deviation.

For *direct FM* methods, the modulating signal directly varies the carrier frequency. Direct FM may also require some frequency multiplication, but it is usually less than in indirect FM. There are a variety of methods for directly controlling the carrier frequency with the modulating signal, including varying the inductance or capacitance of a resonant circuit, using a voltage-controlled oscillator (VCO), or changing the switching frequency of a multivibrator.

To demodulate FM signals, it is necessary to build a device or system which produces an output amplitude that is linearly proportional to the instantaneous frequency of the FM wave. Such devices are called *frequency discriminators*, and there are many different approaches to their implementation. An ideal differentiator has a linear amplitude versus frequency magnitude response and hence should be suitable for FM demodulation.

To see this, consider the FM wave

$$s_{\text{FM}}(t) = A_c \cos\left[\omega_c t + c_f \int m(t)\, dt\right], \tag{6.4.6}$$

which when passed through an ideal differentiator becomes

$$\frac{d}{dt} s_{\text{FM}}(t) = -A_c[\omega_c + c_f m(t)] \sin\left[\omega_c t + c_f \int m(t)\, dt\right]. \tag{6.4.7}$$

Since $c_f|m(t)|_{\max} = \Delta\omega \ll \omega_c$, the envelope of Eq. (6.4.7) is always positive, and hence $m(t)$ can be recovered from Eq. (6.4.7) by envelope detection. For envelope detection to be effective, it is imperative that A_c be constant. To achieve this, the differentiator is usually preceded by a hard limiter that produces a square wave, and then by a bandpass filter centered at ω_c to change the signal back into sinusoidal form. A block diagram of the discriminator is given in Fig. 6.4.3. In practice, the differentiation operation can be achieved by designing a filter such that ω_c falls on a linearly sloping region of the filter magnitude response. Although the development has emphasized FM demodulation, the technique is also valid for PM signals if the envelope detector is followed by an integrator.

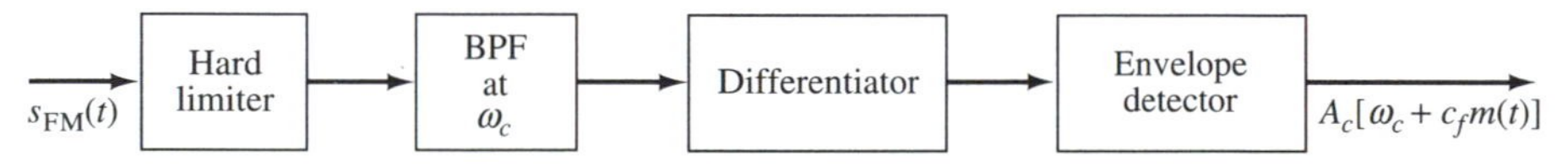

FIGURE 6.4.3 FM discriminator.

Other popular FM demodulators operate by detecting the zero crossings of the FM wave.

6.5 Phase-Locked Loops

For noisy channel applications, feedback demodulators are particularly powerful for recovering the message signal from the received FM waveform. Two types of feedback demodulators are the *FM feedback* (FMFB) *demodulator* and the *phase-locked loop* (PLL). Because of the increasing popularity of the PLL and its availability as an integrated-circuit device, we consider it in more detail here. In this section we investigate the basic operation of PLLs. More details on PLLs are left to the literature.

A block diagram of a phase-locked loop is shown in Fig. 6.5.1. The phase comparator in a PLL can be modeled as a multiplier followed by a unity-gain low-pass filter. As a result, $x(t)$ is given by

$$x(t) = \frac{A_c A_f}{2} \sin\left[(\omega_c - \omega_f)t + \phi_i(t) - \phi_f(t)\right], \tag{6.5.1}$$

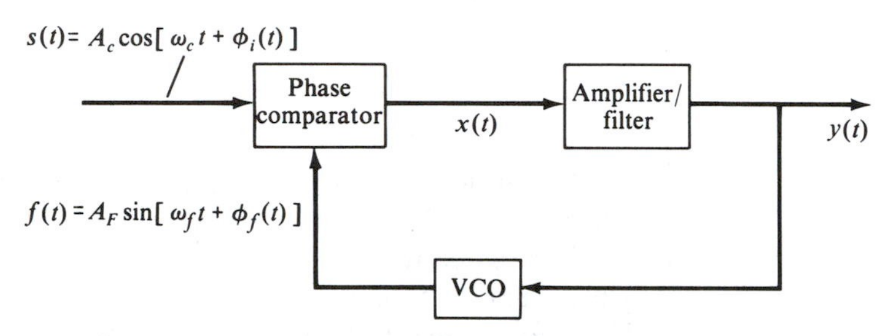

FIGURE 6.5.1 Phase-locked loop.

where the $\omega_c + \omega_f$ term is rejected by the LPF. Generally, we will ignore the frequency acquisition dynamics and assume that we have $\omega_c = \omega_f$, so

$$x(t) = \frac{A_c A_f}{2} \sin\left[\phi_i(t) - \phi_f(t)\right]. \tag{6.5.2}$$

The VCO produces an output signal $f(t)$ with a frequency variation about ω_f that is proportional to its input signal. During ideal operation, we have $\phi_f(t) \cong \phi_i(t)$, so the frequency variation of $f(t)$ is

$$\frac{d}{dt}\phi_f(t) \cong \frac{d}{dt}\phi_i(t),$$

and since

$$\frac{d}{dt}\phi_f(t) = G_f y(t),$$

the PLL output is

$$y(t) = \frac{1}{G_f}\frac{d}{dt}\phi_i(t). \tag{6.5.3}$$

Since $d\phi_i(t)/dt$ is the PLL input signal frequency variation, and since for FM and a message signal $m(t)$ we know that

$$\frac{d}{dt}\phi_i(t) = c_f m(t), \tag{6.5.4}$$

it is evident that we have demodulated the FM waveform $s(t)$.

When $\phi_f(t) \cong \phi_i(t)$ the PLL is said to be operating in *phase lock* with the input. The question remains as to how the PLL in Fig. 6.5.1 acquires and maintains this desirable situation. To begin, we return to Eq. (6.5.2) and note that if the gain of the amplifier/filter in Fig. 6.5.1 is G, then

$$y(t) = \frac{GA_cA_f}{2} \sin[\phi_i(t) - \phi_f(t)]. \tag{6.5.5}$$

Therefore,

$$\frac{d}{dt}\phi_f(t) = \frac{G_fGA_cA_f}{2} \sin[\phi_i(t) - \phi_f(t)], \tag{6.5.6}$$

so letting $G_T = (G_fGA_cA_f)/2$,

$$\phi_f(t) = G_T \int_{t_0}^{t_1} \sin[\phi_i(\tau) - \phi_f(\tau)]\, d\tau. \tag{6.5.7}$$

Defining the phase error $\theta_e(t) = \phi_i(t) - \phi_f(t)$, we then obtain

$$\frac{d}{dt}\theta_e(t) = \frac{d}{dt}\phi_i(t) - \frac{d}{dt}\phi_f(t). \tag{6.5.8}$$

If we assume that the PLL input signal has an abrupt change in frequency variation equal to $\Delta\omega_c = d\phi_i(t)/dt$, then we can write from Eq. (6.5.8) that

$$\frac{d}{dt}\phi_f(t) = \Delta\omega_c - \frac{d}{dt}\theta_e(t) = G_T \sin\theta_e(t), \tag{6.5.9}$$

where the last equality follows from Eq. (6.5.6). Equation (6.5.9) can be rewritten as a differential equation in the phase error as

$$\frac{d}{dt}\theta_e(t) + G_T \sin\theta_e(t) = \Delta\omega_c. \tag{6.5.10}$$

Equation (6.5.10) is sketched in Fig. 6.5.2.

The desired operating point in Fig. 6.5.2 is point O, since at that point the input frequency variation and the frequency variation of $f(t)$ are equal $[d\theta_e(t)/dt = 0]$. Note that for any point on the curve in the upper half plane we have $d\theta_e(t) > 0$ for any positive change in time $dt > 0$. Hence any operating point in the upper half plane must move along the curve in the direction of increasing $\theta_e(t)$. Similarly, for any positive time increment, any operating point in the lower half plane has $d\theta_e(t)$ negative, so this operating point must move along the curve in the negative $\theta_e(t)$ direction. We therefore conclude that the point on the curve labeled "O" in Fig. 6.5.2 is the stable operating point of the PLL. The displacement of O from the origin along the $\theta_e(t)$ axis is the steady-state phase error $(\theta_e)_{ss}$. To make $(\theta_e)_{ss}$ small, the PLL parameters must be chosen appropriately. Further analyses of the PLL are relegated to the problems.

In summary, we see that the PLL operates to force $\theta_e(t) = \phi_i(t) - \phi_f(t) = (\theta_e)_{ss}$, and that for a well-designed loop, $(\theta_e)_{ss}$ will be small.

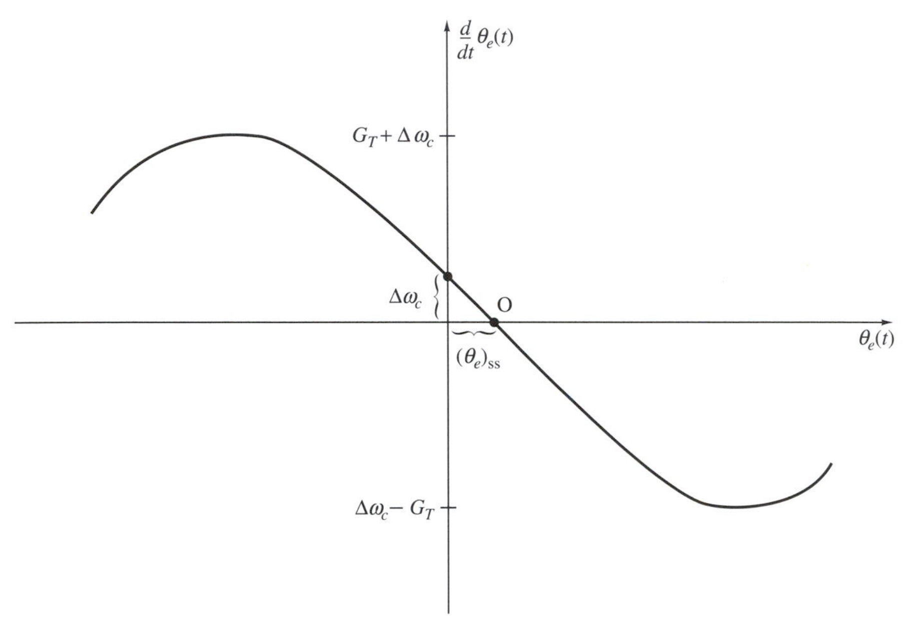

FIGURE 6.5.2 Plot of Eq. (6.5.10).

6.6 Frequency Shift Keying

When frequency modulation is employed to transmit digital messages, particularly binary sequences, the modulation technique is called *frequency shift keying* (FSK). For a binary sequence, FSK simply consists of transmitting a single-frequency sinusoidal pulse for a logic 1 and a different frequency sinusoidal pulse for a logic 0. The pulse shape may vary depending on the application, but if we limit consideration in this section to rectangular pulses of width T, we have for the transmitted signal

$$s(t) = \begin{cases} A_c \cos \omega_1 t, & \text{for } a \ 1, \\ A_c \cos \omega_2 t, & \text{for } a \ 0. \end{cases} \tag{6.6.1}$$

To determine the bandwidth requirements of an FSK signal, we need to model the series of 1's and 0's to be transmitted as a random binary sequence. We present such results shortly. However, to get a rough idea of the bandwidth occupied, we can investigate the alternating infinite sequence of 1's and 0's shown in Fig. 6.6.1(a).

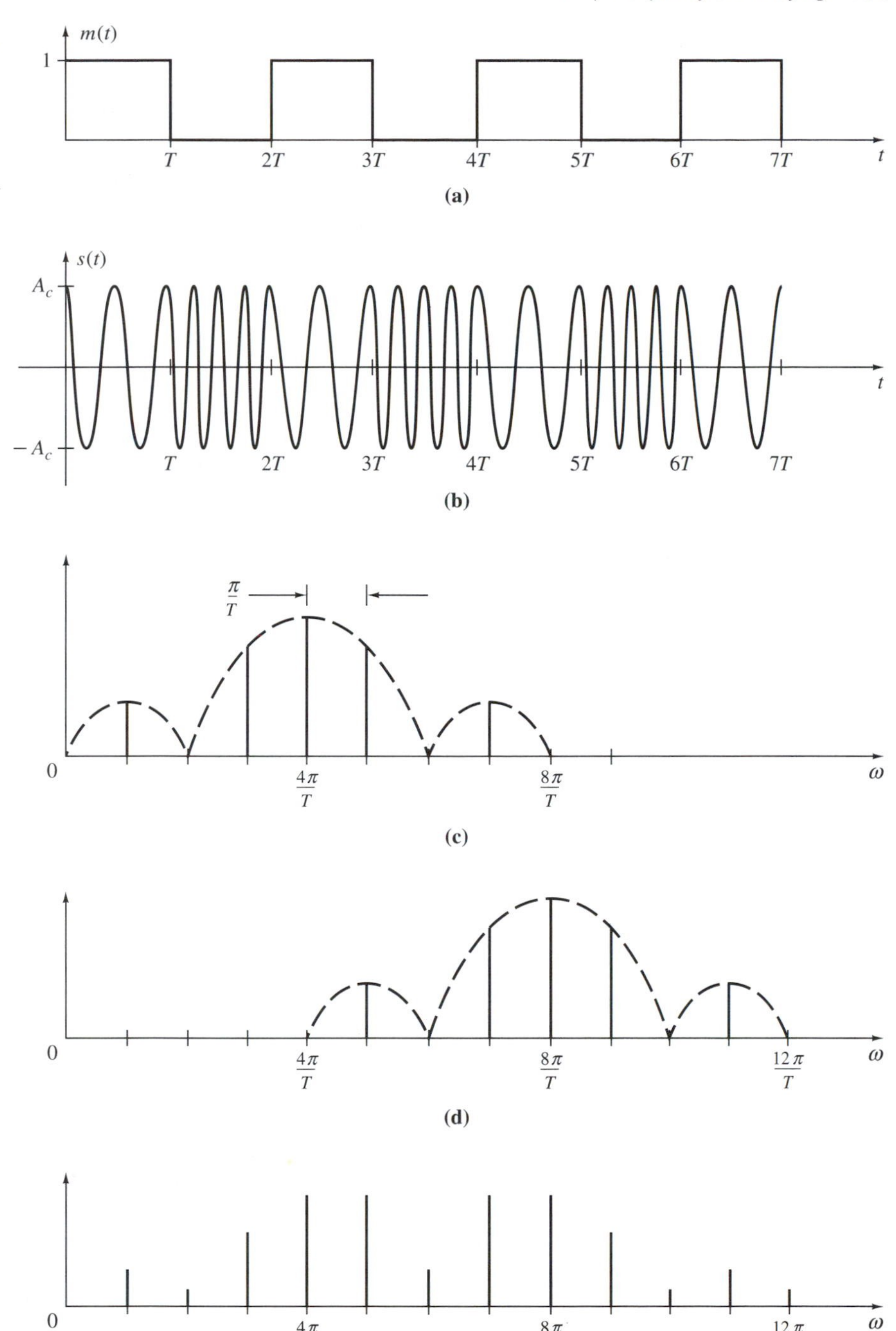

FIGURE 6.6.1 Example 6.6.1, FSK: (a) Message sequence for Example 6.6.1; (b) FSK waveform for $m(t)$ in part (a); (c) magnitude spectrum of $\cos \omega_1 t$ sequence; (d) magnitude spectrum of $\cos \omega_2 t$ sequence; (e) magnitude spectrum of the FSK wave.

EXAMPLE 6.6.1

If we let a pulse of amplitude A denote a 1 and no pulse a 0, the transmitted FSK waveform can be sketched as in Fig. 6.6.1(b), where we have assumed that $\omega_1 = 2\pi/T_1 = 4\pi/T$ and $\omega_2 = 2\pi/T_2 = 8\pi/T$ in Eq. (6.6.1). To evaluate the required bandwidth, we note that the FSK waveform in Fig. 6.6.1(b) is simply the sum of two OOK sequences. The magnitude spectrum corresponding to the cos $\omega_1 t$ carrier is easily calculated to be as shown in Fig. 6.6.1(c), and the magnitude spectrum for the cos $\omega_2 t$ sequence is given in Fig. 6.6.1(d). These two spectra are summed to obtain the magnitude spectrum of the FSK signal sketched in Fig. 6.6.1(e). Depending on which spectral lines we consider significant, we can state the required bandwidth for the FSK waveform.

We can also attack this problem by determining the maximum deviation of the instantaneous frequency. To do this, we assume that the carrier frequency $\omega_c = 6\pi/T$, so that the instantaneous frequency is $\omega_i = \omega_c \pm \Delta\omega = 6\pi/T \pm 2\pi/T$, and thus $\Delta\omega = 2\pi/T$. For Carson's rule we still need the message signal bandwidth. Since in general, $\omega_i = \omega_c + c_f m(t)$, we see that the normalized message signal is a sequence of alternating $+1$ and -1 amplitude rectangular pulses, each of width T. Writing a Fourier series for this sequence and assuming that the highest frequency of importance is at the second sin x/x zero crossing, we have that $\omega_m = 4\pi/T$.

Using Carson's rule, we find that

$$\mathrm{BW} \cong 2(\omega_m + \Delta\omega) = 2\left(\frac{4\pi}{T} + \frac{2\pi}{T}\right) = \frac{12\pi}{T}, \tag{6.6.2}$$

which agrees (under our assumption of significant spectral lines) with Fig. 6.6.1(e).

The exact calculation of the power spectral density of an FSK signal for a random sequence of 1's and 0's is relatively complicated, and hence the derivation is not given in this book. However, plots of the resulting spectral densities for various values of the frequency deviation are presented in Fig. 6.6.2. Note that as the frequency deviation nears $\Delta f = 1/2T$, a sharp peak appears in the spectral density at $f = 1/2T$ hertz. The situation in Example 6.6.1 has $\Delta\omega = 2\pi/T$, which corresponds to $h = 2$. The largest value of h shown in Fig. 6.6.2 is $h = 1.5$ in (d), but the similarity of the spectral shape to the spectral envelope in Fig. 6.6.1(e) is clear. For more details of the power spectral density of FSK, the reader is referred to Anderson and Salz [1965] or Proakis [1989].

Binary FSK signals can be demodulated using synchronous detection or noncoherent detection. We defer the discussion of coherent detection to Chapter 10 and introduce the noncoherent approach here. Noncoherent detection of FSK is particularly popular in data transmission applications. A block diagram for noncoherent detection of FSK is shown in Fig. 6.6.3. The bandpass filters have a narrow passband, so that complete rejection of undesirable frequencies can be achieved.

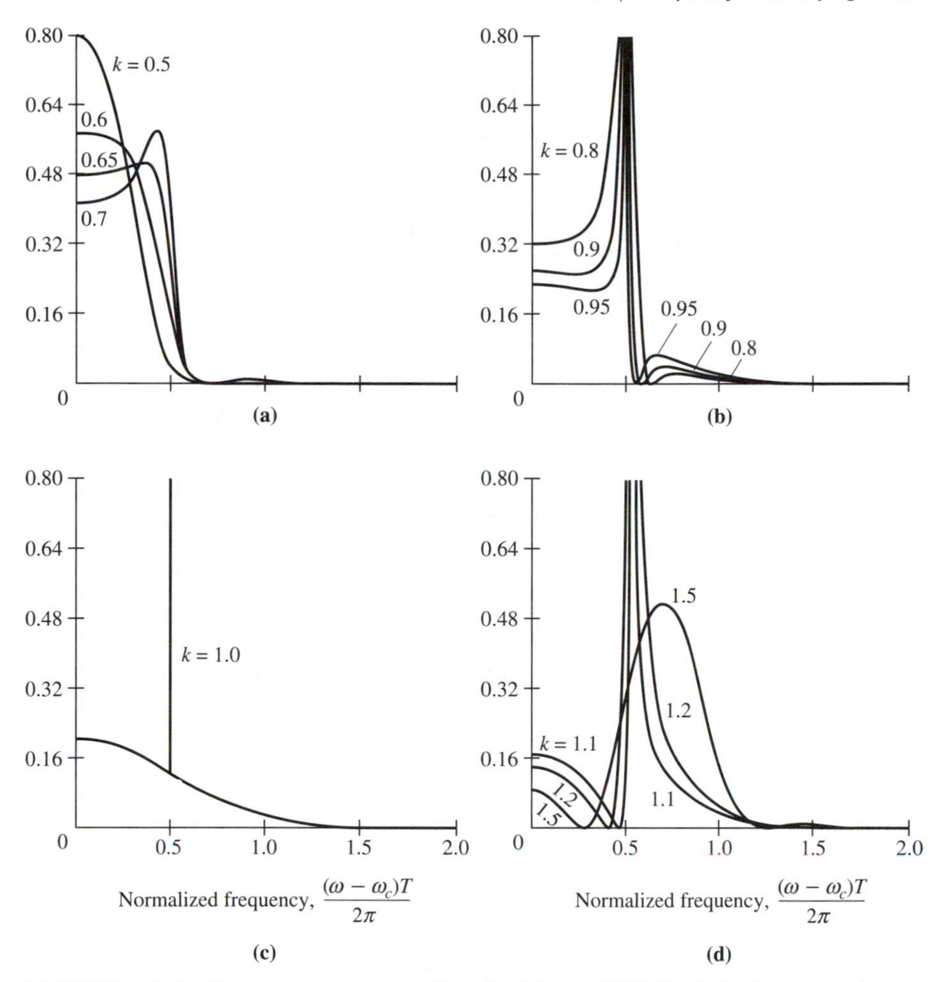

FIGURE 6.6.2 Power spectral densities for binary FSK. R. R. Anderson and J. Salz, "Spectra of Digital FM," *Bell Systems Technical Journal*, Vol. 44, pp. 1165–1189, July–Aug., 1965. Copyright © 1965 AT&T. Reprinted by special permission.

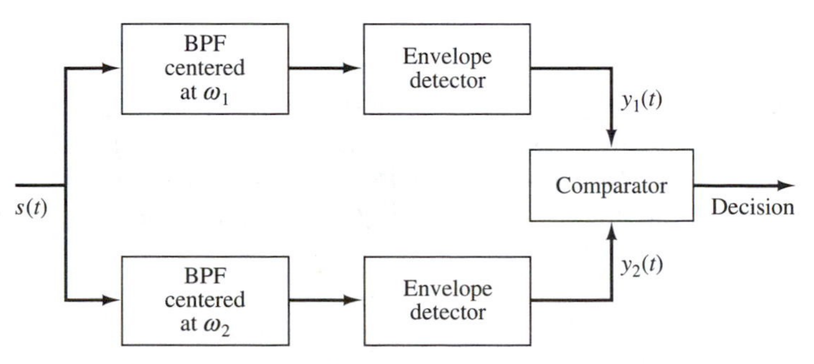

FIGURE 6.6.3 Noncoherent detection of FSK.

When $s(t) = A_c \cos \omega_1(t)$, $y_1(t)$ is a rectangular pulse and $y_2(t)$ is zero, so the comparator output is a logic 1. Similarly, when $s_2(t) = A_c \cos \omega_2 t$, the comparator "decides" that a logic 0 is present. The analysis of this detector in the presence of noise is given in Chapter 10.

6.7 Phase Shift Keying

Phase shift keying (PSK) involves transmitting digital information by shifting the phase of a carrier among several discrete values. When a binary sequence is to be transmitted, the phase is usually switched between 0° and 180°, and the PSK signal is sometimes designated as *phase reversal keying* (PRK). Thus, for PRK, the transmitted signal $s(t)$ is

$$s(t) = \begin{cases} A_c \cos \omega_c t, & \text{for a logic 1} \\ -A_c \cos \omega_c t, & \text{for a logic 0.} \end{cases} \tag{6.7.1}$$

Since for any chosen pulse shaping the envelopes of the two possible transmitted waveforms are the same, and further, they both have the same instantaneous frequency, we are forced to use coherent demodulation to recover the phase information and hence the message.

A block diagram of a detector for PRK is shown in Fig. 6.7.1. As will be discussed in Chapter 10, the filter $H(\omega)$ can be chosen to provide improved noise immunity in addition to rejecting the double-frequency terms. The remaining analysis of Fig. 6.7.1 is left to the reader.

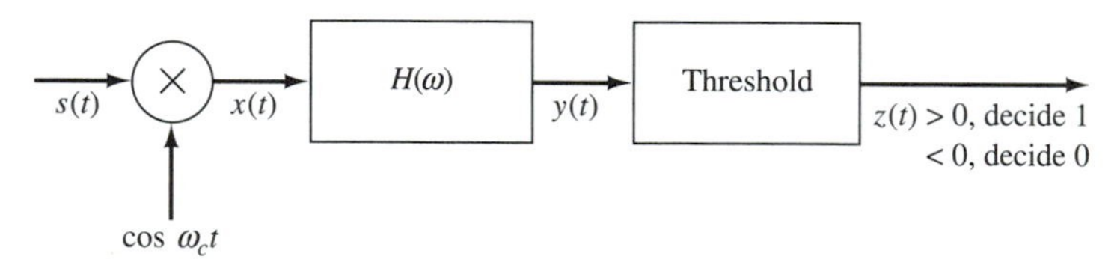

FIGURE 6.7.1 Detector for PRK.

The frequency content of a binary PSK(BPSK) waveform can be obtained using Eq. (A.10.13) from Appendix A and the frequency shifting property, Eq. (3.6.11). In particular, for rectangular pulses of width τ = symbol interval, q = probability of a 1, and $1 - q$ = probability of a 0, the power spectral density of the baseband sequence is

$$S(\omega) = \frac{4q(1-q)}{\tau} \left| \frac{\tau \sin(\omega\tau/2)}{(\omega\tau/2)} \right|^2 + 2\pi(2q-1)^2\delta(\omega). \tag{6.7.2}$$

Note that if the binary symbols are equally likely ($q = 1 - q = \frac{1}{2}$), the spectral line at $\omega = 0$ disappears and the spectral content is the magnitude squared of the familiar sin x/x shape. To obtain the frequency content of BPSK, $S(\omega)$ in Eq. (6.7.2) is shifted to $\pm\omega_c$ according to Eq. (3.6.11).

Since PSK requires coherent demodulation, it is more common to see multiple-phase PSK than PRK. In four-phase PSK, pairs of binary digits are stored and each *pair* is represented by a different transmitted phase. Since there are four possible pairs of binary digits, we must have four different transmitted phases. The phases are usually spaced equally so that the transmitted phases for four-phase PSK are 90° apart. Any set of phases with this spacing will do, although one common choice is the set $\{\pm 45°, \pm 135°\}$.

The transmitted signal $s(t)$ is

$$s(t) = A_c \cos [\omega_c t + \theta_i], \tag{6.7.3}$$

$i = 1, 2, 3, 4$, where we might assign the phases to pairs of binary digits, sometimes called *dibits*, as in Table 6.7.1. The transmitted signal in Eq. (6.7.3) can be rewritten using a trigonometric identity as

$$s(t) = A_c \cos \theta_i \cos \omega_c t - A_c \sin \theta_i \sin \omega_c t. \tag{6.7.4}$$

TABLE 6.7.1 Four-Phase PSK Message Sequence Assignment

Dibit	Carrier Phase
00	+45°
01	−45°
10	+135°
11	−135°

In Eq. (6.7.4) we can clearly distinguish the in-phase and quadrature components, and from this equation we can surmise a possible implementation for the system transmitter.

Following Fig. 6.7.2 and using the phase assignments in Table 6.7.1, we see that for an input binary sequence of 01001110, we would first transmit

$$s(t) = \frac{A_c}{\sqrt{2}} [\cos \omega_c t + \sin \omega_c t], \tag{6.7.5}$$

followed by

$$s(t) = \frac{A_c}{\sqrt{2}} [\cos \omega_c t - \sin \omega_c t]. \tag{6.7.6}$$

The reader should complete the sequence.

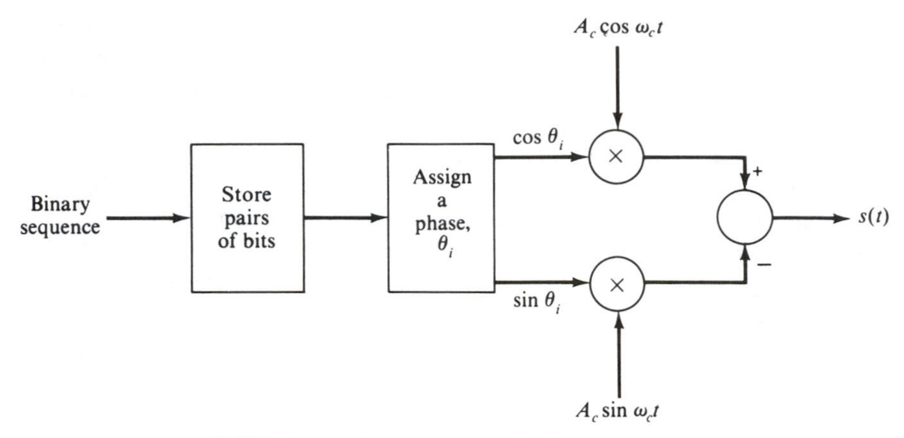

FIGURE 6.7.2 Four-phase PSK transmitter.

A demodulator for the transmitter in Fig. 6.7.2 would take the form shown in Fig. 6.7.3. For the phase assignment in Table 6.7.1, we obtain the decoding rule in Table 6.7.2. It is important to notice that an erroneous phase reference in the local oscillator can cause decoding errors in PSK, and hence a PLL, a transmitted pilot tone, or some other technique must be used to guard against such an eventuality.

TABLE 6.7.2 Decoding Rule for Fig. 6.7.3 and the Phase Assignment in Table 6.7.1

f_i	$-f_q$	θ_i	Decoded Dibit
$\frac{A_c}{\sqrt{2}}$	$\frac{-A_c}{\sqrt{2}}$	45°	00
$\frac{A_c}{\sqrt{2}}$	$\frac{A_c}{\sqrt{2}}$	−45°	01
$\frac{-A_c}{\sqrt{2}}$	$\frac{-A_c}{\sqrt{2}}$	+135°	10
$\frac{-A_c}{\sqrt{2}}$	$\frac{A_c}{\sqrt{2}}$	−135°	11

It is noted that the coherent demodulator in Fig. 6.7.3 is not optimum in terms of minimizing the probability of a bit error. The optimum receiver structures in the presence of noise for PSK and other modulation methods are derived in Chapter 10.

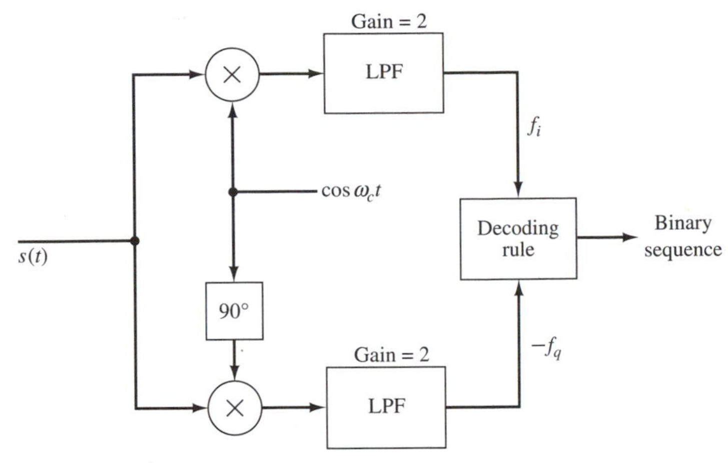

FIGURE 6.7.3 Demodulator for four-phase PSK.

6.8 Differential PSK

Differential phase shift keying (DPSK) avoids the requirement for an accurate local oscillator phase by using the phase during the immediately preceding symbol interval as the phase reference. As long as the preceding phase is received correctly, the phase reference is accurate. Therefore, this approach presupposes that the channel will not produce a serious enough phase error in one symbol interval to cause an erroneous detected phase at the receiver.

There are a variety of encoding techniques for implementing a DPSK system. One technique, which can be used for one-bit-at-a-time transmission, is to obtain a differential binary sequence from the input binary sequence and then assign phases to the bits in the differential sequence. This encoding technique is illustrated in Table 6.8.1. The reference bit is preselected and fixed, but otherwise it is arbitrary. The differential binary sequence is generated by repeating the preceding bit in the differential sequence if the message bit is a 1 or by changing to the opposite bit if the message bit is a 0. Phases are then assigned to the differential binary sequence by transmitting 0° for a 1 and 180° for a 0. The entire procedure is demonstrated in Table 6.8.1.

TABLE 6.8.1 One-Bit-at-a-Time DPSK Scheme

Binary message sequence:										
	0	1	1	0	1	0	0	1	1	1
Differential binary sequence:										
1	0	0	0	1	1	0	1	1	1	1
Reference bit										
Transmitted phase:										
0°	180°	180°	180°	0°	0°	180°	0°	0°	0°	0°

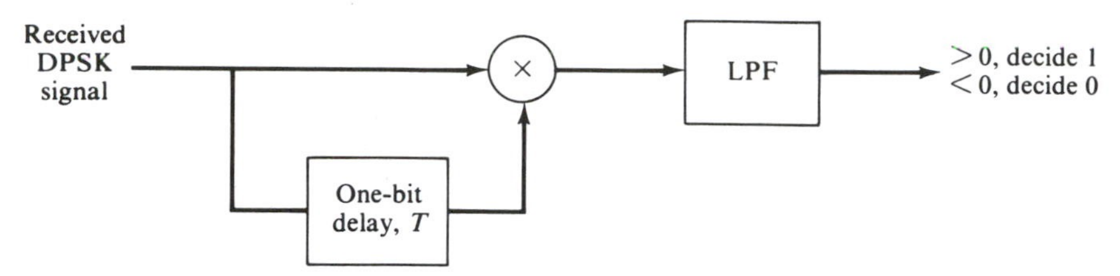

FIGURE 6.8.1 DPSK demodulator for the encoding in Table 6.8.1.

Recovery of the binary message sequence can also be accomplished in several ways. One method is illustrated in Fig. 6.8.1. The received DPSK signal is $\pm\cos \omega_c t$. If the current signal and the immediately preceding received signal agree, the LPF output is positive, and it is decided that the message sequence contains a 1. If the received DPSK signal and its predecessor differ, the LPF output is negative, and the message bit is a 0.

Another common DPSK encoding scheme is to group bits into singles, pairs (dibits), or triples (tribits), and then associate a particular phase *change* with each group. As an example, consider the situation shown in Table 6.8.2, where each pair of binary digits or dibit is represented by the change in transmitted phase indicated. If we wish to transmit the binary message sequence shown in Table 6.8.3, and we assume an initial reference phase of 0°, the transmitted phases are as indicated in the table. Note that the difference between the preceding phase and the present transmitted phase corresponds to the appropriate dibit. Demodulation is accomplished by detecting the received phase sequence, comparing adjacent phases, and outputting the corresponding dibit.

It is common today to use DPSK to transmit tribits, but larger groups of bits are prohibited by phase errors. For many years, DPSK has proven to be a highly effective method for data transmission due to its relative simplicity and reliability.

TABLE 6.8.2 Dibits and Transmitted Phase Changes

Dibit	Phase Change
00	45°
01	135°
10	225°
11	315°

TABLE 6.8.3 Four-Phase DPSK Example

Binary message sequence:		1 0	1 1	0 0	1 0	1 0
Transmitted phase:	0°	225°	180°	225°	90°	315°
Initial phase reference	↑					

SUMMARY

In this chapter we have introduced the important techniques of frequency modulation and phase modulation as well as the concepts of wideband and narrowband FM and their bandwidth requirements. Standard modulation and demodulation methods were also presented, including an introduction to phase-locked loops. The use of phase and frequency modulation for data transmission was also considered, with discussions and developments of FSK, PSK, and DPSK. The importance of these modulation techniques in modem design will become evident in Chapter 8.

PROBLEMS

6.1 Given the waveform

$$s(t) = 5 \cos [5 \times 10^6 t + 100 \sin 200t],$$

(a) If $s(t)$ is a PM wave, what is the message signal $m(t)$?
(b) If $s(t)$ is an FM signal, what is $m(t)$?

6.2 A transmitted angle-modulated waveform is given by

$$s(t) = 20 \cos [10^6 t + 10 \cos 500t].$$

(a) What is the instantaneous frequency of $s(t)$?
(b) What is the approximate bandwidth of $s(t)$?
(c) If $s(t)$ is a PM wave, what is the message signal $m(t)$?
(d) If $s(t)$ is an FM wave, what is the message signal $m(t)$?

6.3 For a message signal given by $m(t) = A_m \cos \omega_m t$, write expressions for the modulation index if:
(a) $m(t)$ is sent by FM.
(b) $m(t)$ is sent by PM. In which case is β independent of ω_m?

6.4 The message $m(t)$ sketched in Fig. P6.4 is to be transmitted using both PM and FM.
(a) For PM let $c_p = \pi/2$ rad/V and assume that $T = 3T_c$, where the carrier frequency $\omega_c = 2\pi/T_c$. Sketch the PM wave.
(b) For FM let $c_f = \pi/T$ rad/sec per volt, with $T = 3T_c$ and $\omega_c = 2\pi/T_c$. Sketch the FM wave.

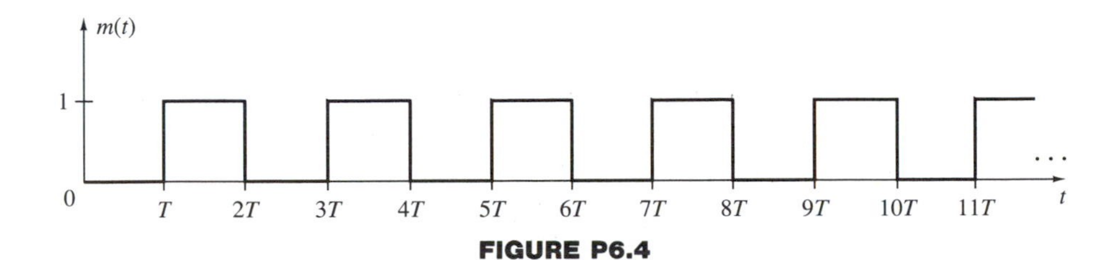

FIGURE P6.4

6.5 Consider the message signal shown in Fig. P6.5.

(a) If $m(t)$ is to be transmitted using FM, let the carrier frequency $\omega_c = 2\pi/T_c$, where $T_c = T/4$ and $c_f = \pi/T$ rad/sec per volt, and sketch the transmitted waveform.

(b) If $m(t)$ is to be transmitted using PM, let $\omega_c = 2\pi/T_c$, $T_c = T/4$, and $c_p = \pi/2$ rad/V, and sketch the PM waveform.

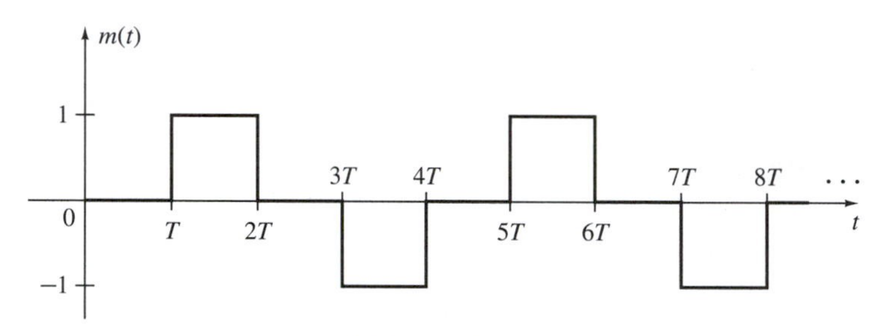

FIGURE P6.5

6.6 If $\theta(t)$ in Eq. (6.2.1) has the form shown in Fig. P6.6, sketch $s_{PM}(t)$. Let $\omega_c = 2\pi/T_c = 8\pi/T$.

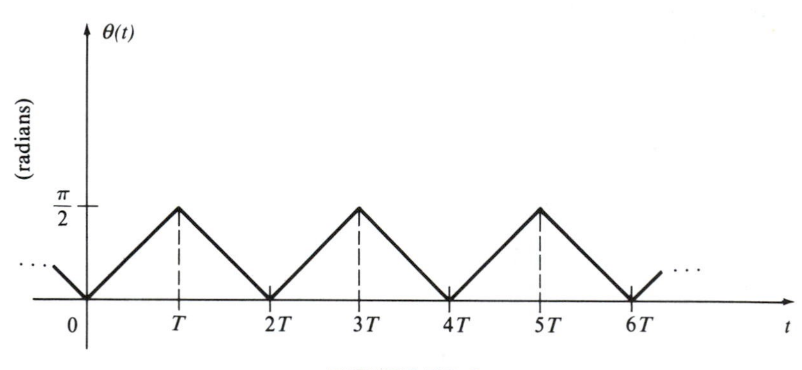

FIGURE P6.6

6.7 If the instantaneous frequency of an FM wave is as shown in Fig. P6.7, sketch the FM waveform. Let $\omega_c = 2\pi/(T/2)$ and $A_c = 1$.

6.8 A 10-MHz carrier signal is frequency modulated by a unit amplitude sinusoid with $\omega_m = 2000\pi$ rad/sec. If $c_f = 10$ rad/sec per volt:

(a) What is the modulation index?

(b) Is this a wideband or a narrowband signal? Why?

(c) What is the bandwidth of the transmitted signal?

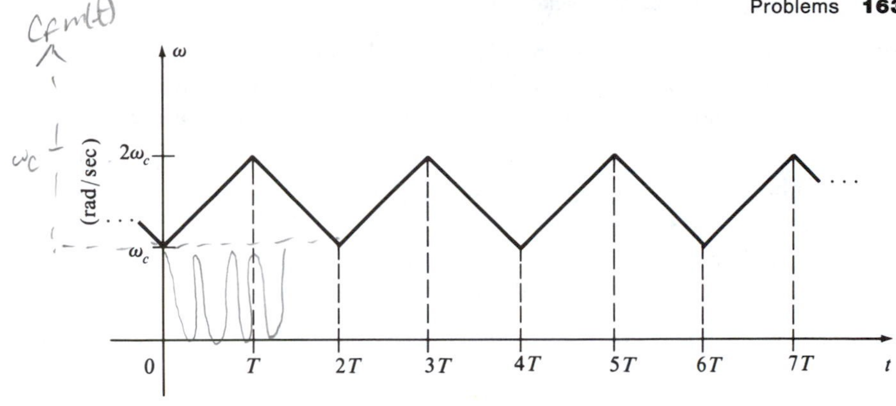

FIGURE P6.7

6.9 Given the angle-modulated waveform

$$s(t) = 10 \cos [90 \times 10^6 t + 200 \sin 4000\pi t].$$

(a) What is the modulation index?
(b) Is $s(t)$ wideband or narrowband? Why?
(c) What is the required bandwidth for $s(t)$?

6.10 Consider a message signal consisting of the sum of two sinusoids,

$$m(t) = A_1 \cos \omega_{m1} t + A_2 \cos \omega_{m2} t.$$

If $m(t)$ is to be transmitted via FM:
(a) Write an expression for the transmitted FM waveform.
(b) What is the maximum frequency deviation?
(c) Obtain an expression for the FM waveform that we can use to determine its bandwidth requirements.

6.11 Show that angle modulation is nonlinear.

6.12 The average power in an FM wave that communicates a pure tone sinusoidal message is $\langle s^2(t) \rangle = (A_c^2/2) \sum_{n=-\infty}^{\infty} J_n^2(\beta)$. For maximum efficiency, we wish all of the power to be in the sidebands. Show how 100% efficiency in this FM signal can be achieved.

6.13 The average power in an angle-modulated signal is $\langle s^2(t) \rangle = A_c^2/2$. Show that $\sum_{n=-\infty}^{\infty} J_n^2(\beta) = 1$.

6.14 Consider an FM wave modulated by a single-tone sinusoid as in Eq. (6.3.4). Sketch the magnitude spectrum of $s_{FM}(t)$ as the amplitude of the modulating signal is increased but ω_m is held fixed. Specifically consider the cases $\beta = 0.1$, $\beta = 1$, and $\beta = 10$.

6.15 For an FM wave modulated by a single-tone sinusoid as in Eq. (6.3.4), sketch the magnitude spectrum of $s_{FM}(t)$ if $\Delta\omega = 60\pi$ rad/sec is held fixed but ω_m is varied. Specifically, let $\omega_m = 60\pi$, 6π, and 2π rad/sec.

6.16 Plot the bandwidth given by Carson's rule versus β when
(a) The bandwidth is normalized by the frequency deviation $\Delta\omega$.
(b) The bandwidth is normalized by the message signal bandwidth ω_m.

6.17 Retaining only those sidebands with an amplitude that is 10% of the unmodulated carrier amplitude or greater, plot the bandwidth of the single-tone modulated FM wave versus β when:
(a) The bandwidth is normalized to $\Delta\omega$.
(b) The bandwidth is normalized to ω_m.

6.18 Repeat Problem 6.17 for sidebands with an amplitude that is 1% of the unmodulated carrier amplitude.

6.19 A narrowband FM wave is given by

$$x(t) = A_c \cos [2\pi \times 10^5 t] - 0.01 A_c \sin [2000\pi t] \sin [2\pi \times 10^5 t].$$

(a) If we write $x(t) = A_c \cos [\omega_c t + \psi(t)]$, what is $\psi(t)$?
(b) Draw a block diagram of an indirect FM transmitter that will generate

$$s(t) = A_c \cos [2\pi \times 10^6 t + \sin 2000\pi t]$$

from $x(t)$.

6.20 A transmitted angle modulated signal is given by $s(t) = A_c \cos [\omega_c t + \theta(t)]$. Unfortunately, an interfering tone is present during transmission, so that the received signal is

$$r(t) = s(t) + A_d \cos [(\omega_c + \omega_d)t].$$

(a) Find an expression for the envelope of $r(t)$.
(b) Find an expression for the phase of $r(t)$.
(c) If $A_c \gg A_d$, what is the approximate instantaneous frequency of $r(t)$?
(d) If $A_d \gg A_c$, what is the approximate instantaneous frequency of $r(t)$?

6.21 A block diagram of a typical broadcast FM receiver is shown in Fig. P6.21. If the IF amplifier center frequency is 10.7 MHz and the IF band-

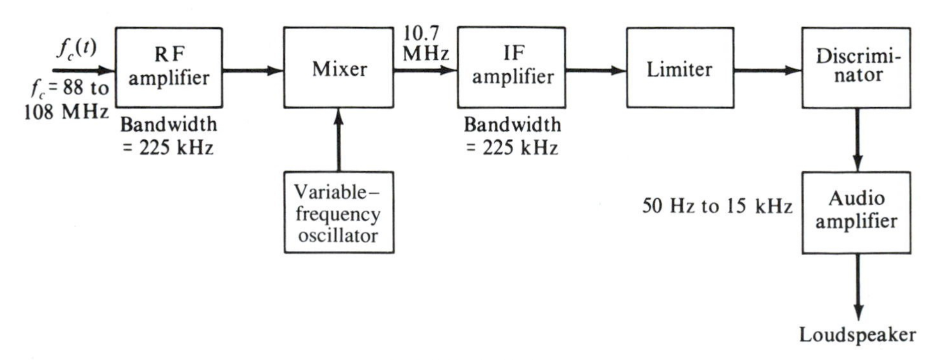

FIGURE P6.21

width is 200 kHz, is there a problem with image stations in broadcast FM?

6.22 For stereo FM, we multiplex the two message signals, along with a pilot tone, and then use narrowband FM. The multiplexing operation is indicated in Fig. P6.22.

(a) If $m_1(t)$ and $m_2(t)$ have the Fourier transforms shown, sketch the Fourier transform of $s(t)$ in Fig. P6.22.

(b) If we use narrowband FM, what is the bandwidth required to transmit $s(t)$?

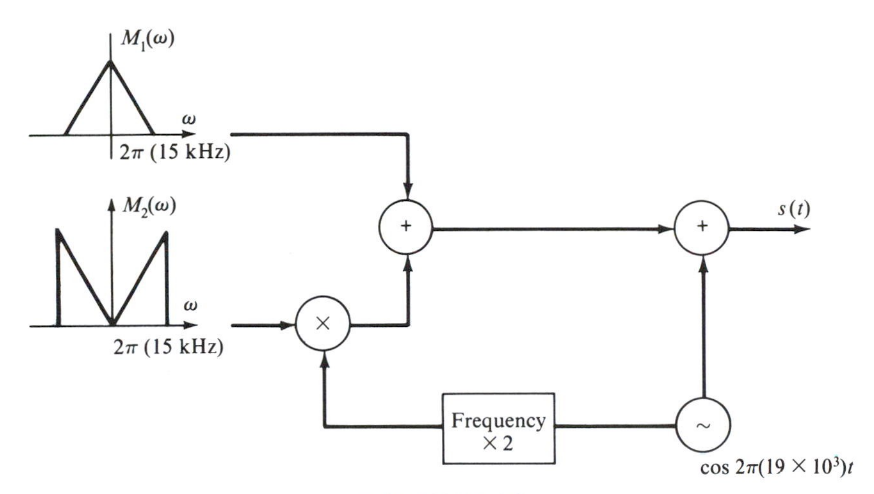

FIGURE P6.22

6.23 A block diagram of an FM stereo receiver is shown in Fig. P6.23(a). If the discriminator output has the Fourier transform shown in Fig. P6.23(b), verify that the LPF outputs are $m_1(t)$ and $m_2(t)$. Note that for monaural FM receivers to be compatible with stereo FM transmission, $m_1(t)$ must represent the sum of the left and right stereo channels. If we then make $m_2(t)$ equal the difference between the left and right channels, we can obtain the left and right channels separately in a stereo receiver by adding and subtracting $m_1(t)$ and $m_2(t)$.

6.24 A nonlinear channel with input $x(t)$ and output $y(t)$ has the following input/output relationship:

$$y(t) = c_1 x(t) + c_2 x^2(t) + c_3 x^3(t).$$

If $x(t) = s_{\mathrm{FM}}(t) = A_c \cos\left[\omega_c t + c_f \int m(t)\, dt\right]$, show how $s_{\mathrm{FM}}(t)$ can be recovered undistorted at the receiver. Compare AMDSB-SC in Problem 5.9.

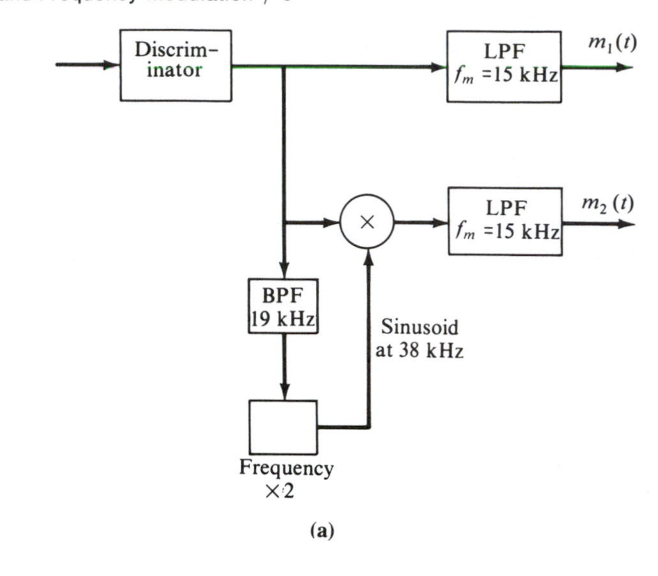

(a)

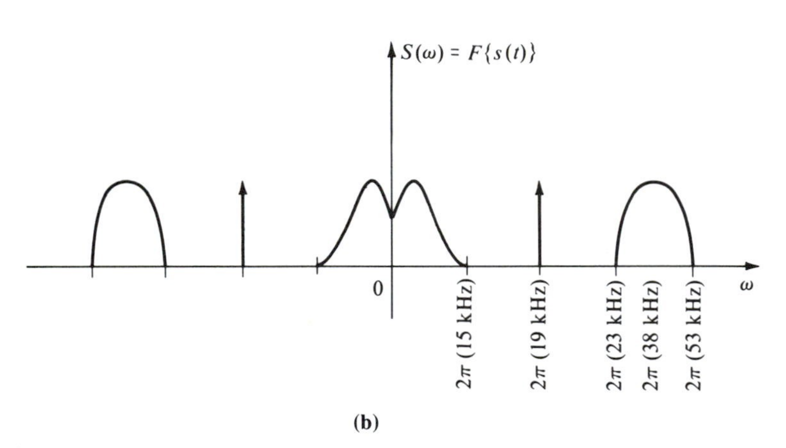

(b)

FIGURE P6.23

6.25 Show that any square-law device with input/output relationship $y(t) = x^2(t)$ can be used as a frequency multiplier.

6.26 Given the narrowband angle-modulated waveform

$$s(t) = A_c \cos \omega_c t - A_c g(t) \sin \omega_c t,$$

where $g(t) = c_p m(t)$ for PM or $g(t) = c_f \int m(t)\, dt$ for FM:

(a) Write $s(t)$ in the envelope-phase form, and thus show that $s(t)$ has amplitude and phase distortion.

(b) Find an infinite series expansion for the instantaneous frequency.

(c) Let $m(t) = A_m \cos \omega_m t$ and assume FM transmission. Retaining only third harmonics and below ($\beta < 1$), write an expression for the instantaneous frequency.

6.27 Use Eq. (6.5.8) to obtain the first-order differential equation for the PLL,

$$\frac{d}{dt}\,\theta_e(t) + G_T\theta_e(t) = \frac{d}{dt}\,\phi_i(t).$$

Hint: Assume that $\theta_e(t)$ is small.

6.28 Letting $r(t) = d\phi_i(t)/dt$, use Laplace transforms to obtain the transfer function of a first-order PLL, $\theta_e(s)/R(s)$, from Problem 6.27.

(a) For a step change in frequency, $R(s) = 1/s$. Show that the steady-state value of $\theta_e(t)$ is $(\theta_e)_{ss} = 1/G_T$.

(b) What is the 3-dB bandwidth of the first-order loop?

(c) Discuss the results of parts (a) and (b).

6.29 Perform the required calculations and sketch the spectra for Example 6.6.1 if $\omega_1 = 20\pi/T$ and $\omega_2 = 24\pi/T$.

6.30 The PLL analysis in Section 6.5 assumes that the amplifier/filter inside the loop in Fig. 6.5.1 has a unity transfer function for all frequencies. Let the transfer function of this filter be $H(\omega)$ and find an expression for $d\theta_e(t)/dt$.

6.31 **(a)** Use the small-angle approximation on Eq. (6.5.6) to obtain a linear model of the PLL.

(b) Repeat part (a) for Problem 6.30.

(c) Find the transfer function of part (b), $\Phi_f(\omega)/\Phi_i(\omega)$.

6.32 Show that an RC high-pass filter as in Fig. P4.20 can be used to obtain Eq. (6.4.7).

Hint: Assume for the frequencies of interest that $\omega \ll 1/RC$.

6.33 Find the transfer function of the first-order, linearized PLL from $\Phi_i(s) = \mathscr{L}\{\phi_i(t)\}$ to $\theta_e(s)$. For a step change in input frequency, find an expression for $\theta_e(t)$.

6.34 For a second-order PLL, the amplifier/filter inside the loop in Fig. 6.5.1 has (approximately)

$$H(s) = \frac{1 + s/\alpha}{s/\beta}.$$

(a) Find the transfer function $\Phi_f(s)/\Phi_i(s)$ of a linearized, second-order PLL.

(b) For a step change in input frequency, $\Phi_i(s) = \Delta\omega/s^2$, find $\theta_e(t)$.

(c) Let $t \to \infty$ in $\theta_e(t)$ and compare to a first-order PLL.

6.35 Can the approach represented by the diagram in Fig. P6.35 be used for FM demodulation? Under what conditions?

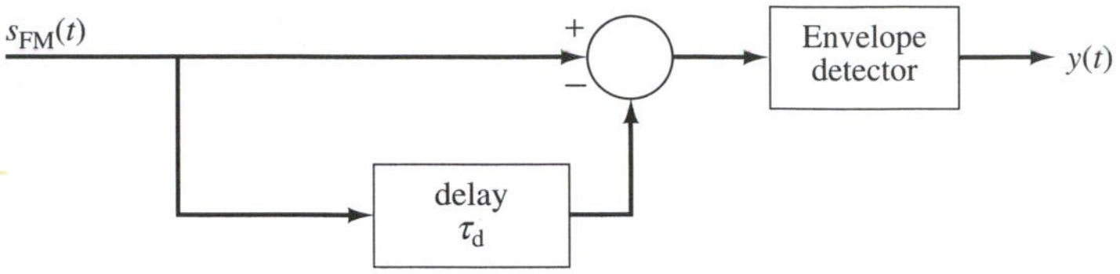

FIGURE P6.35

6.36 Eight-phase PSK is used to transmit binary data sequences by assigning messages (tribits) to phases as shown in Table P6.36. Assume that the carrier frequency is ω_c.

(a) If the sequence to be transmitted is 1 0 1 0 0 1 0 0 0 1 1 1, write expressions for the transmitted waveforms in terms of in-phase and quadrature components.

(b) Specify a decoding table relating the received in-phase and quadrature components and the decoded bits.

TABLE P6.36 Encoding Rule for Problem 6.36

Tribit	000	001	010	011	100	101	110	111
Carrier phase	0°	45°	90°	135°	180°	−45°	−90°	−135°

6.37 A one-bit-at-a-time DPSK encoding scheme does not generate a differential binary sequence but simply transmits no phase change for a binary message 1 and a 180° phase change for a binary message 0. Assuming an initial phase reference of 0°, use this scheme to transmit the binary message sequence in Table 6.8.1. Compare your transmitted phases to the transmitted phase sequence in the table.

6.38 Tribits are to be transmitted using eight-phase DPSK. The tribits and their associated phase changes are listed in Table P6.38. Assuming a 0° initial phase reference, specify the transmitted phase sequence that represents the binary message sequence 1 0 0 0 0 0 1 1 1 1 0 1 0 1 0 0 0 1.

TABLE P6.38 Tribits and Transmitted Phase Changes for Problem 6.38

Tribit	000	001	010	011	100	101	110	111
Phase change	22.5°	67.5°	112.5°	157.5°	202.5°	247.5°	292.5°	337.5°

CHAPTER 7

Noise in Analog Modulation

7.1 Introduction

When designing a communication system, the electrical engineer is usually confronted with limitations on transmitted power and available bandwidth. Required power and bandwidth figured prominently in our discussions of analog modulation techniques, AM, FM, and PM, in Chapters 5 and 6. In this chapter we consider the performance of these analog modulation methods in the presence of noise. Noise is an unavoidable aspect of any communication system. However, if we are willing (and able) to pay the price, say, in terms of excessive transmitted power or system complexity, the noise can be made to have a negligible impact on the performance of a communication system. In most instances, however, we are interested in designing the least expensive, most efficient, highest-performance system possible within the limits of our specifications. Of course, in these situations, analyses of the effects of noise on communication system performance are vitally important. Hence we are led to the purpose of this chapter.

In communication system analyses, there are deterministic impairments such as channels with nonflat amplitude responses and nonlinear phase characteristics, and these deterministic distortions can have a significant impact on communication system performance. However, our interest in this chapter is in random impairments, that is, in distortions that can be modeled as being generated by some random experiment. We call this random distortion *noise*. We also assume throughout the analyses in this chapter that the noise is additive, so that the received signal after transmission over the channel is the undistorted modulated signal plus a random noise waveform. Additive noise is common in practical communication systems, and hence this is a natural place to begin our investigations.

In Section 7.2 the relevant noise models are developed, and in the following sections we use these noise models to obtain objective measures of the various AM, FM, and PM system performances. The principal objective performance indicator is the system output signal-to-noise ratio (SNR). At the end of the chapter we are able to compare the various analog modulation systems on the basis of signal-to-noise ratios, transmitted power, and bandwidth.

7.2 Narrowband Noise

To begin our discussion, we consider a general time function, say $x(t)$, with Fourier transform $X(\omega)$. If $X(\omega)$ is nonzero only over a band of frequencies given by $\omega_0 - \omega_B \leq |\omega| \leq \omega_0 + \omega_B$, where $\omega_0 \gg \omega_B$, $x(t)$ is called a *narrowband waveform* and has a magnitude spectrum of the form shown in Fig. 7.2.1. From our knowledge of amplitude modulation, we see that the spectrum in Fig. 7.2.1 is reminiscent of a carrier wave with a slowly varying envelope, or possibly, from our developments of frequency and phase modulation, a carrier with a slowly varying frequency or phase (narrowband FM). It is thus plausible that $x(t)$ could be expressed as

$$x(t) = \rho(t) \cos [\omega_c t + \theta(t)], \tag{7.2.1}$$

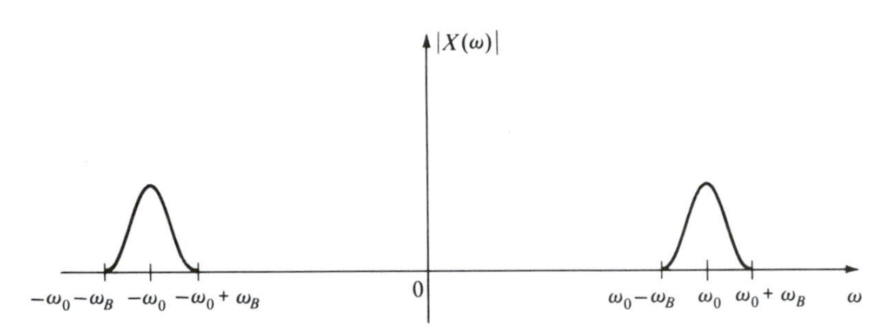

FIGURE 7.2.1 Magnitude spectrum of a narrowband waveform.

where $\rho(t)$ is the slowly varying envelope and $\theta(t)$ is the slowly varying phase. From Problem 5.9 we know that we can use trigonometric identities to expand $x(t)$ as

$$\begin{aligned} x(t) &= \rho(t) \cos \theta(t) \cos \omega_c t - \rho(t) \sin \theta(t) \sin \omega_c t \\ &= x_i(t) \cos \omega_c t - x_q(t) \sin \omega_c t, \end{aligned} \tag{7.2.2}$$

where

$$x_i(t) \triangleq \rho(t) \cos \theta(t) \tag{7.2.3}$$

$$x_q(t) \triangleq \rho(t) \sin \theta(t) \tag{7.2.4}$$

are the in-phase and quadrature components of the signal, respectively. Upon using Eqs. (7.2.3) and (7.2.4), it is immediate that the envelope and phase can be expressed in terms of the in-phase and quadrature components as

$$\rho(t) = \sqrt{x_i^2(t) + x_q^2(t)} \tag{7.2.5}$$

and

$$\theta(t) = \tan^{-1} \frac{x_q(t)}{x_i(t)}. \tag{7.2.6}$$

Now, our interest in this section is noise, and it is common to model the random disturbances contributed by the channel in communication systems as an additive white Gaussian noise process $\{X(t), -\infty < t < \infty\}$ with zero mean and two-sided spectral density $S_X(\omega) = (2\pi)\mathcal{N}_0/2$ watts/rad per second for $-\infty < \omega < \infty$. Clearly, this is a wideband process, and we wonder how narrowband noise is obtained from $\{X(t)\}$. If one examines the analog communication systems in Chapters 5 and 6, the answer becomes quite clear. At the front end of the receivers in analog communication systems, the received signals are passed through at least one bandpass filter whose bandwidth is proportional to the message signal frequency content, say ω_B. Since $\omega_B \ll \omega_c$, the noise process at the output of these filters, denoted by $\{N(t), -\infty < t < \infty\}$, is narrowband with a spectral density $S_N(\omega) = \pi|H(\omega)|^2\mathcal{N}_0$ watts/rad per second [see Eq. (A.9.14)], where $H(\omega)$ is the receiver bandpass filter transfer function. The noise spectral content at the bandpass filter output thus greatly resembles the shape shown in Fig. 7.2.1, and therefore the noise process $\{N(t)\}$ can be written in narrowband form. At this point, to conform with the existing literature and to simplify the notation in subsequent sections, we make a change of notation and let the narrowband noise process be represented by $\{n(t), -\infty < t < \infty\}$ rather than $\{N(t), -\infty < t < \infty\}$. Although this notation differs from that introduced in Appendix A for random processes, it will be used henceforth for narrowband noise processes.

With this change in notation, we can now write the narrowband noise process $\{n(t)\}$ as

$$n(t) = \rho(t) \cos [\omega_c t + \theta(t)] = n_i(t) \cos \omega_c t - n_q(t) \sin \omega_c t, \tag{7.2.7}$$

where $\rho(t)$ and $\theta(t)$ are now random processes and

$$n_i(t) = \rho(t) \cos \theta(t) \tag{7.2.8}$$

and

$$n_q(t) = \rho(t) \sin \theta(t). \tag{7.2.9}$$

Again, the expressions for $\rho(t)$ and $\theta(t)$ follow as

$$\rho(t) = \sqrt{n_i^2(t) + n_q^2(t)} \tag{7.2.10}$$

and

$$\theta(t) = \tan^{-1} \frac{n_q(t)}{n_i(t)}. \tag{7.2.11}$$

Since $\{n(t)\}$ is obtained by linearly filtering the Gaussian process $\{X(t)\}$, $n(t)$ is Gaussian; and further, $n_i(t)$ and $n_q(t)$ can be derived from $n(t)$ by linear transformations, and hence they are also Gaussian processes. We assume for the sequel that $n(t)$ is zero mean.

It is often useful to have available the marginal pdfs of the envelope $\rho(t)$ and the phase $\theta(t)$. Since $n_i(t)$ and $n_q(t)$ are Gaussian processes, and if we assume that they are zero mean (see Property 1 later in this section), then it is straightforward to show via a transformation of variables that the joint pdf of $\rho(t)$ and $\theta(t)$ is

$$f_{\rho,\theta}(\rho, \theta; t) = \frac{\rho(t)}{2\pi\sigma^2} e^{-\rho^2(t)/2\sigma^2} \tag{7.2.12}$$

for $0 \leq \rho(t) < \infty$ and $0 \leq \theta(t) \leq 2\pi$. The marginal pdfs follow directly as

$$\begin{aligned} f_\rho(\rho; t) &= \int_0^{2\pi} f_{\rho,\theta}(\rho, \theta; t)\, d\theta(t) \\ &= \frac{\rho(t)}{\sigma^2} e^{-\rho^2(t)/2\sigma^2} \end{aligned} \tag{7.2.13}$$

for $0 \leq \rho(t) < \infty$, and

$$\begin{aligned} f_\theta(\theta; t) &= \int_0^{\infty} f_{\rho,\theta}(\rho, \theta; t)\, d\rho(t) \\ &= \frac{1}{2\pi} \end{aligned} \tag{7.2.14}$$

for $0 \leq \theta(t) \leq 2\pi$. The pdf in Eq. (7.2.13) is called a *Rayleigh* probability density function, while Eq. (7.2.14) indicates that $\theta(t)$ is uniformly distributed over the interval $[0, 2\pi]$.

We now list several properties of narrowband Gaussian processes without proof, which will be useful in later sections.

Property 1. If $n(t)$ is zero mean, $n_i(t)$ and $n_q(t)$ are zero mean.

Property 2. Since the in-phase and quadrature components of a narrowband process are uncorrelated and since $n_i(t)$ and $n_q(t)$ are jointly Gaussian, then $n_i(t)$ and $n_q(t)$ are independent.

Property 3. The in-phase and quadrature components have the same autocorrelation function, and hence the same spectral density.

Property 4. If $n(t)$ is wide-sense stationary, the $n_i(t)$ and $n_q(t)$ are jointly wide-sense stationary.

We can find the mean-squared value of $n(t)$ starting with Eq. (7.2.7) as

$$E\{n^2(t)\} = E\{n_i^2(t)\} \cos^2 \omega_c t + E\{n_q^2(t)\} \sin^2 \omega_c t, \tag{7.2.15}$$

where we have employed Properties 1 and 2 to eliminate the cross terms. Using trigonometric identities and invoking Property 3, we see that

$$\begin{aligned} E\{n^2(t)\} &= E\{n_i^2(t)\} = E\{n_q^2(t)\} \\ &= \tfrac{1}{2}E\{n_i^2(t)\} + \tfrac{1}{2}E\{n_q^2(t)\}. \end{aligned} \tag{7.2.16}$$

Additional developments concerning narrowband noise are included in the problems.

7.3 AM-Coherent Detection

For the purposes of evaluating the performance of the individual analog modulation systems and comparing the performances among systems, we must select an analytically tractable and physically meaningful quantity. For our developments in this section and the remainder of the chapter, we choose to examine the receiver output signal-to-noise ratio for a specified input signal-to-noise ratio, where we define the signal-to-noise ratio as the ratio of the time average of the squared "message" signal component to the mean-squared value of the noise at the same point. For all the developments, we assume that the channel has added white Gaussian noise which has been bandpass filtered to generate narrowband Gaussian noise at the detector inputs.

For AMDSB-SC signals, the receiver after bandpass filtering can be represented by the block diagram in Fig. 7.3.1. The received waveform after bandpass filtering is given by

$$r(t) = A_1 m(t) \cos \omega_c t + n_i(t) \cos \omega_c t - n_q(t) \sin \omega_c t, \tag{7.3.1}$$

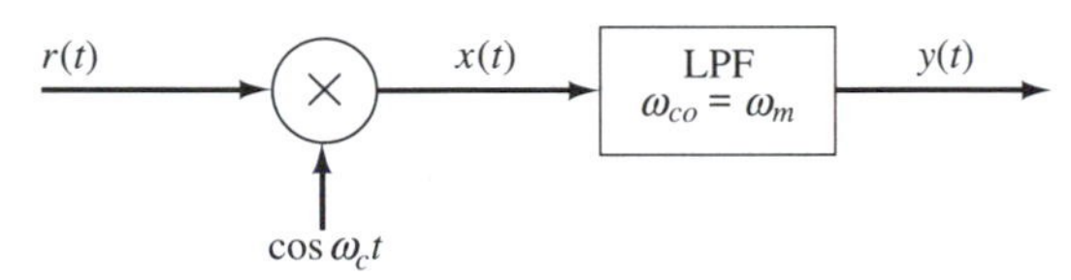

FIGURE 7.3.1 Coherent detector for AMDSB-SC and AMSSB-SC.

where $m(t)$ is the low-pass message signal bandlimited to ω_m. Upon considering $r(t)$ to be the receiver input waveform, it is clear that $A_1 m(t) \cos \omega_c t$ is the desired signal component and the other terms in Eq. (7.3.1) constitute the bandpass Gaussian noise. Therefore, the average power in the desired component is

$$A_1^2 \langle m^2(t) \cos^2 \omega_c t \rangle = \frac{A_1^2}{2} \langle m^2(t) \rangle, \tag{7.3.2}$$

where $\langle \cdot \rangle$ denotes time averaging, and the mean-squared value of the noise is $E\{n^2(t)\}$ as given by Eq. (7.2.15). The input signal-to-noise ratio, SNR_i, is thus

$$\text{SNR}_i = \frac{A_1^2/2 \langle m^2(t) \rangle}{E\{n^2(t)\}}. \tag{7.3.3}$$

To determine the output signal-to-noise ratio, SNR_o, we need an expression for $y(t)$, which follows easily from Fig. 7.3.1 as

$$y(t) = \frac{A_1}{2} m(t) + \frac{1}{2} n_i(t). \tag{7.3.4}$$

The message signal power and mean-squared value of the noise component are thus $A_1^2\langle m^2(t)\rangle/4$ and $E\{n_i^2(t)\}/4$, respectively, so that the output signal-to-noise ratio is

$$\mathrm{SNR}_o = \frac{A_1^2\langle m^2(t)\rangle}{E\{n_i^2(t)\}} = \frac{A_1^2\langle m^2(t)\rangle}{E\{n^2(t)\}}, \tag{7.3.5}$$

where the last equality follows from Eq. (7.2.16). Comparing Eqs. (7.3.3) and (7.3.5), we see that for AMDSB-SC,

$$\mathrm{SNR}_o = 2\mathrm{SNR}_i. \tag{7.3.6}$$

The coherent receiver thus provides a factor of 2 or a 3-dB gain in SNR by eliminating the quadrature component of the narrowband noise.

The coherent receiver block diagram in Fig. 7.3.1 is also valid for AMSSB-SC waveforms; however, in this case the received signal after bandpass filtering is

$$r(t) = A_2 m(t)\cos\omega_c t \pm A_2 m_h(t)\sin\omega_c t + n_i(t)\cos\omega_c t - n_q(t)\sin\omega_c t, \tag{7.3.7}$$

where $m_h(t)$ is the Hilbert transform of $m(t)$. The average power in the desired signal component is

$$\begin{aligned} A_2^2\langle[m(t)\cos\omega_c t \pm m_h(t)\sin\omega_c t]^2\rangle &= \frac{A_2^2}{2}\langle m^2(t)\rangle + \frac{A_2^2}{2}\langle m_h^2(t)\rangle \\ &= A_2^2\langle m^2(t)\rangle, \end{aligned} \tag{7.3.8}$$

where the last equality results from property (c) in Problem 5.15. The input noise power is again just $E\{n^2(t)\}$, so

$$\mathrm{SNR}_i = \frac{A_2^2\langle m^2(t)\rangle}{E\{n^2(t)\}}. \tag{7.3.9}$$

After synchronous detection, we have that

$$y(t) = \frac{A_2}{2}m(t) + \frac{1}{2}n_i(t), \tag{7.3.10}$$

so that the output signal-to-noise ratio follows immediately as

$$\mathrm{SNR}_o = \frac{A_2^2\langle m^2(t)\rangle}{E\{n^2(t)\}}. \tag{7.3.11}$$

Comparing Eqs. (7.3.9) and (7.3.11), we get the result that

$$\mathrm{SNR}_o = \mathrm{SNR}_i \tag{7.3.12}$$

for coherent detection of AMSSB-SC. The receiver does not provide a gain in SNR for SSB as it does for DSB, since the desired signal power in the $m_h(t)$ component is discarded by the receiver along with the quadrature component of the noise.

7.4 AM-Noncoherent Detection

For AMDSB-TC or conventional AM, the receiver simply consists of an envelope detector, and the input to this detector after bandpass filtering is

$$r(t) = A_3[1 + am(t)] \cos \omega_c t + n_i(t) \cos \omega_c t - n_q(t) \sin \omega_c t, \tag{7.4.1}$$

where a is the modulation index and $m(t)$ is assumed to be normalized here. The only term in Eq. (7.4.1) containing the message signal is the component $aA_3 m(t) \cos \omega_c t$, which has an average power

$$\langle a^2 A_3^2 m^2(t) \cos^2 \omega_c t \rangle = \tfrac{1}{2} a^2 A_3^2 \langle m^2(t) \rangle. \tag{7.4.2}$$

Since the mean-squared value of the noise is $E\{n^2(t)\}$, we have that the input SNR is

$$\text{SNR}_i = \frac{a^2 A_3^2 \langle m^2(t) \rangle}{2E\{n^2(t)\}}. \tag{7.4.3}$$

To find the output SNR, we need the envelope of $r(t)$ in Eq. (7.4.1). Although many alternative approaches are possible here, we choose to use Eqs. (7.2.8) and (7.2.9) and write

$$\text{env}[r(t)] = \{A_3^2[1 + am(t)]^2 + 2A_3[1 + am(t)]\rho(t) \cos \theta(t) + \rho^2(t)\}^{1/2}. \tag{7.4.4}$$

To simplify this, we factor $A_3^2[1 + am(t)]^2$ out of the radical and assume that we are in the high carrier-to-noise ratio case,

$$A_3^2 \gg E\{\rho^2(t)\},$$

so that

$$\begin{aligned} \text{env}[r(t)] &= A_3[1 + am(t)]\left\{1 + \frac{2\rho(t) \cos \theta(t)}{A_3[1 + am(t)]} + \frac{\rho^2(t)}{A_3^2[1 + am(t)]^2}\right\}^{1/2} \\ &\cong A_3[1 + am(t)]\left\{1 + \frac{2\rho(t) \cos \theta(t)}{A_3[1 + am(t)]}\right\}^{1/2}. \end{aligned} \tag{7.4.5}$$

Using the approximation for the square root that $\{1 + x\}^{1/2} \cong 1 + \frac{1}{2}x$, we have

$$\begin{aligned} \text{env}[r(t)] &\cong A_3[1 + am(t)] + \rho(t) \cos \theta(t) \\ &= A_3[1 + am(t)] + n_i(t). \end{aligned} \tag{7.4.6}$$

The average power in the term containing the message is thus $\langle A_3^2 a^2 m^2(t) \rangle$, so the output SNR is

$$\text{SNR}_o = \frac{a^2 A_3^2 \langle m^2(t) \rangle}{E\{n^2(t)\}}. \tag{7.4.7}$$

Comparing Eqs. (7.4.3) and (7.4.7), we find that

$$\text{SNR}_o = 2\text{SNR}_i. \tag{7.4.8}$$

Therefore, for high carrier-to-noise ratios, the envelope detector provides a 3-dB gain in output SNR, just as coherent detection does. It is important to note, however, that this result ignores the power contained in the carrier term alone, which is not necessary in AMDSB-SC.

For comparisons to other modulation methods such as FM, it is often convenient to express SNR_i and SNR_o in terms of the input carrier-to-noise ratio. It is evident from Eq. (7.4.1) that the average power in the carrier term alone is $A_3^2/2$, so that the input carrier-to-noise ratio, denoted CNR_i, is given by

$$\text{CNR}_i = \frac{A_3^2}{2E\{n^2(t)\}}. \tag{7.4.9}$$

Thus in terms of CNR_i, Eqs. (7.4.3) and (7.4.7) become

$$\text{SNR}_i = a^2\langle m^2(t)\rangle \text{CNR}_i \tag{7.4.10}$$

and

$$\text{SNR}_o = 2a^2\langle m^2(t)\rangle \text{CNR}_i, \tag{7.4.11}$$

respectively. Of course, Eq. (7.4.8) is unaffected.

For the small carrier-to-noise ratio case ($A_3^2 \ll E\{\rho^2(t)\}$), the envelope in Eq. (7.4.4) can be approximated as

$$\text{env}[r(t)] \cong \rho(t) + A_3[1 + am(t)] \cos \theta(t). \tag{7.4.12}$$

The message signal $m(t)$ is multiplied by the random variable $\cos \theta(t)$ and has added to it a very large amplitude random variable $\rho(t)$. The message is thus completely lost in the case of low carrier-to-noise ratio. There is some threshold value of CNR between the large and small CNR cases above which performance is acceptable and below which performance degrades very quickly. Such a threshold effect is common to nonlinear demodulation methods but is not evident for coherent demodulation of AMDSB-SC.

7.5 Frequency and Phase Modulation

To evaluate the performance of an FM system in the presence of additive narrowband Gaussian noise, we consider a standard FM waveform given by

$$s_{\text{FM}}(t) = A_c \cos\left[\omega_c t + c_f \int m(t)\, dt\right], \tag{7.5.1}$$

which is additively contaminated by the narrowband noise process $n(t)$ in Eq. (7.2.7) to produce the received signal

$$\begin{aligned} r(t) &= s_{\text{FM}}(t) + n(t) \\ &= A_c \cos\left[\omega_c t + c_f \int m(t)\, dt\right] + \rho(t) \cos\left[\omega_c t + \theta(t)\right]. \end{aligned} \tag{7.5.2}$$

The signal $r(t)$ is the waveform present at the IF filter output and at the input to the FM discriminator shown in Fig. 6.4.3. It will be useful to write Eq. (7.5.2) in the form

$$r(t) = A(t) \cos [\omega_c t + \phi(t)], \tag{7.5.3}$$

where $A(t)$ is of no interest due to the presence of the hard limiter in the discriminator. To gain some qualitative insight into the behavior of the phase $\phi(t)$, we let $\alpha(t) = \omega_c t + c_f \int m(t)\,dt$ (only for notational ease) and expand $r(t)$ in Eq. (7.5.2) as

$$\begin{aligned} r(t) &= A_c \cos \alpha(t) + \rho(t) \cos \left[\alpha(t) + \theta(t) - c_f \int m(t)\,dt \right] \\ &= A_c \cos \alpha(t) + \rho(t) \cos \left[\theta(t) - c_f \int m(t)\,dt \right] \cos \alpha(t) \\ &\quad - \rho(t) \sin \left[\theta(t) - c_f \int m(t)\,dt \right] \sin \alpha(t). \end{aligned} \tag{7.5.4}$$

Now, the phase with respect to the in-phase and quadrature components of $\alpha(t)$ can be found in the usual way from Eq. (7.5.4), so that $\phi(t)$ in Eq. (7.5.3) becomes

$$\phi(t) = c_f \int m(t)\,dt + \tan^{-1} \left\{ \frac{\rho(t) \sin [\theta(t) - c_f \int m(t)\,dt]}{A_c + \rho(t) \cos [\theta(t) - c_f \int m(t)\,dt]} \right\}. \tag{7.5.5}$$

For those readers comfortable with phasor notation, Eq. (7.5.5) is tantamount to using the instantaneous phase $c_f \int m(t)\,dt$ as a reference with the second term in Eq. (7.5.5) varying about this reference. Considering the high carrier-to-noise ratio case, where $A_c \gg \rho(t)$, we see from Eq. (7.5.5) that

$$\phi(t) \cong c_f \int m(t)\,dt + \frac{\rho(t)}{A_c} \sin \left[\theta(t) - c_f \int m(t)\,dt \right], \tag{7.5.6}$$

where we have used the small-angle approximation on the arctan. Thus it is evident from Eq. (7.5.6) that in the large CNR case, the noise creates only a small random variation about the desired phase $c_f \int m(t)\,dt$ and is hence not much of a problem.

To examine the small CNR $[A_c \ll \rho(t)]$ behavior of $\phi(t)$, we follow an argument analogous to that employed to obtain Eq. (7.5.5) except that we use $\theta(t)$ as a reference [let $\alpha(t) = \omega_c t + \theta(t)$ and proceed] to find

$$\phi(t) = \theta(t) + \tan^{-1} \left\{ \frac{A_c \sin [c_f \int m(t)\,dt - \theta(t)]}{\rho(t) + A_c \cos [c_f \int m(t)\,dt - \theta(t)]} \right\}. \tag{7.5.7}$$

Invoking the small CNR assumption, we have that

$$\phi(t) \cong \theta(t) + \frac{A_c}{\rho(t)} \sin \left[c_f \int m(t)\,dt - \theta(t) \right]. \tag{7.5.8}$$

Equation (7.5.8) reveals that for $A_c \ll \rho(t)$, the high-noise or weak-signal case, the phase is primarily the phase of the narrowband noise and the message signal cannot be recovered. The noise is sometimes said to *capture* the receiver in this case.

Besides obtaining an intuitive feel for the behavior of an FM receiver, the preceding analysis also makes it clear that our subsequent calculations of output SNR need only focus on the high-CNR case, because in the low-CNR situation, the system will be unusable, and hence of little practical interest.

To determine the output signal-to-noise ratio, we see from Fig. 6.4.3 that we need to differentiate the received signal and then pass it through an envelope detector. For purposes of analysis, this is equivalent to differentiating $\phi(t)$ in Eq. (7.5.6) and then low-pass filtering the result to the message signal bandwidth. Before proceeding, however, we need to simplify Eq. (7.5.6) still further. The term that causes the difficulty is the message component in the argument of the sine function. We know from Section 7.2 that $\theta(t)$ is uniformly distributed over the interval $[0, 2\pi]$, and thus if the message term were not present in the argument of the sine function, we could continue the analysis with little difficulty. Although we will not do so here, it can be shown that the presence of this message term serves only to produce noise outside the message bandwidth and hence can be neglected for the analysis here.

Therefore, returning to Eq. (7.5.6) and ignoring the message component in the sine function argument, we can approximate the additive phase of the carrier (for this analysis) by

$$\begin{aligned}\phi(t) &\cong c_f \int m(t)\, dt + \frac{\rho(t)}{A_c} \sin \theta(t) \\ &= c_f \int m(t)\, dt + \frac{n_q(t)}{A_c},\end{aligned} \tag{7.5.9}$$

where $n_q(t)$ is the white, zero-mean, Gaussian-distributed, quadrature component of the narrowband noise. Differentiating this $\phi(t)$ expression yields

$$\frac{d}{dt}\phi(t) = c_f m(t) + \frac{1}{A_c}\frac{d}{dt} n_q(t). \tag{7.5.10}$$

To find the mean-squared value of the second term in Eq. (7.5.10), we use Eq. (A.9.6) in conjunction with Eq. (A.9.14), where $H(\omega) = j\omega/A_c$ and the power spectral density of $n_q(t)$ is given by

$$S_{n_q}(\omega) = \begin{cases} S_n(\omega - \omega_c) + S_n(\omega + \omega_c), & |\omega| \le (\Delta\omega + \omega_m) \\ 0, & |\omega| > (\Delta\omega + \omega_m), \end{cases} \tag{7.5.11}$$

where ω_m is the bandwidth of the low-pass message $m(t)$, $\Delta\omega$ is the maximum frequency deviation, and $S_n(\omega)$ is the spectral density of the narrowband Gaussian process $n(t)$. If

$$S_n(\omega) = \begin{cases} \mathcal{N}_0/2, & \omega_c - (\Delta\omega + \omega_m) \le |\omega| \le \omega_c + (\Delta\omega + \omega_m) \\ 0, & \text{otherwise}, \end{cases} \tag{7.5.12}$$

then

$$S_{n_q}(\omega) = \mathcal{N}_0 \tag{7.5.13}$$

for $|\omega| \leq (\Delta\omega + \omega_m)$.

We can now use Eq. (A.9.14) to get

$$S_{n_0}(\omega) = \frac{\omega^2}{A_c^2} \mathcal{N}_0, \qquad |\omega| \leq \omega_m, \tag{7.5.14}$$

where we have defined

$$n_o(t) = \frac{1}{A_c} \frac{d}{dt} n_q(t).$$

The output noise power can finally be calculated from Eq. (A.9.6) as

$$E\{n_o^2(t)\} = \frac{1}{2\pi} \int_{-\omega_m}^{\omega_m} \frac{\omega^2 \mathcal{N}_0}{A_c^2} \, d\omega = \frac{\omega_m^3 \mathcal{N}_0}{3\pi A_c^2}. \tag{7.5.15}$$

Clearly, from Eq. (7.5.10), the output message or signal power is $c_f^2\langle m^2(t)\rangle$, so that the output signal-to-noise ratio for FM in the high-CNR case is

$$\mathrm{SNR}_o = \frac{3\pi A_c^2 c_f^2 \langle m^2(t)\rangle}{\omega_m^3 \mathcal{N}_0}. \tag{7.5.16}$$

To determine the detection gain for FM, we need the input signal-to-noise ratio, which is the ratio of signal power to noise power for $r(t)$ in Eq. (7.5.2). Since for FM all of the transmitted power can be in the sidebands, we see that the signal power is from the first term in Eq. (7.5.2) given by $A_c^2/2$. Further, using Eq. (7.5.12) in Eq. (A.9.6), we obtain

$$E\{n^2(t)\} = \frac{\mathcal{N}_0(\Delta\omega + \omega_m)}{\pi}, \tag{7.5.17}$$

so that the input SNR is

$$\mathrm{SNR}_i = \frac{\pi A_c^2}{2(\Delta\omega + \omega_m)\mathcal{N}_0}. \tag{7.5.18}$$

It is common in angle-modulated systems to call SNR_i the CNR. Thus, in terms of the CNR, we can rewrite Eq. (7.5.16) as

$$\mathrm{SNR}_o = \frac{6c_f^2(\Delta\omega + \omega_m)\langle m^2(t)\rangle}{\omega_m^3} (\mathrm{CNR}), \tag{7.5.19}$$

where the CNR $= \mathrm{SNR}_i$ in Eq. (7.5.18).

To draw specific conclusions concerning the detection gain, we need to consider a particular message signal $m(t)$. This is done in the following example.

EXAMPLE 7.5.1

We wish to determine the output signal-to-noise ratio in terms of the CNR for the message signal

$$m(t) = A_m \cos \omega_m t.$$

In this case, Eq. (7.5.19) becomes

$$\begin{aligned} \text{SNR}_o &= \frac{3(\Delta\omega + \omega_m)c_f^2 A_m^2}{\omega_m^3} \text{CNR} \\ &= 3\beta^2(1 + \beta)\text{CNR}, \end{aligned} \tag{7.5.20}$$

where we have used the facts that $\langle m^2(t) \rangle = A_m^2/2$ and that the modulation index $\beta = c_f A_m/\omega_m = \Delta\omega/\omega_m$. If we consider the important special case where $\beta = 5$, then

$$\text{SNR}_o = 450 \text{ CNR}. \tag{7.5.21}$$

Thus the detection gain for FM in the high-CNR case is 450 for a sinusoidal message signal and $\beta = 5$.

Since we know that the bandwidth of an FM signal for sinusoidal modulation is given by $\text{BW} = 2\omega_m(1 + \beta)$, we see that a linear increase in bandwidth results in an exponential improvement in detection gain (for a fixed value of CNR). This last parenthetical remark is extremely important, since it prevents us from concluding that we can increase SNR_o indefinitely by expanding bandwidth. This point is clarified by the following discussion. As the bandwidth is increased, more noise power appears at the detector input. Thus, to keep the CNR constant, the carrier power must be increased. Further, all practical communication systems have a constraint on the maximum available carrier power. Therefore, as we increase β to improve the detection gain, we expand bandwidth and thus admit more noise power into the receiver. To obtain the promised $3\beta^2(1 + \beta)$ improvement in detection gain, we must increase the carrier power in proportion to the noise power. However, we will usually operate at the maximum possible transmitted carrier power, and as a result, our detection gain is limited by this fixed carrier power constraint.

In FM systems, the message signal $m(t)$ is often preemphasized prior to modulation at the transmitter and then the demodulated signal is deemphasized at the receiver. Preemphasis consists of providing a gain for the components of $m(t)$ at the higher frequencies, while deemphasis attenuates the higher frequencies of the detected signal in an inverse proportion to the applied preemphasis. The utility of preemphasis/deemphasis can be made evident as follows. Most message signals, such as voice and music, have decreasing frequency content with increasing frequency. However, from Eq. (7.5.14), it is clear that the detector output noise power spectral density is increasing parabolically with increasing frequency. Therefore, the ratio of message signal power to noise power as a function of frequency decreases as frequency increases. The use of preemphasis boosts the higher frequencies in the message signal to try to compensate for the increased noise power. Of course, the preemphasis distorts the transmitted message signal, and therefore the inverse operation, deemphasis, must be performed on the demodulated message signal at the receiver. The deemphasis also reduces the output noise power at higher frequencies, which yields a better output SNR.

Virtually all of our analyses in this section have been for the high-CNR case, since for the small-CNR case, the demodulated output is due primarily to the noise alone. As the CNR is decreased from a large value to smaller values, there is a point at which there is a precipitous drop in SNR_o if the CNR is decreased still further. This value of CNR is called the *threshold*, and although it varies as a function of modulation index β, the threshold value of CNR is roughly 10 dB. Figure 7.5.1 shows a plot of SNR_o versus CNR as a function of β for a typical FM system. The threshold effect is clearly evident. None of the analyses presented here indicate why the threshold changes with β. However, the effect is explained qualitatively as follows. For small values of β, the FM signal tends to occupy the entire bandwidth $2\omega_m(1 + \beta)$. As β is increased, the bandwidth of the FM signal is also increased, but at any time instant, only a small portion of the FM bandwidth is occupied by the transmitted signal, which sweeps through the wider bandwidth in relation to the message signal. Therefore, at any time instant, the (fixed) signal power occupies a smaller percentage of the

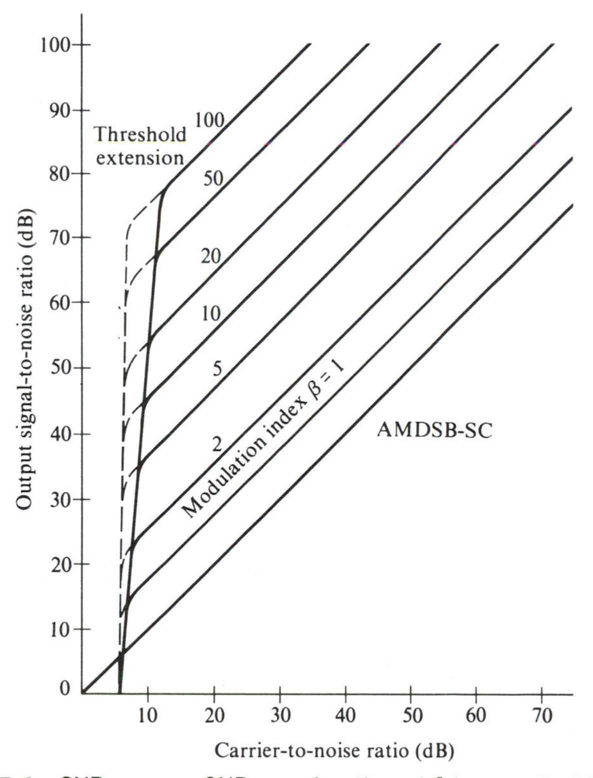

FIGURE 7.5.1 SNR_o versus CNR as a function of β for a typical FM system.

intermediate-frequency bandwidth of $2\omega_m(1 + \beta)$ as β is increased, thus requiring a higher CNR to stay above the threshold.

The threshold can be made independent of β, as indicated by the dashed lines in Fig. 7.5.1, by using what is called FM feedback (FMFB) or a phase-locked loop (PLL). In general, these systems cause the demodulator to track the received signal as a function of time, thus allowing a narrower IF bandwidth to be used. As a result, as β is increased, the IF bandwidth does not increase, and greater noise rejection is achieved.

For phase modulation, the analysis is quite similar to that for FM, but the results are significantly different. The transmitted waveform for PM is

$$s_{\text{PM}}(t) = A_c \cos\left[\omega_c t + c_p m(t)\right] \tag{7.5.22}$$

and the received signal, after contamination with narrowband, white Gaussian noise is

$$\begin{aligned} r(t) &= s_{\text{PM}}(t) + n(t) \\ &= A_c \cos\left[\omega_c t + c_p m(t)\right] + \rho(t) \cos\left[\omega_c t + \theta(t)\right]. \end{aligned} \tag{7.5.23}$$

As before, we can write Eq. (7.5.23) in the form

$$r(t) = A(t) \cos\left[\omega_c t + \phi(t)\right], \tag{7.5.24}$$

where for the large CNR case, $A_c \gg \rho(t)$,

$$\begin{aligned} \phi(t) &\cong c_p m(t) + \frac{\rho(t)}{A_c} \sin\left[\theta(t) - c_p m(t)\right] \\ &\cong c_p m(t) + \frac{\rho(t)}{A_c} \sin\theta(t) \\ &= c_p m(t) + \frac{n_q(t)}{A_c}. \end{aligned} \tag{7.5.25}$$

It is now straightforward to show that the output signal-to-noise ratio for PM is

$$\text{SNR}_o = \frac{\pi A_c^2 c_p^2 \langle m^2(t) \rangle}{\omega_m \mathcal{N}_0}. \tag{7.5.26}$$

The input SNR or CNR is still given by Eq. (7.5.18), so in terms of the CNR we have

$$\text{SNR}_o = \frac{2(\Delta\omega + \omega_m) c_p^2 \langle m^2(t) \rangle}{\omega_m} (\text{CNR}) \tag{7.5.27}$$

for PM.

7.6 SNR Comparisons for Systems

We now attempt to pool the results of the preceding sections to obtain a relative comparison of the several analog communication systems discussed in this chapter. We use the output signal-to-noise ratio as our performance indicator, and for our comparisons to be meaningful, we require that all the systems have the same ratio of *total* transmitted power to received noise power. We denote the latter ratio by SNR_R. Thus the comparisons here are not based on detection gain but on SNR_o for a fixed SNR_R. Of course, bandwidth is another important parameter when comparing communication systems, and we state the bandwidth as some multiple of the message waveform bandwidth, ω_m. Finally, when necessary, we assume a particular message signal, namely single-tone modulation given by $m(t) = A_m \cos \omega_m t$. There is some loss of generality in this last assumption, but in order to calculate $\langle m^2(t) \rangle$, we must have a specific $m(t)$, and a pure tone is a common reference signal.

In Sections 7.3 through 7.5, we found the output signal-to-noise ratio as a function of input SNR, where SNR_i is the ratio of signal power in the *message* component to the noise power. Note the difference between SNR_i and SNR_R, whose numerator is the *total* transmitted power. Thus we shall find the SNR_i in terms of SNR_R for each of the systems of interest, and then substitute the SNR_i value into the appropriate expressions in the preceding sections to obtain SNR_o in terms of SNR_R.

Beginning with AMDSB-SC, the transmitted signal is

$$s_{\text{DSB}}(t) = A_1 m(t) \cos \omega_c t, \tag{7.6.1}$$

where we do not as yet use the assumption that $m(t) = A_m \cos \omega_m t$. For noise with a flat power spectral density of amplitude $\mathcal{N}_0/2$ in $\omega_c - \omega_m \leq |\omega| \leq \omega_c + \omega_m$,

$$\text{SNR}_{R,\text{DSB}} = \frac{(A_1^2/2)\langle m^2(t) \rangle}{(\mathcal{N}_0/2)(4\omega_m)/2\pi} = \frac{2\pi A_1^2 \langle m^2(t) \rangle}{4\mathcal{N}_0 \omega_m}. \tag{7.6.2}$$

Since, for DSB-SC, all of the transmitted power is in the sidebands, it is evident that $\text{SNR}_{i,\text{DSB}} = \text{SNR}_{R,\text{DSB}}$ [see Eq. (7.3.3)], so from Eq. (7.3.6),

$$\text{SNR}_{o,\text{DSB}} = 2\text{SNR}_{R,\text{DSB}} = 2\text{SNR}_R, \tag{7.6.3}$$

where we have defined $\text{SNR}_R \triangleq \text{SNR}_{R,\text{DSB}}$ as our fixed reference.

For AMSSB-SC, the transmitted signal can be written as

$$s_{\text{SSB}}(t) = A_2 m(t) \cos \omega_c t \pm A_2 m_h(t) \sin \omega_c t, \tag{7.6.4}$$

so that the total transmitted power is $A_2^2 \langle m^2(t) \rangle$ [see Eq. (7.3.8)]. We choose A_2 such that the total transmitted power for SSB is the same as for DSB, namely $A_1^2 \langle m^2(t) \rangle / 2$. Only half the bandwidth of DSB is required, so that the received noise power is $(1/2\pi)(\mathcal{N}_0/2)(2\omega_m) = \mathcal{N}_0\omega_m/2\pi$, and therefore,

$$\text{SNR}_{R,\text{SSB}} = \frac{2\pi A_2^2\langle m^2(t)\rangle}{\mathcal{N}_0\omega_m} = \frac{2\pi A_1^2\langle m^2(t)\rangle}{2\mathcal{N}_0\omega_m}$$
$$= 2\text{SNR}_R. \tag{7.6.5}$$

Since for SSB all of the transmitted power is in the sidebands, $\text{SNR}_{i,\text{SSB}} = \text{SNR}_{R,\text{SSB}}$, so using Eq. (7.3.12), we have that

$$\text{SNR}_{o,\text{SSB}} = 2\text{SNR}_R. \tag{7.6.6}$$

The transmitted waveform for AMDSB-TC is

$$s_{\text{AM}}(t) = A_3[1 + am(t)] \cos \omega_c t \tag{7.6.7}$$

$[m(t)$ normalized$]$ with total power $A_3^2[1 + a^2\langle m^2(t)\rangle]/2$, where we have assumed that $\langle m(t)\rangle = 0$. Again, we must keep the total transmitted power the same in all cases, so we adjust A_3 to obtain a total power of $A_1^2\langle m^2(t)\rangle/2$. The received noise power is the same for the suppressed carrier case as for the transmitted carrier case, so

$$\text{SNR}_{\text{AM}} = \frac{A_3^2[1 + a^2\langle m^2(t)\rangle]/2}{2\mathcal{N}_0\omega_m/2\pi}$$
$$= \frac{2\pi A_1^2\langle m^2(t)\rangle}{4\mathcal{N}_0\omega_m} = \text{SNR}_R. \tag{7.6.8}$$

Now, from Eq. (7.4.3),

$$\text{SNR}_{i,\text{AM}} = \frac{2\pi a^2 A_3^2\langle m^2(t)\rangle}{4\mathcal{N}_0\omega_m}$$
$$= \frac{a^2\langle m^2(t)\rangle}{1 + a^2\langle m^2(t)\rangle}\,\text{SNR}_R, \tag{7.6.9}$$

so that using Eq. (7.4.8), we get

$$\text{SNR}_{o,\text{AM}} = \frac{2a^2\langle m^2(t)\rangle}{1 + a^2\langle m^2(t)\rangle}\,\text{SNR}_R. \tag{7.6.10}$$

Using Eqs. (7.6.3), (7.6.6), and (7.6.10), we are now in a position to compare the performance of these three AM systems. We see that AMDSB-SC and AMSSB-SC perform equally well, but since $0 < a \leq 1$ and we have assumed that $m(t)$ is normalized here, $a^2\langle m^2(t)\rangle/[1 + a^2\langle m^2(t)\rangle] < 1$, the performance of AMDSB-TC is poorer than the other two systems. Of course, this is an expected result, since in AMDSB-TC we allocate power to a separate carrier term to simplify the receiver. If we ignore complexity, AMSSB-SC would be the best choice among these three systems, due to its least bandwidth requirements.

To compare FM and PM with the AM systems, we continue to use the assumption that $m(t)$ is normalized and consider Eqs. (7.5.19) and (7.5.27) for

FM and PM, respectively. For FM we substitute Eq. (7.5.18) into (7.5.19), which yields

$$\mathrm{SNR}_{o,\mathrm{FM}} = \frac{3\pi A_c^2 c_f^2 \langle m^2(t)\rangle}{\mathcal{N}_0 \omega_m^3} = \frac{3\beta^2 (2\pi A_c^2)\langle m^2(t)\rangle}{2\mathcal{N}_0 \omega_m}, \tag{7.6.11}$$

where since $|m(t)| \leq 1$, we have let $\Delta\omega = c_f$. With reference to Eq. (7.6.2), we let $A_c = A_1$, so from Eq. (7.6.3) we see that

$$\mathrm{SNR}_{o,\mathrm{FM}} = 3\beta^2 \mathrm{SNR}_{o,\mathrm{DSB}}, \tag{7.6.12}$$

and it follows from Eq. (7.6.6) that

$$\mathrm{SNR}_{o,\mathrm{FM}} = 3\beta^2 \mathrm{SNR}_{o,\mathrm{SSB}}. \tag{7.6.13}$$

Further, it follows directly from Eq. (7.6.10) that

$$\mathrm{SNR}_{o,\mathrm{FM}} = 3\beta^2 \frac{[1 + a^2\langle m^2(t)\rangle]}{a^2\langle m^2(t)\rangle} \mathrm{SNR}_{o,\mathrm{AM}}. \tag{7.6.14}$$

It is often convenient to compare FM and conventional AM in terms of the carrier-to-noise ratio. To do this, we can use Eq. (7.4.9) with $A_3 = A_c$ or Eq. (7.6.9) with $A_3 = A_c$ to produce

$$\mathrm{SNR}_{o,\mathrm{AM}} = a^2\langle m^2(t)\rangle \mathrm{CNR}_{\mathrm{AM}}, \tag{7.6.15}$$

where $\mathrm{CNR}_{\mathrm{AM}} \triangleq 2\pi A_c^2/2\mathcal{N}_0\omega_m$. Thus

$$\mathrm{SNR}_{o,\mathrm{FM}} = 3\beta^2\langle m^2(t)\rangle \mathrm{CNR}_{\mathrm{AM}}. \tag{7.6.16}$$

Under the same assumptions used for FM, it is straightforward to show that for PM,

$$\mathrm{SNR}_{o,\mathrm{PM}} = c_p^2 \mathrm{SNR}_{o,\mathrm{DSB}} = c_p^2 \mathrm{SNR}_{o,\mathrm{SSB}} \tag{7.6.17}$$

and

$$\mathrm{SNR}_{o,\mathrm{PM}} = c_p^2 \frac{[1 + a^2\langle m^2(t)\rangle]}{a^2\langle m^2(t)\rangle} \mathrm{SNR}_{o,\mathrm{AM}}. \tag{7.6.18}$$

It is evident from Eqs. (7.6.12)–(7.6.18) that for $\beta \gg 1$ (wideband FM) that the improvement in SNR provided by FM over AM can be substantial. Of course, the larger β is, the larger the FM or PM bandwidth requirements.

It is difficult to draw general conclusions concerning the relative performance of FM and PM, since their performance varies depending on the characteristics of the message signal. We give one comparison of interest in the following example.

EXAMPLE 7.6.1

In this example we compare the performance of AMDSB-SC, AMDSB-TC, AMSSB-SC, FM, and PM for the particular case of a sinusoidal message signal $m(t) = \cos \omega_m t$. Clearly, Eqs. (7.6.12), (7.6.13), and (7.6.17) still hold and are unaffected by the choice of modulating signal. However, for conventional AM we use the fact that $\langle m^2(t) \rangle = \frac{1}{2}$ and assume that $a = 1$, to obtain

$$\text{SNR}_{o,\text{FM}} = 9\beta^2 \text{SNR}_{o,\text{AM}}. \tag{7.6.19}$$

Similarly for PM,

$$\text{SNR}_{o,\text{PM}} = 3c_p^2 \text{SNR}_{o,\text{AM}}. \tag{7.6.20}$$

To compare FM and PM, we note that here $\beta = c_f/\omega_m$, so that

$$\text{SNR}_{o,\text{FM}} = \frac{3\beta^2}{c_p^2}\,\text{SNR}_{o,\text{PM}} = \frac{3c_f^2}{\omega_m^2 c_p^2}\,\text{SNR}_{o,\text{PM}}. \tag{7.6.21}$$

Thus, if $c_f^2/\omega_m^2 c_p^2 < \frac{1}{3}$, $\text{SNR}_{o,\text{FM}} > \text{SNR}_{o,\text{PM}}$. This is often true for nonbaseband (bandpass) signals.

SUMMARY

In this chapter we have analyzed the performance of analog modulation systems with analog messages first by finding the improvement in signal-to-noise ratio afforded by the demodulation process and then by comparing systems for the same ratio of total transmitted power to noise power. The advantage of coherent detection is clear, and the threshold effects in noncoherent detection of conventional AM and in FM/PM are evident. The performance improvement provided by FM over AM has been demonstrated. Considerably more detailed analyses are possible, but the SNR comparisons in Section 7.6 coupled with a knowledge of complexity and bandwidth requirements for the various methods allow competent design trade-offs to be made.

PROBLEMS

7.1 Derive the joint pdf of $\rho(t)$ and $\theta(t)$ in Eq. (7.2.12).

7.2 Derive expressions for $n_i(t)$ and $n_q(t)$ in terms of $n(t)$ and its Hilbert transform $n_h(t)$.

7.3 Show that if $n(t)$ is a WSS random process, $n(t)$ and its Hilbert transform have the same autocorrelation function.

7.4 Given the WSS narrowband noise $n(t)$ and its Hilbert transform $n_h(t)$, show that:

(a) $R_{nn_h}(\tau) = \mathscr{H}[R_n(\tau)]$.
(b) $R_{n_h n}(\tau) = -\mathscr{H}[R_n(\tau)]$.

7.5 Show that the in-phase and quadrature components of WSS narrowband noise have the same autocorrelation function. This is Property 3.

Hint: See previous problems.

7.6 Derive the cross-correlation function between the in-phase and quadrature components of narrowband noise.

7.7 Derive the spectral densities of the in-phase and quadrature components of WSS narrowband noise.

7.8 Given the power spectral density of narrowband noise $S_n(\omega)$ in Fig. P7.8, sketch $S_{n_i}(\omega)$ and $S_{n_q}(\omega)$.

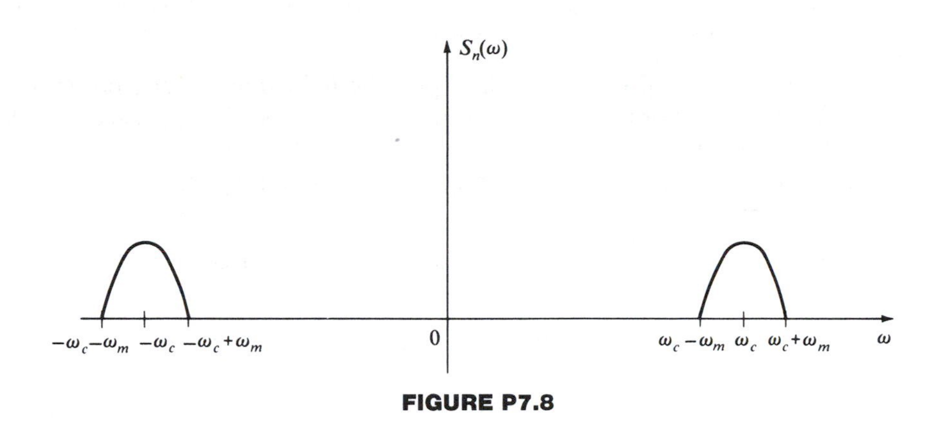

FIGURE P7.8

7.9 Wide-sense stationary white noise with two-sided power spectral density of amplitude $\mathcal{N}_0/2$ is applied to an ideal bandpass filter $H(\omega)$ with $|H(\omega)| = 1$ for $\omega_c - \omega_B \leq |\omega| \leq \omega_c + \omega_B$ and $\angle H(\omega) = 0$. Find the autocorrelation functions of the filter output in-phase and quadrature components.

7.10 Validate Eq. (7.3.8).

7.11 Derive Eq. (7.4.12).

7.12 Validate Eqs. (7.5.6) and (7.5.8).

7.13 Sketch the noise spectral density in Eq. (7.5.14). What does this imply?

7.14 Plot SNR_o in Eq. (7.5.20) versus required bandwidth.

7.15 Plot Eq. (7.6.10) versus a if $\langle m^2(t)\rangle = \frac{1}{2}$.

7.16 Plot SNR_o versus SNR_R for Eqs. (7.6.3), (7.6.6), and (7.6.10) on the same sheet. Let $a = 1$ and $\langle m^2(t)\rangle = \frac{1}{2}$. Develop an expression relating SNR_o and SNR_R for FM and plot it on the same sheet. Let $\beta = 5$.

7.17 For what range of β does FM offer an SNR_o improvement over AMDSB-SC? Over conventional AM if $a = 1$ and $\langle m^2(t)\rangle = \frac{1}{2}$?

7.18 A conventional AM signal can be demodulated by a squarer followed by a low-pass filter. If the input to the square-law detector is a conventional AM signal plus narrowband noise, find an expression for SNR_o in terms of SNR_i. Compare your result to the result for envelope detection in Eq. (7.6.10) by letting $a = 1$ and $\langle m^2(t) \rangle = \frac{1}{2}$.

7.19 Use Eq. (7.6.21) and compare FM and PM systems for the same β. Repeat the comparison if both have the same $\Delta\omega$. Which comparison seems more practical?

7.20 It is mentioned in Section 7.5 that preemphasis/deemphasis is often used to advantage in FM systems. If the preemphasis network has the transfer function $H(\omega) = 1 + j\omega/\omega_0$, find an expression for the output noise power in FM using preemphasis/deemphasis.

7.21 Use the result of Problem 7.20 to plot the reduction in noise power as a function of the ratio ω_m/ω_0, where ω_m is the message signal bandwidth. For commercial FM, $\omega_m = 2\pi f_m = 2\pi(15 \text{ kHz})$ and $\omega_0 = 2\pi f_0 = 2\pi$ (2.1 kHz). What is the reduction in noise power for commercial FM using preemphasis/deemphasis?

7.22 Show that preemphasis/deemphasis can also be used to provide a performance improvement for conventional AM.

7.23 Refer to Fig. 5.3.3 of an envelope detector and the corresponding discussion. Use this development to explain heuristically the threshold effect in an envelope detector for conventional AM in the small carrier-to-noise ratio case.

7.24 The specific definition of the threshold between the large and small carrier-to-noise ratio cases is difficult and has been avoided thus far. Carlson [1975] suggests that it is that value of SNR_i such that the probability of the noise envelope exceeding the carrier amplitude is 0.01. With reference to Eq. (7.4.12), this is $P[\rho(t) > A_3] = 0.01$. Find that SNR_i if $a = 1$ and $\langle m^2(t) \rangle = \frac{1}{2}$.

7.25 Discuss why the output signal-to-noise ratio in stereo FM is worse than in monophonic FM.

7.26 The threshold phenomenon in FM occurs when $A_c \cong \rho(t)$ in Eqs. (7.5.2), (7.5.4), and (7.5.5). In this situation, if we think of A_c and $\rho(t)$ as phasors, $\rho(t)$ can cause a complete 2π phase change in $\phi(t)$ in Eqs. (7.5.3) and (7.5.4), which yields an impulse in $d\phi(t)/dt$. These impulses, called "clicks," have been analyzed by Rice [1963] and are very detrimental to FM system performance [Taub and Schilling, 1971]. The output signal-to-noise ratio, including the effects of clicks for a sinusoidal message signal, is

$$\text{SNR}_o = \frac{\frac{3}{2}\beta^2 \text{SNR}_i}{1 + (12\beta/\pi)\text{SNR}_i e^{-\text{SNR}_i/(1+\beta)2}}.$$

Plot SNR_o versus SNR_i for $\beta = 5$, 10, and 15.

7.27 Use the results of Problem 7.26 to show that:

(a) For a fixed SNR_i, exchanging bandwidth for SNR_o is not always possible.

(b) For a fixed SNR_o, exchanging bandwidth for a reduced SNR_i is not always possible.

7.28 Assume that the narrowband noise has the power spectral density shown in Fig. P7.28. Find $S_{n_i}(\omega)$, $S_{n_q}(\omega)$, $R_{n_q}(\tau)$, $R_{n_i}(\tau)$, $E[n_i^2(t)]$, and $E[n_q^2(t)]$.

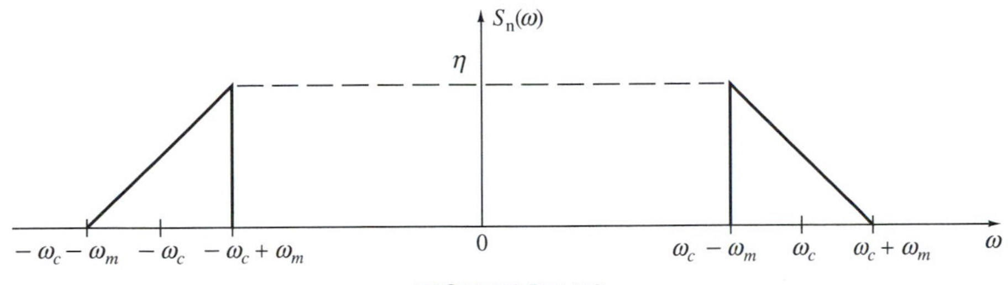

FIGURE P7.28

7.29 Compare $\mathrm{SNR}_{o,\mathrm{DSB}}$ and $\mathrm{SNR}_{o,\mathrm{SSB}}$ for narrowband noise with the power spectral density shown in Fig. P7.28.

7.30 Does a nonflat noise power spectral density as in Fig. P7.28 change the FM SNR_o analysis leading to Eq. (7.5.19)? If so, how?

CHAPTER 8

Data Transmission

8.1 Introduction

This chapter is concerned with transmitting discrete-amplitude (digital) signals over bandwidth-constrained communications channels. To achieve the highest possible transmission rate or data rate in bits per second while minimizing the number of errors in received bits, called the *bit error rate* (BER), requires considerable knowledge, skill, and ingenuity. As a result, this chapter covers what seems to be a wide variety of topics. However, each of these topics plays a fundamental role in the design and analysis of data communication systems.

In previous chapters we have represented digital signals by rectangular (time-domain) pulses of a given amplitude and width. The channel bandwidth limitation precludes the use of rectangular pulses here, since the frequency content of a rectangular pulse has a sin x/x shape. The tails of the sin x/x function decay very slowly, and hence a rectangular pulse requires a wide bandwidth for undistorted transmission. In Section 8.2 we introduce the concept of pulse shaping for restricted bandwidth applications while maintaining zero intersymbol interference with adjacent pulses. Nyquist has shown that there is an upper limit on the rate at which pulses can be transmitted over a bandlimited channel while maintaining zero intersymbol interference. The use of controlled intersymbol interference, called partial response signaling, to achieve the maximum data rate possible in the minimum bandwidth is developed in Section 8.3. For many communications channels, and the telephone channel in particular, impairments are present which are deterministic in that they can be measured for a given transmission path and they do not change perceptibly over a short time period. This deterministic distortion must be well understood, and it is the subject of Section 8.4. A good way to get a subjective idea of the distortion contributed by a transmission path for a particular data transmission system is to display what is called the system eye pattern. Eye patterns are discussed in Section 8.5, followed in Section 8.6 by a development of channel equalization techniques, which attempt to correct any deterministic distortion contributed by the channel. In Section 8.7 we develop data scramblers for randomizing the input message sequence and which thus aid the performance of automatic, adaptive equalizers. The difficult problems of carrier acquisition and symbol synchronization are considered briefly in Section 8.8, followed by Section 8.9, which unifies the chapter by discussing example modem designs.

8.2 Baseband Pulse Shaping

As noted in the preceding section, we have always represented digital signals in earlier chapters by a rectangular time-domain pulse. This pulse shape is convenient in that it is wholly contained within its allocated interval and hence does not interfere with pulses in adjacent intervals. The rectangular time-domain pulse does require excessive bandwidth, however. This is easily seen by considering a unit amplitude, rectangular pulse of width τ centered at the origin,

$$p(t) = \begin{cases} 1, & |t| \leq \dfrac{\tau}{2} \\ 0, & \text{otherwise}, \end{cases} \tag{8.2.1}$$

which has the Fourier transform

$$P(\omega) = \mathscr{F}\{p(t)\} = \tau \frac{\sin(\omega\tau/2)}{\omega\tau/2}. \tag{8.2.2}$$

Since the zero crossings of $P(\omega)$ occur at $\omega = \pm 2n\pi/\tau$, the narrower the pulse, the wider the bandwidth required. Furthermore, the tails of $\sin(\omega\tau/2)/(\omega\tau/2)$ decay very slowly for increasing ω. As a result, the rectangular time-domain pulse is not acceptable for bandwidth-constrained applications.

To begin our search for suitable pulse shapes, we must be more specific concerning our requirements. First, we want the chosen pulse shape to have zero intersymbol interference at the pulse sampling times. Since, at the receiver, pulses are usually sampled at the exact center of their allocated interval, this means that the desired pulse shape is zero at the center of all pulse intervals other than its own. Of course, the rectangular pulse satisfies this requirement, since it is identically zero outside its own interval. However, we are being less restrictive here in that the time-domain pulse may be nonzero in other pulse intervals as long as it is zero at the center of every other pulse interval (the pulse sampling times). Second, the desired pulse shape must have a Fourier transform that is zero for ω greater than some value, say ω_m. This requirement satisfies our bandwidth constraint. Third, we would prefer that the tails of the time-domain pulse shape decay as rapidly as possible, so that jitter in the pulse sequence or sampling times will not cause significant intersymbol interference. This last requirement is a very pragmatic one, but it is necessary, since any data transmission system will have some timing jitter, which will cause pulse sampling to occur at times other than the exact center of each pulse interval. If the pulse tails are not sufficiently damped, sampling in the presence of timing jitter can cause substantial intersymbol interference.

A good starting point in our search for a useful pulse shape is the requirement that its Fourier transform be identically zero for $\omega > \omega_m$. By the symmetry property of Fourier transforms, we know that a $\sin x/x$ time-domain pulse has a rectangular frequency content. In particular, the pulse shape

$$p(t) = \frac{\sin(\pi t/\tau)}{\pi t/\tau} \tag{8.2.3}$$

has the Fourier transform

$$P(\omega) = \begin{cases} \tau, & |\omega| \leq \dfrac{\pi}{\tau} \\ 0, & \text{otherwise.} \end{cases} \tag{8.2.4}$$

Not only does $p(t)$ in Eq. (8.2.3) have a restricted frequency content, but it also is identically zero at the points $t = \pm n\tau$ about its center. Thus, if the pulses are spaced τ seconds apart, this $p(t)$ achieves zero intersymbol interference at the pulse sampling times. Unfortunately, our third requirement on the tails of the pulse is the demise of $p(t)$ in Eq. (8.2.3). The tails of $p(t)$ only decay as $1/t$, and it can be shown that for very long messages even a small error in the sampling times produces substantial intersymbol interference. In fact, even a small error in sampling times can cause the tails of a very long pulse sequence to sum as a divergent series (see Problem 8.3).

Since $p(t)$ in Eq. (8.2.3) satisfies two of our three requirements, we are led to inquire whether this $p(t)$ can be modified to have faster-decaying tails while retaining the same zero crossings and bandlimited feature. Fortunately, Nyquist [1928] has derived a requirement on the Fourier transforms of possible pulse shapes which, if satisfied, guarantees that the zero crossings of $p(t)$ in Eq. (8.2.3) are retained. However, the bandwidth is increased and there may be additional zero crossings. This criterion (called *Nyquist's first criterion*) is that the Fourier transform $G(\omega)$ of a real and even pulse shape $g(t)$ satisfies

$$G\left(\frac{\pi}{\tau} - x\right) + G\left(\frac{\pi}{\tau} + x\right) = \text{constant} \tag{8.2.5}$$

for $0 \leq x \leq \pi/\tau$, where τ is the spacing between pulse sampling instants. The proof of this result can be obtained by allowing the bandwidth to increase over that in Eq. (8.2.4), so that $2\pi B > \pi/\tau$, and writing

$$p(t) = \frac{1}{2\pi} \int_{-2\pi B}^{2\pi B} P(\omega) e^{j\omega t}\, d\omega.$$

We want zero intersymbol interference at the sampling instants, so we let $t = n\tau$,

$$p(n\tau) = \frac{1}{2\pi} \int_{-2\pi B}^{2\pi B} P(\omega) e^{j\omega n\tau}\, d\omega.$$

For $2B\tau = \text{integer} = K > 1$ and since $B > 1/2\tau$,

$$\begin{aligned} p(n\tau) &= \sum_{k=-K}^{K} \frac{1}{2\pi} \int_{(2k-1)\pi/\tau}^{(2k+1)\pi/\tau} P(\omega)e^{j\omega n\tau}\, d\omega \\ &= \sum_{k=-K}^{K} \frac{1}{2\pi} \int_{-\pi/\tau}^{\pi/\tau} P\left(\lambda + \frac{2k\pi}{\tau}\right) e^{jn\tau(\lambda + 2k\pi/\tau)}\, d\lambda \\ &= \frac{1}{2\pi} \int_{-\pi/\tau}^{\pi/\tau} \sum_{k=-K}^{K} P\left(\lambda + \frac{2k\pi}{\tau}\right) e^{jn\tau\lambda}\, d\lambda. \end{aligned}$$

For zero intersymbol interference at the sampling times, $p(0) = 1$, $p(n\tau) = 0$, $n \neq 0$, which requires that

$$\sum_{k=-K}^{K} P\left(\omega + \frac{2k\pi}{\tau}\right) = \begin{cases} \tau, & |\omega| \leq \dfrac{\pi}{\tau} \\ 0, & |\omega| > \dfrac{\pi}{\tau}. \end{cases} \tag{8.2.6}$$

Equation (8.2.6) implies the criterion in Eq. (8.2.5).

Pulse shapes with Fourier transforms that satisfy Eqs. (8.2.5) and (8.2.6) are not guaranteed to have rapidly decaying tails nor are they guaranteed to be bandlimited; however, they are guaranteed to have zero crossings at the nulls of $p(t)$ in Eq. (8.2.3). Thus, by investigating the class of bandlimited Fourier transforms that satisfy Eqs. (8.2.5) and (8.2.6), we may be able to find pulse shapes that satisfy our third requirement as well.

An important pulse shape that satisfies Nyquist's first criterion and our three requirements is called the *raised cosine pulse*, and for real and even pulse shapes $p(t)$, it has a Fourier transform that is even about $\omega = 0$ and that for positive ω is given by

$$P(\omega) = \begin{cases} \tau, & 0 \leq \omega \leq \dfrac{\pi(1-\alpha)}{\tau} \\ \dfrac{\tau[1 + \cos\{\tau[\omega - \pi(1-\alpha)/\tau]/2\alpha\}]}{2}, & \dfrac{\pi(1-\alpha)}{\tau} \leq \omega \leq \dfrac{\pi(1+\alpha)}{\tau} \\ 0, & \omega > \dfrac{\pi(1+\alpha)}{\tau}. \end{cases} \tag{8.2.7}$$

The impulse response corresponding to Eq. (8.2.7) is

$$p(t) = \frac{\sin(\pi t/\tau)}{\pi t/\tau} \cdot \frac{\cos(\alpha\pi t/\tau)}{1 - (4\alpha^2 t^2)/\tau^2}. \tag{8.2.8}$$

It is easy to see that this pulse shape satisfies our three requirements. From Eq. (8.2.7), $p(t)$ is bandlimited, while from Eq. (8.2.8) we observe that it has zero crossings at least at $\pm n\tau$ and its tails decay as $1/t^3$.

A pulse $p(t)$ given by Eq. (8.2.8) or a Fourier transform $P(\omega)$ as shown in Eq. (8.2.7) is said to have *raised cosine spectral shaping with* $100\alpha\%$ *cosine roll-off*. Sketches of $p(t)$ and $P(\omega)$ for various α are shown in Fig. 8.2.1(a) and (b),

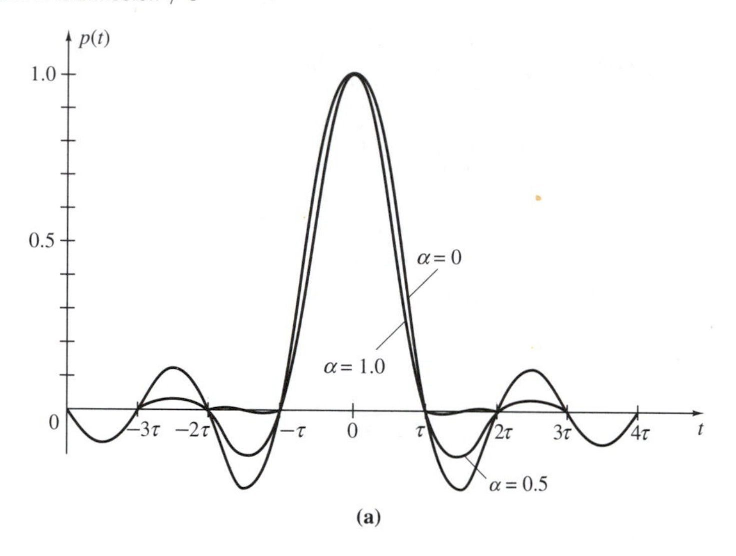

(a)

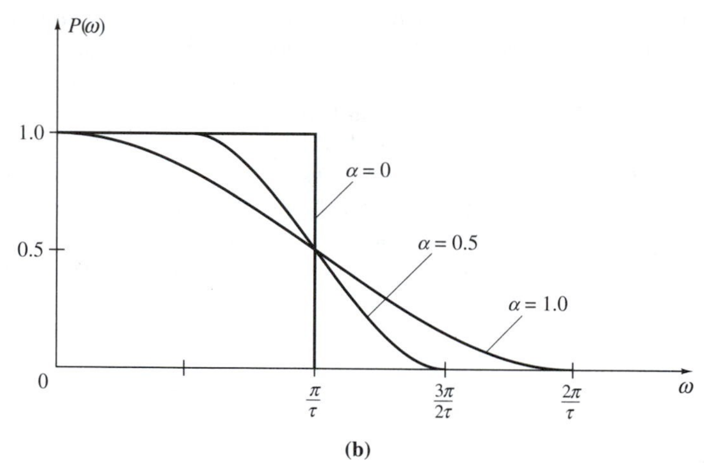

(b)

FIGURE 8.2.1 Raised cosine spectral shaping with 100α% roll-off.

respectively. It is important to note that although $p(t)$ is called a raised cosine pulse, the adjective *raised cosine* refers to the frequency-domain shaping, not the time-domain shaping. Two equivalent forms of Eq. (8.2.7) often appear in the literature,

$$P(\omega) = \begin{cases} \tau, & 0 \le \omega \le \dfrac{\pi(1-\alpha)}{\tau} \\ \tau \cos^2 \left\{ \tau \left[\omega - \dfrac{\pi(1-\alpha)}{\tau} \right] 4\alpha \right\}, & \dfrac{\pi(1-\alpha)}{\tau} \le \omega \le \dfrac{\pi(1+\alpha)}{\tau} \\ 0, & \omega > \dfrac{\pi(1+\alpha)}{\tau} \end{cases} \tag{8.2.9}$$

and

$$P(\omega) = \begin{cases} \tau, & 0 \le \omega \le \dfrac{\pi(1-\alpha)}{\tau} \\ \dfrac{\tau[1 - \sin\{\tau(\omega - \pi/\tau)/2\alpha\}]}{2}, & \dfrac{\pi(1-\alpha)}{\tau} \le \omega \le \dfrac{\pi(1+\alpha)}{\tau} \\ 0, & \omega > \dfrac{\pi(1+\alpha)}{\tau}. \end{cases} \tag{8.2.10}$$

By inspection of Fig. 8.2.1, we see that the parameter α is extremely important, since it affects both the required bandwidth and the "size" of the tails of $p(t)$. For $\alpha = 1$, the tails of $p(t)$ are highly damped, but the required bandwidth is at the maximum value $\omega_m = 2\pi/\tau$. As α is decreased toward zero, the required bandwidth decreases, but the tails of $p(t)$ become increasingly significant. The usual range of α for data transmission is $0.3 \le \alpha \le 1.0$, with α as low as 0.1 for some sophisticated systems.

When raised cosine pulses are used for data transmission, note that the pulses occur every τ seconds; hence the pulse rate, *symbol rate*, or *baud rate*, as it is usually called, is $1/\tau$ symbols/sec. If the transmitted pulses take on only two possible levels, each pulse represents 1 bit, and the bit rate = symbol rate = $1/\tau$ bits/sec.

EXAMPLE 8.2.1

It is desired to send data at a rate of 1000 bits/sec using a positive raised cosine pulse with 100% roll-off to represent a 1 and no pulse to represent a 0.

(a) What is the required symbol rate?
(b) What is the bandwidth of a transmitted positive pulse?
(c) If we wish to send the sequence 1011001, sketch the transmitted pulse stream.

Solution

(a) Since we are using two-level pulses, symbol rate = bit rate = 1000 symbols/sec.
(b) Since $\alpha = 1$, the bandwidth is $2\pi/\tau$, where τ = pulse spacing and $1/\tau$ = symbol rate = 1000 symbols/sec. Thus the pulse bandwidth is $2\pi/\tau = 2000\pi$ rad/sec.
(c) The transmitted pulse stream is sketched in Fig. 8.2.2. To emphasize their individual contributions, the pulses are not added point by point as they would appear on an oscilloscope.

Perhaps we should note at this point that if we wish to determine the bandwidth of a transmitted pulse stream with a given pulse shaping, in general, we need only consider the Fourier transform of the time-domain pulse shape. To

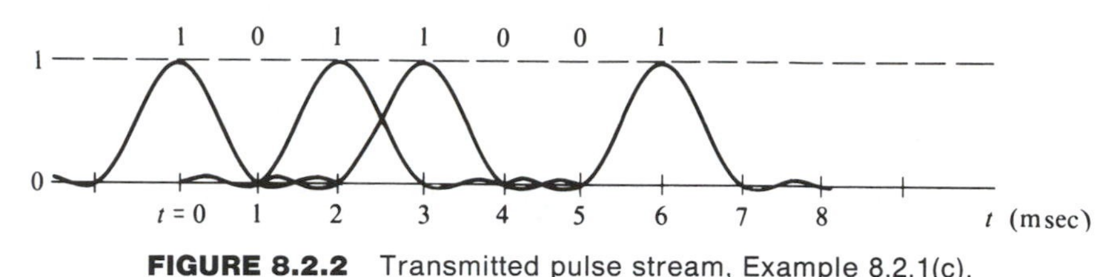

FIGURE 8.2.2 Transmitted pulse stream, Example 8.2.1(c).

show this in the most general situation, we need some expertise in random processes. Deferring this more detailed development, we can still get an idea of what is happening by examining the special case of a periodic sequence of alternating "pulse" and "no pulse." In this case we know from our study of Fourier series that the spacing of the spectral lines depends on the period but that the envelope of the spectral lines depends on the particular pulse shape. We shall thus use the Fourier transform of the specific time-domain pulse being employed as an indicator of data stream bandwidth.

A standard question that arises in data transmission studies is the following. Given a bandlimited channel of bandwidth $2\pi W$ rad/sec, what is the maximum number of noninterfering pulses that can be transmitted per second over this channel? By reconsidering the raised cosine spectral shaping at $\alpha = 0$, we can surmise a possible answer to this question. At $\alpha = 0$, we obtain as our pulse shape $p(t)$ in Eq. (8.2.3) with Fourier transform $P(\omega)$ in Eq. (8.2.4). Thus it seems that the minimum bandwidth within which we can send $1/\tau$ pulses/sec is $\pi/\tau = 2\pi W$ rad/sec or $W = 1/2\tau$ hertz. We are led to guess then that in a bandwidth of W hertz, we can transmit a maximum of $2W$ pulses/sec. We have therefore obtained a special case of Nyquist's [1924] general result that in a bandwidth of W hertz, a maximum of kW pulses/sec can be transmitted without intersymbol interference, where $k \leq 2$ is a proportionality factor depending on the pulse shape and the bandwidth.

Nyquist [1928] also developed two other criteria relating to pulse shaping in data transmission. *Nyquist's second criterion* concerns the required spectral shaping to achieve zero intersymbol interference at the instants halfway between adjacent sampling times, that is, zero intersymbol interference at the edges of the intervals allocated to individual pulses. For real Fourier transforms, the required spectral content is

$$P(\omega) = \begin{cases} \tau \cos \dfrac{\omega\tau}{2}, & 0 \leq |\omega| \leq \dfrac{\pi}{\tau} \\ 0, & |\omega| > \dfrac{\pi}{\tau}. \end{cases} \tag{8.2.11}$$

The impulse response corresponding to $P(\omega)$ in Eq. (8.2.11) is

$$p(t) = \frac{2}{\pi} \cdot \frac{\cos(\pi t/\tau)}{1 - (4t^2/\tau^2)}. \tag{8.2.12}$$

The impulse response in Eq. (8.2.12) is zero at all odd positive or negative multiples of $\tau/2$ except for $t = \pm\tau/2$, where $p(t)$ is $\frac{1}{2}$. Therefore, if we apply an infinite series of impulses spaced τ seconds apart, denoted $\delta_\tau(t)$, to the input of a filter with the transfer function shown in Eq. (8.2.11), measurements of the output taken halfway between adjacent pairs of impulses will be $\frac{1}{2}$ times the sum of the two adjacent impulse weights. The output at these time instants due to other impulses will be zero.

For a binary unit amplitude "pulse" or "no pulse" data stream, the possible outputs at these time instants will be 0, $+\frac{1}{2}$, and $+1$. For a binary sequence of unit amplitude positive and negative pulses, the possible outputs are $+2$, 0, and -2. As a result, simple threshold detectors at these time instants can detect transitions between logic 1's and logic 0's.

$P(\omega)$ in Eq. (8.2.11) is one example of an entire class of spectral shapes that preserves the spacing of these transitions. Members of this class can be obtained by adding any real spectral shaping function $[P_a(\omega)]$ with even symmetry about π/τ to $P(\omega)$. That is, any real function $P_a(\omega)$ that satisfies

$$P_a\left(\frac{\pi}{\tau} - x\right) = P_a\left(\frac{\pi}{\tau} + x\right), \tag{8.2.13}$$

$0 \leq x \leq \pi/\tau$, preserves these transitions. An imaginary function is required to have odd symmetry about π/τ to preserve the transition times.

EXAMPLE 8.2.2

A 100% roll-off raised cosine pulse satisfies Nyquist's second criterion as well as Nyquist's first criterion. To demonstrate that it satisfies Nyquist's second criterion, we need only show that the difference between a 100% roll-off raised cosine spectrum and $P(\omega)$ in Eq. (8.2.11) satisfies Eq. (8.2.13). Letting $\alpha = 1$ in Eq. (8.2.7) and subtracting Eq. (8.2.11), we obtain

$$P_a(\omega) = \begin{cases} \dfrac{\tau[1 - \cos(\omega\tau/2)]}{2}, & 0 \leq |\omega| \leq \dfrac{\pi}{\tau} \\ \dfrac{\tau[1 + \cos(\omega\tau/2)]}{2}, & \dfrac{\pi}{\tau} \leq |\omega| \leq \dfrac{2\pi}{\tau}. \end{cases} \tag{8.2.14}$$

For $0 \leq \omega \leq \pi/\tau$, we replace ω with $\pi/\tau - x$, and manipulate

$$\frac{\tau[1 - \cos(\pi/\tau - x)\tau/2]}{2} = \frac{\tau[1 - \cos(\pi/2 - x\tau/2)]}{2}$$

$$= \frac{\tau[1 - \sin(x\tau/2)]}{2}. \tag{8.2.15}$$

For $\pi/\tau \leq \omega \leq 2\pi/\tau$, we have (letting $\omega = \pi/\tau + x$)

$$\frac{\tau[1 + \cos(\pi/\tau + x)\tau/2]}{2} = \frac{\tau[1 + \cos(\pi/2 + x\tau/2)]}{2}$$

$$= \frac{\tau[1 - \sin(x\tau/2)]}{2}, \tag{8.2.16}$$

which upon comparing Eqs. (8.2.15) and (8.2.16) verifies the property.

The verification that the raised cosine pulse satisfies Nyquist's first criterion is left to the problems.

Nyquist's third criterion involves the area under the received waveform during a symbol time interval. In particular, Nyquist specified a spectral shaping function whose impulse response has an area in its interval proportional to the applied impulse weight and zero area for every other symbol interval. This criterion is developed further in Problem 8.11.

8.3 Partial Response Signaling

In Section 8.2 we stated that it is not possible to send pulses at a rate of $1/\tau$ pulses per second in the minimum bandwidth of $1/2\tau$ hertz without risking substantial intersymbol interference in the presence of timing jitter. The $100\alpha\%$ roll-off raised cosine pulse shaping in Section 8.2 is based on the assumption that the transmitted pulses are independent. Lender [1963, 1964, 1966] showed that by introducing a known dependence or correlation between successive pulse amplitudes, the maximum data rate $1/\tau$ symbols/sec could be achieved in the minimum bandwidth of $1/2\tau$ hertz without excessive sensitivity to timing jitter. The penalty for this performance improvement is that more voltage levels need to be transmitted over the channel. (However, as noted in Section 10.7, this need not result in performance loss for a properly designed receiver.) Lender called his method *duobinary*, while Kretzmer [1965, 1966] introduced the term *partial response* in his papers extending Lender's work. Partial response methods have also been called *correlative techniques* or *correlative coding* [Pasupathy, 1977; Lender, 1981]. In this section we emphasize the two most popular correlative coding schemes, duobinary and class 4 partial response.

For duobinary or class 1 partial response systems, the spectral shaping filter is

$$P(\omega) = \begin{cases} 2\tau e^{-j\omega\tau/2} \cos \dfrac{\omega\tau}{2}, & |\omega| \leq \dfrac{\pi}{\tau} \\ 0, & |\omega| > \dfrac{\pi}{\tau}, \end{cases} \tag{8.3.1}$$

with impulse response

$$p(t) = \frac{4}{\pi} \cdot \frac{\cos\left[\pi(t - \tau/2)/\tau\right]}{1 - 4(t - \tau/2)^2/\tau^2}. \tag{8.3.2}$$

The impulse response, $p(t)$, is sketched in Fig. 8.3.1. If the sampling instant for the present symbol interval is taken to be $t = 0$, it is evident that this pulse will interfere with the pulse in the immediately succeeding symbol interval. Or, in other words, the output at the current sampling instant depends on the present symbol and the preceding symbol.

The transfer function in Eq. (8.3.1) can be expressed as a cascade of a digital filter and an analog ideal LPF, as shown in Fig. 8.3.2. From Figs. 8.3.1 and

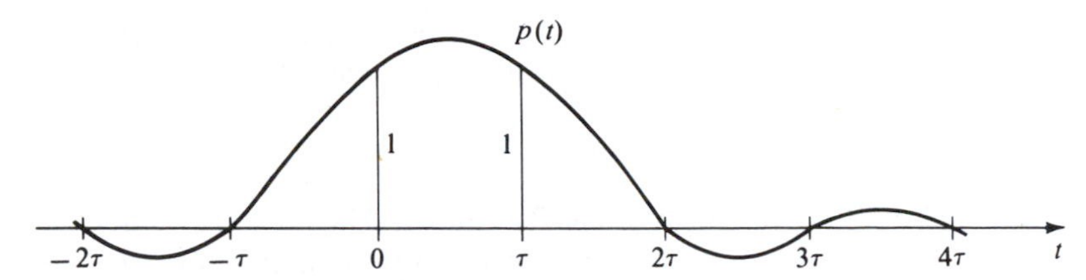

FIGURE 8.3.1 Duobinary impulse response.

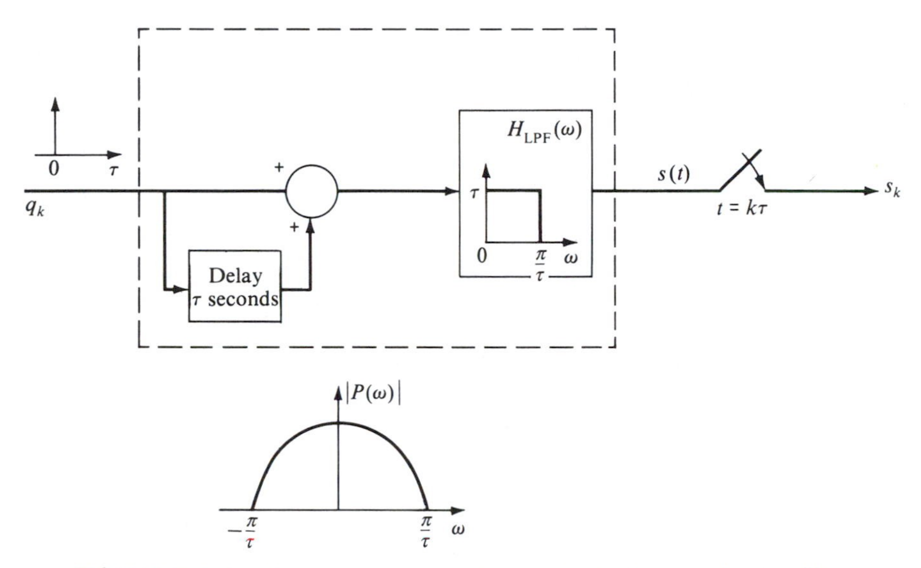

FIGURE 8.3.2 Alternative representation of a duobinary shaping filter.

8.3.2 we know that at the sampling instants,

$$s_k = q_k + q_{k-1}, \tag{8.3.3}$$

and thus if each member of the $\{q_n\}$ sequence takes on the values ± 1, s_k will be $+2$, 0, or -2. The introduction of dependencies between samples thus changes the binary sequence into a three-level sequence (at the sampling times) for transmission. Decoding is possible, since if q_0 is specified, we can find q_1 using Eq. (8.3.3); with q_1 we can find q_2, and so on. We notice immediately from this decoding scheme that errors will tend to propagate, since if q_{k-1} is erroneously decoded, it is likely that q_k will be decoded incorrectly, and so on. Fortunately, Lender [1963] recognized this problem and corrected it by the use of a technique called *precoding*.

Precoding depends on the particular partial response technique being considered, hence we begin by presenting precoding for duobinary or class 1 partial response schemes. We use precoding to generate an appropriate $\{q_k\}$ sequence to serve as the input to the shaping filter in Fig. 8.3.2. Therefore, let the original input data sequence be denoted by $\{a_k\}$, with each a_k taking on the values 0 or 1. We then form a new sequence, denoted $\{b_k\}$, according to

$$b_k = a_k \oplus b_{k-1}, \tag{8.3.4}$$

where the symbol $\oplus$ represents modulo-2 addition. The precoded sequence $\{b_k\}$ still takes on the values 0 or 1, but it is converted into an appropriate sequence $\{q_k\}$ by

$$q_k = 2b_k - 1, \tag{8.3.5}$$

using standard base-10 operations. Equation (8.3.3) still represents the duobinary system encoding rule at each sampling instant, so the decoding rule at the receiver is

$$a_k = \begin{cases} 0, & \text{if } s_k = \pm 2 \\ 1, & \text{if } s_k = 0. \end{cases} \tag{8.3.6}$$

Table 8.3.1 presents an illustrative example of using the precoding operation and of computing the transmitted values at the sampling instants for duobinary signaling. Comparing the $\{a_k\}$ and $\{s_k\}$ sequences in Table 8.3.1 substantiates the validity of the decoding rule in Eq. (8.3.6). The reader should sketch the superposition of the impulse responses to the $\{q_k\}$ sequence in Table 8.3.1 to validate the values shown for $s(t)$ at the sampling instants (see Problem 8.14). To illustrate the effect of transmission errors when precoding is used, we assume that the third value of s_k in Table 8.3.1, s_3, is received as a -2 rather than the correct value 0. It is immediately evident that the only decoding error made is in a_3, which from Eq. (8.3.6) we will erroneously decide was a 0. All other values of a_k are unaffected.

TABLE 8.3.1 Example of Duobinary Signaling with Precoding

k	0	1	2	3	4	5	6	7	8	9	10	11
a_k		0	0	1	1	0	1	1	1	0	0	1
b_k	1	1	1	0	1	1	0	1	0	0	0	1
q_k	1	1	1	-1	1	1	-1	1	-1	-1	-1	1
s_k		2	2	0	0	2	0	0	0	-2	-2	0

Another practically important correlative coding scheme is class 4 partial response, also called modified duobinary, which has the spectral shaping filter

$$P(\omega) = \begin{cases} j2\tau e^{-j\omega\tau} \sin \omega\tau, & |\omega| \le \dfrac{\pi}{\tau} \\ 0, & \text{otherwise,} \end{cases} \tag{8.3.7}$$

which can also be rewritten as

$$P(\omega) = \begin{cases} \tau[1 - e^{-j2\omega\tau}], & |\omega| \leq \dfrac{\pi}{\tau} \\ 0, & \text{otherwise.} \end{cases} \tag{8.3.8}$$

The class 4 partial response filter impulse response follows immediately from Eq. (8.3.8) as

$$p(t) = \frac{\sin(\pi t/\tau)}{\pi t/\tau} - \frac{\sin[\pi(t - 2\tau)/\tau]}{\pi(t - 2\tau)/\tau}. \tag{8.3.9}$$

For $|\omega| \leq \pi/\tau$, we can factor $P(\omega)$ in Eq. (8.3.8) as $P(\omega) = \tau[1 - e^{-j\omega\tau}] \cdot [1 + e^{-j\omega\tau}]$, which is easily shown to yield

$$P(\omega) = \begin{cases} 2\tau[1 - e^{-j\omega\tau}]e^{-j\omega\tau/2} \cos \dfrac{\omega\tau}{2}, & |\omega| \leq \dfrac{\pi}{\tau} \\ 0, & \text{otherwise.} \end{cases} \tag{8.3.10}$$

$P(\omega)$ in Eq. (8.3.10) represents a common implementation of class 4 partial response, which is shown in Fig. 8.3.3.

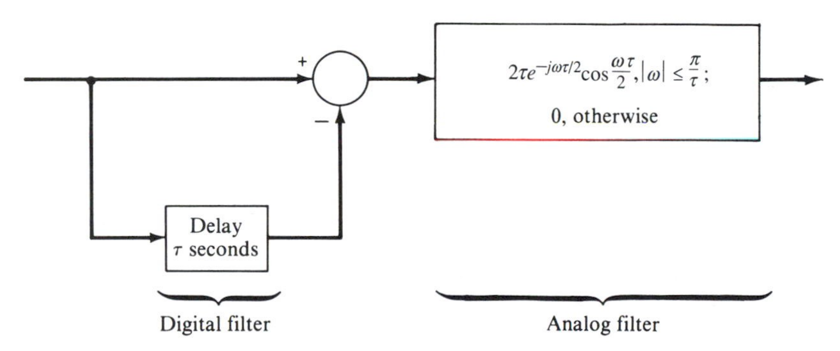

FIGURE 8.3.3 Implementation of class 4 partial response.

If the original data is a sequence of 0's or 1's, denoted $\{a_k\}$, the precoding for class 4 partial response can be expressed as

$$b_k = a_k \oplus b_{k-2}, \tag{8.3.11}$$

and the input to the shaping filter in Fig. 8.3.3 is thus

$$q_k = 2b_k - 1. \tag{8.3.12}$$

The shaping filter output at the sampling instants is the sequence $\{s_k\}$, where

$$s_k = q_k - q_{k-2}, \tag{8.3.13}$$

and the decoding rule in terms of the sample values is

$$a_k = \begin{cases} 0, & \text{if } s_k = 0 \\ 1, & \text{if } s_k = \pm 2. \end{cases} \tag{8.3.14}$$

Table 8.3.2 illustrates the use of class 4 partial response signaling to represent the same $\{a_k\}$ sequence encoded by duobinary in Table 8.3.1. Again, the decoding rule in Eq. (8.3.14) is easily validated by comparing the sequences $\{a_k\}$ and $\{s_k\}$ in Table 8.3.2.

TABLE 8.3.2 Example of Class 4 Partial Response Signaling with Precoding

k	−1	0	1	2	3	4	5	6	7	8	9	10	11
a_k			0	0	1	1	0	1	1	1	0	0	1
b_k	1	1	1	1	0	0	0	1	1	0	1	0	0
q_k	1	1	1	1	−1	−1	−1	1	1	−1	1	−1	−1
s_k			0	0	−2	−2	0	2	2	−2	0	0	−2

A block diagram for generalized partial response signaling is shown in Fig. 8.3.4 [Pasupathy, 1977]. Precoding for generalized partial response consists of

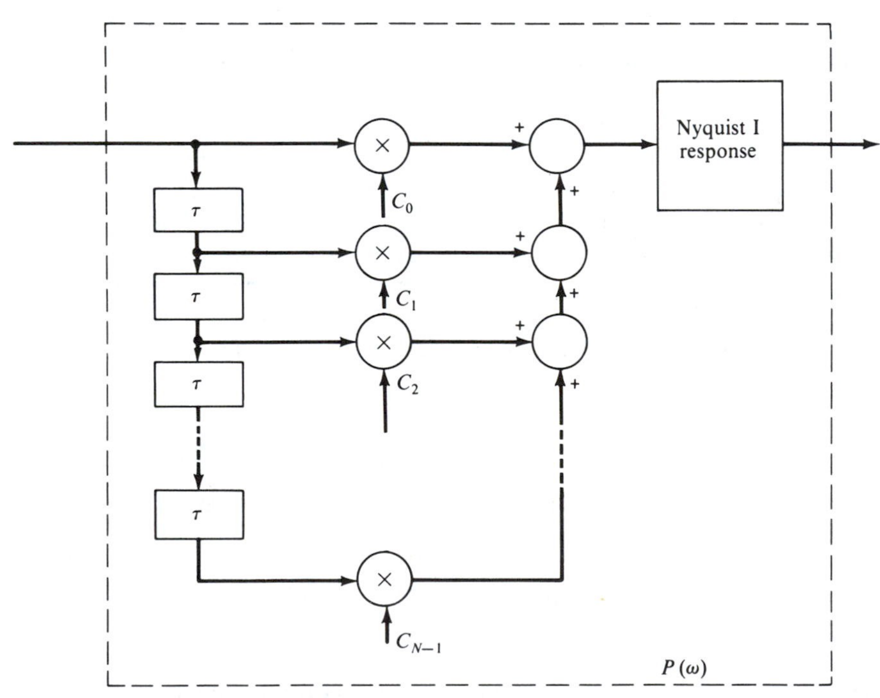

FIGURE 8.3.4 Generalized partial response signaling. From S. Pasupathy, "Correlative Coding: A Bandwidth-Efficient Signaling Scheme," *IEEE Commun. Soc. Mag.*, © 1977 IEEE.

solving the expression

$$a_k = [C_0 b_k + C_1 b_{k-1} + \cdots + C_{N-1} b_{k-N+1}]_{\text{mod } 2} \tag{8.3.15}$$

for b_k, where $\{a_k\}$ is the input data sequence as before. For duobinary $C_0 = 1$ and $C_1 = 1$, while for class 4 partial response $C_0 = 1$, $C_1 = 0$, and $C_2 = -1$. The reader should use Fig. 8.3.4 and Eq. (8.3.15) to derive the previously presented equations for duobinary and modified duobinary (see Problem 8.18). Note from Fig. 8.3.4 that the impulse response of partial response systems is always a weighted superposition of delayed $\sin(\pi t/\tau)/(\pi t/\tau)$ functions. This is clearly evident from Eq. (8.3.9) for class 4 partial response.

To decode, we first note that the transmitted signal $s(t)$ in Fig. 8.3.4 when evaluated at the sampling times is

$$s_k = C_0 q_k + C_1 q_{k-1} + \cdots + C_{N-1} q_{k-N+1}, \tag{8.3.16}$$

so

$$q_k = \frac{1}{C_0}\left[s_k - \sum_{i=1}^{N-1} C_i q_{k-i}\right]. \tag{8.3.17}$$

Using Eq. (8.3.12), we find that

$$b_k = \tfrac{1}{2}[q_k + 1]. \tag{8.3.18}$$

Since all preceding b_j's, $j = k - N + 1, \ldots, k - 1$, are known, we can substitute these values along with b_k from Eq. (8.3.18) into Eq. (8.3.15) to obtain a_k.

Correlative coding systems can also be defined for M-level (rather than binary) input data sequences. For this case, the number of transmitted levels is $2M - 1$ and the operations for precoding and decoding are modulo-M. For more information on these methods, see Lender [1981].

Some partial response spectral shaping functions have nulls at *dc* as well as at $\omega = \pi/\tau$. In particular, as can be seen from Eq. (8.3.7), the class 4 partial response shaping filter has a spectral null at $\omega = 0$. This property can be especially important for data transmission over transformer-coupled telephone lines and when data are to be transmitted using single-sideband modulation.

Since there is correlation among the transmitted levels at sampling instants, this correlation, or memory, can be utilized to detect errors without the insertion of redundant bits at the transmitter. For example, for duobinary signaling the following properties can be discerned: (1) an extreme level (± 2) follows another extreme level of the same polarity only if an even number of zero levels occur in between, (2) an extreme level follows an extreme level of the opposite polarity only if an odd number of zero levels occur in between, and (3) an extreme level of one polarity cannot be followed by an extreme level of the opposite polarity at the next sampling instant. The reader should verify these properties for Table 8.3.1. Similar properties can also be stated for class 4 partial response signaling. Although all errors cannot be detected, these properties allow error monitoring without additional complexity.

8.4 Deterministic Distortion

Impairments that can affect data transmission over telephone channels can be classified as being one of two types: deterministic impairments or random impairments. *Deterministic impairments* include amplitude distortion, delay distortion, nonlinearities, and frequency offset. These impairments are deterministic in that they can be measured for a particular transmission path and they do not change perceptibly over a relatively short period of time. On the other hand, *random impairments* include Gaussian noise, impulsive noise, and phase jitter. Clearly, these random disturbances require a probabilistic description, and hence are treated in a later chapter.

Amplitude distortion and delay distortion can cause intersymbol interference, which decreases the system margin against noise. The magnitude response of a "typical" telephone channel is sketched in Fig. 8.4.1. Note that what is shown is not "gain versus frequency," but "attenuation versus frequency." It is quite evident that the amplitude response as a function of frequency between 200 and 3200 Hz is far from flat, and hence certainly not distortionless.

A demonstration of the detrimental effects of a nonflat frequency response is obtainable by considering the classic example of a filter with ripples in its amplitude response. In particular, we wish to examine the time-domain response of a system with the transfer function

$$H(\omega) = [1 + 2\varepsilon \cos \omega t_0]e^{-j\omega t_d}, \tag{8.4.1}$$

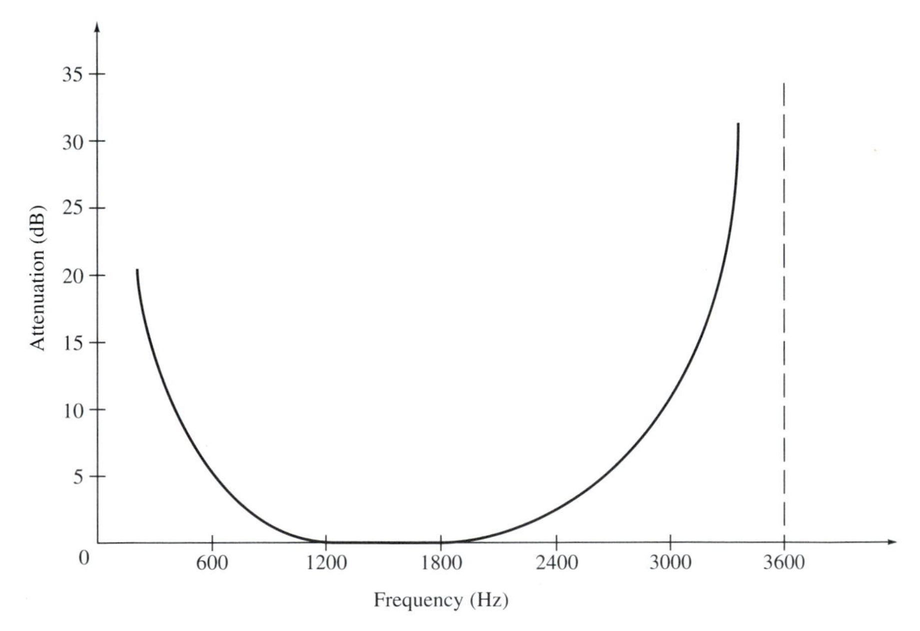

FIGURE 8.4.1 Magnitude response of a typical telephone channel.

where $|\varepsilon| \leq \frac{1}{2}$. It should be evident to the reader that $|H(\omega)|$ has ripples of maximum value $\pm 2\varepsilon$ about 1 and that the phase of $H(\omega)$ is linear. We thus have amplitude distortion only. Rewriting Eq. (8.4.1), we obtain

$$\begin{aligned} H(\omega) &= [1 + \varepsilon(e^{j\omega t_0} + e^{-j\omega t_0})]e^{-j\omega t_d} \\ &= e^{-j\omega t_d} + \varepsilon e^{j\omega(t_0 - t_d)} + \varepsilon e^{-j\omega(t_0 + t_d)}. \end{aligned} \tag{8.4.2}$$

For a general input pulse shape, say $x(t)$, the time-domain response of the filter in Eq. (8.4.2) is

$$y(t) = x(t - t_d) + \varepsilon x(t - t_d + t_0) + \varepsilon x(t - t_d - t_0). \tag{8.4.3}$$

The ripples in the magnitude response of $H(\omega)$ have thus generated scaled replicas of $x(t)$ that both precede and follow an undistorted (but delayed) version of the input. These replicas of $x(t)$ have been called "echoes" in the literature. Depending on the value of t_0, we find that the echoes can interfere with adjacent transmitted pulses, yielding intersymbol interference, and/or the echoes can overlap the main output pulse $x(t - t_d)$, also causing distortion. Sketches of these various cases are left as a problem.

A filter (or channel) with ripples in the phase but a flat magnitude response will also cause echoes. To demonstrate this claim, we consider the transfer function

$$H(\omega) = \exp\{-j[\omega t_d - \varepsilon \sin \omega t_0]\} \tag{8.4.4}$$

with $|\varepsilon| \ll \pi$. To proceed, we note that $e^{-j\omega t_d}$ represents a pure time delay and that

$$e^{j\varepsilon \sin \omega t_0} = \cos[\varepsilon \sin \omega t_0] + j \sin[\varepsilon \sin \omega t_0]. \tag{8.4.5}$$

Since $|\varepsilon| \ll \pi$, we can approximate Eq. (8.4.5) as

$$e^{j\varepsilon \sin \omega t_0} \cong 1 + j\varepsilon \sin \omega t_0, \tag{8.4.6}$$

so

$$H(\omega) \cong e^{-j\omega t_d}\left[1 + \frac{\varepsilon}{2} e^{j\omega t_0} - \frac{\varepsilon}{2} e^{-j\omega t_0}\right]. \tag{8.4.7}$$

For a general input pulse shape $x(t)$, then, the filter output is approximately

$$y(t) \cong x(t - t_d) + \frac{\varepsilon}{2} x(t - t_d + t_0) - \frac{\varepsilon}{2} x(t - t_d - t_0) \tag{8.4.8}$$

and again we have echoes.

We have not established as yet that phase distortion is actually significant over telephone channels. Since the telephone network was originally designed with only voice signals in mind, there is considerable phase nonlinearity over telephone channels. Voice communication is relatively unaffected by the phase nonlinearities present, but as we have seen, data transmission can be impaired significantly. Phase information for a telephone channel is not usually expressed as a system phase response, due to the difficulty of establishing an absolute phase reference and the necessity to count modulo 2π or 360°. To obtain phase information, a related quantity called *envelope delay* is measured instead.

The envelope delay is defined as the rate of change of the phase versus frequency response, and hence can be expressed as

$$t_R = \frac{-d}{d\omega}\,\theta(\omega), \tag{8.4.9}$$

where $\theta(\omega)$ is the channel phase in radians and t_R denotes the envelope delay in seconds. A related quantity is the *phase delay* or *carrier delay*, which is defined as the change in phase versus frequency, and it is expressible as

$$t_c = \frac{-\theta(\omega)}{\omega}, \tag{8.4.10}$$

where t_c denotes the phase or carrier delay. These definitions are instrumental to the measurement of phase distortion over a channel, and the following derivation illustrates the reasoning behind their names as well as the technique used to measure the envelope delay.

We consider a bandpass channel with a constant magnitude response, $|H(\omega)| = K$, but with a nonlinear phase response, $\underline{/H(\omega)} = \theta(\omega)$. If the input to this channel is a narrowband signal

$$x(t) = m(t)\cos\omega_c t, \tag{8.4.11}$$

we wish to write an expression for the channel output, denoted $y(t)$, in terms of the envelope delay and the phase delay. To begin, we assume that the phase nonlinearity is not too severe, and thus we can expand the phase in a Taylor's series about the carrier components $\pm\omega_c$. About $\omega = +\omega_c$, we have

$$\theta(\omega) = \theta(\omega_c) + (\omega - \omega_c)\left.\frac{d\theta(\omega)}{d\omega}\right|_{\omega=\omega_c}, \tag{8.4.12}$$

where higher-order terms are assumed negligible. At the frequency of interest, namely $\omega = +\omega_c$, we know that

$$t_R = -\left.\frac{d\theta(\omega)}{d\omega}\right|_{\omega=\omega_c} \tag{8.4.13}$$

and

$$t_c = -\left.\frac{\theta(\omega)}{\omega}\right|_{\omega=\omega_c}, \tag{8.4.14}$$

so

$$\theta(\omega) = -\omega_c t_c + (\omega - \omega_c)(-t_R) = -\omega_c t_c - (\omega - \omega_c)t_R. \tag{8.4.15}$$

Similarly, about $\omega = -\omega_c$, we find that

$$\theta(\omega) = \omega_c t_c - (\omega + \omega_c)t_R. \tag{8.4.16}$$

Letting $M(\omega) = \mathscr{F}\{m(t)\}$ and $Y(\omega) = \mathscr{F}\{y(t)\}$, we can write the channel response as

$$Y(\omega) = \frac{K}{2}\left[M(\omega - \omega_c) + M(\omega + \omega_c)\right]e^{j\theta(\omega)}$$

$$= \frac{K}{2}\left[M(\omega - \omega_c)e^{j\theta(\omega)} + M(\omega + \omega_c)e^{j\theta(\omega)}\right]. \qquad (8.4.17)$$

Since $M(\omega - \omega_c)$ is located about $\omega = +\omega_c$ and since $M(\omega + \omega_c)$ is located about $\omega = -\omega_c$, we make the appropriate substitutions into Eq. (8.4.17) using Eqs. (8.4.15) and (8.4.16), respectively. Hence

$$Y(\omega) = \frac{K}{2}\left[M(\omega - \omega_c)e^{j[-\omega_c t_c - (\omega - \omega_c)t_R]} + M(\omega + \omega_c)e^{j[\omega_c t_c - (\omega + \omega_c)t_R]}\right]$$

$$= \frac{K}{2}\, e^{-j\omega t_R}\left[M(\omega - \omega_c)e^{-j\omega_c[t_c - t_R]} + M(\omega + \omega_c)e^{j\omega_c[t_c - t_R]}\right]. \qquad (8.4.18)$$

Therefore, using Fourier transform properties we obtain

$$y(t) = \frac{K}{2}\left[m(t)e^{j\omega_c[t - t_c + t_R]} + m(t)e^{-j\omega_c[t - t_c + t_R]}\right]\Big|_{t = t - t_R}$$

$$= Km(t - t_R)\cos\left[\omega_c(t - t_c)\right]. \qquad (8.4.19)$$

Only the envelope $m(t)$ is delayed by t_R, hence t_R is called the *envelope delay*, while only the carrier is delayed by t_c, hence the name *carrier delay*.

The method used to measure the envelope delay over real channels satisfies the assumptions used in the derivation. For example, $m(t)$ is selected to be a low-frequency sinusoid, say $\cos \omega_m t$, so that we have $\omega_c \pm \omega_m \cong \omega_c$ and hence the phase nonlinearity will not be too bad over this region. We can also neglect higher-order terms in the Taylor's series since our range of frequencies of interest is $\omega_c - \omega_m \leq \omega \leq \omega_c + \omega_m$ and ω_m is small. To measure t_R, the derivative is approximated by $\Delta\theta(\omega)/\Delta\omega$, where $\Delta\omega = \omega_m$, and the phase of the received envelope $\cos(\omega_m t - \omega_m t_R)$ is compared to the transmitted envelope to yield $\Delta\theta(\omega) = \omega_m t_R$. Taking the ratio yields the envelope delay.

Results published by Sunde [1961] clearly demonstrate the effects of phase distortion on data transmission. For zero phase distortion, the phase response should be linear, and hence the envelope delay should be constant. Figure 8.4.2 shows plots of 100% roll-off raised cosine pulses subjected to quadratic envelope delay distortion. To obtain these plots, Sunde [1961] inserted delay distortion, which increased quadratically from 0 at $\omega = 0$ to some final value at $\omega = \omega_{\max}$. As the envelope delay increases, the pulse amplitude is reduced and the pulse peak no longer occurs at the desired sampling instant. Furthermore, the zero crossings become shifted and the trailing pulse becomes large enough in amplitude to interfere with adjacent pulses. Obviously, delay distortion or phase distortion can be exceedingly detrimental in data transmission.

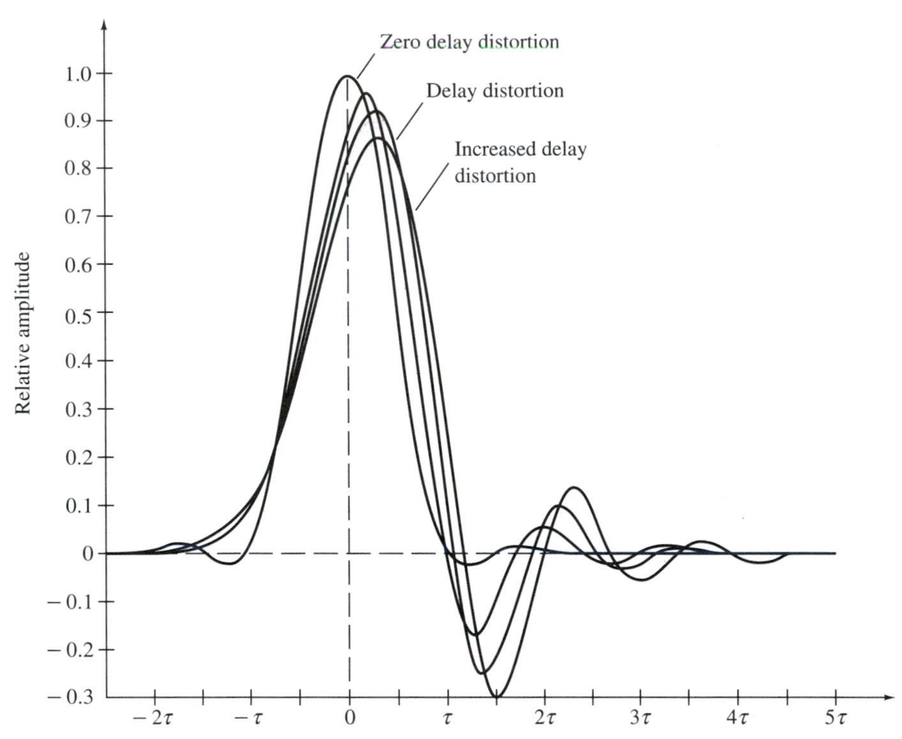

FIGURE 8.4.2 Raised cosine pulses with quadratic delay distortion. From E. D. Sunde, "Pulse Transmission by AM, FM, and PM in the Presence of Phase Distortion," *Bell Syst. Tech. J.*, © 1961 AT&T Bell Laboratories.

8.5 Eye Patterns

A convenient way to see the distortion present on a channel is to display what is called the system *eye pattern* or *eye diagram.* The eye pattern is obtained by displaying the data pulse stream on an oscilloscope, with the pulse stream applied to the vertical input and the sampling clock applied to the external trigger. A drawing of a two-level eye pattern is shown in Fig. 8.5.1, and the source of its name is clearly evident. Typically, one to three pulse (symbol) intervals are displayed and several kinds of distortion are easily observed. For minimum error probability, sampling should occur at the point where the eye is open widest. If all of the traces go through allowable (transmitted) pulse amplitudes only at the sampling instants, the eye is said to be *100% open* or *fully open.* A fully open eye pattern for three-level pulse transmission is sketched in Fig. 8.5.2. The eye pattern is said to be 80% open if, at the sampling instant, the traces deviate by 20% from the fully open eye diagram. This degradation is some-

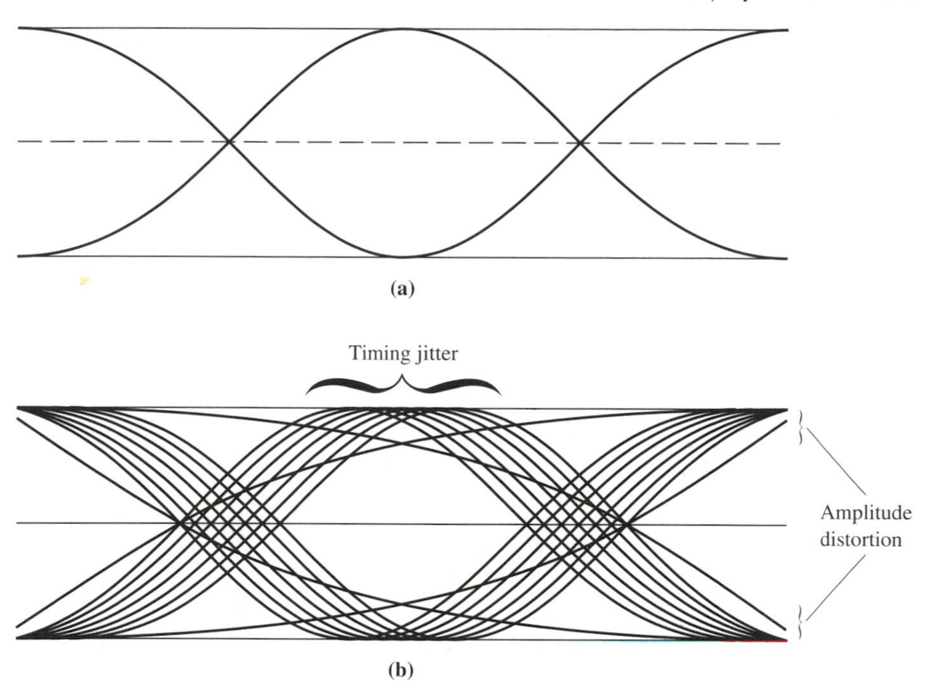

FIGURE 8.5.1 Two-level eye diagrams: (a) Two-level pattern (distortionless); (b) two-level eye pattern with timing jitter.

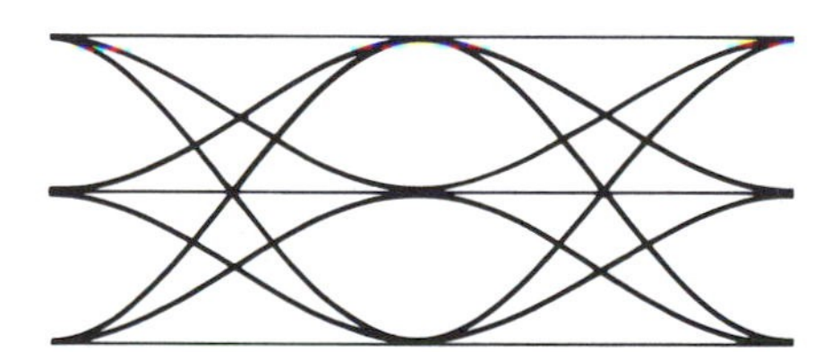

FIGURE 8.5.2 Three-level eye pattern.

times expressed as a loss in signal-to-noise ratio (S/N) by $S/N = 20 \log_{10} 0.8 = -1.9$ dB, where the minus sign indicates a degradation in S/N.

The distance between the decision thresholds and adjacent received pulse traces at the sampling time is the margin of the system against additional noise. As the sampling time is varied about the time instant of maximum eye opening, the eye begins to close. The rate that the eye closes as the sampling instant is varied is an indication of the system's sensitivity to timing error. Jitter in received zero crossings (or threshold crossings) can be particularly insidious, since many receivers extract timing information by averaging zero crossings. The various kinds of distortion are labeled in Fig. 8.5.1.

As can be seen from Fig. 8.5.2, as the number of transmitted levels per pulse is increased for a fixed maximum pulse amplitude, the smaller is the fully open eye's margin against noise. Thus, increasing the number of transmitted levels per pulse, while increasing the transmitted bit rate, also tends to increase the number of bit errors. When the distortion is so severe that the eye is completely filled by pulse traces, the eye is said to be *closed*. Obviously, when the eye is closed, data transmission is extremely unreliable.

8.6 Equalization

For data transmission to be effective, the data must be received accurately. One way to reduce the number of data transmission errors is to correct, or compensate for, the deterministic distortion contributed by the channel. This is easily accomplished if the channel transfer function is known, since if the channel transfer function is $|H_c(\omega)|e^{j\underline{/H_c(\omega)}}$, a cascaded equalizer with transfer function $H_{\text{eq}}(\omega) = |H_c(\omega)|^{-1}e^{-j\underline{/H_c(\omega)}}$ will produce an ideal response. This approach is effective for deterministic distortion, since deterministic impairments are assumed to vary slowly in comparison to a single pulse interval. If the channel transfer function is known and the channel response does not vary significantly with time, the equalizer can be preset and held fixed while the data are transmitted. This situation is called *fixed equalization.*

Of course, the channel transfer function may not be known in advance, in which case the channel characteristics must be measured or "learned." Furthermore, the channel response may be changing slowly but continually with time. If the channel transfer function is unknown, but it can be assumed to be invariant over a relatively long time period, the channel transfer function can be learned during an initialization or startup period by transmitting a previously agreed upon data sequence known to both the transmitter and receiver, and then held fixed during actual data transmission. Equalizers that use this approach are called *automatic equalizers.* If the channel is unknown and slowly time varying, adaptation or learning must continue even after the startup sequence, while the data are being transmitted. Equalizers based on this method are called *adaptive equalizers.* The algorithms used for automatic and adaptive equalization are the same except for the reference signals employed.

A block diagram of a communication system with equalization is shown in Fig. 8.6.1. Modulation and demodulation blocks are not included in this figure

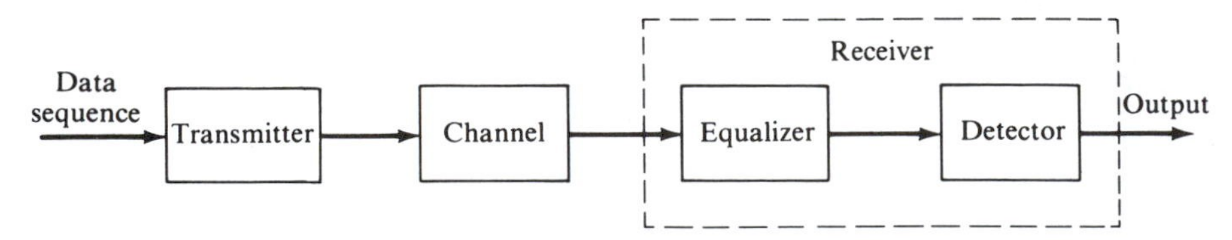

FIGURE 8.6.1 Data communication system with equalization.

and our initial development emphasizes baseband equalization. Passband equalization is discussed at the end of this section. Throughout our development the channel is assumed to be bandlimited to B hertz. Equalizers in data transmission applications typically take the form of a transversal filter. A $2N + 1$ tap transversal filter is illustrated in Fig. 8.6.2. The coefficients $\{c_i, i = -N, -N + 1, \ldots, 0, 1, \ldots, N - 1, N\}$ are called *equalizer tap gains* and the time delays (Δ) are the *tap spacing*. The tap spacing is often fixed at one symbol interval, while the tap gains are adjustable. The equalizer impulse response is

$$h_{\text{eq}}(t) = \sum_{n=-N}^{N} c_n \delta(t - n\Delta), \tag{8.6.1}$$

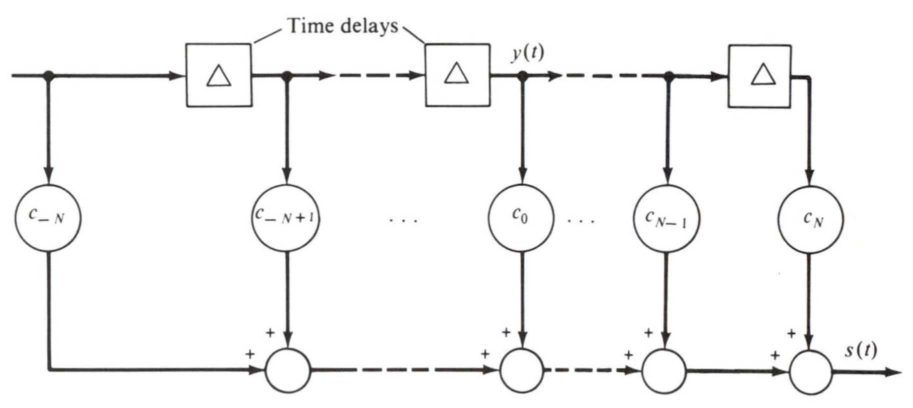

FIGURE 8.6.2 A $2N + 1$ tap equalizer.

which has the transfer function

$$H_{\text{eq}}(\omega) = \sum_{n=-N}^{N} c_n e^{-j\omega n\Delta}. \tag{8.6.2}$$

To illustrate how a transversal filter structure can be used for channel equalization, consider the following example.

EXAMPLE 8.6.1

Consider a data transmission channel with the known transfer function

$$H_c(\omega) = [a_1 + a_2 e^{-j\omega(t_2 - t_1)}]e^{-j\omega t_1}$$

with $t_2 > t_1$ and $(a_2/a_1)^2 \ll 1$. We wish to find $H_{\text{eq}}(\omega) = 1/H_c(\omega)$, and then determine c_{-1}, c_0, c_1, and Δ for a three-tap transversal filter equalizer that approximates $H_{\text{eq}}(\omega)$. Straightforwardly, we have

$$H_{\text{eq}}(\omega) = \frac{1}{H_c(\omega)} = \frac{e^{j\omega t_1}}{a_1} \cdot \frac{1}{1 + a_2 e^{-j\omega(t_2 - t_1)}/a_1}. \tag{8.6.3}$$

Since $(a_2/a_1)^2 \ll 1$, $a_2/a_1 < 1$, and we can use a series expansion to approximate $H_{eq}(\omega)$ as

$$H_{eq}(\omega) \cong \frac{e^{j\omega t_1}}{a_1}\left[1 - \frac{a_2}{a_1} e^{-j\omega(t_2 - t_1)} + \left(\frac{a_2}{a_1}\right)^2 e^{-j2\omega(t_2 - t_1)}\right]. \tag{8.6.4}$$

Letting $N = 1$ in Eq. (8.6.2), we know that for a transversal filter equalizer

$$H_{eq}(\omega) = c_{-1}e^{j\omega\Delta} + c_0 + c_1 e^{-j\omega\Delta} = e^{j\omega\Delta}[c_{-1} + c_0 e^{-j\omega\Delta} + c_1 e^{-j2\omega\Delta}]. \tag{8.6.5}$$

Comparing Eqs. (8.6.4) and (8.6.5), and ignoring the terms in front of the brackets, which involve only gain elements and time shifts, we find that

$$\Delta = t_2 - t_1, \qquad c_{-1} = 1, \qquad c_0 = \frac{-a_2}{a_1}, \qquad c_1 = \left(\frac{a_2}{a_1}\right)^2.$$

Thus a three-tap transversal filter with these parameter values will approximately equalize $H_c(\omega)$.

Let us now proceed to investigate the transversal filter as an automatic equalizer. For $y(t)$ in, as shown in Fig. 8.6.2, the equalizer output is

$$s(t) = y(t) * \sum_{n=-N}^{N} c_n \delta(t - n\Delta) = \sum_{n=-N}^{N} c_n y(t - n\Delta). \tag{8.6.6}$$

If we consider only the sampling times $t = k\Delta$, then Eq. (8.6.6) becomes

$$s(k\Delta) = \sum_{n=-N}^{N} c_n y[(k - n)\Delta]. \tag{8.6.7}$$

It is common notational practice to drop the explicit dependence on Δ and to write

$$s_k = \sum_{n=-N}^{N} c_n y_{k-n} = c_{-N} y_{k+N} + \cdots + c_{-1} y_{k+1} + c_0 y_k + c_1 y_{k-1} + \cdots + c_N y_{k-N}, \tag{8.6.8}$$

where the dependence on Δ is implicit. For a single positive unit-amplitude raised cosine pulse transmitted at time instant (sampling instant) k, the desired sample values at the equalizer output will be $s_k = 1$, $s_{k+j} = 0$ for $j \neq 0$. Thus, if the equalizer input $y(t)$ is received undistorted, we have $y_k = 1$, $y_{k+j} = 0$ for $j \neq 0$, and it is evident that the desired output pulse is centered in time on the equalizer taps.

Although not indicated in Fig. 8.6.2, the equalizer tap gains are adjustable and must be calculated from measurements taken over the channel in question. This is normally accomplished by transmitting a fairly long sequence of pulses, but for purposes of illustration, let us consider the case of a single raised cosine pulse transmitted at time instant $k = 0$. Any other time instant could be used, but $k = 0$ simplifies notation. The following example illustrates how the equalizer tap gains can be calculated from the transmission of this single pulse over the channel and a known reference signal.

EXAMPLE 8.6.2

Consider a three-tap equalizer ($N = 1$) with the input sequence $y_1 = -\frac{1}{2}$, $y_0 = 1$, $y_{-1} = \frac{1}{4}$, and all other $y_k = 0$. Using Eq. (8.6.8), we can find the equalizer output for all values of k as:

$$k \leq -3: \quad s_k = c_{-1}y_{k+1} + c_0 y_k + c_1 y_{k-1} = 0$$

$$k = -2: \quad s_{-2} = c_{-1}y_{-1} + c_0 y_{-2} + c_1 y_{-3} = c_{-1}y_{-1} = \tfrac{1}{4}c_{-1}$$

$$k = -1: \quad s_{-1} = c_{-1}y_0 + c_0 y_{-1} + c_1 y_{-2} = c_{-1}y_0 + c_0 y_{-1} = c_{-1} + \tfrac{1}{4}c_0$$

$$k = 0: \quad s_0 = c_{-1}y_1 + c_0 y_0 + c_1 y_{-1} = -\tfrac{1}{2}c_{-1} + c_0 + \tfrac{1}{4}c_1$$

$$k = 1: \quad s_1 = c_{-1}y_2 + c_0 y_1 + c_1 y_0 = -\tfrac{1}{2}c_0 + c_1$$

$$k = 2: \quad s_2 = c_{-1}y_3 + c_0 y_2 + c_1 y_1 = -\tfrac{1}{2}c_1$$

$$k \geq 3: \quad s_k = 0.$$

If the input to the channel is a single raised cosine pulse at time instant $k = 0$, the desired output of the equalizer is $r_0 = 1$, $r_k = 0$ for all other k. The equations to be solved for the tap gains are thus

$$\tfrac{1}{4}c_{-1} = 0 \tag{8.6.9}$$

$$c_{-1} + \tfrac{1}{4}c_0 = 0 \tag{8.6.10}$$

$$-\tfrac{1}{2}c_{-1} + c_0 + \tfrac{1}{4}c_1 = 1 \tag{8.6.11}$$

$$-\tfrac{1}{2}c_0 + c_1 = 0 \tag{8.6.12}$$

$$-\tfrac{1}{2}c_1 = 0. \tag{8.6.13}$$

We see immediately that these equations are inconsistent.

Since we have only three equalizer coefficients at our disposal, it is unrealistic to expect that we can fix the output at five sampling instants. Hence, to determine the tap gains we retain only the three equations centered on the time instant $k = 0$, Eqs. (8.6.10)–(8.6.12). Solving these equations simultaneously, we find that $c_{-1} = -\frac{1}{5}$, $c_0 = \frac{4}{5}$, and $c_1 = \frac{2}{5}$. For the given equalizer input sequence $\{y_k\}$, we can use Eq. (8.6.8) and the computed tap gains to show that the equalizer output for all k is $s_{k-2} = -\frac{1}{20}$, $s_{k-1} = 0$, $s_k = 1$, $s_{k+1} = 0$, $s_{k+2} = -\frac{1}{5}$, and $s_k = 0$ for all other k. Thus the equalizer has produced an ideal response at time instants $k - 1$, k, and $k + 1$, but the outputs at times $k - 2$ and $k + 2$ are still nonzero, since they cannot be controlled by the equalizer. Since the outputs at the time instants other than k are zeroed, this type of equalizer is called a *zero-forcing* (ZF) equalizer.

Before proceeding with the discussion of equalization, we note that the expression for the equalizer output in Eq. (8.6.8) is a discrete-time convolution, which can also be expressed as a polynomial multiplication. To do this, we write the input sequence $\{y_k\}$ as a polynomial by associating the variable p^j with y_j. Similarly, we associate the variable p^k with c_k to obtain a polynomial from the tap gains. To demonstrate the procedure, we consider the equalizer input sequence from Example 8.6.2, namely $y_1 = -\frac{1}{2}$, $y_0 = 1$, $y_{-1} = \frac{1}{4}$, $y_j = 0$ for all other j, and the resulting computed tap gains $c_{-1} = -\frac{1}{5}$, $c_0 = \frac{4}{5}$, and $c_1 = \frac{2}{5}$. The polynomial multiplication thus becomes

$$\begin{aligned}(y_{-1}p^{-1} + y_0p^0 + y_1p^1)(c_{-1}p^{-1} + c_0p^0 + c_1p^1) \\ = c_{-1}y_{-1}p^{-2} + (c_0y_{-1} + y_0c_{-1})p^{-1} \\ + (y_{-1}c_1 + c_0y_0 + c_{-1}y_1)p^0 + (y_0c_1 + c_0y_1)p^1 + y_1c_1p^2.\end{aligned} \tag{8.6.14}$$

The output at time s_{k+j} is the coefficient of p^j; therefore, upon substituting for the equalizer input and tap gains, we find that $s_{k-2} = -\frac{1}{20}$, $s_{k-1} = 0$, $s_k = 1$, $s_{k+1} = 0$, $s_{k+2} = -\frac{1}{5}$, with all other $s_n = 0$, since these terms are absent from Eq. (8.6.14). Of course, these results are the same as those we obtained using Eq. (8.6.8).

Another type of equalization does not attempt to force the equalizer outputs to zero at all time instants (within its range) except k, but instead, tries to minimize the sum of the squared errors over a set of output samples. Specifically, in minimum mean-squared error (MMSE) equalization, the tap gains are selected to minimize

$$\varepsilon^2 = \sum_{k=-K}^{K} e_k^2 = \sum_{k=-K}^{K} \left[r_k - \sum_{n=-N}^{N} c_n y_{k-n} \right]^2, \tag{8.6.15}$$

where $\{r_k\}$ is the desired equalizer output or reference sequence, c_n, N, and $\{y_k\}$ are as defined previously, and K is to be selected. Taking partial derivatives with respect to the c_m, we have

$$\frac{\partial \varepsilon^2}{\partial c_m} = 0 = -2 \sum_{k=-K}^{K} \left[r_k - \sum_{n=-N}^{N} c_n y_{k-n} \right] y_{k-m} \tag{8.6.16}$$

for $m = -N, -N+1, \ldots, -1, 0, 1, \ldots, N-1, N$, or equivalently,

$$\sum_{n=-N}^{N} c_n \sum_{k=-K}^{K} y_{k-n} y_{k-m} = \sum_{k=-K}^{K} r_k y_{k-m}. \tag{8.6.17}$$

Since the sequences $\{y_j\}$ and $\{r_j\}$ are known, we can compute $\sum_{k=-K}^{K} y_{k-n} y_{k-m}$ and $\sum_{k=-K}^{K} r_k y_{k-m}$ for all m and n, so that from Eq. (8.6.17) we have a set of $2N+1$ simultaneous equations in $2N+1$ unknowns to be solved for the tap gains. To illustrate this procedure, we compute the MMSE equalizer coefficients for the same equalizer input sequence used in Example 8.6.2.

EXAMPLE 8.6.3

The equalizer input sequence is $y_{-1} = \frac{1}{4}$, $y_0 = 1$, $y_1 = -\frac{1}{2}$, with all other $y_j = 0$, and the reference sequence is $r_0 = 1$, $r_j = 0$ for all $j \neq 0$. For $N = 1$ we have to solve the three simultaneous equations represented by [from Eq. (8.6.17)]

$$\sum_{n=-1}^{1} c_n \sum_{k=-K}^{K} y_{k-n} y_{k-m} = \sum_{k=-K}^{K} r_k y_{k-m} \tag{8.6.18}$$

with $m = -1, 0, 1$, for the tap gains $\{c_n\}$. It is instructive to write these equations out in detail. The three simultaneous equations represented by Eq. (8.6.18) are

$m = -1$:

$$\overset{①}{c_{-1} \sum_k y_{k+1}^2} + \overset{②}{c_0 \sum_k y_k y_{k+1}} + \overset{③}{c_1 \sum_k y_{k-1} y_{k+1}} = \sum_k r_k y_{k+1} \qquad (8.6.19)$$

$m = 0$:

$$\overset{④}{c_{-1} \sum_k y_{k+1} y_k} + \overset{⑤}{c_0 \sum_k y_k^2} + \overset{⑥}{c_1 \sum_k y_{k-1} y_k} = \sum_k r_k y_k \qquad (8.6.20)$$

$m = 1$:

$$\overset{⑦}{c_{-1} \sum_k y_{k+1} y_{k-1}} + \overset{⑧}{c_0 \sum_k y_k y_{k-1}} + \overset{⑨}{c_1 \sum_k y_{k-1}^2} = \sum_k r_k y_{k-1}, \qquad (8.6.21)$$

where the summations on k are from $-K$ to K. Normally, the parameter K has to be preselected; however, for our simple example we need only assume that K is large. More will be said on this point later.

To solve Eqs. (8.6.19)–(8.6.21) for the tap gains, we first need to evaluate the terms involving the summations on k. Since $r_j = 0$ for $j \neq 0$, the summations on the right side of the equals sign in these equations are easily evaluated as $\sum_k r_k y_{k+1} = y_1$, $\sum_k r_k y_k = y_0$, and $\sum_k r_k y_{k-1} = y_{-1}$, for $m = -1$, 0, and 1, respectively. For K large it is also easy to see that we do not have to evaluate all nine summations for the terms on the left sides of the equals signs in Eqs. (8.6.19)–(8.6.21), since many of the summations are the same. Specifically, we can show that ① = ⑤ = ⑨, ② = ④ = ⑥ = ⑧, and ③ = ⑦. Thus we only need to calculate three summations, which are given by

$$① = \sum_k y_{k+1}^2 = y_{-1}^2 + y_0^2 + y_1^2 = 1\tfrac{5}{16} = ⑤ = ⑨$$

$$② = \sum_k y_k y_{k+1} = y_{-1} y_0 + y_0 y_1 = -\tfrac{1}{4} = ④ = ⑥ = ⑧$$

$$③ = \sum_k y_{k-1} y_{k+1} = y_{-1} y_1 = -\tfrac{1}{8} = ⑦.$$

Substituting for all of the values of the summations into Eqs. (8.6.19)–(8.6.21), we must solve the three equations

$$\tfrac{21}{16} c_{-1} - \tfrac{1}{4} c_0 - \tfrac{1}{8} c_1 = -\tfrac{1}{2} \qquad (8.6.22)$$

$$-\tfrac{1}{4} c_{-1} + \tfrac{21}{16} c_0 - \tfrac{1}{4} c_1 = 1 \qquad (8.6.23)$$

$$-\tfrac{1}{8} c_{-1} - \tfrac{1}{4} c_0 + \tfrac{21}{16} c_1 = \tfrac{1}{4} \qquad (8.6.24)$$

for c_{-1}, c_0, and c_1. We find that $c_{-1} = -0.2009$, $c_0 = +0.7885$, and $c_1 = 0.3215$.

By substituting for $\{r_k\}$, $\{y_k\}$, and $\{c_k\}$ in Eq. (8.6.15), we can calculate the minimum mean-squared error to be

$$\begin{aligned}\varepsilon^2_{\min} &= [c_{-1}y_{-1}]^2 + [c_{-1}y_0 + c_0 y_{-1}]^2 + [r_0 - c_{-1}y_1 - c_0 y_0 - c_1 y_{-1}]^2 \\ &\quad + [c_0 y_1 + c_1 y_0]^2 + [c_1 y_1]^2 = 0.0352. \end{aligned} \tag{8.6.25}$$

Using the computed tap gains, we can also find the equalizer output sequence to be $s_k = 0$ for $k \le -3$ and $k \ge 3$, $s_{-2} = -0.0502$, $s_{-1} = -0.0038$, $s_0 = 0.9694$, $s_1 = -0.0727$, $s_2 = -0.1608$. Clearly, the MMSE equalizer has not forced any of the output values to zero; however, no other three-tap equalizer can produce a smaller $\varepsilon^2_{\min}$. It should also be noted that output values outside the equalizer's span are not necessarily small, as is seen from the value of s_2.

We have introduced the concepts of ZF equalizers and MMSE equalizers, and we have shown that equalization consists primarily of solving a set of linear simultaneous equations. For automatic and adaptive equalization in data transmission systems, it has proven more advantageous to solve the necessary set of linear simultaneous equations using iterative methods. Generally, iterative methods consist of making incremental adjustments in the equalizer tap gains after each new pulse is received. These iterative methods are best motivated by reconsidering MMSE equalization and Eqs. (8.6.15) and (8.6.16).

The partial derivative in Eq. (8.6.16) is called the gradient of ε^2, and it obviously represents the rate of change of ε^2 with respect to the tap gains $\{c_m\}$. For a minimum of ε^2, we will have a zero rate of change or zero slope, and therefore we equate this rate of change to zero as in Eq. (8.6.16). If, however, we wish to equalize the channel over a long sequence of pulses and we want to do this one sample at a time, rather than wait until after the entire sequence is transmitted, we consider e_k^2 instead of ε^2. From Eq. (8.6.15) we have that

$$e_k^2 = \left[r_k - \sum_{n=-N}^{N} c_n y_{k-n} \right]^2. \tag{8.6.26}$$

We desire to adjust each of the $2N + 1$ equalizer coefficients at each time instant k to minimize e_k^2 in Eq. (8.6.26). To accomplish this goal, we adjust the tap gains by a small amount in a direction opposite to the slope or gradient of e_k^2 with respect to the $\{c_m\}$. Therefore, we are always adjusting the coefficients toward the minimum of e_k^2.

Taking the gradient (partial derivative) of e_k^2 with respect to each of the $\{c_m\}$, we are thus led to construct the iterative MMSE equalizer adjustment algorithm

$$\begin{aligned} c_m(k) &= c_m(k-1) + G\{r_{k-1} - s_{k-1}\}y_{k-m-1} \\ &= c_m(k-1) + G\left\{r_{k-1} - \sum_{n=-N}^{N} c_n(k-1)y_{k-n-1}\right\}y_{k-m-1} \end{aligned} \tag{8.6.27}$$

for $m = -N, \ldots, -1, 0, 1, \ldots, N$, where the time variation of the tap gains is indicated parenthetically and G is a gain constant to be selected. By incrementally changing the tap gains at each time instant k according to Eq. (8.6.27), we minimize e_k^2 in Eq. (8.6.26), and hence approximately solve the set of linear simultaneous equations represented by Eq. (8.6.17). A similar iterative approach can be used for zero-forcing equalization.

As presently written, the equalizer algorithm in Eq. (8.6.27) is applicable to automatic equalization during startup, but not adaptive equalization during data transmission. This is because of the presence of the known reference sequence $\{r_k\}$. Obviously, during actual data transmission, the desired equalizer output is not known. As a result, for adaptive equalization, $\{r_k\}$ is replaced by an estimated reference, say $\{\hat{r}_k\}$, which is derived from the equalizer output sequence $\{s_k\}$. This $\{\hat{r}_k\}$ is always available and equalization can proceed. Since $\hat{r}_k$ is computed from s_k, the equalizer adjustment is called *decision directed*. The reasoning behind using a version of the past equalizer outputs as a reference is that immediately after startup, the equalizer should be well adjusted and the distortion should be small. If the channel response is changing slowly with time, the derived reference should be accurate, and the equalizer can track these slow changes.

A further extension to the decision-directed concept is the use of decision feedback equalization (DFE). In DFE, not only are previously decided symbols used as reference symbols in the adaptive algorithms but they are fed back through a transversal filter and the resulting feedback transversal filter output is subtracted from the equalizer output. The idea here is that once output symbols are detected, their contribution to the equalizer output can be removed. Decision feedback equalizers are nonlinear filters and their development is left to the literature.

When a modulation method such as QAM is used, the equalizer may be implemented in the passband rather than at baseband, as has been discussed here. An advantage of passband equalization is that rapid phase changes can be tracked more easily, since equalizer delay is reduced.

For all of the equalizers discussed in this section, the tap spacing Δ is chosen to be one symbol interval. It is also possible to design fractionally spaced equalizers with tap spacing equal to some fraction of the symbol interval (usually, one-half). Such fractionally spaced equalizers are less sensitive to sampler phase, can accommodate more delay distortion, and can compensate for amplitude distortion with less noise than symbol-spaced equalizers.

The choice among the various equalizer algorithms depends on the requirements of the specific application, particularly desired convergence rate, initial distortion, desired minimum distortion, and complexity. There is a vast literature on equalization, but the original papers by Lucky [1965, 1966] and Lucky and Rudin [1967] and the book by Lucky, Salz, and Weldon [1968] are still useful. Hirsch and Wolf [1970] present a readable performance comparison of some adaptive equalization algorithms, including the ZF and MMSE algorithms. Pahlavan and Holsinger [1988] provide a historical overview of modem evolution, including equalization, and Qureshi [1985] gives an excellent development of adaptive equalization and digital receiver structures.

8.7 Data Scramblers

The performance of data transmission systems must be independent of the specific bit sequence being transmitted. If allowed to occur, repeated bit sequences can cause wide variations in the received power level as well as difficulties for adaptive equalization and clock recovery. Since all these problems are eliminated if the bit sequence is "random" (has no discernible pattern), many modems employ a *data scrambler* to produce a pseudorandom sequence for any given input bit sequence. The scrambler usually takes the form of a shift register with feedback connections, while the unscrambler is a feedforward-connected shift register. The following example clearly illustrates the operation of data scramblers.

EXAMPLE 8.7.1

A data scrambler and unscrambler are shown in Fig. 8.7.1. The scrambler operates in the following fashion. The initial shift register contents are arbitrary but prespecified and fixed to be the same in both the scrambler and unscrambler. The first bit in sequence s_1 (note that the subscript does not indicate a time sequence here but simply denotes bit sequence number 1) is summed modulo-2 with the modulo-2 sum of locations 2 and 5 in the shift register. This sum becomes the first bit in bit sequence s_2. As this bit is presented to the channel, the contents of the shift register are shifted up one stage as follows: $5 \to$ out, $4 \to 5$, $3 \to 4$, $2 \to 3$, $1 \to 2$. The first bit in s_2 is also placed in shift register stage 1. The next bit of sequence s_1 arrives, and the procedure is repeated.

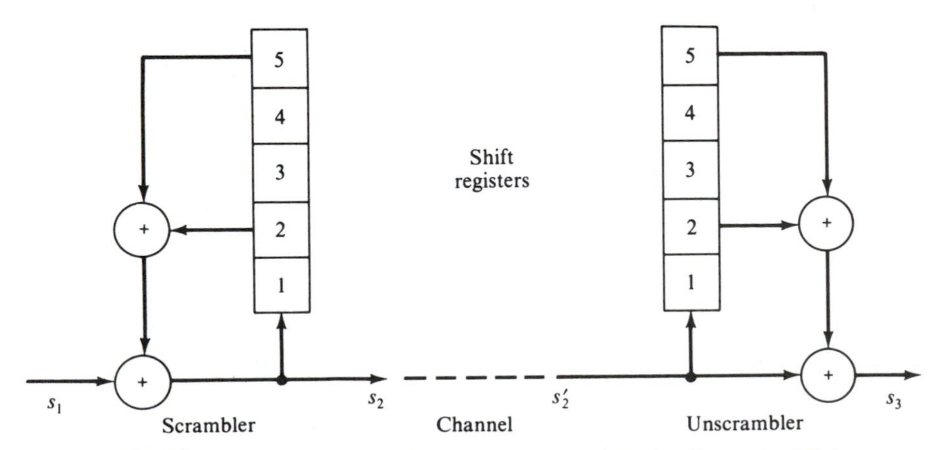

FIGURE 8.7.1 Data scrambler and unscrambler for Example 8.7.1.

TABLE 8.7.1 Scrambled Bit Sequence for Example 8.7.1

Time	1	2	3	4	5	6	7	8	9	10	11	12
s_1	1	0	0	0	0	0	0	0	0	0	0	0
1	0	1	0	1	0	1	1	1	0	1	1	0
2	0	0	1	0	1	0	1	1	1	0	1	1
3	0	0	0	1	0	1	0	1	1	1	0	1
4	0	0	0	0	1	0	1	0	1	1	1	0
5	0	0	0	0	0	1	0	1	0	1	1	1
s_2	1	0	1	0	1	1	1	0	1	1	0	0

The unscrambler operates as follows. The initial contents of the shift register are fixed. When the first bit of sequence s_2' arrives, this bit is summed mod-2 with the mod-2 sum of the initial values of stages 2 and 5. This sum then becomes the first bit of sequence s_3. At this time instant the contents of the shift register are shifted up one as follows: $5 \rightarrow$ out, $4 \rightarrow 5$, $3 \rightarrow 4$, $2 \rightarrow 3$, $1 \rightarrow 2$. The first bit of sequence s_2' is then put in stage 1, and the next bit in s_2' is presented to the unscrambler. The procedure is then repeated.

As an illustration, consider the input sequence s_1 and the initial shift register contents shown in Table 8.7.1 for the scrambler of Fig. 8.7.1. Following the procedure previously outlined for the scrambler, we can compute the scrambler output sequence and shift register contents as shown in the table. Note that in Table 8.7.1, the output bit at any time instant is computed from the current input bit and the current shift register contents. A shift then takes place and the shift register contents at the next time instant are generated. By inspection of s_2, we can see clearly that s_2 is very different from s_1. Whether s_2 has any special randomness properties is not immediately evident.

Assuming that no bit errors occur on the channel, we have $s_2' = s_2$, and Table 8.7.2 illustrates the operation of the unscrambler. We compute s_3 from the current input bit (s_2') and the current shift register contents. A shift is then performed to prepare the shift register for the next bit in s_2'. Note that since $s_1 = s_3$, the data are unscrambled.

TABLE 8.7.2 Unscrambled Bit Sequence for Example 8.7.1

Time	1	2	3	4	5	6	7	8	9	10	11	12
s_2'	1	0	1	0	1	1	1	0	1	1	0	0
1	0	1	0	1	0	1	1	1	0	1	1	0
2	0	0	1	0	1	0	1	1	1	0	1	1
3	0	0	0	1	0	1	0	1	1	1	0	1
4	0	0	0	0	1	0	1	0	1	1	1	0
5	0	0	0	0	0	1	0	1	0	1	1	1
s_3	1	0	0	0	0	0	0	0	0	0	0	0

We have not said anything as yet about how the scrambler and unscrambler circuits are selected. To begin this discussion, we introduce the unit delay operator D, which represents delaying the sequence by one bit. Thus, in this notation, Ds_2 represents the contents of stage 1 in the scrambler shift register, D^2s_2 represents stage 2, and so on. We can thus represent the scrambler circuit in Fig. 8.7.1 as

$$s_2 = D^2s_2 \oplus D^5s_2 \oplus s_1, \tag{8.7.1}$$

where the terms in D, D^3, and D^4 are absent due to the lack of feedback connections at these stages. Taking all terms in s_2 to the left side of the equality, we obtain

$$[1 \oplus D^2 \oplus D^5]s_2 = s_1, \tag{8.7.2}$$

or in the form of a transfer function relationship,

$$s_2 = \frac{1}{1 \oplus D^2 \oplus D^5} s_1. \tag{8.7.3}$$

By inspection of Eq. (8.7.3), we do not get an indication that a single bit produces a much longer sequence, as is evident in Table 8.7.1. To see this effect, we can perform synthetic division on Eq. (8.7.3) to reveal

$$s_2 = [1 \oplus D^2 \oplus D^4 \oplus D^5 \oplus D^6 \oplus D^8 \oplus D^9 \oplus \cdots]s_1. \tag{8.7.4}$$

Hence, from Eq. (8.7.4), we can generate the s_2 sequence from s_1 without the shift register stages.

A representation for the unscrambler can be written similarly as

$$s_3 = [1 \oplus D^2 \oplus D^5]s_2'. \tag{8.7.5}$$

Note that for no channel errors, $s_2' = s_2$, and if we substitute s_2 in Eq. (8.7.3) into Eq. (8.7.5), we find that $s_3 = s_1$.

This is a special case of a more general result that if we have representations for two circuits as

$$s_2 = F(D)s_1 \tag{8.7.6}$$

and

$$s_3 = G(D)s_2, \tag{8.7.7}$$

these two circuits can be used as a scrambler/unscrambler pair whenever

$$F(D)G(D) = 1. \tag{8.7.8}$$

Thus any pair of feedback- and feedforward-connected shift registers that satisfy Eq. (8.7.8) are suitable for use as a scrambler and an unscrambler pair. In Fig. 8.7.1 we chose the feedback connection as the scrambler and the feedforward device as the unscrambler. Equation (8.7.8) indicates that we could have used the feedforward connection for scrambling and the feedback connection for unscrambling.

A primary reason for the choice in Fig. 8.7.1 is bit error propagation. A single bit error into a feedforward connection affects a successive number of bits equal to the shift register length, while for a feedback connection, the effect can be much longer.

For any given shift register length M, there are obviously 2^M possible linear mod-2 sums that can be formed from its contents. All of these connections will not produce a "good" pseudorandom sequence. Furthermore, the output of any linear shift register connection with M stages is periodic with a period of $2^M - 1$ or less. An output sequence with period $2^M - 1$ is a special sequence and is called a *maximal-length linear shift register sequence* [Golomb, 1964]. For the linear feedback shift register connection serving as a scrambler in Fig. 8.7.1, a maximal-length sequence would have a period of $2^5 - 1 = 31$.

Further discussion of scramblers, unscramblers, error propagation, and maximal-length sequences is deferred to the problems and Appendix G.

8.8 Carrier Acquisition and Symbol Synchronization

Carrier acquisition and symbol synchronization or timing extraction are critical to the operation of high-performance modems. In this section we discuss briefly some of the many possible approaches for obtaining this information.

For noncoherent modems, such as those employing FSK, carrier acquisition is unnecessary. Further, we have seen that the very popular DPSK systems circumvent coherent reference difficulties by using the received carrier from the immediately preceding symbol interval. Important modulation methods such as VSB and QAM do require a coherent reference to be acquired in some fashion.

Early modems that employed VSB modulation transmitted a carrier tone in phase quadrature or transmitted pilot tones at the edges of the data spectrum. For the latter method, the received signal can be represented as [Lucky, Salz, and Weldon, 1968]

$$s(t) = m(t)\cos[\omega_c t + \Delta\omega t + \theta(t) + \phi] + \hat{m}(t)\sin[\omega_c t + \Delta\omega t + \theta(t) + \phi] + \alpha\cos[\omega_L t + \Delta\omega t + \theta(t)] + \alpha\cos[\omega_H t + \Delta\omega t + \theta(t)], \tag{8.8.1}$$

where $\Delta\omega$ denotes frequency distortion, $\theta(t)$ denotes phase jitter, and ω_L and ω_H denote the lower and upper pilot tones. The pilot tone components in Eq. (8.8.1) are removed by narrowband filters, multiplied together, and low-pass filtered to yield a cosine term of frequency $\omega_H - \omega_L$. This term is then fed to a phase-locked loop. During the startup time a carrier tone is transmitted and fixed phase differences between the $\cos[\omega_H - \omega_L]$ term and the carrier terms are adjusted out. To complete the carrier extraction, the frequency $\omega_H - \omega_L$ is divided by an integer N and multiplied by the upper pilot tone. The difference frequency component is retained so that

$$\omega_c = \frac{(N-1)\omega_H + \omega_L}{N}. \tag{8.8.2}$$

For more details on this technique, the reader is referred to Lucky, Salz, and Weldon [1968, pp. 184–186] and Holzman and Lawless [1970].

Today, modems rely heavily on suppressed carrier modulation methods such as AMVSB-SC, QAM, and PSK, which have no separate carrier component available. Of course, these modulation methods require a coherent reference for successful operation, and hence the carrier must be regenerated from the received data-carrying signal. There are a variety of techniques for accomplishing this, depending somewhat on the type of modulation being employed. Here we briefly discuss three of these techniques: the squaring loop, the Costas loop, and the Mth power method.

Consider a noise-free, generic AMDSB-SC signal given by

$$s(t) = m(t) \cos [\omega_c t + \phi], \tag{8.8.3}$$

from which we wish to extract a coherent reference signal. For the squaring loop shown in Fig. 8.8.1, $s(t)$ is passed through a square-law device that yields

$$s^2(t) = m^2(t)\{\tfrac{1}{2} + \tfrac{1}{2} \cos [2(\omega_c t + \phi)]\}. \tag{8.8.4}$$

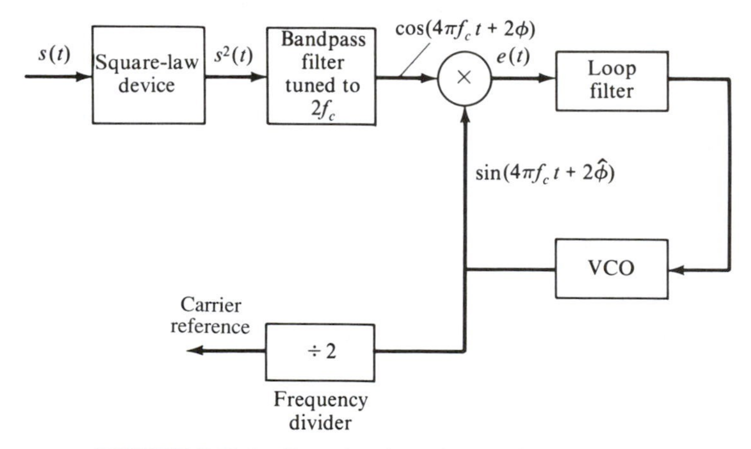

FIGURE 8.8.1 Squaring loop for carrier recovery.

The purpose of the squaring operation is to reduce or eliminate the effects of the modulating signal. To understand this, note that if $m(t) = \pm 1$, then $m^2(t) = 1$ for all data sequences, while if $m(t)$ is multilevel, squaring eliminates phase changes and the remaining amplitude variations can be removed by clipping. Only the double frequency term is retained after bandpass filtering, and this component is fed to a tracking loop (PLL) that locks on to the input frequency and phase. The VCO output is thus a sinusoid at twice the frequency and phase of the received signal, but after frequency division by a factor of 2, an accurate coherent reference is recovered.

The Costas loop in Fig. 8.8.2 differs from the squaring loop in how it eliminates the modulating signal effects and how it generates the input to the tracking

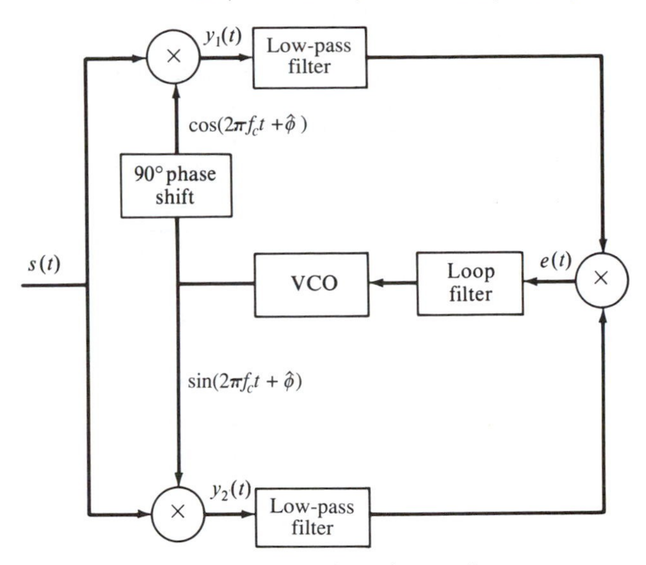

FIGURE 8.8.2 Costas loop for carrier recovery.

loop. If the loop is assumed to be locked on in frequency but with an inaccurate phase $\hat{\phi}$ as shown in Fig. 8.8.2, then for $s(t)$ in Eq. (8.8.3),

$$y_1(t) = \frac{m(t)}{2} \cos (\phi - \hat{\phi}) + \frac{m(t)}{2} \cos (2\omega_c t + \phi + \hat{\phi}) \tag{8.8.5}$$

and

$$y_2(t) = \frac{m(t)}{2} \sin (\hat{\phi} - \phi) + \frac{m(t)}{2} \sin (2\omega_c t + \phi + \hat{\phi}). \tag{8.8.6}$$

Therefore, after filtering and multiplication, we have

$$e(t) = \frac{m^2(t)}{8} \sin [2(\hat{\phi} - \phi)], \tag{8.8.7}$$

which is then passed to the loop filter and VCO. Note that when $\hat{\phi} = \phi$, the output of the low-pass filter in the upper branch of Fig. 8.8.2 is the message or data sequence.

An M-phase (level) PSK waveform can be expressed as

$$s(t) = A_c \cos \left[\omega_c t + \frac{2\pi i}{M} + \phi \right], \tag{8.8.8}$$

$i = 1, 2, \ldots, M$. We need to obtain from $s(t)$ a coherent reference signal of the form $\cos [\omega_c t + \phi]$. The Mth power method, which is similar to the squaring loop, is a technique for doing this. A block diagram of the Mth power method is shown in Fig. 8.8.3. If we pass $s(t)$ in Eq. (8.8.3) through the Mth power device

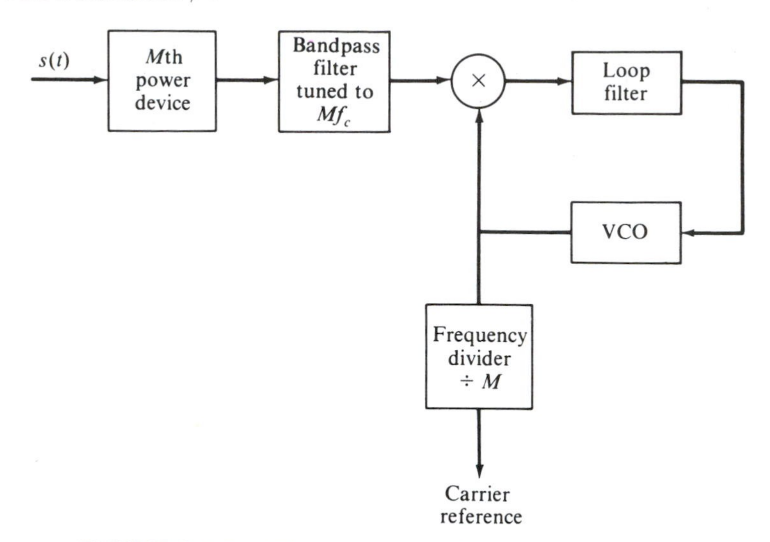

FIGURE 8.8.3 *M*th power method for carrier recovery.

in Fig. 8.8.3, we obtain for M even

$$s^M(t) = A_c^M \left\{ \frac{1}{2^M} \binom{M}{M/2} + \frac{1}{2^{M-1}} \left[\cos M\left(\omega_c t + \frac{2\pi i}{M} + \phi\right) + \binom{M}{1} \cos\left\{(M-2)\left(\omega_c t + \frac{2\pi i}{M} + \phi\right)\right\} + \cdots + \binom{M}{M/2 - 1} \cos 2\left(\omega_c t + \frac{2\pi i}{M} + \phi\right)\right]\right\}. \tag{8.8.9}$$

The bandpass filter tuned to $M\omega_c$ removes all terms except the one containing $\cos M(\omega_c t + 2\pi i/M + \phi) = \cos(M\omega_c t + M\phi)$. The frequency division by M gives the desired coherent reference signal. Additional details on carrier recovery are left to the literature [Franks, 1980; Bhargava et al., 1981; Proakis, 1989].

Symbol synchronization or timing extraction is usually accomplished by operations on the transmitted data sequence. One popular method relies on the threshold crossings of the equalized data sequence and on a highly stable crystal oscillator [Holzman and Lawless, 1970]. The crystal oscillator counts at a rate of about two orders of magnitude above the baud rate, and it effectively counts down from one sampling time to the next. The threshold crossings of the equalized baseband data sequence are used to center the sampling time on the open eye. This is accomplished by what is called an "early–late" decision in every pulse interval. If a transition occurs during the first half of the pulse (symbol) interval, it is desired to delay the sampling time so that it will be centered on the open eye. To accomplish this, we can require an extra "count" or we can effectively delete one of the previous counting pulses, so that this pulse must be counted again. The latter method is used in practice. If a threshold

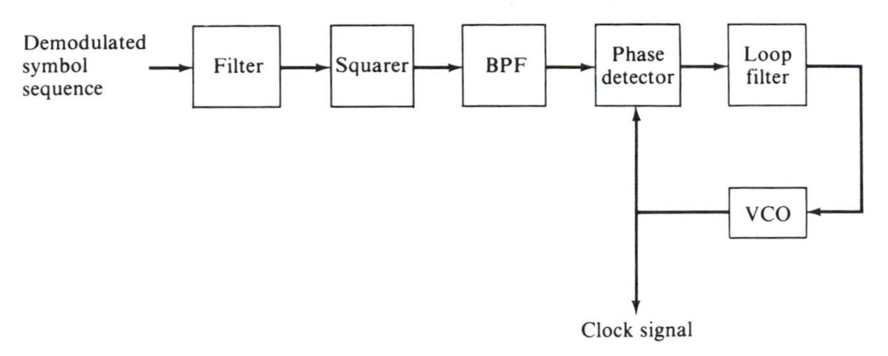

FIGURE 8.8.4 General timing recovery method.

crossing occurs in the last half of the pulse interval, we wish to sample sooner. Hence we add a counting pulse, so that the countdown will be completed sooner. This technique continually centers the sampling instant on the open eye by adjusting the countdown chain in fixed increments equal to the reciprocal of the crystal oscillator counting rate. Accuracies of less than 1% of the baud rate are accomplished easily.

A very common method for symbol synchronization or clock recovery has much the same structure as the squaring loop for carrier recovery in Fig. 8.8.1. A block diagram of a general timing recovery loop is shown in Fig. 8.8.4. The demodulated symbol sequence is passed through a prefilter and then to a squarer or square-law device. As before, the squarer eliminates or reduces the effects of the particular data sequence. The squarer output is then sent to a BPF tuned to the symbol rate. The tracking loop extracts the clock signal from the BPF output. A common example of this approach is shown in block diagram form in Fig. 8.8.5. This latter clock recovery method is similar to that used in regenerative repeaters for pulse code modulation (see Section 9.7). Additional details are left to the references [Franks, 1980; Bhargava et al., 1981; Proakis, 1989].

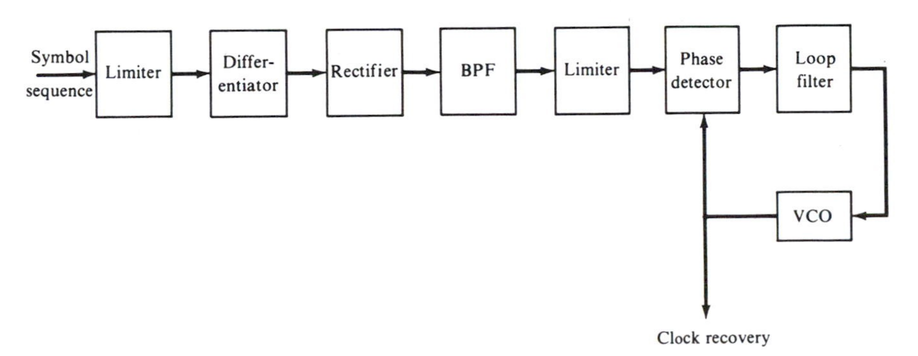

FIGURE 8.8.5 Example clock recovery block diagram.

8.9 Modems

Preceding sections of this chapter develop various aspects of data transmission over voiceband telephone channels. Baseband data signals and voice signals differ substantially in character, and thus devices generically called *modems* (*mod*ulator/*dem*odulator), or previously designated as *data sets* by Bell, are required to prepare data signals for transmission over telephone channels. There is a mind-boggling variety of modem types and configurations available, but in this section we discuss briefly the salient characteristics of the most familiar ones.

For operation at rates of 0 to 300 bits/sec, the type 103 modem is prevalent. This modem transmits data using FSK with a 1070-Hz tone for a “0” (space) and a 1270-Hz tone for a “1” (mark) in one direction, and a 2025-Hz tone for a 0 and a 2225-Hz tone for a 1 in the reverse direction. Since data can be sent in both directions between two modems simultaneously, this modem is said to be *full duplex* or to be capable of operating in the *full-duplex mode*. This is opposed to a modem that can transmit in only one direction at a time, which is called *half-duplex* operation. The type 103 transmits two levels per symbol; hence its symbol rate is 300 baud. These modems are also *asynchronous* in that there is no synchronization between the transmitter and receiver. To let the receiver know that information is being transmitted, a “start” bit is sent first followed by the data bits and a parity bit, and then finally two “stop” bits. This group of bits is sometimes called a *character*, and the bits within this group have a specific timing relationship. However, characters have no fixed synchronism or timing. This asynchronous nature of the 103-type modems make them ideally suited for sending and receiving data to/from teletypes and other character-based terminals.

An important synchronous modem is the 201 type, which can operate in the half-duplex or full-duplex mode and uses four-phase DPSK. As discussed previously, the data sequence to be transmitted is grouped in pairs of bits, called *dibits*, and the carrier phase is shifted by 45° (11), 135° (10), 225° (00), or 315° (01) from the phase in the immediately preceding symbol interval as is appropriate. Since the baud rate of the 201 type is 1200 symbols/sec, the transmitted data rate is 2400 bits/sec. At this low baud rate, only a fixed equalizer is required.

The type 208 modem is a close cousin of type 201 modems. The type 208 groups bits into triples called *tribits* and transmits these 3 bits/symbol using DPSK at a rate of 1600 baud to achieve a data rate of 4800 bits/sec. The differential phases and their associated tribits are 22.5° (001), 67.5° (000), 112.5° (010), 157.5° (011), 202.5° (111), 247.5° (110), 292.5° (100), and 337.5° (101). There is a fixed nominal equalizer, which is adjusted upon installation, at the transmitter, and there is an adaptive passband equalizer at the receiver. There are separate tap adjustments for the in-phase and quadrature components.

One of the early 9600-bits/sec modems was the 203 data set. This modem used eight-level VSB modulation at 3200 baud, 100% roll-off raised cosine pulse shaping, 13- or 17-tap automatic and adaptive equalization, 23-stage shift registers for scrambling and unscrambling, and transmitted pilot tones for carrier frequency recovery. The type 209 data set followed the 203 and achieves a data rate of 9600 bits/sec by using 16-level QAM at 2400 baud.

An important modem design from an independent manufacturer uses a combination of amplitude and phase modulation with two amplitudes per phase for the phases 0°, 90°, 180°, and 270°, and two different amplitudes per phase for the phases 45°, 135°, 225°, and 315°. Thus there are 16 levels, so at 2400 baud a data rate of 9600 bits/sec is attained. This modem uses 31-tap equalization of both the in-phase and quadrature components. Like the 209 data set, no pilot tones are required for carrier recovery.

Table 8.9.1 lists several milestones in the evolution of voiceband data modems. Generally, the modems listed are the first commercially feasible modems at the stated data rate. Increases in transmitted bit rate over the years have come from expanding the bandwidth utilized and by increasing the number of bits transmitted per hertz. These translate into sending more symbols/sec and more bits/symbol, respectively. Improvements in the voiceband channel have allowed both the increased bandwidth utilization and the signal set expansion. In Chapter 10 we develop the ideas of signal sets or constellations beyond the discussions of PSK and QAM encountered thus far.

TABLE 8.9.1 Modem Evolution

Year	Model	Speed (bits/sec)	Bandwidth (Hz)	Modulation	R/W [(bits/sec)/Hz]
1962	Bell 201	2,400	1200	4-PSK	2[a]
1967	Milgo 4400/48	4,800	1600	8-PSK	3[a]
1971	Codex 9600C	9,600	2400	16-QAM	4[a]
1980	Paradyne	14,400	2400	64-QAM	6
1981	Codex/ESE	14,400	2400	64-QAM	6
1984	Codex 2600	14,400	2400	Trellis-coded modulation (TCM)	6[a]
1985	Codex 2680	19,200	2743	Trellis-coded modulation (TCM)	7

Source: G. David Forney, Jr., "Introduction to Modem Technology: Theory and Practice of Bandwidth Efficient Modulation from Shannon and Nyquist to Date," University Video Communications and the IEEE, 1989. Copyright: Motorola.

[a] Standard.

The more recent modems use what is called trellis-coded modulation (TCM) to increase still further the number of bits transmitted per cycle of bandwidth used. TCM is investigated in Chapter 13. The modems that are marked with ([a]) have been accepted as industry standards. For more information on the history of modems, the reader is referred to Pahlavan and Holsinger [1988]. After the reader has digested Chapter 10 and has at least read Section 13.4 on TCM, the videotape by G. David Forney, Jr. [1989] is recommended viewing.

SUMMARY

The basic principles of transmitting data over bandlimited channels have been presented in this chapter. The emphasis has necessarily been on the telephone channel due to the continuing important role of the telephone network in providing data transmission facilities to a wide range of users. As is evident from this chapter, many diverse techniques are required to transmit data reliably over analog voice channels. Today, there is also a significant all-digital, or mostly digital, component in the worldwide telephone network. This all-digital portion of the network is introduced in Chapter 9. The performance analyses of the various techniques discussed in this chapter are included in Chapter 10.

PROBLEMS

8.1 Given a triangular time-domain pulse defined over $[-T/2, T/2]$ with maximum amplitude V, we know its Fourier transform from tables.

(a) Sketch this Fourier transform pair.

(b) Use the symmetry property of Fourier transforms to obtain a time-domain pulse with a triangular spectral content. Sketch this Fourier transform pair.

(c) If we wish to transmit a sequence of pulses occurring every τ seconds using the time-domain pulse shape of part (b), what is the required bandwidth for zero intersymbol interference at sampling times? Compare your result to 100% roll-off raised cosine shaping.

8.2 Given the Fourier transform pair

$$f(t) = \begin{cases} V \cos \dfrac{\pi t}{T}, & |t| \leq \dfrac{T}{2} \\ 0, & \text{otherwise,} \end{cases}$$

$$F(\omega) = \mathscr{F}\{f(t)\} = 2\pi V T \frac{\cos(\omega T/2)}{\pi^2 - \omega^2 T^2},$$

repeat Problem 8.1 for this function and its Fourier transform.

8.3 The input to a receiver is a sequence of positive and negative pulses represented by

$$r(t) = \sum_{n=-N}^{N} b_n p(t - n\tau),$$

where the b_ns are ± 1 and $p(t)$ is given by Eq. (8.2.3). To recover b_0, we wish to sample $y(t)$ at $t = 0$. Assume, instead, that due to timing jitter, we are in error by a small positive amount, say ε, so

$$r(\varepsilon) = \sum_{n=-N}^{N} b_n p(\varepsilon - n\tau).$$

Assume that $\varepsilon < \tau$ and show that as $N \to \infty$ there is a sequence for which $r(\varepsilon)$ diverges.

8.4 Verify that $100\alpha\%$ raised cosine pulse shaping satisfies Eq. (8.2.5). Use any of the three available expressions for $P(\omega)$: Eq. (8.2.7), (8.2.9), or (8.2.10).

8.5 Verify that $100\alpha\%$ raised cosine pulse shaping satisfies Eq. (8.2.6). Use any of the three available expressions for $P(\omega)$: Eq. (8.2.7), (8.2.9), or (8.2.10).

8.6 Rewrite each of the raised cosine spectral shaping expressions in Eqs. (8.2.7), (8.2.9), and (8.2.10) in terms of the transmitted data rate R, thereby obtaining explicit relationships between the data rate and required bandwidth.

8.7 Using a unit amplitude, 50% roll-off raised cosine pulse to indicate a 1 and no pulse to indicate a 0, sketch the pulse sequence corresponding to the binary message sequence 11001.

8.8 Starting with Eq. (8.2.10), demonstrate the equivalence of Eqs. (8.2.7), (8.2.9), and (8.2.10).

8.9 Given a fixed available bandwidth of 4000 Hz:

(a) What is the maximum possible transmitted symbol rate using 100% roll-off raised cosine pulses?

(b) What is the maximum possible transmitted symbol rate using 50% roll-off raised cosine pulses? Compare parts (a) and (b).

8.10 If we generate $100\alpha\%$ roll-off raised cosine pulses starting with rectangular pulses rather than impulses, to maintain the raised cosine shaping of the transmitted pulse, we must modify $P(\omega)$ to account for the rectangular pulse spectral content.

(a) If unit amplitude rectangular pulses of width Δ are to be used to generate $100\alpha\%$ roll-off raised cosine pulses, what should be the transfer function of the shaping filter?

(b) It is desired to generate full cosine roll-off ($\alpha = 1$) raised cosine pulses from rectangular pulses one-tenth of the symbol interval in width. If the symbol rate is 2400 symbols/sec, what is the required shaping filter frequency response?

8.11 A pulse-shaping function that satisfies Nyquist's third criterion is

$$P(\omega) = \begin{cases} \dfrac{\omega\tau/2}{\sin(\omega\tau/2)}, & 0 \le |\omega| \le \dfrac{\pi}{\tau} \\ 0, & |\omega| > \dfrac{\pi}{\tau}, \end{cases}$$

which has the impulse response

$$p(t) = \frac{1}{\pi} \int_0^{\pi/\tau} \frac{\omega\tau/2}{\sin(\omega\tau/2)} \cos \omega t \, d\omega.$$

Show that

$$\int_{(2n-1)\tau/2}^{(2n+1)\tau/2} p(t)\, dt = \begin{cases} 1, & n = 0 \\ 0, & n \neq 0, \end{cases}$$

which demonstrates that $p(t)$ has zero area for every symbol interval other than the one allocated to it, and further, that the response within this interval is directly proportional to the input impulse weight.

8.12 From Fig. 8.3.2, show that the duobinary impulse response can be written as

$$p(t) = \frac{\sin(\pi t/\tau)}{\pi t/\tau} + \frac{\sin[\pi(t-\tau)/\tau]}{[\pi(t-\tau)/\tau]}.$$

Sketch this $p(t)$, thereby verifying the impulse response waveform in Fig. 8.3.1.

8.13 For the $\{a_k\}$ sequence in Table 8.3.1, define a different $\{q_k\}$ sequence by the mapping

a_k		q_k
0	$\Rightarrow$	-1
1	$\Rightarrow$	1

Use Eq. (8.3.3) to represent this new $\{q_k\}$ sequence without precoding and develop a decoding rule. By assuming that s_3 is erroneously received, demonstrate that errors propagate in this scheme.

8.14 Sketch $s(t)$ for Table 8.3.1 and verify the values listed for the sequence $\{s_k\}$.

8.15 It is desired to transmit the binary sequence $\{a_k\}$ given by 01011010001 using duobinary signaling with precoding. Letting $b_0 = 1$, create an encoding table analogous to Table 8.3.1, listing the sequences $\{b_k\}$, $\{q_k\}$, and $\{s_k\}$. Show that an error in s_4 generates only a single decoding error.

8.16 For the input data sequence $\{a_k\}$ given by 01011010001, use class 4 partial response signaling with precoding and create an encoding table analogous to Table 8.3.2. Let $b_0 = 0$ and $b_{-1} = 1$. Demonstrate that if s_3 is received incorrectly, no error propagation occurs.

8.17 Using Eq. (8.3.9), sketch $s(t)$ corresponding to Table 8.3.2.

8.18 Use Fig. 8.3.4 and Eqs. (8.3.15)–(8.3.17) to verify Eqs. (8.3.1), (8.3.3), (8.3.4), and (8.3.6) for duobinary and Eqs. (8.3.7), (8.3.11), (8.3.13), and (8.3.14) for class 4 partial response.

8.19 Let $C_0 = -1$, $C_1 = 0$, $C_2 = 2$, $C_3 = 0$, and $C_4 = -1$ (all other C's zero) in Fig. 8.3.4, and show that

$$|P(\omega)| = 4\tau \sin^2 \omega\tau, \qquad |\omega| \leq \frac{\pi}{\tau}$$

with $|P(\omega)| = 0$ otherwise. Sketch the corresponding impulse response of this shaping filter. This spectral shaping has been denoted class 5 partial response by Kretzmer [1966].

8.20 Double-sideband-suppressed carrier amplitude modulation (AMDSB-SC) is to be used to transmit data over a bandpass channel between 200 and 3200 Hz.

(a) If the signaling rate (symbol rate) is 2000 symbols/sec, what is the maximum raised cosine spectral shaping roll-off that can be used?

(b) If 60% roll-off raised cosine pulses are used, what is the maximum possible signaling rate?

8.21 It is desired to transmit data at 9600 bits/sec over a bandpass channel from 200 to 3200 Hz. If 50% roll-off raised cosine pulses are to be used, compare the symbol (baud) rate and the number of bits/symbol required by AMDSB-SC and AMSSB-SC modulation methods. Discuss the weaknesses of each system.

8.22 Only the range of frequencies from 500 to 3000 Hz are usable on a particular telephone channel. If 50% roll-off raised cosine pulse shaping is used and two levels per symbol are transmitted, calculate the maximum bit rate possible using:

(a) AMVSB-SC with 1.2 times the baseband bandwidth.

(b) Quadrature amplitude modulation.

8.23 Sketch the eye pattern for a data transmission system that sends eight levels in the amplitude range $[-V, V]$ using 100% roll-off raised cosine pulse shaping. Assume that the transmitted levels are $\pm V$, $\pm 5V/7$, $\pm 3V/7$, $\pm 1V/7$, and sketch at least two pulse intervals and three sampling instants.

8.24 Sketch a three-level eye pattern for the pulses with quadratic delay distortion shown in Fig. 8.4.2. Use the pulses with the maximum distortion shown.

8.25 The input to a three-tap equalizer is $y_{-1} = 0.1$, $y_0 = 1.2$, $y_1 = -0.2$, all other $y_i = 0$. The desired equalizer output is $s_0 = 1$, all other $s_i = 0$.

(a) Choose the equalizer tap gains to produce $s_{-1} = 0$, $s_0 = 1$, $s_1 = 0$.

(b) Choose the tap gains to minimize the MSE between the equalizer output and the desired values.

(c) Find the output sequences for the equalizers in parts (a) and (b).

(d) Calculate the MSE for parts (a) and (b). Compare the two values and discuss your results.

8.26 The input to a three-tap equalizer is $y_{-1} = \frac{1}{4}$, $y_0 = 1$, $y_1 = \frac{1}{2}$, all other $y_i = 0$. If the equalizer tap gains are $c_{-1} = -\frac{1}{3}$, $c_0 = 1$, and $c_1 = -\frac{2}{3}$, use polynomial multiplication to find the output sequence $\{s_k\}$ for all k.

8.27 The input to a three-tap equalizer is $y_{-1} = -\frac{1}{4}$, $y_0 = 1$, $y_1 = \frac{1}{4}$, $y_i = 0$ for all other i. The desired output sequence is $s_0 = 1$, $s_i = 0$ for all other i. Find the equalizer tap gains to minimize the mean-squared error between the output and the desired values.

8.28 For the input sequence s_1 and initial shift register contents in Table 8.7.1, use the feedforward device as the scrambler and the feedback connection as the unscrambler.

8.29 In Table 8.7.2, change the third bit in the s_2' sequence and compute s_3. To illustrate the differences in error propagation, flip the third bit in the unscrambler input in Problem 8.28. Compare the results.

8.30 For the scrambler/unscrambler shown in Fig. P8.30, compute the scrambler output s_2 and the unscrambler output s_3 for the input sequence $s_1 =$ 011010011100 and all zero initial shift register contents. Assume that there are no channel errors.

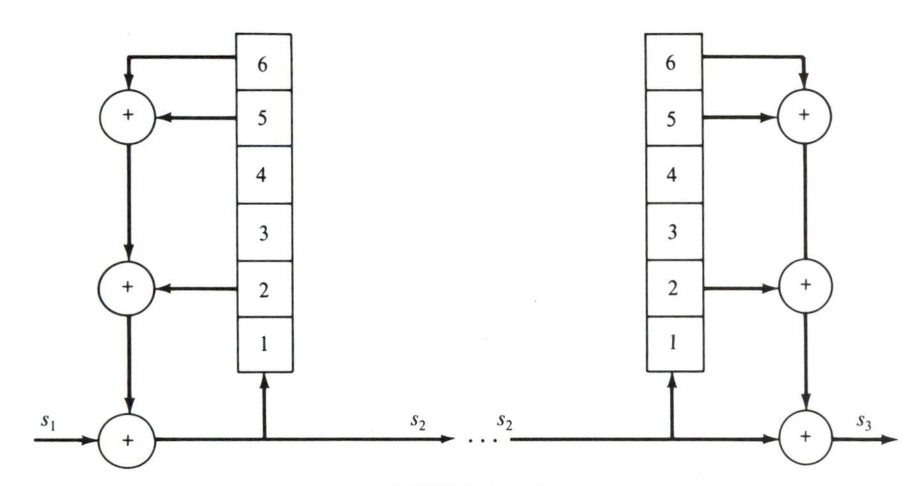

FIGURE P8.30

8.31 Consider a linear, feedback connection, four-stage shift register. Assume that the initial shift register contents are all zero and that the input sequence is a 1 at the first time instant followed by all zeros. Find the output sequence corresponding to each possible feedback connection. Specify the connection that yields a maximal-length sequence.

CHAPTER 9

Pulse Code Modulation

9.1 Introduction

Pulse code modulation (PCM) is a modulation technique unlike the analog modulation methods such as AM, FM, and PM discussed previously. PCM is designed to transmit analog message signals in digital form, as illustrated in Fig. 9.1.1. To accomplish this, a PCM system prefilters the analog message, uses the results of the sampling theorem to discretize in time, discretizes the amplitudes of the previously obtained discrete time pulses (quantization), and assigns a (usually) binary word to each allowable discrete amplitude level. The resulting binary words are then made suitable for transmission over the particular channel of interest by assigning particular pulse shapes and pulse sequences to the binary words (sometimes called line coding). The resulting pulses are then sent directly over the channel rather than passed through an analog modulator as was done in Chapter 8.

Our primary interest in this chapter is in transmitting speech signals using PCM. Now, since we know that present analog telephone channels bandlimit the speech to less than 4 kHz, we know from the sampling theorem that the discretizing in time operation (sampling) will require about 8000 samples/sec to represent the analog speech signal accurately. If we then quantize these samples to one of $2^{12} = 4096$ possible levels, which requires a 12-bit word to represent the chosen level, we see that the single analog voice channel necessitates that we transmit 8000 samples/sec $\times$ 12 bits/sample = 96,000 bits/sec for every

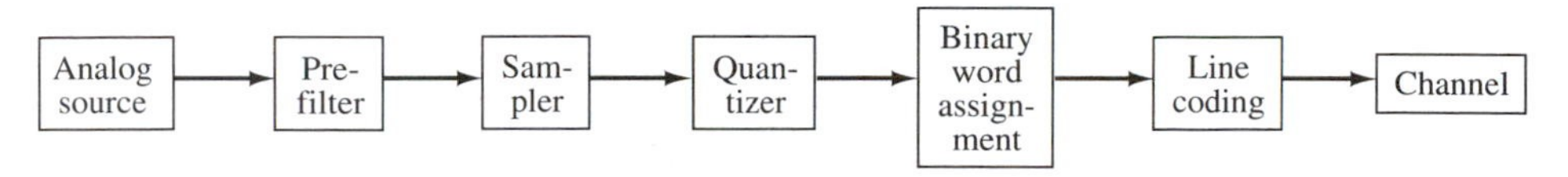

FIGURE 9.1.1 Source representation using PCM.

voice channel. Since required bandwidth is roughly proportional to bit rate, we see that we have expanded the bandwidth of the original 4-kHz voice channel enormously.

Since PCM is widely used, we suspect that there are some advantages to be gleaned from this exorbitant expansion of bandwidth, and indeed there are. We list a few of the main advantages here, not necessarily in their order of importance. First, although quantization introduces irreversible distortion into the signal, we know thereafter that only certain fixed discrete amplitude levels are possible. As a result, we can periodically "clean up" and regenerate the pulses as they become distorted during propagation down the transmission line but before the distortion is sufficiently bad to cause an error. This regeneration of pulses is accomplished by inserting devices called *repeaters* at suitable intervals along the transmission line. A second advantage of PCM is that an analog signal such as speech is changed into a digital signal, which can be handled in communications networks much like data. This is a particular advantage, since data are a constantly increasing percentage of the traffic carrier over the telephone network. A third advantage is that digital switching of signals is faster, cheaper, and generates less noise than analog switching. Another advantage is that digital signals are easily encrypted or scrambled to discourage eavesdropping, and hence provide greater security than present analog methods. A final advantage is that digital circuitry is small, inexpensive, has low power requirements, and seems to become increasingly more so daily.

The preceding advantages of PCM, and others, as well as its applications, will become evident as the chapter unfolds. Since filtering and the sampling theorem are discussed in earlier chapters, we begin our discussion with quantization.

9.2 Uniform Quantization

Quantization is the process of converting a continuous-amplitude signal into one of a finite number of discrete amplitudes. As in Fig. 9.1.1, we assume that time discretization has already occurred via the time sampling operation. An input/output characteristic for a typical uniform quantizer is shown in Fig. 9.2.1. The values -3Δ, -2Δ, $-\Delta$, 0, Δ, 2Δ, and 3Δ marked along the abscissa are called *step points*, and the other two values, -4Δ and $+4\Delta$, are called *overload points*. The values marked along the ordinate, $-7\Delta/2, \ldots, -\Delta/2, \Delta/2, \ldots, 7\Delta/2$, are called *output levels*. The quantizer in Fig. 9.2.1 is said to be *uniform* since the distance between adjacent step points or output levels is the same value Δ, called the *step size*. The quantizer is called a *midriser*, since it does not have a zero level. A quantizer with a zero level is called a *midtread* quantizer. Therefore, since the characteristic in Fig. 9.2.1 has eight output levels and is symmetric, it represents a uniform, symmetric, midriser, eight-level quantizer with step size Δ.

As indicated by the characteristic in Fig. 9.2.1, the input sample amplitude s can take on a continuum of values from $-\infty$ to $+\infty$, but there are only eight

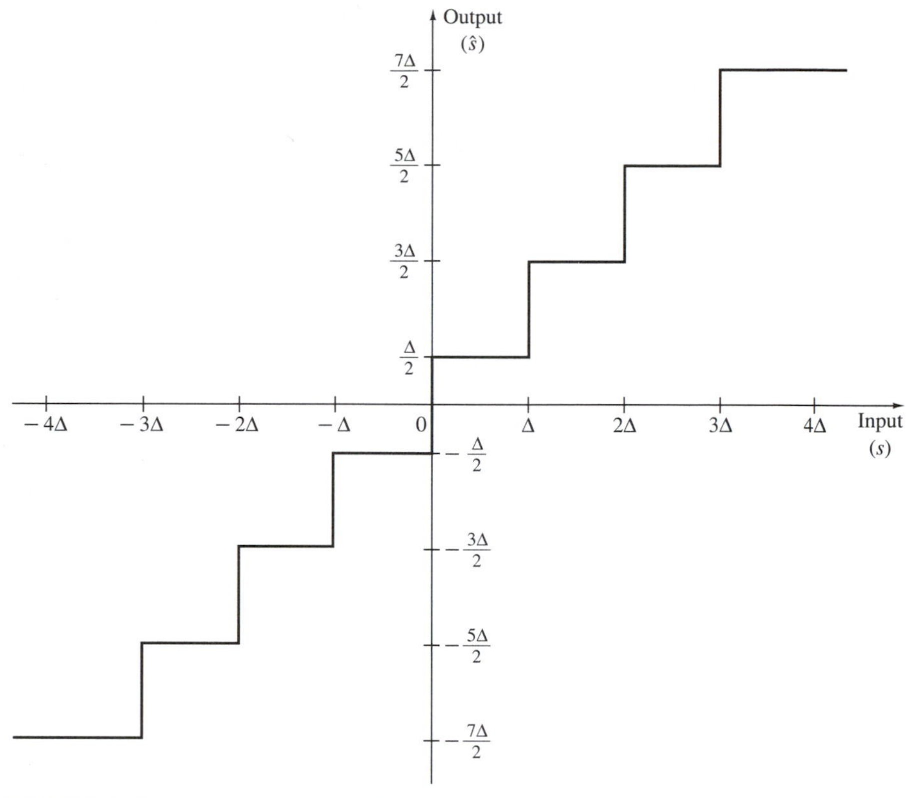

FIGURE 9.2.1 Uniform symmetric midriser eight-level quantizer with step size Δ.

possible values for the output $\hat{s}$. Thus, if an input sample s has an amplitude such that $2\Delta \leq s < 3\Delta$, then $\hat{s} = 5\Delta/2$. As long as the input amplitude s is between -4Δ and 4Δ, the maximum (positive or negative) quantization error is $\pm\Delta/2$. If $s > 4\Delta$ or $s < -4\Delta$, which is called the *overload region*, the quantization error magnitude exceeds $\Delta/2$. Generally, this is an undesirable occurrence and the quantizer is designed, if possible, to avoid overload.

A generic L-level (L even) uniform quantizer for a deterministic input signal with a peak-to-peak amplitude swing of V volts (i.e., $V = s_{\max} - s_{\min}$) can be designed as follows. If we assume that the input signal is symmetric about zero, the overload points are chosen to be $-V/2$ and $V/2$ and the number of output levels in this range is L, so the step size Δ is given by

$$\Delta = \frac{V}{L}. \tag{9.2.1}$$

Thus, in addition to the overload points at $\pm V/2$, we have the step points $\pm\Delta$, $\pm 2\Delta, \ldots, \pm(L/2 - 1)\Delta$, and the output levels $\pm\Delta/2, \pm 3\Delta/2, \ldots, \pm(L-1)\Delta/2$, with Δ given by Eq. (9.2.1).

If we define *quantization error*, sometimes called the *quantization noise*, by

$$q = \hat{s} - s, \tag{9.2.2}$$

we see that q can be plotted as in Fig. 9.2.2 for the eight-level quantizer in Fig. 9.2.1. Again, note that for $-V/2 \le s \le V/2$, $|q| \le \Delta/2$ and we have what is called *granular distortion* or *granular noise*, while for $|s| > V/2$, $|q| > \Delta/2$, and we have *overload distortion*. It should also be noted that although the input signal varies over the range $V = L\Delta$, the quantizer output amplitudes cover only $(L - 1)\Delta = V - \Delta$. Therefore, quantization introduces into the signal an unremovable error, the quantization noise. Of course, this fact is also evident from Fig. 9.2.1, which clearly indicates that quantization is a many-to-one, and hence noninvertible, mapping.

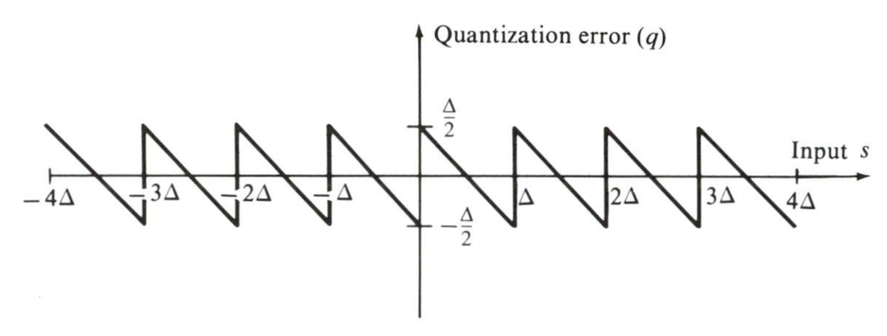

FIGURE 9.2.2 Quantization error for a uniform quantizer.

EXAMPLE 9.2.1

First, we wish to design a 16-level uniform quantizer for an input signal with a dynamic range of ± 10 V. Then we find the quantizer output value and quantization error for an input signal of amplitude 1.2 V.

Solution: To design a uniform, midriser quantizer of the form in Fig. 9.2.1, we need only choose the overload points and the quantizer step size Δ. For no overload, we select the overload point 8Δ to be equal to 10. Therefore, the step size of the uniform quantizer is $\Delta = 1.25$, which agrees with Eq. (9.2.1). Now, an input sample of 1.2 V falls in the range 0 to $\Delta = 1.25$ V, so that the quantizer output is $\Delta/2 = 0.625$ V. From Eq. (9.2.2), the quantization error q is $0.625 - 1.2 = -0.575$ V.

To evaluate and to compare quantizer performance, it is necessary to define an objective distortion measure. One commonly used performance measure is the quantizer signal-to-noise ratio (SQNR) defined as

$$\text{SQNR} = 10 \log_{10} \frac{\sigma^2}{D}, \tag{9.2.3}$$

where σ^2 is the variance (mean-squared value for zero mean) of the input s and D is the mean-squared quantization error. For deterministic input signals, it is often assumed that the quantization error (q) is uniformly distributed in $[-\Delta/2, \Delta/2]$ and that there is no overload distortion. Under these assumptions,

$$D = E\{q^2\} = \int_{-\Delta/2}^{\Delta/2} q^2 \frac{1}{\Delta} \, dq = \frac{\Delta^2}{12}. \tag{9.2.4}$$

To proceed further, we must consider a specific input signal. A common test signal in electrical engineering is the sinusoid; hence if the input to the quantizer is a full-load sine wave with peak value $V/2$, then

$$\sigma^2 = \left[\frac{V/2}{\sqrt{2}}\right]^2 = \frac{V^2}{8} \tag{9.2.5}$$

and from Eq. (9.2.1)

$$D = \frac{\Delta^2}{12} = \frac{V^2}{12L^2}, \tag{9.2.6}$$

so

$$\text{SQNR} = 10 \log_{10} \frac{3L^2}{2} = 20 \log_{10} L + 1.76 \text{ dB}. \tag{9.2.7}$$

Now, if the number of quantization levels $L = 2^n$, we have what is usually called an *n-bit quantizer*, and the SQNR becomes

$$\begin{aligned} \text{SQNR} &= 20n \log_{10} 2 + 1.76 \text{ dB} \\ &= 6.02n + 1.76 \text{ dB}. \end{aligned} \tag{9.2.8}$$

Thus, for an 8-bit quantizer, $L = 2^8 = 256$ levels, and the SQNR for a full-load sine wave is about 50 dB. One particularly important thing to note about Eq. (9.2.8) is that there is a 6-dB gain in SQNR for each additional bit devoted to quantization. This 6-dB/bit increase in SQNR is often stated and shows up in SQNR expressions for other types of inputs also as long as there is no overload.

EXAMPLE 9.2.2

Suppose that we have a uniform n-bit quantizer designed for a full-load sine wave with a peak value of $V/2$, so that the step size $\Delta = V/2^n$. We know that the SQNR obtained at the output of this quantizer for a full-load sine wave input is given by Eq. (9.2.8). However, let us now consider the case where the input to the same quantizer is a sine wave with peak amplitude

$V/8$. Then

$$\text{SQNR} = 10 \log_{10} \frac{[(V/8)/\sqrt{2}]^2}{(V/2^n)^2/12}$$

$$= 6.02n - 10.28 \text{ dB}. \tag{9.2.9}$$

Comparing Eqs. (9.2.8) and (9.2.9), we see that there is a substantial loss in performance (SQNR). This is because the dynamic range of the quantizer (overload points) does not match the dynamic range of the input for this example. Since the quantizer was designed for a wider dynamic range, the step size is too large, thus leading to increased granular noise.

It is often desirable or necessary to design a uniform quantizer for an input modeled by a random variable. If the probability density function (pdf) of the quantizer input is fairly well known, we can choose the step size to minimize some function of the difference between the quantizer input and output (the quantization error).

If we let $g(\hat{s} - s)$ denote the chosen error measure, the expression for the average distortion can be written as

$$D = \int_{-\infty}^{\infty} g[\hat{s} - s] f_S(s)\, ds, \tag{9.2.10}$$

where $f_S(s)$ is the pdf of the input. For a uniform quantizer, the output $\hat{s}$ is $(2i - 1)\Delta/2$ when $(i - 1)\Delta \le s < i\Delta$. Therefore, for a symmetric quantizer with an even number of levels, the distortion in Eq. (9.2.10) can be rewritten as

$$D = 2 \sum_{i=1}^{L/2-1} \int_{(i-1)\Delta}^{i\Delta} g\left[\frac{(2i-1)\Delta}{2} - s\right] f_S(s)\, ds + 2 \int_{(L/2-1)\Delta}^{\infty} g\left[\frac{(L-1)\Delta}{2} - s\right] f_S(s)\, ds, \tag{9.2.11}$$

where Δ is the step size to be chosen. Thus, given an error measure $g(\cdot)$, we must minimize D with respect to Δ. Performing the minimization yields

$$\sum_{i=1}^{L/2-1} (2i-1) \int_{(i-1)\Delta}^{i\Delta} g'\left[\frac{(2i-1)\Delta}{2} - s\right] f_S(s)\, ds + (L-1) \int_{(L/2-1)\Delta}^{\infty} g'\left[\frac{(L-1)\Delta}{2} - s\right] f_S(s)\, ds = 0 \tag{9.2.12}$$

to be solved for Δ_{opt}, where $g'(x) = dg(x)/dx$. For the squared-error distortion measure $g[\hat{s} - s] = (\hat{s} - s)^2$ and a zero-mean, unit-variance Gaussian input pdf, the optimum step sizes (Δ_{opt}), the minimum distortion, and the SQNR for $L = 4$, 8, and 16 levels are presented in Table 9.2.1 [Max, 1960]. (These must be computed numerically.) We see from the table that for this type of quantization we are obtaining about 5 dB for each additional bit devoted to quantization.

TABLE 9.2.1 Optimum Step Sizes for Uniform Quantization of a Gaussian PDF

Number of Levels (L)	Step Size (Δ_{opt})	Minimum Mean-Squared Error (D)	SQNR (dB)
4	0.9957	0.1188	9.25
8	0.5860	0.03744	14.27
16	0.3352	0.01154	19.38

Source: J. Max, "Quantizing for Minimum Distortion," *IRE Trans. Inf. Theory*, © 1960 IEEE.

9.3 Nonuniform Quantization

The most important nonuniform quantization method to date has been the logarithmic quantization used in the telephone network for speech digitization for over 20 years. The general idea behind this type of quantization is that for a fixed, uniform quantizer, an input signal with an amplitude less than full load will have a lower SQNR than a signal whose amplitude occupies the full dynamic range of the quantizer (but without overload). This fact is illustrated by Example 9.2.2. Such a variation in performance (SQNR) as a function of quantizer input signal amplitude is particularly detrimental for speech, since low-amplitude signals can be very important perceptually. There is the additional consideration for speech signals that low amplitudes are more probable than larger amplitudes, since speech is generally stated to have a gamma or Laplacian probability density, which is highly peaked about zero.

Therefore, for speech signals a type of nonlinear quantization was invented called *logarithmic companding*. Initially, this scheme was implemented by passing the analog speech signal through a characteristic of the form

$$F_\mu(s) = \frac{\ln[1 + \mu|s|]}{\ln[1 + \mu]} \operatorname{sgn}(s), \qquad -1 \le s \le 1, \tag{9.3.1}$$

where s is the normalized speech signal and μ is a parameter, usually selected to be $\mu = 100$, or more recently, $\mu = 255$. The function $F_\mu(s)$ is shown in Fig. 9.3.1. Notice that $F_\mu(s)$ tends to amplify small amplitudes more than larger amplitudes whenever $\mu > 0$. The output of $F_\mu(s)$ then served as input to a uniform, n-bit quantizer. To resynthesize the speech signal, the quantizer output $\hat{s}$ was passed through the inverse function of Eq. (9.3.1) given by

$$F_\mu^{-1}(\hat{s}) = \frac{1}{\mu}[(1 + \mu)^{|\hat{s}|} - 1] \operatorname{sgn}(\hat{s}), \tag{9.3.2}$$

where, of course, $-1 \le \hat{s} \le 1$.

The performance in SQNR of this system for $\mu = 100$ and $n = 7$ bits is shown in Fig. 9.3.2. It is evident from this figure that the SQNR is relatively flat over

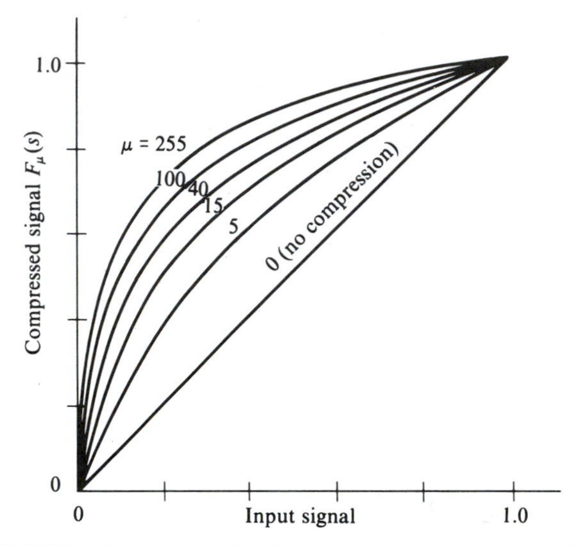

FIGURE 9.3.1 Logarithmic compression characteristics.

a wide dynamic range of input signal power (amplitudes), and hence low-amplitude signals are reproduced almost as well as higher-amplitude signals. In particular, we see from Eqs. (9.2.8) and (9.2.9) that for linear quantization, if we decrease the input signal power by 12 dB (from a peak value of $V/2$ to $V/8$), the output SQNR decreases by 12 dB. However, as we move along the "sine wave" curve in Fig. 9.3.2 from 0 to -12 dB on the input power axis, the SQNR decreases only by about 2 dB. The companding clearly improves the SQNR for low-amplitude signals.

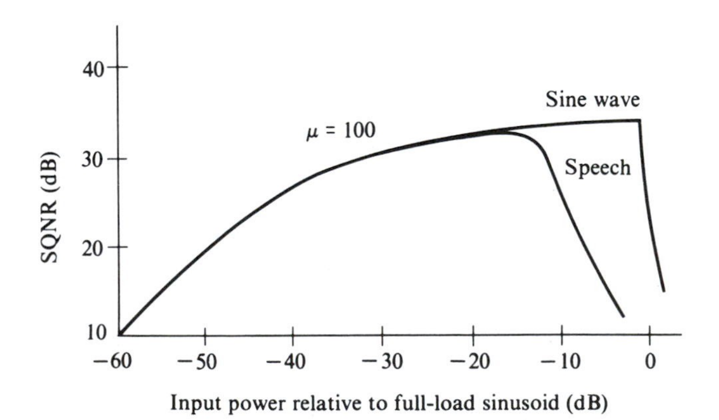

FIGURE 9.3.2 Performance of $\mu = 100$, $n = 7$ bit logarithmic companding.

EXAMPLE 9.3.1

To demonstrate further the utility of the logarithmic companding in Eq. (9.3.1), we consider the uniform 16-level quantizer designed for an input signal dynamic range of ± 10 V in Example 9.2.1. The resulting step size for this quantizer is thus 1.25 V. If we apply an input signal of amplitude 1.2 V to the quantizer, the quantization error is -0.575 V. However, this error should be reduced if we first pass the 1.2-V signal through $F_\mu(s)$ in Eq. (9.3.1), quantize this value, and then reconstruct the nonlinearly quantized output using Eq. (9.3.2).

Therefore, using Eq. (9.3.1) with $\mu = 255$ we have

$$F_\mu\left(\frac{1.2}{10}\right) = \frac{\ln\,[1 + 255|1.2/10|]}{\ln 256}\,\text{sgn}\left(\frac{1.2}{10}\right) = 0.623.$$

The input to the uniform quantizer is thus $10(0.623) = 6.23$, so that the quantizer output is $9\Delta/2 = 5.625$. This value is then substituted as $\hat{s}$ into Eq. (9.3.2) so that the nonlinearly quantized output is 0.848 V for 1.2 V in. The quantization error in this case is $0.848 - 1.2 = -0.352$ V. Note that for the same low-level input voltage of 1.2 V, the quantization error is considerably less for the nonuniform quantizer (the one using companding).

A nonuniform quantizer can also be designed to match a given input probability density function. A general nonuniform midriser quantizer characteristic is shown in Fig. 9.3.3. The step points $x_0 = -\infty, x_1, x_2, \ldots, x_{L-1}, x_L = +\infty$, and the output levels $y_1, y_2, \ldots, y_L$, are constants that can be selected to minimize some function of the quantization error (noise) for a known or assumed input probability distribution. The parameter Δ is simply used for scaling, but it is still often called the step size, even though it cannot be strictly interpreted as such.

Upon letting the input step points be denoted by x_i, $i = 0, 1, 2, \ldots, L$, where $x_0 = -\infty$, $x_L = +\infty$, and letting the output values be denoted by y_i, $i = 1, 2, \ldots, L$, where y_1 is the most negative value, the distortion in Eq. (9.2.10) can be written as

$$D = \sum_{i=1}^{L} \int_{x_{i-1}}^{x_i} g[y_i - s] f_S(s)\, ds. \tag{9.3.3}$$

In Eq. (9.3.3), $\hat{s} = y_i$ if $x_{i-1} \le s < x_i$. For fixed L, we minimize D in Eq. (9.3.3) with respect to both the x_i and y_i. The necessary conditions to be satisfied are

$$g[y_j - x_j] = g[y_{j+1} - x_j], \tag{9.3.4}$$

$j = 1, 2, \ldots, L - 1$, and

$$\int_{x_{j-1}}^{x_j} g'[y_j - s] f_S(s)\, ds = 0, \tag{9.3.5}$$

for $j = 1, 2, \ldots, L$.

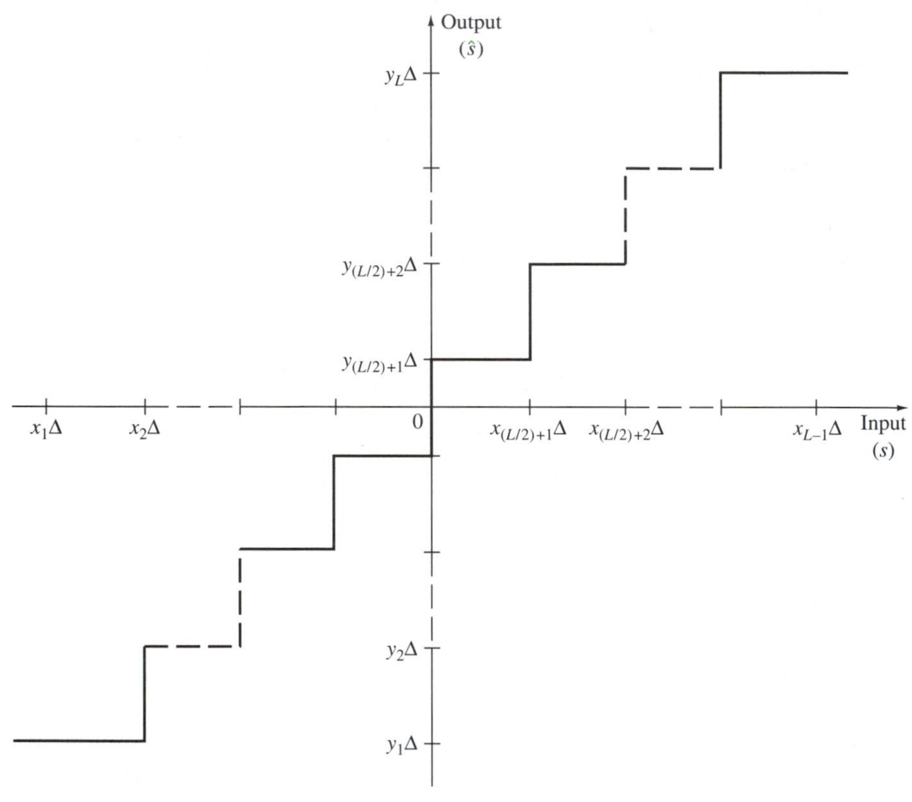

FIGURE 9.3.3 Symmetric nonuniform midriser quantizer with parameter Δ.

For the choice $g[\hat{s} - s] = [\hat{s} - s]^2$, Eqs. (9.3.4) and (9.3.5) become

$$x_j = \frac{y_{j+1} + y_j}{2} \tag{9.3.6}$$

and

$$y_j = \frac{\int_{x_{j-1}}^{x_j} s f_S(s)\, ds}{\int_{x_{j-1}}^{x_j} f_S(s)\, ds}. \tag{9.3.7}$$

Equations (9.3.6) and (9.3.7) define the minimum mean-squared error (MMSE) optimal quantizer or Lloyd–Max quantizer [Lloyd, 1957, 1982; Max, 1960].

Due to the complicated functional relationships, Eqs. (9.3.6) and (9.3.7) must be solved by computer. The approach is to choose a y_1, find x_1 from Eq. (9.3.7) ($x_0 = -\infty$), use Eq. (9.3.6) to find y_2, return to Eq. (9.3.7), and repeat the procedure. If y_L satisfies Eq. (9.3.7) exactly, then y_1 was chosen correctly. If y_L does not satisfy Eq. (9.3.7), a new y_1 must be selected and the computation repeated. Accuracy is at a premium in these calculations. Max [1960] tabulates the x_i and y_i for a Gaussian pdf and 1 to 36 output levels. His results for 4-, 8-, and 16-level quantizers are collected in Tables 9.3.1 to 9.3.3. The values shown are

TABLE 9.3.1 MMSE Four-Level Gaussian Quantizer[a]

i	x_i	y_i
1	−0.9816	−1.510
2	0.0	−0.4528
3	0.9816	0.4528
4	∞	1.510

[a] $D = 0.1175$; SQNR = 9.30 dB.

TABLE 9.3.2 MMSE Eight-Level Gaussian Quantizer[a]

i	x_i	y_i
1	−1.748	−2.152
2	−1.050	−1.344
3	−0.5006	−0.7560
4	0.0	−0.2451
5	0.5006	0.2451
6	1.050	0.7560
7	1.748	1.344
8	∞	2.152

[a] $D = 0.03454$; SQNR = 14.62 dB.

TABLE 9.3.3 MMSE 16-Level Gaussian Quantizer[a]

i	x_i	y_i
1	−2.401	−2.733
2	−1.844	−2.069
3	−1.437	−1.618
4	−1.099	−1.256
5	−0.7996	−0.9424
6	−0.5224	−0.6568
7	−0.2582	−0.3881
8	0.0	−0.1284
9	0.2582	0.1284
10	0.5224	0.3881
11	0.7996	0.6568
12	1.099	0.9424
13	1.437	1.256
14	1.844	1.618
15	2.401	2.069
16	∞	2.733

Source: J. Max, "Quantizing for Minimum Distortion," *IRE Trans. Inf. Theory*, © 1960 IEEE.

[a] $D = 0.009497$; SQNR = 20.22 dB.

normalized and must be multiplied by the input standard deviation (Δ in Fig. 9.3.3) to get the correct step points and output levels. Similarly, the MSE is normalized and must be multiplied by the input variance to obtain the actual mean-squared error.

By comparing the minimum mean-squared errors and SQNRs between the uniform quantizers in Table 9.2.1 and the nonuniform quantizers in Tables 9.3.1 to 9.3.3, it is seen that the nonuniform quantizers generally provide better performance (smaller D, larger SQNR). The nonuniform quantizers may be slightly more complex to implement, however.

9.4 Quantization-Level Coding

Once we have quantized a continuous-amplitude signal sample to one of a finite number of output levels as described in Sections 9.2 and 9.3, we could proceed and transmit a pulse with this amplitude level directly over a communications channel. In practice, however, it is usual to represent each quantizer output level by a binary word, which tends to simplify and standardize interfaces for both communications and storage applications. The assignment of binary words to output quantization levels does not have a generally accepted name in the literature, and here we call it *quantization-level coding.*

Three common binary word assignments for an eight-level quantizer are given in Table 9.4.1 with reference to the quantizer characteristic in Fig. 9.4.1. The *natural binary code* (NBC) simply consists of the binary representations of the base-10 numbers $0, 1, \ldots, 2^N - 1 = L - 1$, where L is the number of quantization levels. It is usual to start with 000 at the most negative level, but this is not mandatory.

TABLE 9.4.1 Binary Representations for an Eight-Level Quantizer

Level Number	Natural Binary Code (NBC)	Folded Binary Code (FBC)	Gray Code (GC)
1	000	011	010
2	001	010	011
3	010	001	001
4	011	000	000
5	100	100	100
6	101	101	101
7	110	110	111
8	111	111	110

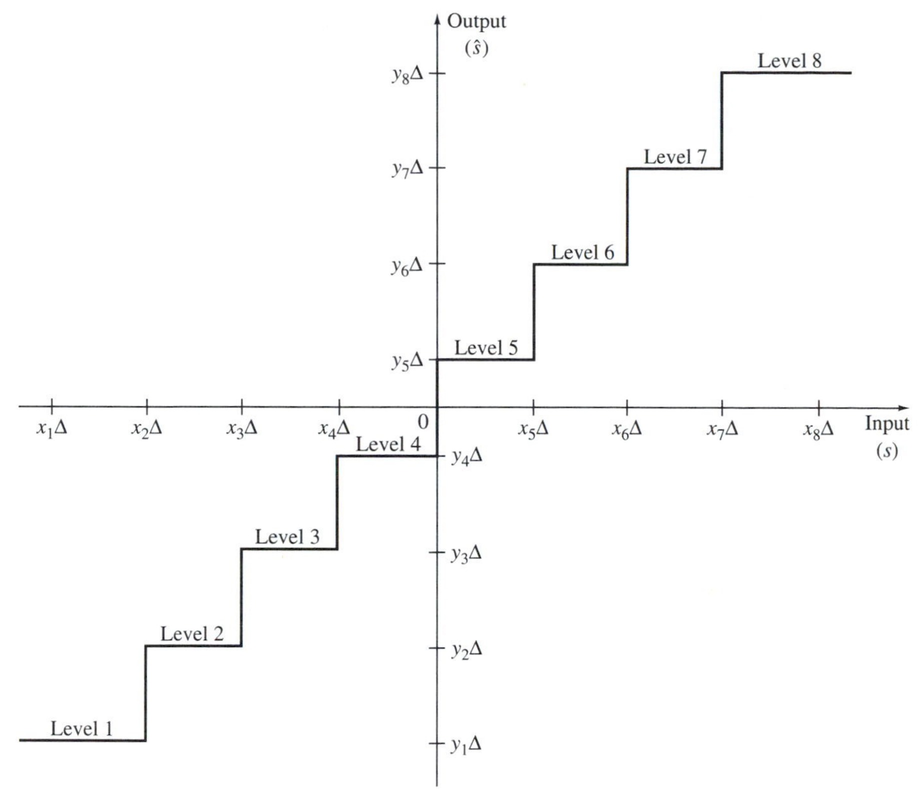

FIGURE 9.4.1 Eight-level symmetric quantizer.

The *folded binary code* (FBC) uses the most significant bit (MSB) or leftmost bit as a sign bit, and then counts out from the origin in both the positive and negative directions with the remaining bits. Since, exclusive of the sign bit, the codewords equidistant from the origin are identical, the FBC is also sometimes called the *sign–magnitude code.*

A third representation shown in Table 9.4.1 is the *Gray code* (GC). In this code, adjacent levels differ by only a single bit, and hence for equally likely bit errors in each position, a single channel bit error is more likely to produce an adjacent output level than in other codes. Of course, an error in the sign bit does not usually result in an adjacent output level (only at the origin).

The discussion thus far has emphasized quantizers with a number of output levels expressible as an integer power of 2. If we have a midtread quantizer (a zero level), there may be an odd number of levels, and of course, there is no rule which says that the number of output levels $L = 2^N$, where N is an integer. In these situations, one or more of the binary codewords are simply not used or they are dedicated to sending synchronization or some other type of signaling information. Which codewords are unused is not important in our present context; however, once we begin to assign symbols to the 1's and 0's, as we do in the next section, the use of one codeword over another may prove beneficial. We defer further discussion until then.

As a final note in this section, we mention that although the NBC and FBC are widely employed, they may not be the most efficient in terms of the average bit rate required. In fact, if the quantization levels are not equally likely (equally probable to occur), variable-length codes may provide a shorter average codeword length, and hence a lower transmitted bit rate. Such techniques are often called *entropy coding*, and the theory and applications of such codes are discussed in Chapter 11.

9.5 Transmission Line Coding

Once we have a sequence of 0's and 1's, whether the 0's and 1's are obtained as in Section 9.4 or are provided to us in some other manner, we must prepare these data for transmission to the receiver. If the data are to be sent over a single bandpass telephone channel, we would proceed as discussed in Chapter 8 (using a modem). However, if the binary information is to be transmitted in baseband or digital form directly, without analog modulation, we must assign a symbol or pulse to each 0 or 1 to be transmitted. This assignment is typically called *line coding*, or as we sometimes call it here, *transmission line coding*. The latter designation is an attempt to be slightly more specific, since many kinds of coding are discussed in this book. When assigning pulses to a sequence of 1's and 0's, we are often willing to work with rectangular pulses, but the key questions are: "How wide are the pulses in relation to the allocated pulse interval?" and "What are the polarities of the pulses?"

An intuitive form of line coding is to assign a pulse of width T seconds and amplitude $+V$ volts to each 1 and an amplitude of 0 V for T seconds to a 0. If the interval allocated to each bit is T seconds, this method is called *unipolar, nonreturn to zero* (NRZ) signaling. The "unipolar" refers to the fact that pulses of only a single polarity are used, and NRZ indicates that the assigned voltage level is maintained throughout the allocated pulse interval. On the other hand, a *polar*, NRZ line code might assign a pulse of amplitude $+V$ volts and width T seconds to a 1 and a pulse of $-V$ volts and width T seconds to a zero. If we continue to enumerate the possibilities, then, a polar, return to zero (RZ) line code might assign a $+V$ volts amplitude pulse of width $T/2$ to a 1 and a $-V$ volts pulse of width $T/2$ to a 0. In this case, the RZ case, the pulse is present only during half of the interval allocated to that particular bit. The voltage level is zero for the other half of the interval. Hence whatever voltage level or polarity a pulse is, it "returns to zero" during the pulse interval of T seconds. The RZ technique described uses pulses one-half the bit interval in width (50% duty cycle), which is the usual case. It is possible to have RZ pulses that occupy any portion of the T-second interval, as long as the pulse width is less than T.

The polar signaling scheme just described is designated as *bipolar* by some writers. However, we reserve the term *bipolar* coding for the alternate mark inversion method used in the digital T-carrier links in the telephone network. Specifically, this bipolar coding scheme uses 0 V for a zero, and alternating polarity, 50% duty cycle RZ pulses for 1's. There are particular advantages to this coding method, which we shall discuss shortly.

EXAMPLE 9.5.1

A binary sequence given by 1 0 1 1 0 0 0 1 is to be represented using (a) unipolar NRZ line coding, (b) polar RZ coding, and (c) bipolar coding. The results are shown in Fig. 9.5.1. In Fig. 9.5.1(a), we used an amplitude of $+V$ for a 1, but we could have also used a $+V$ for a 0 and 0 V for a 1, or we could have used negative polarities. For (a), however, the important things to note are that only one polarity is present and the pulses fill the bit interval.

Polar RZ coding is illustrated in Fig. 9.5.1(b), where $+V$ denotes 1 and $-V$ denotes a 0; of course, the polarities could be reversed. Since RZ pulses are employed, the pulses do not stay at $\pm V$ for the entire bit interval, but instead, drop back to 0 V. Although the pulses are shown as being in the first half of the bit interval, they could be located in the centers of the intervals, and often are.

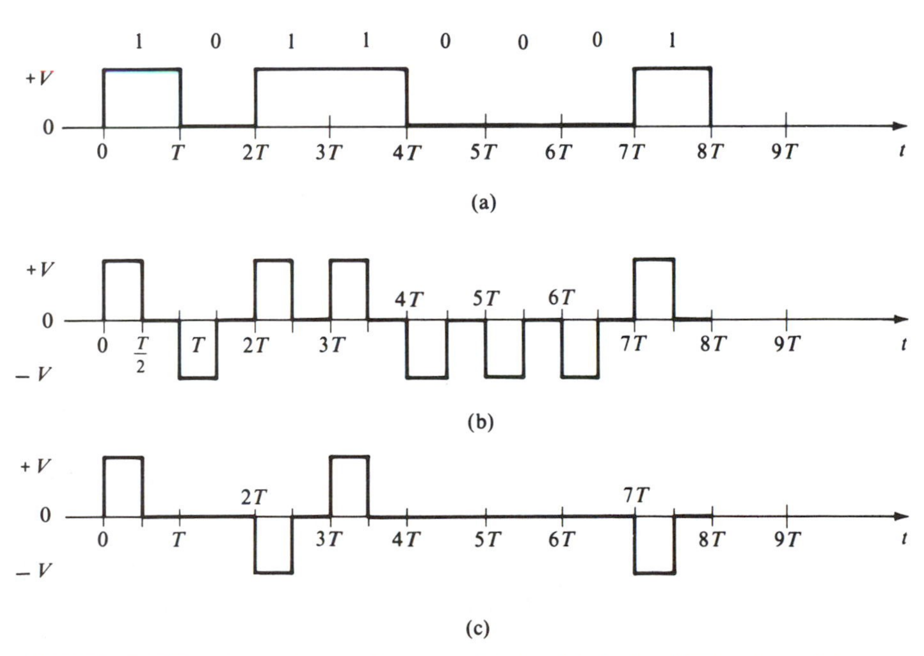

FIGURE 9.5.1 Line codes for Example 9.5.1: (a) Unipolar NRZ coding; (b) polar RZ coding; (c) bipolar (RZ) coding.

Bipolar (RZ) coding of the binary data is demonstrated by Fig. 9.5.1(c). Since there is no preceding 1, the first 1 is arbitrarily shown as a pulse of amplitude $+V$. A 0 is represented by 0 V or no pulse. The next 1 is indicated by a pulse opposite in polarity to the last preceding pulse—no matter how many 0's have occurred in between. Thus the second 1 is sent by a pulse of amplitude $-V$ volts. This process is continued to complete the example.

Clearly, the three line coding methods illustrated in Fig. 9.5.1 yield quite different transmitted signals for the given binary sequence, so the question arises as to how to select an appropriate line coding technique. Three principal considerations in making this selection are the spectrum of the line code, its synchronization properties, and error detection characteristics. Thus the unipolar NRZ code, although conceptually simple, has the disadvantages that there are no pulse transitions for long sequences of 0's or 1's, which are necessary if one wishes to extract timing or synchronization information, and that there is no way to detect when and if an error has occurred from the received pulse sequence. The polar RZ code guarantees the availability of timing information, but there is no error detection capability. The bipolar code has an error detection property, since if two pulses in a row (ignoring intervening 0's) are detected with the same polarity, it is evident that an error has occurred. However, to guarantee that timing data are available for the bipolar code, it is necessary to restrict the allowable number of consecutive 0's.

As we have mentioned, the spectrum of the transmitted line code can also be extremely important. The spectral density of a unipolar sequence, where a 1 is represented by a pulse and a 0 is represented by no pulse, is given by [see Eq. (A.10.13)]

$$S(\omega) = |P(\omega)|^2 \left\{ \frac{p(1-p)}{T} + 2\pi \frac{p^2}{T^2} \sum_{n=-\infty}^{\infty} \delta\left(\omega - \frac{2n\pi}{T}\right) \right\}, \tag{9.5.1}$$

where $P(\omega)$ is the Fourier transform of the pulse shape, p the probability of a pulse (a 1), and $1/T$ the symbol rate. Equation (9.5.1) holds for any $P(\omega)$. A sketch of $S(\omega)$ for rectangular RZ pulses of width $T/2$ and with $p = \frac{1}{2}$ is shown in Fig. 9.5.2. We see from this figure that there is a significant dc component and that most of the energy in the sequence lies at frequencies below twice the symbol rate.

The spectral density of the bipolar code is quite different and can be shown to be (see Problem 9.11)

$$S(\omega) = \frac{2p(1-p)}{T} |P(\omega)|^2 \left\{ \frac{1 - \cos \omega T}{1 + 2(2p-1)\cos \omega T + (2p-1)^2} \right\}, \tag{9.5.2}$$

where $P(\omega)$, p, and T are as given for Eq. (9.5.1). The spectral density in Eq. (9.5.2) is shown in Fig. 9.5.3 for rectangular pulses of width $T/2$ and several values of p. There are significant differences between Figs. 9.5.2 and 9.5.3. First,

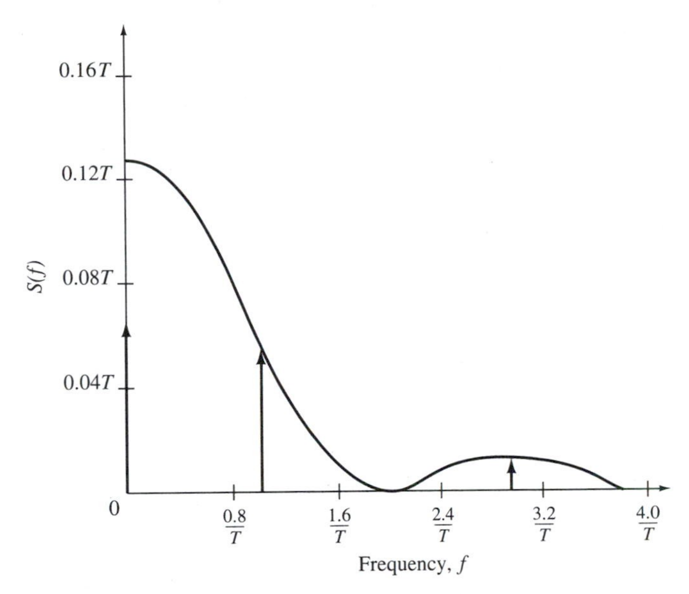

FIGURE 9.5.2 Spectral content of a unipolar RZ Sequence (pulse width = $T/2$ and $p = 1/2$).

the bipolar code has zero frequency content at dc. Second, the bipolar code has its energy concentrated at frequencies below the symbol rate. (Energy above the symbol rate is not shown in Fig. 9.5.3.) Third, there are no discrete components in Fig. 9.5.3. It is thus evident that the selection of a line code can have a considerable impact on the characteristics of the transmitted signal.

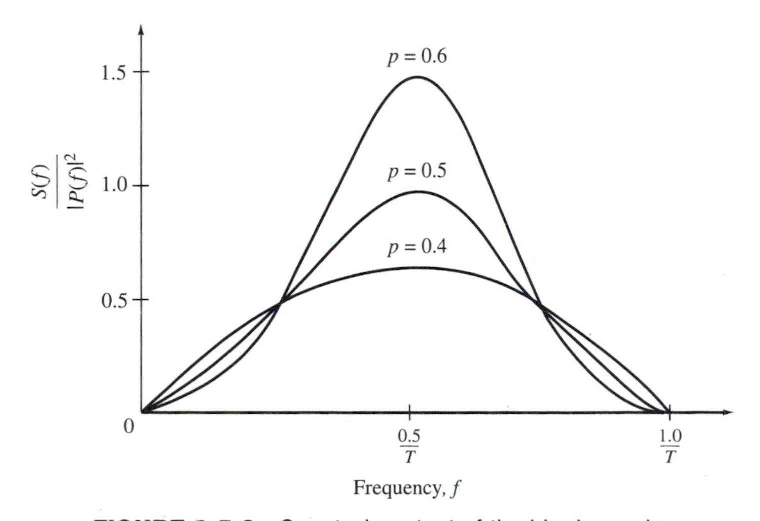

FIGURE 9.5.3 Spectral content of the bipolar code.

In addition to those already discussed, there are several other commonly used line codes. A class of codes currently used in the telephone network is the class of binary with N zero substitution (BNZS) codes. As we noted previously, if the bipolar code is used, care must be taken to prevent long strings of zeros from occurring, since timing information would be unavailable. The BNZS codes are bipolar codes that are modified to replace strings of N zeros with special sequences that contain pulses which cause bipolar violations. The bipolar violations allow the special sequences to be identified at the receiver, so that the N zeros can be reinserted; the presence of the special code during transmission increases the number of pulses, so that timing extraction is easier.

The B6ZS line code is an important specific example of the BNZS line coding procedure. If there are no sequences of 0's longer than five, the B6ZS line code is just the bipolar code described previously. However, when a sequence of six 0's occurs, these six 0's are removed and replaced with one of two possible six bit sequences. If the last pulse before the six 0's is positive, the code substituted for the six 0's is $0 + - 0 - +$, where $+$ denotes a positive 50% duty cycle RZ pulse, $-$ denotes a similar negative pulse, and 0 indicates "no pulse." On the other hand, if the last pulse before the six 0's is negative, the substituted code is $0 - + 0 + -$. Note that these substitutions guarantee (if there are no channel errors) that bipolar violations occur in the second and fifth bit positions, thus allowing the substituted code to be detected at the receiver and replaced with six 0's.

Another important line code is the B3ZS format. In the B3ZS format, the sequence of binary 1's and 0's are encoded by the straight bipolar code except when a string of three consecutive 0's occurs. The three 0's are represented by one of two codes, B0V or 00V, where B denotes a pulse satisfying the bipolar rule, 0 indicates no pulse, and V indicates a pulse that violates the bipolar rule. The choice between these two codes is made such that the number of pulses satisfying the bipolar rule between violations is odd. The bipolar violation after an odd number of pulses satisfying the bipolar rule allows the receiver to detect the substituted code and decode it as three 0's.

EXAMPLE 9.5.2

In this example we wish to illustrate the use of B6ZS and B3ZS line codes and to contrast them with the standard bipolar code introduced previously. Rather than sketch pulses explicitly here, we use a "+" to denote a positive, 50% duty cycle RZ pulse, a "−" to indicate a negative, 50% duty cycle RZ pulse, and a 0 to represent no pulse. We assume that we are given the binary data sequence 10100000000011000110, and we wish to represent this sequence using the bipolar, B6ZS, and B3ZS line codes. The results are shown in Table 9.5.1.

Notice that for all three codes we must specify the polarity of the last pulse representing a 1. Further, for the B3ZS code we must know whether the number of pulses satisfying the bipolar rule since the last violation is odd or even. In Table 9.5.1 we assumed that an even number has occurred.

TABLE 9.5.1 Bipolar, B6ZS, and B3ZS Line Codes for Example 9.5.2

Binary input data	1	0	1	0	0	0	0	0	0	0	1	1	0	0	0	1	1	0
Bipolar code (last 1 a −)	+	0	−	0	0	0	0	0	0	0	+	−	0	0	0	+	−	0
B6ZS (last 1 a −)	+	0	−	0	−	+	0	+	−	0	+	−	0	0	0	+	−	0
B3ZS (last 1 a − and even number of bipolar pulses since last violation)	+	0	−	B +	0 0	V +	B −	0 0	V −	0	+	−	B +	0 0	V +	−	+	0

If we had started with the assumption of an odd number, the B3ZS code in Table 9.5.1 would be different. As a further note, it should be recognized that for the B6ZS format, the nonzero pulses are the same as in a straight bipolar coding of the sequence; however, for B3ZS this may not be true. As an example, compare the final three pulses of the bipolar, B6ZS, and B3ZS codes in Table 9.5.1.

A class of line codes closely related to the binary with N zero substitution codes is the high-density bipolar codes. The most important of these codes is the HDB3 format, which has been adopted as an international standard. The HDB3 code uses bipolar coding whenever possible, but when strings of four 0's occur, they are encoded as either B00V or 000V, where the choice between these two is made (as in B3ZS) such that the number of pulses between bipolar violations is odd.

Another line coding format of some importance, due to its application in some local area networks called Ethernet, is *digital biphase coding* or *Manchester coding*. The primary attractions of a digital biphase code are that it contains a strong timing component and zero dc spectral content. In a digital biphase code, one cycle of a square wave with a particular phase is used to designate a 1 and one cycle of the square wave with the opposite phase is used to indicate a 0. Since there is a transition at the center of every symbol interval, a strong timing signal is available for synchronization. Additionally, since both 1's and 0's are represented by one cycle of a square wave, there is no dc component. Two disadvantages of the digital biphase code are that it has no redundancy for performance monitoring and it requires a wider bandwidth than other line codes, such as bipolar coding. Its principal applications are thus where bandwidth is not at a premium.

There are numerous other line coding formats that have various advantages and have practical applications. Of these, the pair selected ternary (PST) and the 4B3T format come to mind. The PST code maps pairs of binary digits (bits) into two ternary digits, while the 4B3T code maps 4 bits into three ternary digits. Development and discussion of these codes are left to the problems.

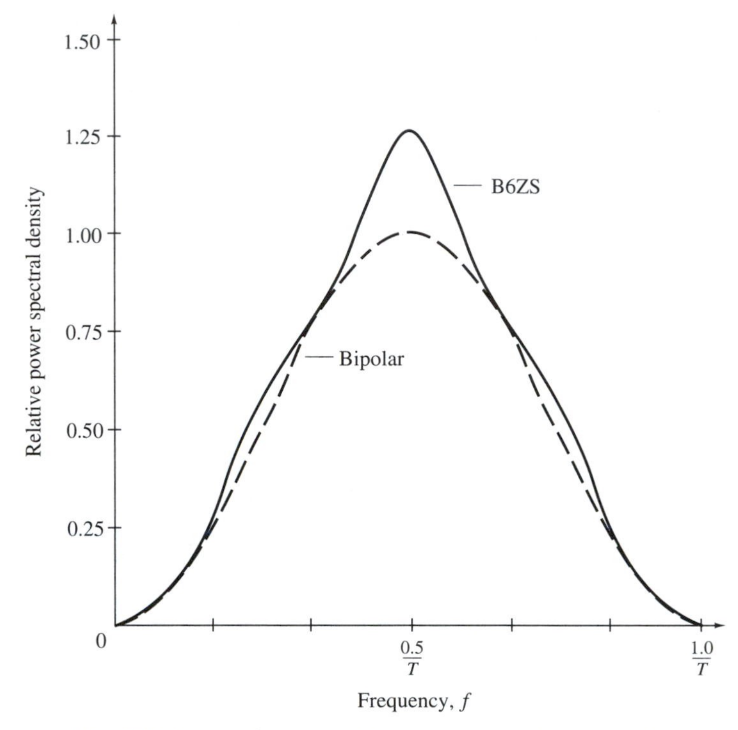

FIGURE 9.5.4 Spectra of bipolar and B6ZS line codes.

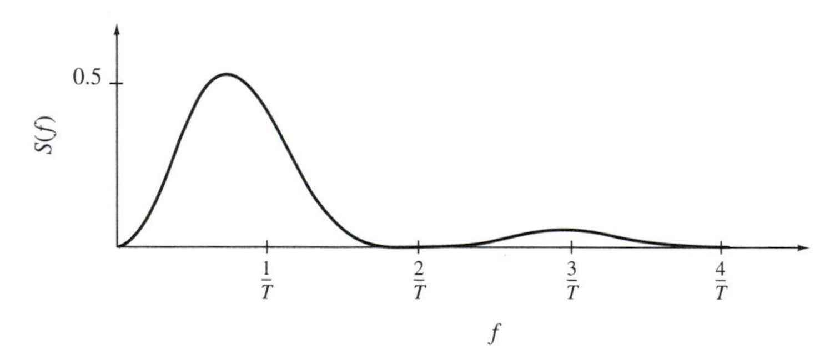

FIGURE 9.5.5 Spectrum of the digital biphase code.

Of course, depending on the transmission medium involved, the spectral content of the various line codes can be of the utmost importance. To close this section, then, we present (without derivation), the spectral densities of several line codes with equally likely 1's and 0's. Figure 9.5.4 shows the spectral content of bipolar, B6ZS, and the PST line codes, and Fig. 9.5.5 illustrates the spectrum of the digital biphase code.

9.6 Codecs and Channel Banks

At the present time, speech digitization is most often accomplished in the telephone network by devices called *codecs* (*co*ders/*dec*oders), which are contained in terminals designated as D-type *channel banks*. The codecs transform the analog speech into 8-bit PCM form, and the channel banks combine the PCM version of numerous voice channels into a single data stream using *time-division multiplexing* (TDM). In this section we briefly describe a few important details of the (U.S.) standard PCM codec and the various channel banks.

We begin by considering the codec used in the D2, D3, and D4 channel banks. This codec bandpass filters the analog voice signal to 200 to 3400 Hz and samples the filtered signal at a rate of 8000 samples/sec. Each of these samples is then quantized to 8-bit accuracy using a nonlinear quantizer based on the μ-law logarithmic characteristic discussed in Section 9.3. In particular, the quantizer is based on a 16-segment piecewise linear approximation to the $\mu = 255$ logarithmic companding characteristic in Eq. (9.3.1). There are eight positive and eight negative segments, but since the two segments around zero are collinear, it is often referred to as a 15-segment approximation. There are 16 equal quantization steps for each segment.

Table 9.6.1 specifies the nonlinear quantizer characteristic assuming that the maximum magnitude is scaled to 8159. The code for each quantization level is also shown in the table. For the 8-bit representation, the first bit is a polarity bit (1 denotes positive values, 0 represents negative values), the next 3 bits indicate the segment number, and the final 4 bits designate the particular step within a segment. Note from Table 9.6.1 that the segment codes and quantization steps for each segment are binary numbers that proceed from largest to smallest as the magnitude increases. Since lower amplitudes occur more often for speech than larger amplitudes, this tends to increase the density (number) of 1's. This is useful because the bipolar transmission line code described in Section 9.5 is employed with these codecs, and thus the number of pulses is increased. The higher density of pulses provides a strong timing component and aids synchronization.

In fact, another constraint is placed on the code assigned to quantizer output levels in Table 9.6.1. If an input sample falls within the most negative range of input amplitudes (quantization bin), the table indicates that the all 0's codeword would be transmitted. However, to guarantee a certain density of 1's, and hence bipolar pulses, the all 0's code is replaced by the codeword 00000010. Although it would cause less of an error to replace the all 0's codeword by 00000001, this is not done, for reasons that will be explained shortly.

Note now that we sampled the analog input voice signal (after filtering) at a rate of 8000 samples/sec, and thus if we pass each sample through the quantizer represented by Table 9.6.1, we get a bit rate of (8000 samples/sec) $\times$ (8 bits/sample) $=$ 64,000 bits/sec or 64 kbits/sec for each voice channel. Referring to Fig. 9.5.4, we see that for the bipolar code and $1/T = 64{,}000$, the required bandwidth is 64 kHz. Thus, by using PCM (the codec), we have expanded

TABLE 9.6.1 Quantizer Characteristic and Code Assignment for D2, D3, and D4 Channel Bank Codecs[a]

Input Amplitude Range:	Step Size:	Polarity Bit:	Quantization Segment Code:	Quantizer Step Code:	Output Value:
0–1	1	1	111	1111	0
1–3				1110	2
3–5	2	1	111	1101	4
⋮				⋮	⋮
29–31				0000	30
31–35				1111	33
⋮	4	1	110	⋮	⋮
91–95				0000	93
95–103				1111	99
⋮	8	1	101	⋮	⋮
215–223				0000	219
223–239				1111	231
⋮	16	1	100	⋮	⋮
463–479				0000	471
479–511				1111	495
⋮	32	1	011	⋮	⋮
959–991				0000	975
991–1055				1111	1023
⋮	64	1	010	⋮	⋮
1951–2015				0000	1983
2015–2143				1111	2079
⋮	128	1	001	⋮	⋮
3935–4063				0000	3999
4063–4319				1111	4191
⋮	256	1	000	⋮	⋮
7903–8159				0000	8031

[a] Positive inputs only; assumed symmetric about zero.

the bandwidth required by a single voice channel from approximately 4 kHz (3.4 kHz at the 3-dB point) to 64 kHz. We must be getting something in return for this extra bandwidth, and in the development in the remainder of this section and in Section 9.7, the primary advantages of PCM transmission are pointed out.

Before proceeding further, it is necessary to define clearly what is meant by time-division multiplexing and how it is used in the digital transmission of speech. In frequency-division multiplexing (FDM), we are given a specified band of frequencies, and we allocate nonoverlapping portions of this band to several different messages or channels. In TDM, we are given a specified time interval, and we allot nonoverlapping subintervals of this larger time slot to binary codewords generated by different codecs.

As a specific example of TDM, we mention what is called the T1 carrier system, which combines 24 PCM voice channels into a single data stream. Since each channel requires 8 bits, we thus have 192 bits for each set of 24 voice channels, usually called a *frame*. One bit is added to this total for synchronization purposes, so that 193 bits are transmitted per frame. Since each voice channel produces a new (8-bit) binary word 8000 times/sec, our frame rate is 8000 frames/sec and the transmitted bit rate for T1 carrier systems is 1.544 Mbits/sec.

Other than synchronization in the usual sense, the added "framing" bit has another purpose. When a telephone call is placed, there is a requirement for what is called *control signaling* information, such as "on-hook" or "off-hook" signals. For T1 carrier systems (with D2, D3, or D4 channel banks) this information is carried in the least significant bit position of each PCM word in every sixth frame. Therefore, in every sixth frame, the samples representing the 24 voice channels have only 7 bits of accuracy, while the intervening five frames use the full 8 bits. As a result, the PCM systems are sometimes said to employ $7\frac{5}{6}$-bit encoding. Since the least significant bit is sometimes "stolen" for signaling, this is why the most negative quantizer output level is encoded as 00000010 rather than 00000001. We would have gained nothing (in terms of a higher density of 1's) if the latter code appeared in the sixth frame.

The T1 systems constitute the first level (or lowest rate) in what is called the *digital hierarchy* in the telephone network. In recent years the various rates (or levels) in the digital hierarchy have been given DS designations, as shown in Table 9.6.2 and Fig. 9.6.1. From Fig. 9.6.1 we see that two DS1 links can be combined to form one DS1C link, four DS1 lines can be combined to form one DS2 link, or 28 DS1 links can be combined to form one DS3 link. Finally, six DS3 links can be combined to form one DS4 signal. Starting with the DS1 rate, which TDMs 24 voice channels in a frame, the method used to combine

TABLE 9.6.2 Data Rates and Line Codes in the Digital Hierarchy

Signal	Bit Rate (Mbits/sec)	Line Code
DS1	1.544	Bipolar
DS1C	3.152	Bipolar
DS2	6.312	B6ZS
DS3	44.736	B3ZS
DS4	274.176	Polar

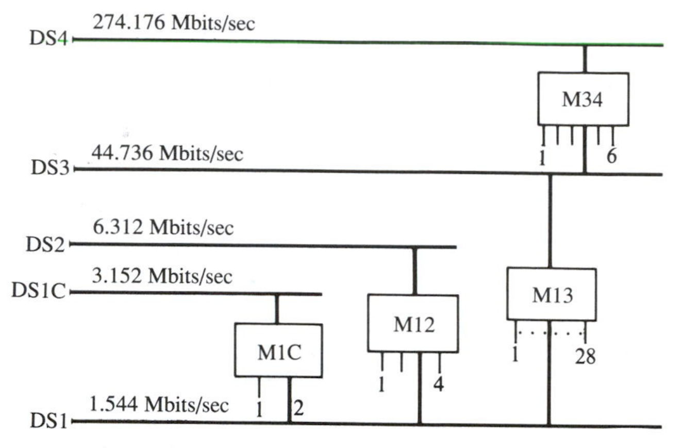

FIGURE 9.6.1 Digital hierarchy multiplexing plan.

all these signals is time-division multiplexing. Note also that in forming the next-highest level of the hierarchy, there seems to be some bits left over; that is, combining four DS1 signals should require 1.544 Mbits/sec $\times$ 4 = 6.176 Mbits/sec, but a DS2 line uses 6.312 Mbits/sec. The "extra" 0.136 Mbit/sec is used for synchronization and framing information, so that the multiplexing can be "undone."

It is also pointed out that as shown in Table 9.6.2, the line codes change as the data rate is increased, except in going from DS1 to DS1C signals. This is because the specifications on timing extraction, dc wander, and hardware change as the bit rate is increased. The line codes designated in Table 9.6.2 should be familiar to the reader from Section 9.5.

9.7 Repeaters

A principal advantage of employing time-division multiplexing of PCM signals to transmit analog waveforms such as speech is that PCM signals can be transmitted, in theory, over any distance without *any* degradation by noise simply by employing devices called *regenerative repeaters.* The concept that forms thebasis for repeaters can be explained as follows. Given a particular line code, only a finite number of fixed levels are allowable. For example, the bipolar line code has allowable levels of $+V$, $-V$, or 0 in each pulse interval, and no other voltage values are used to designate a 1 or 0. As these pulses proceed through a transmission medium (say, a pair of wires), they are smeared by deterministic distortion (amplitude and time delay) and they are subjected to additive noise as well as other random impairments. If the pair of wires is long enough, the pulses will eventually overlap so much and become so distorted that their original identity is irretrievably lost. If, however, we detect these pulses while the three amplitude levels are still distinguishable and they are not too smeared out into adjacent pulse intervals, we can regenerate the original pulse sequence exactly.

Therefore, regenerative repeaters are placed at specified intervals along a time-division multiplexed line to detect and retransmit the desired pulse sequence. The four main functions of these repeaters are equalization, clock recovery, pulse detection, and retransmission. The equalization first corrects for deterministic distortion that is on the line. Next, clock (timing) recovery is accomplished, so that the incoming pulses can be sampled at the center of the pulse intervals, and later, accurate timing is required to resynthesize the pulse sequence. The samples taken from the pulse intervals are used to detect the presence or absence of a pulse, and then the "undistorted" pulse sequence is regenerated and retransmitted. A block diagram of a regenerative repeater is shown in Fig. 9.7.1. The maximum spacing of the repeaters is determined such that the pulses can be reliably detected and the sequence accurately regenerated.

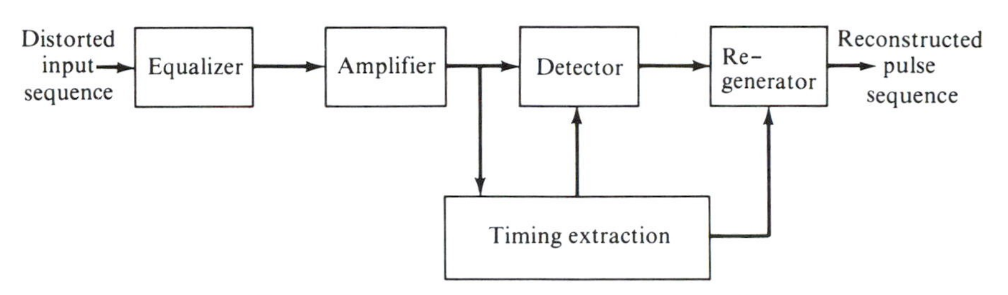

FIGURE 9.7.1 Regenerative repeater block diagram.

Theoretically, it is possible using PCM with TDM over repeatered lines to transmit the original PCM sequence undistorted over any distance. In practice, what occurs is that clock recovery is difficult, and timing inaccuracies from several repeaters begin to accumulate. The resulting distortion, called *timing jitter*, causes bit errors to occur eventually. However, the use of repeaters with PCM is much different from using amplifiers with analog signals. This is because the analog amplifiers not only increase the amplitude of the desired signal, but also amplify in-band noise. The repeaters with PCM are able to eliminate the noise, since only a few amplitude levels are allowable, and thus these levels can be detected and regenerated.

9.8 International Standards

The μ-law quantization characteristic given by Eq. (9.3.1) and the digital (PCM) TDM hierarchy described in Section 9.6 are used in the United States, Japan, and Canada. However, there are other systems implemented in other parts of the world that we mention briefly here. An alternative to the μ-law characteristic discussed in Section 9.3 is the A-law characteristic given by

$$F_A(s) = \begin{cases} \left[\dfrac{A|s|}{1 + \ln A}\right] \operatorname{sgn}(s), & 0 \le |s| \le \dfrac{1}{A} \\ \left[\dfrac{1 + \ln |As|}{1 + \ln A}\right] \operatorname{sgn}(s), & \dfrac{1}{A} \le |s| \le 1, \end{cases} \tag{9.8.1}$$

where $0 \le |s| \le 1$ and $A = 87.6$. The inverse to this characteristic is

$$F_A^{-1}(\hat{s}) = \begin{cases} \dfrac{|\hat{s}|[1 + \ln A]}{A} \operatorname{sgn}(\hat{s}), & 0 \le |\hat{s}| \le \dfrac{1}{1 + \ln A} \\ \dfrac{\exp\{|\hat{s}|[1 + \ln A] - 1\}}{A} \operatorname{sgn}(\hat{s}), & \dfrac{1}{1 + \ln A} \le |\hat{s}| \le 1, \end{cases} \tag{9.8.2}$$

where, if there is no quantization involved, $\hat{s} = F_A(s)$. The primary differences between the A-law and μ-law characteristics are that the A-law has a slightly wider dynamic range but the μ-law has a little better (less) idle channel noise.

The A-law quantizing characteristic is implemented in Europe, Africa, Australia, and South America, and in these countries, a digital hierarchy different from that in Fig. 9.6.1 is used. In particular, thirty 64-kbits/sec voice channels are combined with two 64-kbits/sec channels used for synchronization and signaling to obtain a first-level TDM data rate of 2.048 Mbits/sec. Successive levels of the hierarchy then proceed as shown in Fig. 9.8.1. Just as for the U.S.

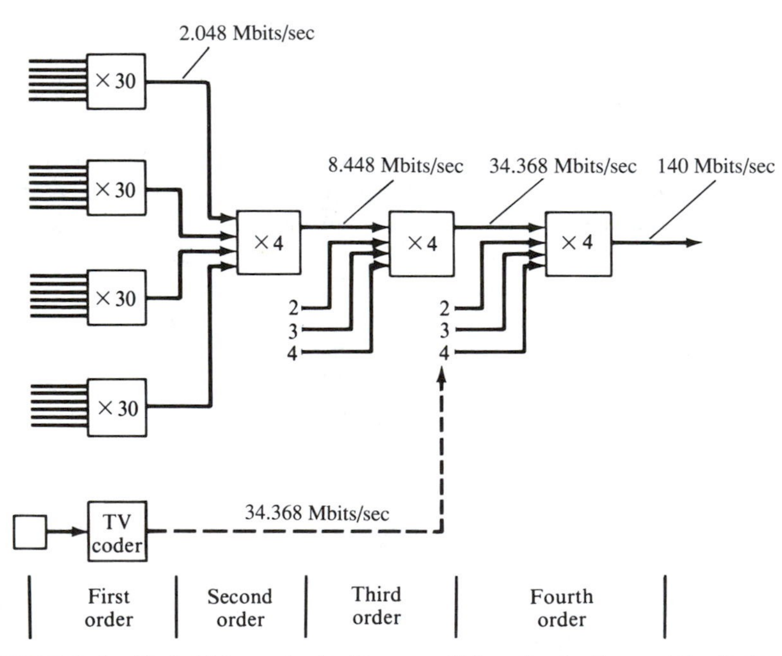

FIGURE 9.8.1 Digital hierarchy for Europe, Africa, Australia, and South America.

digital hierarchy, there are many other interesting details to the functioning of the system, but due to lack of space, they are not covered here. Some elements of A-law codecs and the hierarchy in Fig. 9.8.1 are covered in the problems at the end of the chapter.

SUMMARY

In this chapter we have briefly surveyed the various theoretical, functional, and applied aspects of pulse code modulation. PCM encoding consists of filtering, sampling, quantizing, assigning a level code, assigning a line code, and time-division multiplexing several channels into a chosen frame or block size for transmission. Nonlinear quantization of signals can produce substantial reductions in the required data rate with no loss in voice quality and intelligibility, as is demonstrated by the success of 8 bits/sample μ-law and A-law quantizer characteristics. The assignment of binary codewords to quantization levels and the choice of a line code are affected by several practical considerations, such as synchronization requirements, the spectral content of the line code, and the desired error monitoring capabilities. PCM signals offer the advantages of transmission over long distances with nominal distortion, digital switching, and compatibility with ever-increasing digital traffic. As a result, digital transmission networks based upon time-division multiplexed PCM signals are now an integral part of telecommunications systems throughout the world.

PROBLEMS

9.1 Design a 15-level uniform midtread quantizer for an input signal with a dynamic range of ± 10 V. Find the quantizer output value and the quantization error for an input signal amplitude of 1.2 V.

9.2 Plot the signal-to-quantization noise ratio in Eq. (9.2.7) for $L = 2, 3, 4, 5, 6, 7, 8, 12, 16, 24$, and 32.

9.3 The signal-to-quantization noise ratio expression in Eq. (9.2.7) is exact only for full-load sinusoidal inputs. Derive an equivalent expression if the input signals have the form of $s(t)$ in Fig. P9.3. Assume that the quantization noise is uniformly distributed in $[-\Delta/2, \Delta/2]$. Repeat Problem 9.2 for this new SQNR expression.

9.4 Given a uniform n-bit quantizer designed for a full-load sine wave with a peak value of $V/2$, plot the SQNR of this quantizer for sine-wave inputs with less than a full-load amplitude. What happens to the SQNR expression if the sine-wave input is greater than full load?

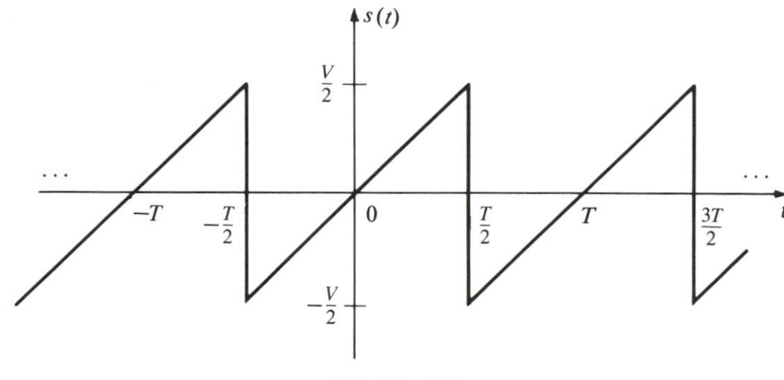

FIGURE P9.3

9.5 Verify the minimum distortion values for the $L = 4$- and 8-level quantizers in Table 9.2.1. Also, verify that Eq. (9.2.12) holds for the four-level quantizer.

9.6 For Example 9.3.1, find the quantization error if the input signal amplitude is $+9.2$ V.

9.7 Consider a uniform eight-level quantizer designed for an input signal dynamic range of ± 10 V. Calculate the quantization error for an input signal of $+1.2$ V. Calculate the quantization error for the same input signal if the quantizer is preceded by a $\mu = 255$ compandor.

9.8 Verify the value of the minimum mean-squared error for the four-level quantizer in Table 9.3.1.

9.9 Specify the natural binary code, the folded binary code, and a Gray code for a 16-level quantizer.

9.10 Given the binary sequence 1 1 0 1 1 1 0, sketch the transmitted pulse sequence for:
(a) Unipolar NRZ line coding.
(b) Polar RZ coding.
(c) Bipolar coding.

9.11 The power spectral density of a line code can be expressed as

$$S(\omega) = \frac{1}{T}\,|P(\omega)|^2\left\{R(0) + 2\sum_{k=1}^{\infty} R(k)\cos k\omega T\right\},$$

where $R(k)$, $k = 0, 1, 2, \ldots$, is the autocorrelation of the code sequence, and $P(\omega)$ is the Fourier transform of the individual pulse shape. Show that for the bipolar code, $R(0) = \frac{1}{2}$, $R(1) = -\frac{1}{4}$, and $R(k) = 0$ for $k \geq 2$. Let $p = \frac{1}{2}$ and derive the power spectral density for the bipolar code in terms of $P(\omega)$ [Davies and Barber, 1973]. Use Eq. (A.10.8).

9.12 One particular line code, called the "2 out of 4" code, restricts adjacent groups of 4 bits to have exactly 2 ones and 2 zeros. If the zeros are represented by a -1 and ones by a $+1$, and $p = \frac{1}{2}$, possible code sequences are shown in Fig. P9.12. Referring to Problem 9.11, show that

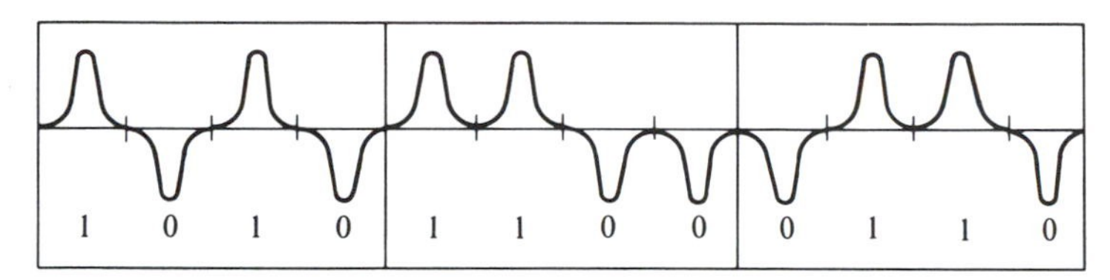

FIGURE P9.12

$R(0) = 1$, $R(1) = -\frac{1}{4}$, $R(2) = -\frac{1}{6}$, $R(3) = -\frac{1}{12}$, and $R(k) = 0$ for $k \geq 4$. Find the power spectral density of this code for a general $P(\omega)$ [Davies and Barber, 1973]. What are the advantages and disadvantages of this code?

9.13 For the input binary sequence 1 1 1 0 0 0 1 0 1 1 1 0 0 1 1 0 0 0 0 0 0 1 1 1, sketch the transmitted pulse sequences for each of the following codes:
(**a**) bipolar.
(**b**) B6ZS.
(**c**) B3ZS.

9.14 Sketch the resulting pulse sequences if **(a)** the HDB3 code and **(b)** a Manchester code are each used to represent the binary sequence in Problem 9.13.

9.15 A pair selected ternary (PST) code maps pairs of binary digits into two ternary digits. The PST code assignments are as shown in Table P9.15(a), where switching between modes is accomplished whenever a 0 1 or 1 0 binary input occurs. The 4B3T line code maps four-bit binary words into three-digit ternary words with $27 - 16 = 11$ codes left over. There is thus flexibility in choosing the codes. One possible 4B3T code assignment is shown in Table P9.15(b). The last six ternary codes are dc balanced, and the other 10 codes can be chosen from one of the two columns to provide a dc balanced transmitted sequence. The alternation between columns is accomplished according to whether there have been fewer negative pulses ($-$ column) or fewer positive pulses ($+$ column) in the entire transmitted sequence [Bellamy, 1982]. Use the PST and 4B3T codes to represent the

TABLE P9.15 (a) PST Coding Rules

Binary	PST	
	+Mode	−Mode
11	+ −	+ −
10	+ 0	− 0
01	0 +	0 −
00	− +	− +

(b) 4B3T Line Code Encoding Table

Binary Word	Ternary Word (Accumulated Disparity)		
	−	0	+
0000	− − −		+ + +
0001	− − 0		+ + 0
0010	− 0 −		+ 0 +
0011	0 − −		0 + +
0100	− − +		+ + −
0101	− + −		+ − +
0110	+ − −		− + +
0111	− 0 0		+ 0 0
1000	0 − 0		0 + 0
1001	0 0 −		0 0 +
1010		0 + −	
1011		0 − +	
1100		+ 0 −	
1101		− 0 +	
1110		+ − 0	
1111		− + 0	

Source: J. Bellamy, *Digital Telephony*, copyright © 1982 John Wiley & Sons, New York. Reprinted by permission of John Wiley & Sons, Inc.

binary sequence 0 1 0 1 1 1 0 0 0 0 0 0 0 0 1 1 0 1 0 0 1 1 1 0 0 1 1 1 0 0 1 1 1 1 1 1 1 0 0 0 1. Do not draw the pulses; use +, −, and 0 to indicate pulse levels. Other 4B3T codes are possible [see Owen, 1982].

9.16 Given the sequence of amplitude samples 12, 37, 108, 220, 392, 601, 150, and 64, use Table 9.6.1 to find the corresponding binary sequence and the decoded output.

9.17 Let $s(t) = e^{-t/\tau} \cos 2\pi f_0 t\, u(t)$, where $\tau = 1$ msec and $f_0 = 4000$ Hz. If $s(t)$ is sampled at a uniform rate of 10,000 samples/sec, specify the sample values, their binary code from Table 9.6.1, and the corresponding decoded values for $1.0 < t \leq 2.0$ msec.

9.18 Calculate the SQNR for a full-load sinusoid in segment 1 of the characteristic in Table 9.6.1. Assume that the quantization noise is uniformly distributed.

9.19 For a full-load sinusoid for the characteristic in Table 9.6.1 (a peak value of 8159), calculate the fraction of time that the amplitude of this sinusoid falls within each segment in the table.

9.20 An approximate value of the SQNR of the characteristic in Table 9.6.1 can be obtained from Eq. (9.2.3) with

$$D = \sum_{i=1}^{8} p_i \frac{\Delta_i^2}{12},$$

where p_i is the probability of the input amplitude falling within segment i and Δ_i is the step size for the ith segment. Use the results of Problem 9.19 to calculate SQNR for the characteristic in Table 9.6.1 with a full-load sinusoid input [Bellamy, 1982].

9.21 Using the SQNR values and input amplitudes from Problems 9.18 and 9.20, show that the coder maintains an SQNR of greater than 30 dB for a range of input powers of 48.4 dB. This quantity is called the coder *dynamic range*. Note that this value is approximately the same as

$$20 \log_{10} \frac{\Delta_{\max}}{\Delta_{\min}} = 20 \log_{10} (1 + \mu).$$

9.22 Message $m_1(t)$ is sampled 8000 times/sec to yield the values 8, 291, 504, 172, -12, -210, -525, and -268. Message $m_2(t)$ is similarly sampled to produce 28, 127, 300, 492, 299, 131, 54, and 6. Encode these samples using Table 9.6.1 and time-division multiplex the two sequences.

9.23 If individual repeaters can regenerate pulses in a T1 sequence only within ± 0.1 μsec, how many repeaters can we go through before this timing jitter causes difficulties?

9.24 The 10 sample values taken from a zero mean, variance 4 Gaussian distribution given by -2, -1.45, -0.2, $+0.15$, $+0.24$, $+0.68$, $+2.2$, $+2.9$, $+3.6$, $+3.9$, $+4.95$ are to be quantized using a 16-level MMSE Gaussian quantizer and encoded using a folded binary code. What are the resulting binary sequence and output amplitudes?

9.25 Sketch the A-law characteristic given by Eq. (9.8.1) with $A = 87.6$. Compare to Fig. 9.3.1.

9.26 Repeat Example 9.3.1 using A-law companding with $A = 87.6$.

9.27 Table P9.27 shows the quantizer characteristic and code assignments for an $A = 87.6$ law codec. Repeat Problems 9.18, 9.19, and 9.20 for this characteristic.

TABLE P9.27 Quantizer Characteristic and Code Assignment for the Segmented A-Law Codec

Input Amplitude Range	Step Size	Quantization Segment Code	Quantizer Step Code	Output Value
0–2			1111	1
2–4		111	1110	3
⋮			⋮	⋮
30–32	2		0000	31
32–34			1111	33
⋮		110	⋮	⋮
62–64			0000	63
64–68			1111	66
⋮	4	101	⋮	⋮
124–128			0000	126
128–136			1111	132
⋮	8	100	⋮	⋮
248–256			0000	252
256–272			1111	264
⋮	16	011	⋮	⋮
496–512			0000	504
512–544			1111	528
⋮	32	010	⋮	⋮
992–1024			0000	1008
1024–1088			1111	1056
⋮	64	001	⋮	⋮
1984–2048			0000	2016
2048–2176			1111	2112
⋮	128	000	⋮	⋮
3968–4096			0000	4032

9.28 Repeat Problem 9.16 for the A-law characteristic in Table P9.27.

9.29 Repeat Problem 9.17 for the A-law characteristic in Table P9.27.

9.30 With reference to Fig. 9.8.1, what are the excess bit rates allocated to frame signaling and synchronization at each stage in the European hierarchy?

CHAPTER 10

Noise in Digital Communications

10.1 Introduction

In this chapter we finally get around to evaluating the performance of the data transmission and digital communication systems developed in Chapters 8 and 9 when the transmitted signals are disturbed by random noise. The transmitted waveforms in Chapters 8 and 9 are really quite different. In Chapter 8 the original digital message sequence is transmitted using analog modulation methods, such as AMVSB-SC, FSK, PSK, and DPSK, while the transmitted signals used in Chapter 9 are actually discrete-amplitude "baseband" pulses. Despite these seemingly drastic physical differences, it turns out that we can express these transmitted signals in a common analytical framework, called signal space, which is developed in Section 10.2. Now, one of the reasons that the transmitted signals discussed in Chapters 8 and 9 are so different is that the channels over which the digital information is to be transmitted are different; that is, in Chapter 8 the channel being considered is a narrowband, bandpass channel (the analog voice channel), while in Chapter 9 the channel of interest is a wideband baseband or low-pass channel.

Because these channels have such different physical characteristics, one feels intuitively that both the deterministic and random distortion present on these two channels would be quite dissimilar. This is indeed the case, and we have discussed briefly in Chapter 8 the various kinds of distortion that affect data transmission over voiceband telephone channels. Further development of several important channel models is included in Appendix C. We are fortunate, however, in that one channel model plays a dominant role in the design and analysis of a major portion of the existing telecommunications systems, even systems as different as those in Chapters 8 and 9. This particular channel, the additive white Gaussian noise channel, is examined in Section 10.3, and there we find that not only does this model have wide applicability, but it is also tractable analytically—a fortuitous situation indeed.

Once we have the analytical framework of Sections 10.2 and 10.3, we are able in Section 10.4 to derive optimum receiver structures for a wide variety of communication systems and to compute the error probability for these optimum systems and other suboptimum communication systems in Section 10.5. Probability of error (or bit error rate, as it is sometimes called) is the basis for our performance evaluations in this chapter, but from previous discussions in the book, we know that there are other physical constraints on transmitted power, bandwidth, and perhaps, complexity, which must be satisfied. Hence, in Section 10.6, we present communication system performance comparisons that illustrate a few of the many trade-offs involved in communication system design. Finally, in Section 10.7 we develop receiver structures that are optimum in the presence of intersymbol interference and additive white Gaussian noise.

10.2 Signal Space and the Gram–Schmidt Procedure

The key to analyzing the performance of many communication systems is to realize that the signals used in these communication systems can be expressed and visualized geometrically. In fact, many signals in today's communication systems are expressible in only two dimensions! Since most undergraduate electrical engineers are quite comfortable with geometric concepts in three dimensions or less and have numerous years of experience with two-dimensional geometry, we are very attracted to a geometrical formulation of the problem.

Let us begin the development by trying to define a two-dimensional representation of a set of time domain waveforms $\{s_1(t), s_2(t), \ldots, s_N(t)\}$ over the time interval $0 \le t \le T$. We know that in standard two-dimensional spaces we have two orthogonal vectors, a concept of the length of a vector, and the idea of the distance between two vectors or two points in the plane. Hence, if we can define similar concepts in the time domain, we can then set up a one-to-one correspondence between the time domain and two-dimensional space. We considered the orthogonality of two time functions over a specified interval in Section 2.2, where we stated that two (real) time functions $f(t)$ and $g(t)$ are said to be orthogonal over the interval $[0, T]$ if

$$\int_0^T f(t)g(t)\,dt = 0 \tag{10.2.1}$$

for $f(t) \neq g(t)$. Thus we already have the needed concept of orthogonality. We obtain the concepts of the length of a signal and the distance between two signals by defining the energy of a signal

$$\int_0^T f^2(t)\,dt = E_f \tag{10.2.2}$$

as its length squared, and the energy of the difference between two signals given by

$$\int_0^T [f(t) - g(t)]^2\,dt = d_{fg}^2 \tag{10.2.3}$$

as the distance between the two signals squared. Thus the geometrical length corresponding to the waveform $f(t)$ is $\sqrt{E_f}$ from Eq. (10.2.2), and the geometrical distance between $f(t)$ and $g(t)$ is d_{fg} from Eq. (10.2.3). As a final step, we should define a pair of ortho*normal* functions to correspond to unit vectors in the two-dimensional space. To do this, we simply let

$$\varphi_1(t) = \frac{f(t)}{\sqrt{E_f}} \quad \text{and} \quad \varphi_2(t) = \frac{g(t)}{\sqrt{E_g}}, \tag{10.2.4}$$

so that

$$\int_0^T \varphi_i(t)\varphi_j(t)\,dt = \begin{cases} 0, & i \neq j \\ 1, & i = j. \end{cases} \tag{10.2.5}$$

Now, if we assume that the set of N signals $\{s_1(t), s_2(t), \ldots, s_N(t)\}$ can be expressed as a linear combination of the orthonormal waveforms $\varphi_1(t)$ and $\varphi_2(t)$, we can write

$$\begin{aligned} s_1(t) &= s_{11}\varphi_1(t) + s_{12}\varphi_2(t) \\ s_2(t) &= s_{21}\varphi_1(t) + s_{22}\varphi_2(t) \\ &\vdots \\ s_N(t) &= s_{N1}\varphi_1(t) + s_{N2}\varphi_2(t). \end{aligned} \tag{10.2.6}$$

Defining the column vectors

$$\varphi_1 = \begin{bmatrix} 1 \\ 0 \end{bmatrix} \quad \text{and} \quad \varphi_2 = \begin{bmatrix} 0 \\ 1 \end{bmatrix}, \tag{10.2.7}$$

we see that we can rewrite the set of equations in Eq. (10.2.6) as

$$\begin{aligned} \mathbf{s}_1 &= s_{11}\varphi_1 + s_{12}\varphi_2 \\ \mathbf{s}_2 &= s_{21}\varphi_1 + s_{22}\varphi_2 \\ &\vdots \\ \mathbf{s}_N &= s_{N1}\varphi_1 + s_{N2}\varphi_2 \end{aligned} \tag{10.2.8}$$

or as row vectors

$$\begin{aligned} \mathbf{s}_1 &= [s_{11} \quad s_{12}] \\ \mathbf{s}_2 &= [s_{21} \quad s_{22}] \\ &\vdots \\ \mathbf{s}_N &= [s_{N1} \quad s_{N2}]. \end{aligned} \tag{10.2.9}$$

In the set of equations denoted by Eq. (10.2.9), it is clear that the first component of the row vector is in the φ_1 direction and the second component is in the φ_2 direction, and we now have a correspondence between the physical, time-domain signal set $\{s_i(t), i = 1, 2, \ldots, N\}$ and points in a two-dimensional space. However, to complete the correspondence, we must find expressions in two-dimensional space analogous to Eqs. (10.2.2), (10.2.3), and (10.2.5).

To do this, we consider the inner product of any two signals in the set given by

$$\begin{aligned}\int_0^T s_i(t)s_j(t)\,dt &= \int_0^T \sum_{n=1}^{2}\sum_{m=1}^{2} s_{in}s_{jm}\varphi_n(t)\varphi_m(t)\,dt \\ &= \sum_{n=1}^{2}\sum_{m=1}^{2} s_{in}s_{jm}\int_0^T \varphi_n(t)\varphi_m(t)\,dt \\ &= \sum_{n=1}^{2} s_{in}s_{jn} = [s_{i1}s_{i2}]\begin{bmatrix} s_{j1} \\ s_{j2}\end{bmatrix} \\ &\triangleq \mathbf{s}_i\cdot\mathbf{s}_j, \end{aligned} \tag{10.2.10}$$

which is a form of Parseval's relations. Equation (10.2.10) is a relationship between the inner product in the time domain and the inner product in our two-dimensional signal space. We consider

$$\begin{aligned}\int_0^T \varphi_i(t)\varphi_j(t)\,dt &= \varphi_i\cdot\varphi_j \\ &= \begin{cases} [1\quad 0]\begin{bmatrix}0\\1\end{bmatrix} = 0, & i \neq j \\ [1\quad 0]\begin{bmatrix}1\\0\end{bmatrix} = 1 \text{ or } [0\quad 1]\begin{bmatrix}0\\1\end{bmatrix} = 1, & i = j,\end{cases}\end{aligned} \tag{10.2.11}$$

which is the same as Eq. (10.2.5). An expression for the energy of any signal in the space is obtained by letting $i = j$ in Eq. (10.2.10), which yields

$$\int_0^T s_i^2(t)\,dt = \sum_{n=1}^{N} s_{in}^2 = \mathbf{s}_i\cdot\mathbf{s}_i, \tag{10.2.12}$$

and distance between any two points (vectors) in the space can be obtained straightforwardly from Eq. (10.2.3) as

$$\begin{aligned}\int_0^T [s_i(t)-s_j(t)]^2\,dt &= \int_0^T \left[\sum_{n=1}^{N}(s_{in}-s_{jn})\varphi_n(t)\right]^2 dt \\ &= \sum_{n=1}^{N}\sum_{m=1}^{N}(s_{in}-s_{jn})(s_{im}-s_{jm})\int_0^T \varphi_n(t)\varphi_m(t)\,dt \\ &= \sum_{n=1}^{N}(s_{in}-s_{jn})^2 \\ &\triangleq d_{ij}^2. \end{aligned} \tag{10.2.13}$$

The correspondence between the original set of time-domain signals and our two-dimensional space, called a *signal space*, is now complete. For any $s_i(t)$, we obtain a vector (point) in the space from Eqs. (10.2.6)–(10.2.9) that has the length (distance from the origin) given by the square root of Eq. (10.2.12). Finally, the distance between any two vectors in the space is d_{ij}, which is available from Eq. (10.2.13).

EXAMPLE 10.2.1

A very important example for actual applications is the set of multilevel quadrature amplitude-modulated signals. In this case we have $\varphi_1(t) = \sqrt{2/T}\cos\omega_c t$ and $\varphi_2(t) = \sqrt{2/T}\sin\omega_c t$ as our orthonormal waveforms for $0 \le t \le T$. So that we can easily enumerate all the signals involved, let us consider initially $N = 4$ with the possible amplitudes of $\pm A_c$ in each direction. Therefore, we have four waveforms defined over $[0, T]$ as

$$\begin{aligned}
s_1(t) &= A_c\varphi_1(t) + A_c\varphi_2(t) \\
&= A_c\sqrt{\frac{2}{T}}\cos\omega_c t + A_c\sqrt{\frac{2}{T}}\sin\omega_c t \\
s_2(t) &= A_c\varphi_1(t) - A_c\varphi_2(t) \\
&= A_c\sqrt{\frac{2}{T}}\cos\omega_c t - A_c\sqrt{\frac{2}{T}}\sin\omega_c t \\
s_3(t) &= -A_c\varphi_1(t) + A_c\varphi_2(t) \\
&= -A_c\sqrt{\frac{2}{T}}\cos\omega_c t + A_c\sqrt{\frac{2}{T}}\sin\omega_c t \\
s_4(t) &= -A_c\varphi_1(t) - A_c\varphi_2(t) \\
&= -A_c\sqrt{\frac{2}{T}}\cos\omega_c t - A_c\sqrt{\frac{2}{T}}\sin\omega_c t.
\end{aligned} \tag{10.2.14}$$

Each of these signals has equal energy given by

$$\int_0^T s_i^2(t)\,dt = 2A_c^2 \triangleq E_s. \tag{10.2.15}$$

We can represent this set of signals by vectors in two-dimensional signal space as

$$\begin{aligned}
\mathbf{s}_1 &= [\ \ A_c \quad\ \ A_c] \\
\mathbf{s}_2 &= [\ \ A_c \quad -A_c] \\
\mathbf{s}_3 &= [-A_c \quad\ \ A_c] \\
\mathbf{s}_4 &= [-A_c \quad -A_c],
\end{aligned} \tag{10.2.16}$$

where from Eq. (10.2.12),

$$\mathbf{s}_i \cdot \mathbf{s}_i = 2A_c^2. \tag{10.2.17}$$

A plot of the signal set in Eq. (10.2.16) is shown in Fig. 10.2.1 and is called a *signal space diagram.* Note that we must know the correspondence between the φ_i and the $\varphi_i(t)$ to go back to the physically meaningful time domain. Furthermore, the points in Fig. 10.2.1 can represent a wide variety of waveforms, depending on this correspondence.

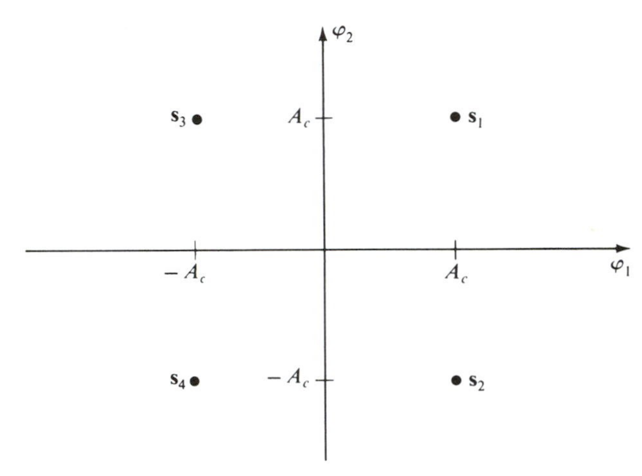

FIGURE 10.2.1 Signal space diagram for four-level QAM in Example 10.2.1 (see also Example 10.2.3).

The distance from the origin to each signal point is obvious as given by Eqs. (10.2.15) and (10.2.17). Furthermore, we can use Eq. (10.2.13) to calculate any d_{ij} of interest, say d_{23}, as

$$d_{23}^2 = \sum_{n=1}^{2} (s_{2n} - s_{3n})^2 = (A_c - (-A_c))^2 + (-A_c - A_c)^2 = 8A_c^2,$$

so $d_{23} = 2\sqrt{2}A_c$, which is obviously true by inspection of Fig. 10.2.1.

If we retain the same orthonormal time functions $\varphi_1(t)$ and $\varphi_2(t)$ and now consider the case where we have four possible amplitudes in each direction, namely, $\pm A_c/3$, $\pm A_c$, then we have $N = 16$, or what is usually called 16-level QAM. We leave it to the reader to write out the set of time-domain waveforms $\{s_i(t),\ i = 1, 2, \ldots, 16\}$ analogous to Eq. (10.2.16), but the signal space corresponding to 16-level QAM is shown in Fig. 10.2.2. Other details concerning the signal space representation of 16-level QAM are left to the problems.

Signal space diagrams are also often called *signal constellations.* To illustrate the generality of signal space diagrams further, we consider two additional examples.

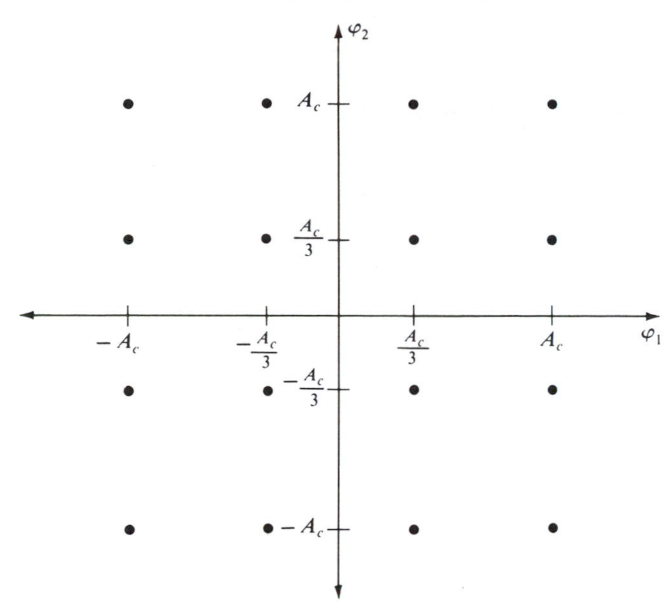

FIGURE 10.2.2 Signal space diagram for 16-level QAM (Example 10.2.1).

EXAMPLE 10.2.2

Phase shift keying (PSK) modulation has been, and continues to be, an important analog modulation technique for transmitting digital information. To begin, we consider binary PSK (BPSK), which has the two transmitted signals

$$s_1(t) = A_c \sqrt{\frac{2}{T}} \cos \omega_c t \tag{10.2.18a}$$

and

$$\begin{aligned} s_2(t) &= A_c \sqrt{\frac{2}{T}} \cos [\omega_c t + 180°] \\ &= -A_c \sqrt{\frac{2}{T}} \cos \omega_c t \end{aligned} \tag{10.2.18b}$$

for $0 \le t \le T$. If we let $\varphi(t) = \sqrt{2/T} \cos \omega_c t$, it is immediately evident that BPSK has a one-dimensional signal space representation as shown in Fig. 10.2.3.

If we now wish to consider four-phase PSK, we might use the four time-domain waveforms

$$s_i(t) = A_c \sqrt{\frac{2}{T}} \cos \left[\omega_c t + \frac{i\pi}{2} \right]$$

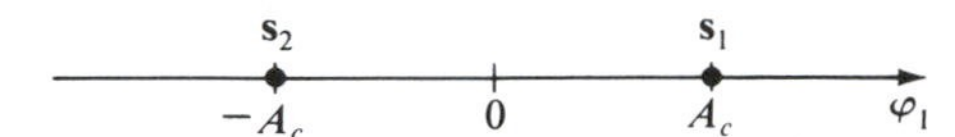

FIGURE 10.2.3 Signal space diagram for BPSK in Example 10.2.2 (see also Example 10.2.3).

for $i = 1, 2, 3, 4$, and $0 \leq t \leq T$. Using a trigonometric identity, we obtain

$$s_i(t) = A_c \sqrt{\frac{2}{T}} \cos \frac{i\pi}{2} \cos \omega_c t - A_c \sqrt{\frac{2}{T}} \sin \frac{i\pi}{2} \sin \omega_c t. \qquad (10.2.19)$$

Letting $\varphi_1(t) = \sqrt{2/T} \cos \omega_c t$ and $\varphi_2(t) = \sqrt{2/T} \sin \omega_c t$, we can rewrite Eq. (10.2.19) as

$$s_i(t) = A_c \cos \frac{i\pi}{2} \varphi_1(t) - A_c \sin \frac{i\pi}{2} \varphi_2(t) \qquad (10.2.20)$$

for $i = 1, 2, 3, 4$. Comparing Eq. (10.2.20) for all i with Fig. 10.2.1, we see that except for a scaling by $\sqrt{2}$ and a renumbering of the signals, the signal space diagram for our four-phase PSK signals has the same form as Fig. 10.2.1, but it is rotated by 45°. Depending on the application, this difference in rotation may or may not be of importance. This point is discussed in more detail in later sections of this chapter.

Eight-phase PSK is an important practical modulation method that can have the transmitted waveforms

$$s_i(t) = A_c \sqrt{\frac{2}{T}} \cos \left[\omega_c t - \frac{i\pi}{4} \right] \qquad (10.2.21)$$

for $i = 1, 2, \ldots, 8$, and $0 \leq t \leq T$. The signal space diagram for this signal set is shown in Fig. 10.2.4, and can be easily surmised by using a trig identity on Eq. (10.2.21).

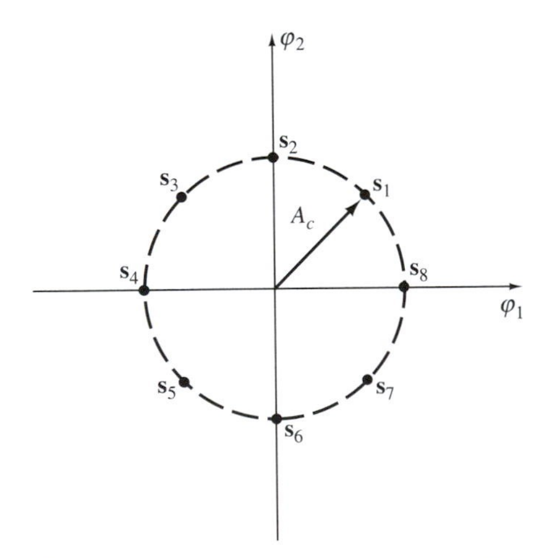

FIGURE 10.2.4 Signal space diagram for 8-phase PSK in Example 10.2.2.

EXAMPLE 10.2.3

For this example we now consider the two orthonormal waveforms over $0 \le t \le T$:

$$\varphi_1(t) = \begin{cases} \sqrt{\dfrac{2}{T}}, & 0 \le t \le \dfrac{T}{2} \\ 0, & \dfrac{T}{2} < t \le T \end{cases} \tag{10.2.22}$$

and

$$\varphi_2(t) = \begin{cases} 0, & 0 \le t \le \dfrac{T}{2} \\ \sqrt{\dfrac{2}{T}}, & \dfrac{T}{2} < t \le T, \end{cases} \tag{10.2.23}$$

where both waveforms are zero outside $[0, T]$. Suppose now that we form the binary signal set (over $0 \le t \le T/2$)

$$s_1(t) = A_c \varphi_1(t) \tag{10.2.24a}$$

and

$$s_2(t) = -A_c \varphi_1(t). \tag{10.2.24b}$$

Then the signal space diagram for Eqs. (10.2.24a,b) is the same as in Fig. 10.2.3. Equations (10.2.24a,b) represent what are often called *binary antipodal signals*.

If we now let $N = 4$ with

$$\begin{aligned} s_1(t) &= A_c\varphi_1(t) + A_c\varphi_2(t) \\ s_2(t) &= A_c\varphi_1(t) - A_c\varphi_2(t) \\ s_3(t) &= -A_c\varphi_1(t) + A_c\varphi_2(t) \\ s_4(t) &= -A_c\varphi_1(t) - A_c\varphi_2(t) \end{aligned} \tag{10.2.25}$$

we discern immediately that the appropriate signal space diagram is identical to the one sketched in Fig. 10.2.1. If we compare the signal sets in Eqs. (10.2.14) and (10.2.25), this result is not surprising, and again points out that a particular signal constellation can represent a variety of physical waveforms depending on the $\varphi_1(t)$ and $\varphi_2(t)$ functions. Can you specify a set of signals using Eqs. (10.2.22) and (10.2.23) that have the signal space diagram in Fig. 10.2.2? See Problem 10.6.

We have considered only two-dimensional signal spaces thus far (because of their physical importance and because of our geometric intuition); however, it is interesting to inquire as to the generality of the approach. That is, if we are presented with a set of N waveforms $\{s_i(t), i = 1, 2, \ldots, N\}$ defined over

$-\infty < t < \infty$ with $\int_{-\infty}^{\infty} s_i^2(t)\,dt < \infty$, can we express each of these signals as a weighted linear combination of orthonormal waveforms

$$s_i(t) = \sum_{j=1}^{M} s_{ij}\varphi_j(t), \tag{10.2.26}$$

where $M \leq N$ and the set $\{\varphi_j(t), j = 1, 2, \ldots, M\}$ satisfies Eq. (10.2.5)? To accomplish an expansion of the form in Eq. (10.2.26), we must first be able to determine a suitable set of orthonormal waveforms. There is a well-defined technique for doing this called the Gram–Schmidt orthogonalization procedure.

Gram–Schmidt Orthogonalization

Here we define a straightforward procedure for generating $M \leq N$ orthonormal functions given any set of N finite-energy waveforms defined over $-\infty < t < \infty$. It is presented in a step-by-step fashion.

STEP 1. For $s_1(t) \not\equiv 0$, find $E_1 = \int_{-\infty}^{\infty} s_1^2(t)\,dt$ and set

$$\varphi_1(t) = \frac{1}{\sqrt{E_1}} s_1(t). \tag{10.2.27}$$

Note that in the form of Eq. (10.2.26),

$$s_1(t) = s_{11}\varphi_1(t) = \sqrt{E_1}\,\varphi_1(t) \tag{10.2.28}$$

and further that we can find s_{11} from $s_1(t)$ as

$$s_{11} = \int_{-\infty}^{\infty} s_1(t)\varphi_1(t)\,dt. \tag{10.2.29}$$

STEP 2. Form

$$f_2(t) = s_2(t) - s_{21}\varphi_1(t), \tag{10.2.30}$$

where

$$s_{21} = \int_{-\infty}^{\infty} s_2(t)\varphi_1(t)\,dt, \tag{10.2.31}$$

so that $f_2(t)$ is orthogonal to $\varphi_1(t)$, that is,

$$\int_{-\infty}^{\infty} f_2(t)\varphi_1(t)\,dt = \int_{-\infty}^{\infty} [s_2(t) - s_{21}\varphi_1(t)]\varphi_1(t)\,dt = 0. \tag{10.2.32}$$

Find $\int_{-\infty}^{\infty} f_2^2(t)\,dt$ and define

$$\varphi_2(t) = \frac{f_2(t)}{\sqrt{\int_{-\infty}^{\infty} f_2^2(t)\,dt}}. \tag{10.2.33}$$

Note that

$$s_{22} = \int_{-\infty}^{\infty} f_2(t)\varphi_2(t)\,dt = \sqrt{\int_{-\infty}^{\infty} f_2^2(t)\,dt}, \tag{10.2.34}$$

so that

$$s_2(t) = s_{21}\varphi_1(t) + s_{22}\varphi_2(t). \tag{10.2.35}$$

Clearly, $\varphi_2(t)$ is orthogonal to $\varphi_1(t)$ from Eqs. (10.2.32) and (10.2.33).

⋮

STEP *k*. The general step in the procedure can be specified by assuming that we have generated $m - 1$ orthonormal waveforms $\{\varphi_j(t), j = 1, 2, \ldots, m - 1\}$ based on $k - 1$ of the original signals $\{s_i(t), i = 1, 2, \ldots, k - 1\}$. We will have $m - 1 \le k - 1$, with strict inequality if at one or more of the preceding steps the $f_i(t) \equiv 0$. This occurrence implies that the original signal set is not linearly independent. Continuing the process, we form

$$f_k(t) = s_k(t) - \sum_{j=1}^{m-1} s_{kj}\varphi_j(t), \tag{10.2.36}$$

where the coefficients are

$$s_{kj} = \int_{-\infty}^{\infty} s_k(t)\varphi_j(t)\, dt. \tag{10.2.37}$$

If $f_k(t) \not\equiv 0$, find $\int_{-\infty}^{\infty} f_k^2(t)\, dt$ and define

$$\varphi_m(t) = \frac{f_k(t)}{\sqrt{\int_{-\infty}^{\infty} f_k^2(t)\, dt}}. \tag{10.2.38}$$

Clearly, $f_k(t)$ is orthogonal to each member of the set, $\{\varphi_j(t), j = 1, 2, \ldots, m - 1\}$, so the set of $\varphi_j(t)$s, including $\varphi_m(t)$, are orthonormal. We have $s_{kj}, j = 1, 2, \ldots, m - 1$, so to write an expression for $s_k(t)$ of the form of Eq. (10.2.26), we need s_{km}. However,

$$s_{km} = \int_{-\infty}^{\infty} \varphi_m(t) f_k(t)\, dt = \sqrt{\int_{-\infty}^{\infty} f_k^2(t)\, dt} \tag{10.2.39}$$

and the expansion is completed.

The process is repeated until all N original signals have been used. The final result is an expansion as in Eq. (10.2.26).

EXAMPLE 10.2.4

To illustrate the Gram–Schmidt procedure, we begin with the four signals $\{s_i(t), i = 1, 2, 3, 4\}$ shown in Fig. 10.2.5. We follow the steps in the procedure as outlined previously.

STEP 1. Calculate

$$E_1 = \int_0^{T/3} (1)^2\, dt = \frac{T}{3},$$

so

$$\varphi_1(t) = \sqrt{\frac{3}{T}}\, s_1(t) \tag{10.2.40}$$

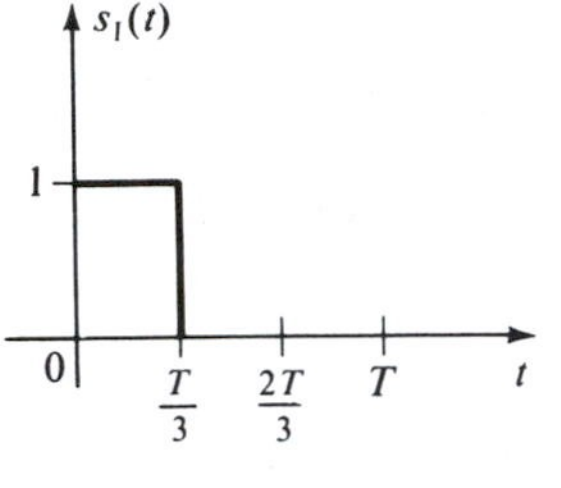

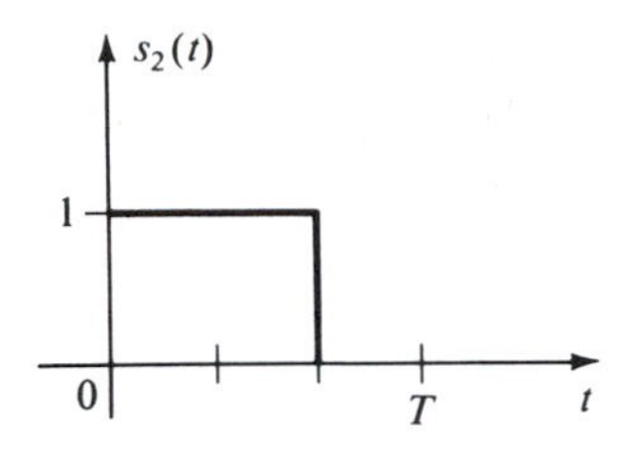

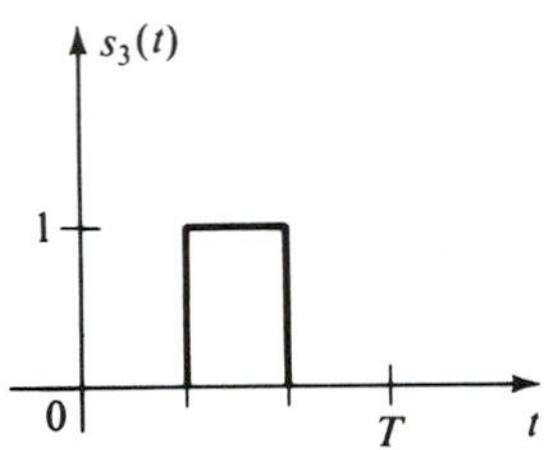

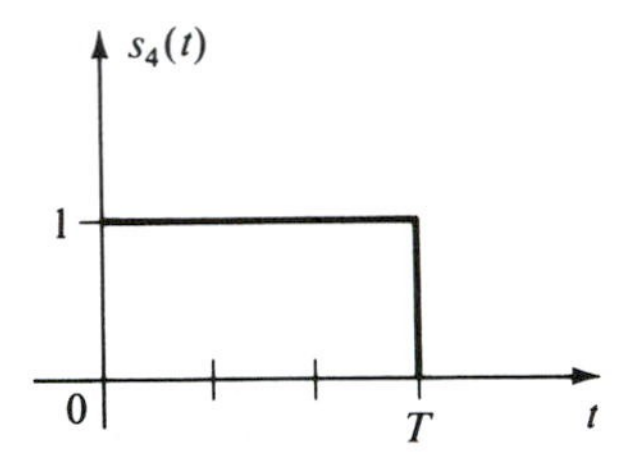

FIGURE 10.2.5 Signal set for Example 10.2.4.

and

$$s_1(t) = s_{11}\varphi_1(t) = \sqrt{\frac{T}{3}}\,\varphi_1(t). \tag{10.2.41}$$

STEP 2. Find

$$s_{21} = \int_{-\infty}^{\infty} s_2(t)\varphi_1(t)\,dt = \int_0^{T/3} \sqrt{\frac{3}{T}}\,dt = \sqrt{\frac{T}{3}}, \tag{10.2.42}$$

then form

$$\begin{aligned} f_2(t) &= s_2(t) - s_{21}\varphi_1(t) \\ &= s_2(t) - s_1(t). \end{aligned} \tag{10.2.43}$$

Now

$$\int_{-\infty}^{\infty} f_2^2(t)\,dt = \int_{T/3}^{2T/3} (1)^2\,dt = \frac{T}{3}, \tag{10.2.44}$$

so we define

$$\varphi_2(t) = \sqrt{\frac{3}{T}}\,[s_2(t) - s_1(t)] \tag{10.2.45}$$

and therefore

$$\begin{aligned} s_2(t) &= s_{21}\varphi_1(t) + s_{22}\varphi_2(t) \\ &= \sqrt{\frac{T}{3}}\,\varphi_1(t) + \sqrt{\frac{T}{3}}\,\varphi_2(t), \end{aligned} \tag{10.2.46}$$

since s_{22} is given by Eq. (10.2.34).

STEP 3. Calculate

$$s_{31} = \int_{-\infty}^{\infty} s_3(t)\varphi_1(t)\,dt = 0, \tag{10.2.47}$$

since $s_3(t)$ and $\varphi_1(t)$ do not overlap, and

$$s_{32} = \int_{-\infty}^{\infty} s_3(t)\varphi_2(t)\,dt = \int_{T/3}^{2T/3} \sqrt{\frac{3}{T}}\,dt = \sqrt{\frac{T}{3}}, \tag{10.2.48}$$

and form

$$\begin{aligned} f_3(t) &= s_3(t) - \sum_{j=1}^{2} s_{3j}\varphi_j(t) \\ &= s_3(t) - \sqrt{\frac{T}{3}}\,\varphi_2(t) \\ &= s_3(t) - \sqrt{\frac{T}{3}} \cdot \sqrt{\frac{3}{T}}\,[s_2(t) - s_1(t)]. \end{aligned} \tag{10.2.49}$$

But from Fig. 10.2.5, we observe that $s_2(t) - s_1(t) = s_3(t)$, so

$$f_3(t) = 0 \tag{10.2.50}$$

and

$$s_3(t) = \sqrt{\frac{T}{3}}\,\varphi_2(t). \tag{10.2.51}$$

From Eq. (10.2.50) we conclude that $s_3(t)$ is a weighted linear combination of $s_1(t)$ and $s_2(t)$, and therefore we proceed to the next step.

STEP 4. Find

$$s_{41} = \int_{-\infty}^{\infty} s_4(t)\varphi_1(t)\,dt = \int_{0}^{T/3} \sqrt{\frac{3}{T}}\,dt = \sqrt{\frac{T}{3}} \tag{10.2.52}$$

$$s_{42} = \int_{-\infty}^{\infty} s_4(t)\varphi_2(t)\,dt = \int_{T/3}^{2T/3} \sqrt{\frac{3}{T}}\,dt = \sqrt{\frac{T}{3}} \tag{10.2.53}$$

and form

$$\begin{aligned} f_4(t) &= s_4(t) - \sum_{j=1}^{2} s_{4j}\varphi_j(t) \\ &= s_4(t) - s_1(t) - [s_2(t) - s_1(t)] \\ &= \begin{cases} 1, & \text{for } \dfrac{2T}{3} \le t \le T \\ 0, & \text{otherwise.} \end{cases} \end{aligned} \tag{10.2.54}$$

So

$$\int_{-\infty}^{\infty} f_4^2(t)\, dt = \frac{T}{3} \tag{10.2.55}$$

and

$$\varphi_3(t) = \sqrt{\frac{3}{T}}\, f_4(t). \tag{10.2.56}$$

Finally,

$$s_{43} = \sqrt{\frac{T}{3}} \tag{10.2.57}$$

and thus the desired orthonormal expansion is

$$s_4(t) = \sqrt{\frac{T}{3}}\, \varphi_1(t) + \sqrt{\frac{T}{3}}\, \varphi_2(t) + \sqrt{\frac{T}{3}}\, \varphi_3(t). \tag{10.2.58}$$

The orthonormal functions $\varphi_1(t)$, $\varphi_2(t)$, and $\varphi_3(t)$ are sketched in Fig. 10.2.6. Note that for this example $M = 3 < N = 4$.

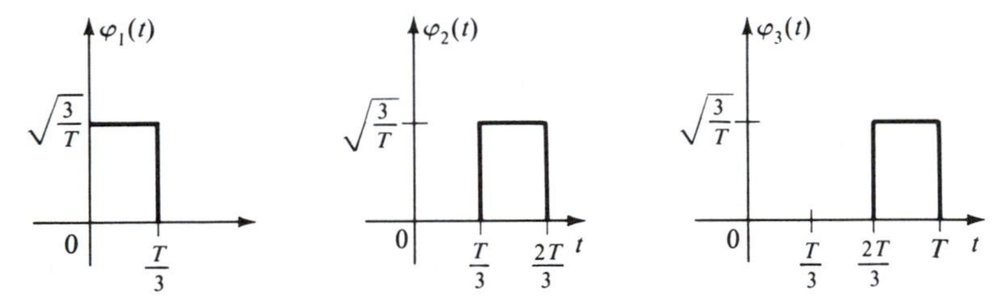

FIGURE 10.2.6 Orthonormal functions for the signal set of Example 10.2.4.

The signal vector expressions for the $s_i(t)$ are

$$\begin{aligned}
\mathbf{s}_1 &= \left[\sqrt{\frac{T}{3}} \quad 0 \quad 0\right] \\
\mathbf{s}_2 &= \left[\sqrt{\frac{T}{3}} \quad \sqrt{\frac{T}{3}} \quad 0\right] \\
\mathbf{s}_3 &= \left[0 \quad \sqrt{\frac{T}{3}} \quad 0\right] \\
\mathbf{s}_4 &= \left[\sqrt{\frac{T}{3}} \quad \sqrt{\frac{T}{3}} \quad \sqrt{\frac{T}{3}}\right].
\end{aligned} \tag{10.2.59}$$

Based on these vectors, we can compute intersignal distances, energies, and inner products in the three-dimensional signal space as needed.

There are several geometrically simple and practically important M-dimensional signal sets, where $M > 2$. We do not pursue these signal sets here; however, several problems are dedicated to the most common ones.

10.3 The Additive White Gaussian Noise Channel

We have discussed in Chapter 8 the effects of deterministic distortion and how we can compensate for this type of distortion by using equalization. We also noted in Chapter 8 that various kinds of random disturbances are present in communication systems, such as impulsive noise, abrupt phase changes (hits), and additive noise, just to name a few. One type of distortion evident in most communication systems can be accurately modeled as additive white Gaussian noise (AWGN). For some communications channels, such as the deep-space channel, AWGN is the principal random disturbance present, while for other channels, such as the telephone channel, other random impairments (impulsive noise, for example) are a greater cause of error events (see Appendix C for brief descriptions of several channel models). Even for channels such as the telephone channel, however, it is common to design systems such that they perform well in the presence of AWGN. There are several reasons usually given for this approach. First, disturbances such as impulsive noise have proven difficult to model. Second, when impulsive noise occurs, it is often of such a large amplitude that a system designed to prevent errors due to this phenomenon would be relatively inefficient when the impulsive noise is not present. Third, for high-speed data transmission, there are so many possible transmitted signals that AWGN does become a significant source of errors. Fourth, practical experience accumulated over many years indicates that communication systems designed to protect against AWGN also perform well in the presence of other disturbances. Fifth, the AWGN channel model is analytically tractable; that is, we can handle it mathematically. For all of the reasons just given, we spend the remainder of this chapter discussing the AWGN channel model and designing communication systems that perform well over such channels.

Given a transmitted time-domain waveform $s_i(t)$, we assume that this signal is contaminated by an additive disturbance $n(t)$, so that the waveform at the receiver input is

$$r(t) = s_i(t) + n(t), \tag{10.3.1}$$

where $n(t)$ is a zero-mean, white Gaussian noise process with power spectral density

$$S_n(\omega) = 2\pi\left(\frac{\mathcal{N}_0}{2}\right) \quad \text{watts/rad/sec}, \qquad -\infty < \omega < \infty. \tag{10.3.2}$$

Of course, we can represent the transmitted waveforms $s(t)$ as some vector in signal space, as discussed in Section 10.2, and we wonder if it is possible to represent $n(t)$, and hence $r(t)$, in the same signal space. Although we do not prove it here, such a representation is in fact possible, and we proceed directly to the vector space formulation of Eq. (10.3.1). If the transmitted waveform $s_i(t)$ can be expressed as

$$s_i(t) = \sum_{j=1}^{M} s_{ij}\varphi_j(t), \tag{10.3.3}$$

where

$$s_{ij} = \int_{-\infty}^{\infty} s_i(t)\varphi_j(t)\,dt, \tag{10.3.4}$$

then we have the vector representation

$$\mathbf{s}_i = [s_{i1} \quad s_{i2} \quad \cdots \quad s_{iM}]. \tag{10.3.5}$$

The vector representation for the noise follows similarly as

$$\mathbf{n} = [n_1 \quad n_2 \quad \cdots \quad n_M], \tag{10.3.6}$$

where

$$n_j = \int_{-\infty}^{\infty} n(t)\varphi_j(t)\,dt, \tag{10.3.7}$$

so

$$\mathbf{r} = [r_1 \quad r_2 \quad \cdots \quad r_M] = \mathbf{s}_i + \mathbf{n} \tag{10.3.8}$$

with

$$r_j = \int_{-\infty}^{\infty} r(t)\varphi_j(t)\,dt. \tag{10.3.9}$$

Based on the given conditions on $n(t)$, it can be demonstrated that the components of $\mathbf{n}$, namely, the $\{n_j, j = 1, 2, \ldots, M\}$, are independent, identically distributed Gaussian random variables with zero mean and variance $\mathcal{N}_0/2$. The pdf of the random vector $\mathbf{n}$ is thus given by

$$f_{\mathbf{n}}(\mathbf{n}) = \frac{1}{[\pi\mathcal{N}_0]^{M/2}} e^{-\sum_{j=1}^{M} n_j^2/\mathcal{N}_0}. \tag{10.3.10}$$

A fact that will be important later is that the pdf in Eq. (10.3.10) is spherically symmetric, which means that the pdf depends only on the magnitude of the noise vector but not its direction.

Equation (10.3.8) thus expresses our signal space model for the received signal, which can be illustrated for two dimensions as shown in Fig. 10.3.1. Therefore, the additive noise vector causes the transmitted signal vector $\mathbf{s}_i$ to be observed at the receiver as another point in signal space, $\mathbf{r}$. The receiver must then make its decision as to which of the N signal vectors $\{\mathbf{s}_j, j = 1, 2, \ldots, N\}$ was transmitted based on the received vector $\mathbf{r}$ and knowledge of the noise vector pdf in Eq. (10.3.10).

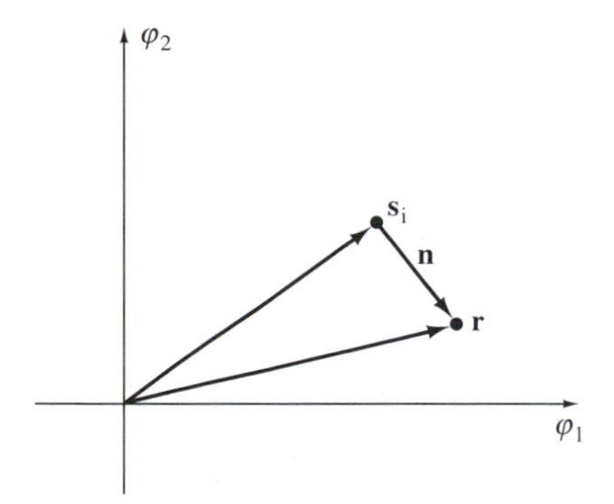

FIGURE 10.3.1 Signal space AWGN channel model.

10.4 Optimum Receivers

Using the signal space representations established in Sections 10.2 and 10.3, we show in this section how to design receivers that are optimum in the sense that they minimize the probability of error, denoted $P[\mathscr{E}]$, for the given transmitted signal set and noise vector pdf. We find that these receivers have an elegant and intuitively pleasing interpretation in signal space. Before proceeding, however, let us consider a block diagram of the overall communication system of interest to us here.

A block diagram of the communication system in terms of our vector space formulation is shown in Fig. 10.4.1. The input to the transmitter m_i is one of a set of N messages $\{m_j, j = 1, 2, \ldots, N\}$. The transmitter assigns a signal vector $\mathbf{s}_i$ (waveform) to the message to be sent from the set of possible transmitted vectors $\{\mathbf{s}_j, j = 1, 2, \ldots, N\}$. The transmitted vector is then disturbed by the additive noise vector $\mathbf{n}$ so that the received vector is $\mathbf{r} = \mathbf{s}_i + \mathbf{n}$. The receiver processes $\mathbf{r}$ to obtain the best possible estimate, $\hat{m}$, of which of the finite number of messages was being communicated. We now turn our attention to the design of the receiver.

For generality, we assume that we know the a priori probabilities of the individual messages, which we denote by $P[m_j]$. We desire to find a receiver that minimizes $P[\mathscr{E}]$, or equivalently, maximizes the probability of a correct

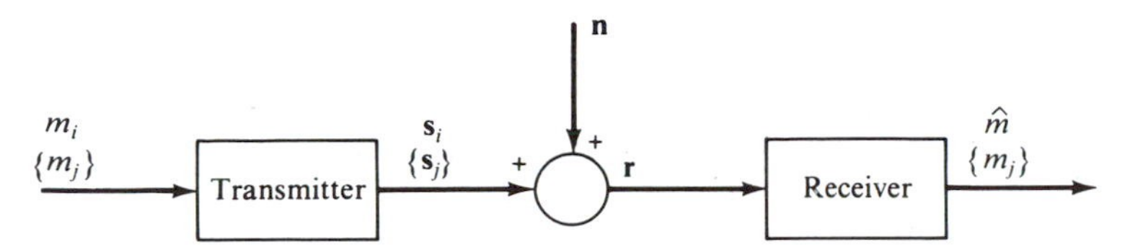

FIGURE 10.4.1 Block diagram of the communication system.

decision, $P[\mathscr{C}]$. If we let $f_{\mathbf{r}}(\mathbf{r})$ denote the pdf of the received vector, the probability of a correct decision can be written as

$$P[\mathscr{C}] = \int_{-\infty}^{\infty} P[\mathscr{C}|\mathbf{r} = \alpha] f_{\mathbf{r}}(\alpha)\, d\alpha. \tag{10.4.1}$$

Since $f_{\mathbf{r}}(\alpha) \geq 0$ and does not depend on the receiver, we see that we maximize $P[\mathscr{C}]$ if we maximize $P[\mathscr{C}|\mathbf{r} = \alpha]$. Now, if $\mathbf{s}_k$ is transmitted, then when the receiver chooses $\hat{m} = m_k$,

$$P[\mathscr{C}|\mathbf{r} = \alpha] = P[m_k|\mathbf{r} = \alpha]. \tag{10.4.2}$$

Therefore, the receiver that maximizes $P[\mathscr{C}]$ sets $\hat{m}$ equal to that m_i which has the maximum a posteriori probability of occurrence $P[m_i|\mathbf{r} = \alpha]$.

Using Bayes' rule, we can expand the a posteriori message probabilities as

$$P[m_i|\mathbf{r} = \alpha] = \frac{f_{\mathbf{r}|m_i}(\alpha|m_i)P[m_i]}{f_{\mathbf{r}}(\alpha)}. \tag{10.4.3}$$

With a little thought, we note that we need not retain the denominator on the right side of Eq. (10.4.3), since it is independent of the index i. Therefore, the optimum receiver sets $\hat{m} = m_k$ if (and only if) for all $i \neq k$,

$$f_{\mathbf{r}|m_k}(\alpha|m_k)P[m_k] > f_{\mathbf{r}|m_i}(\alpha|m_i)P[m_i]. \tag{10.4.4}$$

Since conditioning on m_i is equivalent to conditioning on $\mathbf{s}_i$, the optimum receiver sets $\hat{m} = m_k$ if and only if for all $i \neq k$,

$$f_{\mathbf{r}|\mathbf{s}_k}(\alpha|\mathbf{s}_k)P[m_k] > f_{\mathbf{r}|\mathbf{s}_i}(\alpha|\mathbf{s}_i)P[m_i]. \tag{10.4.5}$$

We can make this result much more explicit by using Eq. (10.3.10) to find the conditional pdfs $f_{\mathbf{r}|\mathbf{s}_i}(\alpha|\mathbf{s}_i)$. We know that $\mathbf{r} = \alpha = \mathbf{s}_i + \mathbf{n}$, so

$$f_{\mathbf{r}|\mathbf{s}_i}(\alpha|\mathbf{s}_i) = f_{\mathbf{n}}(\alpha - \mathbf{s}_i|\mathbf{s}_i). \tag{10.4.6}$$

Assuming that the transmitted signal vector $\mathbf{s}_i$ and the noise are statistically independent, which is a physically reasonable assumption to make, we find that $f_{\mathbf{n}}(\alpha - \mathbf{s}_i|\mathbf{s}_i) = f_{\mathbf{n}}(\alpha - \mathbf{s}_i)$, so

$$f_{\mathbf{r}|\mathbf{s}_i}(\alpha|\mathbf{s}_i) = f_{\mathbf{n}}(\alpha - \mathbf{s}_i). \tag{10.4.7}$$

Upon invoking Eq. (10.3.10) and noting the result in Eq. (10.4.7), the optimum receiver finds that i which maximizes the statistic

$$\begin{aligned} f_{\mathbf{r}|\mathbf{s}_i}(\alpha|\mathbf{s}_i)P[m_i] &= f_{\mathbf{n}}(\alpha - \mathbf{s}_i)P[m_i] \\ &= \frac{1}{[\pi\mathscr{N}_0]^{M/2}} e^{-\sum_{j=1}^{M} (\alpha_j - s_{ij})^2/\mathscr{N}_0} P[m_i]. \end{aligned} \tag{10.4.8}$$

To simplify this, we drop the factor $[\pi\mathscr{N}_0]^{M/2}$, since it is independent of i, and take the natural logarithm of the remaining terms to obtain

$$-\sum_{j=1}^{M} \frac{(\alpha_j - s_{ij})^2}{\mathscr{N}_0} + \ln P[m_i]. \tag{10.4.9}$$

Maximizing Eq. (10.4.9) is the same as minimizing the negative of the quantity, hence the optimum receiver sets $\hat{m} = m_k$ whenever

$$\sum_{j=1}^{M} (\alpha_j - s_{ij})^2 - \mathcal{N}_0 \ln P[m_i] \qquad (10.4.10)$$

is a minimum for $i = k$. We can also rewrite this result in the form of Eq. (10.4.5); that is, the optimum receiver sets $\hat{m} = m_k$ if and only if for all $i \neq k$,

$$\sum_{j=1}^{M} (\alpha_j - s_{kj})^2 - \mathcal{N}_0 \ln P[m_k] < \sum_{j=1}^{M} (\alpha_j - s_{ij})^2 - \mathcal{N}_0 \ln P[m_i]. \qquad (10.4.11)$$

Note that if the messages are all equally likely, then $P[m_i] = P[m_k]$ for all i, k, so Eq. (10.4.11) becomes

$$\sum_{j=1}^{M} (\alpha_j - s_{kj})^2 < \sum_{j=1}^{M} (\alpha_j - s_{ij})^2. \qquad (10.4.12)$$

Therefore, for equal a priori message probabilities, the receiver chooses $\hat{m} = m_k$ if and only if the received vector $\mathbf{r} = \alpha$ is closer to $\mathbf{s}_k$, in terms of Euclidean distance, than to any other $\mathbf{s}_i$, $i \neq k$.

For two dimensions, then, the optimum receiver is simply a partitioning of the two-dimensional signal space into regions, the points of which are closest to a given transmitted signal vector $\mathbf{s}_i$. These regions are called *decision regions.* We illustrate this general result by several examples.

EXAMPLE 10.4.1

As our first illustration, we consider the transmission of four equally likely messages over an AWGN channel with power spectral density $\mathcal{N}_0/2$ watts/Hz using QAM in Example 10.2.1. From Eq. (10.4.12) we see that the optimum decision regions are just the four quadrants; that is, the optimum receiver sets:

$\hat{m} = m_1$ if $\mathbf{r} = \alpha$ is in the first quadrant,

$\hat{m} = m_3$ if $\mathbf{r} = \alpha$ is in the second quadrant,

$\hat{m} = m_4$ if $\mathbf{r} = \alpha$ is in the third quadrant,

$\hat{m} = m_2$ if $\mathbf{r} = \alpha$ is in the fourth quadrant.

The boundaries of the decision regions are the φ_1 and φ_2 axes.

As a second example, consider using the QAM signal set in Fig. 10.2.2 to send 16 equally likely messages over an AWGN channel. The resulting decision region boundaries are shown as dashed lines in Fig. 10.4.2 and include the φ_1 and φ_2 axes. Thus, if $\alpha = [\alpha_1 \ \ \alpha_2]$ is such that $\alpha_1 > 2A_c/3$ and $\alpha_2 > 2A_c/3$, the optimum receiver sets $\hat{m} = m_1$, since α is closest to $\mathbf{s}_1$ in the signal space. Also, if $0 < \alpha_1 < 2A_c/3$ and $0 < \alpha_2 < 2A_c/3$, then the optimum receiver sets $\hat{m} = m_6$. The optimum receiver thus has a simple geometric interpretation for these QAM signal sets.

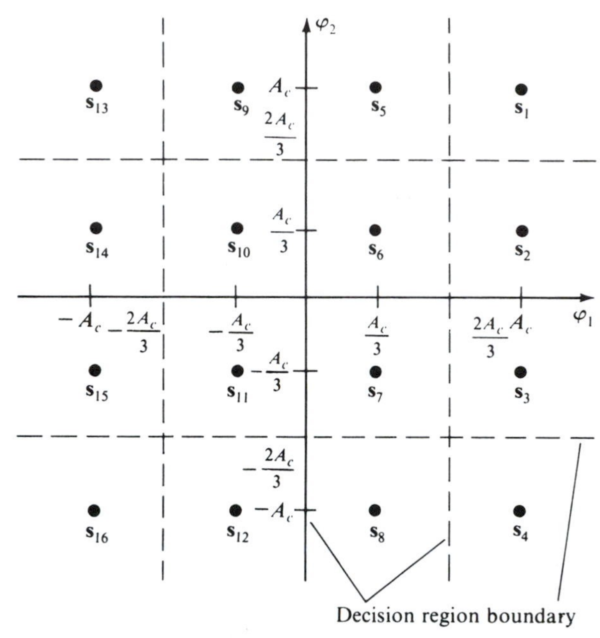

FIGURE 10.4.2 Optimum decision regions for equally likely 16-level QAM (Example 10.4.1).

EXAMPLE 10.4.2

As a second example, we consider using PSK modulation to represent equally likely messages over an AWGN channel. For the binary case, we have the signals in Eqs. (10.2.18a) and (10.2.18b) with the signal space diagram in Fig. 10.2.3. There are only two decision regions for this case, and the decision boundary is the dashed line through 0 shown in Fig. 10.4.3. Thus, if $r = \alpha > 0$, $\hat{m} = m_1$, and if $r = \alpha < 0$, $\hat{m} = m_2$.

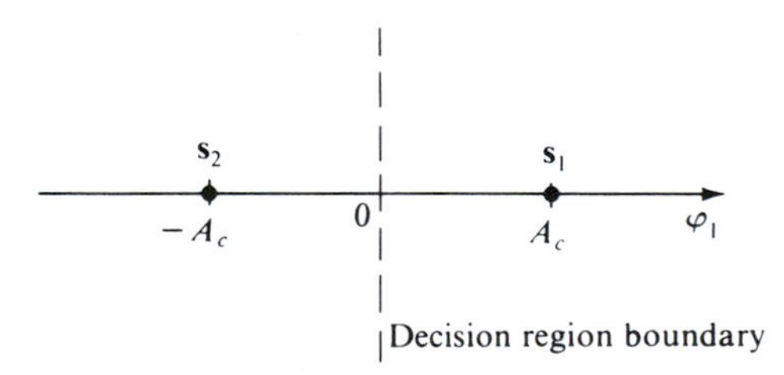

FIGURE 10.4.3 Optimum decision regions for BPSK (Example 10.4.2).

Similar to the other examples, we can sketch the optimum decision boundaries for the eight-phase PSK signal set in Fig. 10.2.4 as shown in Fig. 10.4.4. Of course, we are again assuming equally likely messages and an AWGN channel. Note that the decision regions in Fig. 10.4.4 are not as trivially written as inequalities on the components of α as in the QAM case.

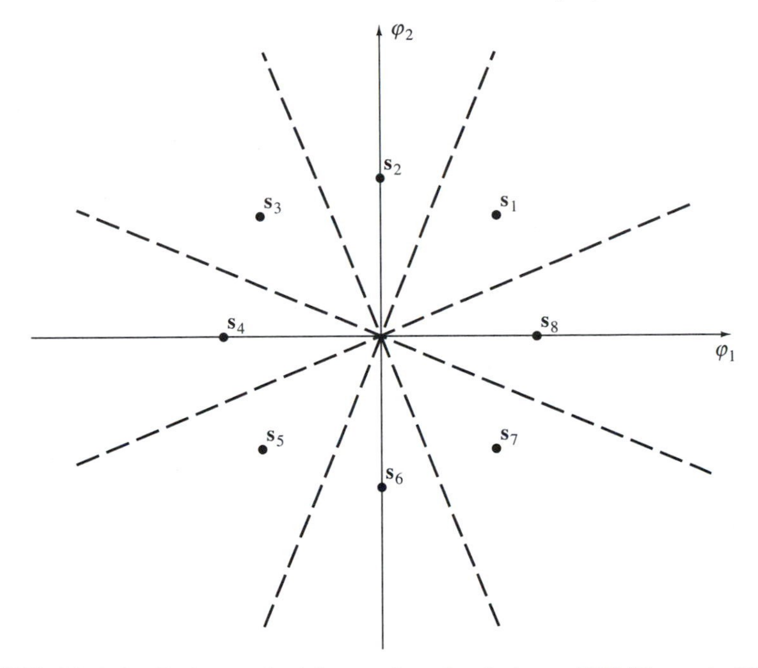

FIGURE 10.4.4 Optimum decision regions for 8-phase PSK (Example 10.4.2).

Before concluding this example, let us return to the BPSK case and assume that $P[m_1] > P[m_2]$. From Eq. (10.4.11), we see that the optimum receiver selects $\hat{m} = m_1$ if

$$\sum_{j=1}^{2} (\alpha_j - s_{1j})^2 - \mathcal{N}_0 \ln P[m_1] < \sum_{j=1}^{2} (\alpha_j - s_{2j})^2 - \mathcal{N}_0 \ln P[m_2] \quad (10.4.13)$$

and $\hat{m} = m_2$ if the inequality is reversed. We can write this optimum decision rule in a compact form as

$$\sum_{j=1}^{2} (\alpha_j - s_{1j})^2 - \mathcal{N}_0 \ln P[m_1] \underset{m_2}{\overset{m_1}{\lessgtr}} \sum_{j=1}^{2} (\alpha_j - s_{2j})^2 - \mathcal{N}_0 \ln P[m_2], \quad (10.4.14)$$

where the labels on the inequality signs indicate the decision if the corresponding inequality is satisfied. Collecting terms involving a priori probabilities on the left side of Eq. (10.4.14) yields

$$\sum_{j=1}^{2} (\alpha_j - s_{1j})^2 - \mathcal{N}_0 \ln \frac{P[m_1]}{P[m_2]} \underset{m_2}{\overset{m_1}{\lessgtr}} \sum_{j=1}^{2} (\alpha_j - s_{2j})^2. \quad (10.4.15)$$

Now, since $P[m_1] > P[m_2]$ and $\mathcal{N}_0 > 0$, the term $-\mathcal{N}_0 \ln(P[m_1]/P[m_2])$ is negative, so the boundary between the two decision regions is shifted away from $\mathbf{s}_1$ nearer to $\mathbf{s}_2$. This shifted boundary is illustrated in Fig. 10.4.5. Note that this result is intuitive, since m_1 is more likely to occur than m_2, and hence should be associated with an increased portion of signal space to minimize $P[\mathscr{E}]$.

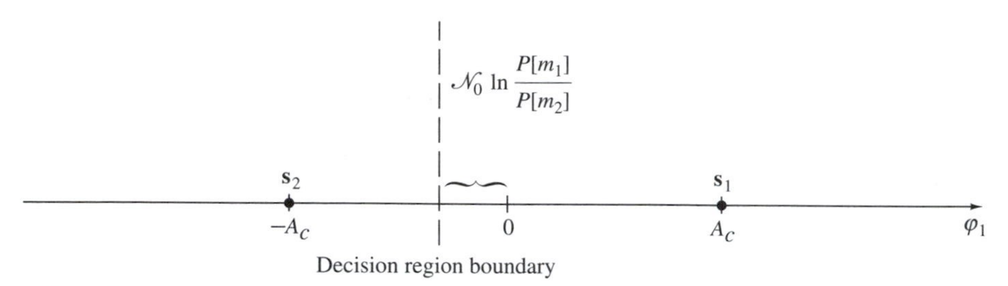

FIGURE 10.4.5 Decision regions for BPSK with $P[m_1] > P[m_2]$ (Example 10.4.2).

Thus far we have only specified our receivers in terms of operations in signal space. We now present receiver implementations that perform the signal space operations. Note from Eq. (10.4.10) that the optimum receiver first calculates the $r_j = \alpha_j$ components according to Eq. (10.3.9) and then computes the quantity in Eq. (10.4.10) for all i. If we expand the first term in Eq. (10.4.10) (letting $r_j = \alpha_j$),

$$\sum_{j=1}^{M} (r_j - s_{ij})^2 = \sum_{j=1}^{M} r_j^2 - 2 \sum_{j=1}^{M} r_j s_{ij} + \sum_{j=1}^{M} s_{ij}^2. \tag{10.4.16}$$

We see that the term $\sum r_j^2$ is independent of i, so that we can rewrite the optimum decision rule as choosing $\hat{m} = m_k$ to yield a minimum of

$$-2 \sum_{j=1}^{M} r_j s_{ij} + \sum_{j=1}^{M} s_{ij}^2 - \mathcal{N}_0 \ln P[m_i] \tag{10.4.17}$$

for $i = k$. Equivalently, we can state the optimum decision rule as setting $\hat{m} = m_k$ if and only if the expression

$$\sum_{j=1}^{M} r_j s_{ij} + \frac{1}{2}\left[\mathcal{N}_0 \ln P(m_i) - \sum_{j=1}^{M} s_{ij}^2\right] \tag{10.4.18}$$

is a maximum for $i = k$.

Using Eqs. (10.4.18) and (10.3.9), we can now draw a block diagram of the optimum receiver as shown in Fig. 10.4.6. In the figure we have defined

$$b_i \triangleq \frac{1}{2}\left[\mathcal{N}_0 \ln P(m_i) - \sum_{j=1}^{M} s_{ij}^2\right] \tag{10.4.19}$$

for $i = 1, 2, \ldots, N$. The reader should note that there are M inputs to the correlator and N outputs. The last block simply compares the N quantities and chooses $\hat{m}$ to be the m_k corresponding to the largest, as indicated by Eq. (10.4.18).

When the orthonormal basis functions $\{\varphi_j(t), j = 1, 2, \ldots, M\}$ are of finite duration, say, limited to be nonzero only over the interval $0 \leq t \leq T$, the multiplications in the implementation of Fig. 10.4.6 can be avoided by using matched filters to generate the components of $\mathbf{r}$. To see this, consider a filter with impulse response $h_j(t), j = 1, 2, \ldots, M$, excited by the random process $r(t)$.

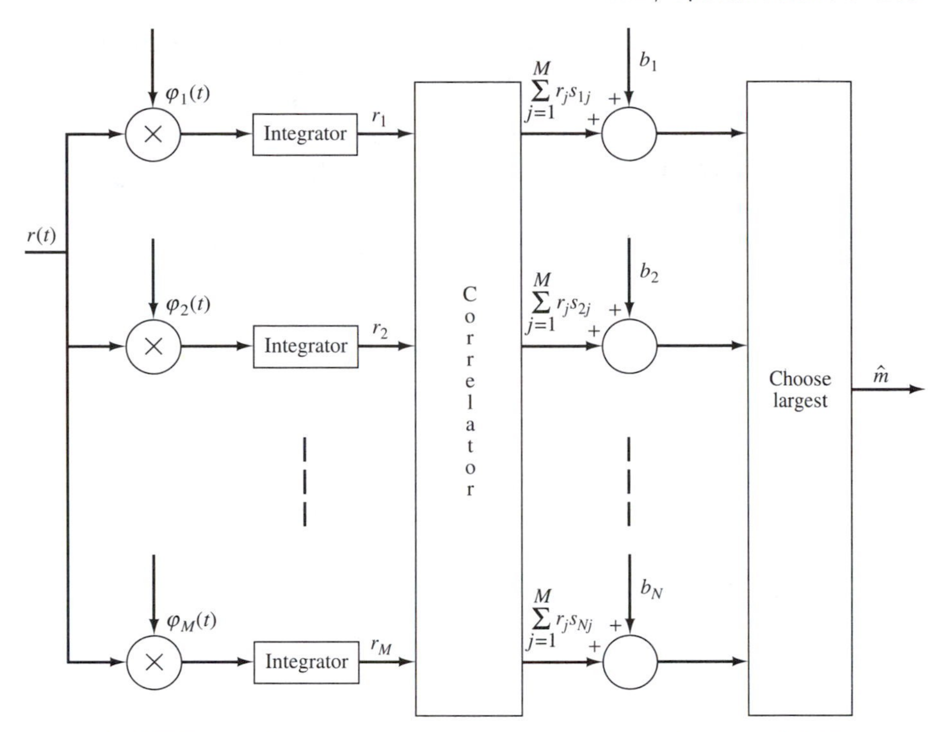

FIGURE 10.4.6 Optimum receiver implementation for Eq. (10.4.18).

The output is then represented by the convolution

$$y_j(t) = \int_{-\infty}^{\infty} r(\lambda)h_j(t - \lambda)\, d\lambda. \tag{10.4.20}$$

Choosing

$$h_j(t) = \varphi_j(T - t), \qquad j = 1, 2, \ldots, M, \tag{10.4.21}$$

we obtain from Eq. (10.4.20) that

$$y_j(t) = \int_{-\infty}^{\infty} r(\lambda)\varphi_j(T - t + \lambda)\, d\lambda. \tag{10.4.22}$$

If the output of the filter is sampled at $t = T$, then

$$y_j(T) = \int_{-\infty}^{\infty} r(\lambda)\varphi_j(\lambda)\, d\lambda, \tag{10.4.23}$$

and we see by comparison with Eq. (10.3.9) that we have produced the components of $\mathbf{r}$, since $y_j(T) = r_j$, $j = 1, 2, \ldots, M$. We thus have an alternative method for generating the $\{r_j\}$ without using multipliers as in Fig. 10.4.6. The remainder of Fig. 10.4.6 is unchanged.

A filter with the impulse response in Eq. (10.4.21) is said to be *matched* to the signal $\varphi_j(t)$, and hence is called a *matched filter* for $\varphi_j(t)$. The matched filter can also be shown to be the filter that maximizes the filter output signal-to-noise ratio when the input is signal plus noise. Since this derivation is available in numerous texts, we do not present it here (see Problem 10.30).

The optimal receiver can also be implemented in terms of the waveforms $\{s_i(t), i = 1, 2, \ldots, N\}$ rather than the orthonormal basis functions $\{\varphi_j(t), j = 1, 2, \ldots, M\}$. To see this, consider the first term in the optimal receiver computation in Eq. (10.4.18) and write

$$\begin{aligned}\sum_{j=1}^{M} r_j s_{ij} &= \sum_{j=1}^{M} s_{ij} \int_{-\infty}^{\infty} r(t)\varphi_j(t)\, dt \\ &= \int_{-\infty}^{\infty} r(t) \sum_{j=1}^{M} s_{ij}\varphi_j(t)\, dt \\ &= \int_{-\infty}^{\infty} r(t)s_i(t)\, dt,\end{aligned} \tag{10.4.24}$$

where we have employed Eqs. (10.3.9) and (10.3.3), respectively. Thus Eq. (10.4.24) implies that the optimal receiver can be implemented by performing N correlations of the received signal $r(t)$ with each of the possible transmitted waveforms. Following an argument similar to that in Eqs. (10.4.20)–(10.4.23), the correlation calculation in Eq. (10.4.24) can be shown to be equivalent to a matched filter implementation if the $s_i(t)$ are constrained to have a time duration T. The reader should draw block diagrams for correlation and matched filter implementations of the optimal receiver in terms of the $\{s_i(t), i = 1, 2, \ldots, N\}$ (see Problem 10.31).

Let us now turn to a consideration of how to use the optimum receiver structures derived in this section, and matched filtering in particular, in conjunction with the pulse shaping requirements in Chapter 8. We consider various degrees of generality and present only the results, leaving the derivations to the literature. To begin, we establish some notation by referring to Fig. 10.4.7, where $H_T(\omega)$, $H_C(\omega)$, and $H_R(\omega)$ are the transfer functions of the transmitter filter, the channel, and the receiver filter, respectively. Suppose now that we wish to utilize a particular pulse shaping characteristic, say $P(\omega)$, and we are sending data over the AWGN channel described by Eqs. (10.3.1) and (10.3.2). In this case, $H_C(\omega) = 1$ for all ω, and the optimum transmitting and receiving filters, which minimize the probability of error and maximize the output peak signal-to-noise ratio, are $H_T(\omega) = \sqrt{P(\omega)}$ and $H_R(\omega) = \sqrt{P(\omega)}$, for $P(\omega)$ purely real. Thus, to

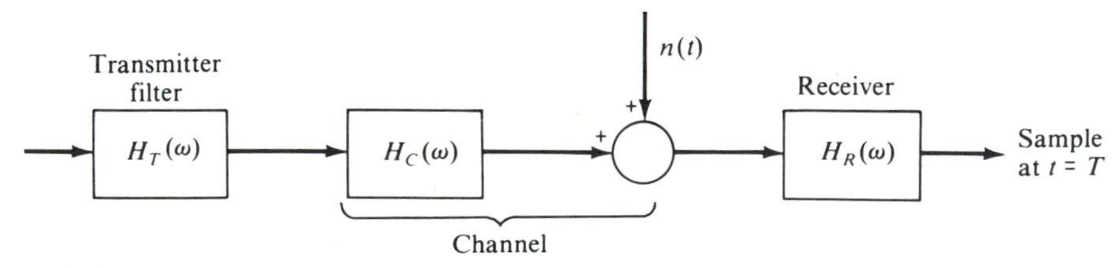

FIGURE 10.4.7 Transfer function representation of the overall communication system.

achieve a pulse shaping $P(\omega)$ at the receiver output and to minimize the symbol error probability, the pulse shaping is evenly split, in a geometric mean sense, between the transmitter and receiver. For simplicity, we are ignoring the pure time delay of T seconds in a matched filter [see Eq. (10.4.21)], and thus the sampling in Fig. 10.4.7 occurs at $t = T = 0$.

The next level of generality is to consider a general $H_C(\omega)$ with additive white Gaussian noise, but only optimize for the case where a single, isolated pulse is transmitted. For $H_T(\omega)$ and $H_C(\omega)$ given, the resulting optimum receiver has the transfer function $H_R(\omega) = H_T^*(\omega)H_C^*(\omega)$. By allowing the possibility of colored noise, so that $S_n(\omega) \neq$ constant but with all other conditions the same, it can be shown that $H_R(\omega) = [H_T^*(\omega)H_C^*(\omega)]/S_n(\omega)$, where $S_n(\omega) \neq 0$ for any ω of interest. A main point to be noted here is that the receiving filter is matched not only to the desired pulse shape, but also to the shaping contributed by the channel. A block diagram of this system is shown in Fig. 10.4.8.

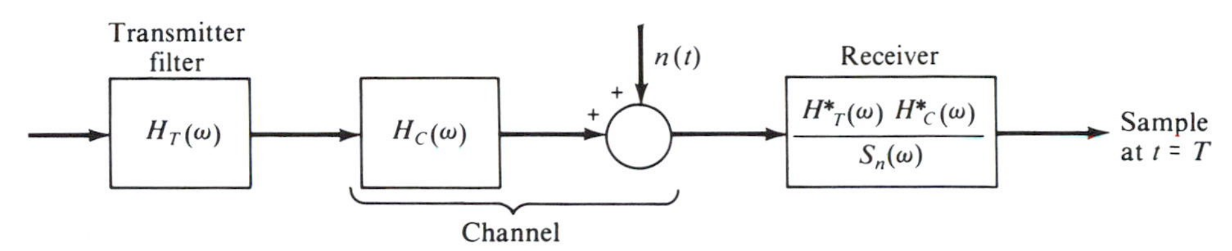

FIGURE 10.4.8 Transfer function representation of an optimal symbol-by-symbol receiver for colored noise and single pulse transmission.

Of course, the treatment for a single isolated pulse is far from reality, since we are virtually always interested in transmitting long sequences of symbols. For $H_C(\omega) \neq$ constant, the pulses overlap and we have intersymbol interference. The optimum symbol-by-symbol receiver for the case of $H_T(\omega)$ and $H_C(\omega)$ given, additive white Gaussian noise, and a sequence of pulses is $H_R(\omega) = H_T^*(\omega)H_C^*(\omega)H_{\text{eq}}(\omega)$, where $H_{\text{eq}}(\omega)$ is the transfer function of a transversal filter equalizer. The communication system thus has the form shown in Fig. 10.4.9. Computing the tap gains of $H_{\text{eq}}(\omega)$ to minimize the error probability is not a

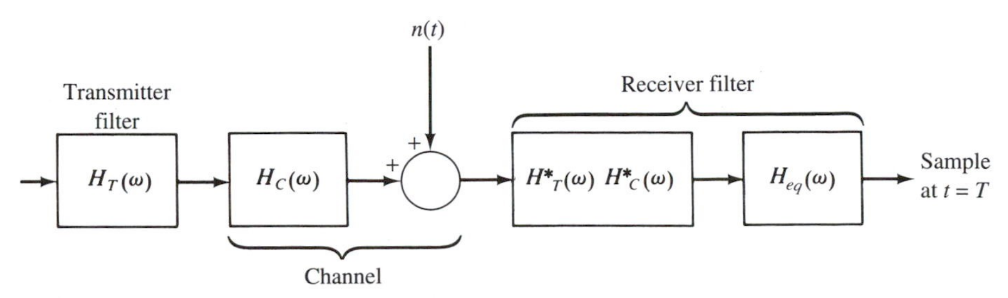

FIGURE 10.4.9 Optimum symbol-by-symbol receiver for white noise and a sequence of transmitted pulses.

simple exercise, and hence it is common to adjust the $H_{eq}(\omega)$ coefficients to minimize peak distortion or mean-squared error as in Chapter 8.

Finally, whenever intersymbol interference is present at the receiver, whether it is due to the channel or a result of inserting controlled intersymbol interference, as in partial response signaling, symbol-by-symbol detection can be improved. Specifically, the optimum receiver in such a case examines a long sequence of symbols and makes a decision concerning the entire transmitted sequence rather than just one symbol at a time. Note that if each symbol has L possible levels and the length of the sequence is K symbols, there are L^K possible (distinct) sequences to be considered by the optimal receiver. Clearly, such receivers can be exceptionally complicated, and perhaps surprisingly, they are becoming common in off-the-shelf communications equipment. The algorithm underlying these receivers is developed in Section 10.7, and several systems that employ receivers based on this concept of optimum sequence detection are treated briefly in Chapter 13. For derivations and much relevant discussion, see Lucky, Salz, and Weldon [1968] and Proakis [1989].

Up to this point in the chapter, our development has emphasized coherent demodulators, although this fact has not been stated explicitly. In retrospect the limitation to coherent receivers is obvious, since we implicitly assume that we have available the components of the received vector $\mathbf{r} = \alpha = [\alpha_1 \; \alpha_2 \; \cdots \; \alpha_M]$. This is possible only if we have coherent references for the orthonormal basis functions $\{\varphi_j(t), j = 1, 2, \ldots, M\}$ at the receiver. Now certainly the basis functions used at the transmitter are known to the receiver; however, from Section 5.2 on AMDSB-SC we know that for coherent demodulation the reference signals (LOs) must include the deterministic effects of the channel. That is, the coherent references at the receiver must be tuned to the $\{\varphi_j(t)\}$ *after* they have been transmitted over the channel. Of course, because of this requirement, coherent demodulators can be complicated, and there are many applications where simpler noncoherent demodulators are appropriate. Thus we now turn our attention to the design of optimum receivers subject to the constraint that noncoherent demodulation is to be used.

If $\mathbf{r}$ is considered to be the output of the noncoherent demodulator rather than the receiver input, the optimum signal processing required after the demodulator is still of the form given by Eq. (10.4.5). To avoid confusion, we denote the output of the noncoherent demodulator (which is a scalar) by z, so that the constrained optimum receiver sets $\hat{m} = m_k$ if and only if for all $i \neq k$,

$$f_{Z|\mathbf{s}_k}(z|\mathbf{s}_k)P[m_k] > f_{Z|\mathbf{s}_i}(z|\mathbf{s}_i)P[m_i]. \tag{10.4.25}$$

Rather than develop this result in its full generality, we now pursue the two important special cases of amplitude shift keying (ASK) and frequency shift keying (FSK). As the reader will see, the geometric interpretation so useful in the coherent receiver development is not as apparent or as important here.

A block diagram for the noncoherent ASK receiver is shown in Fig. 10.4.10. For ASK we have either

$$r(t) = n(t) \text{ for message } m_1, \tag{10.4.26}$$

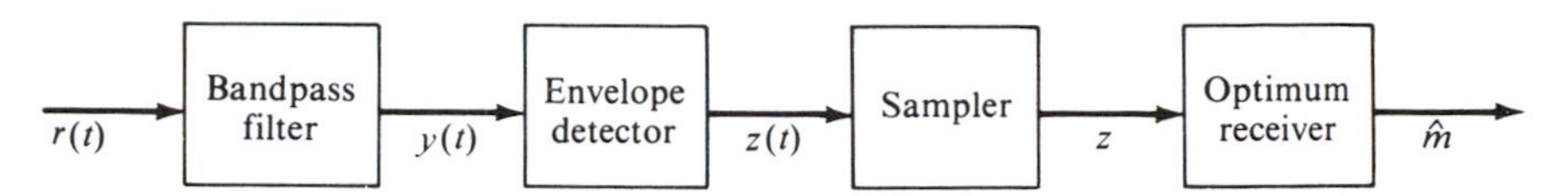

FIGURE 10.4.10 Receiver for noncoherent ASK.

or

$$r(t) = A_c \cos [\omega_c t + \psi] + n(t) \tag{10.4.27}$$

for message m_2 and ψ unknown. For ease of development and without loss of generality, we let $\psi = 0$ in the following. Correspondingly, the bandpass filter output is either

$$y(t) = n_o(t) \tag{10.4.28}$$

or

$$y(t) = A_c \cos \omega_c t + n_o(t), \tag{10.4.29}$$

where $n_o(t)$ is narrowband Gaussian noise as given by Eq. (10.2.7). When $m_1(s_1)$ is sent, the random variable z has a Rayleigh distribution as given by Eq. (10.2.13) or

$$f_{Z|s_1}(z|s_1) = \frac{z}{\mathcal{N}_0/2} e^{-z^2/\mathcal{N}_0}. \tag{10.4.30}$$

The conditional pdf of z given $m_2(s_2)$ requires a little more work. In this case

$$\begin{aligned} z(t) &= \sqrt{[A_c + n_i(t)]^2 + n_q^2(t)} \\ &\triangleq \sqrt{x^2(t) + n_q^2(t)}, \end{aligned} \tag{10.4.31}$$

so to find the pdf of z, we can find the joint pdf of $x = A + n_i$ and n_q and then perform a transformation of variables. Since n_i and n_q are both Gaussian and statistically independent, the desired joint pdf is

$$f[x, n_q] = \frac{1}{\pi \mathcal{N}_0} e^{-[(x - A_c)^2 + n_q^2]/\mathcal{N}_0}. \tag{10.4.32}$$

Defining

$$\theta \triangleq \tan^{-1} \frac{n_q}{x}, \tag{10.4.33}$$

we find that since $z = \sqrt{x^2 + n_q^2}$, $x = z \cos \theta$ and $n_q = z \sin \theta$, we can use a transformation of variables to write the joint pdf in Eq. (10.4.32) as

$$f(z, \theta) = \frac{z}{\pi \mathcal{N}_0} e^{-[z^2 - 2A_c z \cos \theta + A_c^2]/\mathcal{N}_0} \tag{10.4.34}$$

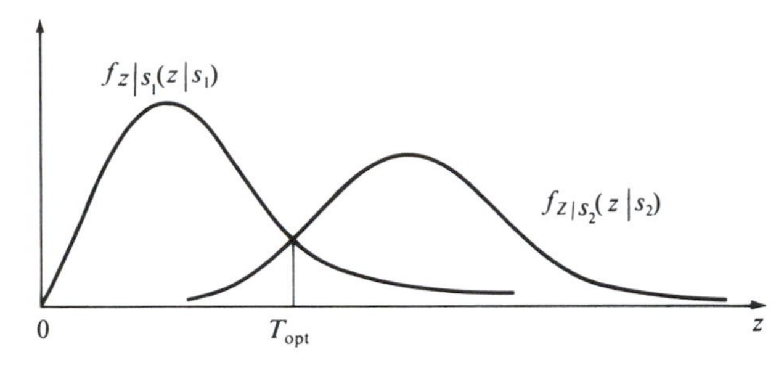

FIGURE 10.4.11 Conditional pdfs and optimum threshold for noncoherent ASK.

for $z \geq 0$ and $-\pi < \theta \leq \pi$. To obtain the desired conditional pdf for z, we must integrate $f(z, \theta)$ over θ. Thus

$$
\begin{aligned}
f_{Z|s_2}(z|s_2) &= \int_{-\pi}^{\pi} f(z, \theta)\, d\theta \\
&= \frac{z}{\pi \mathcal{N}_0} e^{-(z^2 + A_c^2)/\mathcal{N}_0} \int_{-\pi}^{\pi} e^{2A_c z \cos\theta/\mathcal{N}_0}\, d\theta \\
&= \frac{2z}{\mathcal{N}_0} e^{-(z^2 + A_c^2)/\mathcal{N}_0} I_0\left(\frac{2A_c z}{\mathcal{N}_0}\right),
\end{aligned}
\tag{10.4.35}
$$

where $z \geq 0$ and $I_0(\cdot)$ is the zeroth-order modified Bessel function of the first kind. The form of the pdf on the right side of Eq. (10.4.35) is a special one called the Rician pdf.

Using Eqs. (10.4.30) and (10.4.35) in Eq. (10.4.25) with $P(m_1) = P(m_2) = \frac{1}{2}$ yields the optimum receiver for equally likely messages and noncoherent demodulation. The two pdfs and the decision rule are illustrated in Fig. 10.4.11, where T_{opt} is the optimum threshold. If $z > T_{opt}$, $\hat{m} = m_2$, and if $z < T_{opt}$, $\hat{m} = m_1$. The exact determination of T_{opt} is nontrivial. When the error probability is calculated in the next section, an approximate value for T_{opt}, valid in the high signal-to-noise ratio case, is used.

A block diagram of a receiver for noncoherent FSK is shown in Fig. 10.4.12. The transmitted signals for FSK are

$$s_1(t) = A_c \cos \omega_1 t, \qquad 0 \leq t \leq T, \tag{10.4.36a}$$

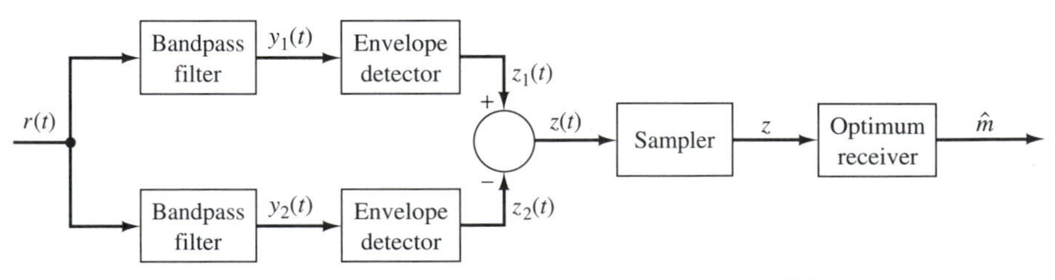

FIGURE 10.4.12 Receiver for noncoherent FSK.

for m_1 and

$$s_2(t) = A_c \cos \omega_2 t, \qquad 0 \le t \le T, \tag{10.4.36b}$$

for m_2, where ω_1 and ω_2 are suitably spaced to minimize spectral overlap. The receiver for noncoherent FSK is just two noncoherent ASK demodulators in parallel followed by the sampler and "optimum" receiver. Hence, if m_1 is sent,

$$z_1(t) = A_c \cos \omega_1 t + n_o(t) \tag{10.4.37}$$

and

$$z_2(t) = n_o(t), \tag{10.4.38}$$

while if m_2 is sent,

$$z_1(t) = n_o(t) \tag{10.4.39}$$

and

$$z_2(t) = A_c \cos \omega_2 t + n_o(t). \tag{10.4.40}$$

In Eqs. (10.4.37)–(10.4.40), $n_o(t)$ is narrowband Gaussian noise. Following the development for noncoherent ASK, the conditional pdfs of $z_1(t)$ and $z_2(t)$ after sampling when m_1 is sent are

$$f_{Z_1|s_1}(z_1 | s_1) = \frac{2z_1}{\mathcal{N}_0} e^{-(z_1^2 + A_c^2)/\mathcal{N}_0} I_0\left(\frac{2A_c z_1}{\mathcal{N}_0}\right) \tag{10.4.41}$$

for $z_1 > 0$, and

$$f_{Z_2|s_1}(z_2 | s_1) = \frac{2z_2}{\mathcal{N}_0} e^{-z_2^2/\mathcal{N}_0} \tag{10.4.42}$$

for $z_2 > 0$. The threshold for $z = z_1 - z_2$ is $T_{\text{opt}} = 0$, so we decide $\hat{m} = m_1$ if $z > 0$ or $z_1 > z_2$. We can write a similar relation for when m_2 is sent. Using these results, the probability of error is easy to calculate.

We conclude this section by noting that we have specified the optimum (minimum $P[\mathscr{E}]$) decision rule for transmitting N messages over an AWGN channel, and have shown how these decision rules might be implemented. We now turn our attention to error probability calculations for performance evaluations and to some aspects of signal design.

10.5 Error Probability and Signal Design

To find expressions for the error probability of an optimum receiver, we consider M-dimensional space and note that Eq. (10.4.11) partitions this space into N regions (decision regions) each one of which is associated with one and only one of the transmitted signal vectors, $\mathbf{s}_i$, $i = 1, 2, \ldots, N$. Due to the correspondence $\mathbf{s}_i \Leftrightarrow m_i$, we see that each of the N regions is also associated with one and only one of the transmitted messages. Denoting the decision region for the ith signal (and message) by $\mathscr{R}_i$, we see that for a received vector $\mathbf{r} = \alpha$, we decide $\hat{m} = m_k$

if $\mathbf{r} = \alpha \in \mathscr{R}_k$, that is, if the received $\mathbf{r}$ vector falls in the region of M-dimensional space associated with m_k.

Therefore, we make an error if $\mathbf{r} = \alpha \notin \mathscr{R}_k$ when $\mathbf{s}_k$ is transmitted. Denoting the probability of error given that $\mathbf{s}_k$, and hence m_k, is sent by $P[\mathscr{E}|m_k]$, we can write

$$\begin{aligned} P[\mathscr{E}|m_k] &= P[\mathbf{r} = \alpha \notin \mathscr{R}_k|m_k] \\ &= 1 - P[\mathbf{r} = \alpha \in \mathscr{R}_k|m_k] \\ &= 1 - P[\mathscr{C}|m_k]. \end{aligned} \tag{10.5.1}$$

Since we are interested in the total probability of error over the entire signal set, we use the a priori message probabilities and write

$$\begin{aligned} P[\mathscr{E}] &= \sum_{i=1}^{N} P[\mathscr{E}|m_i]P[m_i] \\ &= \sum_{i=1}^{N} P[\mathbf{r} = \alpha \notin \mathscr{R}_i|m_i]P[m_i]. \end{aligned} \tag{10.5.2}$$

It is often more convenient to calculate $P[\mathscr{E}]$ by using

$$\begin{aligned} P[\mathscr{E}] &= 1 - P[\mathscr{C}] \\ &= 1 - \sum_{i=1}^{N} P[\mathscr{C}|m_i]P[m_i] \\ &= 1 - \sum_{i=1}^{N} P[\mathbf{r} = \alpha \in \mathscr{R}_i|m_i]P[m_i]. \end{aligned} \tag{10.5.3}$$

Thus, once we calculate the conditional error probabilities or the conditional probabilities of a correct decision, we can find the total error probability from Eq. (10.5.2) or (10.5.3).

The problem is considerably simplified by the fact that we often work with a one- or two-dimensional signal space and that we are assuming an AWGN channel. Rather than continue to pursue this topic in generality, we consider a series of physically important illustrative examples.

EXAMPLE 10.5.1

We reexamine the binary PSK signal set with equally likely messages over an AWGN channel previously demonstrated in Example 10.4.2 to have the optimum decision regions shown in Fig. 10.4.3. From this figure we see that $\mathscr{R}_1 = \{r = \alpha > 0\}$ and $\mathscr{R}_2 = \{r = \alpha < 0\}$. Thus, since $P[m_1] = P[m_2] = \frac{1}{2}$, we have from Eq. (10.5.2) that

$$P[\mathscr{E}] = \tfrac{1}{2}\{P[r < 0|m_1] + P[r > 0|m_2]\}. \tag{10.5.4}$$

Assuming that the noise is zero mean with variance σ^2 (this implies that $\mathscr{N}_0/2 = \sigma^2$), we can write

$$P[r < 0|m_1] = \frac{1}{\sqrt{2\pi}\,\sigma} \int_{-\infty}^{0} e^{-(r - A_c)^2/2\sigma^2}\, dr. \tag{10.5.5}$$

To continue the problem, we must manipulate the integral in Eq. (10.5.5) into the form of the error function defined by Eq. (A.6.11).

Letting $\lambda = -(r - A_c)/\sigma$ produces

$$P[r < 0 | m_1] = \frac{1}{\sqrt{2\pi}} \int_{A_c/\sigma}^{\infty} e^{-\lambda^2/2} \, d\lambda. \tag{10.5.6}$$

Since $\int_0^\infty (1/\sqrt{2\pi}) e^{-y^2/2} \, dy = \frac{1}{2}$, we can rewrite Eq. (10.5.6) as

$$P[r < 0 | m_1] = \frac{1}{2} - \frac{1}{\sqrt{2\pi}} \int_0^{A_c/\sigma} e^{-\lambda^2/2} \, d\lambda$$

$$= \frac{1}{2} - \text{erf}\left[\frac{A_c}{\sigma}\right]. \tag{10.5.7}$$

We leave it to the reader to show that

$$P[r > 0 | m_2] = \frac{1}{2} - \text{erf}\left[\frac{A_c}{\sigma}\right], \tag{10.5.8}$$

so that using Eq. (10.5.4) we obtain

$$P[\mathscr{E}] = \frac{1}{2} - \text{erf}\left[\frac{A_c}{\sigma}\right]. \tag{10.5.9}$$

EXAMPLE 10.5.2

As a second example, we consider the equally likely binary signal set shown in Fig. 10.5.1. This signal set could represent using the time-domain signals $s_1(t) = A_c\sqrt{2/T} \cos \omega_c t$ and $s_2(t) = A_c\sqrt{2/T} \sin \omega_c t$, $0 \leq t \leq T$, for a 1 and 0, among numerous other possibilities. In any event, our goal is to calculate $P[\mathscr{E}]$ for this signal set when it is used over an AWGN channel with zero mean and variance σ^2.

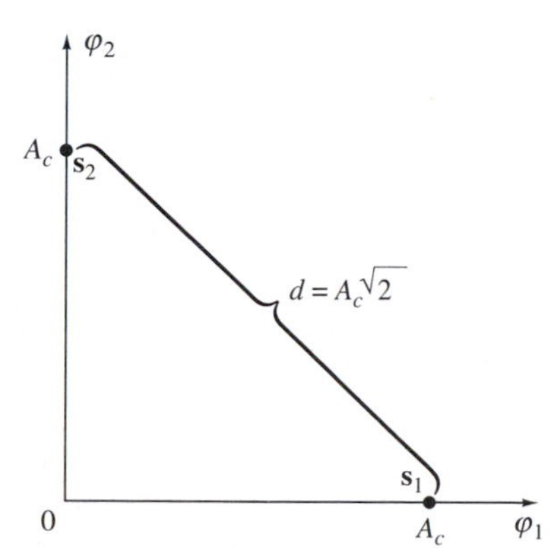

FIGURE 10.5.1 Orthogonal signal set for Example 10.5.2.

To begin, we note from Eq. (10.4.12) that for equally likely signals, the decision rule depends only on the distance between the two signals and not in any way on their orientation in the space. When the signals are not equally likely, it is evident from Eqs. (10.4.10) and (10.4.11) that we can similarly conclude that it is only the distance between the signals plus a bias term which are of importance, and neither depends on where the signals are in space—as long as the distance is preserved.

We thus conclude that we need only compute the distance between $\mathbf{s}_1$ and $\mathbf{s}_2$ in Fig. 10.5.1 and translate this into an equivalent one-dimensional problem as shown in Fig. 10.5.2. The signal constellations in Figs. 10.4.3 and 10.5.2 are identical except for a scale factor of $1/\sqrt{2}$. Hence we use the result from Example 10.5.1 to obtain

$$P[\mathscr{E}] = \frac{1}{2} - \operatorname{erf}\left[\frac{A_c}{\sigma\sqrt{2}}\right]. \tag{10.5.10}$$

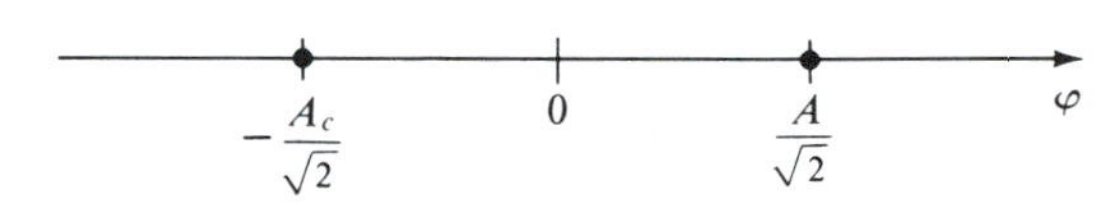

FIGURE 10.5.2 $P[\mathscr{E}]$ equivalent signal set for Example 10.5.2.

Comparing Eqs. (10.5.9) and (10.5.10), we see that since erf(x) is an increasing function of x, equal amplitude binary antipodal signals outperform (have a smaller $P[\mathscr{E}]$) binary orthogonal signals.

In solving Example 10.5.2, we argued that we could rotate and translate the signal set and the decision regions equally without affecting $P[\mathscr{E}]$. This result remains true for more complicated signal constellations and decision regions, and it can be extremely useful in simplifying calculations. In Example 10.5.2 we were able to use this fact to change a two-dimensional problem into a one-dimensional problem. That rotations and translations of the signal set and the corresponding decision regions do not affect $P[\mathscr{E}]$ follows from the decision rules in Eqs. (10.4.10)–(10.4.12), as claimed in the example. In turn, the decision rules were derived based on the assumption of additive white Gaussian noise that is independent of the transmitted signal. Thus the noise is independent of the location of the signal in signal space, and since the Gaussian pdf is spherically symmetric, the rotation of the decision regions does not change $P[\mathscr{E}]$. This invariance of $P[\mathscr{E}]$ to rotations and translations for an AWGN channel is used often in communications.

It should be noted that while translations of signal constellations do not affect $P[\mathscr{E}]$, they can affect both the peak and average energies of the signal set, which are of physical significance. However, for $P[\mathscr{E}]$ analyses we can consider rotated and translated signal sets in order to simplify our calculations.

EXAMPLE 10.5.3

We wish to calculate $P[\mathscr{E}]$ for the signal set in Fig. 10.2.1 when used for equally likely messages over an AWGN channel with zero mean and variance σ^2. This signal set could represent four-level QAM or four-level PSK, among others. Since the signals (messages) are equally likely, we can write

$$P[\mathscr{E}] = \tfrac{1}{4} \sum_{i=1}^{4} P[\mathscr{E}|m_i]. \tag{10.5.11}$$

Thus we need to compute the conditional error probabilities, $P[\mathscr{E}|m_i]$. Starting with m_1, we have

$$\begin{aligned} P[\mathscr{E}|m_1] &= P[\mathbf{r} \notin \mathscr{R}_1|m_1] \\ &= 1 - P[\mathbf{r} \in \mathscr{R}_1|m_1]. \end{aligned} \tag{10.5.12}$$

The region $\mathscr{R}_1$ where we accept m_1 is given by the first quadrant, which is simply the set $r_1 > 0$ and $r_2 > 0$, where $\mathbf{r} = [r_1 \;\; r_2]$. Therefore,

$$P[\mathbf{r} \in \mathscr{R}_1|m_1] = P[r_1 > 0,\, r_2 > 0|m_1]. \tag{10.5.13}$$

Since the components of the noise vector are statistically independent of each other, we can rewrite Eq. (10.5.13) as

$$P[\mathbf{r} \in \mathscr{R}_1|m_1] = P[r_1 > 0|m_1]P[r_2 > 0|m_1]. \tag{10.5.14}$$

Now, given that m_1 is transmitted, $r_1 = s_{11} + n_1$, where n_1 is zero mean, Gaussian with variance σ^2. As a result, r_1 is Gaussian with mean s_{11} and variance σ^2, so

$$P[r_1 > 0|m_1] = \int_0^\infty \frac{1}{\sqrt{2\pi}\sigma} e^{-(r_1 - s_{11})^2/2\sigma^2}\, dr_1, \tag{10.5.15}$$

which becomes upon using the definition of the error function in Eq. (A.6.11) that

$$\begin{aligned} P[r_1 > 0|m_1] &= \frac{1}{2} + \operatorname{erf}\left[\frac{s_{11}}{\sigma}\right] \\ &= \frac{1}{2} + \operatorname{erf}\left[\frac{A_c}{\sigma}\right]. \end{aligned} \tag{10.5.16}$$

The last equality follows from Fig. 10.2.1, since $s_{11} = A_c$. The integral required to compute $P[r_2 > 0|m_1]$ is identical to the one in Eq. (10.5.15), so we have that

$$P[\mathbf{r} \in \mathscr{R}_1|m_1] = \left\{\frac{1}{2} + \operatorname{erf}\left[\frac{A_c}{\sigma}\right]\right\}^2,$$

so

$$P[\mathscr{E}|m_1] = 1 - \left\{\frac{1}{2} + \operatorname{erf}\left[\frac{A_c}{\sigma}\right]\right\}^2$$
$$= \frac{3}{4} - \operatorname{erf}\left[\frac{A_c}{\sigma}\right] - \left\{\operatorname{erf}\left[\frac{A_c}{\sigma}\right]\right\}^2. \tag{10.5.17}$$

Although we can proceed to calculate the other conditional error probabilities directly just as for $P[\mathscr{E}|m_1]$, we take an alternative approach of noting the symmetry of the signal set in Fig. 10.2.1. Furthermore, we have assumed that the noise is Gaussian, and hence circularly symmetric, and that the noise is independent of the particular signal being transmitted. Based on these symmetry arguments, we thus claim that $P[\mathscr{E}|m_2] = P[\mathscr{E}|m_3] = P[\mathscr{E}|m_4] = P[\mathscr{E}|m_1]$ in Eq. (10.5.17). Therefore, from Eq. (10.5.11), the total probability of error is

$$P[\mathscr{E}] = P[\mathscr{E}|m_1] = \frac{3}{4} - \operatorname{erf}\left[\frac{A_c}{\sigma}\right] - \left\{\operatorname{erf}\left[\frac{A_c}{\sigma}\right]\right\}^2. \tag{10.5.18}$$

We now turn our attention to calculating the probability of error for noncoherent ASK and noncoherent FSK. To simplify the development for ASK, we limit consideration to the high signal-to-noise ratio case, where $A_c/\sqrt{\mathscr{N}_0} \gg 1$. In this situation, the conditional pdf given s_2 in Eq. (10.4.34) is approximately Gaussian and is given by

$$f_{Z|s_2}(z|s_2) \cong \frac{1}{\sqrt{\pi\mathscr{N}_0}} e^{-(z-A_c)^2/\mathscr{N}_0} \tag{10.5.19}$$

and the optimum threshold, T_{opt}, is about $A_c/2$. The probability of error for noncoherent ASK is thus

$$P[\mathscr{E}] = \tfrac{1}{2}P[\mathscr{E}|s_1] + \tfrac{1}{2}P[\mathscr{E}|s_2], \tag{10.5.20}$$

where by using Eq. (10.4.30),

$$P[\mathscr{E}|s_1] = \int_{A_c/2}^{\infty} \frac{2z}{\mathscr{N}_0} e^{-z^2/\mathscr{N}_0}\, dz = e^{-A_c^2/4\mathscr{N}_0} \tag{10.5.21}$$

and using Eq. (10.5.19),

$$P[\mathscr{E}|s_2] \cong \int_{-\infty}^{A_c/2} \frac{1}{\sqrt{\pi\mathscr{N}_0}} e^{-(z-A_c)^2/\mathscr{N}_0}\, dz$$
$$\cong \sqrt{\frac{\mathscr{N}_0}{\pi}} \cdot \frac{1}{A_c} e^{-A_c^2/4\mathscr{N}_0}. \tag{10.5.22}$$

To obtain the result in Eq. (10.5.22), we have evaluated the integral in terms of the error function and employed the approximation

$$1 - \operatorname{erf} x \cong \frac{\sqrt{2}}{x\sqrt{\pi}} e^{-x^2/2} \qquad \text{for } x \gg 1.$$

Substituting Eqs. (10.5.21) and (10.5.22) into Eq. (10.5.20) yields

$$P[\mathscr{E}] = \frac{1}{2}e^{-A_c^2/4\mathscr{N}_0} + \frac{1}{2A_c}\sqrt{\frac{\mathscr{N}_0}{\pi}}\,e^{-A_c^2/4\mathscr{N}_0}$$

$$= \left[1 + \frac{1}{A_c}\sqrt{\frac{\mathscr{N}_0}{\pi}}\right]\cdot\frac{1}{2}e^{-A_c^2/4\mathscr{N}_0}. \qquad (10.5.23)$$

Once again invoking the high signal-to-noise ratio assumption, the second term in brackets is small, so

$$P[\mathscr{E}] \cong \tfrac{1}{2}e^{-A_c^2/4\mathscr{N}_0}. \qquad (10.5.24)$$

Although Eq. (10.5.24) is only valid for $A_c/\sqrt{\mathscr{N}_0} \gg 1$, it is often used as the $P[\mathscr{E}]$ expression for noncoherent ASK, and we will follow suit here.

For noncoherent FSK, we need not invoke such approximations, since the threshold is just the variable in the parallel channel. Referring to Fig. 10.4.12 and Eqs. (10.4.41) and (10.4.42), we see that an error is made when m_1 is sent if $z_1 < z_2$ and if m_2 is sent when $z_1 > z_2$. Thus

$$P[\mathscr{E}|s_1] = \int_0^\infty f_{Z_1|s_1}(z_1|s_1)\left[\int_{z_1}^\infty f_{Z_2|s_1}(z_2|s_1)\,dz_2\right]dz_1, \qquad (10.5.25)$$

which upon substituting for $f_{Z_2|s_1}(z_2|s_1)$ and integrating yields

$$P[\mathscr{E}|s_1] = e^{-A_c^2/\mathscr{N}_0}\int_0^\infty \frac{2z_1}{\mathscr{N}_0}\,e^{-z_1^2/(\mathscr{N}_0/2)}I_0\left(\frac{2A_c z_1}{\mathscr{N}_0}\right)dz_1. \qquad (10.5.26)$$

This integral is available in tables, so Eq. (10.5.26) becomes

$$P[\mathscr{E}|s_1] = \tfrac{1}{2}e^{-A_c^2/2\mathscr{N}_0}. \qquad (10.5.27)$$

By symmetry, $P[\mathscr{E}|s_2] = P[\mathscr{E}|s_1]$, so since the messages are equally likely

$$P[\mathscr{E}] = \tfrac{1}{2}e^{-A_c^2/2\mathscr{N}_0}. \qquad (10.5.28)$$

This and other $P[\mathscr{E}]$ results are compared in Section 10.6.

Given a certain transmitted signal set and a particular channel, we know how to design an optimum receiver that minimizes $P[\mathscr{E}]$, and indeed this has been the main thrust of the chapter thus far. The problem can be turned around, however, to obtain an interesting problem in what is called *signal design*. Specifically, given a particular channel and the minimum $P[\mathscr{E}]$ receiver design philosophy, how do we choose the transmitted signal set or signal constellation to minimize $P[\mathscr{E}]$? Here we limit consideration to the coherent receiver case, since this allows us to take maximum advantage of the geometric interpretation and to obtain significant practical results. The problem can be posed with either an average energy or peak energy constraint on the transmitted signals, but in either case, the optimization problem is not a simple one, since conditional pdfs must be integrated over peculiarly shaped decision regions and the sum of these integrals must be minimized. In the high-SNR case, only adjacent decision regions need to be considered, but the problem remains

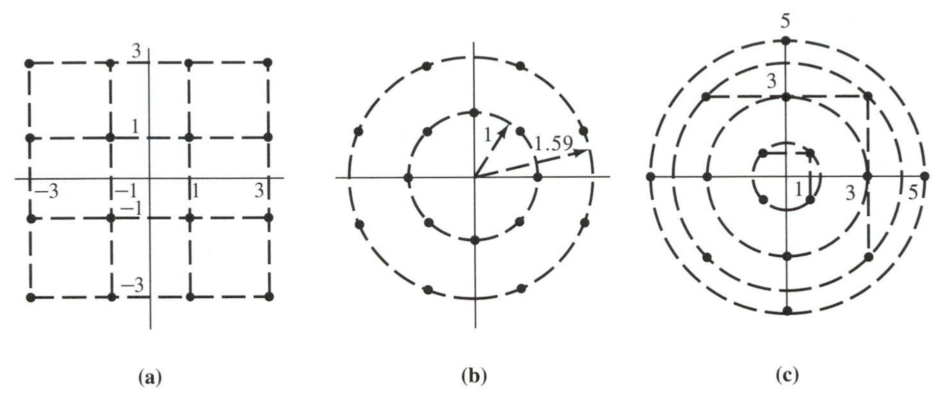

FIGURE 10.5.3 Common 16-point signal constellations: (a) Four-level QAM; (b) modified 8–8 AM/PM; (c) circular (4, 90°) constellation.

difficult. On the other hand, the problem is relatively easy to state verbally as one of maximizing the spacing between adjacent points in signal space subject to the appropriate energy constraint. Three well-known 16-point signal constellations not obtained via this optimization procedure are shown in Fig. 10.5.3 for benchmark purposes. The optimum 16-point constellation under an average energy constraint for an AWGN channel is shown in Fig. 10.5.4 [Foschini, Gitlin, and Weinstein, 1974]. Note the lack of symmetry in comparison to the signal constellations in Fig. 10.5.3. The optimum signal constellation subject to a peak energy constraint for an AWGN channel is presented in Fig. 10.5.5 [Kernighan and Lin, 1973]. Notice that this constellation is less "random looking" than the constellation in Fig. 10.5.4, but not nearly as symmetric as those in Fig. 10.5.3. If the channel causes random carrier phase jitter in addition to injecting AWGN, the optimum signal set subject to a peak energy constraint is that shown in Fig. 10.5.6 [Kernighan and Lin, 1973]. When phase jitter is included, points farther from the origin suffer increased phase distortion, and hence one of the outer signal points in Fig. 10.5.5 is moved to the origin in Fig. 10.5.6. Calculations pertaining to these signal sets and signal design in particular are left to the references.

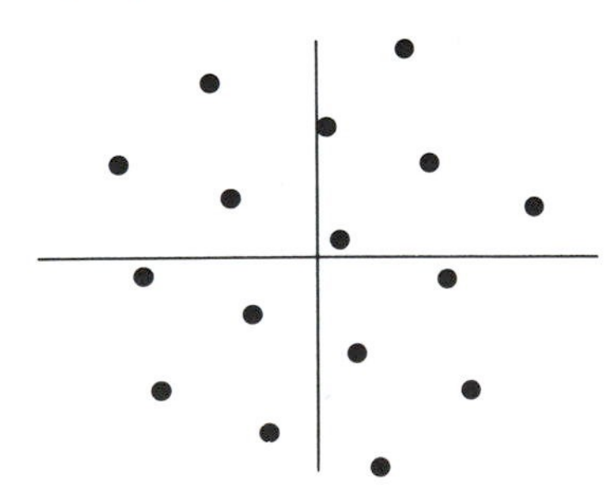

FIGURE 10.5.4 Optimum 16-point signal constellation for an average energy constraint. From G. J. Foschini, R. D. Gitlin, and S. B. Weinstein, "Optimization of Two-Dimensional Signal Constellations in the Presence of Gaussian Noise," *IEEE Trans. Commun.*, © 1974 IEEE.

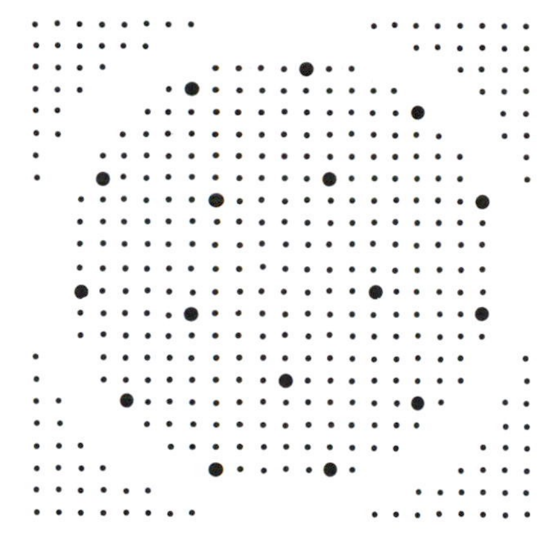

FIGURE 10.5.5 Optimum 16-point signal constellation for a peak energy constraint (peak SNR = 27 dB). From B. W. Kernighan and S. Lin, "Heuristic Solution of a Signal Design Optimization Problem," Proc. 7th Annual Princeton Conference on Information Science and Systems, Mar. 1973.

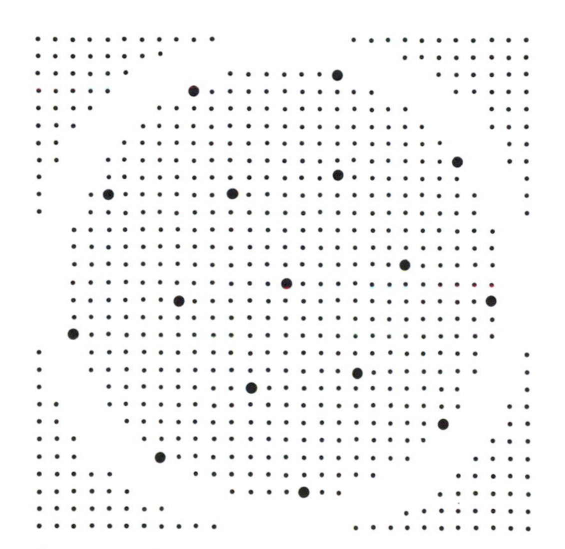

FIGURE 10.5.6 Optimum 16-point signal constellation for a peak energy constraint (SNR = 27 dB, RMS phase jitter = 3°). From B. W. Kernighan and S. Lin, "Heuristic Solution of a Signal Design Optimization Problem," Proc. 7th Annual Princeton Conference on Information Science and Systems, Mar. 1973.

10.6 System Performance Comparisons

The first question that arises when an engineer is designing a digital communication system is: Which signaling scheme is best? The answer to this question involves many considerations such as probability of error, energy constraints, complexity, bandwidth, channel nonlinearities, and others, and these details are usually known only for very specific communications problems. Here we briefly provide some comparisons based primarily on $P[\mathscr{E}]$, with some discussion concerning energy constraints, bandwidth, and complexity. The signaling methods examined are coherent and noncoherent ASK, coherent and noncoherent FSK, coherent BPSK, and differential PSK(DPSK). We first focus on the bit error probabilities of these schemes, which are usually plotted versus the ratio of energy per bit to one-sided noise power spectral density, $E_b/\mathscr{N}_0$.

We begin by writing the previously derived $P[\mathscr{E}]$ expressions in terms of E_b and $\mathscr{N}_0$, which we accomplish by noting that $A_c = \sqrt{E_b}$, since we are transmitting one bit per dimension and the basis functions are orthonormal, and that $\sigma^2 = \mathscr{N}_0/2$. Using these definitions, we have from Eq. (10.5.9) that for coherent BPSK,

$$P[\mathscr{E}] = \frac{1}{2} - \text{erf}\left[\sqrt{\frac{2E_b}{\mathscr{N}_0}}\right]. \tag{10.6.1}$$

For coherent FSK where the signals are sufficiently separated in frequency to be considered orthogonal, Eq. (10.5.10) yields

$$P[\mathscr{E}] = \frac{1}{2} - \text{erf}\left[\sqrt{\frac{E_b}{\mathscr{N}_0}}\right]. \tag{10.6.2}$$

It is straightforward to show that for coherent ASK with a peak power constraint that

$$P[\mathscr{E}] = \frac{1}{2} - \text{erf}\left[\sqrt{\frac{E_b}{2\mathscr{N}_0}}\right], \tag{10.6.3}$$

while for an average power constraint coherent ASK has a $P[\mathscr{E}]$ given by Eq. (10.6.2). For noncoherent ASK subject to a peak power constraint, we have from Eq. (10.5.24),

$$P[\mathscr{E}] \cong \tfrac{1}{2}e^{-E_b/4\mathscr{N}_0}, \tag{10.6.4}$$

while with an average power constraint,

$$P[\mathscr{E}] \cong \tfrac{1}{2}e^{-E_b/2\mathscr{N}_0}. \tag{10.6.5}$$

The error probability for noncoherent FSK can be obtained from Eq. (10.5.28) as

$$P[\mathscr{E}] = \tfrac{1}{2}e^{-E_b/2\mathscr{N}_0}. \tag{10.6.6}$$

Finally, we display some error probability results for a binary DPSK system as described in Chapter 5. Although the derivation is somewhat involved, and hence not given here, the bit error probability for DPSK is simple and is given by

$$P[\mathscr{E}] = \tfrac{1}{2}e^{-E_b/\mathscr{N}_0}. \tag{10.6.7}$$

Plots of these error probabilities subject to an average energy constraint are shown in Fig. 10.6.1. For a peak power constraint, the coherent and noncoherent ASK curves are shifted to the right by 3 dB.

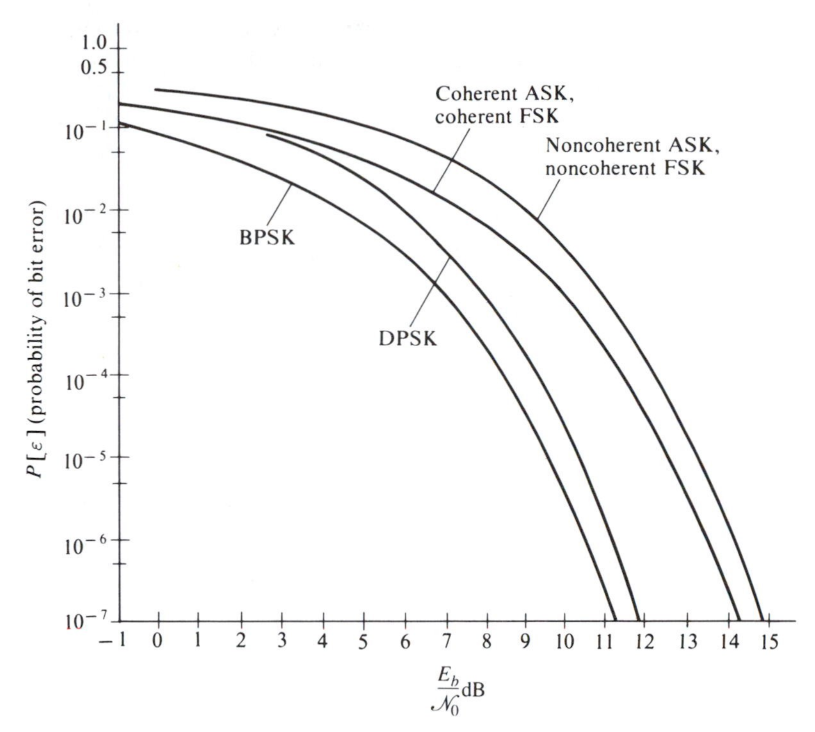

FIGURE 10.6.1 Bit error probabilities for binary signaling methods subject to an average power constraint.

Since there are many possible definitions of bandwidth for phase- and frequency-modulated carriers and random pulse sequence messages, comparative bandwidth statements are necessarily imprecise. In general, FSK methods require a greater bandwidth than ASK, DPSK, and BPSK, which have similar bandwidth requirements. In terms of complexity, noncoherent ASK is the least complex, followed by noncoherent FSK, DPSK, BPSK, and coherent ASK and FSK. Note that for high signal-to-noise ratios ($E_b/\mathscr{N}_0$), noncoherent ASK and FSK perform within 1 dB of their coherent counterparts and hence are quite attractive as a compromise between complexity and performance. Similarly, DPSK has performance close to BPSK at high $E_b/\mathscr{N}_0$ and is simpler to implement, since a coherent reference is not required. These statements provide some indication as to why noncoherent FSK and DPSK have found widespread acceptance in many practical communication systems.

10.7 Maximum Likelihood Sequence Estimation

In this section we develop receiver structures that are truly optimum in the presence of intersymbol interference and additive white Gaussian noise. As noted in Section 10.4, the intersymbol interference may be caused by the channel or it may have been inserted purposely at the transmitter by using partial response signaling. The underlying algorithm is the same in both cases, and the improvement in performance over optimum symbol-by-symbol detection is due to the receiver making decisions for an entire sequence of symbols rather than just on individual symbols.

We must now include a time index in our notation, so that if the ith message is input to the transmitter at time k, we denote this as $m_i(k)$, and the corresponding transmitted signal vector at time k is $\mathbf{s}_i(k)$. If each transmitted symbol interferes with J other symbols, the received vector at time k for a general transmitted signal sequence $\{\mathbf{s}(k)\}$ can be written as

$$\mathbf{r}(k) = \sum_{j=0}^{J} d_j \mathbf{s}(k-j) + \mathbf{n}(k), \tag{10.7.1}$$

where the coefficients d_j, $j = 0, 1, \ldots, J$, account for the intersymbol interference and $\mathbf{n}(k)$ is the Gaussian noise vector at time k. We assume that all messages are equally likely and consider a transmitted sequence K symbols long, $K > J$. If we denote the received and transmitted sequences, respectively, as $\mathbf{R}_K = \{\mathbf{r}(1), \mathbf{r}(2), \ldots, \mathbf{r}(K)\}$ and $\mathbf{S}_K = \{\mathbf{s}(1), \mathbf{s}(2), \ldots, \mathbf{s}(K)\}$, following Section 10.4, the optimum receiver for the sequence maximizes

$$f_{\mathbf{R}|\mathbf{S}}(\mathbf{R}_K | \mathbf{S}_K) \triangleq f_{\mathbf{R}|\mathbf{S}}(\mathbf{r}(K), \ldots, \mathbf{r}(1) | \mathbf{s}(K), \ldots, \mathbf{s}(1)). \tag{10.7.2}$$

As before, the additive Gaussian noise is assumed to be white, so that Eq. (10.7.2) can be expressed as

$$f_{\mathbf{R}|\mathbf{S}}(\mathbf{R}_K | \mathbf{S}_K) = \prod_{k=1}^{K} f_{\mathbf{r}|\mathbf{S}_J}(\mathbf{r}(k) | \mathbf{s}(k), \ldots, \mathbf{s}(k-J)). \tag{10.7.3}$$

The optimum sequence estimator in Eq. (10.7.3) clearly subsumes the optimum symbol-by-symbol receiver, since if there is no intersymbol interference, the conditional pdfs on the right side of Eq. (10.7.3) reduce to $f_{\mathbf{r}|\mathbf{s}}(\mathbf{r}(k)|\mathbf{s}(k))$, as in Section 10.4. It is common to take (natural) logarithms of both sides of Eq. (10.7.3), to yield

$$\ln f_{\mathbf{R}|\mathbf{S}}(\mathbf{R}_K | \mathbf{S}_K) = \sum_{k=1}^{K} \ln f_{\mathbf{r}|\mathbf{S}_J}(\mathbf{r}(k) | \mathbf{s}(k), \ldots, \mathbf{s}(k-J)) \tag{10.7.4}$$

as the quantity to be maximized by the selection of the decoded sequence $\mathbf{S}_K$.

To perform the maximization, we note that $\mathbf{s}(k) = 0$ for $k \le 0$, and write out a few terms of the summation explicitly to find

$$\ln f_{\mathbf{R}|\mathbf{S}}(\mathbf{R}_K | \mathbf{S}_K) = \sum_{k=3}^{K} \ln f_{\mathbf{r}|\mathbf{S}_J}(\mathbf{r}(k) | \mathbf{s}(k), \ldots, \mathbf{s}(k-J))$$
$$+ \ln f_{\mathbf{r}|\mathbf{S}_J}(\mathbf{r}(2) | \mathbf{s}(2), \mathbf{s}(1)) + \ln f_{\mathbf{r}|\mathbf{S}_J}(\mathbf{r}(1) | \mathbf{s}(1)), \tag{10.7.5}$$

which we wish to maximize with respect to the selection of $\mathbf{s}(1), \ldots, \mathbf{s}(K)$. Thus we can write

$$
\begin{aligned}
\max_{\mathbf{S}_K} \ln f_{\mathbf{R}|\mathbf{S}}(\mathbf{R}_K|\mathbf{S}_K) &= \max_{\mathbf{s}(1)} \Bigg\{ \ln f_{\mathbf{r}|\mathbf{S}_J}(\mathbf{r}(1)|\mathbf{s}(1)) + \max_{\mathbf{s}(2)} \Bigg\{ \ln f_{\mathbf{r}|\mathbf{S}_J}(\mathbf{r}(2)|\mathbf{s}(2), \mathbf{s}(1)) + \cdots \\
&\quad + \max_{\mathbf{s}(K)} \{ \ln f_{\mathbf{r}|\mathbf{S}_J}(\mathbf{r}(K)|\mathbf{s}(K), \ldots, \mathbf{s}(K-J)) \} \underbrace{\cdots \Bigg\}\Bigg\}}_{K \text{ braces}} \\
&= \max_{\mathbf{s}(1)} \ln f_{\mathbf{r}|\mathbf{S}_J}(\mathbf{r}(1)|\mathbf{s}(1)) + \max_{\mathbf{s}(1),\mathbf{s}(2)} \ln f_{\mathbf{r}|\mathbf{S}_J}(\mathbf{r}(2)|\mathbf{s}(2), \mathbf{s}(1)) + \cdots \\
&\quad + \max_{\mathbf{s}(1), \ldots, \mathbf{s}(K)} \ln f_{\mathbf{r}|\mathbf{S}_J}(\mathbf{r}(K)|\mathbf{s}(K), \ldots, \mathbf{s}(K-J)). \qquad (10.7.6)
\end{aligned}
$$

There are N possible transmitted signal vectors $\mathbf{s}(k)$ at each time instant k, so that for a sequence of length K, there are N^K possible sequences that must be considered in the most general case. However, for our current situation where there is intersymbol interference with $J < K$ symbols, the computations required are somewhat less. To see this, we expand the last part of Eq. (10.7.6) as

$$
\begin{aligned}
\max_{\mathbf{S}_K} \ln f_{\mathbf{R}|\mathbf{S}}(\mathbf{R}_K|\mathbf{S}_K) &= \max_{\mathbf{s}(1)} \ln f_{\mathbf{r}|\mathbf{S}_J}(\mathbf{r}(1)|\mathbf{s}(1)) + \max_{\mathbf{s}(1),\mathbf{s}(2)} \ln f_{\mathbf{r}|\mathbf{S}_J}(\mathbf{r}(2)|\mathbf{s}(2), \mathbf{s}(1)) + \cdots \\
&\quad + \max_{\mathbf{s}(1), \ldots, \mathbf{s}(J+1)} \ln f_{\mathbf{r}|\mathbf{S}_J}(\mathbf{r}(J+1)|\mathbf{s}(J+1), \ldots, \mathbf{s}(1)) \\
&\quad + \max_{\mathbf{s}(1),\mathbf{s}(2), \ldots, \mathbf{s}(J+2)} \ln f_{\mathbf{r}|\mathbf{S}_J}(\mathbf{r}(J+2)|\mathbf{s}(J+2), \ldots, \mathbf{s}(2)) + \cdots \\
&\quad + \max_{\mathbf{s}(1), \ldots, \mathbf{s}(K)} \ln f_{\mathbf{r}|\mathbf{S}_J}(\mathbf{r}(K)|\mathbf{s}(K), \ldots, \mathbf{s}(K-J)). \qquad (10.7.7)
\end{aligned}
$$

After time instant $k = J$, each term requires the computation of N^{J+1} probabilities, so that the number of probabilities needed for Eq. (10.7.7) is upper bounded by KN^{J+1}. If $J \ll K$, this upper bound is fairly tight.

Let us now consider all the terms in Eq. (10.7.7) with explicit conditioning on $\mathbf{s}(1)$. If these $J + 1$ maximizations yield the same value for $\mathbf{s}(1)$, that is the decoded value corresponding to $\mathbf{r}(1)$. If, however, these maximizations yield different values for $\mathbf{s}(1)$, the decision on $\mathbf{s}(1)$ must be deferred. At the next time instant, $k = J + 2$, we try to make a decision on $\mathbf{s}(1)$ and $\mathbf{s}(2)$. Again, if all the maximizations yield the same $\mathbf{s}(1)$ or $\mathbf{s}(1)$ and $\mathbf{s}(2)$, then $\mathbf{s}(1)$ or $\mathbf{s}(1)$ and $\mathbf{s}(2)$, respectively, can be definitely decided. If not, the decision is again deferred. Now the question becomes: How long can the decision on $\mathbf{s}(1)$ and following symbols be delayed? The answer is that since we are making decisions on sequences and each sequence overlaps the immediately preceding sequence by J symbols, the memory is essentially infinite, so the decision on $\mathbf{s}(1)$ could be delayed until the end of the sequence, time instant $k = K$. However, since K is typically very large, an upper limit is usually chosen on the delay in deciding each particular symbol, and it has been determined that if this delay is greater than or equal to $5J$, there is negligible loss in performance. If at some time instant k, all candidate sequences disagree on the symbol $\mathbf{s}(k - 5J)$, then the value of $\mathbf{s}(k - 5J)$ in the most probable sequence is picked as the decoded value.

It is important to notice that implicit in the optimization in Eq. (10.7.7) is the fact that at some time instant k, only the N best sequences through time instant $k-J$ need to remain under consideration. This is a manifestation of what is called *the principle of optimality*, which states that the optimal sequence including some value of $\mathbf{s}(k-J)$ has as a subset the optimal sequence from the beginning of the decoding process (time 0) to the candidate output $\mathbf{s}(k-J)$. The explicit statement of this concept is an aid when one is solving specific problems.

EXAMPLE 10.7.1

To illustrate the use of this optimum sequence estimation algorithm, we consider the use of duobinary or class I partial response signaling to transmit binary (± 1) data (see Section 8.3 for details on duobinary). The received values at time instants 1 and 2 are thus

$$r(1) = s(1) + n(1) \tag{10.7.8}$$

and

$$r(2) = s(2) + s(1) + n(2). \tag{10.7.9}$$

We assume here that K is very large, and we know for duobinary that $J = 1$. The pertinent optimization is thus summarized as

$$\begin{aligned}\max_{\mathbf{S}_K} \ln f_{\mathbf{R}|\mathbf{S}}(\mathbf{R}_K|\mathbf{S}_K) &= \max_{s(1)} \ln f_{r|\mathbf{S}_1}(r(1)|s(1)) + \max_{s(1),s(2)} \ln f_{r|\mathbf{S}_1}(r(2)|s(2), s(1)) \\ &+ \max_{s(1),s(2),s(3)} \ln f_{r|\mathbf{S}_1}(r(3)|s(3), s(2)) + \cdots \\ &+ \max_{s(1),\ldots,s(k)} \ln f_{r|\mathbf{S}_1}(r(k)|s(k), s(k-1)) + \cdots \\ &+ \max_{s(1),\ldots,s(K)} \ln f_{r|\mathbf{S}_1}(r(K)|s(K), s(K-1)).\end{aligned} \tag{10.7.10}$$

Since for any k, $s(k) = \pm 1$, we can represent all possible output sequences from time 0 to time k by a tree diagram as shown in Fig. 10.7.1. Noting that the noise is zero-mean Gaussian and the inputs are equally likely, we have that an individual term in Eq. (10.7.10) is given by

$$\max_{s(1),\ldots,s(k)} \ln f_{r|\mathbf{S}_1}(r(k)|s(k), s(k-1)) = \max_{s(1),\ldots,s(k)} \{-[r(k) - s(k) - s(k-1)]^2\}. \tag{10.7.11}$$

The first maximization to be examined is therefore

$$\max_{s(1),s(2)} \{-[r(1) - s(1)]^2 - [r(2) - s(2) - s(1)]^2\}, \tag{10.7.12}$$

and it has four possible values corresponding to the four candidate paths to depth (time instant) 2 in the tree of Fig. 10.7.1. Now, at the next time instant, the received value does not involve $s(1)$, so that we need only retain at time instant 2, $N = 2$ paths, consisting of the best path terminating with $s(2) = +1$

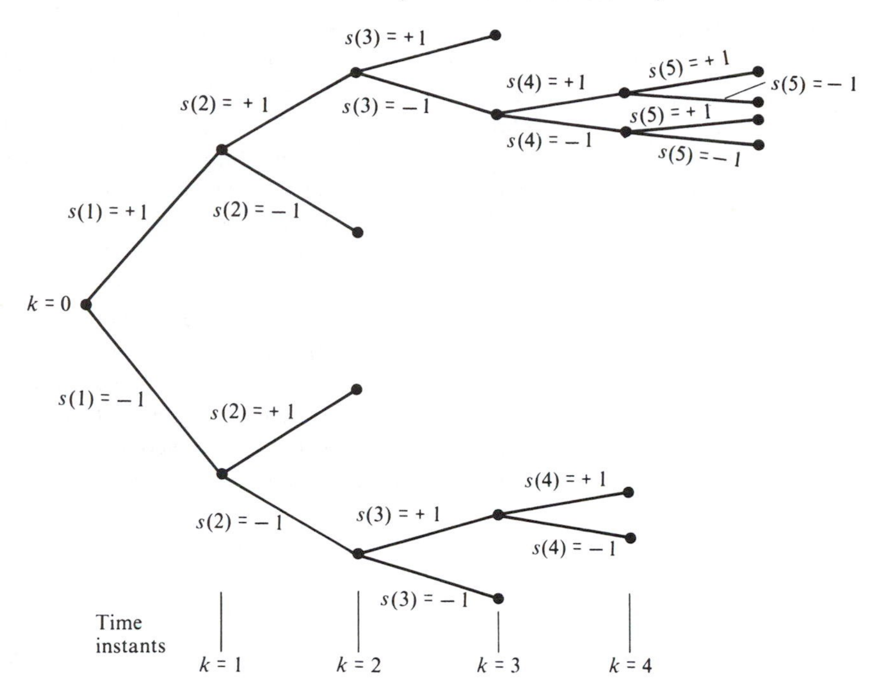

FIGURE 10.7.1 Tree diagram corresponding to maximum likelihood sequence estimation for binary-driven duobinary.

and the best path ending at $s(2) = -1$. Since we do not pick explicit values for the received values in this example, we have arbitrarily indicated the extension of two paths at $k = 2$ in Fig. 10.7.1. Note that these paths do not have the same value for $s(1)$, so this output cannot be decided as yet. At time instant 3, the maximization is

$$\max_{s(1),s(2),s(3)} \Big\{ -[r(3) - s(3) - s(2)]^2 + \max_{s(1),s(2)} \{-[r(1) - s(1)]^2 - [r(2) - s(2) - s(1)]^2\} \Big\}, \qquad (10.7.13)$$

so that we only need to extend the two best paths as shown in Fig. 10.7.1. Again, we note that the next received value at time instant 4 does not depend on $s(3)$, so we retain only the $N = 2$ best sequences leading to $s(3) = -1$ and $s(3) = +1$. A possible pruning of paths is indicated in the figure, and the two best paths still do not have the same value of $s(1)$.

Continuing this process as indicated in Fig. 10.7.1, we see that at time instant 4, the best paths to $s(4) = +1$ and $s(4) = -1$ have the same value of $s(1)$; indeed, they have the same values for $s(1)$, $s(2)$, and $s(3)$, namely, $+1$, $+1$, and -1, respectively. Therefore, these values can all be definitely decided at time instant 4. The procedure continues until the end of the sequence is reached.

The optimum algorithm presented in this section is called the *maximum likelihood sequence estimator*, since it maximizes the conditional pdf in Eq. (10.7.2), called the *likelihood function*, which is the likelihood that the observed sequence would be received for all candidate input sequences. This algorithm not only outlines the optimum receiver for partial response signaling and channels with intersymbol interference, it is also important in the decoding of convolutional codes (see Section 12.4) and for the reception of sequences transmitted with continuous phase modulation and coded modulation (see Chapter 13).

It has often been stated that there is a loss in signal-to-noise ratio (SNR) when partial response signaling is used in comparison to independent symbol transmission at the maximum rate [Lucky, Salz, and Weldon, 1968]. However, this loss in SNR occurs only when optimum symbol-by-symbol detection is used at the receiver, and this loss does not occur when the receiver employs maximum likelihood sequence estimation to decode the received symbols (see Proakis [1989] for details concerning these performance comparisons).

SUMMARY

The effects of channel noise in digital communications, including data transmission using analog modulation methods such as ASK, FSK, and PSK, have been developed based on the geometric approach introduced by Kotel'nikov [1947] and advanced by Wozencraft and Jacobs [1965]. This approach unifies the analyses of digital communication systems and thus avoids the impression that communication theory is simply a collection of analytical techniques with little or no structure. Using this geometric approach, we have also been able to derive expressions for receivers that minimize the bit error probability and even gain insight into how one might choose the transmitted signals in order to optimize system performance. Particularly important signal constellations have been presented and comparisons of bit error probabilities given. The material presented in this chapter has had a profound effect upon digital communication system design for the last 20 years and is fundamental to understanding systems being developed today.

PROBLEMS

10.1 Use Eq. (10.2.13) to calculate all d_{ij} of interest for the signal set given by Eq. (10.2.16). Verify these distances by comparison with Fig. 10.2.1.

10.2 Write out the time-domain waveforms for the 16-level QAM signal set in Example 10.2.1. Specify all signal energies and the distances between each signal and its neighbors. Check by comparison with Fig. 10.2.2.

10.3 Verify that the signal set given by Eq. (10.2.21) can be represented by the signal space diagram in Fig. 10.2.4.

10.4 Specify an eight-phase PSK signal set in which one of the signals is $s_1(t) = A_c\sqrt{2/T}\cos[\omega_c t + 22.5°]$. Draw the corresponding signal space diagram.

10.5 Using $\varphi_1(t)$ in Eq. (10.2.22), specify an eight-level signal set over $0 \le t \le T/2$. Draw its signal space diagram.

10.6 Use Eqs. (10.2.22) and (10.2.23) to define a signal set corresponding to the signal space diagram in Fig. 10.2.2.

10.7 Given the waveforms $\{s_i(t),\ i = 1, 2, 3, 4\}$ shown in Fig. P10.7, use the Gram–Schmidt procedure to find appropriate signal vector representations.

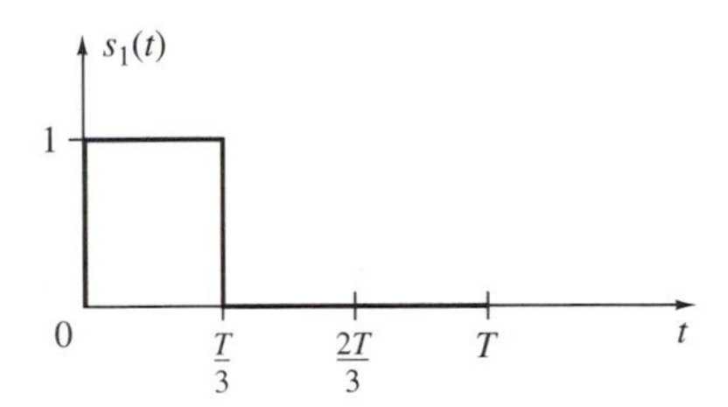

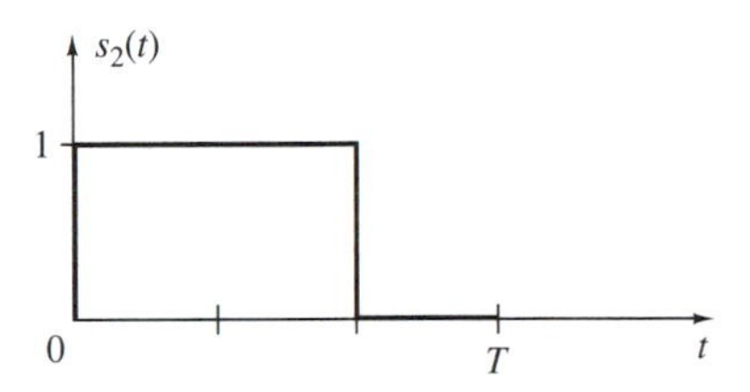

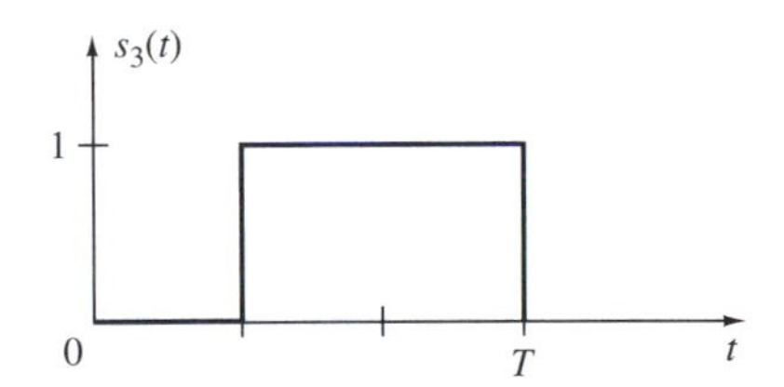

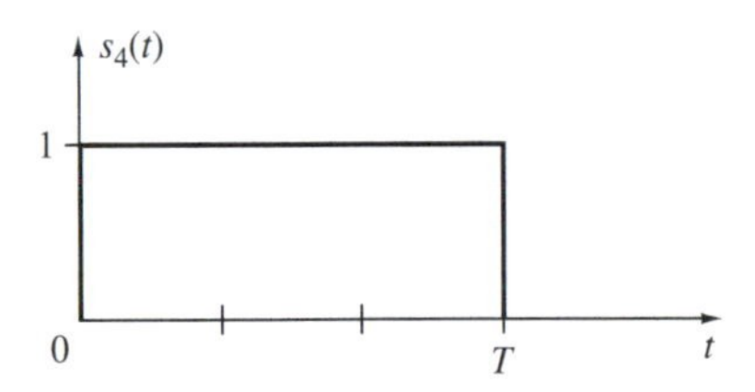

FIGURE P10.7

10.8 Renumber the signals in Fig. 10.2.5 as $s_1'(t) = s_4(t)$, $s_2'(t) = s_3(t)$, $s_3'(t) = s_2(t)$, and $s_4'(t) = s_1(t)$. Use the Gram–Schmidt procedure on this new signal set $\{s_i'(t),\ i = 1, 2, 3, 4\}$. Compare the results to Example 10.2.4.

10.9 Redefine the signals in Fig. P10.7 as $s_1'(t) = s_3(t)$, $s_2'(t) = s_2(t)$, $s_3'(t) = s_4(t)$, and $s_4'(t) = s_1(t)$. Use the Gram–Schmidt procedure on this new "primed" signal set $\{s_i'(t),\ i = 1, 2, 3, 4\}$.

10.10 Consider the M-dimensional signal set given by the vectors

$$\mathbf{s}_i = [s_{i1}\ s_{i2}\ \cdots\ s_{iM}], \qquad i = 1, 2, \ldots, N,$$

where

$$s_{ij} = \begin{cases} +d \\ \text{or} \\ -d \end{cases} \qquad \text{for all } i, j.$$

Sketch this signal set in two ($M = 2$) and three ($M = 3$) dimensions. How many signal vectors are there in M dimensions? Can you specify a set of time-domain waveforms that correspond to these signal vectors?

10.11 Specify the M-dimensional vectors for an orthogonal signal set. Sketch the signal sets in two and three dimensions. How many vectors are there in M dimensions?

10.12 Write signal vectors for the M-dimensional generalization of binary antipodal signals (called *biorthogonal* signals). Sketch these signal vectors in two and three dimensions. How many signals are there in M dimensions?

10.13 If each of the signal vectors in Problems 10.10, 10.11, and 10.12 have equal energy, say E_s, compare the distance between nearest neighbors for the three signal sets when $M = 3$.

10.14 Each member of the set of $N = M$ orthogonal signal vectors, denoted $\{\mathbf{s}_i, i = 1, 2, \ldots, M\}$, is translated by $\mu = (1/M)\sum_{i=1}^{M} \mathbf{s}_i$ to produce what is called the *simplex* signal set given by $\mathbf{s}_i' = \mathbf{s}_i - \mu, i = 1, 2, \ldots, M$. Sketch the simplex signal set in two and three dimensions. If $\mathbf{s}_i \cdot \mathbf{s}_j = E_s$ for all i, j, for the orthogonal signal set, find $\mathbf{s}_i' \cdot \mathbf{s}_j'$ for all i, j, for the simplex signal set.

10.15 In this problem and Problems 10.16 through 10.19, we prove that the vector $\mathbf{r}$ in Eq. (10.3.8) has all the "information" in $r(t)$ necessary to design an optimum receiver and that the pdf of the noise vector is given by Eq. (10.3.10). We begin by noting that the time-domain expression corresponding to Eq. (10.3.8) is

$$r'(t) \triangleq \sum_{j=1}^{M} r_j\varphi_j(t) = s_i(t) + n_o(t)$$

for some i, where $n_o(t) \triangleq \sum_{j=1}^{M} n_j\varphi_j(t)$ corresponds to $\mathbf{n}$. Now consider the difference signal $e(t) \triangleq r(t) - r'(t)$. Show that since $n(t)$ is a zero-mean Gaussian process, then $e(t)$ and $n_o(t)$ are zero-mean, jointly Gaussian processes.

10.16 Using the results of Problem 10.15, show that $e(t)$ and $n_o(t)$ are statistically independent.

10.17 Based on Eq. (10.4.3), show that if a receiver has $\mathbf{r}$ and $\mathbf{e}$ [a vector of time samples corresponding to $e(t)$] available, the conditional pdf $f_{\mathbf{e}|\mathbf{r},\mathbf{s}_i}$ is needed in the optimum receiver (see Problem 10.15).

10.18 Use Bayes' rule and Problem 10.16 to show that $\mathbf{e}$ is statistically independent of $\mathbf{r}$ and $\mathbf{s}_i$, and hence only $\mathbf{r}$ is needed in the optimum receiver.

10.19 Use the results of Problems 10.15 and 10.16 to prove that $\mathbf{n}$ has the pdf in Eq. (10.3.10).

10.20 We consider the transmission of four messages over an AWGN channel with spectral density $\mathcal{N}_0/2$ watts/Hz using the signal set specified by

Eqs. (10.2.14) and (10.2.16). If $P[m_1] = \frac{1}{2}$, $P[m_2] = \frac{1}{4}$, and $P[m_3] = P[m_4] = \frac{1}{8}$, sketch the optimum decision regions.

10.21 The signal constellation in Fig. P10.21 is used to send eight equally likely messages over a zero-mean AWGN channel with power spectral density $\mathcal{N}_0/2$ watts/Hz. Determine and sketch the optimum decision regions.

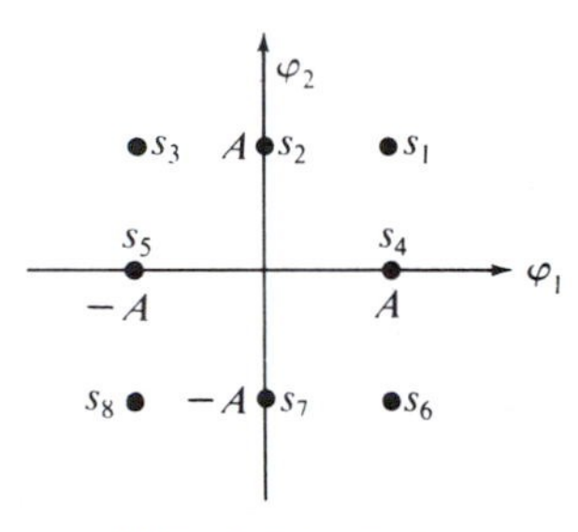

FIGURE P10.21

10.22 Repeat Problem 10.21 for the signal set in Fig. P10.22.

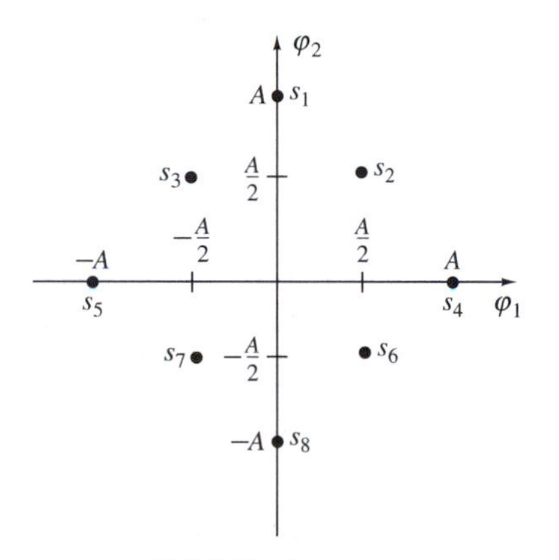

FIGURE P10.22

10.23 Derive the pdf in Eq. (10.4.34).

10.24 Use the approximation for $A_c/(\mathcal{N}_0/2) \gg 1$,

$$I_0\left(\frac{2A_cz}{\mathcal{N}_0}\right) \cong \frac{e^{2A_cz/\mathcal{N}_0}}{\sqrt{2\pi(2A_cz/\mathcal{N}_0)}},$$

to show that the pdf in Eq. (10.4.35) can be approximated as

$$f_{Z|s_2}(z|s_2) \cong \sqrt{\frac{z}{\pi A_c\mathcal{N}_0}}\, e^{-(z-A_c)^2/\mathcal{N}_0}.$$

Plot this pdf as a function of z, and hence validate the approximation in Eq. (10.5.19).

10.25 If the signal set in Problem 10.10, with equally likely signals, is used over an AWGN channel with power spectral density $\mathcal{N}_0/2$ watts/Hz, what is $P[\mathscr{E}]$ for an optimum receiver?

10.26 Write an expression for $P[\mathscr{E}]$ if the M orthogonal signals in Problem 10.11 are used over an AWGN channel with power spectral density $\mathcal{N}_0/2$ watts/Hz. Assume equally likely signals, each with energy E_s, and an optimum receiver.

10.27 Repeat Problem 10.26 for the simplex signal set.

10.28 Find an expression for the probability of error when the biorthogonal signals in Problem 10.12 are used to transmit equally likely messages over an AWGN channel with spectral density $\mathcal{N}_0/2$ watts/Hz. Assume that each signal has energy E_s and an optimum receiver.

10.29 Plot $P[\mathscr{E}]$ for coherent and noncoherent ASK subject to a peak power constraint.

10.30 A received waveform $r(t)$ given by Eq. (10.3.1) is applied to the input of a linear filter with impulse response $h(t)$. Show that the filter which maximizes the signal-to-noise ratio (SNR) at its output is a matched filter. Let $\text{SNR} \triangleq s_i^2(t)/E[n^2(t)]$.

Hint: Use the Schwarz inequality,

$$\left|\int_{-\infty}^{\infty} A(\omega)B(\omega)\, d\omega\right|^2 \leq \int_{-\infty}^{\infty} |A(\omega)|^2\, d\omega \int_{-\infty}^{\infty} |B(\omega)|^2\, d\omega.$$

10.31 Draw block diagrams of the optimal correlation and matched filter receiver structure involving the set of possible transmitted waveforms $\{s_i(t), i = 1, 2, \ldots, N\}$. Compare to Fig. 10.4.6.

10.32 Repeat Example 10.7.1 for the case where the inputs to the duobinary system (after precoding) take on the four levels, ± 1, $\pm\frac{1}{3}$.

10.33 Repeat Example 10.7.1 for binary-driven class 4 partial response.

CHAPTER 11

Information Theory and Rate Distortion Theory

11.1 Introduction

Thus far in the book, the term *information* has been used sparingly and when it has been used, we have purposely been imprecise as to its meaning. Although everyone has an intuitive feeling for what information is, it is difficult to attach a meaningful quantitative definition to the term. In the context of communication systems, Claude Shannon was able to do exactly this, and as a result, opened up an entirely new view of communication systems analysis and design [1948, 1949, 1959]. The principal contribution of Shannon's information theory to date has been to allow communication theorists to establish absolute bounds on communication systems performance that cannot be exceeded no matter how ingeniously designed or complex our communication systems are. Fundamental physical limitations on communication systems performance is another topic that has been largely ignored in the preceding chapters, but it is a subject of exceptional practical importance. For example, for any of the numerous communication systems developed thus far in the book, we could decide to design a new system that would outperform the accepted standard for a particular application. The first question that we should ask is: How close is the present system to achieving theoretically optimum performance? If the existing communication system operates at or near the fundamental physical limit on performance, our task may be difficult or impossible. However, if the existing system is far away from the absolute performance bound, this might be an area for fruitful work.

Of course, in specifying the particular communication system under investigation, we must know the important physical parameters, such as transmitted power, bandwidth, type(s) of noise present, and so on, and information theory allows these constraints to be incorporated. However, information theory does not provide a way for communication system complexity to be explicitly included. Although this is something of a drawback, information theory itself provides a way around this difficulty, since it is generally true that as we approach the fundamental limit on the performance of a communication system, the system complexity increases, sometimes quite drastically. Therefore, for a simple communication system operating far from its performance bound, we may be able to improve the performance with a relatively modest increase in complexity. On the other hand, if we have a rather complicated communication system operating near its fundamental limit, any performance improvement may be possible only with an extremely complicated system.

In this chapter we are concerned with the rather general block diagram shown in Fig. 11.1.1. Most of the early work by Shannon and others ignored the source encoder/decoder blocks and concentrated on bounding the performance of the channel encoder/decoder pair. Subsequently, the source encoder/decoder blocks have attracted much research attention. In this chapter we consider both topics and expose the reader to the nomenclature used in the information theory literature. Quantitative definitions of information are presented in Section 11.2 that lay the foundation for the remaining sections. In Sections 11.3 and 11.4 we present the fundamental source and channel coding theorems, give some examples, and state the implications of these theorems. Section 11.5 contains a brief development of rate distortion theory, which is the mathematical basis for data compression. A few applications of the theory in this chapter are presented in Section 11.6, and a technique for variable-length source coding is given in Section 11.7.

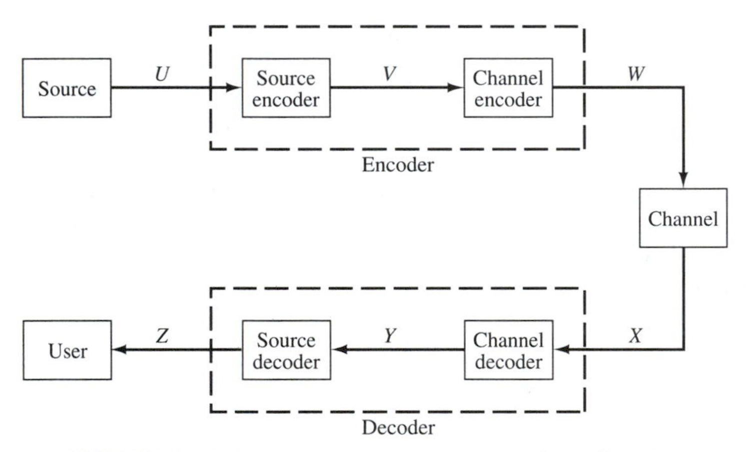

FIGURE 11.1.1 Communication system block diagram.

11.2 Entropy and Average Mutual Information

Consider a discrete random variable U that takes on the values $\{u_1, u_2, \ldots, u_M\}$, where the set of possible values of U is often called the *alphabet* and the elements of the set are called *letters* of the alphabet. Let $P_U(u)$ denote the probability assignment over the alphabet, then we can define the *self-information* of the event $u = u_j$ by

$$I_U(u_j) = \log \frac{1}{P_U(u_j)} = -\log P_U(u_j). \tag{11.2.1}$$

The quantity $I_U(u_j)$ is a measure of the information contained in the event $u = u_j$. Note that the base of the logarithm in Eq. (11.2.1) is unspecified. It is common to use base e, in which case $I_U(\cdot)$ is in natural units (nats), or base 2, in which case $I_U(\cdot)$ is in binary units (bits). Either base is acceptable since the difference in the two bases is just a scaling operation. We will use base 2 in all of our work, and hence $I_U(\cdot)$ and related quantities will be in bits. The average or expected value of the self-information is called the *entropy*, also discrete entropy or absolute entropy, and is given by

$$H(U) = -\sum_{j=1}^{M} P_U(u_j) \log P_U(u_j). \tag{11.2.2}$$

The following example illustrates the calculation of entropy and how it is affected by probability assignments.

EXAMPLE 11.2.1

Given a random variable U with four equally likely letters in its alphabet, we wish to find $H(U)$. Clearly, $M = 4$ and $P_U(u_i) = \frac{1}{4}$ for $i = 1, 2, 3, 4$. Thus, from Eq. (11.2.2),

$$\begin{aligned} H(U) &= -\sum_{j=1}^{4} \tfrac{1}{4} \log \tfrac{1}{4} \\ &= -\log \tfrac{1}{4} = 2 \text{ bits}. \end{aligned} \tag{11.2.3}$$

We now consider a discrete random variable X with four letters such that $P_X(x_1) = \frac{1}{2}$, $P_X(x_2) = \frac{1}{4}$, $P_X(x_3) = P_X(x_4) = \frac{1}{8}$. Again, we wish to find the entropy of this random variable. Directly,

$$H(X) = -\tfrac{1}{2} \log \tfrac{1}{2} - \tfrac{1}{4} \log \tfrac{1}{4} - \tfrac{1}{8} \log \tfrac{1}{8} - \tfrac{1}{8} \log \tfrac{1}{8} = 1.75 \text{ bits}. \tag{11.2.4}$$

Comparing Eqs. (11.2.3) and (11.2.4), we see that equally likely letters produce a larger entropy. This result is true in general, as we will see shortly.

We now consider two jointly distributed discrete random variables W and X with the probability assignment $P_{WX}(w_j, x_k), j = 1, 2, \ldots, M, k = 1, 2, \ldots, N$. We are particularly interested in the interpretation that w is an input letter to a noisy channel and x is the corresponding output. The information provided about the event $w = w_j$ by the occurrence of the event $x = x_k$ is

$$I_{W;X}(w_j; x_k) = \log \frac{P_{W|X}(w_j | x_k)}{P_W(w_j)}. \tag{11.2.5}$$

Further, the information provided about the event $x = x_k$ by the occurrence of $w = w_j$ is

$$I_{X;W}(x_k; w_j) = \log \frac{P_{X|W}(x_k|w_j)}{P_X(x_k)}. \tag{11.2.6}$$

We can show that $I_{W;X}(w_j; x_k) = I_{X;W}(x_k; w_j)$ by starting with Eq. (11.2.5) and using conditional probability as follows:

$$\begin{aligned} I_{W;X}(w_j; x_k) &= \log \frac{P_{WX}(w_j, x_k)}{P_W(w_j)P_X(x_k)} \\ &= \log \frac{P_{X|W}(x_k|w_j)}{P_X(x_k)} = I_{X;W}(x_k; w_j). \end{aligned} \tag{11.2.7}$$

Because of this symmetry, Eqs. (11.2.5) and (11.2.6) are called the *mutual information* between the events $w = w_j$ and $x = x_k$. The *average mutual information* over the joint ensemble is an important quantity defined by

$$\begin{aligned} I(W; X) &= \sum_{j=1}^{M} \sum_{k=1}^{N} P_{WX}(w_j, x_k) I_{W;X}(w_j; x_k) \\ &= \sum_{j=1}^{M} \sum_{k=1}^{N} P_{WX}(w_j, x_k) \log \frac{P_{W|X}(w_j|x_k)}{P_W(w_j)}. \end{aligned} \tag{11.2.8}$$

By a straightforward manipulation of Eq. (11.2.5),

$$\begin{aligned} I_{W;X}(w_j; x_k) &= -\log P_W(w_j) + \log P_{W|X}(w_j|x_k) \\ &= I_W(w_j) - I_{W|X}(w_j|x_k), \end{aligned} \tag{11.2.9}$$

where

$$I_{W|X}(w_j|x_k) \triangleq -\log P_{W|X}(w_j|x_k) \tag{11.2.10}$$

is called the *conditional self-information*, and is interpreted as the information that must be supplied to an observer to specify $w = w_j$ after the occurrence of $x = x_k$. Substituting Eq. (11.2.9) into Eq. (11.2.8), we find that

$$I(W; X) = H(W) - H(W|X), \tag{11.2.11}$$

where $H(W|X)$ is the *average conditional self-information*. Since entropy is a measure of uncertainty, we see from Eq. (11.2.11) that the average mutual information can be interpreted as the average amount of uncertainty remaining after the observation of X.

EXAMPLE 11.2.2

Here we wish to calculate the mutual information and the average mutual information for the probability assignments (with $M = 2$ and $N = 2$)

$$P_W(w_1) = P_W(w_2) = \tfrac{1}{2} \tag{11.2.12}$$

and

$$P_{X|W}(x_1|w_1) = P_{X|W}(x_2|w_2) = 1 - p \tag{11.2.13}$$

$$P_{X|W}(x_1|w_2) = P_{X|W}(x_2|w_1) = p. \tag{11.2.14}$$

If we interpret W as the input to a channel and X as the output, the transition probabilities in Eqs. (11.2.13) and (11.2.14) are representative of what is called a *binary symmetric channel* (BSC).

To calculate the mutual information, we could use either Eq. (11.2.5) or (11.2.6). From the probabilities given, Eq. (11.2.6) seems to be slightly simpler. Thus we need $P_X(x_1)$ and $P_X(x_2)$. Directly,

$$\begin{aligned} P_{WX}(w_1, x_1) &= P_{X|W}(x_1|w_1)P_W(w_1) = \frac{1-p}{2} \\ &= P_{WX}(w_2, x_2) \end{aligned} \tag{11.2.15}$$

and

$$\begin{aligned} P_{WX}(w_1, x_2) &= P_{X|W}(x_2|w_1)P_W(w_1) \\ &= \frac{p}{2} = P_{WX}(w_2, x_1). \end{aligned} \tag{11.2.16}$$

Thus

$$\begin{aligned} P_X(x_1) &= P_{WX}(w_1, x_1) + P_{WX}(w_2, x_1) \\ &= \tfrac{1}{2} = P_X(x_2), \end{aligned} \tag{11.2.17}$$

so the four mutual information values are

$$I_{X;W}(x_1; w_1) = \log 2(1 - p) = I_{X;W}(x_2; w_2) \tag{11.2.18}$$

and

$$I_{X;W}(x_1; w_2) = \log 2p = I_{X;W}(x_2; w_1). \tag{11.2.19}$$

The average mutual information follows in a straightforward fashion from Eq. (11.2.8) as

$$\begin{aligned} I(W; X) = I(X; W) &= \sum_{j=1}^{2} \sum_{k=1}^{2} P_{WX}(w_j, x_k) \log \frac{P_{X|W}(x_k|w_j)}{P_X(x_k)} \\ &= \frac{1-p}{2} \log 2(1-p) + \frac{p}{2} \log 2p + \frac{p}{2} \log 2p + \frac{1-p}{2} \log 2(1-p) \\ &= 1 + (1-p) \log (1-p) + p \log p. \end{aligned} \tag{11.2.20}$$

The average mutual information given by Eq. (11.2.20) is plotted versus p in Fig. 11.2.1. The results of this example are discussed more in Section 11.4 in a communications context.

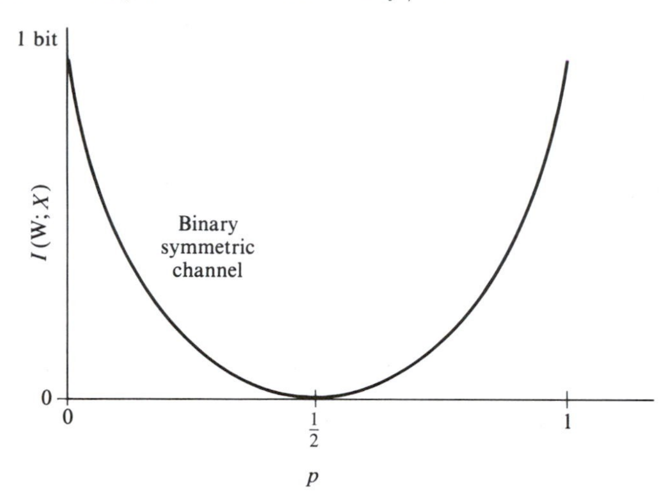

FIGURE 11.2.1 Average mutual information for a BSC with equally likely inputs (Example 11.2.2).

There are numerous useful properties of entropy and average mutual information. We state two of these properties here without proof.

Property 1. Let U be a random variable with possible values $\{u_1, u_2, \ldots, u_M\}$. Then

$$H(U) \leq \log M \tag{11.2.21}$$

with equality if and only if the values of U are equally likely to occur.

Example 11.2.1 illustrates Property 1.

Property 2. Let W and X be jointly distributed random variables. The average mutual information between W and X satisfies

$$I(W; X) \geq 0 \tag{11.2.22}$$

with equality if and only if W and X are statistically independent.

Thus far we have defined the entropy and average mutual information only for discrete random variables. Given an absolutely continuous random variable U with pdf $f_U(u)$, we define the *differential entropy* of U as

$$h(U) = -\int_{-\infty}^{\infty} f_U(u) \log f_U(u)\, du. \tag{11.2.23}$$

Although Eqs. (11.2.2) and (11.2.23) are functionally quite similar, there is a significant difference between the interpretations of absolute or discrete entropy and differential entropy. While $H(U)$ is an absolute indicator of "randomness," $h(U)$ is only an indicator of randomness with respect to a coordinate system:

hence the names "absolute entropy" for $H(U)$ and "differential entropy" for $h(U)$. The following example illustrates the calculation of differential entropy and its property of indicating randomness with respect to a coordinate system.

EXAMPLE 11.2.3

Consider an absolutely continuous random variable with uniform pdf

$$f_U(u) = \begin{cases} \frac{1}{a}, & \frac{-a}{2} \le u \le \frac{a}{2} \\ 0, & \text{elsewhere.} \end{cases} \tag{11.2.24}$$

(1) Let $a = 1$ in Eq. (11.2.24) and find $h(U)$. Then

$$h(U) = -\int_{-1/2}^{1/2} \log 1 \, du = 0. \tag{11.2.25}$$

(2) Let $a = 32$ and find $h(U)$. We have

$$h(U) = -\int_{-16}^{16} \tfrac{1}{32} \log\left(\tfrac{1}{32}\right) du = 5. \tag{11.2.26}$$

(3) Finally, let $a = \frac{1}{32}$ and find $h(U)$. Here

$$h(U) = -\int_{-1/64}^{1/64} 32 \log(32) \, du = -5. \tag{11.2.27}$$

The fact that differential entropy is a relative indicator of randomness is evident from these three special cases of the uniform distribution. Clearly, $h(U)$ is not an absolute indicator of randomness, since in case (3) $h(U)$ is negative, and negative randomness is difficult to interpret physically! The "reference" distribution is the uniform distribution over a unit interval, with "broader" distributions having a positive entropy and "narrower" distributions having a negative differential entropy.

Differential entropies are unlike absolute entropy in that differential entropy is not always positive, not necessarily finite, not invariant to a one-to-one transformation of the random variable, and not subject to interpretation as an average self-information.

The average mutual information of two jointly distributed continuous random variables, say W and X, can also be defined as

$$\begin{aligned} I(W; X) &= \int_{-\infty}^{\infty} \int_{-\infty}^{\infty} f_{WX}(w, x) \log \frac{f_{WX}(w, x)}{f_W(w) f_X(x)} \, dw \, dx \\ &= I(X; W). \end{aligned} \tag{11.2.28}$$

As in the discrete case, the average mutual information can be expressed in terms of entropies as

$$\begin{aligned} I(W; X) &= h(W) - h(W \mid X) \\ &= h(X) - h(X \mid W), \end{aligned} \tag{11.2.29}$$

where

$$h(W|X) = -\int_{-\infty}^{\infty}\int_{-\infty}^{\infty} f_{WX}(w, x) \log f_{W|X}(w|x)\, dw\, dx. \qquad (11.2.30)$$

Fortunately for our subsequent uses of $I(W; X)$, the average mutual information is invariant under any one-to-one transformation of the variables, even though the individual differential entropies are not.

EXAMPLE 11.2.4

In this example we compute the differential entropy of the input and the average mutual information between the input and output of an additive white Gaussian noise channel with zero mean and variance σ_c^2. The input is also assumed to be Gaussian with zero mean and variance σ_s^2. Representing the channel as shown in Fig. 11.2.2, we thus have that

$$f_W(w) = \frac{1}{\sqrt{2\pi}\,\sigma_s} e^{-w^2/2\sigma_s^2} \qquad (11.2.31)$$

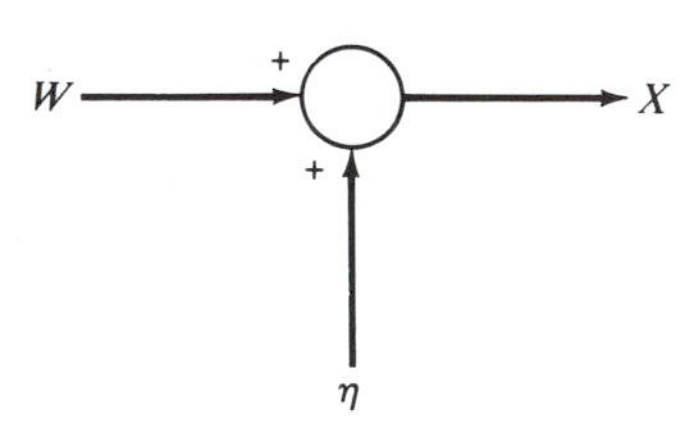

FIGURE 11.2.2 Additive noise channel.

and

$$f_{X|W}(x|w) = \frac{1}{\sqrt{2\pi}\,\sigma_c} e^{-(x-w)^2/2\sigma_c^2}. \qquad (11.2.32)$$

From Eq. (11.2.23) the differential entropy of the input is

$$\begin{aligned} h(W) &= -\int_{-\infty}^{\infty} f_W(w) \log f_W(w)\, dw \\ &= \int_{-\infty}^{\infty} f_W(w) \left\{ \log \sqrt{2\pi}\,\sigma_s + \frac{w^2}{2\sigma_s^2} \log e \right\} dw \\ &= \log \sqrt{2\pi}\,\sigma_s + \tfrac{1}{2} \log e = \tfrac{1}{2} \log 2\pi e \sigma_s^2. \end{aligned} \qquad (11.2.33)$$

To calculate the average mutual information, we choose to employ the expression $I(W; X) = h(X) - h(X|W)$. We already have $f_{X|W}(x|w)$, but we need $f_X(x)$. This follows directly as

$$f_X(x) = \frac{1}{\sqrt{2\pi}[\sigma_s^2 + \sigma_c^2]^{1/2}} e^{-x^2/2(\sigma_s^2 + \sigma_c^2)}, \qquad (11.2.34)$$

since the input and the noise are independent, zero-mean Gaussian processes. By analogy with Eq. (11.2.33), we have from Eqs. (11.2.32) and (11.2.34) that

$$h(X) = \tfrac{1}{2} \log 2\pi e[\sigma_s^2 + \sigma_c^2] \tag{11.2.35}$$

and

$$h(X \mid W) = \tfrac{1}{2} \log 2\pi e \sigma_c^2. \tag{11.2.36}$$

We thus find the average mutual information to be

$$\begin{aligned} I(W; X) &= \tfrac{1}{2} \log 2\pi e[\sigma_s^2 + \sigma_c^2] - \tfrac{1}{2} \log 2\pi e \sigma_c^2 \\ &= \frac{1}{2} \log \left(1 + \frac{\sigma_s^2}{\sigma_c^2}\right). \end{aligned} \tag{11.2.37}$$

In Eq. (11.2.37), we see that for $\sigma_s^2 \ll \sigma_c^2$, $I(W; X) \cong 0$, while if $\sigma_c^2 \to 0$, then $I(W; X) \to \infty$.

11.3 Source Coding Theorem

We now begin to interpret the quantities developed in Section 11.2 within a communications context. For this particular section, we consider the communication system block diagram in Fig. 11.1.1 under the assumptions that the channel is ideal (no noise or deterministic distortion) and the channel encoder/decoder blocks are identity mappings (a straight wire connection from input to output of each of these blocks). We are left with a communication system block diagram consisting of a source, source encoder, ideal channel, source decoder, and user. Since the channel is ideal and the channel encoder and decoder are identities, we have, with reference to Fig. 11.1.1, that $V = W = X = Y$. A block diagram of the simplified communication system is shown in Fig. 11.3.1.

Just what are the physical meanings of each of the components in Fig. 11.3.1? The source is some kind of data generation device, such as a computer or computer peripheral, which generates a discrete-valued random variable U with

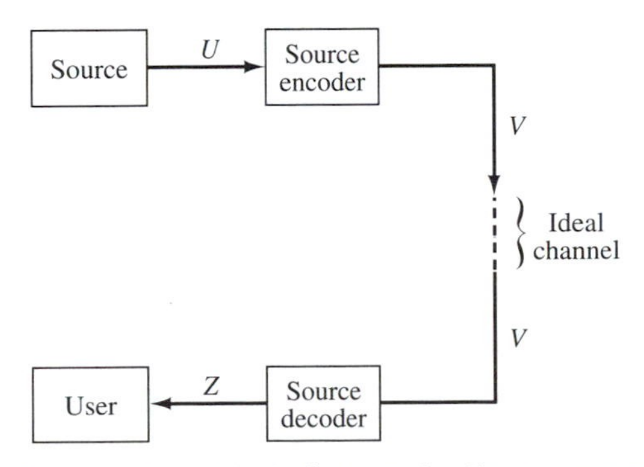

FIGURE 11.3.1 Simplified block diagram for the source coding problem.

alphabet $\{u_1, u_2, \ldots, u_M\}$ and probability assignment $P_U(\cdot)$. For the development in this section, we assume that the successive letters produced by the source are statistically independent. Such a source is called a *discrete memoryless source* (DMS). The ideal channel can be thought of as some perfectly operating modem or a mass storage device that is error-free. The user is a machine or individual that requires the data to accomplish a task. With these descriptions thus far, it is unclear why we need the source encoder/decoder blocks or what utility they might be. The answer is that data at the output of any given discrete source may not be in a form that yields the minimum required transmitted bit rate. The following example illustrates this point.

EXAMPLE 11.3.1 [Wyner, 1981]

The source output is a ternary-valued random variable that takes on the values $\{u_1, u_2, u_3\}$ with probabilities $P(u_1) = 0.7$, $P(u_2) = 0.15 = P(u_3)$. The source letter produced at each time instant is assumed to be independent of the letter produced at any other time instant, so that we have a DMS. We wish to find a binary representation of any sequence of ternary source letters such that the source sequence can be recovered exactly (by the decoder) and such that the average number of binary digits per source letter is a minimum.

A straightforward assignment of binary words, called a source code, is to let u_1 be represented by 00, u_2 by 10, and u_3 by 11. This code transmits an average of 2 bits per source letter. Since u_1 occurs much more often than the other two letters, it seems intuitive that we should use a shorter sequence to represent u_1 than those used for u_2 and u_3. One such code assignment is

$$u_1 \to 0$$

$$u_2 \to 10$$

$$u_3 \to 11.$$

Since this code has the special property (called the *prefix condition*) that no binary word assignment (codeword) is a prefix of any other codeword, the ternary source data can be uniquely recovered from its binary encoding. The average number of bits required per source letter, denoted here by $\bar{n}$, is thus

$$\begin{aligned}\bar{n} &= 1 \cdot P(u_1) + 2 \cdot P(u_2) + 2 \cdot P(u_3) \\ &= 0.7 + 0.3 + 0.3 = 1.3 \text{ bits/source letter.}\end{aligned} \tag{11.3.1}$$

This is a clear improvement over the original 2-bits/source letter code.

To try and reduce $\bar{n}$ further, we encode pairs of source letters, which are listed in Table 11.3.1 together with the probability of each pair. If we now assign a binary word to each pair of source letters as shown in the column labeled "codeword" in Table 11.3.1, we find that the average binary codeword length per source letter is

$$\begin{aligned}\bar{n} &= \tfrac{1}{2}\{1(0.49) + 3(0.105) + 3(0.105) + 3(0.105) + 4(0.105) + 6(0.0225) \\ &\quad + 6(0.0225) + 6(0.0225) + 6(0.0225)\} \\ &= 1.1975 \text{ bits/source letter.}\end{aligned} \tag{11.3.2}$$

TABLE 11.3.1 Pairs of Ternary Letters and a Source Code for Example 11.3.1

Source Letters	Probability	Codeword
u_1u_1	0.49	0
u_1u_2	0.105	111
u_1u_3	0.105	100
u_2u_1	0.105	101
u_3u_1	0.105	1100
u_2u_2	0.0225	110110
u_2u_3	0.0225	110111
u_3u_2	0.0225	110100
u_3u_3	0.0225	110101

Source: A. D. Wyner, "Fundamental Limits in Information Theory," *Proc. IEEE*, © 1981 IEEE.

The value in Eq. (11.3.2) is slightly better than the 1.3 achieved by our second code. The code in Table 11.3.1 is also uniquely decodable back into the original ternary sequence.

Although we have not described how the binary codes were selected, it is clear that at least in this particular case, source coding allows the ternary data to be represented with a smaller number of bits than one might originally think. Thus the utility of the source encoder/decoder blocks in Fig. 11.3.1 is demonstrated.

Example 11.3.1 demonstrates that source coding techniques can be useful for reducing the bit rate required to represent (exactly) a discrete source, and in Section 11.7 we present a constructive technique for designing source codes that is due to Huffman [1952]. It is also important to note that in Example 11.3.1 rates are expressed in terms of bits/source letter (or bits/letter). This may be confusing, since prior chapters have discussed data rates primarily in terms of bits/sec. There is no difficulty, however, since bits/letter is just bits/symbol, and hence if we multiply by the number of letters or symbols transmitted per second, the rate in bits/sec results. We thus state all of our rates in this chapter in bits/letter with this understanding in mind. A fundamental question raised by Example 11.3.1 is: What is the minimum bit rate required to represent a given discrete source? The answer is provided by the source coding theorem, which we state in a nonmathematical form here.

Theorem 11.3.1 Source Coding Theorem *For a DMS with entropy $H(U)$, the minimum average codeword length per source letter ($\bar{n}$) for any code is lower bounded by $H(U)$, that is, $\bar{n} \geq H(U)$, and further, $\bar{n}$ can be made as close to $H(U)$ as desired for some suitably chosen code.*

The (absolute) entropy of a discrete source is thus a very important physical quantity, since it specifies the minimum bit rate required to yield a perfect replication of the original source sequence. Therefore, for the DMS in Example 11.3.1, since

$$\begin{aligned} H(U) &= -0.7 \log 0.7 - 0.15 \log 0.15 - 0.15 \log 0.15 \\ &= 1.18129\ldots, \end{aligned} \tag{11.3.3}$$

we know that $\bar{n} \geq 1.18129\ldots$, and hence only a slight further reduction in rate (from 1.1975 bits/source letter) can be achieved by designing additional source codes (see Section 11.7).

11.4 Channel Coding Theorem

We now turn our attention to the problem of communicating source information over a noisy channel. With respect to the general communication system block diagram in Fig. 11.1.1, we are presently interested in the channel encoder, channel, and channel decoder blocks. For our current purposes, it is of no interest whether the source encoder/decoder blocks are present or whether the source output is connected directly to the channel encoder input and the channel decoder output is passed directly to the user. What we are interested in here is the transmission of information over a noisy channel. More specifically, we would like to address the question: Given the characterization of a communications channel, what is the maximum bit rate that can be sent over this channel with negligibly small error probability? We find that the average mutual information between the channel input and output random variables plays an important role in providing the answer to this question.

Because of the scope of this topic, we limit consideration to *discrete memoryless channels* that have finite input and output alphabets and for which the output letter at any given time depends only on the channel input letter at the same time instant. Therefore, with reference to Fig. 11.1.1, we define a *discrete memoryless channel* (DMC) with input alphabet $\{w_1, w_2, \ldots, w_M\}$ and probability assignment $P_W(w_j)$, $j = 1, 2, \ldots, M$, and with output alphabet $\{x_1, x_2, \ldots, x_N\}$ and *transition probabilities* $P_{X|W}(x_k|w_j)$, $j = 1, 2, \ldots, M$, and $k = 1, 2, \ldots, N$. From these quantities we can calculate the average mutual information between the channel input W and output X according to Eq. (11.2.8). We define the capacity of a DMC with input W and output X by

$$\begin{aligned} C &\triangleq \max_{\text{all } P_W(\cdot)} I(W; X) \\ &= \max_{\text{all } P_W(\cdot)} \sum_{j=1}^{M} \sum_{k=1}^{N} P_{WX}(w_j, x_k) \log \frac{P_{WX}(w_j, x_k)}{P_W(w_j) P_X(x_k)}, \end{aligned} \tag{11.4.1}$$

where the maximum is taken over all channel input probability assignments. We note that $I(W; X)$ is a function of the input probabilities and the transition probabilities, whereas the channel capacity C is a function of the input probabilities only. We have no control over the channel transition probabilities. In words, Eq. (11.4.1) says that the capacity of a DMC is the largest average mutual information that can be transmitted over the channel in one use. The operational significance of the channel capacity is illustrated further by the following theorem.

Theorem 11.4.1 Channel Coding Theorem [Berger, 1971] *Given a DMS with entropy H bits/source letter and a DMC with capacity C bits/source letter, if $H \le C$, the source output can be encoded for transmission over the channel with an arbitrarily small bit error probability. Further, if $H > C$, the bit error probability is bounded away from* 0.

To calculate the channel capacity, it is evident from Eq. (11.4.1) that we must perform a maximization over M variables, the $P_W(w_j)$, subject to the constraints that $P_W(w_j) \ge 0$ for all j and $\sum_{j=1}^{M} P_W(w_j) = 1$. In general, this is a difficult task. The following example illustrates the calculation of capacity for the special case of what is called a *binary symmetric channel* (BSC).

EXAMPLE 11.4.1

A special case of a DMC is the BSC with binary input and output alphabets $\{0, 1\}$ and transition probabilities $P_{X|W}(0|0) = P_{X|W}(1|1) = 1 - p$ and $P_{X|W}(0|1) = P_{X|W}(1|0) = p$, where W is the input random variable and X is the output random variable. A standard diagram for the BSC is shown in Fig. 11.4.1. We can rewrite the average mutual information in Eq. (11.2.8) as

$$I(W; X) = \sum_{k=1}^{N} \sum_{j=1}^{M} P_{X|W}(x_k|w_j)P_W(w_j) \log \frac{P_{X|W}(x_k|w_j)}{\sum_{l=1}^{M} P_{X|W}(x_k|w_l)P_W(w_l)}. \tag{11.4.2}$$

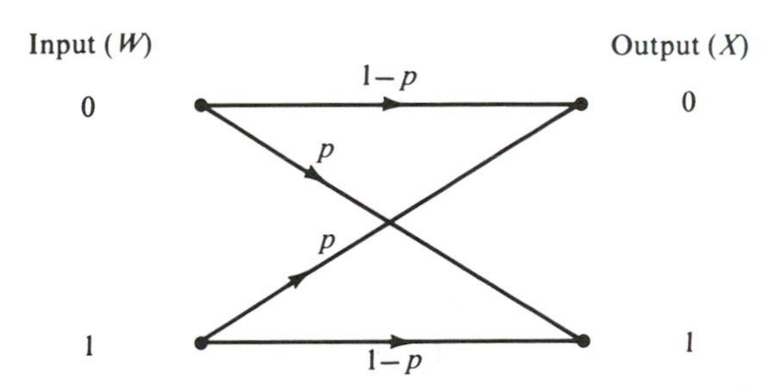

FIGURE 11.4.1 Binary symmetric channel (see Example 11.4.1).

For the current example $M = N = 2$, $\{x_1, x_2\} = \{0, 1\}$, and $\{w_1, w_2\} = \{0, 1\}$. We need to write $I(W; X)$ in terms of $P_W(0)$ and $P_W(1) = 1 - P_W(0)$ so that we can perform the maximization. First, we evaluate the denominators of the argument of the logarithm as

$$x_k = 0: \quad \sum_{l=1}^{2} P_{X|W}(0|w_l)P_W(w_l) = (1-p)P_W(0) + pP_W(1)$$

$$= (1-2p)P_W(0) + p \triangleq \text{den}_0$$

$$x_k = 1: \quad \sum_{l=1}^{2} P_{X|W}(1|w_l)P_W(w_l) = pP_W(0) + (1-p)P_W(1)$$

$$= 1 - p - (1-2p)P_W(0) \triangleq \text{den}_1,$$

since $P_W(1) = 1 - P_W(0)$. Now, expanding Eq. (11.4.2), we have

$$I(W; X) = (1-p)P_W(0) \log \frac{1-p}{\text{den}_0} + pP_W(1) \log \frac{p}{\text{den}_0}$$

$$+ pP_W(0) \log \frac{p}{\text{den}_1} + (1-p)P_W(1) \log \frac{1-p}{\text{den}_1}$$

$$= p \log p + (1-p) \log (1-p) + [2pP_W(0) - P_W(0) - p] \log (\text{den}_0)$$

$$+ [p + P_W(0) - 2pP_W(0) - 1] \log (\text{den}_1), \tag{11.4.3}$$

where the last equality results from letting $P_W(1) = 1 - P_W(0)$ and simplifying.

Taking the partial derivative of $I(W; X)$ with respect to $P_W(0)$ yields

$$\frac{\partial}{\partial P_W(0)} I(W; X) = (2p-1) \log \text{den}_0 + \frac{2pP_W(0) - P_W(0) - p}{\text{den}_0}(1-2p)$$

$$+ (1-2p) \log \text{den}_1 + \frac{p + P_W(0) - 2pP_W(0) - 1}{\text{den}_1}(2p-1)$$

$$= -(1-2p) \log \text{den}_0 + (1-2p) \log \text{den}_1, \tag{11.4.4}$$

where the last simplification follows after using the definitions of den_0 and den_1. Equating the partial derivative to 0, we find that $P_W(0) = \frac{1}{2}$. Upon substituting this value back into Eq. (11.4.3), the capacity of the BSC is found to be

$$C = 1 + p \log p + (1-p) \log (1-p) \tag{11.4.5}$$

and is achieved with equally likely inputs. Although we have not shown that the partial derivative yields a maximum as opposed to a minimum, $I(W; X)$ is a convex $\cap$ (read "cap") function of $P_W(\cdot)$, and hence we have found a maximum. Proofs of such properties of $I(W; X)$ are available elsewhere [McEliece, 1977].

We now turn our attention to finding the capacity of discrete-time, memoryless channels with input and output alphabets that consist of the entire set of real numbers. In other words, the input (W) and output (X) are absolutely continuous random variables here, as opposed to discrete random variables previously. Since the calculation of capacity can be difficult in general for continuous-valued random variables, we limit ourselves to the physically important case of the additive Gaussian noise channel with an average power constraint. The channel being considered can thus be represented by the diagram in Fig. 11.4.2, where the noise ζ has the pdf

$$f_\zeta(\zeta) = \frac{1}{\sqrt{2\pi}\,\sigma} e^{-\zeta^2/2\sigma^2}, \tag{11.4.6}$$

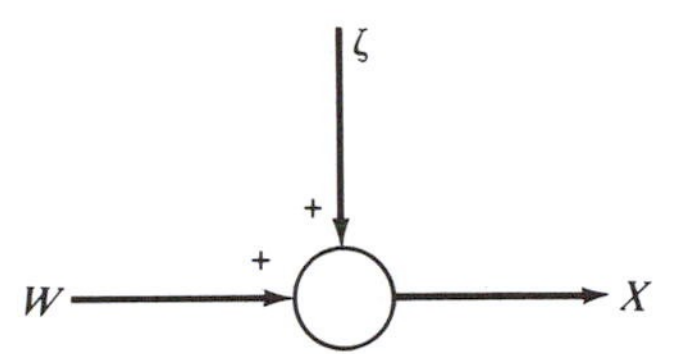

FIGURE 11.4.2 Additive noise channel.

$-\infty < \zeta < \infty$, $E[W] = 0$ and $E[W^2] \leq S$ (the average power constraint on the input), and each output letter is probabilistically dependent only on the current input letter (the memoryless assumption).

The average mutual information is given by Eq. (11.2.29). However, for any independent, additive noise channel

$$f_{X|W}(x|w) = f_\zeta(x - w), \tag{11.4.7}$$

so

$$\begin{aligned} h(X|W) &= -\int_{-\infty}^{\infty}\int_{-\infty}^{\infty} f_W(w) f_\zeta(x-w) \log f_\zeta(x-w)\, dx\, dw \\ &= -\int_{-\infty}^{\infty} f_W(w)\left\{\int_{-\infty}^{\infty} f_\zeta(\zeta)\log f_\zeta(\zeta)\, d\zeta\right\} dw \\ &= \int_{-\infty}^{\infty} f_W(w) h(\zeta)\, dw = h(\zeta). \end{aligned} \tag{11.4.8}$$

Therefore, for an additive noise channel,

$$I(W; X) = h(X) - h(\zeta). \tag{11.4.9}$$

To find the channel capacity, we wish to maximize $I(W; X)$ with respect to the choice of the input pdf $f_W(w)$ subject to the constraint $E[W^2] \leq S$. From Eq. (11.4.9) it is evident that $h(\zeta)$ is independent of the input, and hence only $h(X)$ is affected by the input pdf. Thus, to maximize $I(W; X)$, it is equivalent to maximize $h(X)$.

To complete the calculation of capacity for the additive Gaussian noise channel, the following intermediate results are needed.

Theorem 11.4.2 [Shannon, 1948; Gallager, 1968] *For any absolutely continuous random variable* ξ, *the pdf that maximizes the differential entropy*

$$h(\xi) = -\int f_\xi(\xi) \log f_\xi(\xi)\, d\xi \tag{11.4.10}$$

subject to the constraint that

$$\int_{-\infty}^{\infty} \xi^2 f_\xi(\xi)\, d\xi \le \sigma_{\max}^2 \tag{11.4.11}$$

is

$$f_\xi(\xi) = \frac{1}{\sqrt{2\pi}\,\sigma_{\max}} e^{-\xi^2/2\sigma_{\max}^2}. \tag{11.4.12}$$

Proof: This result can be proved in several ways, including calculus of variations [Shannon, 1948] and Jensen's inequality [McEliece, 1977]; however, an alternative method is used here [Gallager, 1968]. Let $f_\eta(\eta)$ be an arbitrary pdf that satisfies the constraint in Eq. (11.4.11), and let $f_\xi(\xi)$ be given by Eq. (11.4.12). Then

$$\begin{aligned} -\int_{-\infty}^{\infty} f_\eta(\alpha) \log f_\xi(\alpha)\, d\alpha &= \int_{-\infty}^{\infty} f_\eta(\alpha)\left\{\log \sqrt{2\pi}\,\sigma_{\max} + \frac{\alpha^2}{2\sigma_{\max}^2} \log e\right\} d\alpha \\ &= \frac{1}{2} \log 2\pi e \sigma_{\max}^2. \end{aligned} \tag{11.4.13}$$

Now consider

$$\begin{aligned} h(\eta) - \frac{1}{2} \log 2\pi e \sigma_{\max}^2 &= \int_{-\infty}^{\infty} f_\eta(\alpha) \log \frac{f_\xi(\alpha)}{f_\eta(\alpha)}\, d\alpha \\ &\le \log e \int_{-\infty}^{\infty} f_\eta(\alpha)\left[\frac{f_\xi(\alpha)}{f_\eta(\alpha)} - 1\right] d\alpha = 0, \end{aligned} \tag{11.4.14}$$

where the inequality follows from the fact that $\log \beta \le (\beta - 1) \log e$. Thus

$$h(\eta) \le \tfrac{1}{2} \log 2\pi e \sigma_{\max}^2 \tag{11.4.15}$$

with equality if and only if $f_\xi(\alpha)/f_\eta(\alpha) = 1$ for all α. Hence the theorem follows.

From Theorem 11.4.2, $I(W; X)$ in Eq. (11.4.9) is maximized if X is Gaussian with $E[X^2] = S + \sigma^2$ (since $X = W + \zeta$ and $E[W^2] \le S$); but if X and ζ are Gaussian, then W must be Gaussian. Therefore, the input pdf that achieves channel capacity is

$$f_W(w) = \frac{1}{\sqrt{2\pi S}} e^{-w^2/2S}, \tag{11.4.16}$$

so from Eqs. (11.4.6) and (11.4.9), the channel capacity of the discrete time, memoryless, additive Gaussian noise channel with an average power constraint on the input is

$$C = \tfrac{1}{2}\log 2\pi e(S + \sigma^2) - \tfrac{1}{2}\log 2\pi e\sigma^2$$
$$= \frac{1}{2}\log\left(1 + \frac{S}{\sigma^2}\right) \quad \text{bits/source letter.} \tag{11.4.17}$$

The channel capacity given in Eq. (11.4.17) is a classical, often-quoted result that has considerable intuitive appeal. For instance, if $S/\sigma^2 \ll 1$, then $C \cong 0$, while as $S/\sigma^2 \to \infty$, $C \to \infty$. Of course, in practical applications, the signal power is constrained, as we assumed in the derivation of Eq. (11.4.17).

We have only calculated channel capacity for the two special cases of a binary symmetric channel and an additive Gaussian noise channel. The calculation of capacity for other channels can be a tedious and difficult task. A variety of theorems and techniques have been developed to aid in the calculation of capacity and several possibilities are examined in the problems.

11.5 Rate Distortion Theory

In Section 11.3 the transmitted data rate required to reproduce a discrete-time, discrete-amplitude source *exactly* (with no error) is considered, and the minimum rate necessary is shown to be the absolute or discrete entropy. The process of exactly representing a discrete-amplitude source with a reduced or minimum number of binary digits is called *noiseless source coding*, or simply, *source coding*. If the source to be transmitted is a continuous-amplitude random variable or random process, the source has an infinite number of possible amplitudes, and hence the number of bits required to reproduce the source *exactly* at the receiver is infinite. This is indicated by the fact that continuous-amplitude sources have infinite *absolute* entropy. Therefore, to represent continuous-amplitude sources in terms of a finite number of bits/source letter, we must accept the inevitability of some amount of reconstruction error or distortion. We are thus led to the problem of representing a source with a minimum number of bits/source letter subject to a constraint on allowable distortion. This problem is usually called *source coding with a fidelity criterion*. Source coding with respect to some distortion measure may also be necessary for a discrete-amplitude source. For instance, if we are given a DMS with entropy H such that $H > C$ (see Theorem 11.4.1), it may be necessary to accept some amount of distortion in the reproduced version of the DMS in order to reduce the required number of bits/source letter below C.

For source coding with a fidelity criterion, the function of interest is no longer H, but the rate distortion function, denoted $R(D)$. The rate distortion function $R(D)$ with respect to a fidelity criterion is the minimum information rate necessary to represent the source with an average distortion less than or equal to D. To be more specific, we again focus on the block diagram in Fig. 11.3.1, where the channel is ideal and the channel encoder/decoder blocks are identities, and on discrete memoryless sources. We must choose or be given a meaningful measure of distortion for the source/user pair. If the source generates the output letter u_j and this letter is reproduced at the source decoder output as z_k, we denote the distortion incurred by this reproduction as $d(u_j, z_k)$. The quantity $d(u_j, z_k)$ is sometimes called a *single-letter* distortion measure or fidelity criterion. The average value of $d(\cdot, \cdot)$ over all possible source outputs and user inputs is

$$\bar{d}(P_{Z|U}) = \sum_{j=1}^{J} \sum_{k=1}^{K} P_U(u_j) P_{Z|U}(z_k|u_j) d(u_j, z_k), \tag{11.5.1}$$

where the source outputs are $\{u_1, u_2, \ldots, u_J\}$ and the user inputs are $\{z_1, z_2, \ldots, z_K\}$. The average distortion in Eq. (11.5.1) is a function of the transition probabilities $\{P_{Z|U}(z_k|u_j),\ j = 1, 2, \ldots, J,\ k = 1, 2, \ldots, K\}$, which are determined by the source encoder/decoder pair. To find the rate distortion function, we wish only to consider those conditional probability assignments $\{P_{Z|U}\}$ that yield an average distortion less than or equal to some acceptable value D, called *D-admissible* transition probabilities, and denoted by

$$\mathscr{P}_D = \{P_{Z|U}(z_k|u_j) : \bar{d}(P_{Z|U}) \leq D\}. \tag{11.5.2}$$

For each set of transition probabilities, we have an average mutual information

$$I(U; Z) = \sum_{j=1}^{J} \sum_{k=1}^{K} P_U(u_j) P_{Z|U}(z_k|u_j) \log \frac{P_{Z|U}(z_k|u_j)}{P_Z(z_k)}. \tag{11.5.3}$$

We are now able to define the rate distortion function of the source with respect to the fidelity criterion $d(\cdot, \cdot)$ as

$$R(D) = \min_{P_{Z|U} \in \mathscr{P}_D} I(U; Z) \tag{11.5.4}$$

for a chosen or given fixed value D. The importance of the rate distortion function is attested to by the fact that for a channel of capacity C, it is possible to reproduce the source at the receiver with an average distortion D if and only if $R(D) \leq C$.

A useful property of $R(D)$ is stated in the following theorem.

Theorem 11.5.1 *For a DMS with J output letters,*

$$0 \leq R(D) \leq \log J. \tag{11.5.5}$$

Proof: Using Eqs. (11.2.11), (11.2.21), and (11.2.22),

$$\begin{aligned} 0 \leq I(U; Z) &= H(U) - H(U|Z) \\ &\leq H(U) \leq \log J. \end{aligned} \tag{11.5.6}$$

Now, $R(D)$ is the minimum of $I(U; Z)$ over the admissible conditional probabilities; hence Eq. (11.5.5) follows.

The evaluation of the rate distortion function is not straightforward, even for discrete memoryless sources, and therefore we present the following example of a rate distortion function without indicating how it is derived.

EXAMPLE 11.5.1 [Berger, 1971]

Here we examine a special case of a DMS called a binary symmetric source (BSS) that produces a 0 with probability p and a 1 with probability $1 - p$. If we define $u_1 = z_1 = 0$ and $u_2 = z_2 = 1$, the single-letter distortion measure is specified to be

$$d(u_j, z_k) = \begin{cases} 0, & j = k \\ 1, & j \neq k. \end{cases} \tag{11.5.7}$$

Then, for $p \leq \frac{1}{2}$,

$$\begin{aligned} R(D) = & -p \log p - (1 - p) \log (1 - p) + D \log D \\ & + (1 - D) \log (1 - D), \qquad 0 \leq D \leq p. \end{aligned} \tag{11.5.8}$$

This $R(D)$ is plotted in Fig. 11.5.1 for $p = 0.1$, 0.2, 0.3, and 0.5. The reader should verify that $R(D) \leq H(U)$ for each p and that $R(p) = 0$.

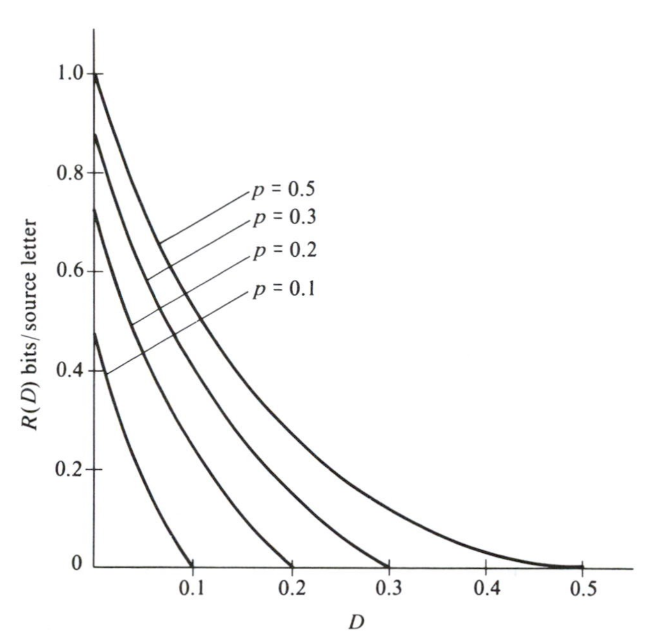

FIGURE 11.5.1 $R(D)$ for a BSS with $p \leq \frac{1}{2}$ (Example 11.5.1). From T. Berger, 1971. *Rate Distortion Theory: A Mathematical Basis for Data Compression*, 1971, Englewood Cliffs, N.J.: Prentice Hall, Inc. Reprinted by permission of the author.

For a discrete-time continuous amplitude source with single-letter distortion measure $d(u, z)$, each conditional pdf relating the source output to the user input produces an average distortion given by

$$\bar{d}(f_{Z|U}) = \int_{-\infty}^{\infty} \int_{-\infty}^{\infty} f_U(u) f_{Z|U}(z|u) d(u, z)\, du\, dz \tag{11.5.9}$$

and an average mutual information

$$I(U; Z) = \int_{-\infty}^{\infty} \int_{-\infty}^{\infty} f_U(u) f_{Z|U}(z|u) \log \frac{f_{Z|U}(z|u)}{f_Z(z)}\, du\, dz. \tag{11.5.10}$$

The admissible pdfs are described by the set

$$\mathscr{P}_D = \{f_{Z|U}(z|u) : \bar{d}(f_{Z|U}) \le D\}.$$

The rate distortion function is then defined as[1]

$$R(D) = \min_{f_{Z|U} \in \mathscr{P}_D} I(U; Z). \tag{11.5.11}$$

A significant difference between the rate distortion functions for discrete-amplitude and continuous-amplitude sources is that for $R(D)$ in Eq. (11.5.11), as $D \to 0$, $R(D) \to \infty$.

Analytical calculation of the rate distortion function for continuous-amplitude sources often is extremely difficult, and relatively few such calculations have been accomplished. We present the results of one such calculation in the following example.

EXAMPLE 11.5.2

For the squared-error distortion measure

$$d(u - z) = (u - z)^2, \tag{11.5.12}$$

a discrete-time, memoryless Gaussian source with zero mean and variance σ_s^2 has the rate distortion function

$$R(D) = \begin{cases} \dfrac{1}{2} \log \dfrac{\sigma_s^2}{D}, & 0 \le D \le \sigma_s^2 \\ 0, & D \ge \sigma_s^2, \end{cases} \tag{11.5.13}$$

which is sketched in Fig. 11.5.2. Note that as $D \to 0$, $R(D) \to \infty$.

---

[1] Strictly speaking, the "min" in Eq. (11.5.11) should be replaced with "inf," denoting infimum or greatest lower bound.

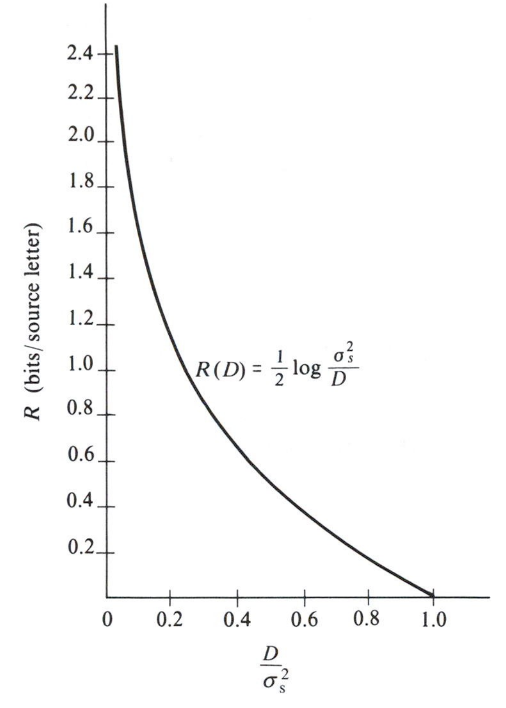

FIGURE 11.5.2 *R*(*D*) for a memoryless Gaussian source and squared-error distortion (Example 11.5.2).

Equation (11.5.13) is a relatively simple result that is given added importance by the following theorem.

Theorem 11.5.2 *Any memoryless, zero-mean, continuous-amplitude source with variance σ_s^2 has a rate distortion function R(D) with respect to the squared-error distortion measure that is upper bounded as*

$$R(D) \leq \frac{1}{2} \log \frac{\sigma_s^2}{D}, \qquad 0 \leq D \leq \sigma_s^2. \tag{11.5.14}$$

Proof: [Berger, 1971].

Theorem 11.5.2 thus implies that the Gaussian source is a worst-case source in the sense that it requires the maximum rate of all possible sources to achieve a specified mean squared-error distortion. The rate distortion function, when it can be evaluated, provides an absolute lower bound on the performance achievable by these systems.

11.6 Applications to Source and Receiver Quantization

In this section we present a series of examples that demonstrate how the material in Sections 11.2 through 11.5 affects realistic communication systems. The first example concerns the coarse quantization of a discrete-time, continuous-amplitude source.

EXAMPLE 11.6.1

A memoryless, zero-mean, unit-variance Gaussian source is quantized using a four-level MMSE Gaussian quantizer with the characteristic summarized in Table 9.3.1. This process produces a discrete-time, discrete-amplitude memoryless source with output letters $\{u_1 = -1.510,\ u_2 = -0.4528,\ u_3 = 0.4528,\ u_4 = 1.510\}$. The probability assigned to each of these values must be calculated from the given Gaussian pdf and the quantizer step points in Table 9.3.1. That is, since all values of the source from $-\infty$ to -0.9816 are assigned to the output level $-1.510 = u_1$, then

$$P_U(u_1) = \int_{-\infty}^{-0.9816} \frac{1}{\sqrt{2\pi}} e^{-u^2/2}\, du = 0.1635. \tag{11.6.1}$$

Similarly, we find that

$$P_U(u_2) = \int_{-0.9816}^{0} \frac{1}{\sqrt{2\pi}} e^{-u^2/2}\, du = \frac{1}{2} - \int_{-\infty}^{-0.9816} \frac{1}{\sqrt{2\pi}} e^{-u^2/2}\, du$$

$$= 0.3365 \tag{11.6.2}$$

$$P_U(u_3) = \int_{0}^{0.9816} \frac{1}{\sqrt{2\pi}} e^{-u^2/2}\, du = 0.3365 \tag{11.6.3}$$

and

$$P_U(u_4) = \int_{0.9816}^{\infty} \frac{1}{\sqrt{2\pi}} e^{-u^2/2}\, du = 0.1635. \tag{11.6.4}$$

Using Eqs. (11.6.1)–(11.6.4), we can find the absolute entropy of this newly created DMS from Eq. (11.2.2) as 1.911 bits/source letter. Three interesting observations can be made concerning these results. First, the quantizer has accomplished entropy reduction in that it has transformed a continuous amplitude source with infinite absolute entropy into a DMS with a finite entropy of 1.911 bits/source letter. Second, the minimum bit rate required to represent the quantizer outputs exactly is 1.911 bits/source letter, which does not seem to be significantly less than the 2 bits/source letter needed by the NBC or FBC. Hence the extra effort required to achieve the minimum bit rate using coding procedures as illustrated in Example 11.3.1 or in Section 11.7

may not be worthwhile. Third, from Table 9.3.1, the mean-squared error distortion achieved by this quantizer is $D = 0.1175$ at a minimum rate of 1.911 bits/source letter. This result can be compared to the rate distortion function of a memoryless Gaussian source to ascertain how far away from optimum this quantizer operates.

As a second example, we illustrate the calculation of channel capacity for binary antipodal signals transmitted over an AWGN with a particular receiver.

EXAMPLE 11.6.2

Here we examine the transmission of binary messages over an AWGN channel with zero mean and variance $\sigma_c^2 = 1$ using BPSK signals as given by Eq. (11.2.18a) for message m_1 and Eq. (11.2.18b) for message m_2. The optimum receiver in terms of minimizing the probability of error has a threshold at 0 and decides m_1 if the received signal is positive and decides m_2 if the received signal is negative. Thinking of the binary messages as channel inputs, say w_1 and w_2, and the binary decisions of the receiver as channel outputs, say x_1 and x_2, we have a discrete memoryless channel. To complete the specification of the channel, we must determine the input–output transition probabilities.

If we let $A_c = 1$ in Eqs. (11.2.18a) and (11.2.18b), the received signal when m_1 is transmitted is $r = 1 + \eta$, where η is the Gaussian noise variable, and when m_2 is transmitted the received signal is $r = -1 + \eta$. The transition probabilities for the DMS are thus

$$P_{X|M}(x_1|m_1) = \int_0^{\infty} \frac{1}{\sqrt{2\pi}} e^{-(r-1)^2/2}\, dr = 0.8413 \tag{11.6.5}$$

$$P_{X|M}(x_1|m_2) = \int_0^{\infty} \frac{1}{\sqrt{2\pi}} e^{-(r+1)^2/2}\, dr = 0.1587 \tag{11.6.6}$$

$$P_{X|M}(x_2|m_1) = \int_{-\infty}^{0} \frac{1}{\sqrt{2\pi}} e^{-(r-1)^2/2}\, dr = 0.1587 \tag{11.6.7}$$

and

$$P_{X|M}(x_2|m_2) = \int_{-\infty}^{0} \frac{1}{\sqrt{2\pi}} e^{-(r+1)^2/2}\, dr = 0.8413. \tag{11.6.8}$$

From Eqs. (11.6.5)–(11.6.8) it is evident that we have a BSC with $p = 0.1587$. It is known from Example 11.4.1 that the capacity for this channel is given by Eq. (11.4.5), so

$$\begin{aligned} C &= 1 + (0.1587) \log (0.1587) + (0.8413) \log (0.8413) \\ &= 0.369 \text{ bit/source letter} \end{aligned} \tag{11.6.9}$$

and is achieved by equally likely channel inputs.

In this third example, the rate distortion functions of several memoryless continuous-amplitude sources are presented and the performance of optimum quantizers in comparison to their rate distortion bound is discussed.

EXAMPLE 11.6.3

Memoryless, discrete-time, continuous-amplitude sources with uniform, Gaussian, Laplacian, and gamma distributions are of importance in a variety of practical applications. Typical pdfs for these sources are listed below.

Uniform pdf:

$$f_U(u) = \begin{cases} \frac{1}{2}, & -1 \leq u \leq 1 \\ 0 & \text{otherwise} \end{cases}. \tag{11.6.10}$$

Gaussian pdf:

$$f_U(u) = \frac{1}{\sqrt{2\pi}} e^{-u^2/2}, \qquad -\infty < u < \infty. \tag{11.6.11}$$

Laplacian pdf:

$$f_U(u) = \frac{1}{\sqrt{2}} e^{-\sqrt{2}|u|}, \qquad -\infty < u < \infty. \tag{11.6.12}$$

Gamma pdf:

$$f_U(u) = \frac{\sqrt{3}}{4\sqrt{\pi|u|}} e^{-\sqrt{3}|u|/2}, \qquad -\infty < u < \infty. \tag{11.6.13}$$

The rate distortion functions for sources with the pdfs in Eqs. (11.6.10)–(11.6.13) are shown in Fig. 11.6.1. In Table 11.6.1, the minimum-distortion Gaussian and Laplacian quantizer performance at rates 1, 2, and 3 bits/source letter are compared to the rate distortion functions for Gaussian and Laplacian sources in terms of SQNR $= 10 \log_{10} \sigma_s^2/D$. The quantizer outputs are assumed to be noiselessly encoded at their minimum possible rate (see Section 11.7), which is the absolute entropy of the output letters. To close the performance gap further requires more exotic coding techniques. Except for the Gaussian case, the results in Fig. 11.6.1 and Table 11.6.1 were obtained numerically.

These three examples demonstrate the utility of the theory in Sections 11.2 to 11.5 to familiar communications problems from earlier chapters. As can be seen, information theory and rate distortion theory can be extremely useful for obtaining bounds on communication system performance, and hence can indicate whether further improvements in system design are likely to be worth the time, effort, and/or complexity involved.

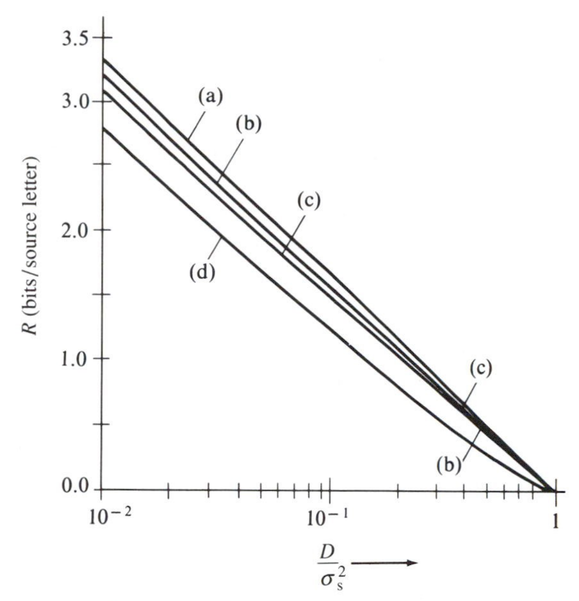

FIGURE 11.6.1 Rate Distortion Functions for (a) Gaussian, (b) Laplacian, (c) Uniform, and (d) Gamma Distributed Sources. From P. Noll and R. Zelinski, "Bounds on Quantizer Performance in the Low Bit-Rate Region," *IEEE Trans. Commun.*, © 1978 IEEE.

TABLE 11.6.1 SQNR (dB) Comparison of $R(D)$ and Optimum Quantizer Performance for Gaussian and Laplacian Sources

Rate (bits/source letter)	Gaussian Source		Laplacian Source	
	$R(D)$	Optimum Quantizer	$R(D)$	Optimum Quantizer
1	6.02	4.64	6.62	5.76
2	12.04	10.55	12.66	11.31
3	18.06	16.56	18.68	17.20

Source: N. Farvardin and J. W. Modestino, "Optimum Quantizer Performance for a Class of Non-Gaussian Memoryless Sources," *IEEE Trans. Inf. Theory*, © 1984 IEEE.

11.7 Variable-Length Source Coding

We mention in Section 9.5 that if the output quantization levels are not equally likely to occur, the average bit rate required may be reduced in comparison to the NBC or FBC by using a variable-length code. Theorem 11.3.1 states that a lower bound on the average codeword length per source letter, denoted by $\bar{n}$, is the source entropy. Furthermore, Example 11.2.1 illustrates that a DMS with letters that are not equally likely has a smaller entropy than a source with the same number of letters and equally probable outputs. Finally, in Example 11.3.1, a variable-length code is constructed for a DMS with nonequally likely output letters.

The design of variable-length codes with an $\bar{n}$ that approaches the entropy of the DMS is generically referred to as *entropy coding*, and there are several procedures for finding such codes. In this section we present the most familiar and most straightforward of the available techniques for entropy coding, which is due to Huffman [1952] and is thus called *Huffman coding*. To specify the encoding procedure, we consider a DMS U with M output letters $\{u_1, u_2, \ldots, u_M\}$ and probabilities $P_U(u_j), j = 1, 2, \ldots, M$. We also assume for simplicity that the letters are numbered such that $P_U(u_1) \geq P_U(u_2) \geq \cdots \geq P_U(u_M)$. Of course, if this property does not hold at the outset, we can always renumber the letters to produce it. The constructive procedure for designing the variable-length code can be described as follows. The letters and their probabilities are listed in two columns in the order of decreasing probability. The two lowest-probability letters are combined by drawing a straight line out from each and connecting them. The probabilities of these two letters are then summed, and this sum is considered to be the probability of a new letter denoted u'_{M-1}. The next two lowest-probability letters, among $u_1, u_2, \ldots, u_{M-2}$, and u'_{M-1}, are combined to create another letter with probability equal to their sum. This process is continued until only two letters remain and a type of "tree" is generated. Binary codewords are then assigned by moving from right to left in the tree, assigning a 0 to the upper branch and a 1 to the lower branch, where each pair of letters has been combined. The codeword for each letter is read off the tree from right to left. An example will greatly clarify the procedure.

EXAMPLE 11.7.1

Given a DMS U with five letters and probabilities $P_U(u_1) = 0.5$, $P_U(u_2) = P_U(u_3) = 0.2$, and $P_U(u_4) = P_U(u_5) = 0.05$, we wish to design a binary variable-length code for this source. With reference to Fig. 11.7.1, we combine the

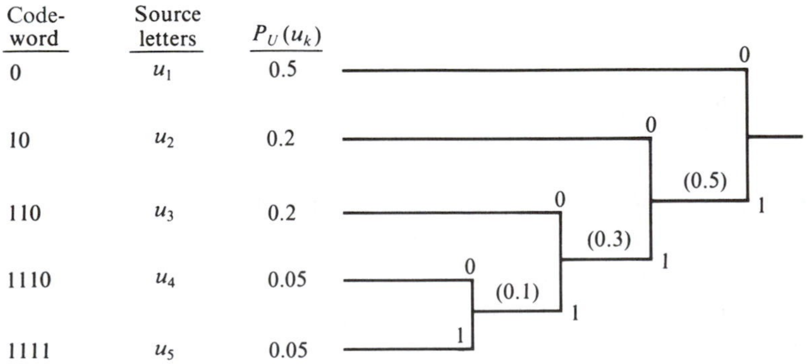

Codeword	Source letters	$P_U(u_k)$
0	u_1	0.5
10	u_2	0.2
110	u_3	0.2
1110	u_4	0.05
1111	u_5	0.05

FIGURE 11.7.1 Huffman encoding for the DMS in Example 11.7.1.

two least likely letters and sum their probabilities to create a new "letter." The least likely letters are combined again, and the procedure is continued until only two letters remain. Codewords are then assigned by moving right to left and assigning a 0 to an upper branch and a 1 to a lower branch. Codewords are also read off in a right-to-left fashion and are shown in the leftmost column of Fig. 11.7.1. The average codeword length ($\bar{n}$) for this variable-length code is

$$\bar{n} = (4)(0.05) + (4)(0.05) + (3)(0.2) + (2)(0.2) + (1)(0.5)$$
$$= 1.9 \text{ bits/source letter.} \qquad (11.7.1)$$

We know from Eq. (11.2.21) that $H(U)$ is upper bounded by $\log_2 M = \log_2 5 \cong 2.322$ bits/source letter, and we can show that the entropy of U is $H(U) \cong 1.86$ bits/source letter. Thus $H(U) < \bar{n} < \log_2 M$.

The codewords in Example 11.7.1 are *uniquely decodable* in that each source letter has a codeword that differs from that assigned to any other letter. *The Huffman procedure yields the smallest average codeword length of any uniquely decodable set of codewords.* Although another code can do as well as the Huffman code, none can be better.

Recall that in Example 11.3.1, a code is designed for single-letter encoding of the source, and then another code is designed to represent pairs of source letters. Should we use Huffman coding on pairs of letters for the source in Example 11.7.1? Probably not, since $\bar{n}$ is already very close to the entropy. However, Example 11.3.1 demonstrates that by encoding blocks of source letters, an average codeword length nearer the source entropy can be obtained (with added complexity). A more general illustration of this property is provided by the following two inequalities. The Huffman encoding procedure generates a code for the DMS U with an $\bar{n}$ that satisfies

$$H(U) \leq \bar{n} < H(U) + 1 \text{ bits/source letter} \qquad (11.7.2)$$

for letter-by-letter (or symbol-by-symbol) encoding. If blocks of L letters are combined before using the Huffman technique, $\bar{n}$ is bounded by

$$H(U) \leq \bar{n} < H(U) + \frac{1}{L} \text{ bits/source letter.} \qquad (11.7.3)$$

Thus encoding pairs of letters is at least as good as single-letter encoding, and for large L, $\bar{n}$ can be made arbitrarily close to $H(U)$. Whether block encoding makes sense depends on the particular DMS and its letter probabilities (and perhaps, the application of interest).

Only the binary Huffman procedure has been described here, but nonbinary codes can be designed using the Huffman method. The details are somewhat more complicated and nonbinary codes are less commonly encountered than binary ones, so further discussion is left to the problems and the literature.

SUMMARY

In this chapter we have discussed very briefly some of the salient results from information theory and rate distortion theory and have indicated how these results can be used to bound communication system performance. It is perhaps surprising to the reader that physically meaningful quantitative measures of information can be defined, but such is the case. It has been impossible to present and to develop the many elegant and insightful results available from information theory, and hence a clear view of the importance of this field may not be available to the reader. However, information theory and rate distortion theory provide the theoretical basis and performance bounds for the practical source coding and channel coding systems in use today.

PROBLEMS

11.1 A random variable U has a sample space consisting of the set of all possible binary sequences of length N, denoted $\{u_j, j = 1, 2, \ldots, 2^N\}$. If each of these sequences is equally probable, so that $P[u_j] = 2^{-N}$ for all j, what is the self-information of any event $u = u_j$?

11.2 Given a random variable U with the alphabet $\{u_1, u_2, u_3, u_4\}$ and probability assignments $P(u_1) = 0.8$, $P(u_2) = 0.1$, $P(u_3) = 0.05$, $P(u_4) = 0.05$, calculate the entropy of U. Compare your result to a random variable with equally likely values.

11.3 Given the binary erasure channel (BEC) shown in Fig. P11.3, find an expression for the average mutual information between the input and output $I(W; X)$ if $P_W(w_1) = P_W(w_2) = \frac{1}{2}$. The BEC might be a good channel model for the physical situation where binary antipodal signals are transmitted and the receiver makes a decision if the received signal is much greater than or much less than the threshold, but asks for a retransmission of the received signal if the received signal is very near the threshold.

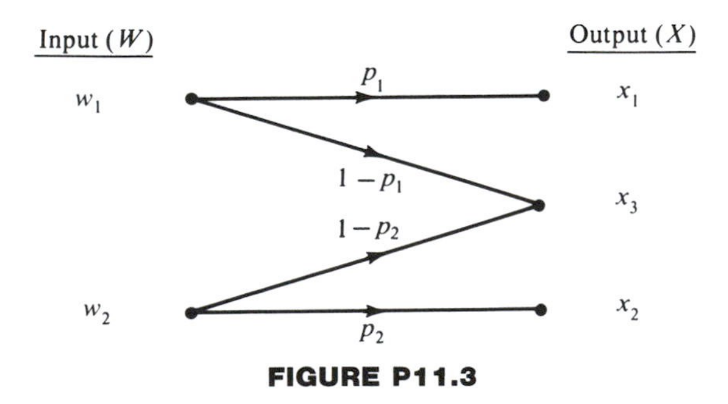

FIGURE P11.3

11.4 For the DMC in Fig. P11.4 with $P_W(0) = \frac{1}{3}$ and $P_W(1) = \frac{2}{3}$, find $H(W)$ and $H(W|X)$. What is the average mutual information for this channel and input probability assignment?

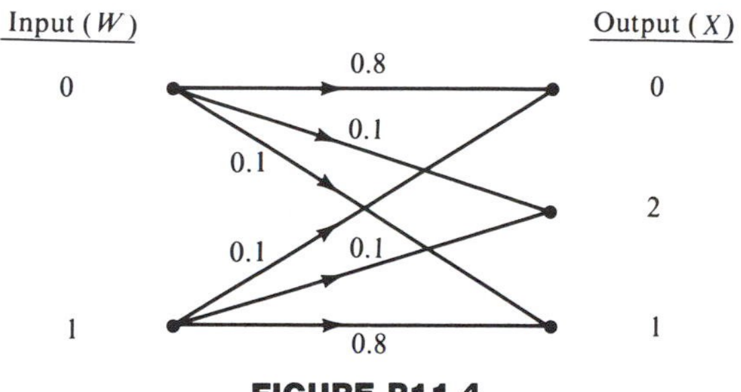

FIGURE P11.4

11.5 Calculate the differential entropy for an absolutely continuous random variable with pdf $f_U(u) = (1/\alpha)e^{-u/\alpha}$, $0 < u < \infty$, and $f_U(u) = 0$ for $u \leq 0$.

11.6 Show that the differential entropy for a random variable U with the Laplacian pdf $f_U(u) = (1/\sqrt{2})e^{-\sqrt{2}|u|}$ for $-\infty < u < \infty$ is given by $h(U) = \log(e\sqrt{2})$.

11.7 Given an absolutely continuous random variable X with pdf $f_X(x)$ and the transformation $Y = aX + b$, find an expression for the entropy of Y in terms of $h(X)$. How has the transformation affected the result?

11.8 Consider the discrete random variable X in Example 11.2.1 with $P_X(x_1) = \frac{1}{2}$, $P_X(x_2) = \frac{1}{4}$, $P_X(x_3) = P_X(x_4) = \frac{1}{8}$, and the linear transformation $Y = aX + b$. Find $H(Y)$. What effect has the transformation had on the entropy of the discrete random variable X?

11.9 For a general multivariate pdf of absolutely continuous random variables denoted by $f_X(x_1, x_2, \ldots, x_N)$, consider the one-to-one transformation represented by $y_i = g_i(X)$, $i = 1, 2, \ldots, N$. Find an expression for the differential entropy of the joint pdf of Y, $f_Y(y_1, y_2, \ldots, y_N)$ in terms of the Jacobian of the transformation. See Eqs. (A.5.4) and (A.5.5).

11.10 Use Eq. (11.2.29) and the result of Problem 11.9 to show that average mutual information is invariant to one-to-one transformations. That is, given two continuous random vectors X and Z with average mutual information $I(X; Z)$ and a one-to-one transformation $y_i = g_i(X)$, $i = 1, 2, \ldots, N$, find $I(Y; Z)$.

11.11 Given a Gaussian source U with mean μ_s and variance σ_s^2, find an expression for its differential entropy $h(U)$.

11.12 A DMS U has $P_U(u_1) = 0.4$, $P_U(u_2) = 0.3$, and $P_U(u_3) = P_U(u_4) = 0.15$. Construct a source code for U by mimicking the procedure in Example 11.3.1 and Table 11.3.1.

11.13 For the BSC in Example 11.4.1 plot $I(W; X)$ as a function of p when $P_W(0) = \frac{3}{4}$ and $P_W(1) = \frac{1}{4}$ and compare to a plot of C given by Eq. (11.4.5).

11.14 For a one-sided pdf $f_X(x)$ such that $f_X(x) = 0$ for $x \leq 0$ with mean

$$\mu = \int_0^\infty x f_X(x)\, dx,$$

show that the maximum differential entropy is achieved when

$$f_X(x) = \frac{1}{\mu} e^{-x/\mu} u(x)$$

and that $h(X) = \log \mu e$ [Shannon, 1948].

11.15 An often useful approach to finding channel capacity for discrete memoryless channels relies on the following theorem.

Theorem *A DMC has input W and output X with transition probabilities $P_{X|W}(x_k|w_j)$, $j = 1, \ldots, M$, $k = 1, 2, \ldots, N$. Necessary and sufficient conditions for a set of input probabilities $P_W(w_j), j = 1, 2, \ldots, M$, to achieve capacity is for*

$$I(w = w_j; X) = C \qquad \text{for all } w_j \text{ with } P_W(w_j) > 0$$

and

$$I(w = w_j; X) \leq C \qquad \text{for all } w_j \text{ with } P_W(w_j) = 0,$$

for some number C, where

$$I(w = w_j; X) = \sum_{k=1}^{N} P_{X|W}(x_k|w_j) \log \frac{P_{X|W}(x_k|w_j)}{\sum_{j=1}^{M} P_W(w_j) P_{X|W}(x_k|w_j)}.$$

The number C is the channel capacity.

The main use of this theorem is to check the validity of some hypothesized set of input probabilities. Thus, for the BSC in Example 11.4.1, we might guess by symmetry that $P_W(0) = P_W(1) = \frac{1}{2}$ achieves capacity. Substantiate this claim and find the capacity in Eq. (11.4.5) by using this theorem.

11.16 Use the theorem in Problem 11.15 to show that the capacity of the binary erasure channel in Fig. P11.3 with $p_1 = p_2$, called the binary *symmetric* erasure channel (BSEC), is p_1.

Hint: Guess equally likely inputs.

11.17 For the channel in Fig. P11.4, find channel capacity using the theorem in Problem 11.15.

11.18 We have only calculated the capacity of channels with continuous inputs and outputs when the noise is Gaussian. The following theorem expands the utility of the Gaussian result.

Theorem *Consider the additive noise channel in Fig.* 11.4.2, *where* $E[W^2] \le S$ *and* $\mathrm{var}(\zeta) = \sigma^2$. *The capacity of this channel is bounded by*

$$\frac{1}{2}\log\left(1+\frac{S}{\sigma^2}\right) \le C \le \frac{1}{2}\log\left[2\pi e(S+\sigma^2)\right] - h(\zeta).$$

In essence, this theorem says that for a fixed noise variance, Gaussian noise is the worst since it lower bounds the channel capacity.

(a) Use Eq. (11.4.9) to prove the right inequality.
(b) Follow the proof of Theorem 11.4.2 to prove the left inequality.

Hint: See Gallager [1968].

11.19 Jensen's inequality states that for a random variable W with a distribution defined on an appropriate interval and for a convex $\cup$ function, say $g(x)$, then

$$E[g(W)] \ge g[E(W)]$$

if $E[W]$ exists. If $g(x)$ is convex $\cap$, the inequality is reversed to yield

$$E[g(W)] \le g[E(W)].$$

Use Jensen's inequality to prove Theorem 11.4.2. Jensen's inequality is particularly useful when working with average mutual information, since $I(W; X)$ is a convex $\cup$ function of the transition probabilities, and $I(W; X)$ is a convex $\cap$ function of the input probabilities.

11.20 The rate distortion function for a Laplacian source U with pdf

$$f_U(u) = \frac{\lambda}{2} e^{-\lambda|u|}, \qquad -\infty < u < \infty$$

subject to the absolute value of the error distortion measure $d(u-z) = |u-z|$ is

$$R(D) = -\log \lambda D, \qquad 0 \le D \le \frac{1}{\lambda}$$

[Berger, 1971]. Plot this rate distortion function.

11.21 A memoryless, zero-mean, unit-variance Gaussian source is quantized using the MMSE Gaussian quantizer characteristic in Table 9.3.2. Find the equivalent DMS and its entropy. Compare the performance of this quantizer to the rate distortion bound.

11.22 Use the eight-level uniform quantizer in Table 9.2.1 to quantize a memoryless, zero-mean, unit-variance Gaussian source. Find the output entropy and compare its performance to the rate distortion bound.

11.23 Compare the results of Problems 11.21 and 11.22.

11.24 Binary messages are transmitted over an AWGN channel with zero mean and variance $\sigma_c^2 = \frac{1}{2}$ using BPSK signals as given by Eq. (11.2.18a) for message m_1 and Eq. (11.2.18b) for message m_2. The receiver decides m_1 if the received signal is positive and m_2 if the received signal is negative. Considering the binary messages as channel inputs, w_1 and w_2, and the binary receiver decisions as outputs, x_1 and x_2, specify the equivalent DMC if $A_c = 1$ in Eqs. (11.2.18a,b). Calculate the channel capacity.

11.25 Repeat Problem 11.24 for the case where the receiver has the four output values $\{x_1, x_2, x_3, x_4\}$ and where x_1 occurs if the received value $r > 1.032$, x_2 occurs if $0 < r < 1.032$, x_3 occurs if $-1.032 < r < 0$, and x_4 occurs if $r < -1.032$.

11.26 Given a DMS U with four letters and probabilities $P_U(u_1) = 0.5$, $P_U(u_2) = 0.25$, and $P_U(u_3) = P_U(u_4) = 0.125$, use the Huffman procedure to design a variable-length code. Find $\bar{n}$ and compare to $H(U)$.

11.27 Use the Huffman procedure to design variable-length codes for the single-letter and paired-letter sources in Example 11.3.1.

11.28 Plot the upper bound on $\bar{n}$ in Eq. (11.7.3) as a function of L for the sources in Examples 11.3.1 and 11.7.1.

CHAPTER 12

Channel Coding

12.1 Introduction

Our focus in this chapter is on designing codes for the reliable transmission of digital information over a noisy channel. The pertinent block diagram for this particular problem is shown in Fig. 12.1.1. The binary message sequence presented to the input of the channel encoder may be the output of a source encoder or the output of a source directly. For the development in this chapter, we are only interested in the fact that the channel encoder input is a binary sequence at a rate of R_s bits/sec. The channel encoder introduces redundancy into the data stream by adding bits to the (input) message bits in such a way as to facilitate the detection and/or correction of bit errors in the original binary message sequence at the receiver. Perhaps the most familiar example of channel coding (for error detection) to the reader is the addition of a parity check bit to a block of message bits. For each k bits into the channel encoder, $n > k$ bits are produced at the channel encoder output. Thus the channel coding process adds $n - k$ bits to each k-bit input sequence. As a result, a dimensionless parameter called the *code rate*, denoted by R_c and defined by $R_c = k/n =$ (number of bits in/number of bits out), is associated with the code. Since $k < n$, and hence $R_c < 1$, the data rate at the channel encoder output is $R_T = R_s/R_c > R_s$ bps. Thus the channel coding operation has increased the transmitted bit rate.

For the purposes of this chapter we assume that the modulator maps the coded bits in a one-to-one fashion onto an analog waveform suitable for propagation over the channel, and that the modulation process is totally separate from the coding operation. The modulator output is then transmitted over the channel and demodulated to produce a binary sequence that serves as the input to the channel decoder. The channel decoder then generates a decoded binary sequence which hopefully is an accurate replica of the input binary message sequence at the transmitter. The fact that the demodulator output is a binary sequence means that the demodulator is making "hard decisions," and the subsequent decoding process is called *hard decision decoding.* The alternative to hard decisions is for the demodulator to pass on three or more level versions of its output to the decoder, which improves decoding accuracy over hard decisions (two-level information). The latter procedure is called *soft decision*

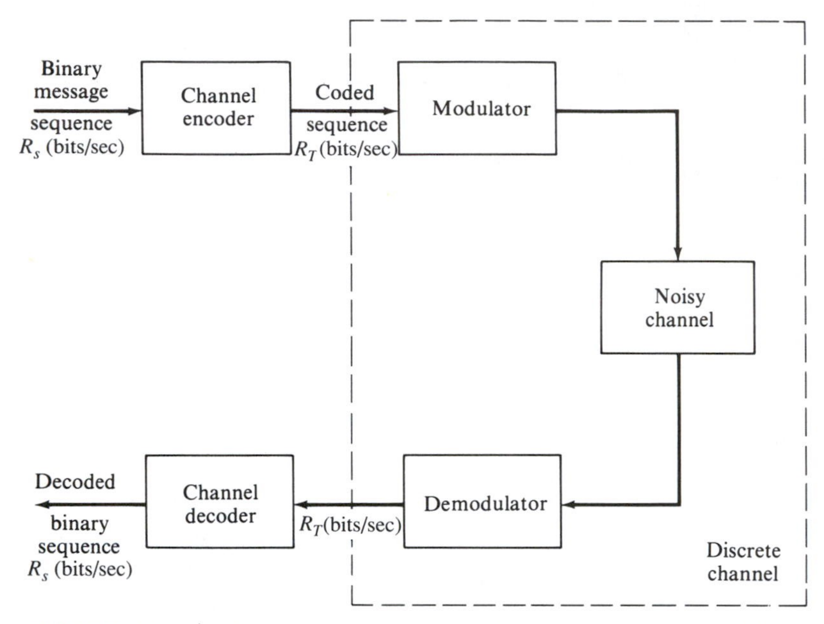

FIGURE 12.1.1 Basic block diagram for the channel coding problem.

decoding and generally increases receiver complexity. Soft decision decoding can substantially improve communication system performance; however, for our introductory treatment, we focus on hard decisions.

Since the input to the modulator is an R_T-bits/sec binary sequence and the demodulator output is an R_T-bits/sec binary sequence, it is common to lump the modulator and demodulator with the noisy channel and model the entire three block series as a discrete channel with binary inputs and binary outputs (shown by dashed lines in Fig. 12.1.1). Working with the lumped discrete (binary) channel, we now only need to consider channel encoders and decoders that have binary inputs and outputs. If this binary channel is memoryless and the probability of error for a 0 and a 1 are the same, we have the familiar binary symmetric channel (BSC).

The two types of codes most often studied and used in communication systems are block codes and convolutional codes. These codes are discussed in Sections 12.2 and 12.4, respectively, with an emphasis on a few special cases and on forward error correction (as opposed to error detection). In Section 12.3 we discuss a special class of linear block codes, called *cyclic codes*, that are particularly easy to encode and decode and hence have found a number of practical applications. In Section 12.5 we describe an error control strategy called *automatic repeat-request* (ARQ), which relies on error detection and retransmission rather than forward error correction. Some qualitative and quantitative comparisons of code performance and implementation complexity are presented in Section 12.6.

12.2 Block Codes

For the purposes of our current discussion, we refer to the block diagram in Fig. 11.1.1, where the input to the channel encoder is denoted by V and the channel encoder output is denoted by W. We will call $V = \{v_1, v_2, \ldots\}$ the message sequence. For block codes, the message sequence is partitioned into blocks of fixed length k, indicated here by $\mathbf{v} = (v_1, v_2, \ldots, v_k)$. The channel encoder maps each input message block $\mathbf{v}$ into an n-component output block called a *codeword* and denoted by $\mathbf{w} = (w_1, w_2, \ldots, w_n)$, where $n > k$. Since we are only considering binary sequences, there are 2^k possible messages and 2^k possible codewords, and the code is called an (n,k) block code. We limit our attention to *linear* block codes, since linearity gives us the structure needed to reduce encoding and decoding complexity.

A linear *systematic* block code has the additional structure that $n - k$ digits of a codeword are a linear combination of the components of the message block, while the other k digits of the codeword are the k message digits themselves. Often, the two sets of digits are grouped such that the first $n - k$ digits are a linear combination of the components of the message block, while the last k digits are the k message digits themselves. Thus

$$\begin{aligned}
w_1 &= v_1 g_{11} + v_2 g_{21} + \cdots + v_k g_{k1} \\
w_2 &= v_1 g_{12} + v_2 g_{22} + \cdots + v_k g_{k2} \\
&\vdots \\
w_{n-k} &= v_1 g_{1,n-k} + v_2 g_{2,n-k} + \cdots + v_k g_{k,n-k} \\
w_{n-k+1} &= v_1 \\
w_{n-k+2} &= v_2 \\
&\vdots \\
w_n &= v_k,
\end{aligned} \tag{12.2.1}$$

where all additions in this chapter are considered to be modulo-2. Based on the set of equations in Eq. (12.2.1), we can construct a *generator matrix* given by

$$\begin{aligned}
\mathbf{G} &= \begin{bmatrix} g_{11} & g_{12} & \cdots & g_{1,n-k} & 1 & 0 & \cdots & \cdots & 0 \\ g_{21} & g_{22} & \cdots & g_{2,n-k} & 0 & 1 & 0 & \cdots & 0 \\ \vdots & & & & \vdots & & & & \\ g_{k1} & g_{k2} & \cdots & g_{k,n-k} & 0 & \cdots & \cdots & 0 & 1 \end{bmatrix} \\
&\triangleq [\mathbf{P} \quad \mathbf{I}_k],
\end{aligned} \tag{12.2.2}$$

where the $k \times n - k$ matrix $\mathbf{P}$ is defined as shown and $\mathbf{I}_k$ is a $k \times k$ identity matrix. (*Note:* Some authors may reverse $\mathbf{I}_k$ and $\mathbf{P}$ in $\mathbf{G}$. The performance is equivalent when used over a memoryless channel. In fact, any column permutation does not alter the code performance over a memoryless channel.) Thus the components of the codeword $\mathbf{w}$ in Eq. (12.2.1) can be produced by considering the k-tuple $\mathbf{v}$ as a row vector and performing the multiplication

$$\mathbf{w} = \mathbf{vG}. \tag{12.2.3}$$

Thus, for a given generator matrix, we can obtain the codewords corresponding to all possible length-k sequences in a message block. We do not discuss how one tries to find a good $\mathbf{G}$ matrix here. It is not easy.

In addition to the $\mathbf{G}$ matrix, every linear block code has a related matrix called the *parity check matrix* $\mathbf{H}$, which for $\mathbf{G}$ in Eq. (12.2.2) is given by

$$\mathbf{H} = [\mathbf{I}_{n-k} \quad \mathbf{P}^T], \tag{12.2.4}$$

where $\mathbf{I}_{n-k}$ is the $n - k \times n - k$ identity matrix and $\mathbf{P}$ is as defined previously. The importance of $\mathbf{H}$ stems from the fact that

$$\begin{aligned} \mathbf{wH}^T &= \mathbf{v}[\mathbf{P} \quad \mathbf{I}_k]\begin{bmatrix} \mathbf{I}_{n-k} \\ \mathbf{P} \end{bmatrix} \\ &= \mathbf{vP} + \mathbf{vP} = \mathbf{0}, \end{aligned} \tag{12.2.5}$$

which simply says that any codeword multiplied by $\mathbf{H}^T$ is the $(n - k)$-length zero vector. This property is extremely useful for error detection and correction. In particular with reference to Fig. 12.1.1, let $\mathbf{e}$ be an n-dimensional binary error vector added to $\mathbf{w}$ by the channel so that the channel output, and hence the channel decoder input is

$$\mathbf{x} = \mathbf{w} + \mathbf{e}. \tag{12.2.6}$$

Now the *syndrome* of the received vector $\mathbf{x}$ is defined as

$$\mathbf{s} = \mathbf{xH}^T. \tag{12.2.7}$$

Clearly, if $\mathbf{x}$ is a codeword, then $\mathbf{s} = \mathbf{0}$, but if $\mathbf{x}$ is not a codeword, $\mathbf{s} \neq \mathbf{0}$. Thus an error is detected if $\mathbf{s} \neq \mathbf{0}$. The syndrome $\mathbf{s}$ is not dependent on which codeword is transmitted, and in fact, $\mathbf{s}$ depends only on the error vector $\mathbf{e}$. This can be demonstrated straightforwardly by substituting Eq. (12.2.6) into Eq. (12.2.7) and then using Eq. (12.2.5) to find that

$$\mathbf{s} = \mathbf{xH}^T = \mathbf{wH}^T + \mathbf{eH}^T = \mathbf{eH}^T. \tag{12.2.8}$$

Equation (12.2.8) implies that there are some *undetectable error patterns* that occur whenever $\mathbf{e}$ is a codeword. Since there are $2^k - 1$ nonzero codewords, there are $2^k - 1$ undetectable error patterns.

It would seem that Eq. (12.2.8) would allow us to find the error pattern $\mathbf{e}$ and hence implement the decoding operation at the receiver. However, Eq. (12.2.8) gives $n - k$ equations in n unknowns, the components of $\mathbf{e}$. Thus Eq. (12.2.8) has 2^k solutions, which implies that there are 2^k error patterns for each syndrome. How then does the decoder choose a single-error vector out of the 2^k possible error patterns for a given syndrome? The answer is that we select that error pattern which minimizes the probability of error as in Chapter 10. For independent, identically distributed channel inputs and a binary symmetric channel, the error vector that minimizes the probability of error is the error pattern with the smallest number of 1's. Note from Eq. (12.2.6) that $\mathbf{x}$ is the received codeword and once $\mathbf{e}$ is known, we can calculate the transmitted codeword as

$$\mathbf{w} = \mathbf{x} + \mathbf{e}, \tag{12.2.9}$$

since subtraction and addition modulo-2 are identical. Thus for each syndrome the decoder must have available the error vector with the fewest number of 1's out of the corresponding 2^k error patterns. The operation of the channel decoder therefore proceeds as follows:

STEP 1. Calculate the syndrome of the received vector, that is, $\mathbf{s} = \mathbf{x}\mathbf{H}^T$.

STEP 2. Find the error vector $\mathbf{e}$ with the fewest 1's corresponding to $\mathbf{s}$.

STEP 3. Compute the decoder output as $\mathbf{w}' = \mathbf{x} + \mathbf{e}$.

The following example illustrates the many concepts presented thus far.

EXAMPLE 12.2.1

A systematic linear block code called a (7,4) Hamming code has the generator matrix

$$\mathbf{G} = \begin{bmatrix} 1 & 1 & 0 & 1 & 0 & 0 & 0 \\ 0 & 1 & 1 & 0 & 1 & 0 & 0 \\ 1 & 1 & 1 & 0 & 0 & 1 & 0 \\ 1 & 0 & 1 & 0 & 0 & 0 & 1 \end{bmatrix}, \tag{12.2.10}$$

so that the $n = 7$ bit codewords for all possible $k = 4$-bit message blocks are shown in Table 12.2.1 and are calculated via Eq. (12.2.3). By inspection of

TABLE 12.2.1 The (7,4) Hamming Code

Message Bits (**v**)	Codewords (**w**)
0 0 0 0	0 0 0 0 0 0 0
0 0 0 1	1 0 1 0 0 0 1
0 0 1 0	1 1 1 0 0 1 0
0 0 1 1	0 1 0 0 0 1 1
0 1 0 0	0 1 1 0 1 0 0
0 1 0 1	1 1 0 0 1 0 1
0 1 1 0	1 0 0 0 1 1 0
0 1 1 1	0 0 1 0 1 1 1
1 0 0 0	1 1 0 1 0 0 0
1 0 0 1	0 1 1 1 0 0 1
1 0 1 0	0 0 1 1 0 1 0
1 0 1 1	1 0 0 1 0 1 1
1 1 0 0	1 0 1 1 1 0 0
1 1 0 1	0 0 0 1 1 0 1
1 1 1 0	0 1 0 1 1 1 0
1 1 1 1	1 1 1 1 1 1 1

Eq. (12.2.10),

$$\mathbf{P} = \begin{bmatrix} 1 & 1 & 0 \\ 0 & 1 & 1 \\ 1 & 1 & 1 \\ 1 & 0 & 1 \end{bmatrix}, \tag{12.2.11}$$

so that from Eq. (12.2.4), the parity check matrix of this code is

$$\mathbf{H} = \begin{bmatrix} 1 & 0 & 0 & 1 & 0 & 1 & 1 \\ 0 & 1 & 0 & 1 & 1 & 1 & 0 \\ 0 & 0 & 1 & 0 & 1 & 1 & 1 \end{bmatrix}. \tag{12.2.12}$$

Thus the validity of Eq. (12.2.5) can be demonstrated for any codeword in Table 12.2.1 using $\mathbf{H}$ in Eq. (12.2.12).

If we assume that the input message $\mathbf{v} = (1\ 0\ 1\ 0)$ is to be transmitted, the transmitted codeword is $\mathbf{w} = (0\ 0\ 1\ 1\ 0\ 1\ 0)$. For no channel errors, $\mathbf{e} = \mathbf{0}$ and

$$\mathbf{s} = \mathbf{w}\mathbf{H}^T = [0 \quad 0 \quad 1 \quad 1 \quad 0 \quad 1 \quad 0] \begin{bmatrix} 1 & 0 & 0 \\ 0 & 1 & 0 \\ 0 & 0 & 1 \\ 1 & 1 & 0 \\ 0 & 1 & 1 \\ 1 & 1 & 1 \\ 1 & 0 & 1 \end{bmatrix}$$

$$= (0 \quad 0 \quad 0) = \mathbf{0}. \tag{12.2.13}$$

If the channel produces the error pattern $\mathbf{e} = (1\ 0\ 0\ 0\ 0\ 0\ 0)$, then

$$\mathbf{x} = \mathbf{w} + \mathbf{e} = (1 \quad 0 \quad 1 \quad 1 \quad 0 \quad 1 \quad 0) \tag{12.2.14}$$

and

$$\mathbf{s} = \mathbf{x}\mathbf{H}^T = [1 \quad 0 \quad 1 \quad 1 \quad 0 \quad 1 \quad 0] \begin{bmatrix} 1 & 0 & 0 \\ 0 & 1 & 0 \\ 0 & 0 & 1 \\ 1 & 1 & 0 \\ 0 & 1 & 1 \\ 1 & 1 & 1 \\ 1 & 0 & 1 \end{bmatrix}$$

$$= (1 \quad 0 \quad 0). \tag{12.2.15}$$

To finish the decoding procedure, we need to find the error vector out of the $2^k = 2^4 = 16$ error vectors associated with the $\mathbf{s}$ that has the fewest 1's. We do this by letting $\mathbf{e} = (e_1\ e_2\ e_3\ e_4\ e_5\ e_6\ e_7)$ and writing out Eq. (12.2.8)

TABLE 12.2.2 Error Patterns That Satisfy Eqs. (12.2.16)

e_1	e_2	e_3	e_4	e_5	e_6	e_7	
1	1	0	1	1	1	0	
1	0	0	0	0	0	0	←
1	0	1	0	1	1	1	
1	1	0	1	1	1	0	
0	1	1	1	1	1	1	
0	0	0	0	1	1	0	
0	0	1	0	0	0	1	
1	1	1	1	0	0	1	
0	1	0	0	1	0	1	
0	1	0	1	0	0	0	
0	0	1	1	1	0	0	
0	0	0	1	0	1	1	
0	0	1	0	0	0	1	
1	1	1	0	1	0	0	
1	0	1	1	0	1	0	
1	1	0	0	0	1	1	

as

$$1 = e_1 + e_4 + e_6 + e_7 \tag{12.2.16a}$$

$$0 = e_2 + e_4 + e_5 + e_6 \tag{12.2.16b}$$

$$0 = e_3 + e_5 + e_6 + e_7. \tag{12.2.16c}$$

The 16 error patterns that satisfy these equations are shown in Table 12.2.2. Note that only one vector has a single 1, namely, $\mathbf{e} = (1\ 0\ 0\ 0\ 0\ 0\ 0)$, marked by an arrow. Thus for a BSC this is the most likely error pattern and we compute the decoder output as [using $\mathbf{x}$ in Eq. (12.2.13)]

$$\mathbf{w}' = \mathbf{x} + \mathbf{e} = (0\quad 0\quad 1\quad 1\quad 0\quad 1\quad 0),$$

which is the actual transmitted codeword. The corresponding message block from Table 12.2.1 is $\mathbf{v} = (1\ 0\ 1\ 0)$, and the decoder has correctly recovered the transmitted message block.

When choosing a channel code, important questions that must be answered are: (1) How many errors does the code correct? and (2) How many errors can the code detect? The answers revolve around a parameter of the code called the minimum distance. To begin we define the *Hamming weight* or just weight of a codeword $\mathbf{w}$ as the total number of 1's in the codeword, and we denote the weight by $w(\mathbf{w})$. Therefore, the weight of the codeword $\mathbf{w} = (1\ 1\ 0\ 0\ 1\ 0\ 0)$ is $w(\mathbf{w}) = 3$. Using this idea, we next define the Hamming distance or just distance between any two codewords $\mathbf{w}$ and $\mathbf{v}$ as $w(\mathbf{w} + \mathbf{v})$, which is denoted by $d(\mathbf{w}, \mathbf{v})$. Thus, for $\mathbf{w}$ as just given and $\mathbf{v} = (0\ 0\ 1\ 0\ 0\ 1\ 1)$, $d(\mathbf{w}, \mathbf{v}) = w(\mathbf{w} + \mathbf{v}) = w[(1\ 1\ 1\ 0\ 1\ 1\ 1)] = 6$. The *minimum distance* for a code is finally given by

$$d_{\min} = \min\{d(\mathbf{w}, \mathbf{v}) = w(\mathbf{w} + \mathbf{v}), \text{ over all codewords } \mathbf{w}, \mathbf{v} \text{ with } \mathbf{w} \neq \mathbf{v}\}. \quad (12.2.17)$$

Now for a linear block code, the sum of any two codewords is also a codeword, so Eq. (12.2.17) can be rewritten as

$$\begin{aligned} d_{\min} &= \min\{w(\mathbf{u}) \text{ for all codewords } \mathbf{u} \neq \mathbf{0}\} \\ &\triangleq w_{\min}. \end{aligned} \quad (12.2.18)$$

The quantity $w_{\min}$ is called the *minimum weight* of the code. From Eq. (12.2.18) we see that $d_{\min}$ for a code is simply the minimum weight of all nonzero codewords. Scanning the codewords in Table 12.2.1, it is evident that $d_{\min} = 3$ for the (7,4) Hamming code.

Our minimum probability of error decoding scheme relies very heavily on the concept of how close a received vector is to a transmitted codeword. In fact, our decoder decodes the received vector as that codeword which is closest to the received vector in terms of Hamming distance. For this decoding to be unique, we need for the distance between any pair of codewords to be such that no received vector is the same distance from two or more codewords. This is achieved if for a t-error *correcting* code we have

$$d_{\min} \geq 2t + 1. \quad (12.2.19)$$

The number of detectable errors for a given code also depends explicitly on $d_{\min}$. Since $d_{\min}$ is the minimum distance between any pair of codewords, if the channel error pattern causes $d_{\min} - 1$ or fewer errors, the received vector will not be a codeword, and hence the error is detectable. Thus a t-error *detecting* code only requires that

$$d_{\min} \geq t + 1 \quad (12.2.20)$$

and comparing to Eq. (12.2.19), $d_{\min}$ can be smaller for t-error detecting codes than for t-error correcting codes. Since there are $2^k - 1$ nonzero codewords, there are $2^k - 1$ undetectable error patterns. (Why?) Further, since there are $2^n - 1$ nonzero error patterns, there are $2^n - 1 - (2^k - 1) = 2^n - 2^k$ detectable error patterns for an (n,k) linear block code. This is in contrast to the fact that the same linear block code has only $2^{n-k} - 1$ correctable error patterns.

EXAMPLE 12.2.2

For the (7,4) Hamming code in Example 12.2.1, we know that $d_{\min} = 3$. Thus this code can detect error patterns of $d_{\min} - 1 = 2$ or fewer errors and can correct ($d_{\min} = 3 = 2t + 1 \Rightarrow t = 1$) single errors. The total number of detectable error patterns is $2^n - 2^k = 2^7 - 2^4 = 112$. However, it can correct only seven single-error patterns.

Thus far we have avoided calculating the probability of error for a linear block code, and in general, this is not a trivial task. We can write an upper bound on the probability of a decoding error (block error) for a t-error-correcting block code used over a BSC with independent probability of a bit error p by

$$P_B(\mathscr{E}) \leq \sum_{j=t+1}^{n} \binom{n}{j} p^j (1-p)^{n-j}, \tag{12.2.21}$$

where the right side simply sums all possible combinations of $t+1$ or more errors. Further consideration of error probabilities for linear block codes are not pursued here because of their complexity.

12.3 Cyclic Codes

A key issue in any application of error control codes is implementation complexity. A subset of linear codes, called *cyclic codes*, have the important property that encoders and decoders can be implemented easily using linearly connected shift registers. Given the code vector $\mathbf{w} = (w_1 \ w_2 \ \cdots \ w_n)$, we define a cyclic shift to the right of this code vector by one place as [Lin and Costello, 1983]

$$\mathbf{w}^{(1)} = (w_n \quad w_1 \quad w_2 \quad \cdots \quad w_{n-1}),$$

and a cyclic shift of i places by

$$\mathbf{w}^{(i)} = (w_{n-i+1} \quad w_{n-i+2} \quad \cdots \quad w_n \quad w_1 \quad \cdots \quad w_{n-i}).$$

A linear (n,k) code is called a cyclic code if every cyclic shift of a codeword is another codeword. Thus, for a cyclic code, if 1 0 0 1 0 1 1 is a codeword, so is 1 1 0 0 1 0 1.

The generator matrix of a cyclic code can take on a special form as illustrated by the generator matrix of a (7,4) cyclic code, which is

$$G' = \begin{bmatrix} 1 & 1 & 0 & 1 & 0 & 0 & 0 \\ 0 & 1 & 1 & 0 & 1 & 0 & 0 \\ 0 & 0 & 1 & 1 & 0 & 1 & 0 \\ 0 & 0 & 0 & 1 & 1 & 0 & 1 \end{bmatrix}. \tag{12.3.1}$$

Note that each row is a cyclic shift of an adjacent row. Exactly how this matrix is formed for a particular cyclic code is described in more detail later. Although the generator matrix is not in systematic form, it can be put in systematic form by elementary row operations. For example, keeping rows 1 and 2 intact, we replace row 3 by the sum of rows 1 and 3, and then replace row 4 by the sum of rows 1, 2, and 4, which yields the systematic generator matrix in Eq. (12.2.10). All sums are taken modulo-2.

We return to the specification of generator matrices and their corresponding parity check matrices later. To facilitate the development of encoder and decoder shift register implementations for cyclic codes, we present an alternative development of cyclic codes in terms of a polynomial representation. The development here is adapted from Lin and Costello [1983] and Bertsekas and Gallager [1987]. Let

$$w(D) = w_n D^{n-1} + w_{n-1} D^{n-2} + \cdots + w_2 D + w_1 \tag{12.3.2}$$

be the polynomial in D corresponding to the transmitted codeword, and let

$$v(D) = v_k D^{k-1} + v_{k-1} D^{k-2} + \cdots + v_2 D + v_1 \tag{12.3.3}$$

be the polynomial corresponding to the message bits. The codeword polynomial is expressible as the product of $v(D)$ and the generator polynomial $g(D)$, so

$$w(D) = v(D)g(D). \tag{12.3.4}$$

The generator polynomial of an (n,k) cyclic code is a factor of the polynomial $D^n + 1$; in fact, any polynomial of degree $n - k$ that is a factor of $D^n + 1$ generates an (n,k) cyclic code. For example, for the (7,4) linear code, we have

$$D^7 + 1 = (1 + D)(1 + D^2 + D^3)(1 + D + D^3). \tag{12.3.5}$$

There are two polynomials of degree $n - k = 3$ that generate cyclic codes. The factor $1 + D + D^3$ is known to generate a single error-correcting cyclic code of minimum distance 3. The generator matrix for this generator polynomial is the one in Eq. (12.3.1), where the coefficients of increasing powers of D appear in the first row of the matrix starting with the D^0. Subsequent rows are cyclic shifts of the first row.

To get the systematic form of the generator matrix starting with the generator polynomial, we divide $D^{n-k+i-1}$ by $g(D)$ for $i = 1, 2, \ldots, k$, which yields

$$D^{n-k+i-1} = a_i(D)g(D) + \mathbf{g}_i(D) \tag{12.3.6}$$

or

$$D^{n-k+i-1} + \mathbf{g}_i(D) = a_i(D)g(D), \tag{12.3.7}$$

where we write

$$\mathbf{g}_i(D) = g_{i1} + g_{i2}D + \cdots + g_{i,n-k}D^{n-k-1}. \tag{12.3.8}$$

The coefficients $\{g_{ij};\ i = 1, 2, \ldots, k, j = 1, \ldots, n - k\}$ are the coefficients in the systematic generator matrix of Eq. (12.2.2).

To encode in systematic form using the polynomial notation, we observe that the code vector polynomial must equal the message vector polynomial multiplied by D^{n-k} and summed with a parity check polynomial of order $n - k - 1$. Thus

$$w(D) = D^{n-k}v(D) + p(D), \tag{12.3.9}$$

where

$$p(D) = p_1 + p_2 D + \cdots + p_{n-k}D^{n-k-1}. \tag{12.3.10}$$

Since $w(D) = a(D)g(D)$, $p(D)$ is just the remainder when we divide $D^{n-k}v(D)$ by $g(D)$. Expanding Eq. (12.3.9), we obtain

$$\begin{aligned} w(D) &= D^{n-k}v(D) + p(D) \\ &= v_k D^{n-1} + v_{k-1} D^{n-2} + \cdots + v_1 D^{n-k} \\ &\quad + p_{n-k} D^{n-k-1} + \cdots + p_2 D + p_1, \end{aligned} \tag{12.3.11}$$

so that

$$\mathbf{w} = (p_1 \quad p_2 \quad \cdots \quad p_{n-k} \quad v_1 \quad v_2 \quad \cdots \quad v_k). \tag{12.3.12}$$

The first $n - k$ symbols are parity check bits and the last k digits are the information bits.

EXAMPLE 12.3.1

Starting with the generator polynomial $g(D) = 1 + D + D^3$ of the (7,4) linear code, we wish to encode the message vector $\mathbf{v} = (1 \; 0 \; 1 \; 0)$ using the systematic version of the code. Thus we need to find $p(D)$ as the remainder when dividing $D^3[D^2 + 1] = D^5 + D^3$ by $g(D)$. Using synthetic division, we get $p(D) = D^2$ and $w(D) = D^5 + D^3 + D^2$ and the transmitted codeword from Eq. (12.3.12) is $\mathbf{w} = (0 \; 0 \; 1 \; 1 \; 0 \; 1 \; 0)$, which agrees with the result in Section 12.2.

We can also start with $g(D)$ and form the systematic generator matrix based on the approach in Eqs. (12.3.6)–(12.3.8). $D^{n-k+i-1} = D^3, D^4, D^5, D^6$, for $i = 1, 2, 3$, and 4, respectively, so

$$\mathbf{g}_1(D) = 1 + D$$

$$\mathbf{g}_2(D) = D + D^2$$

$$\mathbf{g}_3(D) = 1 + D + D^2$$

$$\mathbf{g}_4(D) = 1 + D^2$$

and the generator matrix in Eq. (12.2.10) follows from Eq. (12.2.2).

The polynomial operations for decoding a cyclic code are rather straightforward. Let the polynomial corresponding to the received vector be

$$x(D) = w(D) + e(D), \tag{12.3.13}$$

where $e(D)$ is the error vector polynomial. Dividing $x(D)$ by $g(D)$ gives

$$x(D) = b(D)g(D) + s(D) \tag{12.3.14}$$

with $s(D)$ the syndrome polynomial. Since $w(D) = a(D)g(D)$, we can combine Eqs. (12.3.13) and (12.3.14) to produce a relationship between the syndrome polynomial and the error pattern polynomial,

$$e(D) = (a(D) + b(D))g(D) + s(D). \tag{12.3.15}$$

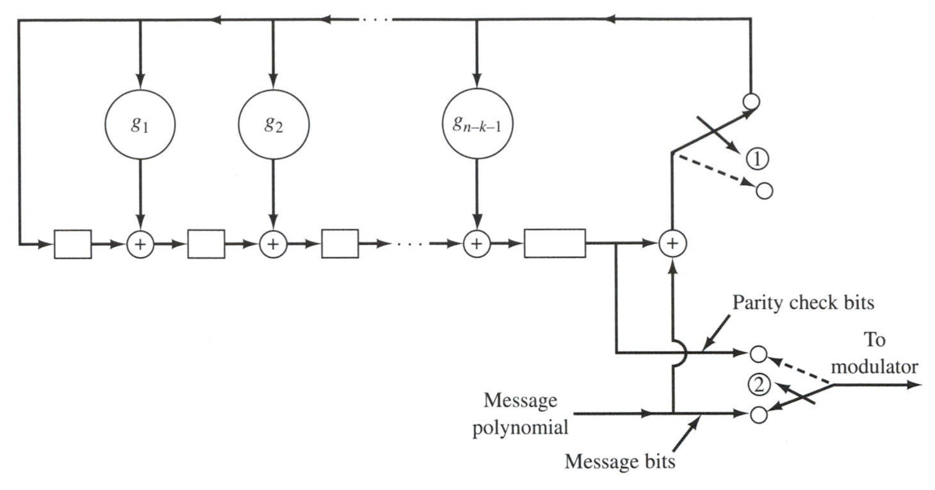

FIGURE 12.3.1 Encoding of a cyclic code based on the generator polynomial. From J. G. Proakis, *Digital Communications*, 2nd ed., New York: McGraw-Hill, Inc., 1989. Reproduced with permission of McGraw-Hill, Inc.

If there are no errors, $e(D) = 0$. Since we know that $g(D)$ divides $w(D)$, we assume that $b(D) = a(D)$. Thus, from Eq. (12.3.15), this yields $s(D) = 0$.

Referring to Eqs. (12.3.9)–(12.3.11) for encoding and Eqs. (12.3.13) and (12.3.14) for decoding, we see that both of these operations require the division of one polynomial by another polynomial. Fortunately, this operation is implementable with a linear feedback shift register circuit. To be more specific, recall that encoding requires multiplication of $v(D)$ by D^{n-k}, division of $D^{n-k}v(D)$ by $g(D)$ to get the remainder, and the formation of $D^{n-k}v(D) + p(D)$. Figure 12.3.1 shows a general circuit for accomplishing this. With switch ① in the up position and switch ② in the down position, the message bits are shifted in. Immediately, after the last of these bits have been shifted in, the shift register contents are the coefficients of $p(D)$. Switch ① is moved to the lower position and switch ② to the up position, and the contents of the shift register are sent to the modulator. In this manner, $\mathbf{w}$ in Eq. (12.3.12) is formed.

EXAMPLE 12.3.2

A linear feedback shift register connection for the (7,4) linear code is shown in Fig. 12.3.2, where the shift register contents are for $v(D) = D^2 + 1$. Table 12.3.1 shows the contents of the shift register as the message is shifted in. The final output of the circuit is the 7-bit codeword found in Example 12.3.1.

A linear shift register feedback connection for calculating the syndrome based on $g(D)$ is shown in Fig. 12.3.3. The initial contents of the shift register is all zeros. The components of $x(D)$ are shifted in one bit at a time, and after the last bit is shifted in, the contents of the shift register specify the syndrome.

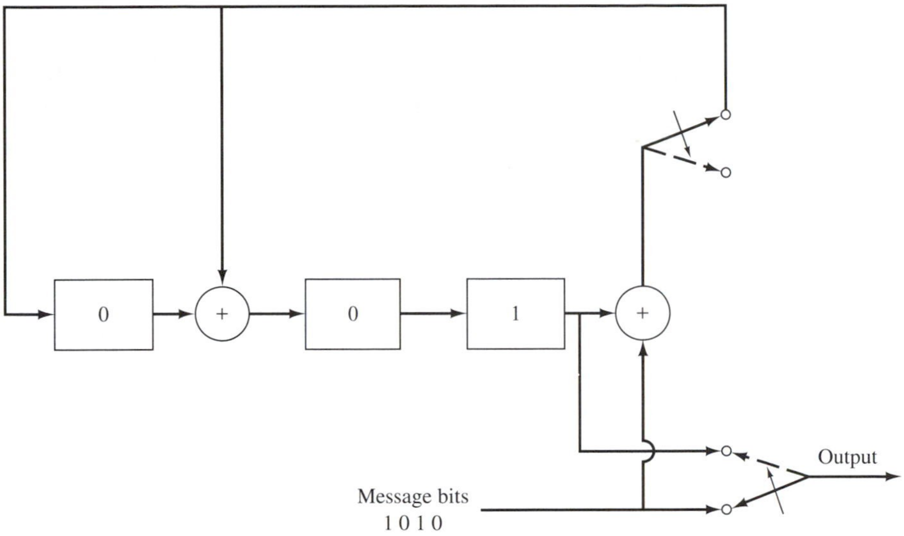

FIGURE 12.3.2 Encoding circuit for a (7,4) cyclic code. From J. G. Proakis, *Digital Communications*, 2nd ed., New York: McGraw-Hill, Inc., 1989. Reproduced with permission of McGraw-Hill, Inc.

TABLE 12.3.1 Shift Register Contents for the Encoder of Figure 12.3.2

Input	Shift Register Contents
	0 0 0
0	0 0 0
1	1 1 0
0	0 1 1
1	0 0 1 = $(p_1 p_2 p_3)$

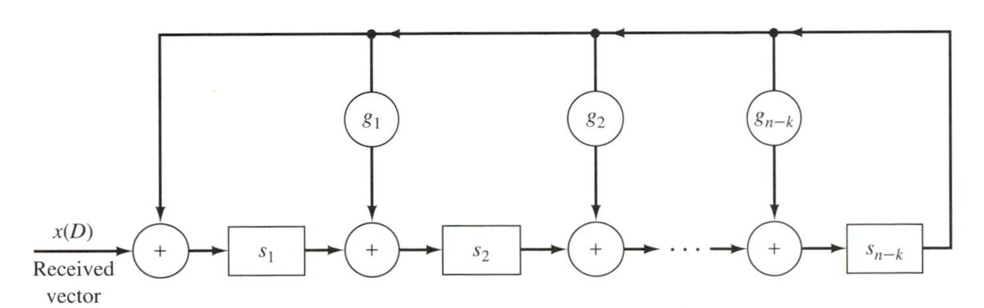

FIGURE 12.3.3 Shift register connection for obtaining the syndrome.

EXAMPLE 12.3.3

A syndrome calculation circuit for the (7,4) cyclic code based on $g(D) = D^3 + D + 1$ is shown in Fig. 12.3.4. The shift register contents as $x(D) = 1 + D^2 + D^3 + D^5$ is shifted in are listed in Table 12.3.2.

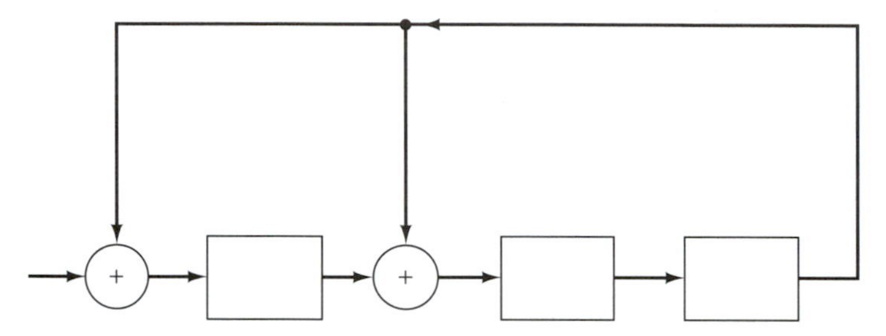

FIGURE 12.3.4 Circuit for calculating the syndrome of the (7,4) code with $g(D) = D^3 + D + 1$.

TABLE 12.3.2 Syndrome Calculation Corresponding to Figure 12.3.4

Received	Shift Register Contents
	0 0 0
0	0 0 0
1	1 0 0
0	0 1 0
1	1 0 1
1	0 0 0
0	0 0 0
1	1 0 0 = $(s_1 s_2 s_3)$

The simple fact that $g(D)$ divides $w(D)$ and the assumption that if $g(D)$ divides $x(D)$ the received codeword is accepted as being correct, allow a number of conclusions to be reached concerning the error detecting capability of a cyclic code. If there is a single bit error in position i, then $e(D) = D^i$. For this error to not be detected, $g(D)$ must divide $e(D)$. However, if $g(D)$ has more than one term, it cannot divide D^i, and all single errors are detectable.

For a double bit error in the ith and jth positions, we have

$$e(D) = D^i + D^j = D^i(1 + D^{j-i}),$$

assuming that $i < j$. If $g(D)$ has three or more terms, neither component will be divisible by $g(D)$ and all double errors will be detected.

Some statements can also be made concerning burst error detection, that is, a sequence of consecutive bit errors. If the burst is of length $n - k$ or less, we can write $e(D)$ as

$$e(D) = D^i B(D), \tag{12.3.16}$$

where $B(D)$ is of degree $n - k - 1$ or less and $0 \leq i \leq n - 1$. The generator polynomial is of degree $n - k$, and if it has more than one term, it cannot divide either factor of Eq. (12.3.16). Therefore, all bursts of length $n - k$ or less are detectable. Other burst error-detecting capabilities are left to the problems (Problems 12.21 and 12.22).

Some cyclic codes are called *cyclic redundancy check* (CRC) *codes* and a few have been selected as standards for networking applications. The generator polynomials for these codes are given in Table 12.3.3. These codes are discussed in Chapter 15.

TABLE 12.3.3 Generator Polynomials for CRC Code Standards

Code	Generator Polynomial
CRC-12	$1 + D + D^2 + D^3 + D^{11} + D^{12}$
CRC-16	$1 + D^2 + D^{15} + D^{16}$
CRC-CCITT	$1 + D^5 + D^{12} + D^{16}$
CRC-32	$1 + D + D^2 + D^4 + D^5 + D^7 + D^8 + D^{10}$ $+ D^{11} + D^{12} + D^{16} + D^{22} + D^{23} + D^{26} + D^{32}$

12.4 Convolutional Codes

Convolutional codes are important channel codes for many applications, and the encoding and decoding operations for convolutional codes differ quite substantially from those described for linear block codes in the preceding section. The nomenclature and structure of convolutional codes is most simply (and most commonly) developed by example. Thus we begin by considering the convolutional code in Fig. 12.4.1 with binary input sequence $\mathbf{v} = (v_1 \ v_2 \ v_3 \ \cdots)$ and binary output sequence $\mathbf{w} = (w_1 \ w_2 \ w_3 \ \cdots)$. The blocks labeled "$D$" represent unit delays and the $\oplus$ represent summations. The output sequence is obtained by alternately sampling the upper and lower summer outputs. The encoder inputs and outputs are not blocked as in Section 12.2 but are "semi-infinite" binary sequences. For the input sequence $\mathbf{v} = (1 \ 0 \ 1 \ 1 \ 0 \ 1 \ 0 \ 0 \ \cdots)$, where the three dots indicate a repetition of zeros, the output sequence can be found from Fig. 12.4.1 to be (assuming zero initial conditions)

$$\mathbf{w} = (1 \ 1 \ 0 \ 1 \ 0 \ 0 \ 1 \ 0 \ 1 \ 0 \ 0 \ 0 \ 0 \ 1 \ 1 \ 1 \ 0 \ 0 \ \cdots).$$

For each input bit in $\mathbf{v}$ there are two output bits in $\mathbf{w}$, so the code is said to have rate $R = k/n = \frac{1}{2}$. Our development in this section is limited to rate $1/n$ convolutional codes (see Problem 12.35 for an example of a higher rate code).

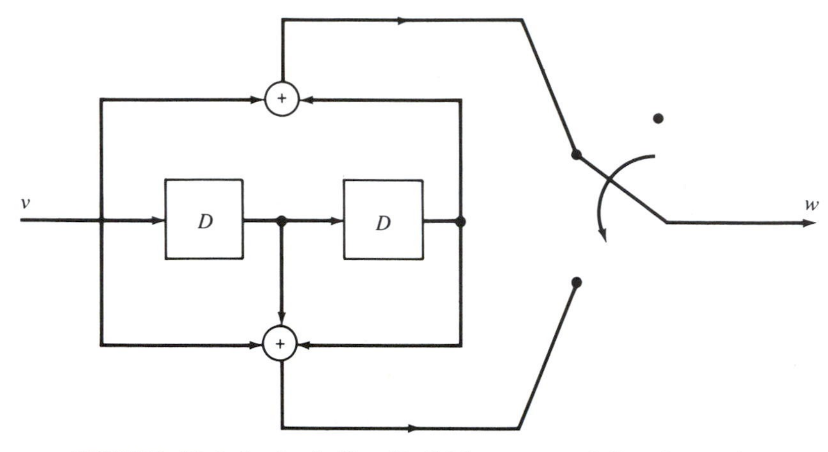

FIGURE 12.4.1 An $(n,k) = (2, 1)$ binary convolutional encoder.

There are several alternative ways of drawing a convolutional encoder that appear in the literature. One alternative is illustrated in Fig. 12.4.2, where delay elements are not indicated, but storage locations are shown. Some authors say that the diagram in Fig. 12.4.1 is a two-stage shift register (by counting delays), while others claim that Fig. 12.4.2 represents a three-stage shift register by counting storage locations. This is only nomenclature, however, and either diagram is easy to work with. Another point of confusion in convolutional coding terminology is the definition of *constraint length.* We define the constraint length of a convolutional code to be equal to the number of storage locations or digits available as inputs to the modulo-2 summers, and denote it by ν. Several other definitions of constraint length are employed in the literature, all of which are related to our definition of ν in some manner.

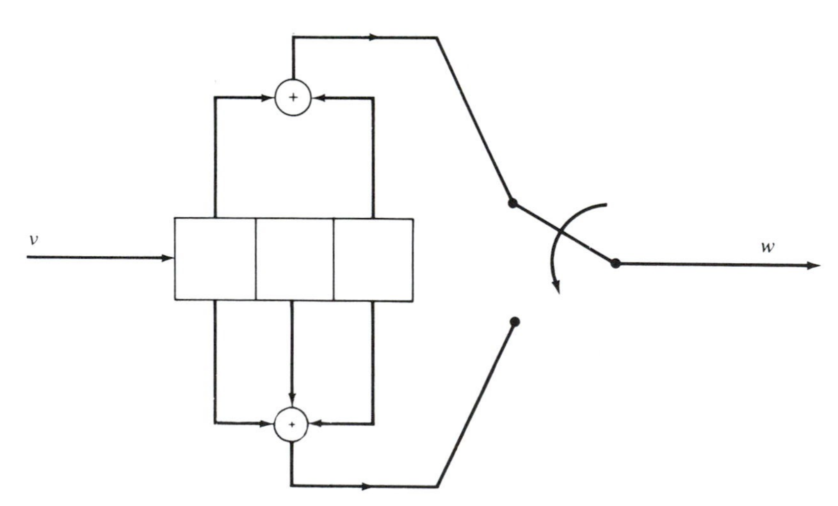

FIGURE 12.4.2 Alternative diagram for an $(n,k) = (2, 1)$ binary convolutional encoder.

The encoder output is an interleaving of the outputs of the upper and lower summer outputs. If the encoder input is an impulse, $\mathbf{v} = (1\ 0\ 0\ \cdots)$, the impulse responses of the upper and lower branches are

$$\mathbf{w}^{(1)} = (1 \quad 0 \quad 1 \quad 0 \quad 0 \quad \cdots) \tag{12.4.1}$$

and

$$\mathbf{w}^{(2)} = (1 \quad 1 \quad 1 \quad 0 \quad 0 \quad \cdots). \tag{12.4.2}$$

These impulse responses are usually given the special name *generator sequences* and denoted by $\mathbf{g}^{(1)} = \mathbf{w}^{(1)}$ in Eq. (12.4.1) and $\mathbf{g}^{(2)} = \mathbf{w}^{(2)}$ in Eq. (12.4.2). For any given encoder input $\mathbf{v}$, the two outputs $\mathbf{w}^{(i)}$, $i = 1, 2$, can be found by discrete-time convolution for all time instants $m \geq 1$ as

$$\begin{aligned}\mathbf{w}^{(i)} &= \mathbf{v} * \mathbf{g}^{(i)} \\ &= \sum_{j=0}^{\nu-1} v_{m-j} g_{j+1}^{(i)} = v_m g_1^{(i)} + v_{m-1} g_2^{(i)} + \cdots + v_{m-\nu+1} g_\nu^{(i)}\end{aligned} \tag{12.4.3}$$

with $v_{m-j} \triangleq 0$ for $m - j \leq 0$. The encoder output sequence is then $\mathbf{w} = (w_1^{(1)}\ w_1^{(2)}\ w_2^{(1)}\ w_2^{(2)}\ w_3^{(1)}\ w_3^{(2)}\ \cdots)$. A generator matrix for convolutional codes can be obtained by interleaving the generator sequences to form a row and then creating subsequent rows by shifting n columns to the right. All unspecified components are zero. The generator matrix for the code in Figs. 12.4.1 and 12.4.2 is

$$\mathbf{G} = \begin{bmatrix} g_1^{(1)} & g_1^{(2)} & g_2^{(1)} & g_2^{(2)} & g_3^{(1)} & g_3^{(2)} & \cdots & g_\nu^{(1)} & g_\nu^{(2)} & & \\ \cdots & \cdots & g_1^{(1)} & g_1^{(2)} & g_2^{(1)} & g_2^{(2)} & \cdots & \cdots & \cdots & g_\nu^{(1)} & g_\nu^{(2)} \\ & & & & \ddots & & & & & & \end{bmatrix}$$

$$= \begin{bmatrix} 1 & 1 & 0 & 1 & 1 & 1 & 0 & 0 & \cdots & \cdots & \cdots & \cdots & \cdots \\ & & 1 & 1 & 0 & 1 & 1 & 1 & 0 & 0 & \cdots & \cdots & \cdots \\ & & & & 1 & 1 & 0 & 1 & 1 & 1 & 0 & 0 & \cdots \\ & & & & & \ddots & & & & & & & \end{bmatrix}. \tag{12.4.4}$$

The encoder output can be written in terms of this generator matrix and the encoder input treated as a row vector as

$$\mathbf{w} = \mathbf{v}\mathbf{G}, \tag{12.4.5}$$

so for $\mathbf{v} = (1\ 0\ 1\ 1\ 0\ 1\ 0\ 0\ \cdots)$ as before,

$$\mathbf{w} = \mathbf{v}\mathbf{G} = (1\ 1\ 0\ 1\ 0\ 0\ 1\ 0\ 1\ 0\ 0\ 0\ 0\ 1\ 1\ 1\ 0\ 0\ \cdots).$$

Of course, $\mathbf{G}$ is different from the linear block code case since it is a semi-infinite matrix here.

Graphical displays of convolutional codes have proven invaluable over the years for their understanding and analysis. A particularly useful graphical presentation is the *code tree*. A code tree is created by assuming zero initial conditions for the encoder and considering all possible encoder input sequences. For a 0 input digit a tree branch is drawn by moving upward, drawing a horizontal line and labeling the line with the encoder output. For a 1 input a tree branch

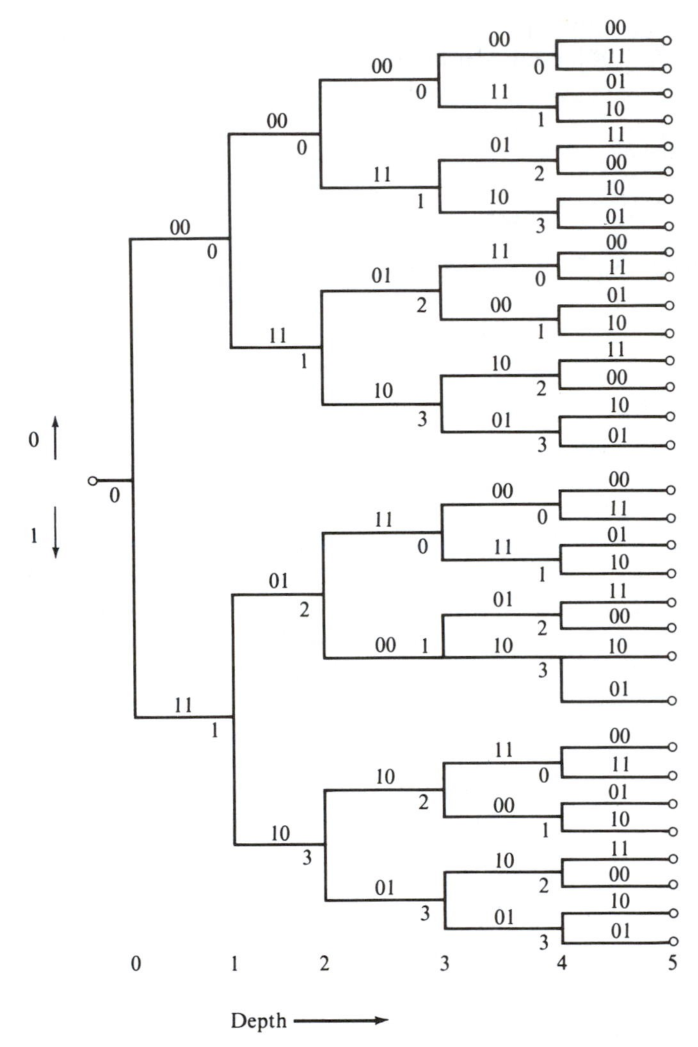

FIGURE 12.4.3 Code tree for the binary convolutional encoder in Figs. 12.4.1 and 12.4.2.

is drawn by moving down and drawing a horizontal line labeled with the encoder output. The same process is repeated at the end of each horizontal line indefinitely. The code tree for the convolutional code in Figs. 12.4.1 and 12.4.2 is shown in Fig. 12.4.3. Note that the output **w** in Eq. (12.4.5) can be obtained using the code tree and the given encoder input by stepping up for an input 0 and down for an input 1 and reading off the two-digit branch labels. To find all of **w** in Eq. (12.4.5), the code tree needs to be extended. The tree diagram (code tree) thus allows the encoder output to be found very easily for a given input sequence. Unfortunately, for l input bits, the tree has 2^l branches, so that as l, the number of input bits, gets large, the code tree has an exponentially growing number of branches.

The code tree can be redrawn into a more manageable diagram called a *trellis* by noting that after v input bits, the tree becomes repetitive. Each of the nodes (or branching points indicated by black dots) in Fig. 12.4.3 is labeled by one of the numbers 0, 1, 2, or 3. These numbers indicate the two most recent encoder input bits by the correspondence $0 \leftrightarrow 00$, $1 \leftrightarrow 01$, $2 \leftrightarrow 10$, and $3 \leftrightarrow 11$, and are usually called *states*. All like-numbered nodes after depth $2(v - 1)$ in the tree have identical output digits on the branches emanating from them. Therefore, these nodes can be merged. By merging like-numbered nodes after depth $2(v - 1)$ in the tree, we obtain a trellis, as shown in Fig. 12.4.4. As in the tree, an upper branch out of a node is taken for a 0 input digit and the lower branch is taken for a 1 input bit. The branch labels are the output bits. Note that after all states are reached, the trellis is repetitive. For a given input sequence, we can find the output by tracing through the trellis, but without the exponential growth in branches as in the tree.

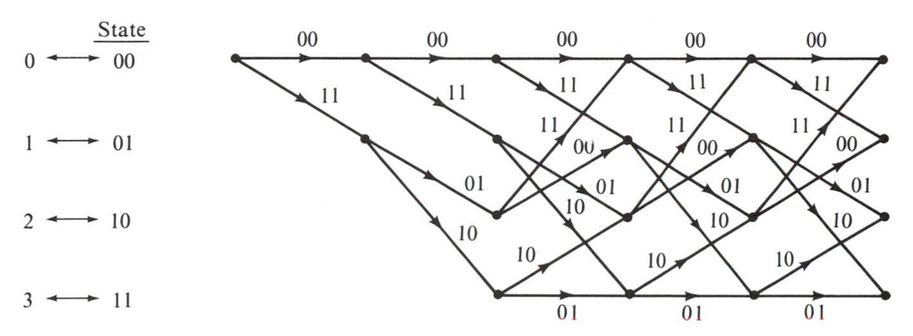

FIGURE 12.4.4 Trellis corresponding to the tree in Fig. 12.4.3.

Just as the encoding operation for convolutional codes is quite different from the encoding operation for linear block codes, the decoding process for convolutional codes proceeds quite differently. Since we can represent a transmitted codeword for a convolutional code as a path through a trellis, the decoding operation consists of finding that path through the trellis which is "most like" the received binary sequence. As with the linear block codes, we are interested in hard decision decoding and a decoder that minimizes the probability of error, and therefore, for a given received binary vector, the decoder finds that path through the trellis which has minimum Hamming distance from the received sequence. Given a long received sequence of binary digits and a trellis similar to that in Fig. 12.4.4, it would seem quite a formidable task to search all possible paths for the best path. However, there exists an iterative procedure called the Viterbi algorithm that greatly simplifies matters. This algorithm is a special case of *forward dynamic programming* and relies on the "principle of optimality" (see Section 10.7). As applied to our particular problem, this principle states that the best (smallest Hamming distance) path through the trellis that includes a particular node necessarily includes the best path from the beginning of the trellis to this node. What this means to us is that for each node in the trellis we need only retain the single best path to a node, thus limiting the number of retained paths at any time instant to the number of nodes in the trellis at that time. Therefore, for the trellis in Fig. 12.4.4, no more than four paths are retained at any time instant. The retained paths are called *survivors*, and the nodes correspond to *states* as shown in the figure. The basic steps in the Viterbi algorithm are as follows:

STEP 1. At time instant j, find the Hamming distance of each path entering a node by adding the Hamming distance required in going from the node(s) at time $j - 1$ to the node at time j with the accumulated Hamming distance of the survivor into the node at time $j - 1$.

STEP 2. For each node in the trellis, discard all paths into the node except for the one with the minimum total Hamming distance—this is the survivor.

STEP 3. Go to time instant $j + 1$ and repeat steps (1) and (2).

Use of the Viterbi algorithm can be clarified by a concrete example.

EXAMPLE 12.4.1

We assume that the rate-$\frac{1}{2}$ binary convolutional code represented by the encoders in Figs. 12.4.1 and 12.4.2 and the trellis in Fig. 12.4.4 is used and that the transmitted codeword is the all-zero sequence $\mathbf{w} = (0\ 0\ 0\ 0\ 0\ 0\ 0\ \cdots)$. The received codeword is assumed to be $\mathbf{x} = (1\ 0\ 1\ 0\ 0\ 0\ \cdots)$ and we wish to use the Viterbi algorithm to decode this received sequence.

The decoding operation is indicated by a series of incomplete trellises in Fig. 12.4.5. Through time instant 2, no paths need be discarded, since the trellis is just "fanning out." The four possible paths with the Hamming distance into each node listed above the node are shown in Fig. 12.4.5(a). These distances are calculated by comparing the received bits with the path labels. Thus the Hamming distance associated with state 2 at time instant 2 is $\mathbf{w}[(1\ 0\ 1\ 0) + (1\ 1\ 0\ 1)] = \mathbf{w}[(0\ 1\ 1\ 1)] = 3$, as shown. We now extend each of the paths to the next level and calculate Hamming distances as demonstrated in Fig. 12.4.5(b). The paths to be discarded are shown dashed and their corresponding Hamming distances are given in parentheses below the node. The minimum Hamming distance to each node is written above the node and the retained paths (survivors) are the solid lines. The survivors are extended and the calculations repeated in Fig. 12.4.5(c) and (d). After time instant 5 in Fig. 12.4.5(d), it is evident that the decoder output will be $\mathbf{y} = (0\ 0\ 0\ 0\ 0\ 0\ \cdots)$, since the total Hamming distance to state 0 at time instant 5 is 2, while the distance to all other nodes is 3. Since subsequent decisions cannot reduce the path distance, the all-zero sequence is the decoder output. In this instance, the decoder correctly decoded the received codeword. This is not always the case, since uncorrectable error patterns can occur for convolutional codes just as for block codes.

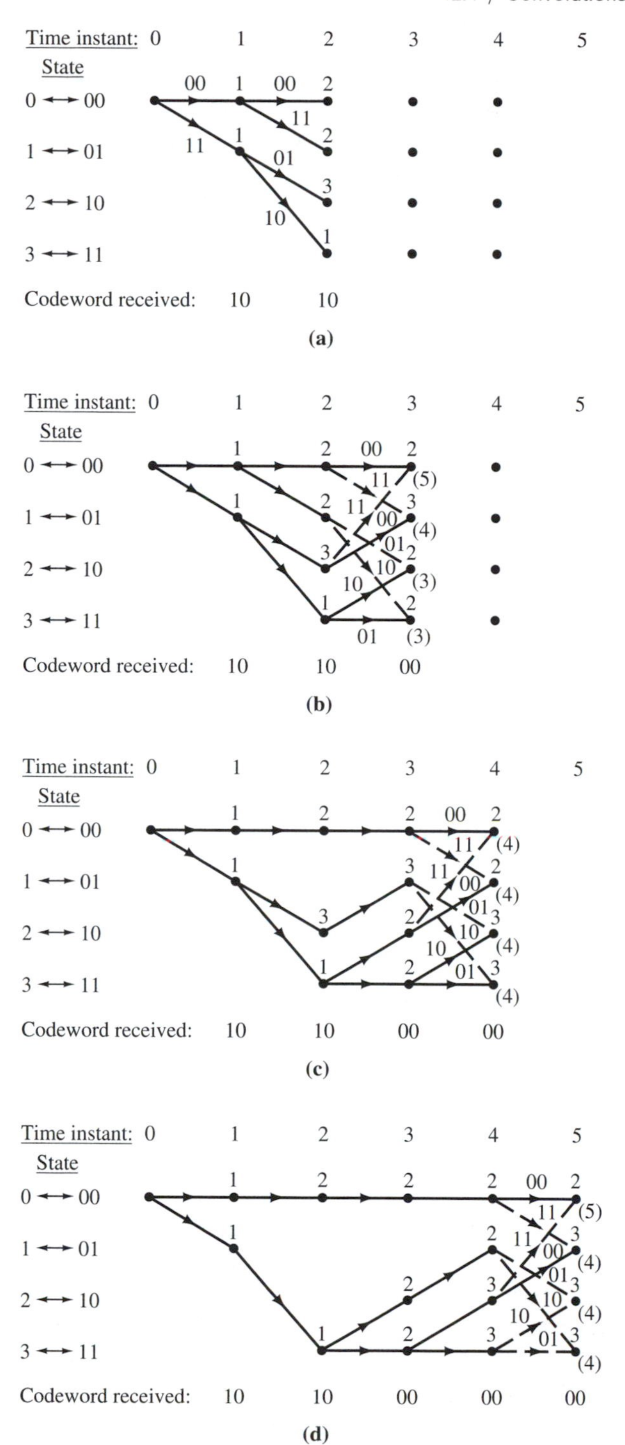

FIGURE 12.4.5 Viterbi decoding for Example 12.4.1.

The Viterbi algorithm has the drawback that the number of decoding calculations is proportional to $2^{\nu-1}$, where ν is the constraint length. As ν gets large, the computational load of Viterbi decoding becomes prohibitive. Additionally, the Viterbi decoder requires the same decoding effort, whether the channel is very noisy or relatively error-free. For these reasons other decoders for convolutional codes have been developed. Examination of these other decoding algorithms is beyond the intended scope of this book. Other important topics not covered here are probability of error calculations for convolutional codes and soft decision decoding, which is very useful with convolutional codes. The reader should consult the references for additional information [Lin and Costello, 1983].

12.5 Automatic Repeat-Request Systems

Sections 12.2 and 12.4 are concerned primarily with what is called *forward error correction* (FEC), that is, codes which try to detect *and* to correct channel errors. The possibility of using linear block codes for error detection alone is also mentioned briefly in Sections 12.2 and 12.3, where it is noted that error detection is easier than error correction. This is because for each $(n - k)$-length syndrome there are 2^k error patterns that could have generated this syndrome, but only one of the error patterns can be corrected. Thus there are $2^{n-k} - 1$ correctable error patterns, but as demonstrated in Section 12.2, there are $2^n - 2^k$ detectable error patterns. From Example 12.2.2 for the (7,4) Hamming code, there are 112 detectable error patterns but only $2^{7-4} - 1 = 7$ correctable error patterns. If error detection is simpler than error correction, how do we make use of the fact that we have detected an error? *Automatic repeat-request* (ARQ) *systems* employ strategies that use error detection information to improve communication system performance.

Here we discuss three basic ARQ strategies: (1) the stop-and-wait ARQ, (2) the go-back-N ARQ, and (3) the selective repeat ARQ. As for any communication system, these strategies differ in complexity and performance. Generally, ARQ systems employ a good error detecting code and compute the syndrome of the received vector. If the syndrome is zero, the received vector is taken to be correct, and the receiver notifies the transmitter over a reverse or return channel that the transmitted vector has been correctly received by sending an acknowledgment (ACK). The reverse channel is assumed to be low speed and accurate (no errors). When the syndrome is nonzero, an error is detected in the received vector, and the transmitter is informed of this occurrence by the receiver sending a negative acknowledgment (NAK) through the return channel. When the transmitter receives the NAK, it retransmits the erroneously received vector. The procedure is repeated until all received vectors have a zero syndrome (no error is detected). The specific ARQ strategies are variations on this theme.

The stop-and-wait ARQ method has the transmitter send a codeword to the receiver and then not send another codeword until a positive acknowledgment, ACK, is returned by the receiver. If NAK is returned to the transmitter, the codeword is retransmitted until ACK is received by the transmitter. After reception of an ACK, the next codeword is transmitted. This stop-and-wait ARQ scheme is obviously inefficient because of the time spent waiting for an acknowledgment. One suggestion for overcoming this idle time inefficiency is to use very long block length codes, but this has the disadvantage of increasing retransmissions, since a longer block is more likely to contain errors.

The go-back-N ARQ system continuously transmits codewords rather than waiting for an acknowledgment after each received vector. The quantity $N - 1$ is the number of codewords transmitted during the time it takes for a vector to reach the receiver and for the acknowledgment of the vector to be returned to the transmitter. This time is called the *round-trip delay*. When an NAK is returned, the transmitter retransmits the erroneously received codeword *and* the $N - 1$ immediately following codewords. Thus a buffer is not needed at the receiver since the $N - 1$ vectors following an erroneous received vector are discarded irrespective of whether they are correctly received. This go-back-N ARQ method efficiently utilizes the forward channel because of its continuous transmission, but it may not be effective at delivering usable data to the receiver if the transmitted bit rate is high and the round-trip delay relatively large. This is because N will be large in this case, and too many error-free codewords are retransmitted after an error is detected at the receiver.

The selective-repeat ARQ scheme improves upon the preceding ARQ systems by continuously transmitting codewords and retransmitting only erroneously received codewords. Therefore, as for the go-back-N system, codewords are continuously transmitted and those codewords for which an NAK is returned are retransmitted. However, the selective-repeat ARQ method *only* resends codewords for which an NAK is returned. For received vectors to be released to the user in the correct order, a buffer is now required at the receiver. Correctly received vectors are continuously released until a codeword is received in error. Subsequently received correct codewords are then stored at the receiver until the incorrect codeword is received without error. The entire string of codewords is then released from the receiver buffer to the user in the correct order. The receiver buffer must be sufficiently long to prevent buffer overflow for this system.

The primary indicator of ARQ performance is what is called the *throughput*, which is defined to be the ratio of the average number of message digits delivered to the user per unit time to the maximum number of digits that could be transmitted per unit time. For an (n,k) linear block code and with P being the probability of a received vector being accepted, the throughput of the selective repeat system is [Lin and Costello, 1983]

$$\eta_{\text{sr}} = \frac{k}{n} P \tag{12.5.1}$$

and the throughput of the go-back-N scheme is

$$\eta_{\text{gb}N} = \frac{P}{P + (1 - P)N} \frac{k}{n}. \tag{12.5.2}$$

To write the expression for the throughput of the stop-and-wait method, we define the time spent waiting for an acknowledgment as τ_d and the transmitted data rate in bits/sec as r. The throughput of the stop-and-wait scheme can now be written as

$$\eta_{sw} = \frac{P}{1 + (r\tau_d/n)} \frac{k}{n}. \tag{12.5.3}$$

Clearly, η_{sr} is greater than both η_{gbN} and η_{sw}. Note, however, that the expression for η_{sr} in Eq. (12.5.1) assumes an infinite buffer at the receiver. More detailed analysis with a reasonable finite buffer size of say, N, reveals that the throughput of the selective-repeat system is still significantly better than the other two [Lin and Costello, 1983].

12.6 Code Comparisons

The linear block codes in Section 12.2 and the convolutional codes in Section 12.4, along with the decoding algorithms presented in these sections, are but a very few of a host of channel codes and their decoding algorithms that have been investigated and described in the literature. For a variety of reasons, the performance evaluation of channel codes can be a difficult task that is certainly beyond the scope of this book. We collect here some performance data concerning both block and convolutional codes to give the reader an idea of what can be expected.

Of course, an often used indicator of relative performance for linear block codes is the minimum distance d_{min}, since the number of errors that can be detected and corrected is proportional to d_{min}. There is a related quantity that can be defined for convolutional codes called the *free distance*, d_{free}, which is the minimum distance between any two codewords in a code. Because of the exponential growth with constraint length v of the number of states and possible codewords in a convolutional code, the calculation of d_{free} for a code can be a difficult task. In any event, when d_{free} is known, the larger d_{free}, the better the code performance.

Another shortcut often taken to ascertaining how well a class of codes performs is to devise coding bounds. These bounds are primarily of two types, bounds on d_{min} and bounds on performance. Two bounds on d_{min}, the Hamming bound and the Plotkin bound, give the maximum possible d_{min} for a specific code rate and code block length. Another bound, called the Gilbert bound, gives a lower bound on d_{min} for the best code. The other type of bounds, performance bounds, are generally of the type called *random coding bounds*. Bounds in this class bound the *average* performance of all codes with a certain structure, and typically are a demonstration that the average probability of error decreases exponentially with code "block" length. Such random coding bounds only imply that codes which perform better than the average exist and do not indicate how good codes can be found.

Clearly, the coding bounds are not helpful if we need to evaluate the performance of a specific code, and the performance of particular codes is often

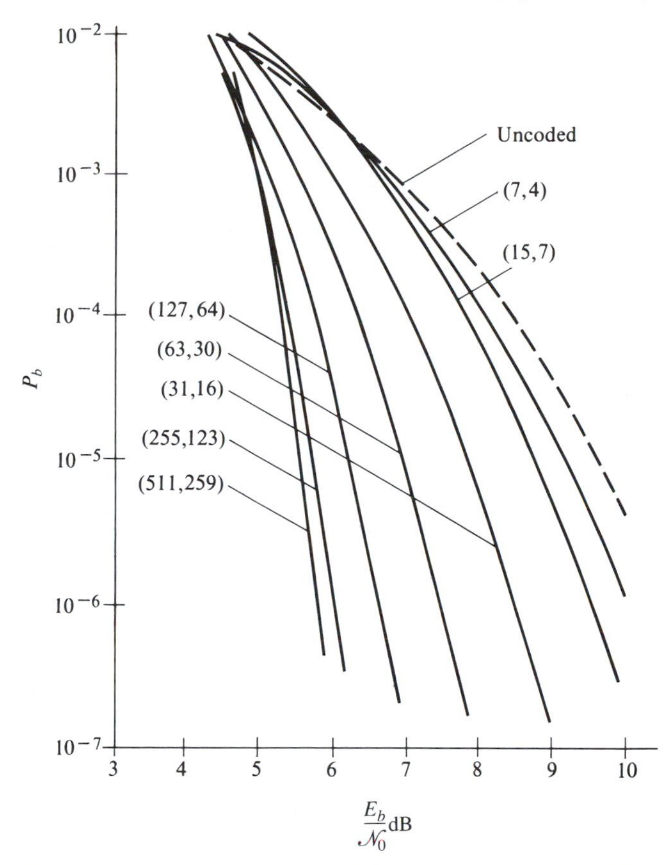

FIGURE 12.6.1 Bit error probability for BCH codes using binary PSK modulation and hard decisions. From G. C. Clark, Jr., and J. B. Cain, *Error-Correction Coding for Digital Communications*, New York: Plenum Press, 1981. © 1981 Plenum Press.

difficult to calculate. An important generalization of the Hamming codes for multiple-error correction is the class of codes called BCH (Bose, Chaudhuri, Hocquenghem) codes. Figure 12.6.1 compares the bit error probability (P_b) of several (n,k) BCH codes as a function of $E_b/\mathcal{N}_0$ for binary PSK with hard decisions where the channel bit error probability $p = Q[\sqrt{2kE_b/n\mathcal{N}_0}]$. Note that the (7,4) code shown is a Hamming code. It is evident from this figure that the BCH codes can significantly improve performance in comparison to uncoded binary PSK (dashed line). It is noted that P_b in Fig. 12.6.1 is not exact but is an upper bound on the bit error probability of the codes [Clark and Cain, 1981]. Despite this, considerable performance improvement is evident.

Some computer simulation results for binary convolutional codes with Viterbi decoding and hard decisions are presented in Fig. 12.6.2 as the constraint length is varied [Heller and Jacobs, 1971]. The improvement over uncoded transmission can be seen by comparison with the uncoded curve in Fig. 12.6.1. The reduction in bit error probability as the constraint length is increased is clear

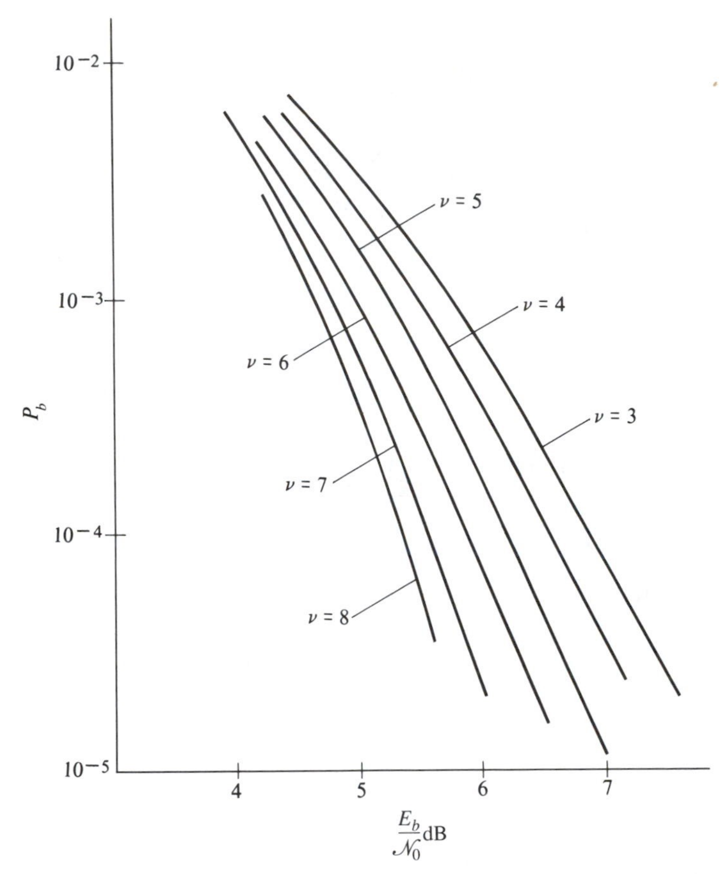

FIGURE 12.6.2 Bit error probability for convolutional codes using PSK and hard decisions. From J. A. Heller and I. M. Jacobs, "Viterbi Decoding for Satellite and Space Communications," *IEEE Trans. Commun. Technol.*, © 1971 IEEE.

from Fig. 12.6.2. By referring back to Example 12.4.1, the reader can see that the Viterbi algorithm must search to some (as yet unspecified) depth in the trellis before the best path is found, or more generally, before all retained paths spring from the same first step. When implementing the Viterbi algorithm, this depth must be selected. If this depth is chosen too small, there is ambiguity in the decoding decision [see, e.g., Fig. 12.4.5(c) where two paths have distance 2 and a different first step]. If this depth is chosen large, excessive storage is required. This "search depth" is generally selected to be several times (say, 5 to 10 times) the constraint length of the code as a compromise. The simulations for Fig. 12.6.2 used a search depth of 32. Figure 12.6.3 presents comparative performance results as the search depth or decoder memory, denoted by λ, is varied for a $\nu = 5$, rate-$\frac{1}{2}$ convolutional code. The three curves labeled $Q = 2$

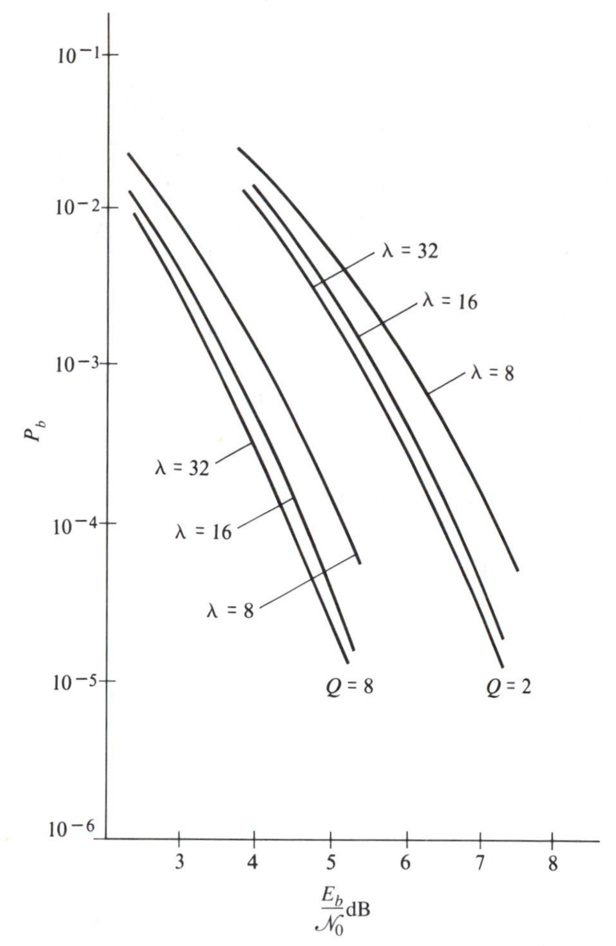

FIGURE 12.6.3 Convolutional code performance as a function of decoder memory or search depth (λ). From J. A. Heller and I. M. Jacobs, "Viterbi Decoding for Satellite and Space Communications," *IEEE Trans. Commun. Technol.*, © 1971 IEEE.

are for the hard decision case. The $\lambda = 32$ choice yields a slight improvement over the $\lambda = 16$ depth.

Thus far in this chapter we have limited consideration to hard decision decoding, where the output of the demodulator is simply a binary sequence. In this case, optimum (also maximum a posteriori and minimum probability of error) decoding simplifies to calculating the appropriate Hamming distances. Hard decision decoding corresponds to a two-level quantization of the decoder input. We know that we could better implement the optimum decision rule for an AWGN channel if the demodulator output were unquantized, often called infinite quantization, or at least quantized more finely than to two levels. The reader should review optimum receivers for unquantized outputs in Section 10.4

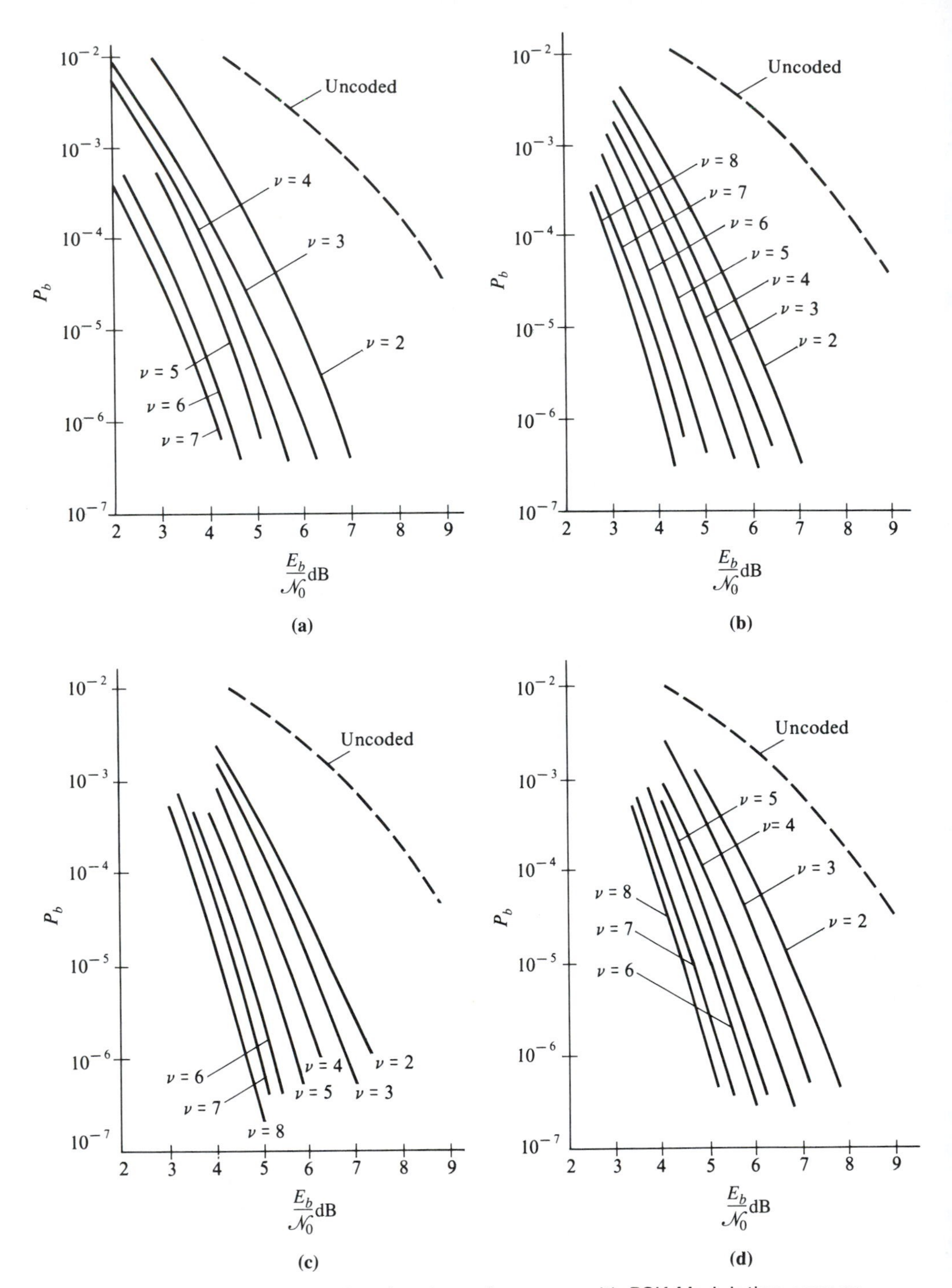

FIGURE 12.6.4 Convolutional code performance with PSK Modulation over an AWGN channel with soft decision Viterbi decoding: (a) $R_c = \frac{1}{3}$; (b) $R_c = \frac{1}{2}$; (c) $R_c = \frac{2}{3}$; (d) $R_c = \frac{3}{4}$. From G. C. Clark, Jr., and J. B. Cain, *Error-Correction Coding for Digital Communications*, New York: Plenum Press, 1981. © 1981 Plenum Press.

if this notion seems fuzzy and see Example 11.6.2 and Problems 11.24 and 11.25 for examples of quantizing the channel or demodulator output. When the number of levels (Q) at the demodulator output is greater than 2 for binary transmitted symbols, the system is said to be using soft decision demodulation or soft decision decoding. Soft decision decoding can be used with either block or convolutional codes, although it is much simpler to use with convolutional codes. Figure 12.6.3 contains performance results for eight-level ($Q = 8$) soft decision decoding as well as for hard decision decoding. The soft decisions have a clear performance advantage. When using soft decisions, there is always a trade-off between receiver complexity and performance, both of which increase with the number of output quantization levels. Figure 12.6.4(a) to (d) present the performance of rate-$\frac{1}{3}$, -$\frac{1}{2}$, -$\frac{2}{3}$, and -$\frac{3}{4}$ convolutional codes for various constraint lengths (v) when used with binary PSK over an AWGN channel and infinitely quantized soft decision Viterbi decoding [Clark and Cain, 1981]. The lower rate codes and the longer constraint lengths for a given rate code provide better performance.

The question often arises as to which type of code, a block code or a convolutional code, is better for a particular application. Whether block codes or convolutional codes are better in general is a point of continuing debate among coding theorists. McEliece [1977] states that the BCH (block) codes are to be preferred for use over a BSC and other symmetric DMCs, since these are the channels the BCH codes were designed for, whereas for the AWGN channel, soft decision Viterbi decoded convolutional codes are preferable. This is primarily because of the ease with which soft decisions can be combined with Viterbi decoding. He also states that convolutional codes show up in more practical applications than block codes because the AWGN channel occurs more often in practical communication situations. Obviously, we cannot give a clear answer here as to whether one should use a block code or a convolutional code. The choice depends on the particular application of interest.

SUMMARY

We have touched on a few of the most important concepts involved in forward error correction and error detection. Both block codes and convolutional codes play a fundamental role in many communication systems today and their analysis and design continues to be an area of ongoing research interest. Automatic repeat-request systems implement strategies that allow the powerful error detection capabilities of codes to be used to great advantage and new ARQ strategies will certainly be developed. Performance studies of block and convolutional codes are of great importance, and much research remains to be performed here. The proper integration of error control codes into practical communication systems has lagged the development of the codes and coding theory, with the result that many existing communication systems do not perform as well as they could. For the application of coding to spread, it is important that undergraduate engineers, as a minimum, grasp the material in this chapter.

PROBLEMS

12.1 With reference to Fig. 12.1.1, assume that the rate of the binary message sequence is $R_s = 20{,}000$ bits/sec. If the rate of the channel code is $R_c = \frac{2}{3}$, what is the rate transmitted over the discrete channel?

12.2 Which of the following generator matrices are systematic? What is the rate of the corresponding codes?

(a)
$$\mathbf{G}_1 = \begin{bmatrix} 1 & 0 & 0 & 0 & 1 & 1 & 0 \\ 0 & 1 & 0 & 0 & 0 & 1 & 1 \\ 0 & 0 & 1 & 0 & 1 & 1 & 1 \\ 0 & 0 & 0 & 1 & 1 & 0 & 1 \end{bmatrix}$$

(b)
$$\mathbf{G}_2 = \begin{bmatrix} 1 & 1 & 1 & 0 & 0 & 0 & 0 \\ 0 & 0 & 1 & 1 & 0 & 1 & 0 \\ 0 & 1 & 1 & 0 & 1 & 1 & 0 \\ 0 & 1 & 0 & 0 & 0 & 1 & 1 \end{bmatrix}$$

(c)
$$\mathbf{G}_3 = \begin{bmatrix} 1 & 0 & 0 & 1 & 1 \\ 0 & 1 & 0 & 0 & 1 \\ 0 & 0 & 1 & 0 & 1 \end{bmatrix}$$

12.3 Assume that the codes in Problem 12.2 are to be used over a memoryless channel and write each generator matrix in the form of Eq. (12.2.2). Construct the corresponding parity check matrices.

12.4 For each of the generator matrices in Problem 12.2, construct codeword tables analogous to Table 12.2.1. Compare the codes generated by $\mathbf{G}_1$ and $\mathbf{G}_2$ with the (7,4) Hamming code in Table 12.2.1. This may be easier to do symbolically using Eq. (12.2.3).

12.5 Find the generator matrix for the parity check matrix

$$\mathbf{H} = \begin{bmatrix} 1 & 1 & 0 & 1 & 0 & 0 \\ 0 & 1 & 1 & 0 & 1 & 0 \\ 1 & 1 & 1 & 0 & 0 & 1 \end{bmatrix}.$$

If the message $\mathbf{v} = (1\ \ 0\ \ 1)$ is to be transmitted, find the corresponding codeword and show that its syndrome is $\mathbf{0}$.

12.6 For the (7,4) Hamming code in Example 12.2.1, decode the received sequence $\mathbf{x} = (0\ 1\ 0\ 0\ 1\ 1\ 1)$.

12.7 Decoding by hand can be simplified and linear block decoding can be better understood by creating what is called the *standard array* for a code. For an (n,k) linear block code, the standard array is constructed as follows. The all-zero vector followed by the 2^k codewords are placed in a row. An error pattern is placed under the all-zero vector and a second row is generated by adding this error pattern to the codeword immediately above in the first row. To begin the third row, an error pattern that has not yet appeared anywhere in the table is selected and placed in the first column. This error pattern is then used to fill out the third row by adding this error pattern to the codeword at the top of each column. The process is continued until the table contains all $2^n - 2^k$ detectable error patterns below the codewords, and there will be 2^{n-k} rows corresponding to the correctable error patterns. The error patterns in the first column of each row should be chosen as the remaining vector of least Hamming weight. We now form the standard array for the code with generator matrix

$$\mathbf{G} = \begin{bmatrix} 1 & 1 & 1 & 0 & 0 \\ 0 & 0 & 1 & 1 & 0 \\ 1 & 1 & 1 & 1 & 1 \end{bmatrix}.$$

The possible codewords are listed in the first row of Table P12.7 preceded by the zero vector. A minimum weight error pattern, in this case, 0 0 0 0 1, is placed in the first position of the second row and added to the codewords at the top of each column to fill out the row. Note that 0 0 0 1 0 and 0 0 1 0 0 appear in the second row, and hence the minimum weight vector chosen to begin row 3 is 0 1 0 0 0. The process is continued until there are $2^{n-k} = 2^{5-3} = 4$ rows, as given in Table P12.7. The standard array contains all 24 detectable error patterns. To use this table for decoding, we need to involve the syndrome. It is interesting that all binary sequences in a row have the same syndrome (see Problem 12.10). Thus we need only calculate the syndrome for each word in the first column, which is called the *coset leader*. These are shown in Table P12.7. Decoding using the standard array therefore consists of the following steps: (1) The syndrome of the received vector is calculated; (2) the coset leader corresponding to the calculated syndrome is assumed to be the error pattern; (3) the error pattern is added to the received vector to get the decoded sequence. Given the received vector $\mathbf{x} = (1\ 0\ 1\ 1\ 1)$, use the standard array to decode this sequence.

TABLE P12.7 Standard Array for the (5,3) Code in Problem 12.7

Syndrome	Coset Leader							
00	00000	11100	11010	00110	11111	11001	00101	00011
01	00001	11101	11011	00111	11110	11000	00100	00010
10	01000	10100	10010	01110	10111	10001	01101	01011
11	10000	01100	01010	10110	01111	01001	10101	10011

12.8 Generate the standard array for the code in Problem 12.7 using 0 0 0 1 0 as a coset leader. Is this a good code? This illustrates why the coset leaders are called correctable error patterns.

12.9 Generate the standard array for the (7,4) Hamming code and use it to decode the received vector in Example 12.2.1.

12.10 Prove that all binary sequences in a row of the standard array have the same syndrome.

12.11 Let A_i, $i = 0, 1, 2, \ldots, n$, be the number of code vectors with Hamming weight i for a particular code. The A_i are called the *weight distribution* of the code and are very useful for probability of error calculations. Find the weight distribution of the (7,4) Hamming code in Table 12.2.1.

12.12 Find the weight distribution for the code with the generator matrix

$$\mathbf{G} = \begin{bmatrix} 0 & 1 & 1 & 1 & 0 & 0 \\ 1 & 0 & 1 & 0 & 1 & 0 \\ 1 & 1 & 0 & 0 & 0 & 1 \end{bmatrix}.$$

12.13 There are $2^k - 1$ undetectable error patterns for an (n,k) linear block code that occur when an error pattern is the same as a nonzero codeword. Write the probability of an undetected error, $P_{\text{ud}}(\mathscr{E})$, in terms of the weight distribution of the (7,4) Hamming code and the BSC bit error probability p. Evaluate $P_{\text{ud}}(\mathscr{E})$ for $p = 10^{-1}, 10^{-2}, 10^{-3}, 10^{-4}$, and 10^{-5}.

12.14 Evaluate Eq. (12.2.21) for $n = 7$, $t = 1$, and $p = 10^{-1}$, 10^{-2}, 10^{-3}, and 10^{-4}.

12.15 Use Eq. (12.3.4) and $g(D) = D^3 + D + 1$ to generate a (7,4) cyclic code. Compare to the Hamming code in Table 12.2.1.

12.16 **(a)** Write the polynomial corresponding to the code vector cyclically shifted by i places,

$$\mathbf{w}^{(i)} = (w_{n-i+1} \cdots w_n w_1 \cdots w_{n-i}).$$

(b) Form $D^i w(D)$ and show that $D^i w(D) = f(D)(D^n + 1) + w^{(i)}(D)$, where $w^{(i)}(D)$ is the polynomial in part (a). Specify $f(D)$.

12.17 For the (7,3) cyclic code with generator polynomial $g(D) = (1 + D) \cdot (1 + D + D^3)$, find the generator matrix and the corresponding parity check matrix in systematic form.

12.18 Repeat Problem 12.17 for the (15,11) cyclic code with generator polynomial $g(D) = 1 + D + D^4$.

12.19 Draw encoder and decoder circuits for the (7,3) cyclic code with generator polynomial $g(D) = (1 + D)(1 + D + D^3)$.

12.20 **(a)** Using the encoding circuit in Fig. 12.3.2, find the codeword for $\mathbf{v} = (1\ 1\ 0\ 1)$. Show the shift register contents for all shifts of the circuit.

(b) If the error pattern, $\mathbf{e} = (0\ 0\ 1\ 0\ 0\ 0\ 0)$, is added to the code vector of part (a), use Fig. 12.3.4 to find the syndrome and decode.

12.21 An error burst of i bits at the beginning of the codeword and an error burst of j bits at the end of a codeword is called an *end–around* burst of length $i + j$. Show that an (n,k) cyclic code can detect end–around bursts of length $n - k$ by showing that $s(D) \neq 0$.

12.22 Show that the fraction of undetectable bursts of length $n - k + 1$ is $2^{-(n-k-1)}$.

12.23 What can you say about the burst error-correcting capability of the cyclic (7,4) code generated by $g(D) = 1 + D + D^3$? The (15,11) code generated by $g(D) = 1 + D + D^4$? See also Problem 12.22.

12.24 A polynomial $x(D)$ of degree m is said to be irreducible if $x(D)$ cannot be divided by any polynomial of degree m or less ($m \neq 0$). Show that D^2, $D^2 + 1$, and $D^2 + D$ are not irreducible but $1 + D + D^2$ is irreducible.

12.25 A primitive polynomial $x(D)$ can be defined as an irreducible polynomial of degree m, where the smallest integer n for which $x(D)$ divides $D^n + 1$ is $n = 2^m - 1$. Show that $1 + D + D^3$ divides $1 + D^7$ and that $1 + D + D^4$ divides $1 + D^{15}$. To show that these polynomials are primitive would require checking $1 + D^n$ for all $1 \leq n \leq 2^m - 1$ in each case [Lin and Costello, 1983]. A binary cyclic code with block length $n = 2^m - 1$ is called a primitive cyclic code [Blahut, 1983].

12.26 If the input sequence to the (2,1) binary convolutional encoder in Fig. 12.4.1 is $\mathbf{v} = (1\ 1\ 0\ 0\ 1\ 0\ 1\ 0\ 0\ \cdots)$, find the encoder output sequence $\mathbf{w}$ from this figure.

12.27 Repeat Problem 12.26 using Eq. (12.4.5).

12.28 The rate-$\frac{1}{3}$ convolutional encoder in Fig. P12.28 is a modification of the rate-$\frac{1}{2}$ convolutional encoder in Fig. 12.4.1, obtained by feeding the input

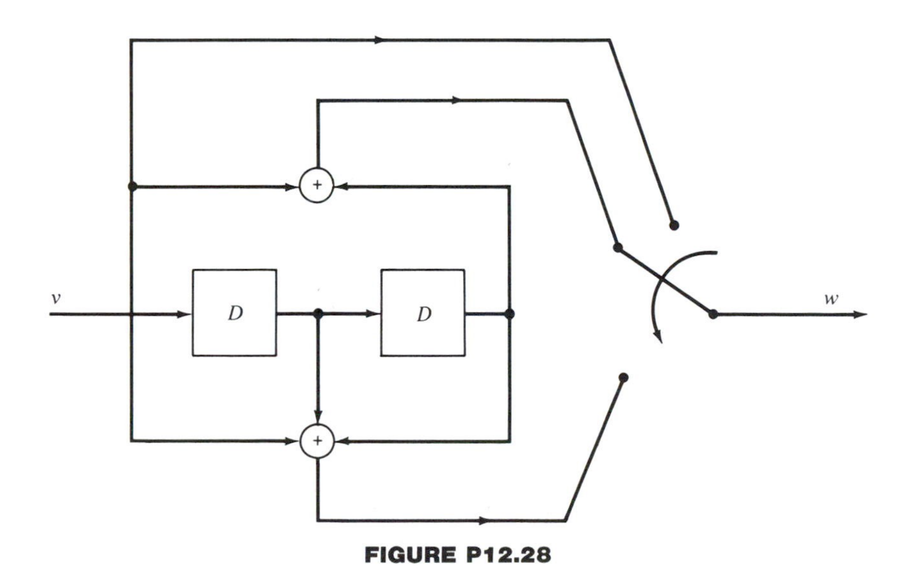

FIGURE P12.28

bit directly to the output. For the input sequence

$$\mathbf{v} = (1\ 0\ 1\ 1\ 0\ 1\ 0\ 0\ \cdots),$$

find the encoder output sequence $\mathbf{w}$. Sketch the alternative block diagram for this encoder analogous to that shown in Fig. 12.4.2.

12.29 Specify the generator sequences for the encoder in Fig. P12.28 and find the output sequence $\mathbf{w}$ for $\mathbf{v} = (1\ 0\ 1\ 1\ 0\ 1\ 0\ 0\ \cdots)$ using Eq. (12.4.3).

12.30 For the convolutional encoder in Fig. P12.28, write the generator matrix and find the output $\mathbf{w}$ for $\mathbf{v} = (1\ 0\ 1\ 1\ 0\ 1\ 0\ 0\ \cdots)$ using Eq. (12.4.5).

12.31 Specify the constraint length of the convolutional encoder in Fig. P12.28 and draw and label the code tree for this encoder to depth 5.

12.32 Starting with the code tree in Problem 12.31, draw the trellis for the convolutional encoder in Fig. P12.28.

12.33 The movement through a trellis as demonstrated in Fig. 12.4.4 is simply a set of transitions between a finite number of states, where all states may not be reachable from any other state. A convolutional encoder thus has an interpretation as what is called a *finite state machine.* A finite state machine can be represented by a *state diagram* that shows the states as circles and the possible state transitions. State diagrams play an important role in the analysis and understanding of convolutional encoders. The state diagram for the trellis in Fig. 12.4.4 is shown in Fig. P12.33, where the states are shown as appropriately labeled circles and the state transitions by directed arrows, with the message bit above the arrow and the encoder output bits in parentheses below the arrows. Convince yourself of the validity of this state diagram and use it to find the encoder output $\mathbf{w}$ for the input message sequence $\mathbf{v} = (1\ 0\ 1\ 1\ 0\ 1\ 0\ 0\ \cdots)$.

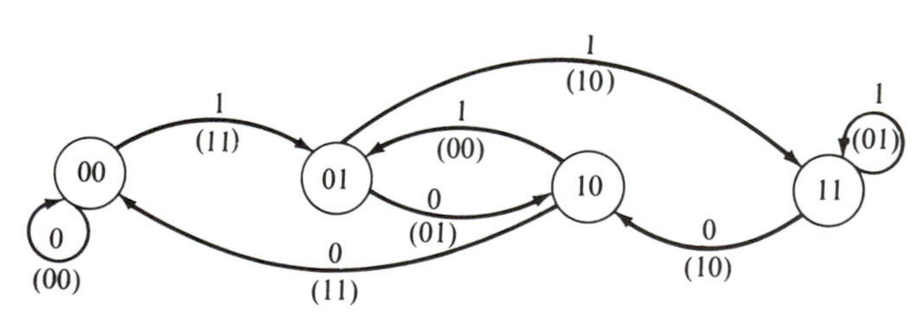

FIGURE P12.33

12.34 Draw the state diagram (see Problem 12.33) for the trellis in Problem 12.32.

12.35 Figure P12.35 shows the convolutional encoder for a rate-$\frac{2}{3}$ code. The sequence of message bits is now used two at a time to generate encoder

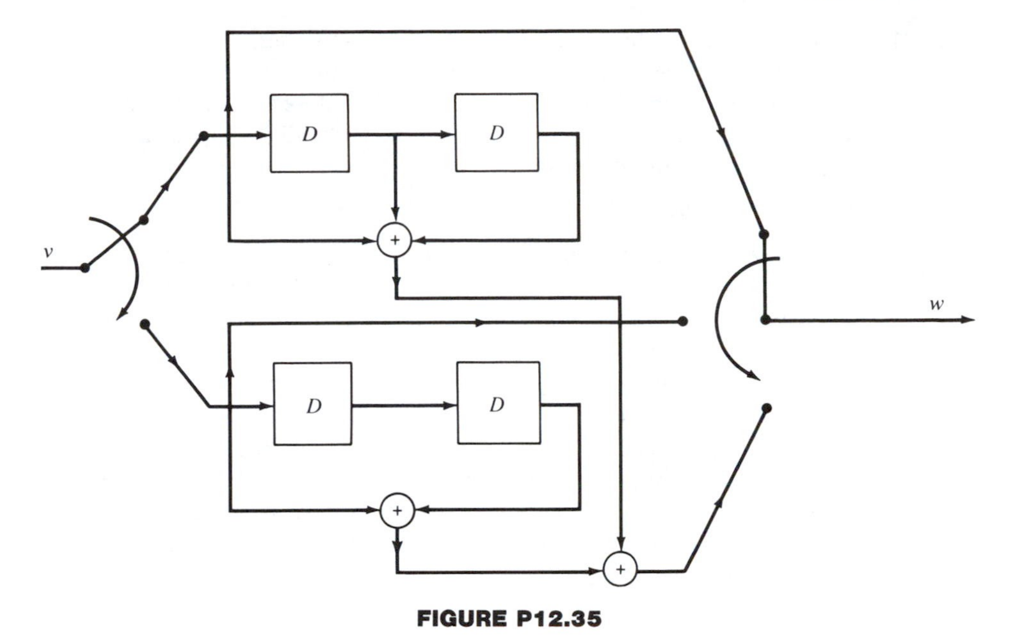

FIGURE P12.35

outputs by alternately applying them to the upper and lower inputs. For the input sequence $\mathbf{v} = (1\ 1\ 0\ 0\ 1\ 0\ 1\ 1\ 0\ 0\ 1\ 0\ 0\ 0\ \cdots)$, find the output sequence for this encoder. The constraint length for such a code is sometimes said to be $v + k$.

12.36 Draw the tree corresponding to the encoder in Fig. P12.35.

12.37 Use the results of Problem 12.36 to draw the trellis and state diagram (see Problem 12.33 for the definition) for the convolutional encoder in Fig. P12.35.

12.38 The rate-$\frac{1}{2}$ binary convolutional code in Fig. 12.4.1 with the trellis in Fig. 12.4.4 is used and the transmitted codeword is the all-zero sequence $\mathbf{w} = (0\ 0\ 0\ 0\ 0\ 0\ 0\ 0\ \cdots)$. The received vector is

$$\mathbf{x} = (1\ 1\ 0\ 0\ 0\ 0\ 0\ 0\ \cdots).$$

Use the Viterbi decoder to decode the received sequence.

12.39 Repeat Problem 12.38 if the received sequence is

$$\mathbf{x} = (1\ 1\ 1\ 0\ 0\ 0\ 0\ 0\ \cdots).$$

12.40 For the rate-$\frac{1}{3}$ convolutional encoder in Fig. P12.28, the all-zero codeword is transmitted and the received vector is

$$\mathbf{x} = (1\ 1\ 0\ 1\ 0\ 0\ 0\ 0\ 0\ \cdots).$$

Use Viterbi decoding to decode this received sequence.

12.41 In this chapter we have discussed only linear codes. Further, in our studies of Viterbi decoding, we have always assumed that the transmitted codeword is all zeros. It turns out that in all analyses of Viterbi decoder performance, we can assume without loss of generality that the all-zeros sequence is transmitted. Why?

Hint: See McEliece [1977].

12.42 Although exact calculation of the exact error probability for various codes as shown in Figs. 12.6.1 through 12.6.4 is difficult, in general, it is often possible to determine bounds on the error probability that are sufficient to evaluate the code performance. For example, consider the use of an (n,k) block code to represent one of M messages for transmission with BPSK modulation over an additive white Gaussian noise channel with a two-sided power spectral density of $\mathcal{N}_0/2$ watts/Hz. Let the energy per symbol be E_s so that the energy per information bit is $E_b = E_s/k$. If we define E_c to be the energy per codeword bit and $R_c = k/n$, then also $E_b = E_c/R_c$. The optimal coherent receiver observes the n demodulator outputs given by $r_j = \pm E_c + n_j$, $j = 1, 2, \ldots, n$, when n_j is the sampled additive white Gaussian noise, and forms the test statistic

$$Z_i = \sum_{j=1}^{n} (2c_{ij} - 1)r_j, \qquad i = 1, 2, \ldots, M,$$

where c_{ij} is the jth bit of the ith codeword.

(a) Assume that the all-zeros codeword was transmitted, with $\mathbf{c}_1 = [c_{11}\ c_{12} \cdots c_{1n}] = [0\ 0\ \cdots\ 0]$; find $E[Z_1|\mathbf{c}_1]$, $E[Z_i|\mathbf{c}_1]$, $i = 2, 3, \ldots, M$, and var $[Z_i|\mathbf{c}]$ for all i.

(b) Show that

$$P[Z_j > Z_1|\mathbf{c}_1] = \frac{1}{2} - \operatorname{erf}\left[\sqrt{\frac{E_s w_j}{2n\mathcal{N}_0}}\right]$$

for all $j \neq 1$, where w_j is the Hamming weight of the jth codeword.

(c) Since each of the Z_i's is equally likely, use the union bound in Appendix H to obtain

$$P_e \leq \sum_{j=2}^{M} \left[\frac{1}{2} - \operatorname{erf}\left(\frac{E_b}{2\mathcal{N}_0} w_j R_c\right)\right].$$

12.43 Note that the error probability in part (c) of Problem 12.42 requires knowledge of the code weight distribution.

(a) Show that another bound on error probability is given by

$$P_e \leq (M - 1)\left[\frac{1}{2} - \operatorname{erf}\left(\frac{E_b}{2\mathcal{N}_0} R_c d_{\min}\right)\right],$$

where $d_{\min}$ is the code minimum distance.

(b) A quantity called the *coding gain* is often defined, which is the difference in SNR required without and with coding to achieve a given error probability. For BPSK without coding,

$$P_e = \frac{1}{2} - \text{erf}\left[\sqrt{\frac{2E_b}{\mathcal{N}_0}}\right] < \frac{1}{2}e^{-2E_b/\mathcal{N}_0},$$

where the last inequality is a commonly used bound [Proakis, 1989]. Show that the result from part (a) with coding can be bounded as

$$P_e < \frac{M}{2}\, e^{(-E_b/2\mathcal{N}_0)R_c d_{\min}}.$$

(c) By comparing the two bounds in part (b), show that the coding gain for an (n,k) block code is given by

$$10 \log_{10}\left(\frac{R_c d_{\min}}{4} - \frac{k\mathcal{N}_0}{2E_b}\ln 2\right) \qquad \text{dB}.$$

CHAPTER 13

Coded Modulation

13.1 Introduction

For the digital communications systems discussed thus far in the book, we have assumed that coding (operations on the binary data stream) and modulation are independent in the sense that the modulator simply accepts the binary sequence presented to it and maps a set of one or more bits, in a one-to-one fashion, onto a waveform appropriate for transmission over the channel. This is the most widely accepted approach to digital communications today, and with reference to the signal space diagrams in Chapter 10, it implies that only two parameters can be adjusted to improve communication system performance: transmitted power and channel bandwidth. Specifically, for a fixed data rate, the symbol error probability can be reduced by increasing the transmitted power, or for a fixed error probability, the data rate can be increased if the available bandwidth is increased. A second implication of this separation of coding and modulation is that if channel coding (as described in Chapter 12) is used, a higher data rate results, which necessitates a wider bandwidth. Thus it seems that whatever we do to improve digital communication system performance either increases the transmitted power or the required bandwidth.

Fortunately, since transmitted power and channel bandwidth are precious commodities today, there is a way out of this dilemma. The basic concept involved is that of observing an entire sequence or block of data before making a decision at the receiver, rather than operating on a symbol-by-symbol basis as is done now. To be more exact, consider a sequence of symbols of length N. If each symbol is a point in a two-dimensional signal constellation as in Chapter 10, the sequence of symbols is a point in $2N$-dimensional space. The idea is thus to consider the Euclidean distance between *sequences* in this $2N$-dimensional space, and the signal design problem is to select the allowable *sequences* such that they are as far apart as possible. As long as the message bit rate, or bit rate to be transmitted, is small enough that some of the points in the two-dimensional signal constellation are unused, by letting N get large, the error probability can be reduced. Coding becomes intertwined with modulation, since channel codes are employed to specify the set of allowable transmitted sequences in the $2N$-dimensional space. This combined modulation and coding approach fits within the theory provided 20 years ago by Wozencraft and Jacobs [1965], and when convolutional codes (in contrast to block codes) are employed, the approach is called *trellis-coded modulation* (TCM).

Spectrally efficient constant envelope modulation techniques are related to trellis-coded modulation methods, although they are not the same, and there is some debate as to whether spectrally efficient constant envelope modulation falls within the theoretical framework described by Wozencraft and Jacobs [1965]. However, the combination of coding and modulation is evident in these techniques. The requirement for a constant envelope in the transmitted waveform follows from the fact that nonlinearities present in some communications channels, such as satellite channels, can distort as well as translate envelope variations into other types of distortion. Holding the envelope constant naturally implies that we are using either phase or frequency modulation. We know that FM and PM can require relatively large bandwidths, primarily because of rapid changes in transmitted phase. Spectrally efficient constant envelope methods only allow "smooth" changes in the phase, thus reducing bandwidth requirements, and accommodate an input message that requires a large total phase change by spreading this phase change out over several symbol intervals. Thus, to recover the total phase change corresponding to a particular message in a particular symbol interval, the receiver must examine a sequence of phases and compare it to all allowable sequences of phase changes. Combined coding and modulation is thus again evident in these schemes, although not as explicitly as in TCM.

The common denominator in both types of coded modulation, TCM and constant envelope methods, is the necessity of comparing received sequences to allowable transmitted sequences and finding the "nearest" one. Therefore, the Viterbi algorithm and related trellis search algorithms are an integral part of the receivers for such modulations.[1] Thus, although we are getting improved performance without increasing transmitted power or channel bandwidth, we are not getting something for nothing. The price we are paying is in terms of increased receiver complexity or increased signal processing. However, this is not a major problem in many communication systems today because of the success of microelectronics and VLSI, which allow very complicated devices to be constructed in very small packages with lower power consumption. Furthermore, these devices may be quite inexpensive, particularly if there is a large commercial market, as is possible in telecommunications.

Section 13.2 illustrates the idea that comparing sequences can provide a performance improvement over symbol-by-symbol decoding/demodulation by considering binary FSK, which is a constant envelope technique. Spectrally efficient constant envelope modulation is then developed in Section 13.3, followed by a treatment of trellis-coded modulation in Section 13.4. The summary in Section 13.5 tries to place these techniques in perspective and to whet the reader's appetite for a more detailed study of the ideas involved.

---

[1] As a result, the reader should probably be familiar with Section 10.7 and Chapter 12 before proceeding here.

13.2 Coherent Binary FSK

In this section we consider a particular example of constant envelope modulation due to de Buda [1972], which is included to demonstrate explicitly that observing more than one symbol interval at the receiver can improve communication system performance. This example is not crucial to the developments in subsequent sections but is provided for motivational purposes. Following de Buda [1972], we examine coherent binary frequency shift keying (FSK), and we represent the transmitted waveform at any time instant by its pre-envelope

$$s(t) = e^{j\{(\omega_1 + \omega_2)t/2 + \phi(t)\}}, \tag{13.2.1}$$

where ω_1 corresponds to a 1 being transmitted, ω_2 corresponds to a 0, and $\phi(t)$ is varied to yield the desired transmitted signal. We use the analytic signal representation because it is simple to manipulate. The actual transmitted waveform is the real part of $s(t)$, and the pre-envelope representation only requires the assumption that the signal is bandpass (see Problem 13.3 and Rice [1982]).

Letting $s_0(t)$ denote the pre-envelope when a 0 is transmitted, then we find that

$$s_0(t) = e^{j[\omega_2 t + \phi(0)]}, \qquad 0 \le t \le T, \tag{13.2.2}$$

which upon comparing Eqs. (13.2.1) and (13.2.2) implies that

$$\phi(t) - \phi(0) = \frac{\omega_2 - \omega_1}{2}\, t. \tag{13.2.3}$$

Thus, at time $t = T$,

$$\begin{aligned}\phi(T) - \phi(0) &= \frac{\omega_2 - \omega_1}{2}\, T \\ &= \pi(f_2 - f_1)T,\end{aligned} \tag{13.2.4}$$

so if $\omega_2 > \omega_1$, the phase increases by $\pi(f_2 - f_1)T$ in T seconds. Defining $s_1(t)$ as the pre-envelope when a 1 is transmitted, we obtain

$$s_1(t) = e^{j[\omega_1 t + \phi(0)]}, \qquad 0 \le t \le T, \tag{13.2.5}$$

and it is straightforward to show that the phase decreases by $\pi(f_2 - f_1)T$ when a 1 is sent. It is common to define a parameter h called the *modulation index* by

$$h = (f_2 - f_1)T. \tag{13.2.6}$$

If at time $t = (j + k)T$, j 0's and k 1's have been sent, the phase is given by

$$\phi[(j + k)T] - \phi[0] = (j - k)h\pi. \tag{13.2.7}$$

The possible phase differences $\phi(t) - \phi(0)$ can be sketched as a function of time as shown in Fig. 13.2.1.

It is of great interest to find the value of the modulation index h that minimizes the probability of error. If we limit consideration to the AWGN channel, we

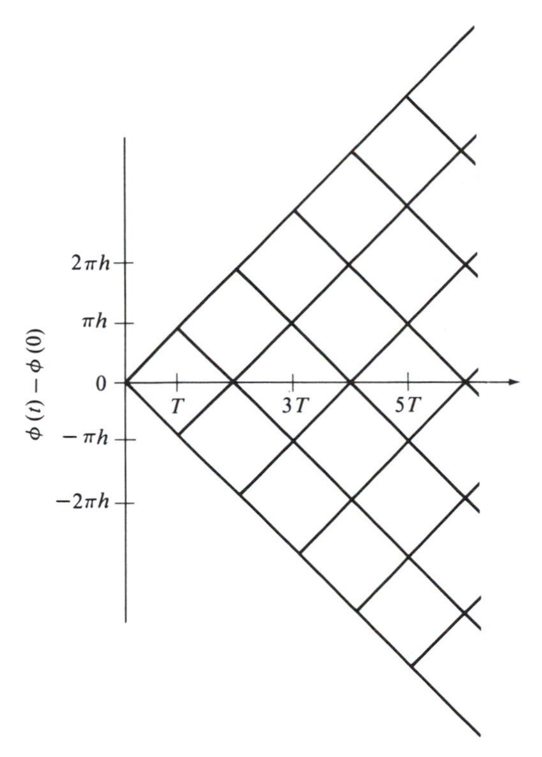

FIGURE 13.2.1 Possible phase differences $\phi(t) - \phi(0)$ as a function of time.

know from Chapter 10 that we need to select h to maximize the distance between $s_0(t)$ and $s_1(t)$. Thus we need to maximize

$$d^2(s_0, s_1) = \int_0^T |s_0(t) - s_1(t)|^2 \, dt \tag{13.2.8}$$

with respect to h. It can be shown by straightforward manipulations that

$$|s_0(t) - s_1(t)| = 2 \sin \frac{\pi h t}{T}, \tag{13.2.9}$$

so Eq. (13.2.8) yields

$$d^2(s_0, s_1) = 4 \int_0^T \sin^2 \frac{\pi h t}{T} \, dt = 2T\left[1 - \frac{\sin(2\pi h/T)}{(2\pi h/T)}\right], \tag{13.2.10}$$

the second term of which has a minimum of -0.217 when $\tan 2\pi h = 2\pi h$, or when $h = 0.714$. Therefore, h_{opt} seems to be 0.714.

The reason that we say "seems to be" in this last sentence is that in deriving this value for h, we have made an incorrect implicit assumption concerning the integration limits in Eq. (13.2.8). Although integrating over one symbol interval seems intuitive, we show in what follows that using a different value of h and integrating over two symbol intervals yields improved performance. We begin with Eq. (13.2.1) and write for $0 \leq t \leq T$,

$$\begin{aligned} s(t) &= e^{j\{(\omega_1+\omega_2)t/2+\phi(t)-\phi(0)+\phi(0)\}} \\ &= e^{j\{(\omega_1+\omega_2)t/2 \pm \pi h t/T + \phi(0)\}}, \end{aligned} \tag{13.2.11}$$

where we have substituted for $\phi(t) - \phi(0)$, with the $+$ sign for a 0 and the $-$ sign for a 1. For $-T \leq t \leq 0$, $s(t)$ is the same as in Eq. (13.2.11) except that the signs on the $\phi(t) - \phi(0)$ term are reversed. To recover the phase (message) sequence, we multiply $s(t)$ by $e^{-j(\omega_1+\omega_2)t/2}$ and take the real part, so that

$$\begin{aligned} r(t) &\triangleq \mathrm{Re}\left[s(t)e^{-j(\omega_1+\omega_2)t/2}\right] \\ &= \begin{cases} \cos\left[\pm\left(\dfrac{\pi h t}{T}\right) + \phi(0)\right], & 0 \leq t \leq T, \\ \cos\left[\mp\left(\dfrac{\pi h t}{T}\right) + \phi(0)\right], & -T \leq t \leq 0. \end{cases} \end{aligned} \tag{13.2.12}$$

Now, according to Eq. (13.2.7) with $h = 0.5$, the phase difference is either 0 or π at even multiples of T; hence, considering 0 an even multiple of T, we let $\phi(0) = 0$ or π. Equation (13.2.12) thus becomes

$$r(t) = \begin{cases} \cos\left[\pm\dfrac{\pi t}{2T}\right], & 0 \leq t \leq T, & \phi(0) = 0 \\ -\cos\left[\pm\dfrac{\pi t}{2T}\right], & 0 \leq t \leq T, & \phi(0) = \pi \\ \cos\left[\mp\dfrac{\pi t}{2T}\right], & -T \leq t \leq 0, & \phi(0) = 0 \\ -\cos\left[\mp\dfrac{\pi t}{2T}\right], & -T \leq t \leq 0, & \phi(0) = \pi, \end{cases}$$

which, since $\cos\left[\pm\pi t/2T\right] = \cos\left[\mp\pi t/2T\right]$, collapses to

$$r(t) = \begin{cases} \cos\dfrac{\pi t}{2T}, & \phi(0) = 0 \\ -\cos\dfrac{\pi t}{2T}, & \phi(0) = \pi \end{cases} \tag{13.2.13}$$

for $-T \leq t \leq T$. From this equation it is evident that we can decide what phase was transmitted at $t = 0$, either $\phi(0) = 0$ or $\phi(0) = \pi$, by observing $r(t)$ over $-T \leq t \leq T$. Thus $r(t)$ consists of binary antipodal signals over $-T \leq t \leq T$, and has an error probability in terms of signal-to-noise ratio (SNR) given by [see Eq. (10.5.9)]

$$P_{\mathrm{opt}}[\mathscr{E}] = \frac{1}{2} - \mathrm{erf}\left[\sqrt{\mathrm{SNR}}\right], \tag{13.2.14}$$

which is 3 dB better than coherently orthogonal (over $0 \leq t \leq T$) FSK, which has [see Eq. (10.5.10)]

$$P_{\mathrm{FSK}}[\mathscr{E}] = \frac{1}{2} - \mathrm{erf}\left[\sqrt{\frac{\mathrm{SNR}}{2}}\right]. \tag{13.2.15}$$

Observing more than one symbol interval thus provides improved system performance. The following section delves further into these constant envelope modulation methods.

13.3 Constant Envelope Modulation

We begin by considering a constant envelope, phase-varying sinusoid of the form

$$s(t) = \sqrt{\frac{2E}{T}} \cos\left[\omega_c t + \phi(t)\right], \tag{13.3.1}$$

where T is the symbol interval, E the symbol energy, ω_c the carrier frequency, and $\phi(t)$ the time-varying phase. We are presented with a sequence of underlying phase changes, corresponding to the message sequence to be transmitted, that is represented by $\{a_i, i = \ldots, -2, -1, 0, 1, 2, \ldots\}$, where each a_i is a member of the M-ary alphabet $\pm 2\pi h, \pm 3(2\pi h), \ldots, \pm(M-1)(2\pi h)$ for M even, or $0, \pm 2(2\pi h), \ldots, \pm(M-1)(2\pi h)$ for M odd. The constant parameter h is called the *modulation index*, and the phase $\phi(t)$ in Eq. (13.3.1) changes in accordance with the data sequence $\{a_i\}$ through the relationship

$$\phi(t) = \sum_{i=-\infty}^{\infty} a_i f(t - iT), \qquad -\infty < t < \infty, \tag{13.3.2}$$

where $f(t)$ is a phase smoothing function to be chosen. If $\phi(t)$ is a continuous function of time, we obtain a spectrally efficient form of constant envelope modulation called *continuous phase modulation* (CPM). The function $f(t)$ is called a *phase pulse* and is sometimes defined in terms of a frequency pulse $g(t)$ as

$$f(t) = \int_{-\infty}^{t} g(\tau)\, d\tau, \qquad -\infty < t < \infty. \tag{13.3.3}$$

Let a_n be the phase change associated with the interval $[(n-1)T, nT]$, where the change occurs just after time $(n-1)T$. The smoothing function $f(t)$ causes this change to be spread over the next L symbol intervals. The phase pulse $f(t)$ can be any function that is zero before some (possibly negative) value of t and is $\frac{1}{2}$ for some sufficiently large value of t. Thus, if we consider two possible amplitudes of the nth phase change, denoted by $a_n(1)$ and $a_n(2)$ such that $a_n(1) - a_n(2) = 2k(2\pi h)$, the carrier phases corresponding to $a_n(1)$ and $a_n(2)$ will eventually differ by $k(2\pi h)$ radians if all other phase changes a_i, $i \neq n$, in the sequences are the same. In terms of the frequency pulse $g(t)$, we have a *causal* CPM system if

$$\begin{aligned} g(t) &\equiv 0, && \text{for } t < 0 \text{ and } t > LT \\ g(t) &\not\equiv 0, && \text{for } 0 \leq t \leq LT. \end{aligned} \tag{13.3.4}$$

Figure 13.3.1 shows four possible phase smoothing functions $f(t)$ and their corresponding instantaneous frequency pulses $g(t)$ [Anderson et al., 1981]. The $f(t)$ in Fig. 13.3.1(a) represents standard PSK, where there is a step change in the phase. The $f(t)$ in (b) causes the phase to change linearly over one interval,

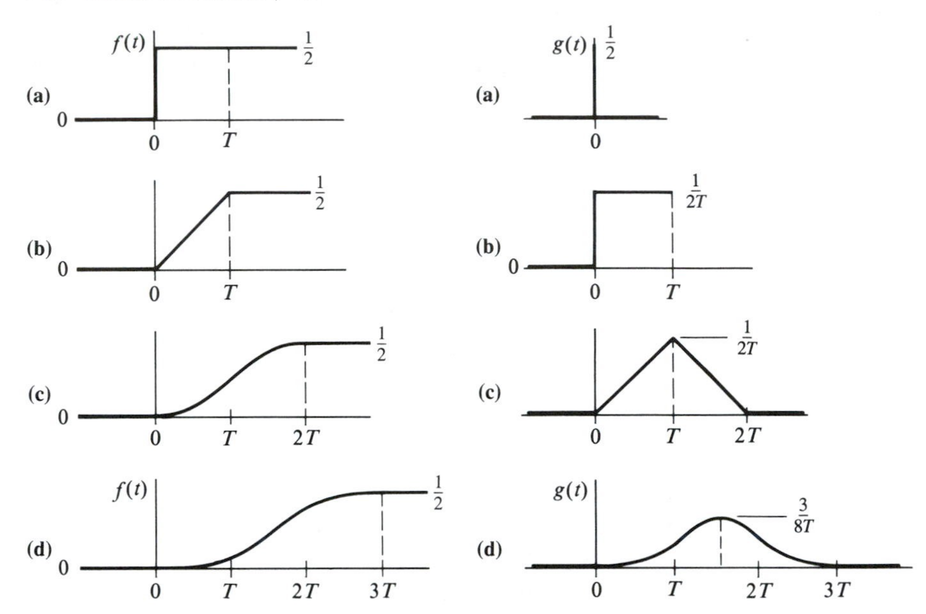

FIGURE 13.3.1 Four possible phase smoothing functions and their corresponding frequency pulses. From J. B. Anderson, C.-E. W. Sundberg, T. Aulin, and N. Rydbeck, "Power-Bandwidth Performance of Smoothed Phase Modulation Codes," *IEEE Trans. Commun.*, © 1981 IEEE.

which, as is evident from $g(t)$, is the same as standard FSK. Upon letting $\alpha = t/T$, the third phase smoothing function can be written as

$$f(\alpha) = \begin{cases} 0, & \alpha < 0 \\ \dfrac{\alpha^2}{4}, & 0 \le \alpha \le 1 \\ -\dfrac{\alpha^2}{4} + \alpha - \dfrac{1}{2}, & 1 \le \alpha \le 2 \\ \dfrac{1}{2}, & \alpha \ge 2 \end{cases} \tag{13.3.5}$$

and the function in (d) is given by

$$f(\alpha) = \begin{cases} 0, & \alpha < 0 \\ \dfrac{\alpha^3}{12}, & 0 \le \alpha \le 1 \\ -\dfrac{\alpha^3}{6} + \dfrac{3\alpha^2}{4} - \dfrac{3\alpha}{4} + \dfrac{1}{4}, & 1 \le \alpha \le 2 \\ \dfrac{\alpha^3}{12} - \dfrac{3\alpha^2}{4} + \dfrac{9\alpha}{4} - \dfrac{7}{4}, & 2 \le \alpha \le 3 \\ \dfrac{1}{2}, & \alpha \ge 3. \end{cases} \tag{13.3.6}$$

Note that all of the smoothing functions shown are causal, and that $L = 0$, 1, 2, and 3 for (a), (b), (c), and (d), respectively. Another popular choice for the smoothing function is the raised cosine class given by

$$f(\alpha) = \begin{cases} 0, & \alpha < 0 \\ \dfrac{\alpha}{2L} - \dfrac{\sin(2\pi\alpha/L)}{4\pi}, & 0 \le \alpha \le L \\ \dfrac{1}{2}, & \alpha \ge L, \end{cases} \tag{13.3.7}$$

where L is the number of intervals over which the phase is smoothed. In conjunction with CPM, $L = 1$ is called *full response* signaling and $L > 1$ is called *partial response* signaling. This nomenclature simply indicates that the phase change is ($L > 1$, partial response) or is not ($L = 1$, full response) spread over more than one symbol interval, and is not necessarily the same as the baseband spectral shaping discussed in Chapter 7.

The most studied case of continuous-phase, constant envelope modulation is the scheme with $L = 1$, which is variously known by the label of *fast FSK* (FFSK), *minimum shift keying* (MSK), or as a special case of continuous phase FSK (CPFSK). We investigate this modulation method in the following example.

EXAMPLE 13.3.1

We are interested in studying MSK, which has $L = 1$ and $f(t)$ as in Fig. 13.3.1(b) expressible by

$$f(t) = \begin{cases} 0, & t \le 0 \\ \dfrac{t}{2T}, & 0 \le t \le T \\ \dfrac{1}{2}, & t \ge T. \end{cases} \tag{13.3.8}$$

With $h = \frac{1}{2}$ and $M = 2$, the input phase changes are $\pm\pi$, so that since $f(t) = \frac{1}{2}$ for $t \ge T$, the total phase change per input symbol is $\pm\pi/2$. From $g(t)$ in Fig. 13.3.1(b), we see that what we are really doing is switching back and forth according to the input data sequence between two frequencies spaced $1/2T$ hertz apart. Since this spacing is one-half the spacing required for noncoherent orthogonality and used in most noncoherent FSK systems, MSK is often labeled fast FSK or FFSK. Furthermore, $1/2T$ is the minimum frequency spacing allowable for two FSK signals to be coherently orthogonal, which explains the title MSK.

If we assume that $\phi(0) = 0$, all possible phase trajectories starting at time $t = 0$ can be displayed by the tree diagram shown in Fig. 13.3.2. Suppose that the input binary message sequence consists of a series of $+1$s or -1s and that we associate $+1$ with $a_i = +\pi$ and -1 with $a_i = -\pi$. Then for any given binary input message sequence, we can trace through the tree diagram in Fig. 13.3.2 and write the transmitted phase at any time instant. Thus,

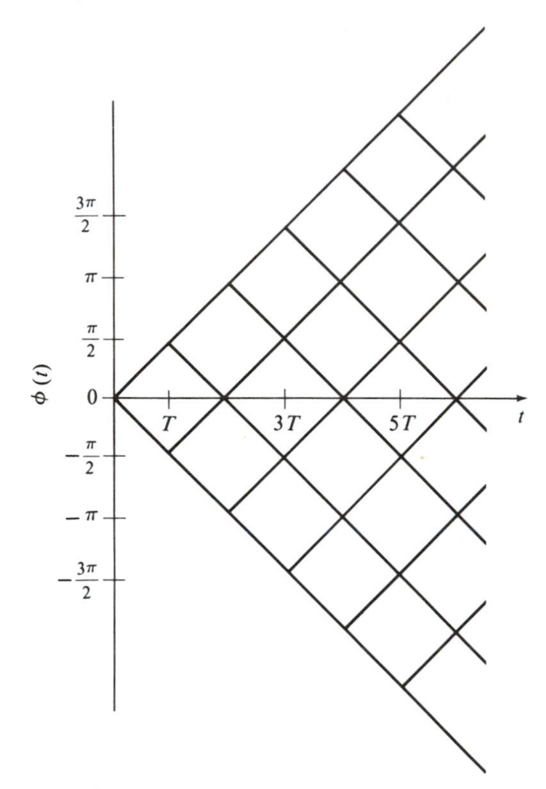

FIGURE 13.3.2 MSK phase diagram.

for the input binary sequence $-1, -1, +1, -1, -1, +1, +1, +1$, the corresponding transmitted phases at time instants nT, $n \geq 1$, are $-\pi/2$, $-\pi$, $-\pi/2$, $-\pi$, $-3\pi/2$, $-\pi$, $-\pi/2$, and 0.

Recovery of the input message sequence requires that the received phases be compared to all possible transmitted phase sequences using a tree search procedure, such as the Viterbi algorithm described in Section 10.7 and Chapter 12, and the "closest" path to the received sequence decoded as the input message sequence. Thus, although classical channel encoders are not used here, certainly, familiar decoders are being employed, and the boundary between coding and modulation is blurred. The complexity of the receiver over classical noncoherent FSK has clearly increased, but improved spectral efficiency is evident from Fig. 13.3.3, which compares the power spectral densities of binary PSK (BPSK), quadrature PSK (QPSK), and MSK. MSK has a narrower main lobe and much lower sidelobes than BPSK, and in comparison to QPSK, MSK has lower sidelobes but a wider main lobe. Thus, in terms of spectral efficiency, there is a trade-off between QPSK and MSK that depends on the particular application.

It is interesting to note that MSK can also be expressed and implemented as a shaped pulse version of offset or staggered QPSK. This connection is pursued in the problems.

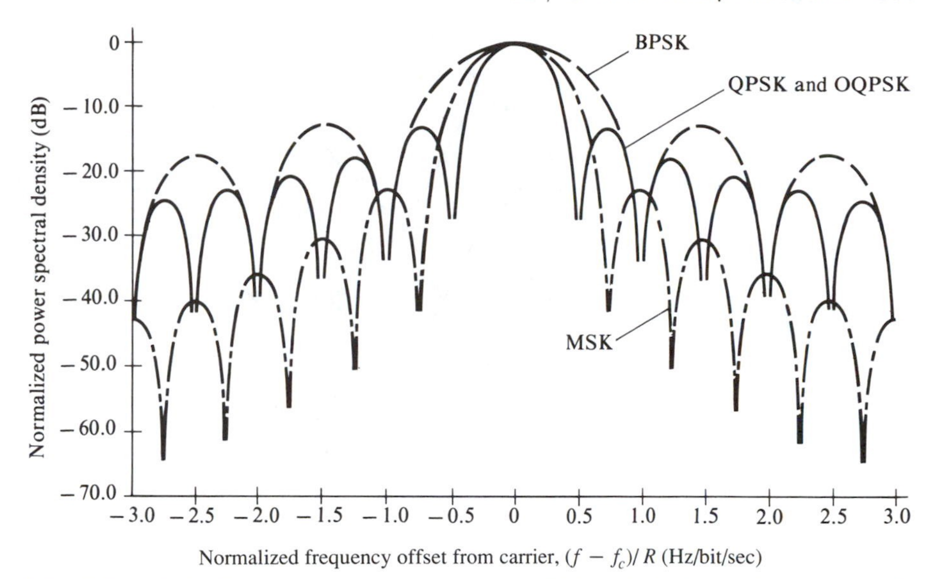

FIGURE 13.3.3 Power spectral densities of BPSK, QPSK, and MSK. From F. Amoroso, "The Bandwidth of Digital Data Signals," *IEEE Commun. Mag.*, © 1980 IEEE.

Another variation on spectrally efficient constant envelope modulation is what is called *multi-h CPM*. The transmitted waveform for a multi-h CPM system has the same form as $s(t)$ in Eq. (13.3.1), but now $\phi(t)$ is given by

$$\phi(t) = \sum_{i=-\infty}^{\infty} a_i h_i f(t - iT), \qquad -\infty < t < \infty, \tag{13.3.9}$$

where the a_i and $f(t)$ are as before and the set $\{h_i, i = 1, 2, \ldots, K\}$ represents ordered modulation indices that are cyclically used proceeding as $h_1, h_2, \ldots, h_K, h_1, h_2, \ldots$. Just as in the constant-h case, any phase change is spread over L symbol intervals depending on $f(t)$, but h_i is changing from symbol to symbol. The h_i are chosen to be rational numbers. The multi-h CPM schemes produce phase trees just like the constant-h case. A phase tree for an $\{h_1, h_2\} = \{\frac{1}{4}, \frac{2}{4}\}$ CPM system with $L = 1$ [see Fig. 13.3.1(b)] is shown in Fig. 13.3.4(a). Since phase is a modulo-2π quantity, the tree in part (a) can be drawn as a trellis as shown in Fig. 13.3.4(b), where the dashed lines represent "wrap-around" state transitions [Mazur and Taylor, 1981]. Note that for a binary input sequence, the total transmitted phase changes by $\pm\pi/4$ radians when $h_1 = \frac{1}{4}$ is used and by $\pm\pi/2$ radians when $h_2 = \frac{2}{4}$ is used. The binary sequence $+1$, -1, -1, -1, $+1$, $+1$, -1, starting at $t = 0$, would thus yield transmitted phases at time nT of $\pi/4$, $-\pi/4$, $-\pi/2$, $-\pi$, $-3\pi/4$, $-\pi/4$, and $-\pi/2$.

Decoding of multi-h codes still consists of comparing the received phase sequence to all possible paths in the phase tree and selecting that possible path which is closest to the received sequence. However, calculating distances can be more difficult, since the transmitted signals need not be orthogonal, as they

(a)

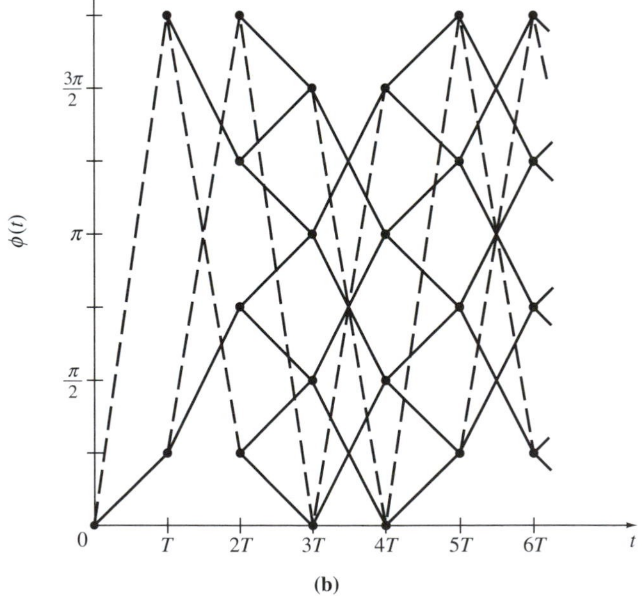

(b)

FIGURE 13.3.4 (a) Tree and (b) trellis structures for a multi-h $\{\frac{1}{4}, \frac{2}{4}\}$ $L = 1$ CPM system. From B. A. Mazur, and D. P. Taylor, "Demodulation and Carrier Synchronization of Multi-h Phase Codes," *IEEE Trans. Commun.*, © 1981 IEEE.

are, for instance, for MSK. This topic is beyond our intended development and is not pursued further here. The choice of sets of modulation indices must be done carefully and different choices can yield quite different performances and spectra. Reduced spectral requirements for multi-h codes compared to MSK have been reported in the literature [Anderson and Lesh, 1981; Ziemer and Peterson, 1985]. Research is continuing on these CPM or spectrally efficient constant envelope methods, and the reader should consult the references and additional literature for more details.

13.4 Trellis-Coded Modulation

In this section we do not restrict ourselves to constant envelope modulations, and the use of coding as described in Chapter 12 will be seen to be more explicit. As noted in the introduction to this chapter and as in Sections 13.2 and 13.3, the basic operational change over more familiar modulation methods is that the receiver bases its decision on an entire sequence of symbols rather than on a symbol-by-symbol calculation. The concept that allows such an operational technique to provide a performance improvement is described in Wozencraft and Jacobs [1965] and is explained geometrically as follows. Let each transmitted symbol be a point in a two-dimensional signal constellation, and let us assume that some of the points in the two-dimensional constellation are unused. To be more specific, let the number of available signal points in the two-dimensional constellation be A, but only $B < A$ of these points are actually used for transmitted signals. The ratio $B/A < 1$ remains fixed if the number of message bits/symbol is held fixed. Now we consider a sequence of two of these symbols. There will be A^2 available signal points, but only B^2 need to be used. If we consider a sequence of N symbols, these numbers become A^N and B^N, and the ratio of used signal points to unused signal points is $(B/A)^N$. Since $B/A < 1$, $(B/A)^N \to 0$ as $N \to \infty$; that is, the fraction of the possible transmitted symbols actually used becomes smaller as the length of the sequence increases. This implies that the distance between transmitted signals can be made larger with increasing N, and hence that the error probability can be reduced by increasing N. Not all choices of the B^N out of the A^N points are good ones (far apart), and coding combined with what is called "mapping by set partitioning" helps us select a good set of B^N points. Either block codes or convolutional codes can be employed; however, our development emphasizes convolutional codes, since they have received the most attention and seem to be better for coded modulation implementations. Note that channel coding alone cannot provide the available gain, and the mapping of the channel encoder output bits into symbols via the mapping by set partitioning procedure is critical to the success of the method.

To be more specific, we assume that we wish to transmit m bits/symbol, thus requiring a signal constellation of at least 2^m signal points. For coded modulation this signal set is expanded to 2^{m+1} (other expansions greater than 2^m will work) signals, so that in terms of our previous discussion, $B = 2^m$ and $A = 2^{m+1}$, yielding $B/A = \frac{1}{2}$. We now consider N symbol sequences and select the 2^{mN} best transmitted sequences out of the $2^{(m+1)N}$ possible transmitted sequences. In this manner we are able to keep the bit rate at m bits/symbol and still spread out the points in $2N$-dimensional space.

To expand the signal set from 2^m to 2^{m+1}, we use a rate $m/(m+1)$ convolutional code with constraint length $m+1$. Note that a block code will also do, but we stick with convolutional codes, for the reasons noted previously. The mapping by set partitioning procedure is then used to map these bits into channel signals. Since convolutional codes can be represented by a trellis structure, the rules for this mapping can be stated as follows [Ungerboeck, 1982]:

1. All parallel transitions in the trellis structure are allocated the maximum possible Euclidean distance in the signal constellation;
2. All transitions diverging from or merging with a trellis state receive the next maximum possible Euclidean distance.

These rules guarantee that all single and multiple errors exceed the Euclidean distance of the uncoded m-bit/symbol constellation. Although this all seems mysterious at this point, the following example will help clarify the procedure.

EXAMPLE 13.4.1

We consider the transmission of 2 bits/symbol using a rate-$\frac{2}{3}$, constraint length 3 convolutional code and eight-phase PSK from Ungerboeck [1982]. The partitioning of the eight-phase PSK signal constellation into subsets with increasing Euclidean distance is shown in Fig. 13.4.1. The trellis structure of the chosen convolutional code is given in Fig. 13.4.2, where the digits

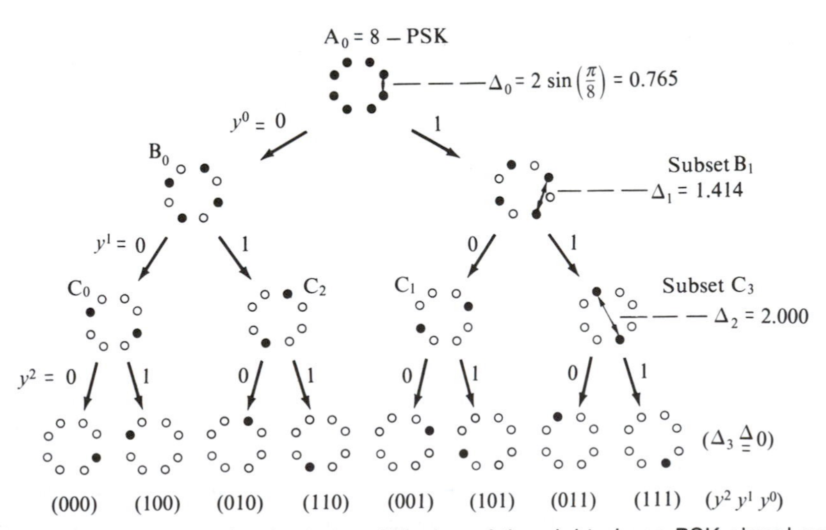

FIGURE 13.4.1 Mapping by set partitioning of the eight-phase PSK signal set. From G. Ungerboeck, "Channel Coding with Multilevel/Phase Signals," *IEEE Trans. Inf. Theory*, © 1982 IEEE.

State ($s_n^3 s_n^2 s_n^1$)	Subset transmitted
0 0 0	0 4 2 6
0 1 0	1 5 3 7
1 0 0	4 0 6 2
1 1 0	5 1 7 3
0 0 1	2 6 0 4
0 1 1	3 7 1 5
1 0 1	6 2 4 0
1 1 1	7 3 5 1

$d_{\text{free}} = \sqrt{d_1^2 + d_0^2 + d_1^2}$
(3.6-dB gain over 4-PSK)

FIGURE 13.4.2 Trellis diagram of the rate-$\frac{2}{3}$ constraint length 3 convolutional code indicating allowable phase transitions for 2 bits/symbol. From G. Ungerboeck, "Channel Coding with Multilevel/Phase Signals," *IEEE Trans. Inf. Theory*, © 1982 IEEE.

in the column "subset transmitted" indicate the transitions out of each node in top-to-bottom order. The parameter d_{free} is called the *maximum free distance* and is an indicator of the performance of the coded modulation scheme. The larger d_{free}, the better the performance. The convolutional encoder that generates the trellis in Fig. 13.4.2 is shown in Fig. 13.4.3. The subscript n denotes the time instant, x_n^1 and x_n^2 are the input bits, y_n^0, y_n^1, and y_n^2 are the output bits, s_n^1, s_n^2, and s_n^3 are bits representing the state, and a_n is the transmitted symbol after mapping by set partitioning. For a review of convolutional encoding, see Chapter 12. Note that the particular diagram for this convolutional encoder as shown in Fig. 13.4.3 is called a minimal realization, since it is simpler than other realizations.

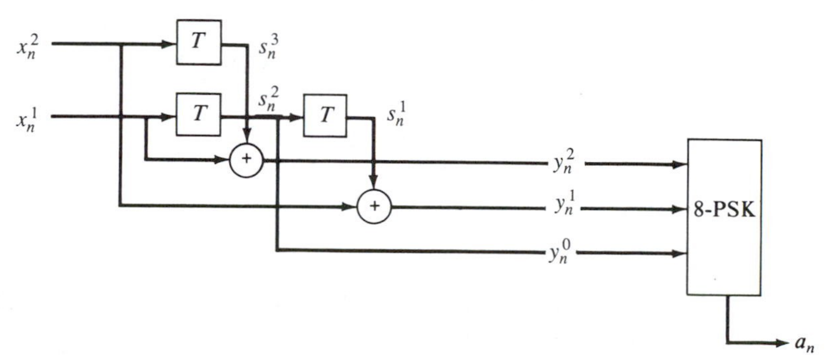

FIGURE 13.4.3 Minimal realization of the convolutional encoder for the trellis in Fig. 13.4.2. From G. Ungerboeck, "Channel Coding with Multilevel/Phase Signals," *IEEE Trans. Inf. Theory*, © 1982 IEEE.

TABLE 13.4.1 Coding and Modulation Process for Example 13.4.1

Time (n)	Input ($x_n^2 x_n^1$)	State ($s_n^3 s_n^2 s_n^1$)	Coder Output ($y_n^2 y_n^1 {}^2_n$)	Next State ($s_{n+1}^3 s_{n+1}^2 s_{n+1}^1$)	Transition Path and Transmitted Signal (i)
1	11	000	110	110	6
2	00	110	101	001	5
3	01	001	110	010	6
4	01	010	101	011	5
5	10	011	001	101	1
6	11	101	000	110	0

Using Figs. 13.4.2 and 13.4.3, we can find the encoder output and the transmitted symbol sequences for a given binary input sequence. For example, let the initial encoder state be $(s_0^3\ s_0^2\ s_0^1) = (0\ \ 0\ \ 0)$ and assume that the input sequence is 1 1 0 0 0 1 0 1 1 0 1 1. Table 13.4.1 shows the input, current state, coder output, path transition, transmitted symbol, and next state as a function of time n. By comparing the numbers in the transmitted signal column in Table 13.4.1 with the numbers in Fig. 13.4.1, the specific phases being transmitted can be determined.

To calculate the distance between two paths, we use the distances (d_i) shown in Fig. 13.4.1 and take the square root of the sum of the squared distances between each transmitted signal in the two paths. For instance, we calculate the distance between the two labeled paths in Fig. 13.4.2, namely (0, 0, 0) and (6, 7, 6), by first finding the distances between signals 0 and 6, 0 and 7, and 0 and 6, which are d_1, d_0, and d_1, respectively. Then the total distance between the two paths (or sequences) is $\sqrt{d_1^2 + d_0^2 + d_1^2} = 2.141$. Similarly, it can be found that the distance between the path sequence (4, 1, 2) and (6, 5, 6) is $\sqrt{d_1^2 + d_0^2 + d_2^2} = 2.566$.

The quantity called *free distance* and denoted by d_{free}, is important, since the probability of an error event (deciding on a path diverging from the true path) for optimal soft decision decoding and an AWGN channel is lower bounded at high signal-to-noise ratios by

$$P[\text{error event}] \geq K(d_{\text{free}})\left\{\frac{1}{2} - \text{erf}\left[\frac{d_{\text{free}}}{2\sigma}\right]\right\}, \tag{13.4.1}$$

where $K(d_{\text{free}})$ is the average number of error events at distance d_{free}. The calculation of d_{free} generally requires an exhaustive search over all possible error events, which we do not describe here. The comparison of two modulation/coding schemes involves the calculation of what is called the *coding gain*, defined as

$$\text{coding gain} \triangleq 10 \log_{10} \frac{(d_{\text{free}}^2/P_{\text{av}})_{\text{coded}}}{(d_{\text{min}}^2/P_{\text{av}})_{\text{uncoded}}}, \tag{13.4.2}$$

where $d_{\min}$ is the minimum distance between transmitted signal points in the uncoded constellation and P_{av} is the average power for each signal set. If we fix P_{av} to be the same for both coded and uncoded signal sets, as is usually done, we can set $P_{\text{av}} = 1$, so Eq. (13.4.2) becomes

$$\text{coding gain} \triangleq 10 \log_{10} \frac{(d^2_{\text{free}})_{\text{coded}}}{(d^2_{\min})_{\text{uncoded}}}. \tag{13.4.3}$$

Thus we can compute the coding gain for the coded modulation method in Example 13.4.1 with respect to four-phase PSK by noting that for four-phase PSK $d_{\min} = \sqrt{2}$, and since we know from Fig. 13.4.2 and our calculations that $d_{\text{free}} = 2.141$, then

$$\text{coding gain} = 10 \log_{10} \frac{(2.141)^2}{(\sqrt{2})^2} = 3.6 \text{ dB}, \tag{13.4.4}$$

as specified in Fig. 13.4.2. The maximum practical improvement available is about 6 dB, although we cannot delve into this topic in this book.

EXAMPLE 13.4.2

We consider the transmission of 3 bits/symbol using a 16-point QAM signal constellation with coding [Ungerboeck, 1982]. The partitioning of the 16-point QAM constellation into subsets with increasing Euclidean distance is demonstrated in Fig. 13.4.4. The convolutional encoder chosen and its

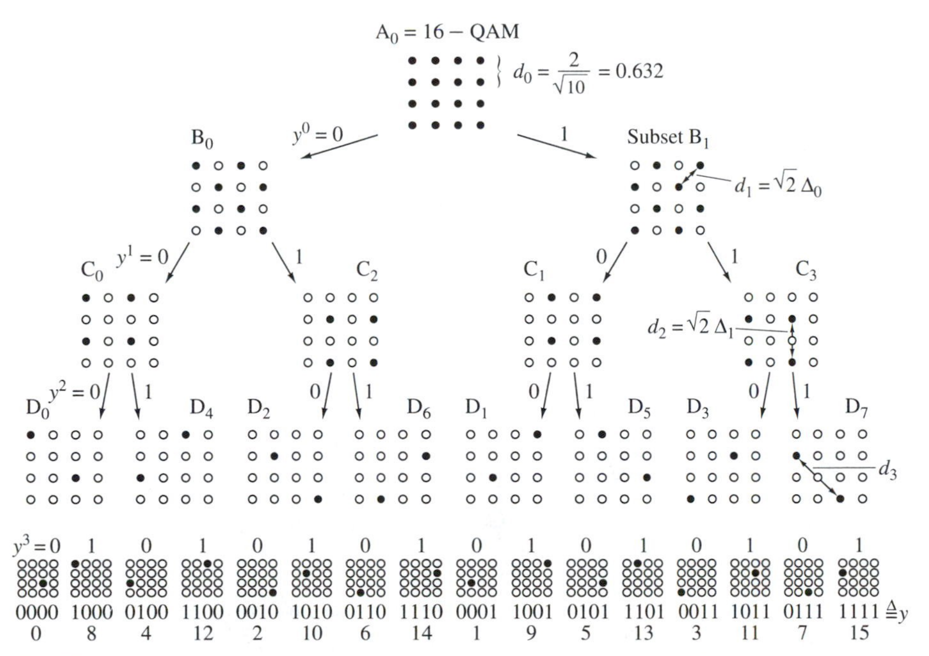

FIGURE 13.4.4 Mapping by set partitioning of the 16-point QAM constellation. From G. Ungerboeck, "Channel Coding with Multilevel/Phase Signals," *IEEE Trans. Inf. Theory*, © 1982 IEEE.

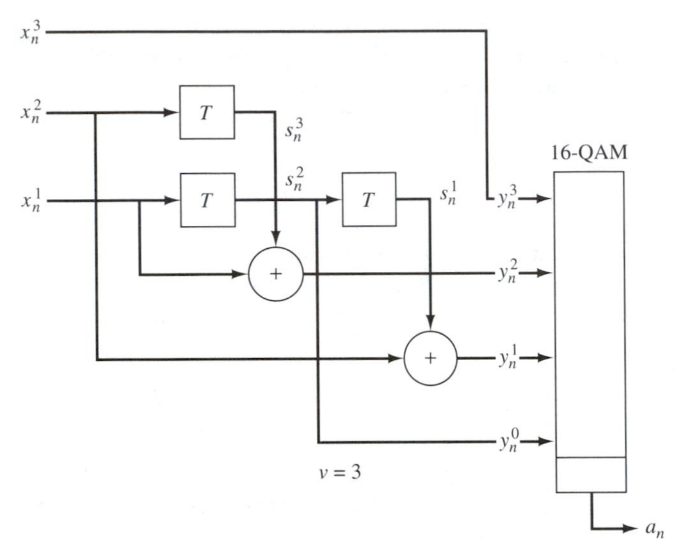

FIGURE 13.4.5 Rate-$\frac{3}{4}$ convolutional encoder for Example 13.4.2. From G. Ungerboeck, "Channel Coding with Multilevel/Phase Signals," *IEEE Trans. Inf. Theory*, © 1982 IEEE.

associated trellis are shown in Figs. 13.4.5 and 13.4.6, respectively. Note that the rate-$\frac{3}{4}$ convolutional code in Fig. 13.4.5 is just the rate-$\frac{2}{3}$ convolutional code used in Example 13.4.1 with an additional uncoded bit. The coded bits select the specific subset and the uncoded bit chooses the particular signal within that subset. Note also that the trellis diagram in Fig. 13.4.6 has double lines drawn between states. These double lines are called *parallel transitions* and are due to the uncoded bit being either 0 or 1.

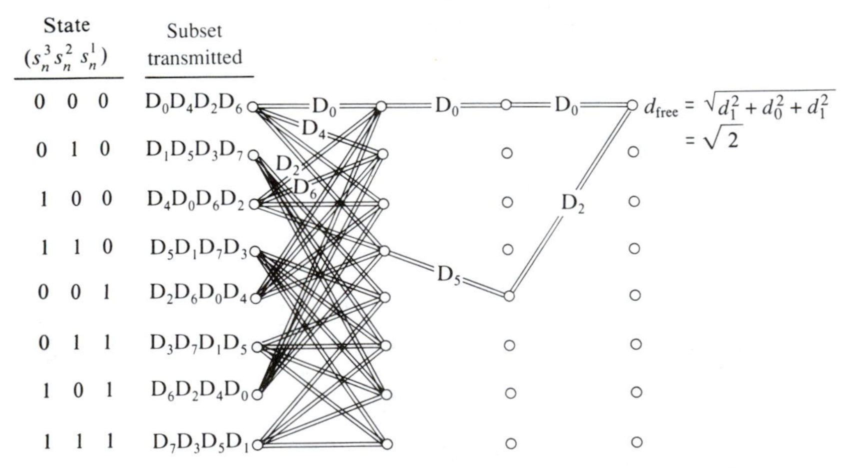

FIGURE 13.4.6 Trellis diagram for convolutional code indicating allowable phase transitions for 3 bits/symbol in Example 13.4.2. From G. Ungerboeck, "Channel Coding with Multilevel/Phase Signals," *IEEE Trans. Inf. Theory*, © 1982 IEEE.

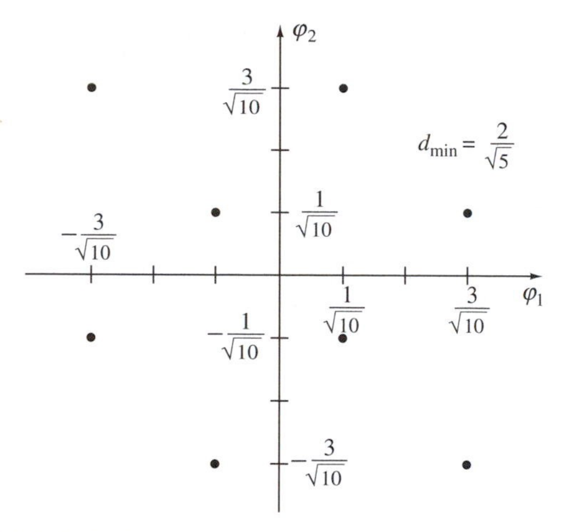

FIGURE 13.4.7 Eight-point AM/PM constellation with $P_{av} = 1$.

For uncoded eight-phase PSK, $d_{\min} = 0.765$, so the coding gain of the coded 16-point QAM scheme (with $d_{\text{free}} = \sqrt{2}$) over eight-phase PSK is 5.3 dB. If we compare the coded QAM method with the eight-point AM/PM scheme shown in Fig. 13.4.7, we find that

$$\begin{aligned} \text{coding gain} &= 10 \log_{10} \frac{2}{4/5} \\ &= 3.98 \text{ dB}. \end{aligned} \tag{13.4.5}$$

The problems investigate additional aspects of this example.

We have given heuristic arguments for how trellis-coded modulation provides a performance improvement, we have described two TCM systems in the examples and have demonstrated their operation, and we have introduced the concept of coding gain. It is beyond the scope of this book to develop the techniques for designing TCM systems and to provide the asymptotic theory that bounds TCM system performance. However, with the material in this section and in Chapter 12, the reader should be equipped to understand existing TCM systems and to begin to read the literature on the subject.

SUMMARY

We have described in this chapter some of the earliest and, at present, most successful systems that combine the operations of coding and modulation. These topics are the subject of much current research and development, and the coded modulation techniques are already available in off-the-shelf hardware (see Section 8.9). Coded modulation is a dynamic and exciting field, and the reader is referred to more advanced texts and technical papers for a more complete discussion.

PROBLEMS

13.1 A transmitted signal $s(t)$ is a weighted linear combination of nonoverlapping time shifted pulses $p(t)$ with

$$s(t) = \sum_{j=1}^{M} s_j p(t - jT),$$

where $p(t) \neq 0$ for $-T \leq t \leq 0$, $p(t) = 0$ for all other t, and

$$\int_{-T}^{0} p^2(t)\, dt = E_b$$

is the energy transmitted per bit. We assign the s_j according to

$$s_j = \begin{cases} +1, & \text{if the } j\text{th message bit is 1} \\ -1, & \text{if the } j\text{th message bit is 0.} \end{cases}$$

If $s(t)$ is sent over an AWGN channel with power spectral density $\mathcal{N}_0/2$ watts/Hz, show that $P[\mathscr{E}] = 1 - (1 - p)^M$, where $p = \frac{1}{2} - \text{erf}[\sqrt{2E_b/\mathcal{N}_0}]$. This has been called "bit-by-bit" signaling [Wozencraft and Jacobs, 1965]. The signal $s(t)$ can be represented geometrically as 2^M signal points at the vertices of a hypercube in M dimensions. What happens to $P[\mathscr{E}]$ as $M \to \infty$?

13.2 An M-bit binary message sequence is to be transmitted in a time interval τ seconds long. The transmitted signals are equally likely and of the form

$$s_i(t) = \sqrt{E}\, \varphi(t - iT),$$

where $\varphi(t) \neq 0$ for $0 \leq t \leq T$, $\varphi(t) = 0$ for t otherwise, $\int_0^T \varphi^2(t)\, dt = 1$, $T = \tau/2^M$, and $i = 0, 1, 2, \ldots, 2^M - 1$ is the number corresponding to the binary message sequence. Show that $P[\mathscr{E}] \leq (2^M - 1)p$, where $p = \frac{1}{2} - \text{erf}[\sqrt{E/\mathcal{N}_0}]$, if these signals are transmitted over an AWGN channel with spectral density $\mathcal{N}_0/2$ watts/Hz. This has been called "block orthogonal" signaling [Wozencraft and Jacobs, 1965]. Note that the energy transmitted per bit $E_b = E/M$, and consider $P[\mathscr{E}]$ as $M \to \infty$. Compare to the result in Problem 13.1.

13.3 The pre-envelope $s(t)$ of a real bandpass signal $x(t)$ is defined as [Rice, 1982]

$$s(t) = \hat{r}(t)e^{j\omega_0 t + j\hat{\varphi}(t)} = x(t) + jx_h(t),$$

where ω_0 is arbitrary and $x_h(t)$ is the Hilbert transform of $x(t)$. Clearly, $\text{Re}[s(t)] = x(t)$ and $\hat{r}(t) = \sqrt{x^2(t) + x_h^2(t)}$ is the magnitude of $s(t)$.

(a) Let $x(t) = m(t) \cos \omega_c t$ and show that $\hat{r}(t)$ is the envelope of $x(t)$.

(b) Given a signal of the form

$$z(t) = x(t) \cos \omega_0 t - y(t) \sin \omega_0 t,$$

where the highest frequency in $x(t)$ and $y(t)$ is less than ω_0, find an expression for the analytic signal corresponding to $z(t)$ and expressions for $x(t)$ and $y(t)$.

13.4 Derive Eqs. (13.2.9) and (13.2.10).

13.5 Obtain $r(t)$ in Eq. (13.2.12) for the case where $-T \leq t \leq 0$.

13.6 Derive Eqs. (13.2.14) and (13.2.15).

13.7 Obtain the frequency pulses corresponding to the smoothing functions in Eqs. (13.3.5) and (13.3.6).

13.8 Sketch the smoothing function in Eq. (13.3.7) and its frequency pulse for $L = 2$ and $L = 3$.

13.9 Show that the minimum spacing of FSK frequencies for noncoherent orthogonality is the symbol rate, $1/T$.

13.10 Given a binary data sequence $\{b_k, k = 0, 1, 2, \ldots\}$ with $b_k = \pm 1$, a quadrature PSK (QPSK) signal can be written as

$$s(t) = \frac{b_i(t)}{\sqrt{2}} \cos\left[\omega_c t + \frac{\pi}{4}\right] + \frac{b_q(t)}{\sqrt{2}} \sin\left[\omega_c t + \frac{\pi}{4}\right],$$

where $b_i(t)$ is a rectangular pulse $2T$ seconds wide with an amplitude (± 1) determined by the even-subscripted bits, $k = 0, 2, 4, \ldots$. Similarly, $b_q(t)$ is a rectangular pulse whose amplitude is specified by the odd-subscripted bits, $k = 1, 3, 5, \ldots$ [Pasupathy, 1979].

(a) Show that the possible transmitted phases are 0; $\pm 90°$; and $180°$ every $2T$ seconds.

(b) For the input binary sequence $\{b_k\} = \{1, -1, 1, -1, -1, -1, 1, 1\}$, specify the transmitted phase sequence and sketch the transmitted waveform.

13.11 Offset QPSK (OQPSK), also called staggered QPSK, can be generated from the formulation in Problem 13.10 by delaying $b_q(t)$ with respect to $b_i(t)$ by T seconds [Pasupathy, 1979].

(a) Show that the possible transmitted phases are 0 and $\pm 90°$ every T seconds.

(b) For the same binary input sequence as in Problem 13.10(b), find the transmitted phase sequence and sketch the transmitted waveform for OQPSK.

(c) Compare part (b) of Problems 13.10 and 13.11.

13.12 MSK can be written as a version of OQPSK described in Problem 13.11 in which the quadrature components have cosine and sine pulse shaping.

In particular, an MSK waveform can be expressed as

$$s(t) = b_i(t) \cos \frac{\pi t}{2T} \cos \omega_c t + b_q(t) \sin \frac{\pi t}{2T} \sin \omega_c t,$$

where $b_i(t)$ and $b_q(t)$ are as defined in Problems 13.10 and 13.11.

(a) Show that $s(t)$ can be rewritten as

$$s(t) = \cos\left[\omega_c t - b_i(t) b_q(t) \frac{\pi t}{2T} + \phi_k\right],$$

where

$$\phi_k = \begin{cases} 0, & \text{if } b_i(t) = +1 \\ \pi, & \text{if } b_i(t) = -1. \end{cases}$$

(b) Use $s(t)$ in part (a) to demonstrate that MSK is also an FSK signal transmitting the two frequencies $\omega_+ = \omega_c + \pi/2T$ and $\omega_- = \omega_c - \pi/2T$ [Pasupathy, 1979].

13.13 For the binary input message sequence $\{+1, +1, +1, +1, -1, -1, +1, -1\}$, find the transmitted phases at the time instants nT, $n \geq 1$ for MSK using the trellis diagram in Fig. 13.3.2.

13.14 Draw a trellis diagram corresponding to the phase smoothing function $f(t)$ in Fig. 13.3.1(c). Let $h = \frac{1}{2}$ and $M = 2$.

13.15 Draw a phase tree for multi-h CPM for $f(t)$ in Fig. 13.3.1(b), $M = 2$, and $\{h_1, h_2\} = \{\frac{3}{8}, \frac{4}{8}\}$ [Ziemer and Peterson, 1985].

13.16 Find the sequence of transmitted phases for the multi-h CPM system in Problem 13.15 if the input binary message string is

$$\{+1, -1, -1, -1, +1, +1, +1, -1\}.$$

13.17 For the coded modulation scheme in Example 13.4.1, let the initial encoder state be $(s_0^3\ s_0^2\ s_0^1) = (0\ 0\ 0)$ and assume that the input sequence is 1 0 1 1 0 1 1 1 0 1 0 1. Specify the current state, coder output, path transition, transmitted symbol, and next state as a function of time n.

13.18 Find the distance between the path (0, 0, 0) and (6, 5, 2) in Problem 13.17.

13.19 Three eight-phase PSK coded modulation methods for sending 2 bits/symbol have the d_{free} shown.

Code:	1	2	3
d_{free}:	1.608	2.0	2.274

Calculate the coding gain of each of these techniques with respect to four-phase PSK.

13.20 A rate-$\frac{2}{3}$ convolutional coder is shown in Fig. P13.20(a) along with its associated trellis in Fig. P13.20(b). A bit-to-symbol mapping for a one-dimensional signal set is shown in Fig. P13.20(c). For this coded modulation method, $d_{\min}^2/P_{\text{av}} = 36/21$ [Thapar, 1984].

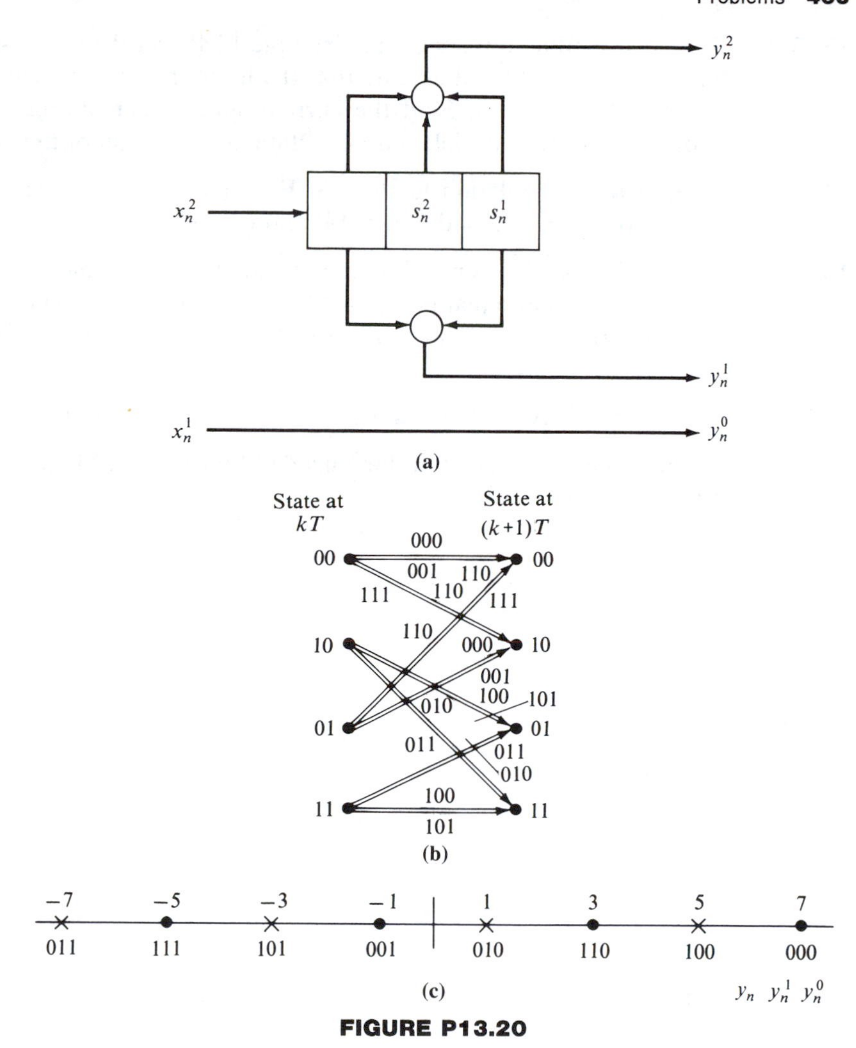

FIGURE P13.20

(a) For an initial encoder state of $(s_0^2\ s_0^1) = (0\ 0)$ and the input sequence 1 1 0 1 1 0 1 0, find the current state, coder output, path transition, transmitted symbol, and next state as a function of time.

(b) Find the coding gain with respect to a four-level one-dimensional signal set.

13.21 For the trellis in Fig. P13.20(b), find the distance between the paths (7, 7, 7) and (3, −3, −1).

13.22 Redraw the one-dimensional mapping in Fig. P13.20(c) as a circle, thus generating a two-dimensional mapping. Calculate the coding gain of this new code with respect to four-phase PSK [Thapar, 1984].

13.23 For the coded modulation system in Example 13.4.2, let the initial state be $(s_n^3\ s_n^2\ s_n^1) = (0\ 0\ 0)$ and assume that the input message sequence is 1 0 1 1 1 0 1 1 1 0 0 1. Find the current state, coder output, path transition, transmitted symbol, and next state as a function of time n.

13.24 Given the coded modulation method in Example 13.4.2, calculate the distance between the two paths (8, 6, 14) and (12, 14, 9).

13.25 Parallel transitions allow single-bit-error events to occur, which implies that d_{free} is less than or equal to d_{min} for the signal subset corresponding to the parallel transition. What is d_{min} for the parallel transition subsets in Fig. 13.4.6?

13.26 Find the distance between parallel transitions for the code in Fig. P13.20.

13.27 Draw a one-state trellis diagram for uncoded four-phase PSK at a rate of 2 bits/symbol. Specify d_{free}.

CHAPTER 14

Spread Spectrum Communications

14.1 Introduction

Spread spectrum communication systems have the characteristic attributes that the required transmission bandwidth is greater (often much greater) than the baseband message signal bandwidth and that the transmission bandwidth is determined by a spreading signal that is independent of the message. Furthermore, the receiver recovers the message by using a replica of the spreading signal. Spread spectrum systems as just defined do not include some familiar communication techniques such as wideband FM and PCM that occupy a relatively large transmission bandwidth, since the transmission bandwidth is a function of the message signal for these systems.

The primary advantage obtained by spreading the bandwidth is interference rejection, both intentional and unintentional interference, and thus spread spectrum systems have found a number of applications in military communications. In addition to interference rejection, spread spectrum techniques also offer low probability of intercept, multiuser random access, and high-resolution ranging. Although applications outside the military have been limited to date, it has been studied for mobile radio applications, positioning systems, and satellite communications. The benefits of spread spectrum and the possible applications will become clearer as the chapter unfolds. Spread spectrum systems have a long history, and for background, details, and origins of original work beyond that reported in this chapter, the reader should refer to the literature [Scholtz, 1982; Dixon, 1976; Simon et al., 1985].

In Section 14.2 we provide a quantitative demonstration of the interference rejection capability afforded by spread spectrum and define a commonly quoted figure of merit called processing gain. The two principal approaches to spread spectrum are direct sequence and frequency hopping techniques, and these methods are developed in Sections 14.3 and 14.4, respectively. The application and advantages of using error control coding in conjunction with spread spectrum are investigated in Section 14.5. Spread spectrum techniques for code-division multiple-access communications, which have received considerable research attention in the last 10 years, are presented in Section 14.6. In Sections 14.7 and 14.8 we examine the problems of code acquisition and code tracking, which are essential to the recovery of the message at the receiver.

14.2 Processing Gain

The interference rejection capability of spread spectrum communications is the property that has motivated most applications to date. Here we present two examples that illustrate this capability, one of which relies only on deterministic spectral arguments similar to those in Chapters 5 and 6, while the other utilizes a digital communications development similar to that in Chapter 10. Along the way, the quantity called processing gain is defined, which is a widely quoted performance indicator for spread spectrum systems.

The transmitted waveform in a spread spectrum system can be written as [Blahut, 1990]

$$s(t) = m(t) \cos [\omega_0 t + \phi(t)], \tag{14.2.1}$$

where $m(t)$ is the message signal with bandwidth B hertz and $\phi(t)$ is wideband phase modulation induced by the spreading signal, independent of $m(t)$, so that the bandwidth occupied by $s(t)$ is $W \gg B$. During transmission, an interference signal denoted by $v(t)$ is added, so that the received signal is

$$\begin{aligned} r(t) &= s(t) + v(t) \\ &= m(t) \cos [\omega_0 t + \phi(t)] + v(t). \end{aligned} \tag{14.2.2}$$

The additive interference signal $v(t)$ may be intentional or unintentional and may be wideband or narrowband. For the present discussion it is assumed that the interference is intentional, that is, caused by a jammer, since this represents a common application and allows us to choose the interference to yield worst-case performance.

To demodulate the message, the receiver multiplies $r(t)$ by $\cos [\omega_0 t + \phi(t)]$ to produce

$$\begin{aligned} r(t) \cos [\omega_0 t + \phi(t)] &= m(t) \cos^2 [\omega_0 t + \phi(t)] + v(t) \cos [\omega_0 t + \phi(t)] \\ &= m(t)[\tfrac{1}{2} + \tfrac{1}{2} \cos (2\omega_0 t + 2\phi(t))] + v(t) \cos [\omega_0 t + \phi(t)]. \end{aligned} \tag{14.2.3}$$

The interference signal can be expanded in in-phase and quadrature form about f_0 as (see Chapter 7)

$$v(t) = v_i(t) \cos \omega_0 t - v_q(t) \sin \omega_0 t, \tag{14.2.4}$$

which upon substituting into Eq. (14.2.3) and ignoring components at $2f_0$ yields

$$s_0(t) = m(t) + v_i(t) \cos \phi(t) - v_q(t) \sin \phi(t). \tag{14.2.5}$$

The message signal has been returned to its original baseband form. However, the in-phase and quadrature components of the interference are each multiplied by wideband signals, so that the products are wideband (with a bandwidth of approximately W hertz). The average power of the interference in Eq. (14.2.5) is approximately (see Section 4.9) $v_i^2(t) + v_q^2(t) \triangleq P_J$, the total jammer power, but it is now spread over a bandwidth $W \gg B$. Therefore, lowpass filtering of $s_0(t)$ to retrieve $m(t)$ includes an interference signal with power proportional to $(B/W)P_J$, which since $B \ll W$, is very small. The ratio W/B is sometimes called the *processing gain* and is often cited as a figure of merit for a spread spectrum system.

Another example based on a digital communications development similar to that used in Chapter 10 is now offered [Pickholtz, Schilling, and Milstein, 1982]. A data sequence is transmitted using M equiprobable, equal-energy, orthogonal signals represented by an N-dimensional orthonormal basis set $\{\vartheta_j(t),\ j = 1, 2, \ldots, N\}$ as

$$s_i(t) = \sum_{j=1}^{N} s_{ij}\vartheta_j(t), \tag{14.2.6}$$

where $0 \le t \le T$, $1 \le i \le M < N$, and

$$s_{ij} = \int_0^T s_i(t)\vartheta_j(t)\,dt. \tag{14.2.7}$$

For the purposes of illustration, rather than use a pseudorandom sequence to modulate the data, it is assumed that the components s_{ij} are selected randomly and independently such that they have zero mean and

$$E\{s_{ij}s_{ik}\} = \begin{cases} \dfrac{E_s}{N}, & j = k \\ 0, & j \ne k, \end{cases} \tag{14.2.8}$$

where $1 \le i \le M$, with

$$\int_0^T E\{s_i^2(t)\}\,dt = \sum_{j=1}^{N} E\{s_{ij}^2\} \triangleq E_s. \tag{14.2.9}$$

Of course, the receiver is informed of the sequences $\{s_{ij}\}$, so that the data can be demodulated, but the sequences are unknown to the jammer who is generating the intentional interference.

The jammer signal is expressed as

$$s_J(t) = \sum_{k=1}^{N} J_k\vartheta_k(t), \tag{14.2.10}$$

$0 \le t \le T$, with total energy

$$\int_0^T s_J^2(t)\,dt = \sum_{k=1}^{N} J_k^2 \triangleq E_J. \tag{14.2.11}$$

The jamming is additive, so that the received signal is

$$r(t) = s_i(t) + s_J(t) \tag{14.2.12}$$

for the ith message. To recover the message, $r(t)$ is correlated with all of the known random sequences; hence the output of the ith correlator when $s_i(t)$ is transmitted is

$$Y_i = \int_0^T r(t)s_i(t)\,dt = \sum_{j=1}^{N} [s_{ij}^2 + J_j s_{ij}] \tag{14.2.13}$$

and the output of the ith correlator when $s_l(t)$, $l \neq i$, is transmitted is

$$Y_i = \int_0^T r(t)s_l(t)\,dt = \sum_{j=1}^{N} [s_{lj}s_{ij} + J_j s_{ij}], \tag{14.2.14}$$

$l \neq i$.

A performance indicator for this random spread spectrum system is the signal-to-noise ratio defined by

$$\text{SNR} \triangleq \sum_{l=1}^{M} \frac{E^2(Y_i|s_l)}{\text{var}(Y_i|s_l)} P(s_l), \tag{14.2.15}$$

which is the average signal-to-noise ratio at the ith correlator output. To find the numerator of Eq. (14.2.15), we first calculate the conditional expectations given that $s_i(t)$ is transmitted

$$E[Y_i|s_i] = \sum_{j=1}^{N} E[s_{ij}^2] = E_s \tag{14.2.16}$$

and given that $s_l(t)$, $l \neq i$, is transmitted

$$\begin{aligned} E[Y_i|s_l] &= \sum_{j=1}^{N} \{E[s_{lj}s_{ij}] + E[J_j s_{lj}]\} \\ &= \sum_{j=1}^{N} \{E[s_{lj}]E[s_{ij}] + E[J_j]E[s_{lj}]\} = 0, \end{aligned} \tag{14.2.17}$$

so

$$E(Y_i) = \frac{E_s}{M}. \tag{14.2.18}$$

To calculate the variance at the ith correlator output, we first assume that $s_i(t)$ is transmitted to obtain (see Pickholtz, Schilling, and Milstein, 1984)

$$\text{var}(Y_i|s_i) = \sum_{j=1}^{N}\sum_{l=1}^{N} J_j J_l E\{s_{ij}s_{il}\} = \sum_{l=1}^{N} J_l^2 \frac{E_s}{N} = \frac{E_s}{N} E_J, \tag{14.2.19}$$

where we have used Eqs. (14.2.8) and (14.2.9), and then we assume that $s_l(t)$, $l \neq i$, is transmitted, which yields

$$\begin{aligned} \text{var}(Y_i|s_l) &= E(Y_i^2|s_l) \\ &= \sum_{j=1}^{N} \{E[s_{ij}^2 s_{lj}^2] + E[J_j^2 s_{ij}^2]\} \\ &= \frac{E_s^2}{N} + \frac{E_s}{N} E_J. \end{aligned} \tag{14.2.20}$$

Substituting Eqs. (14.2.16), (14.2.17), (14.2.19), and (14.2.20) into Eq. (14.2.15) gives

$$\text{SNR} = \frac{E_s}{E_J}\frac{N}{M}. \tag{14.2.21}$$

The ratio $N/M > 1$ is also called the *processing gain* and is the ratio of the number of available dimensions to the actual number of dimensions used. Since the jammer does not know which of the N dimensions are used, it must cover them all. This ratio can be related back to the bandwidth of the spread spectrum signal W hertz and the message signal bandwidth by noting that $N \cong 2WT$ and $M \cong 2BT$ [Wozencraft and Jacobs, 1965], so

$$\frac{N}{M} \cong \frac{2WT}{2BT} = \frac{W}{B} \tag{14.2.22}$$

as before.

The two examples in this section demonstrate the effectiveness of spread spectrum techniques for interference rejection and the origin of processing gain as a performance indicator. The next two sections describe the two most common spread spectrum techniques.

14.3 Direct Sequence Systems

A direct sequence (DS) spread spectrum system takes a message-data-modulated carrier waveform and modulates it a second time using a pseudorandom sequence of ± 1s. Pseudorandom sequences and their properties are developed in Appendix G, and that information is a prerequisite for a full understanding of the present section. For this beginning discussion, the modulation method used for the message data is assumed to be BPSK. Since in BPSK the phase is switching between $\pm 180°$ every T seconds, this can be represented equivalently by multiplying the carrier by a sequence of ± 1s, each with duration T seconds. Therefore, in Fig. 14.3.1, with $m(t)$ representing the message data, we write the message-data-modulated carrier as

$$s_m(t) = Am(t)\cos\omega_0 t, \tag{14.3.1}$$

where $f_0 = \omega_0/2\pi$ is the carrier frequency.

The pseudorandom spreading signal then multiplies $s_m(t)$ to give

$$s_m(t)s_c(t) = As_c(t)m(t)\cos\omega_0 t, \tag{14.3.2}$$

which is the signal transmitted over the channel. The spreading signal is a sequence of ± 1s, each of duration T_c with $T_c \ll T$, so that $f_c = 1/T_c \gg f_b \triangleq 1/T$. Thus the bandwidth of the spreading signal is much greater than the message bandwidth. Further, the transmitted signal bandwidth is much greater than that of the message and is (virtually) independent of the message. The spreading signal is often called the *chip sequence*, a signal $+1$ or -1 is called a *chip*, and T_c is the chip duration (length).

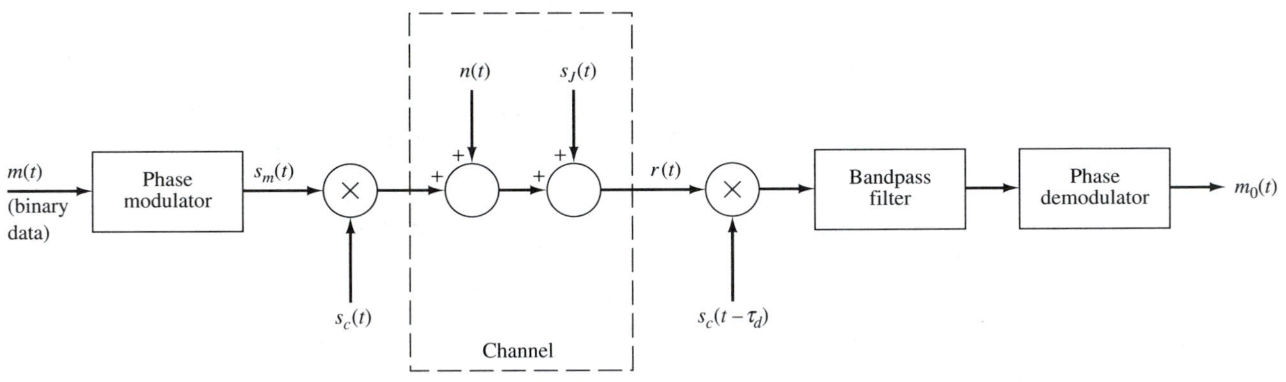

FIGURE 14.3.1 BPSK direct sequence spread spectrum system.

During transmission over the channel, the signal is contaminated by thermal noise $n(t)$ with two-sided spectral density $\mathcal{N}_0/2$ watts/Hz and by an interfering signal that we call a jammer, given by $s_J(t)$ with total power P_J watts spread over the entire bandwidth of the transmitted signal. The received signal is

$$r(t) = As_c(t)m(t) \cos \omega_0 t + n(t) + s_J(t), \tag{14.3.3}$$

and hence if the chip waveform at the receiver is perfectly synchronized ($\tau_d = 0$ in Fig. 14.3.1), the input to the bandpass filter is

$$r(t)s_c(t) = As_c^2(t)m(t) \cos \omega_0 t + n(t)s_c(t) + s_J(t)s_c(t). \tag{14.3.4}$$

Since $s_c(t) = \pm 1$, $s_c^2(t) = 1$ drops out of the first term and the power spectral density of $n(t)s_c(t)$ is the same as $n(t)$.

For the purposes of this analysis, we assume that $s_J(t)$ is a zero-mean, white Gaussian random process. Multiplying $s_J(t)$ by $s_c(t)$, as in Eq. (14.3.4), results in the total jammer power being spread over a bandwidth of approximately $2f_c$ hertz, so that the one-sided power spectral density of the interference multiplied by $s_c(t)$ is $\mathcal{N}_J = P_J/2f_c$ watts/Hz. For coherent matched filter detection of the product in Eq. (14.3.4), the probability of bit error is given by (see Chapter 10)

$$P_e = \frac{1}{2} - \operatorname{erf}\left[\sqrt{\frac{2E_s}{\mathcal{N}_0 + \mathcal{N}_J}}\right], \tag{14.3.5}$$

where $E_s = A^2T/2$. From Eq. (14.3.5) it is evident that the effect of the jammer can be made negligible by choosing f_c sufficiently large. However, the spreading operation provides no performance improvement for thermal noise, since the total power in $n(t)$ increases with bandwidth.

Under the assumption that the jammer power spectral density $\mathcal{N}_J$ is much greater than the thermal noise spectral density $\mathcal{N}_0$ (i.e., $\mathcal{N}_J \gg \mathcal{N}_0$), Eq. (14.3.5) becomes

$$P_e \cong \frac{1}{2} - \operatorname{erf}\left[\sqrt{\frac{2E_s}{\mathcal{N}_J}}\right]. \tag{14.3.6}$$

This last expression can be written in terms of the processing gain by noting that $E_s = TP_s = P_s/B$, where P_s is the total signal power per bit, and that $\mathcal{N}_J = P_J/W$, so upon substituting into Eq. (14.3.6) gives

$$P_e \cong \frac{1}{2} - \operatorname{erf}\left[\sqrt{\frac{2W/B}{P_J/P_s}}\right]. \tag{14.3.7}$$

Thus, from Eq. (14.3.7), the bit error probability decreases with increasing processing gain.

The other ratio in the argument of the error function in Eq. (14.3.7), namely P_J/P_s, is often called the *jamming margin*. It is so-named because if $E_s/\mathcal{N}_J$ is fixed to give a desired error probability, say P_e^*, and the processing gain is specified, the jamming margin is the largest value of the ratio P_J/P_s such that the error probability is less than or equal to P_e^*. For example, let $P_e^* = 10^{-5}$

and $W/B = 1000$. Then $10 \log_{10}(E_s/\mathcal{N}_J) \cong 9.6$ dB, and $10 \log_{10}(P_J/P_s) = 30 - 9.6 = 20.4$ dB. Any larger value of P_J/P_s will cause P_e to be greater than 10^{-5}.

EXAMPLE 14.3.1

We can examine the effect of increasing processing gain on P_e if P_J/P_s is held fixed and W/B is varied. Thus we let $P_J/P_s = 100$ and find P_e for $W/B = 1$, 10, 100, and 1000. From Eq. (14.3.7) we find that

$$P_e \cong \begin{cases} 0.444, & \dfrac{W}{B} = 1 \\ 0.328, & \dfrac{W}{B} = 10 \\ 0.08, & \dfrac{W}{B} = 100 \\ 3.4 \times 10^{-6}, & \dfrac{W}{B} = 1000. \end{cases}$$

The dependence of P_e on processing gain is clear from these results (it is asymptotically exponential) and this example provides reinforcement for the discussion in Section 14.2.

Equations (14.3.5)–(14.3.7) demonstrate that DS spread spectrum can be very effective against continuous jamming of the communications channel. However, there is a particular jamming strategy that can cause serious problems for a DS spread spectrum system. Consider a noise jammer with total average power P_J as before, but now the jammer transmits pulses of wideband noise for only a fraction, say γ, of the time. Thus when the jammer is not transmitting, the transmitted bits are received in the presence of thermal noise only, but when the jammer is transmitting, the probability of bit error has the same form as in Eqs. (14.3.5)–(14.3.7), but with $\mathcal{N}_J$ and P_J replaced by $\mathcal{N}_J/\gamma$ and P_J/γ, respectively. The average probability of bit error for the DS system with this pulsed jamming is thus

$$P_e = (1 - \gamma)\left[\frac{1}{2} - \operatorname{erf}\left(\sqrt{\frac{2E_s}{\mathcal{N}_0}}\right)\right] + \gamma\left[\frac{1}{2} - \operatorname{erf}\left(\sqrt{\frac{2E_s}{\mathcal{N}_0 + \mathcal{N}_J/\gamma}}\right)\right]. \tag{14.3.8}$$

For simplicity we have assumed that the pulsed jammer affects an integer number of bits. When $\mathcal{N}_0$ is small compared to $\mathcal{N}_J$, the first term on the right of Eq. (14.3.8) is small compared to the second term, and furthermore, $\mathcal{N}_0$ can be neglected in the second term. Equation (14.3.8) therefore becomes

$$P_e \cong \gamma\left[\frac{1}{2} - \operatorname{erf}\left(\sqrt{\frac{2\gamma E_s}{\mathcal{N}_J}}\right)\right]. \tag{14.3.9}$$

If the jammer selects γ to maximize this bit error probability, it can be shown by differentiating Eq. (14.3.9) with respect to γ that [Houston, 1975; Viterbi and

Jacobs, 1975; Proakis, 1989; Blahut, 1990]

$$\gamma_{\text{opt}} = \begin{cases} \dfrac{0.71}{E_s/\mathcal{N}_J}, & \dfrac{E_s}{\mathcal{N}_J} \geq 0.71 \\ 1, & \dfrac{E_s}{\mathcal{N}_J} < 0.71, \end{cases} \tag{14.3.10}$$

with

$$P_e = \begin{cases} \dfrac{0.083}{E_s/\mathcal{N}_J}, & \dfrac{E_s}{\mathcal{N}_J} \geq 0.71 \\ \dfrac{1}{2} - \text{erf}\left[\sqrt{\dfrac{2E_s}{\mathcal{N}_J}}\right], & \dfrac{E_s}{\mathcal{N}_J} < 0.71. \end{cases} \tag{14.3.11}$$

Plots of Eq. (14.3.9) for various values of γ are given in Fig. 14.3.2. Note for $P_e = 10^{-4}$ that there is about a 20-dB loss in performance for BPSK with

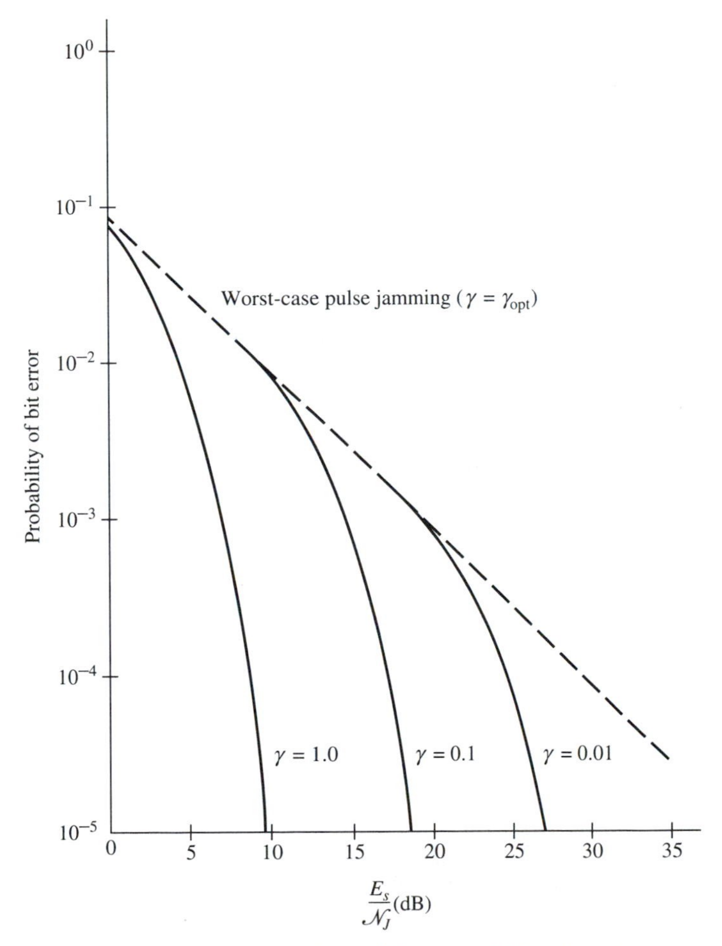

FIGURE 14.3.2 Bit error probability of DS binary PSK with pulsed jamming. From J. G. Proakis, *Digital Communications*, 2nd ed., New York: McGraw-Hill, Inc., 1989. Reproduced with permission of McGraw-Hill, Inc.

pulsed jamming (γ_{opt}) as compared to BPSK in thermal noise of the same average power. As we shall see in Section 14.5, the very detrimental effects of pulsed jamming can be compensated for by the judicious use of error control coding. Although these results are for BPSK message data modulation and noise jamming, similar expressions can be derived for other modulation methods and tone jamming [Ziemer and Peterson, 1985; Schilling et al., 1980].

14.4 Frequency Hopping Systems

Frequency hopping (FH) spread spectrum communication systems operate by changing the frequency of the message-data-modulated carrier from time to time according to a pseudorandom sequence. The possible carrier frequencies are usually spaced as far apart as the width of the message spectrum. A block diagram of a FH spread spectrum communication system is shown in Fig. 14.4.1.

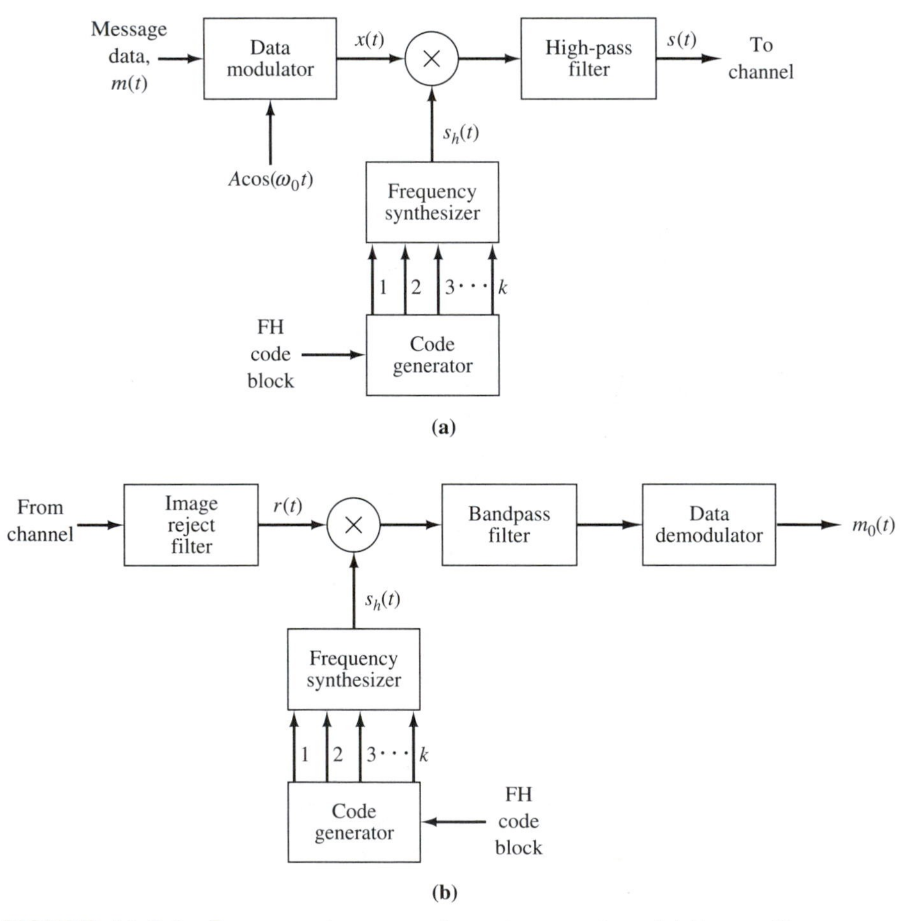

FIGURE 14.4.1 Frequency hop spread spectrum system: (a) Transmitter; (b) receiver.

By letting

$$s_h(t) = \sum_{n=-\infty}^{\infty} g(t - nT_c) \cos \omega_n t, \tag{14.4.1}$$

where $g(t)$ is a rectangular pulse of duration T_c seconds and the period of the pseudorandom sequence is considered infinitely long (because of the infinite summation), the signal transmitted over the channel is

$$\begin{aligned} s(t) &= [x(t)s_h(t)]_+ \\ &= \left[A \cos \left\{ \omega_0 t + c_f \int m(\lambda)\, d\lambda \right\} s_h(t) \right]_+ \\ &= \sum_{n=-\infty}^{\infty} Ag(t - nT_c) \cos \left\{ (\omega_0 + \omega_n)t + c_f \int m(\lambda)\, d\lambda \right\}. \end{aligned} \tag{14.4.2}$$

The subscript "+" on the brackets in Eq. (14.4.2) denotes taking the sum frequency components from the product operation. The modulation method shown is FM, which is M-ary FSK for an M-level message data sequence, but it is also often DPSK. Noncoherent detection is commonly used in the receiver to avoid the difficulties of achieving coherence with the frequency synthesizers.

Both slow FH and fast FH systems are possible, where for binary FSK, slow FH has $T_c \geq T$ and fast FH has $T_c < T$. Therefore, for slow FH the carrier frequency changes no more than once per data symbol, while for fast FH there may be several carrier frequency changes during a data symbol. In a fast FH system there are several demodulated chips per symbol and the detector may combine these using a majority vote or may employ some other decision rule such as maximum likelihood.

Instead of jamming the entire frequency band spanned by the complete set of carrier frequencies, a jammer may utilize *partial band jamming* against FH systems [Pickholtz, Schilling, and Milstein, 1982; Proakis, 1989], which consists of additive white Gaussian noise over only a fraction of the total frequency-hopped band. For example, the jammer may corrupt only a fraction γ of the band, but if the jammer's average power spectral density over the entire band is $\mathcal{N}_J$, the power spectral density in the jammed frequency range is $\mathcal{N}_J/\gamma$, $0 < \gamma \leq 1$. The average probability of bit error over the band is thus

$$P_e = \frac{1-\gamma}{2} \exp\left\{\frac{-E_s}{2\mathcal{N}_0}\right\} + \frac{\gamma}{2} \exp\left\{\frac{-E_s}{(\mathcal{N}_0 + \mathcal{N}_J/\gamma)2}\right\}, \tag{14.4.3}$$

where we have used Eq. (10.6.6) for noncoherent FSK. If we assume that the jammer power spectral density dominates and that $E_s/\mathcal{N}_0 \gg 1$, the first term in Eq. (14.4.3) is near zero and $\mathcal{N}_0$ can be neglected in the second term, so that [Viterbi and Jacobs, 1975; Proakis, 1989]

$$P_e \cong \frac{\gamma}{2} \exp\left\{\frac{-\gamma E_s}{2\mathcal{N}_J}\right\}. \tag{14.4.4}$$

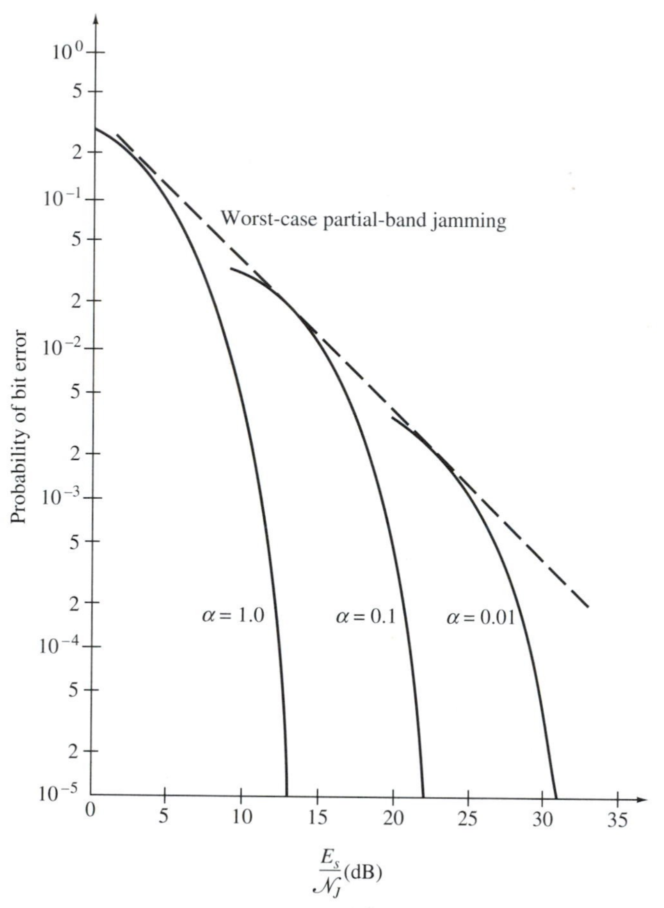

FIGURE 14.4.2 Probability of bit error for noncoherent FSK in partial band jamming. From J. G. Proakis, *Digital Communications*, 2nd ed., New York: McGraw-Hill, Inc., 1989. Reproduced with permission of McGraw-Hill, Inc.

The worst-case γ can be found to satisfy

$$\gamma_{\text{opt}} = \begin{cases} \dfrac{1}{E_s/2\mathcal{N}_J}, & \dfrac{E_s}{\mathcal{N}_J} \geq 2 \\ 1, & \dfrac{E_s}{\mathcal{N}_J} < 2, \end{cases} \tag{14.4.5}$$

which yields for $E_s/\mathcal{N}_J \geq 2$,

$$P_e \cong \frac{e^{-1}}{E_s/\mathcal{N}_J}. \tag{14.4.6}$$

Plots of Eq. (14.4.4) for various γ are presented in Fig 14.4.2. Note that at $P_e = 10^{-4}$, the partial band jammer with γ_{opt} causes a loss in performance greater than 20 dB.

In the following section we show how diversity and error control coding can be employed to improve this undesirable situation.

14.5 Diversity and Coding

The rather dramatic performance degradations caused by a pulsed jammer for DS systems and by partial band jamming of FH systems clearly necessitates some modifications. In this section we describe the use of interleaving, diversity, and coding to provide significant performance improvements in these situations.

We begin by considering pulsed jamming of a DS system to which we have added ideal interleaving and diversity transmission. It is specifically assumed that each bit is repeated L times, the resulting bit stream is interleaved, and the interleaved bits are transmitted via BPSK. To simplify the analysis, we make the possibly idealized assumption that the receiver can determine which bits have been jammed and which bits are not jammed. This *side information* may be obtainable from noise power measurements in nearby frequency bands or otherwise, but in any event, the knowledge of which bits are jammed greatly simplifies the analysis and yields a useful expression for performance comparisons.

The receiver forms the weighted sum of the L received pulses representing the L copies of a transmitted bit and performs coherent matched filter detection on this composite waveform to obtain the test statistic [Blahut, 1990]

$$Z(T) = \sum_{l=1}^{L} a_l r(k_l T), \tag{14.5.1}$$

where the L repetitions have a total pulse length of T seconds. The weights are chosen using the perfect side information according to whether or not an individual pulse is jammed, that is,

$$a_l = \begin{cases} \dfrac{1}{\mathcal{N}_0}, & \text{not jammed} \\[2ex] \dfrac{1}{\mathcal{N}_0 + \mathcal{N}_J/\gamma}, & \text{jammed.} \end{cases} \tag{14.5.2}$$

Furthermore, in those situations where only a portion, say L_J, of the total number of pulses are jammed, $L_J < L$, we assume that $\mathcal{N}_J/\gamma$ is sufficiently large that the a_l for the jammed pulses are effectively zero. Therefore, for L_J jammed pulses, the test statistic in Eq. (14.5.1) becomes

$$Z(T) = \sum_{l=1}^{L-L_J} a_l r(k_l T). \tag{14.5.3}$$

This implies that the received bit energy corresponding to Eq. (14.5.3) is $E_s(L - L_J)/L$.

With the fraction of jammed pulses equal to γ, the probability that l pulses are jammed is $\binom{L}{l}\gamma^l(1-\gamma)^{L-l}$, so

$$P_e \cong \sum_{l=0}^{L-1} \binom{L}{l}\gamma^l(1-\gamma)^{L-l}\left\{\frac{1}{2} - \operatorname{erf}\left(\sqrt{\frac{2E_s(L-l)}{L\mathcal{N}_0}}\right)\right\} + \gamma^L\left\{\frac{1}{2} - \operatorname{erf}\left(\sqrt{\frac{2E_s}{\mathcal{N}_0 + \mathcal{N}_J/\gamma}}\right)\right\}. \tag{14.5.4}$$

If we assume that $\mathcal{N}_0$ is negligible, only the last term in Eq. (14.5.4) persists, which can be written as

$$P_e \cong \gamma^L\left\{\frac{1}{2} - \operatorname{erf}\left(\sqrt{\frac{2\gamma E_s}{\mathcal{N}_J}}\right)\right\}. \tag{14.5.5}$$

Plots of this expression for $L = 1, 2,$ and 3, in comparison to a full message jammer, are shown in Fig. 14.5.1, where γ is optimized for each L. As little as

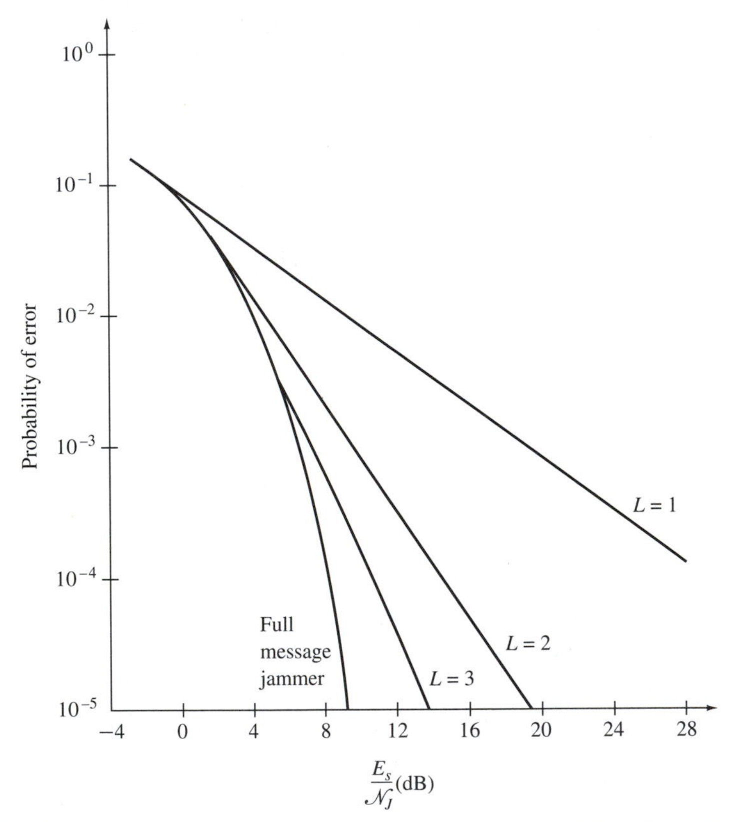

FIGURE 14.5.1 Bit error probability for L-fold diversity in DS BPSK with pulsed jamming. From R. E. Blahut, *Digital Transmission of Information* © 1990 by Addison-Wesley Publishing Company, Inc. Reprinted with permission of the publisher.

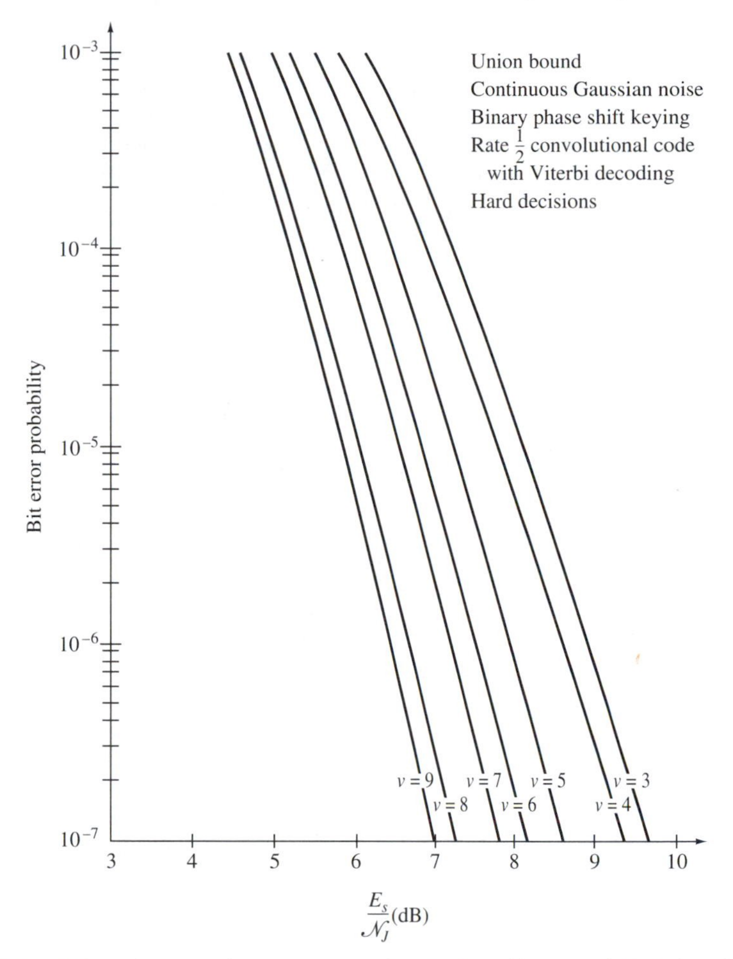

FIGURE 14.5.2 DS spread spectrum performance with convolutional codes and hard decisions against continuous jamming. From D. R. Martin and P. L. McAdam, "Convolutional Code Performance with Optimal Jamming," *Conf. Rec.*, 1980 IEEE International Conference on Communications, Seattle, Wash., June 8–12.

threefold diversity significantly reduces the effectiveness of the pulsed jamming strategy on a DS system.

The diversity method just described is a repetition code or an $(n,1)$ block code. Martin and McAdam [1980] present results for rate-$\frac{1}{2}$ convolutional coding of the interleaved bit stream without ideal side information (i.e., which bits are jammed is not known perfectly). They approximate the bit error probability using the union bound and verify the efficacy of the approach with simulations. Figure 14.5.2 shows the bit error probability for continuous jamming of a DS spread spectrum system with interleaving and rate-$\frac{1}{2}$ convolutional codes decoded using hard decisions and Viterbi decoding. The bit error probability for the same system against continuous jamming but with soft

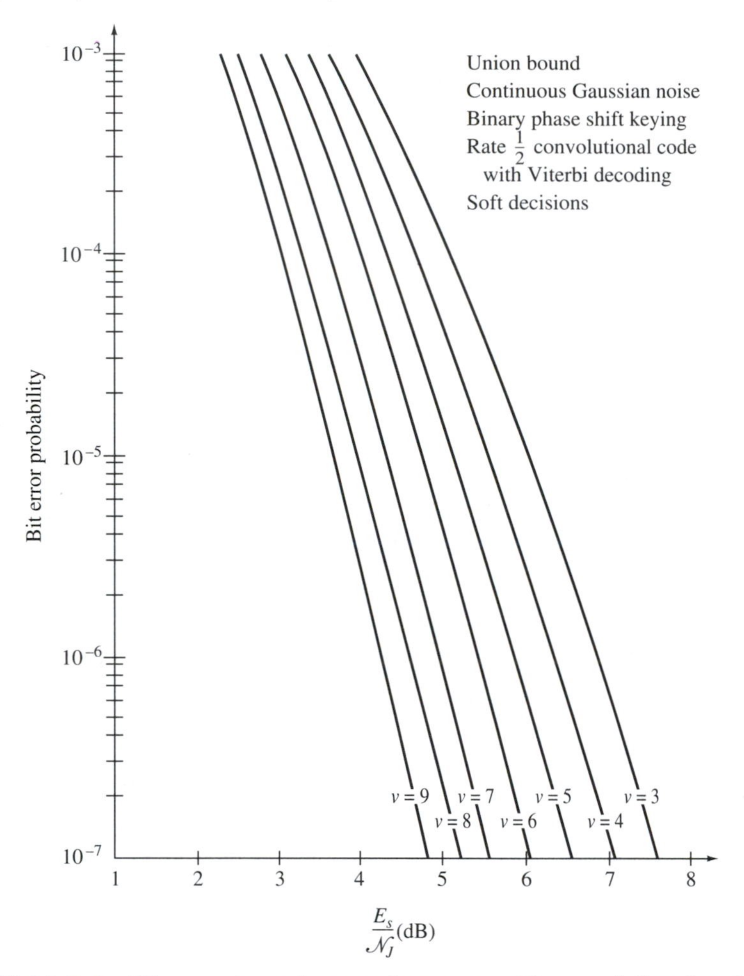

FIGURE 14.5.3 DS spread spectrum performance with convolutional codes and soft decisions against continuous jamming. From D. R. Martin and P. L. McAdam, "Convolutional Code Performance with Optimal Jamming," *Conf. Rec.*, 1980 IEEE International Conference on Communications, Seattle, Wash., June 8–12.

decision decoding is given in Fig. 14.5.3. From Fig. 14.5.2 we see that increasing the constraint length v from 3 to 9 provides a 2-dB performance improvement for $P_e = 10^{-6}$. Comparing the two figures, we also see the advantage of using soft decisions over hard decisions is about 2 dB at $P_e \leq 10^{-5}$.

Figures 14.5.4 and 14.5.5 show the performance of the same system against an optimized pulsed jammer. Clearly, the pulsed jammer is a more difficult threat, since for hard decisions and $v = 9$, performance at $P_e = 10^{-7}$ requires an additional 4 dB over continuous jamming. It is also interesting to note that soft decisions suffer a degradation of only about 2.5 dB ($v = 9$, $P_e = 10^{-7}$) in changing from continuous to pulsed jamming. However, the most important observation is that rate-$\frac{1}{2}$ convolutional coding, $v = 9$, and hard decisions have reduced the required $E_s/\mathcal{N}_J$ for $P_e = 10^{-4}$ from about 30 dB to less than

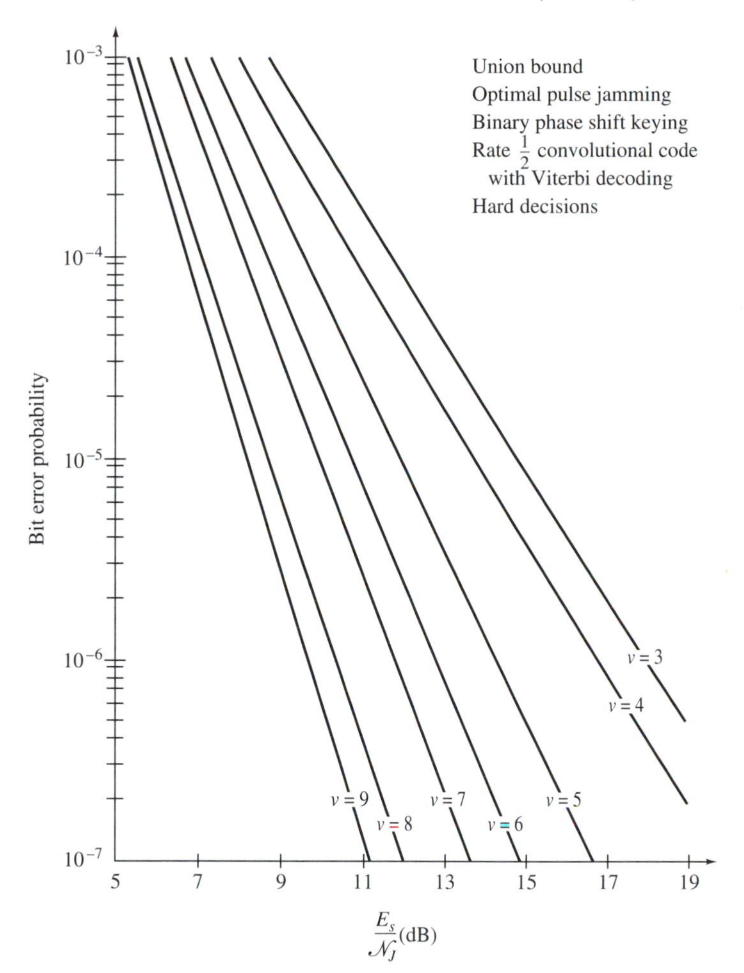

FIGURE 14.5.4 DS spread spectrum performance with convolutional codes and hard decisions against optimal pulsed jamming. From D. R. Martin and P. L. McAdam, "Convolutional Code Performance with Optimal Jamming," *Conf. Rec.*, 1980 IEEE International Conference on Communications, Seattle, Wash., June 8–12.

7 dB, and soft decisions drop this to less than 4 dB. To see this, compare Figs. 14.5.4 and 14.5.5 to Fig. 14.3.2.

Diversity and coding can also reduce the very serious effects of partial band jamming on FH spread spectrum systems. The type of diversity referred to here consists of transmitting each message symbol L times, with each transmission occurring at a different frequency. Thus the chip rate or hopping rate R_h is L times the symbol rate R, $R_h = LR$. The received signals for each chip or transmission frequency must be combined to produce a decision for each symbol, and there are a variety of ways to do this [Viterbi and Jacobs, 1975; Ziemer and Peterson, 1985; Levitt, 1985; Viswanathan and Taghizadeh, 1988].

For M-ary FSK modulation, hard decisions on the individual chips, and majority logic diversity combining of the hard decisions, the performance of

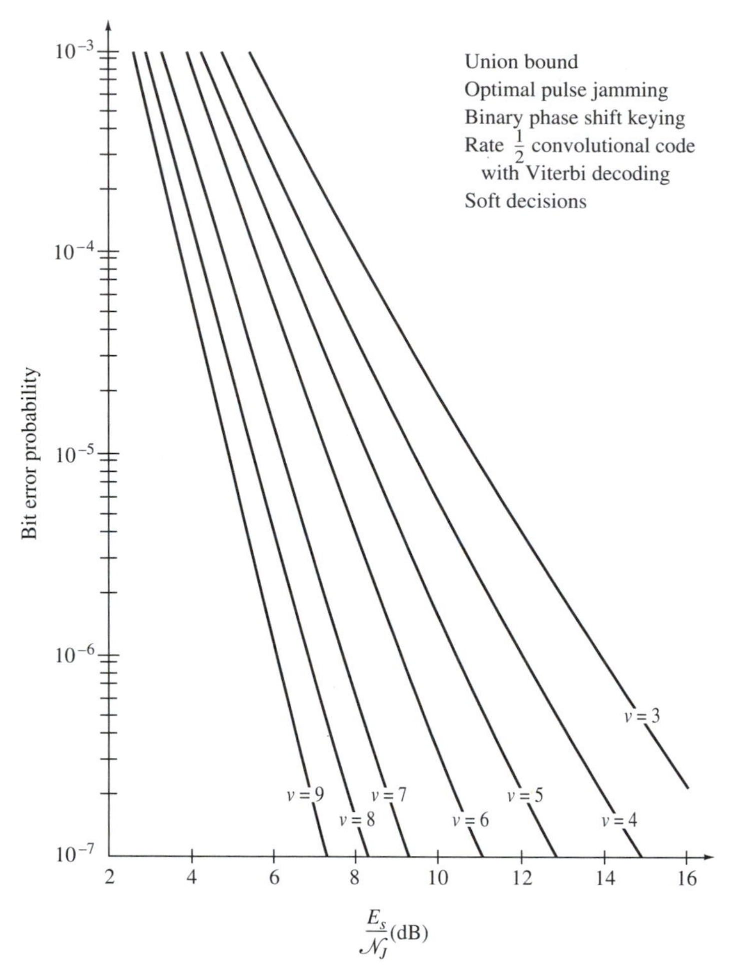

FIGURE 14.5.5 DS spread spectrum performance with convolutional codes and soft decisions against optimal pulsed jamming. From D. R. Martin and P. L. McAdam, "Convolutional Code Performance with Optimal Jamming," *Conf. Rec.*, 1980 IEEE International Conference on Communications, Seattle, Wash., June 8–12.

the FH/MFSK system with worst-case partial band jamming is as shown in Fig. 14.5.6. For a given $E_b/\mathcal{N}_J$, diversity provides a clear performance improvement against partial band jamming. For example, fivefold diversity provides over a 20-dB performance improvement for $P_e = 10^{-5}$.

Assuming soft decisions on individual chips, Viterbi and Jacobs [1975] obtain Chernoff bounds [Wozencraft and Jacobs, 1965] for a variety of diversity and coding configurations with partial band jamming. For uncoded binary, orthogonal FSK using frequency hopping without diversity in partial band jamming, the bit error probability for a worst-case γ is bounded as [Viterbi and Jacobs, 1975]

$$\max_{0<\gamma<1} P_e(\gamma) > \frac{e^{-1}}{E_b/\mathcal{N}_J}, \tag{14.5.6}$$

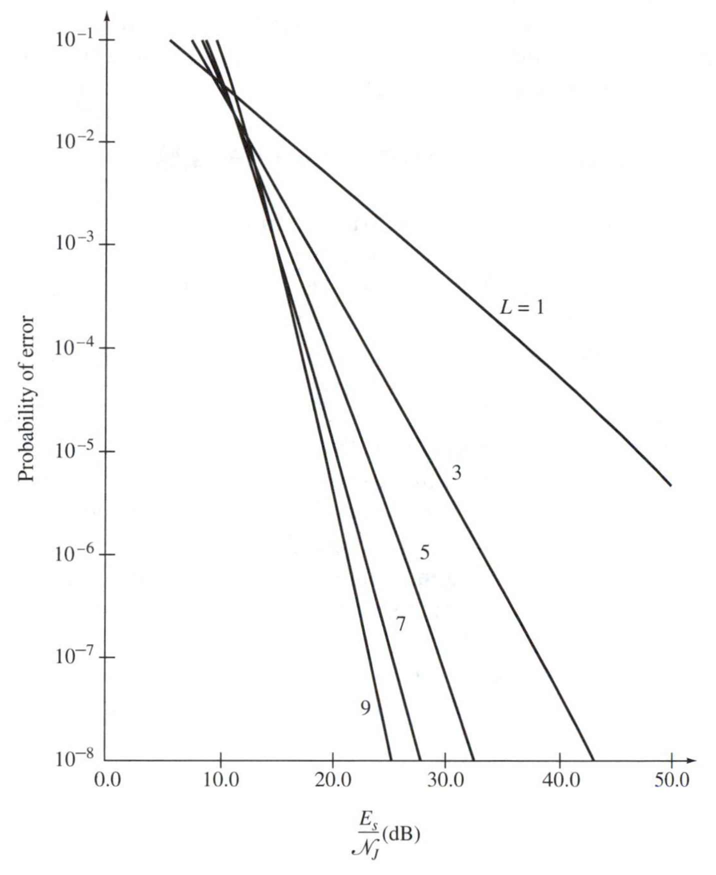

FIGURE 14.5.6 FH/MFSK performance in worst-case partial-band jamming with L-fold diversity. From R. E. Ziemer and R. L. Peterson, *Digital Communications and Spread Spectrum Systems*, 1985. Copyright © 1985 by Macmillan Publishing Company. Reprinted with the permission of Macmillan Publishing Company.

with equality when $E_b/\mathcal{N}_J \geq 2$. The interesting point to be made is that to achieve $P_e = 10^{-5}$ requires $E_b/\mathcal{N}_J = 45.7$ dB for worst-case partial band jamming, compared to only 13.4 dB for noncoherent detection without jamming. With the L repetition code diversity discussed previously,

$$\max_{0<\gamma<1} P_e(\gamma) < \frac{1}{2}\left[\frac{4e^{-1}L}{E_b/\mathcal{N}_J}\right]^L, \tag{14.5.7}$$

if $E_b/L\mathcal{N}_J \geq 3$. Minimizing Eq. (14.5.7) over L gives

$$\min_L \max_\gamma P_e(\gamma) < \frac{1}{2}\exp\left(-\frac{E_b}{4\mathcal{N}_J}\right) \tag{14.5.8}$$

for $E_b/L\mathcal{N}_J = 4$. Thus to achieve $P_e = 10^{-5}$ requires $E_b/\mathcal{N}_J = 16.4$ dB and L approximately equal to 11. Diversity provides an almost 30-dB reduction in the signal-to-noise ratio required to achieve $P_e = 10^{-5}$ in the presence of jamming.

Other results illustrating the utility of diversity and coding in ameliorating the effects of partial band jamming are available in Viterbi and Jacobs [1975]; Proakis [1989]; Schilling et al. [1982].

14.6 Code-Division Multiple Access

A long-held attraction of spread spectrum communications has been, and continues to be, the possibility of frequency "reuse" using code-division multiple-access (CDMA) systems [Schilling, Pickholtz, and Milstein, 1990]. Furthermore, there are other applications where CDMA may prove a feasible alternative to traditional multiple-access schemes such as TDMA and FDMA. We introduce the concept of CDMA here and then point out some of its advantages and disadvantages.

The key idea in a CDMA system is that each user is assigned a distinct PN sequence, so that even when L users transmit in the same frequency band simultaneously, the data from a particular user can be extracted with low bit error probability by correlating the received signal with the particular user's PN sequence. To provide the supporting analysis for this statement, we consider binary PSK modulation with coherent detection over a channel of total (spread) bandwidth W hertz with additive thermal noise of two-sided power spectral density $\mathcal{N}_0/2$ watts/Hz. Each of the L users transmits a message $m_i(t)$ with power P_i watts at the carrier frequency f_0 hertz and has an assigned PN signal $s_{ci}(t)$, $i = 1, 2, \ldots, L$. The transmitted signal due to all L users in the frequency band W is thus

$$s(t) = \sum_{i=1}^{L} \sqrt{P_i}\, s_{ci}(t) m_i(t) \cos[\omega_0 t + \theta_i], \tag{14.6.1}$$

where a random phase θ_i has been included, which is assumed statistically independent of all θ_j, $j \neq i$. The signal at the receiver front end is thus

$$r(t) = s(t) + n(t), \tag{14.6.2}$$

where $n(t)$ is the thermal noise.

The receiver consists of L parallel receivers, each of which multiplies the received signal by the carrier with an appropriate phase and by the appropriate PN sequence. Thus, in the kth parallel receiver,

$$r(t)s_{ck}(t) \cos[\omega_0 t + \theta_k] = \sum_{i=1}^{L} \sqrt{P_i}\, s_{ci}(t) s_{ck}(t) m_i(t) \cos[\omega_0 t + \theta_i] \times \cos[\omega_0 t + \theta_k] + s_{ck}(t) n(t) \cos[\omega_0 t + \theta_k]. \tag{14.6.3}$$

For the purposes of the present development, the PN sequences are assumed white and independent for all users, so after low-pass filtering, the product in Eq. (14.6.3) becomes

$$v(t) = \sqrt{P_k}\, m_k(t) + \sum_{\substack{i=1 \\ i \neq k}}^{L} \sqrt{P_i}\, s_{ci}(t) s_{ck}(t) m_i(t) \cos[\theta_i - \theta_k] + n_0(t), \tag{14.6.4}$$

where $n_0(t)$ is white Gaussian noise, since $n(t)$ is independent of all transmitted signals.

In Eq. (14.6.4), $\sqrt{P_k}m_k(t)$ is the desired signal and the remaining terms are considered to be noise. We assume that the transitions of $s_{ck}(t)$ and $s_{ci}(t)$ are aligned so that the product $s_{ck}(t)s_{ci}(t)$ is another PN sequence, say $s'_{ci}(t)$, such that the $s'_{ci}(t)$ are mutually orthogonal for all i. Thus, when computing the power in the unwanted signals in Eq. (14.6.4), all cross terms fall out, and total noise power is

$$P_n = \tfrac{1}{2} \sum_{\substack{i=1 \\ i \neq k}}^{L} P_i + \mathcal{N}_0 W. \tag{14.6.5}$$

To complete the analysis, we assume that the received signal power from each user is the same, $P_i = P_s$ for all i, so the bit error probability for the kth user is

$$P_{e,k} = \frac{1}{2} - \operatorname{erf}\left[\frac{2E_s}{\mathcal{N}_0 + (L-1)P_s/2W}\right]. \tag{14.6.6}$$

If we neglect thermal noise ($\mathcal{N}_0$ small), this simplifies to

$$P_{e,k} = \frac{1}{2} - \operatorname{erf}\left[\frac{2E_s}{(L-1)P_s/2W}\right], \tag{14.6.7}$$

or in terms of processing gain

$$P_{e,k} = \frac{1}{2} - \operatorname{erf}\left[\frac{4(W/B)}{(L-1)}\right]. \tag{14.6.8}$$

Note by comparing Eq. (14.6.7) with (14.3.6) that the other $L - 1$ users simply look like a noise jammer with power spectral density $(L-1)P_s/2W$, or by comparing Eq. (14.6.8) with (14.3.7) that the other $L - 1$ users represent a jamming margin of $(L-1)/2$. Therefore, the processing gain is reduced in direct proportion to $L - 1$.

EXAMPLE 14.6.1

In a CDMA system, we have a processing gain of 1000 and we desire that each user have a bit error probability of 10^{-5} or less. If we neglect thermal noise effects, how many users can the system support?

From Eq. (14.6.8) with $P_{e,k} = 10^{-5}$, we find that

$$\frac{4(W/B)}{L-1} = 18.2, \tag{14.6.9}$$

so

$$L = \lfloor 220.8 \rfloor = 220 \text{ users}, \tag{14.6.10}$$

where $\lfloor \cdot \rfloor$ represents "the integer part of."

In arriving at Eqs. (14.6.6)–(14.6.8), it was assumed that all users have the same received power level and that the PN sequences are mutually orthogonal. Of course, the problem of finding good PN sequences is a long-standing one, and mutual orthogonality is not easy to achieve in practice.

The assumption of equal received power levels for all users is not a good one, in general, without special system modifications. For ground-based systems such as cellular mobile radio, the received power levels from different users are often unequal since some users are closer to the receiver than the others. Furthermore, in other applications, the various users may actually transmit at different power levels. The consequences of variations in received power can be seen if we use Eq. (14.6.5) to write the bit error probability for the kth user,

$$P_{e,k} = \frac{1}{2} - \operatorname{erf}\left[\frac{P_k/B}{\mathcal{N}_0 + \frac{1}{2}\sum_{i=1,\, i\neq k}^{L} P_i/W}\right], \tag{14.6.11}$$

or neglecting thermal noise,

$$P_{e,k} = \frac{1}{2} - \operatorname{erf}\left[\frac{P_k(W/B)}{\frac{1}{2}\sum_{i=1,\, i\neq k}^{L} P_i}\right]. \tag{14.6.12}$$

If $P_k \ll P_i$ for all $i \neq k$ (or even some i), then $P_{e,k} > P_{e,i}$; that is, the bit error probability will vary as a function of the user.

This problem of unequal received power levels for different users is called the "near–far" problem in the literature and is the subject of much research effort. One system modification that compensates for unequal received power levels is to transmit a pilot tone from the receiver to the user so that the user can determine the path loss and hence the required transmitted power level. This technique is used by Gilhousen et al. [1990] in a mobile satellite application.

There is an interesting contrast between CDMA systems and FDMA or TDMA systems. In FDMA or TDMA, the number of users is constrained by the number of "slots" available in frequency or time, respectively. Once these slots are full, additional users can be accommodated only by a system redesign. For CDMA, however, the number of users is constrained by performance. New users can always be added, but the performance for all users will degrade according to Eqs. (14.6.6)–(14.6.8).

The assumption of mutual orthogonality used in deriving Eqs. (14.6.6)–(14.6.8) is difficult to attain for easily generated PN sequences. However, to minimize the "code noise" in a CDMA system, it is necessary that the cross-correlation between PN sequences be as low as possible. For two periodic sequences of ± 1s with period N, the periodic cross-correlation is defined as [Sarwate and Pursley, 1980]

$$\theta_{ab}(l) = \frac{1}{N} \sum_{k=1}^{N} a_k b_{k+l}. \tag{14.6.13}$$

Sarwate and Pursley [1980] have tabulated the periodic cross-correlation for maximal-length shift register sequences with $N = 2^n - 1$ and $3 \leq n \leq 12$. These

TABLE 14.6.1 Maximum Cross-Correlations for *m*-Sequences and Gold Sequences

		$\lvert\theta_{ab}(l)\rvert$	
n	$N = 2^n - 1$	m-Sequences	Gold Sequences
3	7	0.71	0.71
4	15	0.60	0.60
5	31	0.35	0.29
6	63	0.36	0.27
7	127	0.32	0.13
8	255	0.37	0.13
9	511	0.22	0.06
10	1023	0.37	0.06
11	2047	0.14	0.03
12	4095	0.34	0.03

Source: D. V. Sarwate and M. B. Pursley, "Crosscorrelation Properties of Pseudorandum and Related Sequences," *Proc. IEEE*, Vol. 68, May, pp. 593–619., © 1980 IEEE.

values are shown in Table 14.6.1, where it is evident that $\lvert\theta_{ab}(l)\rvert$ can be relatively large. Gold [1967, 1968] proposed a set of periodic sequences with good cross-correlation properties that are not maximal-length shift register sequences of length N, but can be derived from m-sequences. Their cross-correlations are also shown in Table 14.6.1. Note the reduction in $\lvert\theta_{ab}(l)\rvert$ for equivalent Gold sequences in comparison to m-sequences.

14.7 Code Acquisition

In our previous discussions of spread spectrum systems, such as the direct sequence systems in Section 14.3, we have always assumed that the received carrier frequency and phase are known exactly and that the despreading PN code is perfectly synchronized with the received PN code sequence. In reality, neither the carrier frequency/phase nor the correct PN code alignment (phase) of the received signal can be known a priori, and hence this information must be extracted by the receiver. The uncertainty in carrier frequency and phase can occur due to frequency drift in the transmitter local oscillator or as a result of Doppler shift caused by relative motion between the transmitter and receiver. Inaccurate PN code synchronization may result from the particular application, such as ranging (see Problem 14.6), or from drift in the code clock. Receiver synchronization with the received carrier frequency/phase and the received PN code sequence is one of the major challenges in spread spectrum system design, and in this section and Section 14.8 we touch on a few of the common approaches to obtaining this information for DS spread spectrum.

Figure 14.7.1 is a block diagram of a noncoherent demodulator for a DS spread spectrum system. There are three principal components of this receiver.

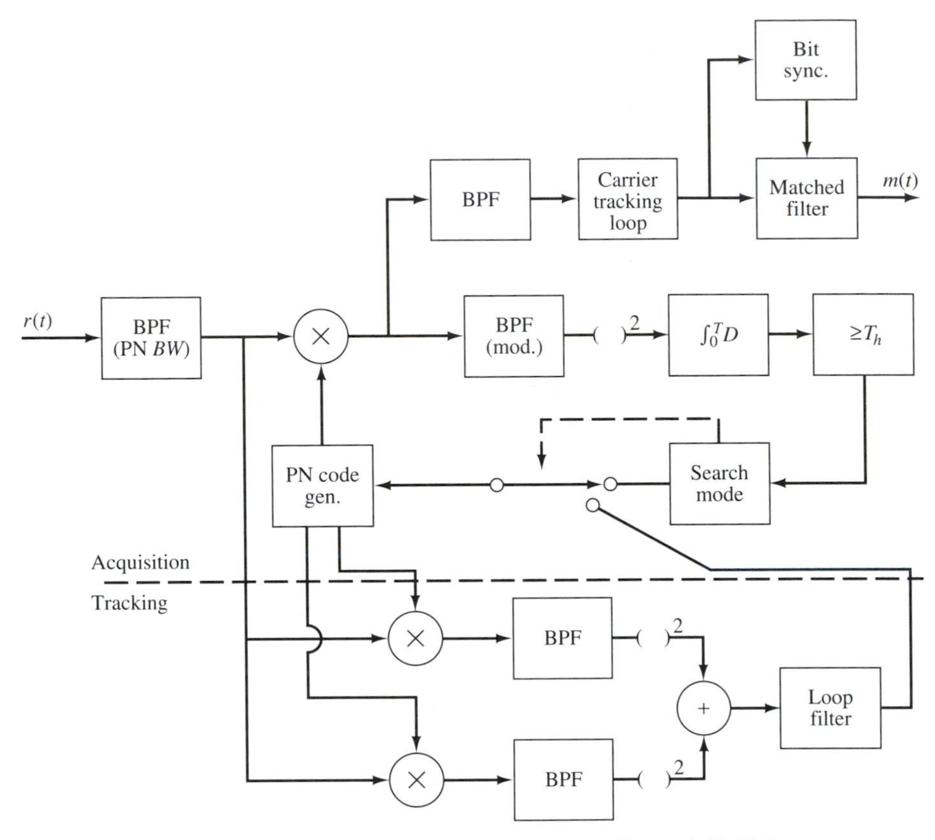

FIGURE 14.7.1 Noncoherent PN code demodulator. From J. K. Holmes, *Coherent Spread Spectrum Systems*, New York: John Wiley & Sons, Inc., © 1982. Reprinted by permission of John Wiley & Sons, Inc.

The uppermost branch of the figure is concerned with tracking the received carrier frequency and phase. The center part of the diagram, between the dashed lines, is dedicated to the initial acquisition of (coarse) PN code synchronization, while the bottom part of the figure, below the lower dashed line, achieves fine PN code synchronization and tracking. In this section we emphasize the initial acquisition stage of carrier and code synchronization, while in Section 14.8 we discuss the tracking stage.

The difficulties that arise as a result of inaccurate carrier phase and code synchronization can be demonstrated straightforwardly by letting the received signal $r(t)$ be given by Eq. (14.3.3), but now assuming that $r(t)$ is multiplied by an unsynchronized PN sequence, $s_c(t - \tau_d)$, and by a receiver local oscillator with an incorrect phase, $\cos(\omega_0 t + \phi)$, so that

$$\begin{aligned} r(t)s_c(t - \tau_d)\cos(\omega_0 t + \phi) = {} & As_c(t)s_c(t - \tau_d)m(t)\cos\omega_0 t\cos(\omega_0 t + \phi) \\ & + n(t)s_c(t - \tau_d)\cos(\omega_0 t + \phi) \\ & + s_J(t)s_c(t - \tau_d)\cos(\omega_0 t + \phi). \end{aligned} \tag{14.7.1}$$

To achieve coherent demodulation, we would normally low-pass filter this product, which in this situation gives

$$[r(t)s_c(t - \tau_d) \cos(\omega_0 t + \phi)]_{\text{message component}} = As_c(t)s_c(t - \tau_d)m(t) \cos \phi. \quad (14.7.2)$$

If the quantity in Eq. (14.7.2) is averaged over a PN sequence of length N chips, $Am(t) \cos \phi$ will be multiplied by $-1/N$ (see Problem 14.5) unless the PN sequence is perfectly synchronized ($\tau_d = 0$). Furthermore, if $\phi \neq 0$, the power in the message component is diminished by the factor $\cos^2 \phi$. Clearly, $\tau_d \neq 0$ or $\phi \neq 0$ can reduce the effectiveness (perhaps substantially) of the spread spectrum system.

The acquisition stage of PN synchronization can be very time consuming. Typically, the acquisition portion of the receiver in Fig. 14.7.1 operates as follows. For a given carrier frequency and phase, a particular PN code sequence time alignment is chosen and multiplied times the received signal. The resulting product is passed through the bandpass filter and to the integrator, where it is integrated over T_i seconds and compared to a threshold. If the threshold is exceeded, acquisition is achieved and the tracking stage is begun. If the threshold is not exceeded, the reference PN sequence is shifted by half (usually) a chip interval, and the process is repeated.

The search rate is thus $1/T_i$ half chip intervals ($T_c/2$) per second, so that for an initial disparity in PN sequence alignment of τ_d seconds, the time required to check all possible alignments is

$$T_{\text{acq}} = \frac{\tau_d}{(T_c/2)(1/T_i)} = \frac{2T_i\tau_d}{T_c} \qquad \text{seconds.} \quad (14.7.3)$$

Since the integration time is often some multiple, say K, of the PN sequence length, NT_c, Eq. (14.7.3) gives

$$T_{\text{acq}} = 2KN\tau_d \qquad \text{seconds.} \quad (14.7.4)$$

This result is for a given carrier frequency and phase. The search is noncoherent because of the squarer before the envelope detector, so the phase is not critical. However, if the frequency is not known to some accuracy, the preceding process may have to be repeated for several other frequencies because of the BPF. Certainly, it is evident from Eq. (14.7.4) that the acquisition time can be relatively long, since N is often large.

One way to reduce the acquisition time is to integrate over only one sequence length [$K = 1$ in Eq. (14.7.4)] or to integrate over only a portion of the PN sequence. In the latter case there are no longer any guarantees about the orthogonality of the PN sequences over this shorter interval, and the sequence autocorrelations become partial correlations and are random variables (see Appendix G and Problems 14.21 through 14.24). As a result, the probability of finding the correct synchronization is reduced for short integration times.

A method for achieving shorter acquisition times is to use a matched filter, where the filter is matched to all or part of the PN sequence being received. As the received signal $r(t)$ is shifted through the matched filter, the matched

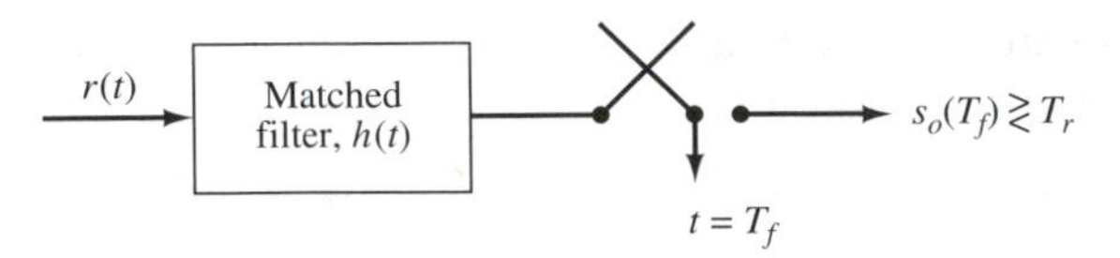

FIGURE 14.7.2 Matched filter acquisition.

filter output is monitored to see if it exceeds a threshold. If it does, acquisition is declared. The matched filter approach is depicted in Fig. 14.7.2.

To illustrate the method, let the matched filter impulse response be

$$h(t) = \begin{cases} s(T_f - t), & 0 \leq t \leq T_f \\ 0, & \text{elsewhere.} \end{cases} \tag{14.7.5}$$

Then, with $r(t)$ as in Eq. (14.3.3), the matched filter output evaluated at $t = T_f$ is

$$\begin{aligned} s_0(T_f) &= \int_0^{T_f} h(\lambda)r(T_f - \lambda)\, d\lambda \\ &= \int_0^{T_f} s(T_f - \lambda)[As_c(T_f - \lambda)m(T_f - \lambda) \cos \omega_0(T_f - \lambda) \\ &\quad + n(T_f - \lambda) + s_J(T_f - \lambda)]\, d\lambda \\ &= \int_0^{T_f} As(T_f - \lambda)s_c(T_f - \lambda)m(T_f - \lambda) \cos \omega_0(T_f - \lambda)\, d\lambda \\ &\quad + \int_0^{T_f} s(T_f - \lambda)n(T_f - \lambda)\, d\lambda + \int_0^{T_f} s(T_f - \lambda)s_J(T_f - \lambda)\, d\lambda. \end{aligned} \tag{14.7.6}$$

If the filter is matched to the signal component, that is,

$$s(T_f - t) = As_c(T_f - t) \cos \omega_0(T_f - t),$$

then the last two terms in Eq. (14.7.6) are still "noise" components, but the first term simplifies greatly, so

$$s_0(T_f) = \int_0^{T_f} A^2 m(T_f - \lambda) \cos^2 \omega_0(T_f - \lambda)\, d\lambda + P_n(T_f) + P_J(T_f). \tag{14.7.7}$$

Thus the first term is the desired signal component and $s_0(T_f)$ is compared to a threshold.

There are two difficulties with the matched filter approach, both of which are illustrated by Eq. (14.7.7). First, the desired signal component will be a maximum if T_f is an integer multiple of NT_c. Since N can be very large, the matched filter length can be quite long. Second, where to set the threshold to declare acquisition is difficult, since the energy in the remaining two terms of Eq. (14.7.7) can be large. To correct the latter problem, a two-path filtering process, sometimes called a self-referencing matched filter, is used. A block diagram of such a matched filter is shown in Fig. 14.7.3 [Cooper and McGillem, 1986]. The impulse response of the reference channel filter has the two requirements that

$$\int_0^{T_f} g(t)h(t)\, dt = 0$$

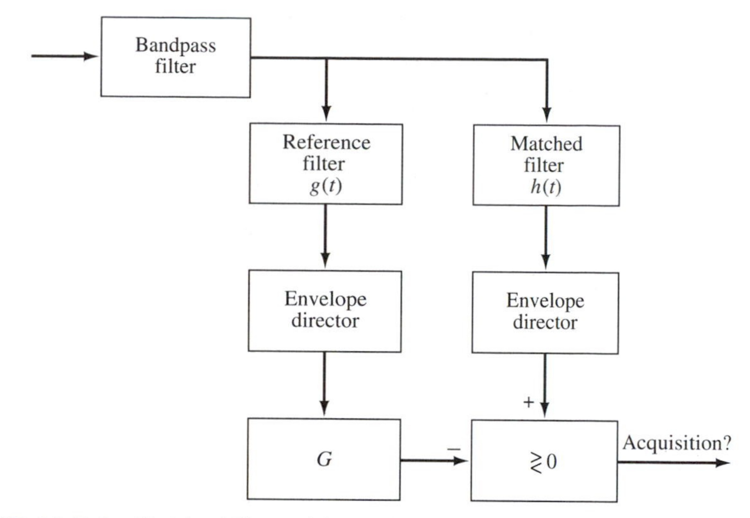

FIGURE 14.7.3 Matched filter with a reference channel. From G. R. Cooper and C. D. McGillem, *Modern Communications and Spread Spectrum*, New York: McGraw-Hill, Inc., 1986, p. 368. Reproduced with permission of McGraw-Hill, Inc.

and

$$\int_0^{T_f} g^2(t)\, dt = 1.$$

The reference branch thus gives a reference "noise" power level that can be adjusted by the gain G to produce the desired performance.

Another technique for reducing acquisition time is to transmit a short, periodic PN sequence called a *preamble* [Dixon, 1984]. A short PN sequence allows either the sliding correlation or the matched filter approach to be used to obtain synchronization before the message data are sent. This method is useful for commercial applications, as in Gilhousen et al. [1990], but for secure communications, the preamble can reduce system effectiveness, since they are more susceptible to jamming and detection by a jammer.

The discussion here has emphasized direct sequence spread spectrum. The treatment of acquisition methods for frequency hopping systems is left to the literature [Dixon, 1984; Proakis, 1989].

14.8 Code Tracking

Once the acquisition step brings the reference PN sequence within $T_c/2$ seconds of being synchronized, the switch in Fig. 14.7.1 is moved to change from the acquisition mode to the tracking mode. The tracking loop shown in the figure is noncoherent (because of the squaring operations), which allows for the possibility that the carrier frequency and phase are not known accurately due, perhaps, to Doppler effects. If the carrier frequency and phase are known exactly, a coherent loop or baseband loop can be used. To motivate the basic concepts, we begin the discussion with a baseband tracking loop.

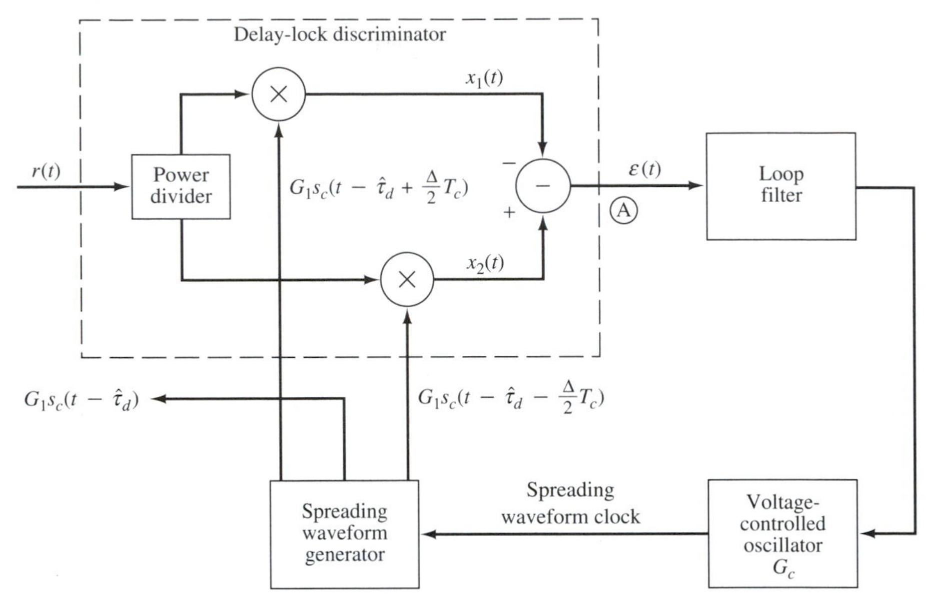

FIGURE 14.8.1 Baseband code tracking loop. From R. E. Ziemer and R. L. Peterson, *Digital Communications and Spread Spectrum Systems*, 1985. Copyright © 1985 by Macmillan Publishing Company. Reprinted with the permission of Macmillan Publishing Company.

A block diagram of a baseband code tracking loop is shown in Fig. 14.8.1. The loop is assumed to be preceded by coherent demodulation with accurate carrier frequency and phase, so that the signal presented to the loop input is

$$r(t) = As_c(t - \tau_d) + n(t), \tag{14.8.1}$$

where $s_c(t - \tau_d)$ is the PN sequence with an unknown delay τ_d, and $n(t)$ is additive noise. However, in the subsequent development, we neglect the noise, $n(t)$. The acquisition loop has produced an estimate of τ_d, denoted $\hat{\tau}_d$, that is within $\pm T_c/2$ seconds of the true value (τ_d). The function of the tracking loop is to keep the estimate of τ_d within this range and reduce it to zero if possible. To achieve this, an early–late tracking technique is employed, where after power division the received signal is passed to two branches, one of which is multiplied by an early version of the chip sequence, $s_c(t - \hat{\tau}_d + (\Delta/2)T_c)$, and the other which is multiplied by a late version of the chip sequence, $s_c(t - \hat{\tau}_d - (\Delta/2)T_c)$. Therefore, the output of the early multiplier is

$$x_1(t) = G_1 \frac{A}{\sqrt{2}} s_c(t - \tau_d) s_c\left(t - \hat{\tau}_d + \frac{\Delta}{2} T_c\right) \tag{14.8.2}$$

and the output of the late multiplier is

$$x_2(t) = G_1 \frac{A}{\sqrt{2}} s_c(t - \tau_d) s_c\left(t - \hat{\tau}_d - \frac{\Delta}{2} T_c\right), \tag{14.8.3}$$

where Δ is the time difference between early and late sequences divided by T_c.

From Fig. 14.8.1, the error signal becomes

$$\varepsilon(t) = G_1 \frac{A}{\sqrt{2}} s_c(t - \tau_d)\left[s_c\left(t - \hat{\tau}_d - \frac{\Delta}{2} T_c\right) - s_c\left(t - \hat{\tau}_d + \frac{\Delta}{2} T_c\right)\right], \quad (14.8.4)$$

which serves as input to the loop filter. The average value of $\varepsilon(t)$ over one sequence period, NT_c, is the quantity that is used for tracking. Thus

$$\begin{aligned} \varepsilon_{\text{avg}} &= \frac{1}{NT_c} \int_{-NT_c/2}^{NT_c/2} \varepsilon(t)\, dt \\ &= \frac{1}{NT_c} \int_{-NT_c/2}^{NT_c/2} G_1 \frac{A}{\sqrt{2}} s_c(t - \tau_d)\left[s_c\left(t - \hat{\tau}_d - \frac{\Delta}{2} T_c\right) - s_c\left(t - \hat{\tau}_d + \frac{\Delta}{2} T_c\right)\right] dt \\ &= G_1 \frac{A}{\sqrt{2}}\left[R_{s_c}\left(\tau_d - \hat{\tau}_d - \frac{\Delta}{2} T_c\right) - R_{s_c}\left(\tau_d - \hat{\tau}_d + \frac{\Delta}{2} T_c\right)\right]. \quad (14.8.5) \end{aligned}$$

If the chip pulses are rectangular with amplitudes ± 1, the autocorrelations, $R_{s_c}(\cdot)$, have the shape shown in Fig. P14.5. Plots of ε_{avg} from Eq. (14.8.5), appropriately normalized, are shown in Fig. 14.8.2. The tracking loop acts to drive ε_{avg} toward 0 and Δ is a design parameter. Note from Fig. 14.8.2(a) and (b) that as Δ is decreased from 2.0 to 1.0, the slope of the discriminator characteristic changes, so that $\Delta = 1.0$ yields a higher slope around the origin, but simultaneously, the tracking range [the distance along the $(\tau_d - \hat{\tau}_d)/T_c$ axis between positive and negative peaks] decreases. As Δ is reduced below 1.0, the slope remains the same, but the tracking range decreases still further. However, it can be shown that the variance of the tracking error decreases as Δ decreases for $\Delta \leq 1.0$ [Ziemer and Peterson, 1985, p. 432]. Thus, choosing Δ requires trade-offs among tracking range, rate of response, and tracking error variance.

The noncoherent tracking loop version of the baseband loop in Fig. 14.8.1 has bandpass filters and squarers added to each of the early and late branches, but its operation is essentially identical. These types of loops work well but they suffer from at least two drawbacks. First, the loop is relatively complex, requiring duplicate circuitry for the early and late branches and the generation of three PN sequences. Second, any mismatch or imbalance between the early and the late branch circuitry can degrade tracking performance. The *tau-dither code tracking loop* in Fig. 14.8.3 was invented to address both of these problems.

The tau-dither loop operates by dithering back and forth (according to a square wave) between the early and late sequences, so that we are working with only one autocorrelation function (see Fig. P14.5). Figure 14.8.4 shows how the dithering is used to generate a tracking error signal [Holmes, 1982]. In this figure, the correlations from both the early and late sequences lag the input code. This fact is available from the "correlated envelope" in the upper right portion of the figure because the equal positive and negative swings of the original dither signal are multiplied by different gains. If the tracking error is zero, the dithering takes place about $\tau = 0$ in $R(\tau)$ and the correlated envelope has equal positive and negative swings.

The discriminator characteristic of the tau-dither loop for various dither amplitudes (Δ) is shown in Fig. 14.8.5. Note that the characteristic looks much

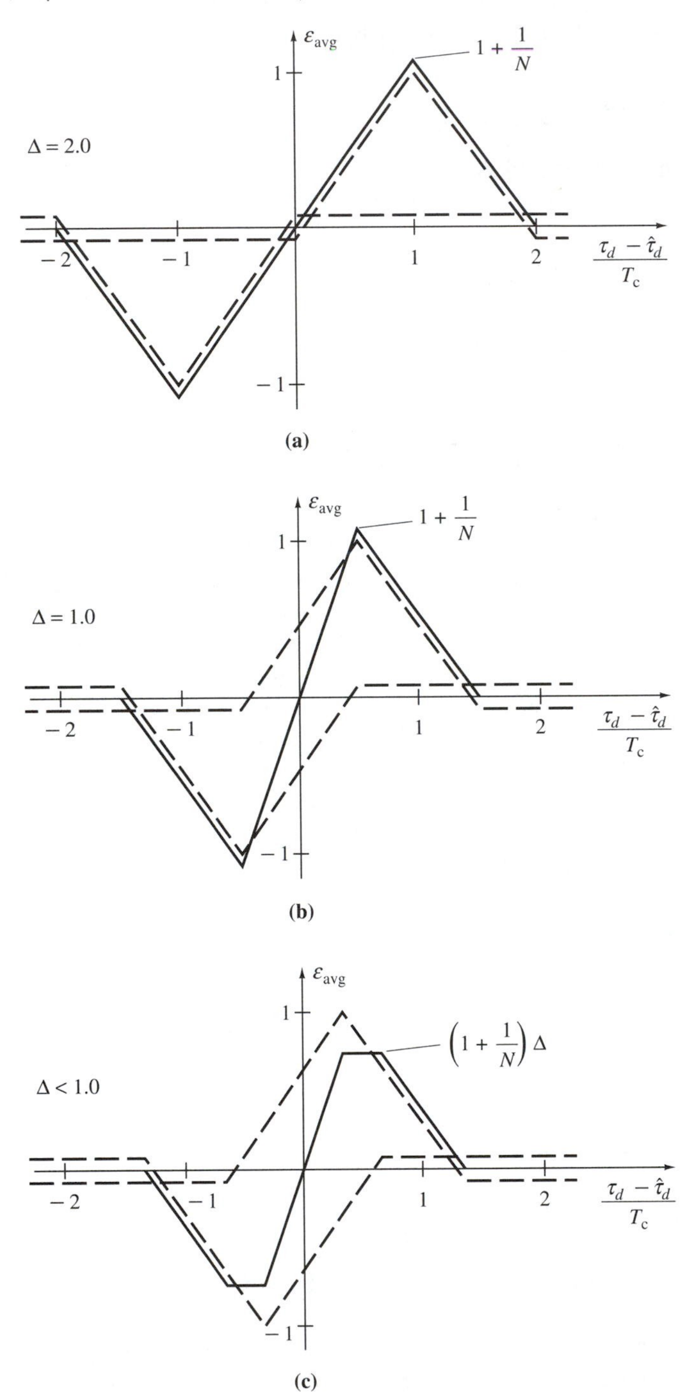

FIGURE 14.8.2 Baseband tracking loop discriminator characteristic.

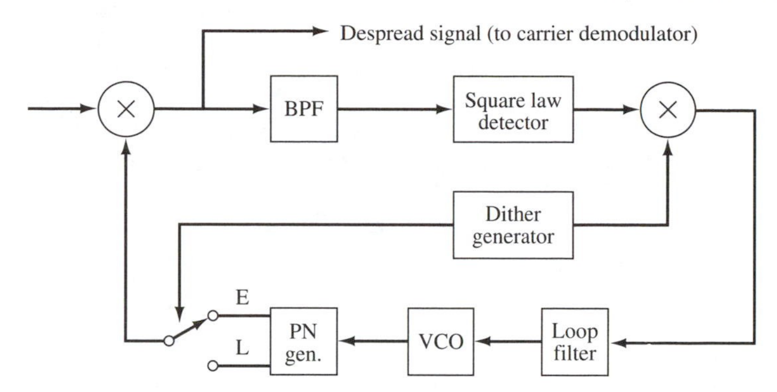

FIGURE 14.8.3 Tau-dither code tracking loop. From J. K. Holmes, *Coherent Spread Spectrum Systems*, New York: John Wiley & Sons, Inc., © 1982. Reprinted by permission of John Wiley & Sons, Inc.

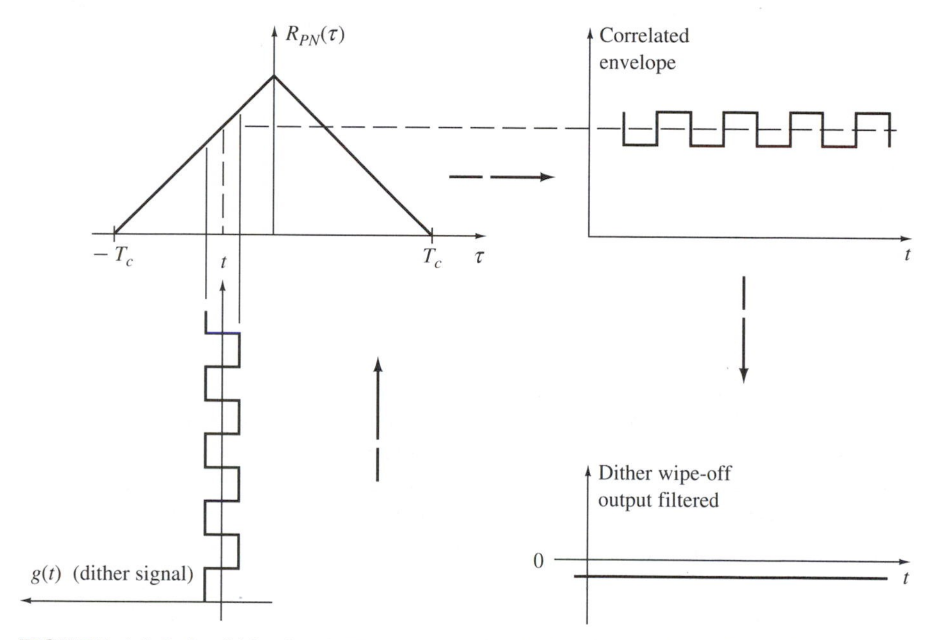

FIGURE 14.8.4 Dithering between early and late correlations. From J. K. Holmes, *Coherent Spread Spectrum Systems*, New York: John Wiley & Sons, Inc., © 1982. Reprinted by permission of John Wiley & Sons, Inc.

the same as those for the nondithered (called full-time) loop in Fig. 14.8.2. Our discussion has ignored noise, however, and the tau-dither loop requires a slightly higher signal-to-noise ratio for performance equivalent to the full-time loop [Ziemer and Peterson, 1985]. In fact, the tau-dither loop is often operated with $\Delta = 0.2 = 0.1T_c$, so that the tracking range is quite small [Holmes, 1982].

This concludes our discussion on code synchronization in DS spread spectrum systems. For more details and for information on frequency-hopping synchronization, the reader is referred to the literature [Ziemer and Peterson, 1985; Holmes, 1982; Dixon, 1984; Cooper and McGillem, 1986].

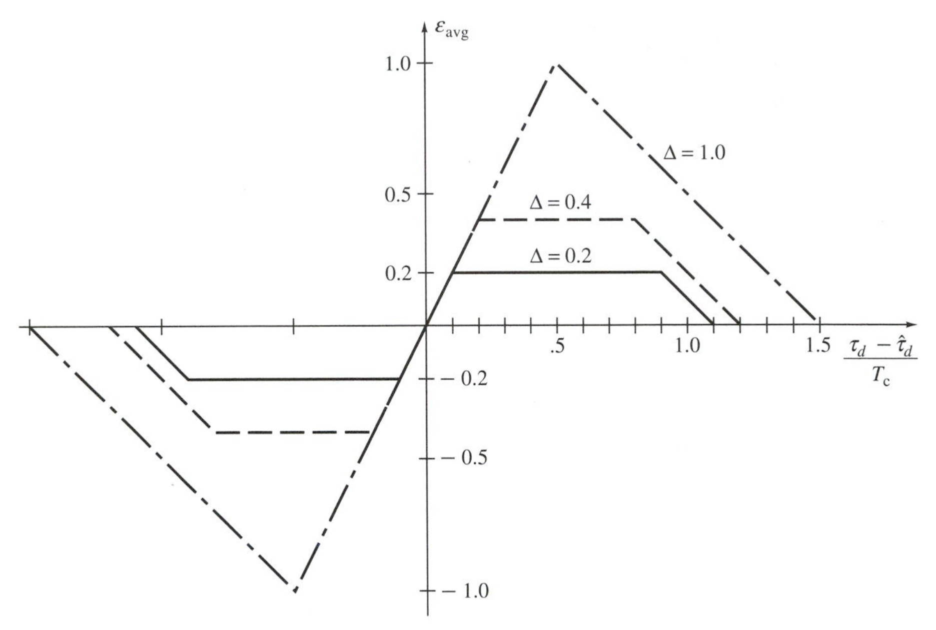

FIGURE 14.8.5 Discriminator characteristic for tau-dither loop. From J. K. Holmes, *Coherent Spread Spectrum Systems*, New York: John Wiley & Sons, Inc., © 1982. Reprinted by permission of John Wiley & Sons, Inc.

SUMMARY

We have presented the basic concepts underlying direct sequence and frequency hopping spread spectrum communications. The jammer-resistant properties of such systems were demonstrated, and applications to secure communications and multiple-access communications were discussed. The importance of diversity and error control coding to system performance was demonstrated. The critical issues of carrier phase and PN code sequence synchronization were examined and methods for acquisition and tracking of these quantities were described. Interest in spread spectrum communications, particularly for commercial systems, continues to grow [Schilling et al., 1990].

PROBLEMS

14.1 The connection between $s(t)$ in Eq. (14.2.1) and a DS spread spectrum signal is made explicit by letting $\phi(t)$ take the values $\pm 180°$ according to some pseudorandom sequence, so we can write

$$s(t) = m(t) \cos [\omega_0 t + \phi(t)] = m(t) \cos \phi(t) \cos \omega_0 t = m(t)s_c(t) \cos \omega_0 t,$$

where $s_c(t)$ is a sequence of pulses with amplitudes ± 1 and duration $T_c \ll T$ and T is the message symbol duration. Of course, the pseudorandom sequence of ± 1s, which constitute the amplitudes of the $s_c(t)$ pulses, is the chip sequence or the spreading signal. Note that $s_c(t)$ is independent of $m(t)$. Let the $s_c(t)$ pulses be rectangular and the interfering signal in Eq. (14.2.2) be a pure tone at the carrier frequency, so $v(t) = A \cos \omega_0 t$. Verify the claim that the power in the interference signal at the receiver output is proportional to $(B/W)P_J$.

14.2 Draw a block diagram of the receiver that produces the statistics in Eqs. (14.2.13) and (14.2.14). Why is the SNR in Eq. (14.2.15) a valid performance indicator?

14.3 Show that spread spectrum provides no performance improvement against additive white thermal noise.

14.4 For a processing gain of 1000, what is the jamming margin (in dB) of a DS spread spectrum system against a continuous noise jammer if the desired $P_e = 10^{-6}$? How does this change for operation against a pulsed noise jammer that jams $\gamma = 0.01$ of the DS pulses?

14.5 Let a DS pseudonoise spreading signal be given by

$$s_c(t) = \sum_{n=-\infty}^{\infty} b_n p(t - nT_c),$$

where we assume here that $p(t)$ is rectangular,

$$p(t) = \begin{cases} 1, & 0 \le t \le T_c \\ 0, & \text{otherwise,} \end{cases}$$

and the b_n are a sequence of ± 1s with periodic autocorrelation given by Eqs. (G.2.1) and (G.2.2). Show that

$$R(\tau) = \frac{1}{NT_c} \int_0^{NT_c} s_c(t) s_c(t + \tau)\, dt$$

is as shown in Fig. P14.5.

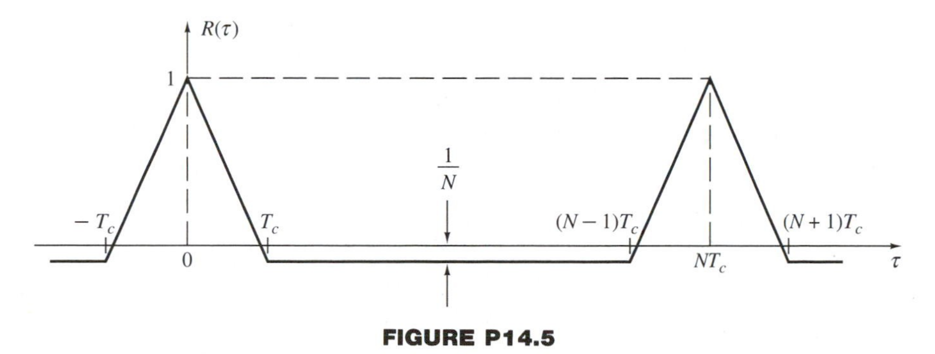

FIGURE P14.5

14.6 A common application of PN signals is for ranging measurements. Here the transmitted PN signal is reflected from a target and the received (reflected) signal is correlated with the transmitted signal to determine the target range. Let the PN signal that is transmitted (using appropriate modulation) be $s_c(t)$, as given in Problem 14.5. Let the received PN sequence (after appropriate demodulation) be given by $r_c(t) = s_c(t - \tau)$. Calculate

$$R(\tau_d) = \int_0^{NT_c} s_c(t - \tau)s_c(t - \tau_d)\, dt$$

and hence show how this technique works. What is a limit on the accuracy of this ranging method?

14.7 Compute the jamming margin to achieve $P_e \cong 10^{-6}$ for the cases where the processing gain is 100, 1000, and 10,000.

14.8 Suppose that a DS spread spectrum system uses binary DPSK to transmit the data sequences. The channel adds thermal noise with two-sided power spectral density $\mathcal{N}_0/2$ watts/Hz and the transmitted signal is also contaminated by an additive continuous jamming signal modeled as white Gaussian noise with a two-sided power spectral density of $\mathcal{N}_J/2$ watts/Hz.

(a) Write an expression for the bit error probability of this system.

(b) Assume that $\mathcal{N}_J \gg \mathcal{N}_0$, and write P_e in terms of the processing gain and jamming margin.

(c) Plot P_e from part (b) versus $E_s/\mathcal{N}_J$.

14.9 Rework parts (a) and (b) of Problem 14.8 if the jammer transmits pulses of wideband noise for only a fraction γ of the time. Plot the expression corresponding to part (b) versus $E_s/\mathcal{N}_J$ for $\gamma = 1.0$, 0.1, and 0.01.

14.10 Using Eq. (14.3.5), plot P_e versus $E_s/\mathcal{N}_J$ for $E_s/\mathcal{N}_0 = 5$, 10, and 15 dB. Compare to Fig. 14.3.2 with $\gamma = 1$.

14.11 Find an expression for the bit error probability of a slow FH spread spectrum system using binary DPSK modulation against a partial band jammer. Plot P_e versus $E_s/\mathcal{N}_J$ assuming that thermal noise is negligible for several values of the fraction of the band jammed, γ. Find the optimum γ and the resulting P_e.

14.12 Plot Eq. (14.5.4), that is, P_e versus $E_s/\mathcal{N}_J$, for $L = 1$, 2, and 3, $E_s/\mathcal{N}_0 = 5$, 10, and 15 dB, and $\gamma = 0.01$.

14.13 In a CDMA spread spectrum system, we desire that each user has a bit error probability of 10^{-6} or less. Neglecting thermal noise effects, calculate the number of users for processing gains of 1, 10, 100, and 1000.

14.14 Plot $P_{e,k}$ in Eq. (14.6.6) versus $E_s/[(L-1)P_s/2W]$ as a function of $E_s/\mathcal{N}_0 = 5$, 10, and 15 dB, with $L = 101$.

14.15 To illustrate the near–far problem, let $P_i = P^*$ for $i \neq k$, and $P_k = 50P^*$.

(a) Write an expression for the bit error probability of user k. Repeat for user $j \neq k$.

(b) If we desire a bit error probability of at least 10^{-5} for all users, how many users can this system support if the processing gain is 1000?

14.16 An often-used measure of distinguishability between two signals, $s_1(t)$ and $s_2(t+\tau)$, is the average power over some interval T seconds of the difference between the two; that is,

$$d = \frac{1}{T}\int_0^T [s_1(t) \pm s_2(t+\tau)]^2\, dt,$$

where the + sign is included to allow the possibility of the signal $-s_2(t+\tau)$.

(a) Show that d will be large, in general, only if

$$R_{12}(\tau) = \frac{1}{T}\int_0^T s_1(t)s_2(t+\tau)\, dt$$

is small.

(b) Let

$$s_1(t) = \sum_{k=-\infty}^{\infty} s_{1k}p(t - kT_c)$$

and

$$s_2(t) = \sum_{k=-\infty}^{\infty} s_{2k}p(t - kT_c).$$

If $p(t)$ is a unit rectangular pulse of length $T_c = T/N$, show that

$$R_{xy}(\tau) = T_c\theta_{xy}(l') + (\tau - l'T_c)[\theta_{xy}(l'+1) - \theta_{xy}(l')], \qquad 0 \le \tau \le T,$$

where l' is the largest integer such that $l'T_c \le \tau$ [Sarwate and Pursley, 1980].

14.17 Note from Table 14.6.1 that there can be a relatively large number of m-sequences and that the number available must be at least as great as the number of users in a CDMA system. Assume that the number of m-sequences needed is relatively large and that their maximum cross-correlation is relatively high. Show how these assumptions affect the derivation of $P_{e,k}$ in Eq. (14.6.6).

14.18 The aperiodic cross-correlation function defined by [Sarwate and Pursley, 1980]

$$C_{xy}(l) = \begin{cases} \sum_{k=1}^{N-l} x_k y_{k+l}, & 1 \le l \le N \\ \sum_{k=1}^{N+l} x_{k-l} y_k, & -N \le l \le 0 \\ 0, & |l| \ge N, \end{cases}$$

is also of importance in direct sequence spread spectrum systems. To see this, we consider the following situation. A user wishes to transmit the binary message sequence $\{c_k\} = \{\ldots, c_{-1}, c_0, c_1, \ldots\}$ and multiplies this sequence by the unique PN sequence $\mathbf{a} = \{a_1, a_2, \ldots, a_N\}$ to get the transmitted sequence $\{\ldots, c_{-1}\mathbf{a}, c_0\mathbf{a}, c_1\mathbf{a}, \ldots\}$. Another user is simultaneously transmitting the data sequence $\{\ldots, d_{-1}, d_0, d_1, \ldots\}$ using a different PN sequence $\mathbf{b} = \{b_1, b_2, \ldots, b_N\}$. Show that the output of the first user's correlation receiver is

$$z_k = c_k\theta_a(0) + \left[d_{k-1}\sum_{i=1}^{l} b_{N-l+i}a_i + d_k\sum_{i=l+1}^{N} b_{i-l}a_i\right].$$

Note the presence of the aperiodic cross-correlations [Sarwate and Pursley, 1980].

14.19 The union bound technique can be employed to obtain a bound on the error probability of a DS spread spectrum system in the presence of jamming. Furthermore, it is possible to include the effect of an (n,k) block code on system performance by using the approach in Problem 12.42.

The system operates in the following fashion. One information bit arrives at the encoder every T seconds. The encoder uses an (n,k) block code to take k information bits and map them into n coded bits that occupy the time interval kT seconds. As in Section 14.3, the spreading (PN) sequence is a sequence of ± 1s, each of duration T_c seconds, $T_c \ll T$, and where we assume that $L = T/T_c$ is an integer. The PN sequence multiplies the output of the encoder to produce a binary sequence to be presented to the BPSK modulator.

To make this specific, let $\{b_k\}$ be the PN sequence of $\{0,1\}$ (binary) symbols, $\mathbf{c}_i = [c_{i1} \quad c_{i2} \quad \cdots \quad c_{in}]$ be the ith n-bit codeword of the (n,k) code, and $p(t)$ be the basic pulse shape, so that the jth waveform supplied to the BPSK modulator is

$$s_j(t) = (2b_j - 1)(2c_{ij} - 1)p(t - jT_c).$$

For coherent PSK demodulation, the received waveform for the jth bit is

$$r_j(t) = (2b_j - 1)(2c_{ij} - 1)p(t - jT_c) + \eta(t).$$

A soft decision decoder forms the statistic

$$Z_j = \sum_{i=1}^{n} (2c_{ij} - 1)r_j, \qquad j = 1, 2, \ldots, M.$$

(a) Let $\mathbf{c}_1 = [0 \quad 0 \quad \cdots \quad 0]$, and assuming that $\mathbf{c}_1$ is transmitted, specify $Z_j, j = 1, 2, \ldots, M$.

(b) Assume that the additive interference terms are Gaussian and show that the probability of deciding codeword l given that codeword 1 is transmitted is

$$P[Z_l > Z_1 | \mathbf{c}_1 \text{ transmitted}] = \frac{1}{2} - \text{erf}\left(\sqrt{\frac{E_b}{2\mathcal{N}_J}}\, R_c w_l\right),$$

where $\mathcal{N}_J$ is the interference power spectral density and w_l is the Hamming weight of the lth codeword.

(c) Use the union bound and the assumption that the codewords are all equally likely to find

$$P_e \le \sum_{l=2}^{M} \left[\frac{1}{2} - \mathrm{erf}\left(\sqrt{\frac{E_b}{2\mathcal{N}_J} R_c w_l}\right)\right] < M\left[\frac{1}{2} - \mathrm{erf}\left(\sqrt{\frac{E_b}{2\mathcal{N}_J} R_c d_{\min}}\right)\right].$$

14.20 Express the bounds in Problem 14.19 in terms of processing gain and jamming margin. Note that there is a third factor, the coding gain. If an $(n,1)$ binary repetition code is used, what is the processing gain? What is the coding gain [Proakis, 1989]?

14.21 Consider the partial correlation of a PN sequence as given by Eq. (G.3.1),

$$\theta_{b'}(l, l', N_w) = \frac{1}{N_w} \sum_{k=l'}^{l'+N_w} b'_k b'_{k+l}.$$

Assume that the b'_k take on the values ± 1 with equal probability.

(a) Calculate $E\{\theta_{b'}(l, l', N_w)\}$.

(b) Calculate $\mathrm{var}\{\theta_{b'}(l, l', N_w)\}$.

Compare with Eqs. (G.3.2) and (G.3.3).

14.22 Derive the mean and variance of the partial correlations for m-sequences given by Eqs. (G.3.2) and (G.3.3).

14.23 Define the partial autocorrelation function as

$$R_{N_w}(\tau) = \frac{1}{N_w T_c} \int_0^{N_w T_c} s_c(t) s_c(t + \tau)\, dt,$$

$N_w < N = 2^n - 1$, the m-sequence length. Letting $s_c(t)$ be given as in Problem 14.5, show that

$$R_{N_w}(\tau) = \begin{cases} R_p(\tau) + \dfrac{\theta_{b'}(1, l', N_2)}{N_w} R_p(\tau - T_c), & 0 \le \tau \le T_c \\ \dfrac{1}{N_w}[\theta_{b'}(1, l', N_w) R_p(\tau - T_c) & T_c \le \tau \le 2T_c, \\ \quad + \theta_{b'}(2, l', N_w) R_p(\tau - 2T_c)], & \end{cases}$$

where

$$R_p(\tau) = \frac{1}{T_c} \int_{-\infty}^{\infty} p(t) p(t + \tau)\, dt.$$

14.24 Calculate and plot the mean and variance of $R_{N_w}(\tau)$ in Problem 14.23.

14.25 A more quantitative analysis of the sliding correlator for PN code sync acquisition is possible. To do this, let the received signal be $r(t) = A s_c(t) m(t) \cos \omega_0 t + n(t)$ (no jamming), and let the receiver be of the form shown in Fig. P14.25. Assume that the additive noise is Gaussian and zero mean with two-sided power spectral density $\mathcal{N}_0/2$ watts/Hz.

(a) Find an expression for the output signal as a function of τ_d and ϕ.

(b) Show that $\tau_d = 0$ and $\phi = 0$ maximize the output signal-to-noise ratio.

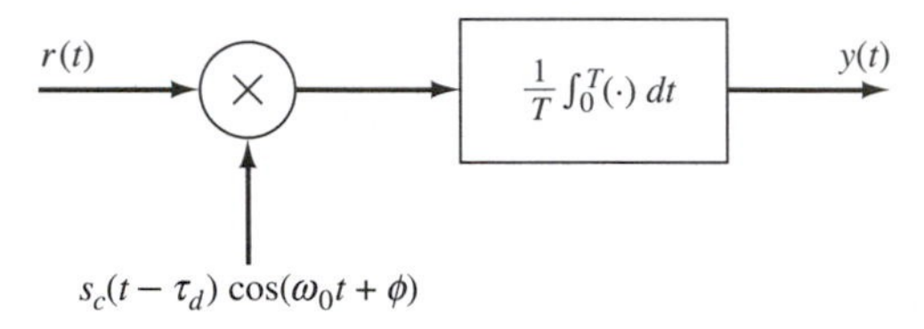

FIGURE P14.25

14.26 Consider the 7-bit PN sequence sketched in Fig. P14.26.

(a) Sketch the impulse response of a matched filter for this sequence.

(b) If the sequence in Fig. P14.26 is input to the matched filter in part (a), calculate and plot the matched filter output for various shifts. Assume that the input is periodic and integrate the matched filter over $7T_c$ seconds.

(c) Repeat part (b) if the matched filter integration time is $5T_c$ seconds.

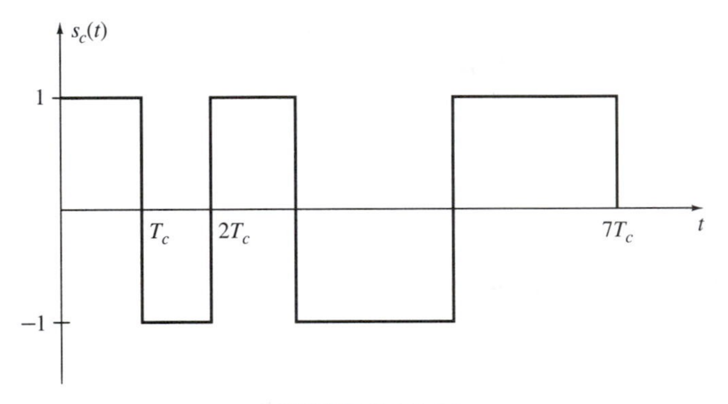

FIGURE P14.26

14.27 Sketch the discriminator characteristic for the baseband code tracking loop in Fig. 14.8.1 if $\Delta = 0.5$.

14.28 Calculate the power spectral density of the PN sequence in Problem 14.5.

14.29 Consider a message $m(t)$ consisting of an equally likely, random sequence of rectangular pulses with amplitudes ± 1 and width T. Use the result of Problem 14.28 to find the power spectral density of the modulated PN sequence for different relationships between T and NT_c.

14.30 With reference to Eq. (14.7.2), consider the case where $T \ll NT_c$, and note that the desired message signal component is reduced on the average by $(1/NT_c)\int_0^{NT_c} s_c(t)s_c(t-\tau_d)\,dt$. Let $\phi = 0$, and plot P_e versus signal-to-noise ratio for $\tau_d \in [-T_c, T_c]$.

CHAPTER 15

Computer Communication Networks

15.1 Introduction

Telecommunications networks are an unavoidable part of everyday life in the 1990s, from the ubiquitous telephone network to the many networks that connect computers, computer peripherals, and computer terminals in our businesses and on our campuses. The need for communications networks is evident—in order to transfer a message from one physical location to another, we need to provide some sort of connection between these two points. Equally obvious is the fact that we cannot provide permanent connections between all possible pairs of users. Instead, we provide a limited number of connections between various special locations (say, nodes), and these several nodes can be connected or disconnected (switched) to yield a diversity of possible communication paths. One of the principal ways of categorizing communications networks is according to how this switching function is performed.

Communications networks are usually classified as using circuit switching, packet switching, or message switching. Circuit switching has the property that once a connection is established, a fixed maximum transmission rate of (say) R_t bits/sec is allocated to this connection for the entire duration of the session. This is true whether messages are being transmitted continuously or just intermittently. In message switching, sometimes called store-and-forward switching, an established connection is used at its maximum rate when a message needs to be transmitted and the connection is not already in use. After the transmission of one message is complete, the connection is available for transmission of another message, perhaps for a completely different pair of users than the first message. Thus the transmission capacity of the connection can be fully utilized, even if each of the several different messages require transmission only intermittently. This switching technique is often called store-and-forward, since once a message has been transmitted from one connection point, called a node, to the next connection point, the message may have to be stored until a connection is available to another node in the network.

Packet switching is a version of store-and-forward switching, where the message length is limited to relatively short blocks of data called packets. Each packet must have extra bits added to it, called a header, in order to specify routing information for the packet. Packet-switched networks usually employ either virtual circuit (connection-oriented) routing or datagram (connectionless) routing. In virtual circuit routing, a fixed, specific path through the network is initially set up, and all packets are sent over this path for the duration of the session. Thus, although this method is similar to circuit switching in that a fixed path is used, the transmission capacity of a particular connection or link is used only as required and is not dedicated to a particular session between users. In datagram routing, each packet may take a different route through the network than adjacent packets or any other packet. Packets in a virtual circuit network require less header information than for datagrams, and datagram packets are not guaranteed to arrive in any order, and in fact, packets may be lost. Datagram routing allows more flexibility for the user.

Circuit switching is commonly used in the telephone network that is so familiar to us, and a concise discussion of circuit switching is given in Appendix E. The present chapter is devoted to providing a development of packet-switched networks, with the goal of introducing the user to the nomenclature and technical issues involved with such networks.

Of course, one of the first questions to be asked is why do we need a packet-switched network in addition to the already available, and easily accessible, circuit-switched telephone network? That is, why not use FDM or TDM over the existing telephone network to send the data? This question is answered in some detail later in the chapter, but we can provide a somewhat heuristic motivation for packet-switched networks here [Bertsekas and Gallager, 1987]. A packet-switched network operates as a statistical multiplexor that combines all packets arriving at a node into a single queue, and the packets are transmitted over the available link on a first-in first-out basis. The efficiency of the approach comes from the fact that the total transmission rate of the link is devoted to each packet when it moves to the front of the queue. So if the link data rate is C bits/sec and the packet is m bits long, the time to transmit this packet is m/C seconds. If we now consider a TDM link of capacity C bits/sec serving n users, the data rate available to any single user is C/n bits/sec, and the time required to send a message m bits long is mn/C seconds (assuming the TDM time slots are a small fraction of the length of a message—an identical result is obtained if the time slots are large enough to accommodate the entire m-bit message, since the wait between message blocks is $n - 1$ times the length of the time slot). Thus the delay for an individual packet in TDM is n times as long as the delay in statistical multiplexing, on the average. A similar result holds for FDM (see Problem 15.1).

In addition to classification according to virtual circuit or datagram routing, packet-switched networks may also be classified according to geographical coverage. Local area networks (LANs) cover an area with a diameter up to a few kilometers and may be limited to a single building or include an entire university campus. Metropolitan area networks (MANs), which is a relatively new term, have a geographical coverage of a few tens of kilometers. Wide area networks (WANs) may have up to worldwide coverage [Spragins et al., 1991]. There are other inherent differences among these network types other than geographic coverage. LANs typically have wide bandwidth, since they are locally owned and the owners lay their own cable, whereas WANs generally use the public telephone network in some way, with its limited bandwidth transmission media originally designed for another purpose. Thus LANs operate at megabit per second or higher data rates, while WANs operate at lower data rates and must be very efficient with bandwidth usage, since there is much less available. Further, the WANs rely more heavily on point-to-point communication links than LANs, which typically employ multiple access techniques [Tanenbaum, 1988].

The routing of packets through a network and the interconnection of networks require the establishment of rules that specify the content and format of the data in the packets. These rules are *protocols.* There can be as many different protocols as there are networks since the same information can be sent within a network in many different ways. However, in order for different networks to be interconnected and for the equipment from several manufacturers to be installed interchangeably, it is desirable to establish standards wherever possible. We begin our discussion in Section 15.2 with an overview of the Open Systems Interconnection (OSI) standard, other protocols, and common network architectures. Section 15.3 becomes much more quantitative, with a presentation of background material on queueing theory needed for the rest of the chapter. Multiple access schemes are described and compared in Section 15.4, and routing and flow control issues and algorithms are examined in Section 15.5.

15.2 Network Architectures

The idea of having available for our use a nationwide, or perhaps a worldwide, communications network that can handle data files or just about any type of message has extraordinary appeal. Anyone who has used Internet or a local area network for electronic mail or data file transfer has experienced the connectivity and convenience that such networks can provide. However, a network that is widely distributed geographically, and which cuts across institutional and political lines, is far from trivial to design. Certainly, each individual user will have his or her own special needs and will want to choose which equipment satisfies these requirements at an affordable cost. Thus the network must be capable of connecting a vast array of terminals, computers, and even smaller networks, all of which consist of equipment purchased from possibly different manufacturers with different hardware and software interfaces.

The International Standards Organization (ISO) began to address this issue in 1978 by developing the Open Systems Interconnection (OSI) Reference Model as a guideline for standardizing the interconnection of networks. The adjective "open" in the model title is intended to convey that any two systems in the world that conform to the OSI Reference Model standards can communicate with each other, irrespective of their differences. To achieve this standardization, the OSI model uses a layered approach that separates the several functions required and allows considerable flexibility.

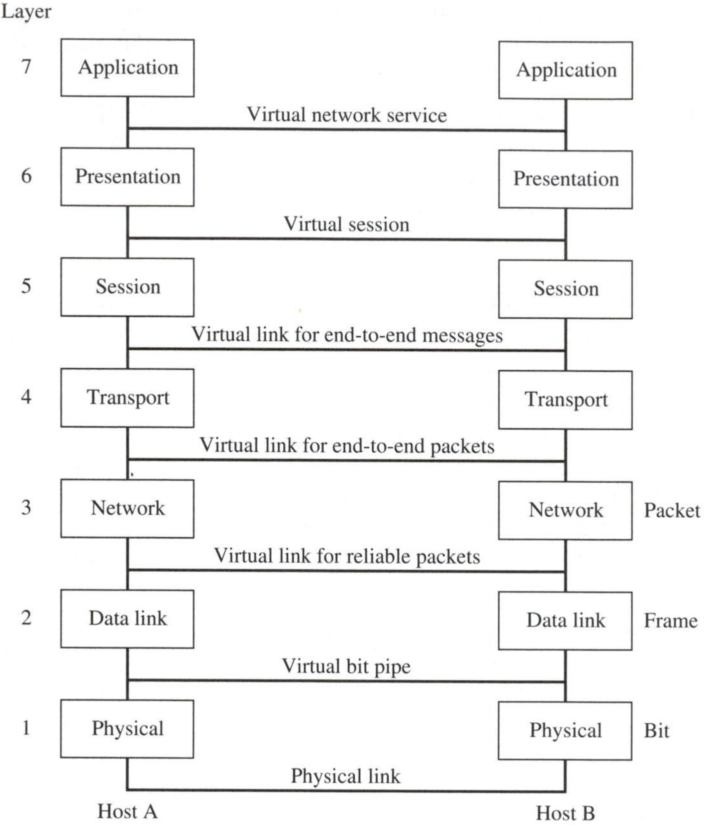

FIGURE 15.2.1 Seven-layer OSI network architecture. Adapted from Dimitri Bertsekas and Robert Gallager, *Data Networks*, © 1987. Reprinted by permission of Prentice Hall, Englewood Cliffs, New Jersey.

The seven layers of the OSI model are depicted in Fig. 15.2.1. Layers 1 to 3, the bottom three layers, are (usually) the responsibility of the network, while the top four layers, 4 to 7, reside in the host system. The physical layer transmits the bit stream over the communications channel, called the physical link in Fig. 15.2.1, and is concerned with issues of the type we have discussed in Chapters 8 and 9. The data link layer attempts to present to the network layer data that are error-free. To accomplish this, the data link layer structures the data into frames and adds header bits to each frame to delineate the beginning and ending of a frame and to detect errors. Additionally, it must send and receive an acknowledgment if the frame is received correctly.

The network layer and the data link layer operate asynchronously with respect to each other in that the data link layer handles frames of data created at that layer, but the network layer handles the basic network data unit, the packet. The network layer is responsible for routing the packets through the network and controlling congestion in the network. These issues are discussed in Section 15.5. The transport layer is the true interface layer between the network services and the host-provided services. The transport layer resides in the host system and it breaks the data received from the session layer into packets and then puts the packets back together to form messages at the receiver. It also performs multiplexing, and it sets up and tears down connections through the network.

The session layer allows the various services to be set up between host machines and performs some functions such as token exchange and accounting. The presentation layer handles data formatting or code conversion and may also provide services such as data compression and encryption. The application layer provides certain services, such as file transfer, remote job entry, and electronic mail, to name a few. It is sometimes stated that the presentation layer is concerned with syntax while the application layer is concerned with semantics [Schwartz, 1987]. The actual physical connection is between physical layers, and if there is an intervening node or nodes between hosts, the node contains only layers 1 to 3, the physical, data link, and network layers, as illustrated in Fig. 15.2.2.

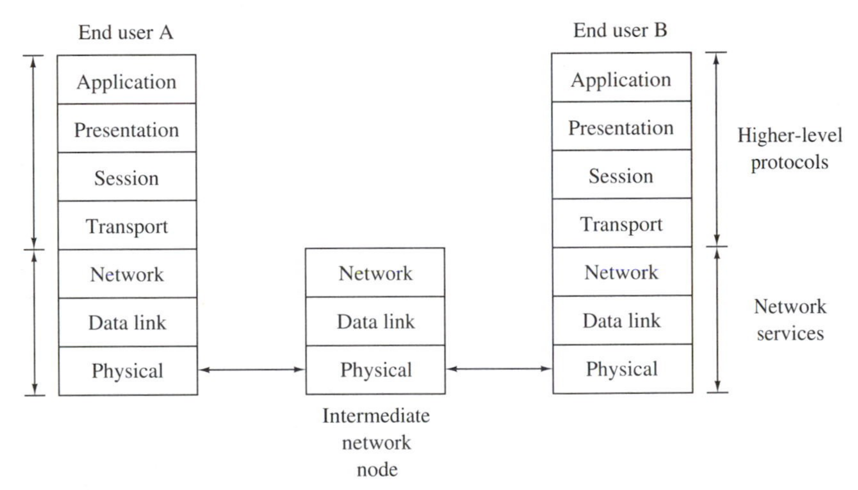

FIGURE 15.2.2 Functional diagram of the layered network architecture. Adapted from M. Schwartz, *Telecommunications Networks: Protocols, Modeling, and Analysis*, © 1987, Addison-Wesley Publishing Company, Inc. Reprinted with permission of the publisher.

If we have data to send at host A in Fig. 15.2.1, these data are presented to the application layer, which then appends a header or protocol, denoted by AH, to the data. The presentation layer then accepts the data with the appended header AH and attaches a header of its own, PH. Figure 15.2.1 makes the point that there is peer-to-peer communication; that is, layer N in host A communicates with layer N in host B, by showing the several virtual connections that are established. Layer N in a host does not communicate with adjacent layers $N + 1$ and $N - 1$, it simply treats what passes between them as data. Thus the physical layer provides a virtual bit pipe, the data link layer provides a virtual link for reliable packets, and so on.

It is evident that layered architectures allow a great deal of flexibility and hence can be useful for standardization purposes. However, how many layers are needed and which functions to associate with which layer are open for debate. The OSI model is but one possibility arrived at by committee. Other layered architectures, such as the Xerox Network Systems (XNS) architecture, IBM's System Network Architecture (SNA), and the Digital Equipment Corporation (DEC) Digital Network Architecture (DNA), exist and may or may not conform to the OSI model at each layer [Keiser, 1989]. There is also a set of standards called the IEEE 802 family that covers local area networks. Figure 15.2.3 shows the IEEE 802 standards and their relationship to the ISO standard. The data link layer is subdivided into a media access control (MAC) layer and a logical link control (LLC) layer. The IEEE 802.1 standard provides an overview of the other 802 standards and describes the relationship of these standards with the OSI model and other higher-level protocols. The MAC layer also takes on certain functions covered by the physical layer in the OSI model.

A key point is that the IEEE 802 standards do not specify a network layer or a layer corresponding to the network layer. The reason for this is that the physical, MAC, and LLC layers provide the capability for accessing a network, but they do not consider switching and routing functions. However, local area networks do not require these functions since users have a single communication link.

Many architectures and protocols have grown up outside the OSI Reference Model, and along with these architectures and protocols, there is a vast array of nomenclature and jargon. We briefly survey some of the more common terms here.

The physical layer protocol specifies the electrical, mechanical, and procedural interface between the data communications equipment (DCE) and the data terminal equipment (DTE). The most familiar physical layer standards are RS-232-C, RS-449, and X.21. The RS-232-C standard has existed for a long time and has the restrictions of a data rate of 20 kbits/sec or less and a maximum cable length of 15 meters. The RS-449 standard replaces RS-232-C but has two electrical standards, one that is similar to RS-232-C, designated RS-423-A, and one that allows higher speeds and longer cables, denoted as RS-422-A. The X.21 standard was adopted by CCITT to specify a common interface with digital lines such as T1 links. X.21 is part of the CCITT standard X.25 that specifies protocols for the three lowest OSI layers, layers 1 to 3.

Protocols for the data link control (DLC) layer have slightly different names, depending on the network or the standards-setting organization. IBM's data link protocol for SNA, called synchronous data link control (SDLC), was modified by the ISO to get the high-level data link control (HDLC) protocol, which was then modified by CCITT to generate the link access procedure (LAP) protocol that is part of X.25 [Tanenbaum, 1988]. To provide error-free data to the network layer, the DLC layer in these standards implements an error detecting code that is a version of the CRC-CCITT polynomial code mentioned in Section 12.3. The LLC layer in the IEEE 802 standard uses the 32-bit polynomial error control code, CRC-32, given in Section 12.3.

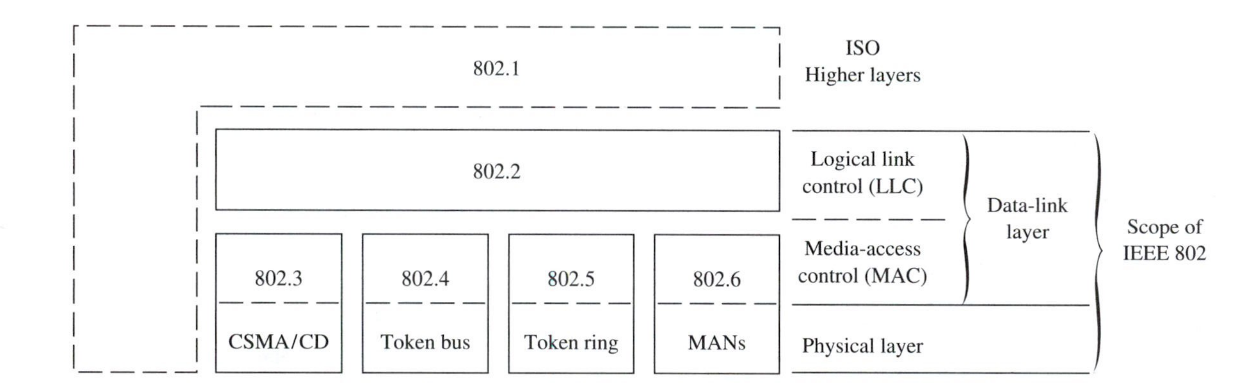

IEEE family of protocols:
IEEE 802.1 Relationship of 802.X standards to ISO reference model, higher-layer protocols, internetworking, and management and control
IEEE 802.2 Logical link control protocol standard
IEEE 802.3 CSMA/CD bus access method
IEEE 802.4 Token passing bus access method
IEEE 802.5 Token passing ring access method
IEEE 802.6 Metropolitan area network (MAN) access method

FIGURE 15.2.3 The IEEE 802 family and ISO model. From G. E. Keiser, *Local Area Networks*, New York: McGraw-Hill, Inc., 1989. Reproduced with permission of McGraw-Hill, Inc.

The network layer provides routing, flow control, and internetworking, and there are considerable differences in what this layer does depending on whether the network is a virtual circuit or datagram network. Routing and congestion control are discussed in detail in Section 15.5. As mentioned earlier, X.25 is the CCITT network layer interface standard, and X.25 is a virtual circuit or connection-oriented protocol. On the other hand, Internet is a datagram network, and its network layer protocol is Internet Protocol (IP).

The transport layer generally views the network layer and below as subnets (subnetworks) and must provide a variety of protocols if it is to deal with the different subnets. A common transport-level protocol is the ARPA Internet (ARPANET) Transmission Control Protocol (TCP), which along with its associated network layer protocol IP, has found a relatively broad application. Berkeley UNIX supports TCP/IP.

The presentation layer handles code conversion between machines, data compression, and data encryption among other duties. The code conversion problem arises since some machines use two's complement arithmetic while others use one's complement; similarly, some machines use ASCII code while others use EBCDIC. As part of OSI, ISO has developed what is called the abstract syntax notation 1 (ASN.1), which can be used by the presentation layer of different machines as a common representation of data structures. The data compression function is often used to keep costs down, since usage charges may be on a per byte basis. Commonly used data compression techniques are Huffman coding (Section 11.7), arithmetic coding, run-length coding, and Lempel–Ziv coding (Gersho and Gray, 1991). Data encryption may also be desirable and there are a variety of methods for achieving different levels of security [Newman and Pickholtz, 1987; Tanenbaum, 1988]. A particular encryption method, called the Data Encryption Standard (DES), has been standardized by the National Bureau of Standards, and the widespread availability of DES implementations make it attractive.

The most popular LAN is Ethernet, originally developed at Xerox in 1974. Ethernet, which corresponds to a version of the IEEE 802.3 specification, has a maximum data rate of 10 Mbits/sec and was designed for use over coaxial cable, although optical fiber versions exist [Keiser, 1989]. Since it is an IEEE 802.3 specification, the Ethernet protocols are concerned with the physical, MAC, and LLC network layers.

In the following sections we provide background material, descriptions, and analyses of several of the important functions required in the layered architectures.

15.3 Queueing Theory

We wish to model the situation of a single node in a network with a buffer to receive incoming packets on a first-in first-out basis. The packets are assumed to arrive at a rate of λ per second and all packets have the same length: L bits. If the transmission capacity over the link is C bits/sec, the average service time is $L/C = 1/\mu$, where μ is the service rate in packets/second. Two important parameters are the average delay in transmitting a packet, which we denote as $\bar{T}$, and the average number of packets in the system (or the average number of customers in the queue and being serviced), denoted by $\bar{N}$. Throughout our development, we assume that time averages are equal to ensemble averages.

We begin with a derivation of Little's formula or Little's theorem:

$$\bar{N}_t = \lambda \bar{T}_t, \tag{15.3.1}$$

where $\bar{N}_t$ is the average number of packets in the system at time t and $\bar{T}_t$ is the average time a packet spends in the system at time t. Referring to Fig. 15.3.1, we let $\alpha(t)$ = the number of arriving packets in the time interval $[0, t]$ and $\beta(t)$ = the number of departing (transmitted) packets in $[0, t]$. The number of packets in the system at time t is therefore

$$N(t) = \alpha(t) - \beta(t)$$

and the shaded region in Fig. 15.3.1 is given by the integral, $\int_0^t N(\tau)\, d\tau$. Letting the arrival time of the ith packet be t_i and the time this packet spends in the system be T_i, the shaded region in Fig. 15.3.1 is also expressible as

$$\sum_{i=1}^{\beta(t)} T_i + \sum_{i=\beta(t)+1}^{\alpha(t)} (t - t_i).$$

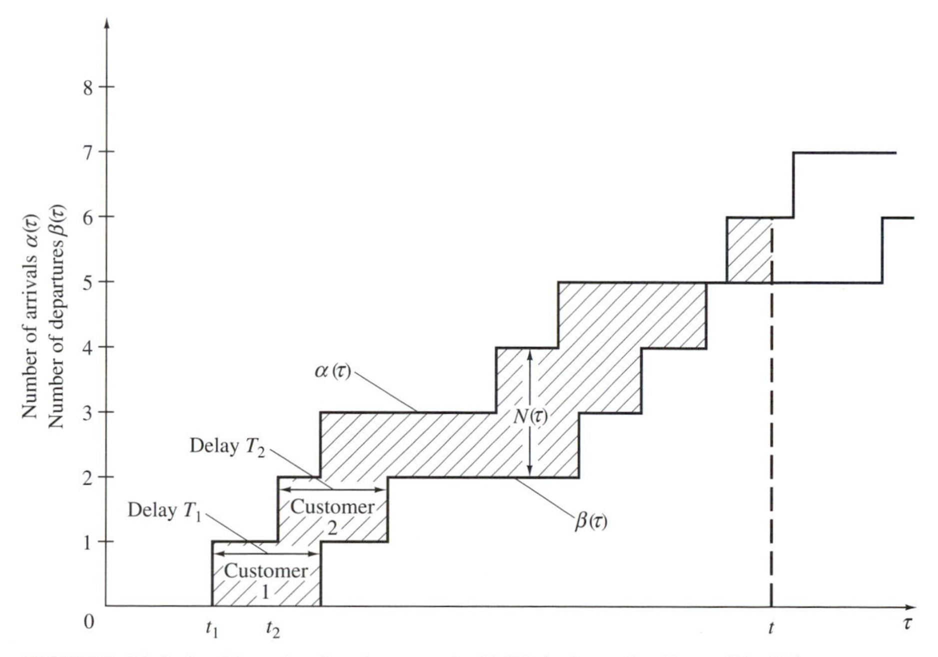

FIGURE 15.3.1 Diagram for the proof of Little's formula. From Dimitri Bertsekas and Robert Gallager, *Data Networks*, © 1987. Reprinted by permission of Prentice Hall, Englewood Cliffs, New Jersey.

Dividing both expressions by the length of the interval (t) and equating them, we have

$$\bar{N}_t = \frac{\int_0^t N(\tau)\, d\tau}{t} = \frac{\sum_{i=1}^{\beta(t)} T_i + \sum_{i=\beta(t)+1}^{\alpha(t)} (t - t_i)}{t}$$

$$= \frac{\alpha(t)}{t} \cdot \frac{\sum_{i=1}^{\beta(t)} T_i + \sum_{i=\beta(t)+1}^{\alpha(t)} (t - t_i)}{\alpha(t)} = \lambda \bar{T}_t, \qquad (15.3.2)$$

where

$$\bar{N}_t = \frac{\int_0^t N(\tau)\, d\tau}{t} = \text{average number of packets in the system in } [0, t]$$

$$\lambda = \frac{\alpha(t)}{t} = \text{average packet arrival rate in } [0, t]$$

$$\bar{T}_t = \frac{\sum_{i=1}^{\beta(t)} T_i + \sum_{i=\beta(t)+1}^{\alpha(t)} (t - t_i)}{\alpha(t)}$$

$$= \text{average time a packet spends in the system in } [0, t].$$

If we assume that the system is in steady state, $\bar{N}_t \to \bar{N}$ and $T_t = \bar{T}$; we write

$$\bar{N} = \lambda \bar{T}. \qquad (15.3.3)$$

Thus we see that Little's formula is intuitive and states simply that the average number of packets in the system is equal to the average arrival rate multiplied by the average time a packet spends in the system. This is a very useful formula.

To study packet-switched systems further, we model a single node in the network by the single-server queue shown in Fig. 15.3.2. Both the arrival of the messages (packets) and the service time of a message are random, with the average rate of arrivals given by λ and the average service time denoted by μ. We adopt the Poisson model of message arrivals, which is based on the following three assumptions [Schwartz, 1990]:

1. The probability of an arrival in the time interval $(t, t + \Delta t)$ is $\lambda \, \Delta t \ll 1$ (proportional to its length and independent of t).
2. The probability of no arrival in the time interval $(t, t + \Delta t)$ is $1 - \lambda \, \Delta t$.
3. The number of arrivals in nonoverlapping time intervals are independent.

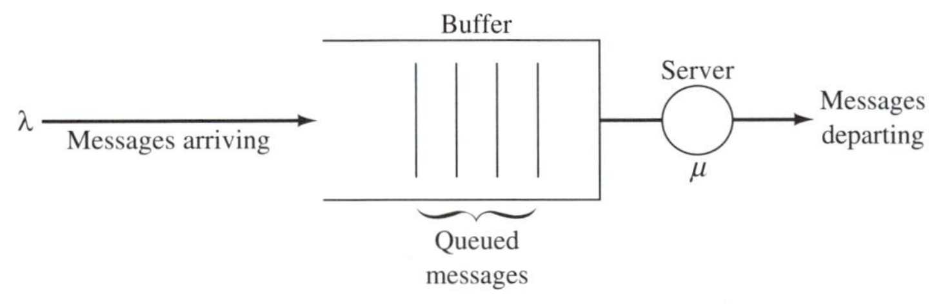

FIGURE 15.3.2 Model of a single-server queue. From G. E. Keiser, *Local Area Networks*, New York: McGraw-Hill, Inc., 1989. Reproduced with permission of McGraw-Hill, Inc.

Based on these three assumptions, and the additional one that there are a large number of independently arriving messages, it can be shown that the probability of n messages arriving during a time interval t seconds long is

$$P_n(t) = \frac{(\lambda t)^n e^{-\lambda t}}{n!},$$

for $n = 0, 1, 2, \ldots$. Using the results for a Poisson random variable from Appendix A, we know that in an interval T seconds long, the average number of arrivals is

$$E[N] = \sum_{n=0}^{\infty} nP_n(t) = \lambda T \tag{15.3.4}$$

and the variance of the number of arrivals is

$$\text{var}(N) = \lambda T. \tag{15.3.5}$$

Another important fact needed for later analyses is the probability density function of the arrival times between the messages. Letting t_i, $i = 0, 1, 2, \ldots$, be the arrival times of independent messages, we define the interarrival times $\tau_k = t_k - t_{k-1}$, $k \geq 1$, that are assumed to be independent, identically distributed random variables. It can then be shown that the interarrival time τ has the exponential pdf

$$f_\tau(\tau) = \lambda e^{-\lambda \tau} u(\tau), \tag{15.3.6}$$

where $u(\cdot)$ is the unit step function. It is simple to show that the average time between message arrivals is

$$E(\tau) = \frac{1}{\lambda}. \tag{15.3.7}$$

To this point we have let the message lengths all be the same and fixed at L bits. To get an expression for the service times, we assume that the message lengths have an exponential pdf

$$f_L(l) = \bar{L} e^{-\bar{L} l} u(l), \tag{15.3.8}$$

where $\bar{L}$ is the average message length in bits. We can convert this to seconds by dividing by the transmission rate of C bits/sec, so the pdf of the message length in seconds is

$$f_m(m) = \mu e^{-\mu m} u(m), \tag{15.3.9}$$

where $\mu = C/\bar{L}$.

The utility of the exponential assumption in Eqs. (15.3.8) and (15.3.9) is that it greatly simplifies the analysis. By analogy with the Poisson arrival times of the messages, which leads to exponential interarrival times, we see that exponentially distributed message lengths imply independent, Poisson distributed departures from the server.

For a particular queue as in Fig. 15.3.2, one of the quantities of great importance to us is the average time that a message (or packet) spends in the system. That is, we need to know the average time it takes for a message to be transmitted (depart) from the time it enters the queue. This quantity is the sum of two components, the average time spent waiting in the queue and the average service time. For a system with exponentially distributed message lengths as in Eq. (15.3.9) and a fixed transmission capacity of C bits/sec, the average service time is [Keiser, 1989]

$$E[S(t)] = \frac{\bar{L}}{C} = \frac{1}{\mu} \text{ sec/packet}, \tag{15.3.10}$$

and the service time has the exponential pdf,

$$f_S(s) = \mu e^{-\mu s} u(s). \tag{15.3.11}$$

To calculate the average time waiting in the queue, we need to find the pdf of the number of messages in the queue.

To continue our analysis, we focus on the queueing system shown in Fig. 15.3.2, with the stated assumptions, which is known as an $M/M/1$ queue. This notation is standard in queueing theory and $M/M/1$ has the following meaning. The first letter M indicates that the arrival process is memoryless, which implies Poisson arrivals and an exponential distribution for the interarrival times. The second letter denotes the service time distribution and the M indicates that it is exponential. The last symbol is a digit that specifies the number of servers, which is 1 for this case. We wish to find an expression for the probability of exactly n users in the system when the $M/M/1$ queue is in equilibrium or steady state.

To do this, we write equations for the flow of messages through the system according to the state diagram in Fig. 15.3.3, letting the system states $n = 0, 1, 2, \ldots,$ be the number of messages in the queue and p_n be the probability of state n. We assume that each state can only be reached from adjacent states and that the joint probability of an arrival and a departure in Δt is negligible. If there are no messages in the queue, we are in state 0 with probability p_0. The only possible transition is to state 1, and since the system is in equilibrium,

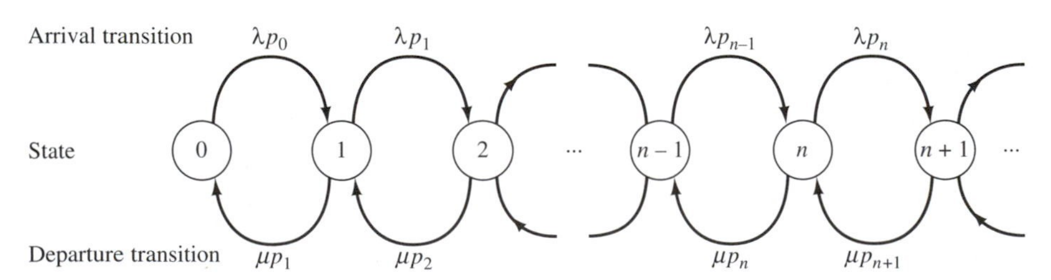

FIGURE 15.3.3 State diagram for the $M/M/1$ queue. From G. E. Keiser, *Local Area Networks*, New York: McGraw-Hill, Inc., 1989. Reproduced with permission of McGraw-Hill, Inc.

p_0 is a constant. Thus there is a transition to state 1 with average rate λp_0, but there is an equal transition rate back into state 0 of μp_1, so

$$\lambda p_0 = \mu p_1. \tag{15.3.12}$$

Performing a similar analysis for state 1, we have an average flow of messages into state 1 of $\lambda p_0 + \mu p_2$ and an average outward flow of $(\mu + \lambda)p_1$. In equilibrium, these must be equal, so

$$(\mu + \lambda)p_1 = \lambda p_0 + \mu p_2. \tag{15.3.13}$$

Using Eq. (15.3.12), this simplifies to

$$\lambda p_1 = \mu p_2.$$

The queue is assumed to have an unbounded length, so the analysis for state 1 holds for any general state $n \geq 1$, giving

$$\lambda p_{n-1} = \mu p_n. \tag{15.3.14}$$

Rewriting Eq. (15.3.14), we have

$$p_n = \frac{\lambda}{\mu} p_{n-1} = \rho p_{n-1}, \tag{15.3.15}$$

where $\rho \triangleq \lambda/\mu$ is called the traffic intensity. Iterating backward on Eq. (15.3.15) yields

$$p_n = \rho^n p_0. \tag{15.3.16}$$

Since probability must sum to 1 and the queue is infinite,

$$1 = \sum_{n=0}^{\infty} p_n = \sum_{n=0}^{\infty} \rho^n p_0 = \frac{p_0}{1 - \rho}, \tag{15.3.17}$$

where we were able to sum the geometric series, since $\rho = \lambda/\mu < 1$. Solving Eq. (15.3.17) gives us

$$p_0 = 1 - \rho \tag{15.3.18}$$

and

$$p_n = (1 - \rho)\rho^n, \qquad n \geq 0. \tag{15.3.19}$$

With Eq. (15.3.19), it is possible to find the average number of messages in the queueing system as

$$\bar{N} \triangleq E[N] = \sum_{n=0}^{\infty} np_n = \sum_{n=0}^{\infty} (1 - \rho)\rho^n = \frac{\rho}{1 - \rho}. \tag{15.3.20}$$

Equation (15.3.20) is plotted in Fig. 15.3.4 versus the traffic intensity ρ. Clearly, as λ approaches μ, the average number of messages in the system grows rapidly.

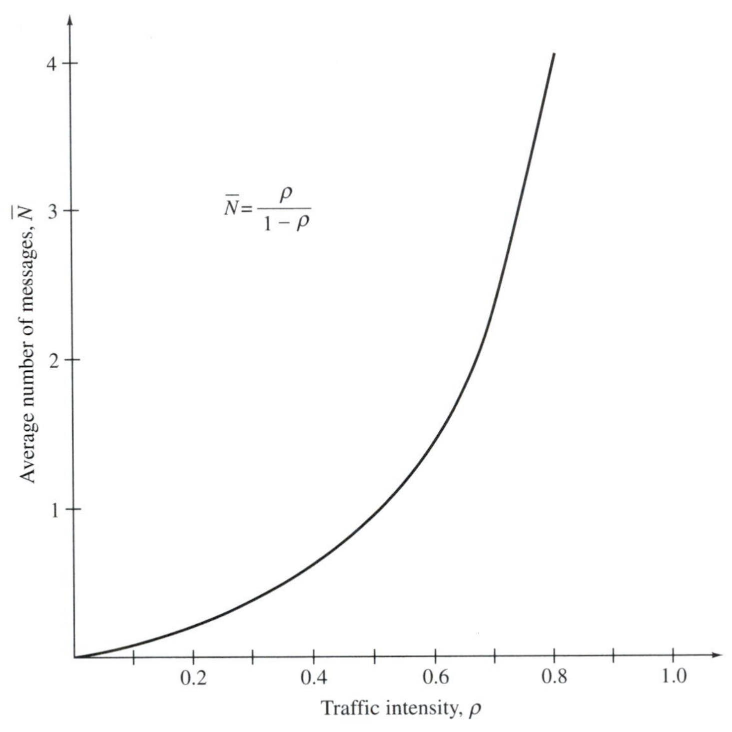

FIGURE 15.3.4 Average number of messages in an *M/M/*1 queue as a function of the traffic intensity. From G. E. Keiser, *Local Area Networks*, New York: McGraw-Hill, Inc., 1989. Reproduced with permission of McGraw-Hill, Inc.

Since we now have $\bar{N}$, we can use Little's formula from Eq. (15.3.3) to find the average time spent in the system by a message,

$$\bar{T} = \frac{\bar{N}}{\lambda} = \frac{1}{\mu(1-\rho)}, \tag{15.3.21}$$

which is plotted in Fig. 15.3.5.

We can now return to a comparison of statistical multiplexing with TDM or FDM that was first addressed in Section 15.1. If we consider a single packet-switched node attached to an output link with capacity C bits/sec, and the average packet length is $\bar{L}$ bits/packet, then $C/\bar{L} = \mu =$ packets/sec, and $\bar{T} = 1/(\mu - \lambda)$ sec/packet is the average waiting time for a packet to be transmitted from Eq. (15.3.21). For a TDM link, we find the difference between the possible packet transmission rate and the packet arrival rate and then take the reciprocal to get the waiting time. With K users the total capacity C is divided into subchannels of capacity C/K, each subchannel being allocated time slots K slots apart. Thus the transmission rate in packets/second is $(C/K)/\bar{L} = \mu/K$ for each subchannel. The packet arrival rate is equally distributed among subchannels, so that the average arrival rate at each subchannel is λ/K packets/sec.

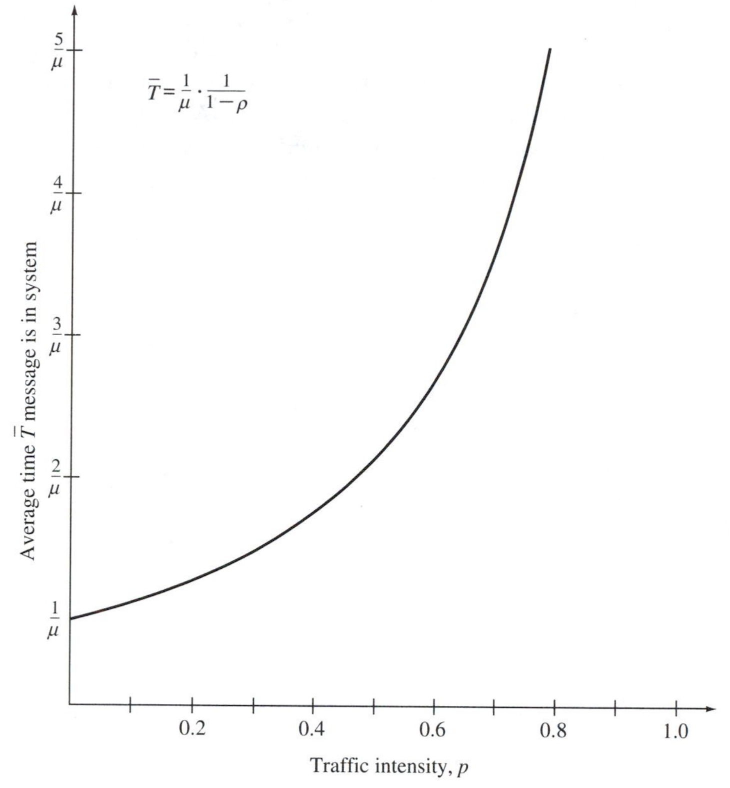

FIGURE 15.3.5 Average time a message spends in an *M/M/1* system as a function of the traffic intensity. From G. E. Keiser, *Local Area Networks*, New York: McGraw-Hill, Inc., 1989. Reproduced with permission of McGraw-Hill, Inc.

The average delay per packet for TDM is therefore

$$\bar{T}_{\text{TDM}} = \frac{1}{\mu/K - \lambda/K} = \frac{K}{\mu - \lambda}, \tag{15.3.22}$$

which is K times the delay of statistical multiplexing. A similar calculation results for FDM. The key advantage of statistical multiplexing is that the multiplexing method matches the arrival process of the messages and is not fixed as in TDM and FDM.

15.4 Multiple Access Methods

As has been pointed out in Section 15.3, a fixed multiplexing scheme such as synchronous TDM or FDM does not make efficient use of the available transmission capacity for messages that arrive intermittently and do not continuously use the allocated capacity. However, to achieve the promised performance of statistical multiplexing, we must devise techniques that allow users at several different nodes (locations) in the networks to access the available transmission capacity with minimum interference among users. In this section we present an overview of the most common multiple access schemes, including ALOHA, slotted ALOHA, carrier-sense multiple access (CSMA), CSMA/CD (collision detection), and token ring. The important performance curves for evaluating the various methods are throughput versus offered traffic and average packet delay versus throughput.

The predecessor of many multiple access schemes in use today is the ALOHA algorithm, or as it is sometimes called to distinguish it from later variations, the *pure* ALOHA method. With the pure ALOHA method, each node (or user) transmits a message whenever the node has a message to send. The node transmits a packet of data, waits for an acknowledgment to be returned that the packet was received, and then transmits another packet. This approach is clearly quite simple, but it suffers from the problem of collisions. That is, packets from two or more users may overlap (called a *collision*), which causes the data in both packets to be lost. The transmitting nodes wait for an acknowledgment that their packet has been received correctly, which in the case of a collision will not be returned, or they wait for some maximum time interval, after which the nodes retransmit the preceding packet. To reduce the likelihood of a collision upon retransmission, each node waits a random amount of time before transmitting again.

To analyze the performance of the pure ALOHA method, we assume that packets are transmitted immediately upon arrival and that arriving packets, including both new messages and retransmissions, have the Poisson distribution

$$P_n(t) = \frac{(\lambda t)^n}{n!} e^{-\lambda t}. \tag{15.4.1}$$

For offered traffic with a mean of G transmission attempts/packet, which includes the effect of all users, and a packet length of τ_p seconds, the parameter λ in Eq. (15.4.1) is $\lambda = G/\tau_p$. In pure ALOHA, the transmission of a packet is successful (no collisions) if no other station transmits within one packet length of the packet's start time. This is because any packet transmitted less than τ_p seconds before the packet's start will overlap the beginning of the packet, and any packet transmitted less than τ_p seconds after the packet is transmitted collides with the end of the packet. Thus, in pure ALOHA, the vulnerable period for a collision is $2\tau_p$ seconds long.

There will be no collision if there are zero arrivals in this period of $2\tau_p$ seconds, and the probability of this event is [using Eq. (15.4.1)] $e^{-2\lambda\tau_p} = e^{-2G}$. The throughput S is then the product of the offered traffic G and the probability of no collisions, so

$$S = Ge^{-2G}. \tag{15.4.2}$$

Equation (15.4.2) is plotted in Fig. 15.4.1, where it is observed that the maximum throughput occurs at $G = 0.5$ and is only 0.184. This is extraordinarily low, but it is not surprising due to the total lack of coordination among the transmitting stations in pure ALOHA. Further, for bursty, interactive data, the number of users accommodated can be relatively large (see Problem 15.16).

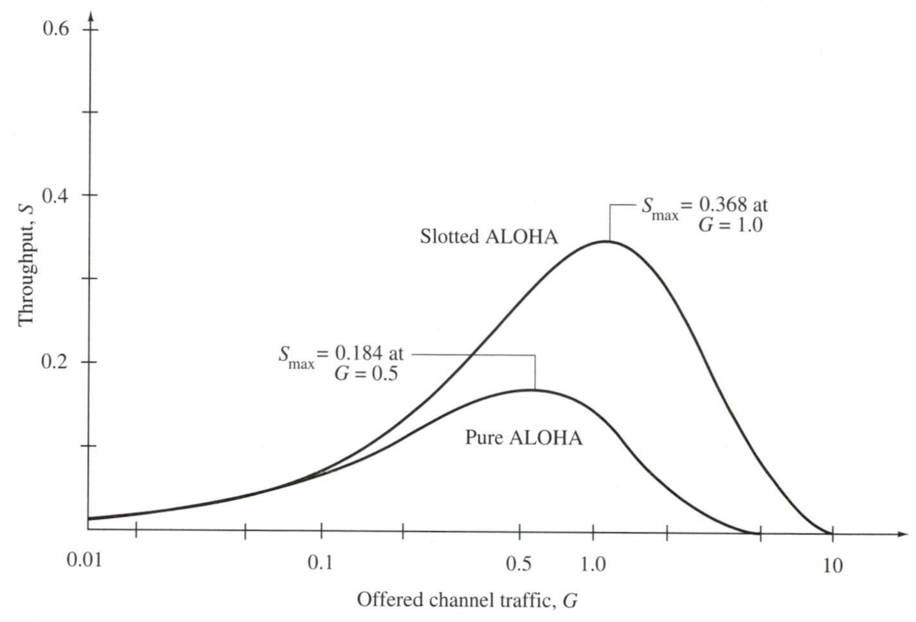

FIGURE 15.4.1 Comparison of the throughput as a function of offered load for pure and slotted ALOHA. From G. E. Keiser, *Local Area Networks*, New York: McGraw-Hill, Inc., 1989. Reproduced with permission of McGraw-Hill, Inc.

A technique proposed to improve the efficiency of pure ALOHA, called *slotted* ALOHA, divides the channel into fixed-length time slots and allows a station to transmit only at the beginning of a time slot. This method guarantees that collisions can occur only when two or more users transmit at the beginning of a time slot, and thus the vulnerable period for collisions is reduced to one time slot. Letting a time slot equal one packet length τ_p, we obtain the throughput for slotted ALOHA:

$$S = Ge^{-G}, \tag{15.4.3}$$

which is also plotted in Fig. 15.4.1. Note that the maximum throughput has been doubled and the maximum occurs at $G = 1.0$. If a collision occurs, slotted ALOHA retransmits a collided packet after a random delay corresponding to some integral number of time slots.

Since the offered traffic G includes both new arrivals and retransmissions, it is evident that G varies in proportion to the number of backlogged packets n_b, which is reflected in the number of retransmissions. Upon letting the total number of nodes be N, the probability of a new arrival in a time slot be p_a, and the probability of retransmission be p_r, the offered traffic varies as

$$G(n_b) = (N - n_b)p_a + n_b p_r. \tag{15.4.4}$$

The first term in Eq. (15.4.4) is the average number of new packets in the system and it can be plotted versus n_b as shown in Fig. 15.4.2. We can also plot the average throughput of slotted ALOHA in Eq. (15.4.3) versus n_b by using $G(n_b)$ in Eq. (15.4.4) for G in Eq. (15.4.3). This plot is also shown in Fig. 15.4.2. Possible

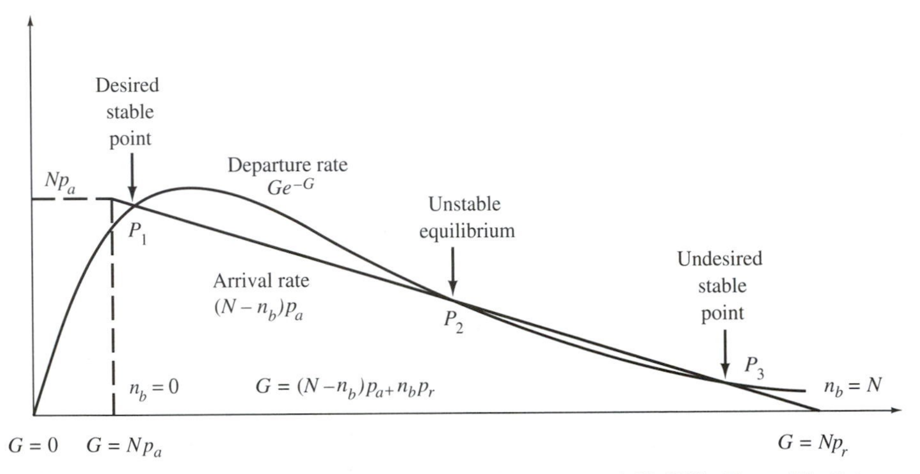

FIGURE 15.4.2 Stability, instability, and drift in slotted ALOHA. From Dimitri Bertsekas and Robert Gallager, *Data Networks*, © 1987. Reprinted by permission of Prentice Hall, Englewood Cliffs, New Jersey.

points of equilibrium between throughput and offered traffic for the network occur where the two functions intersect.

The discussion as to whether each of the equilibrium points shown is stable is facilitated by defining the drift in state n_b as the expected change in the number of backlogged packets in one time slot given that there are n_b packets in the current backlog [Bertsekas and Gallager, 1987]. The drift is therefore

$$D_{n_b} = (N - n_b)p_a - \bar{S}(n_b), \tag{15.4.5}$$

where $\bar{S}(n_b)$ is approximately given by Eq. (15.4.3) with G replaced by $G(n_b)$ in Eq. (15.4.4). We thus see that the drift is the difference between the straight line in Fig. 15.4.2 and the throughput curve Ge^{-G}.

From the figure we see that for n_b to the left of P_1, the drift is positive, and for n_b to the right of P_1, the drift is negative. Thus P_1 is a stable equilibrium point. Around P_2, if n_b is to the left of P_2, the drift is negative (toward P_1), and for n_b to the right of P_2, the drift is positive toward P_3. Hence P_2 is unstable. To the right of P_3, the drift is negative toward P_3, so P_3 is stable.

The equilibrium point P_3 is undesirable, since the throughput is low and the backlog is high. A network that behaves according to Fig. 15.4.2 will operate around either P_1 or P_3 with infrequent switching between the two operating points. Such operation is unacceptable, since if the network ever switches to P_3, network efficiency is low and it takes an unusual situation for the network to return to P_1.

The analysis here assumes that there is no queueing of packets at the individual nodes, so if a packet must be retransmitted at a node, incoming packets at this node are discarded until retransmission is successful. The retransmission probability is a parameter that is adjustable by the system designer, so let us investigate the behavior of Fig. 15.4.2 as p_r is varied. Note that changing p_r changes the relationship between n_b and $G(n_b)$. Generally, decreasing p_r (increasing the time interval for retransmissions) tends to stabilize the network.

That is, decreasing p_r expands Ge^{-G} horizontally, so that the number of backlogged packets required to exceed P_2, and hence cause the network to operate at P_3, is increased. Thus the network operates at P_1 but with an increased delay in retransmitted packets. If p_r is increased, the curve Ge^{-G} is contracted horizontally, and a smaller value of n_b causes the network to drift toward P_3, but the retransmission delay is smaller.

It is also interesting to examine the effect of the number of users on network stability. If N is increased sufficiently, the straight line will intersect Ge^{-G} only at high n_b and low throughput. In this situation, the network is seriously overloaded.

Kleinrock and Lam [1975] have defined a network to be stable when the drift line intersects the throughput curve at one and only one point and not at a point of tangency. The motivation for this definition is evident from the preceding discussion. Unfortunately, operating at this point (reducing p_r so that the curves only intersect at P_1) can cause large retransmission delays. What is usually done in networks is to estimate n_b in some way and use this estimated n_b to adjust $G(n_b)$ to maintain maximum throughput [$G(n_b) = 1$ here].

For wide-bandwidth networks, which is often the case for LANs, the efficiency of slotted ALOHA can be improved upon by using carrier-sense multiple access (CSMA) or listen-before-talk (LBT) schemes. As the latter name implies, in these schemes each station listens to the transmission medium before sending a packet and transmits only if the line is sensed to be idle. Three CSMA schemes are usually distinguished: nonpersistent CSMA, 1-persistent CSMA, and p-persistent CSMA. In nonpersistent CSMA, the station transmits if the channel is sensed to be idle, or if the channel is sensed to be busy, it waits for a random length of time and retransmits if the channel becomes idle. For 1-persistent CSMA, the station again transmits if the channel is sensed idle, but if the channel is busy, the channel is monitored and a packet is transmitted as soon as an idle channel is detected. With $0 < p \leq 1$, p-persistent CSMA transmits with probability p when the channel is sensed idle, and if the channel is busy, the channel is monitored until it is sensed idle. When an idle channel is sensed, a packet is transmitted with probability p.

To see how CSMA is an improvement over ALOHA, let the maximum propagation delay between any two stations be t_d, which is assumed much less than a packet length τ_p. The vulnerable period in CSMA is equal to t_d, since this is the time it can take a packet that has been transmitted at one station to propagate to the most distant station. During this propagation time, another station may sense that the channel is idle and transmit even though another station has already transmitted at some time $t < t_d$ seconds before. Note, however, that if the propagation time is small compared to the packet length ($t_d \ll \tau_p$) as we have assumed, this vulnerable period is much less than in pure ALOHA and slotted ALOHA, where the vulnerable period is $2\tau_p$ or τ_p, respectively.

The performance differences among nonpersistent, 1-persistent, and p-persistent CSMA can be significant. To begin our comparisons, we derive an expression for the throughput of nonpersistent CSMA by considering the situations shown in Fig. 15.4.3. The first busy period is successful, since packet 1 is

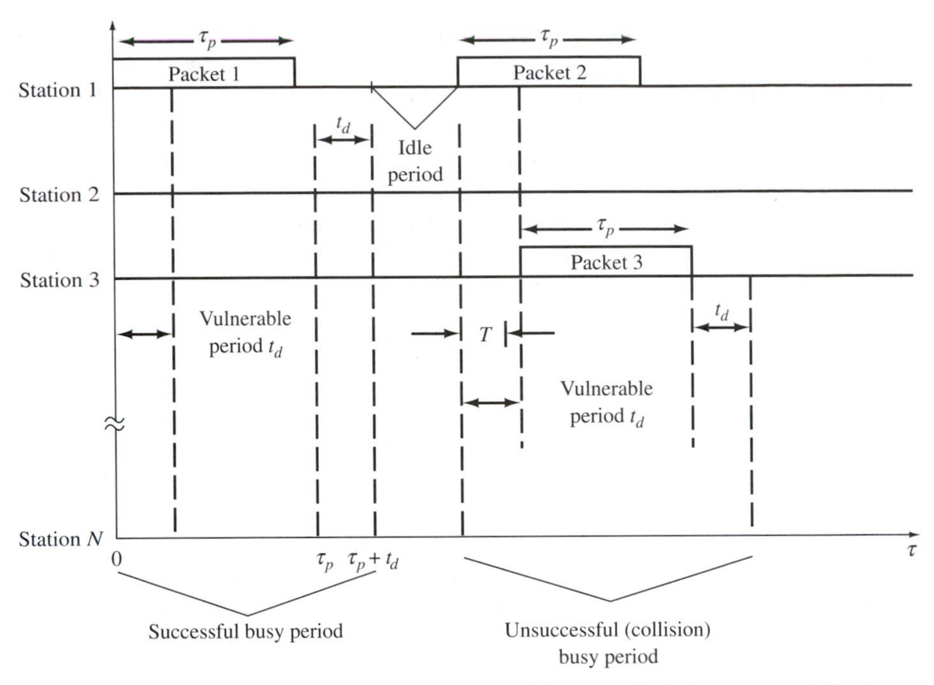

FIGURE 15.4.3 Successful and unsuccessful busy periods for nonpersistent CSMA. From G. E. Keiser, *Local Area Networks*, New York: McGraw-Hill, Inc., 1989. Reproduced with permission of McGraw-Hill, Inc.

sent without a collision. In the second busy period, however, packet 2 has not yet arrived at station 3, so station 3 senses that the line is idle and transmits packet 3. This leads to a collision. A common definition of throughput is the average portion of a cycle with a successful transmission ($\bar{P}$) divided by the average length of a cycle, where the average cycle length equals the average busy period ($\bar{B}$) plus the average idle period ($\bar{I}$) [Keiser, 1989]. Thus

$$S = \frac{\bar{P}}{\bar{B} + \bar{I}}. \tag{15.4.6}$$

This particular definition assumes that the system is in steady state and that there is little difference statistically between cycles.

With reference to Fig. 15.4.3, we can calculate the various quantities in Eq. (15.4.6). If G is the offered traffic in packets/sec, the probability of successful transmission is the probability of no new arrivals in the vulnerable period $0 \leq t \leq t_d$, or from Section 15.3 (assuming Poisson arrivals),

$$\bar{P} = e^{-t_d G}. \tag{15.4.7}$$

Still working with Poisson arrivals, we find that the average length of the idle period is just the average length of time after the busy period with no arrivals, so from Eq. (15.3.7),

$$\bar{I} = \frac{1}{G}. \tag{15.4.8}$$

Again with reference to Fig. 15.4.3, the length of a busy period is represented by the random variable

$$B = \tau_p + t_d + T, \tag{15.4.9}$$

where τ_p and t_d are fixed and T is a random variable. The random variable T represents the time after transmission of packet 2 that the last packet collision occurs. The cumulative distribution function of this random variable is equal to the probability of no arrivals in the interval $t_d - T$, so

$$\begin{aligned} F_T(\tau) &= P[0 \text{ arrivals in } (t_d - \tau)] \\ &= e^{-G(t_d - \tau)}, \qquad 0 < \tau \le t_d. \end{aligned} \tag{15.4.10}$$

The pdf can be obtained by differentiation to be $Ge^{-G(t_d-\tau)}$, so

$$\begin{aligned} \bar{T} &= \int_0^{t_d} \tau G e^{-G(t_d - \tau)}\, d\tau \\ &= t_d - \frac{1}{G}[1 - e^{-Gt_d}]. \end{aligned} \tag{15.4.11}$$

Therefore, directly

$$\bar{B} = \tau_p + 2t_d - \frac{1}{G}[1 - e^{-Gt_d}], \tag{15.4.12}$$

and upon substitution of Eqs. (15.4.7), (15.4.8), and (15.4.12) into Eq. (15.4.6), the throughput is found to be

$$S = \frac{Ge^{-Gt_d}}{G(\tau_p + 2t_d) + e^{-Gt_d}}. \tag{15.4.13}$$

This result is plotted in Fig. 15.4.4 as a function of the normalized variable $a = t_d/\tau_p$, where it is clear that for t_d small, nonpersistent CSMA is very effective, but as t_d approaches the packet length, the throughput is drastically reduced.

The throughput versus offered load expression is more complicated to derive for 1-persistent CSMA, so we simply present the result here as [Keiser, 1989; Hammond and O'Reilly, 1986]

$$S = \frac{G[1 + G + aG(1 + G + aG/2)]e^{-G(1+2a)}}{G(1 + 2a) - (1 - e^{-aG}) + (1 + aG)e^{-G(1+a)}}, \tag{15.4.14}$$

where $a = t_d/\tau_p$. Equation (15.4.14) is plotted in Fig. 15.4.5. From the figure it is evident that nonpersistent CSMA has a much higher throughput than 1-persistent CSMA at higher offered loads. However, the saving grace of 1-persistent CSMA is that it has a smaller average packet delay.

Figure 15.4.5 has a plot of the throughput versus offered load for nonpersistent CSMA/CD, and rather than discuss p-persistent CSMA, we now turn our attention to collision detection schemes. Collision detection methods are listen-while-talk (LWT) methods, which means that not only does a station sense the line before transmitting, it also monitors the line during transmission. If a collision is sensed, the station stops its transmission immediately and inserts a

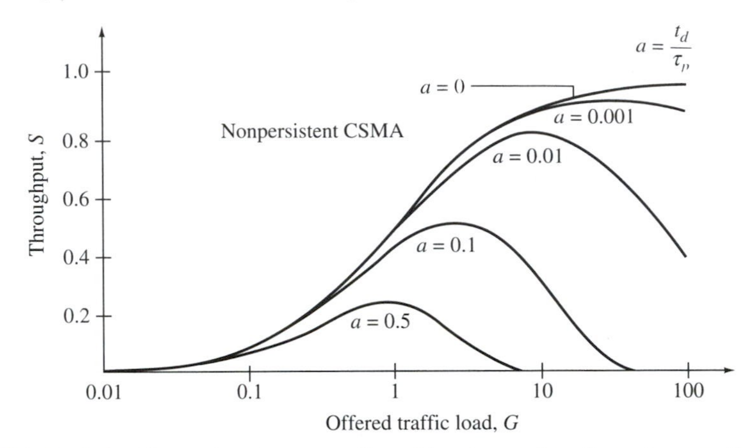

FIGURE 15.4.4 Throughput versus offered load as a function of normalized propagation delay for nonpersistent CSMA. From G. E. Keiser, *Local Area Networks*, New York: McGraw-Hill, Inc., 1989. Reproduced with permission of McGraw-Hill, Inc.

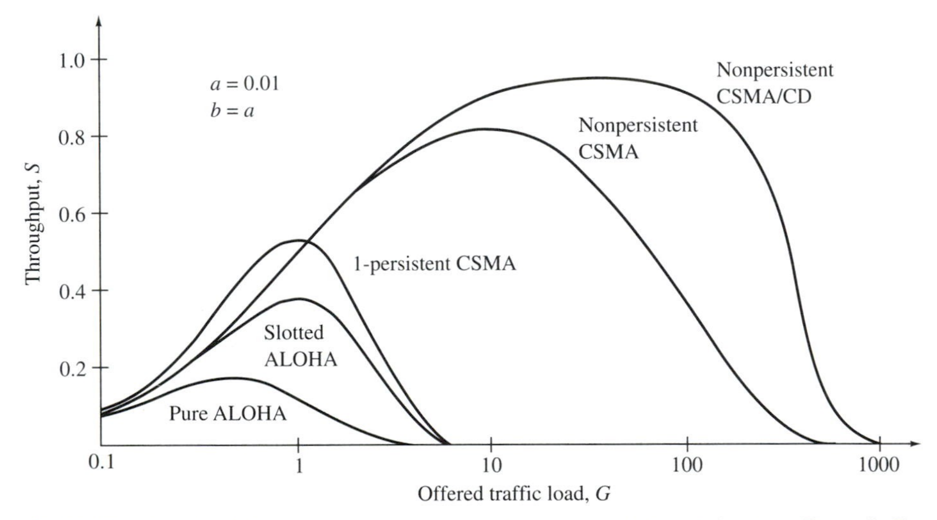

FIGURE 15.4.5 Throughput comparison of random access schemes. From G. E. Keiser, *Local Area Networks*, New York: McGraw-Hill, Inc., 1989. Reproduced with permission of McGraw-Hill, Inc.

jamming signal to indicate to other users that a collision has occurred. The collision detection thus reduces the time spent on unsuccessful transmissions.

The throughput expression for nonpersistent CSMA/CD is given by

$$S = \frac{Ge^{-aG}}{Ge^{-aG} + bG(1 - e^{-aG}) + 2aG(1 - e^{-aG}) + (2 - e^{-aG})}, \quad (15.4.15)$$

where b is the length of the jamming signal normalized by the packet length. Equation (15.4.15) is plotted in Fig. 15.4.5, where it is obvious that collision detection improves throughput even more for higher offered loads. Note that the plot in Fig. 15.4.5 of S for CSMA/CD sets $b = a = 0.01$. For collision detection to offer a performance increment, the length of the jamming signal must be small compared to the packet length. The performance analysis for 1-persistent CSMA/CD is rather complicated, as is the final result, so we present neither one here. We do note, however, that 1-persistent CSMA/CD has a throughput versus offered load curve that falls between those of nonpersistent CSMA and CSMA/CD.

Our discussions and analyses in this section have employed the infinite node network model with a single customer per node. Under this model, CSMA and CSMA/CD are both unstable with the random retransmission of blocked packets. Both protocols can be stabilized, however, by choosing appropriate packet retransmission strategies [Bertsekas and Gallager, 1987; Keiser, 1989].

An important standard for LAN random access is the IEEE 802.3 Standard, which is based on the Ethernet specification developed by Xerox, Digital Equipment Corporation, and Intel. The IEEE 802.3 standard uses 1-persistent CSMA/CD with a stabilized retransmission policy. The stations wait a random amount of time after receiving the jamming signal to retransmit. Stability is maintained by using what is called the truncated binary exponential backoff retransmission scheme. After each unsuccessful transmission, the random delay before transmission is an integer n_d times a base backoff time, such that $0 \leq n_d \leq 2^k$, where $k = \min(n, 10)$, with n being the number of unsuccessful transmission attempts. The base backoff time is often twice the end-to-end propagation delay. After 16 unsuccessful transmissions, a packet is discarded [Keiser, 1989; Hammond and O'Reilly, 1986].

Another important medium access approach is token passing, where a token is a single bit or group of bits and the token circulates from user to user. When a user has possession of the token, that user is allowed to transmit data over the link. When the user is finished transmitting data, the token is passed on to the next user. The token ring is a prominent local area network configuration and has the idealized form shown in Fig. 15.4.6. In the idle state, the token typically circulates around the ring in one direction (assumed counterclockwise here). The token, which might be an 8-bit all-one's pattern 1 1 1 1 1 1 1 1, is taken in by each station and passed on to the next station, after a delay of m bits, $m \geq 1$ if there are no data waiting to be transmitted. If the token arrives at a station that has data awaiting transmission, the token is changed to a busy status, perhaps by flipping the last bit to get 1 1 1 1 1 1 1 0, and the station transmits its data. The transmitted data follow immediately after the busy token, but exactly when the token is changed back to idle status can vary.

The differences in which the token is switched to the idle state correspond to one of three methods: multiple-token, single-token, or single-packet operation. In multiple-token operation, the idle token immediately follows the last bit of transmitted data, so that there can be several busy tokens on the ring at a time, but only one idle token. This is because station 1 may send a busy token followed by data and an idle token, but a subsequent station may change the idle token to a busy token and transmit its data followed by an idle token. Of course, another station could also change the idle token and transmit data followed by an idle token: thus the name *multiple-token operation.*

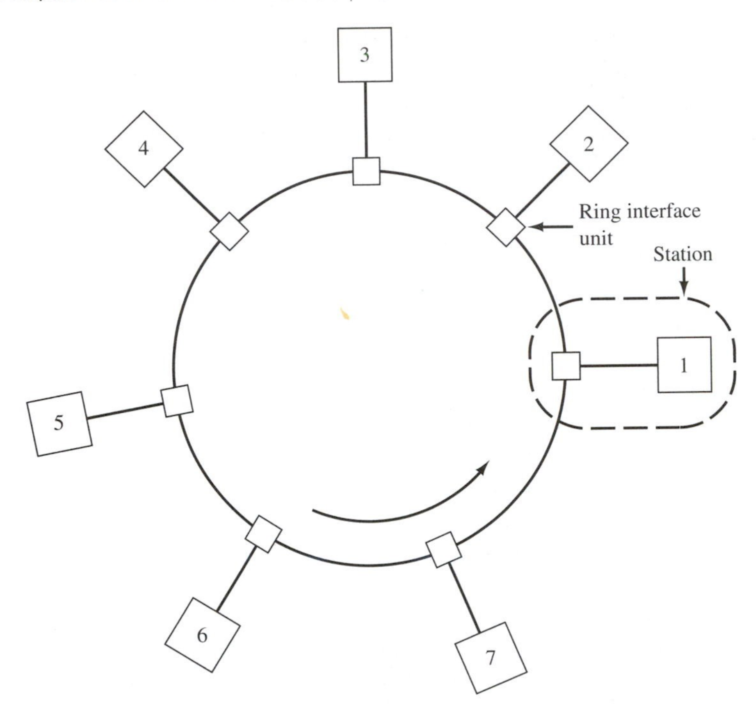

FIGURE 15.4.6 Token ring network. From J. L. Hammond and P. J. P. O'Reilly, *Performance Analysis of Local Computer Networks*, © 1986 by Addison-Wesley Publishing Company, Inc., p. 224. Reprinted by permission of the publisher.

In single-token operation, the transmitting station waits to generate an idle token until it receives its own busy token. However, if the data packet is longer than the propagation time around the ring, the station must wait until after the last bit of the packet is transmitted to insert the idle token. Note that the latter case corresponds to multiple-token operation, so that multiple-token and single-token operation differ only when the packet length is less than the propagation time around the ring. The IEEE 802.5 standard uses single-token operation. A third mode of operation, called single-packet, does not generate an idle token until the last transmitted bit is received by the transmitting station.

The utilization factor or the throughput, defined as the total transmitted data rate on the ring divided by the maximum data rate possible on the ring, is straightforward to calculate under certain assumptions. For this development, we follow Stallings [1984]. Letting the maximum possible data rate on the ring be R bits/sec, the number of stations be N, and the total propagation time around the entire ring be τ_d, we assume that the average delay between adjacent stations is τ_d/N and consider single-token operation. We further assume that each station always has data awaiting transmission, and note that the average delay in passing a token between stations is τ_d/N. The throughput S is different depending on whether the packet length τ_p is greater than or less than τ_d. For $\tau_d < \tau_p$, it takes τ_p seconds to transmit a packet, immediately after which an idle token is generated. Since the token needs τ_d/N seconds to reach the next station, the total delay from the beginning of the transmission until the next station can transmit is $\tau_p + \tau_d/N$, so the actual transmitted data rate is $1/(\tau_p + \tau_d/N)$. Normalizing this by the maximum transmission rate $1/\tau_p$, we get the throughput S as

$$S = \frac{\tau_p}{\tau_p + \tau_d/N}, \qquad \tau_p > \tau_d. \tag{15.4.16}$$

For $\tau_p < \tau_d$, the idle token is generated when the transmitting station receives the first bit in its transmitted packet, so the total delay from the beginning of the transmission to when the next station can transmit is $\tau_d + \tau_d/N$. The throughput is thus $1/(\tau_d + \tau_d/N)$ normalized by $1/\tau_p$, so

$$S = \frac{\tau_p}{\tau_d(1 + 1/N)}, \qquad \tau_p < \tau_d. \tag{15.4.17}$$

If we define $a = \tau_d/\tau_p$, we can write the preceding expressions for throughput as [Stallings, 1984]

$$S = \begin{cases} \dfrac{1}{1 + a/N}, & a < 1 \\ \dfrac{1}{a(1 + 1/N)}, & a > 1. \end{cases} \tag{15.4.18}$$

The normalized delay a usually falls between 0.01 and 0.1.

The expression for throughput in Eq. (15.4.18) is effectively for the heavily loaded case, since we assumed that each station has data awaiting transmission at all times. Certainly, the throughput is quite dependent on the normalized propagation delay a; more interesting, however, is the fact that for a small, the throughput S approaches 1 under heavy loads.

15.5 Routing and Flow Control

In a communications network there are often many possible paths from a source node to a destination node, with (possibly) many nodes along any path. The delivery of data packets from the source to the destination thus requires the selection of a path through the network for the data packets. This is the task of routing algorithms. Routing algorithms find the best path through a network by minimizing some cost function. Typical cost criteria for communications networks include minimizing the number of hops, minimizing the delay, minimizing distance, or minimizing the cost of a connection.

Routing algorithms may be centralized or distributed and may use a static or an adaptive strategy in finding the best routes. A centralized algorithm collects all of the necessary information at a single location or node, computes the optimum routes for all paths between nodes, and distributes this routing information throughout the network. Distributed algorithms perform routing calculations at various nodes, if not all, in the network, and thus require that the nodes interchange information.

Static routing strategies find the optimum routes between nodes and do not modify these routes in response to dynamic changes in network traffic. On the other hand, adaptive strategies attempt to keep track of the network traffic status and adapt the optimal routes to these changes.

If too many packets are presented to any node in a network, the network becomes congested at this node. It is the job of flow control algorithms to limit the offered traffic at the nodes so that such congestion can be avoided. Routing and flow control are usually studied separately even though there is a close interaction between them. This interaction is illustrated by Fig. 15.5.1. If a routing algorithm performs poorly, packet delay increases, which then requires flow control to reduce the number of new packets presented to the particular node. The term *congestion* is usually considered to be different from *flow control* in that congestion control algorithms address network congestion as a whole, whereas flow control is a local issue.

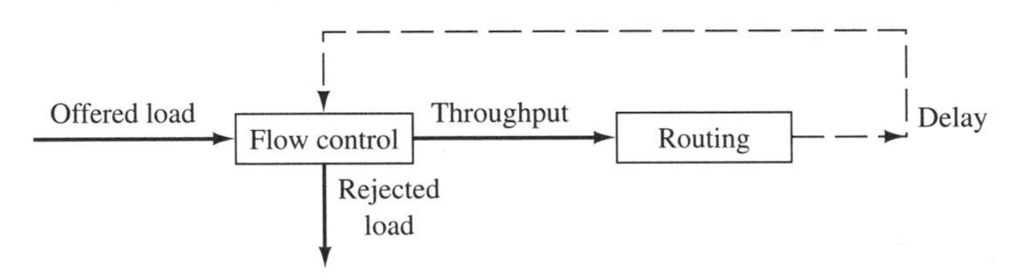

FIGURE 15.5.1 Interaction between routing and flow control. From Dimitri Bertsekas and Robert Gallager, *Data Networks*, © 1987. Reprinted by permission of Prentice Hall, Englewood Cliffs, New Jersey.

We now begin a more detailed discussion of routing algorithms. The end result of a routing algorithm is a routing table for each node that specifies, for every destination, the next node in the network. Thus, when a packet arrives at a node, the route to be taken is found from the routing table by looking up the packet's destination node. Routing in a virtual circuit network is somewhat simpler than in a datagram network, since the route is maintained for the entire length of the session, whereas for datagrams, each packet in a message may take a different route.

The first routing algorithm that we discuss is called the Bellman–Ford algorithm [Bertsekas and Gallager, 1987], the Bellman–Ford–Moore algorithm [Spragins, Hammond, and Pawlikowski, 1991] or the Ford–Fulkerson algorithm [Schwartz, 1987]. This algorithm finds the minimum-cost path between nodes subject to the constraint that the path contains h or fewer hops, or links. Such algorithms are usually called "shortest path" algorithms, irrespective of whether the cost per link is specified in distance, delay, or actual dollar cost. Letting $D_i^{(h)}$ denote the minimum cost or length of the shortest path from node 1 to node i with h or fewer hops, we define $D_1^{(h)} = 0$ for all h and $D_i^{(0)} = \infty$ for all $i \neq 1$. If the cost in going from node i to node j is c_{ji}, we can write the Bellman–Ford algorithm as, for each $h = 1, 2, \ldots,$

$$D_i^{(h+1)} = \min_j \left[D_j^{(h)} + c_{ji}\right], \qquad i \neq 1. \tag{15.5.1}$$

We do not demonstrate optimality here; this is left to the problems. The algorithm is probably best illustrated by an example.

EXAMPLE 15.5.1

Consider the network in Fig. 15.5.2, where node 1 is the source node and node 6 is the destination node. The cost per hop is listed adjacent to each link. Starting with $h = 1$, we find the shortest paths with a length of 1 hop, as shown in Fig. 15.5.3(a). In the next step we find the shortest path to each

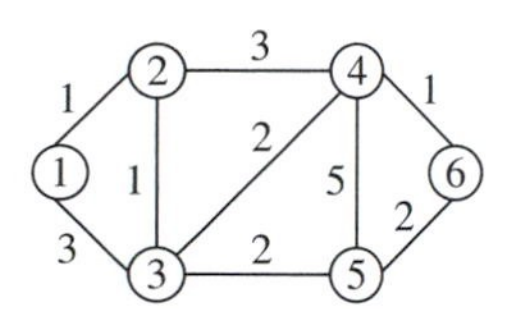

FIGURE 15.5.2 Network with specified cost per hop.

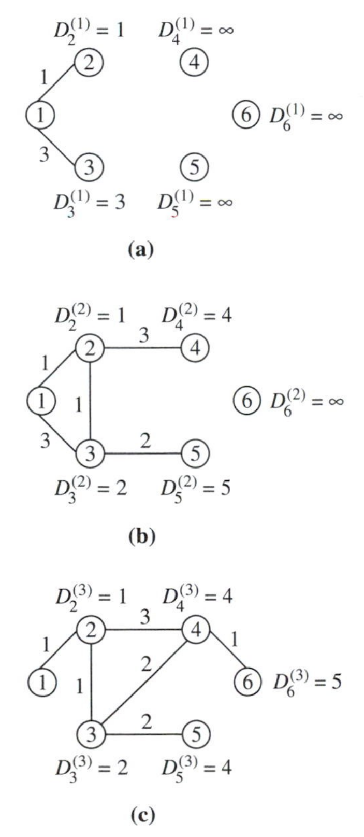

FIGURE 15.5.3 (a) Shortest paths with $h = 1$; (b) shortest paths with $h \leq 2$; (c) shortest paths with $h \leq 3$.

of the nodes with $h \leq 2$ hops. This is the same as evaluating

$$D_i^{(2)} = \min_j \left[D_j^{(1)} + c_{ji}\right] \tag{15.5.2}$$

for $i = 2, 3, \ldots, 6$. The end result is shown in Fig. 15.5.3(b). In these figures the nodes are labeled with the appropriate $D_i^{(h)}$. The shortest paths with three or fewer hops are shown in part (c) of the figure. By comparing (b) and (c), we see that allowing three hops yields a small minimum cost to node 5 and admits another route with the same minimum cost to node 4. Letting $h = 4$ produces another path to node 6 with $D_6^{(4)} = 5$, but no other changes.

A second important routing algorithm, called Dijkstra's algorithm, searches for the shortest path from a source node to all other nodes by extending only the lowest-cost path. Let the source node be node 1 and initially define the set $S = \{1\}$, the set containing only node 1. The single link cost from node 1 to node j is $D_j = d_{j1}$, so $D_1 = 0$. The algorithm consists of the following steps:

STEP 1. Initialization. With $S = \{1\}$ set $D_j = d_{j1}$.

STEP 2. Find the closest node. Find $i \notin S$ such that

$$D_i = \min_{j \notin S} D_j.$$

Set $S = S \cup \{i\}$. The algorithm terminates when all nodes belong to S.

STEP 3. Update costs. For all $j \notin S$, find

$$D_j = \min[D_j, D_i + d_{ji}]. \tag{15.5.3}$$

Go to step 2.

We illustrate the algorithm by reconsidering the network in Fig. 15.5.2.

EXAMPLE 15.5.2

We initialize with $S = \{1\}$, $D_1 = 0$, and the distortions shown in Table 15.5.1. The successive steps are listed in Table 15.5.1, and the steps are illustrated graphically in Fig. 15.5.4. Comparing Figs. 15.5.3(c) and 15.5.4(f), we see a slight difference in the routing tables obtained by the Bellman–Ford algorithm and Dijkstra's algorithm. The Bellman–Ford algorithm finds two

TABLE 15.5.1 Dijkstra's Algorithm for Figure 15.5.2

Step	S	D_2	D_3	D_4	D_5	D_6
Initial	{1}	1	3	∞	∞	∞
1	{1, 2}	1	1	3	∞	∞
2	{1, 2, 3}	1	1	2	2	∞
3	{1, 2, 3, 4}	1	1	2	2	1
4	{1, 2, 3, 4, 6}	1	1	2	2	1
5	{1, 2, 3, 4, 5, 6}	1	1	2	2	1

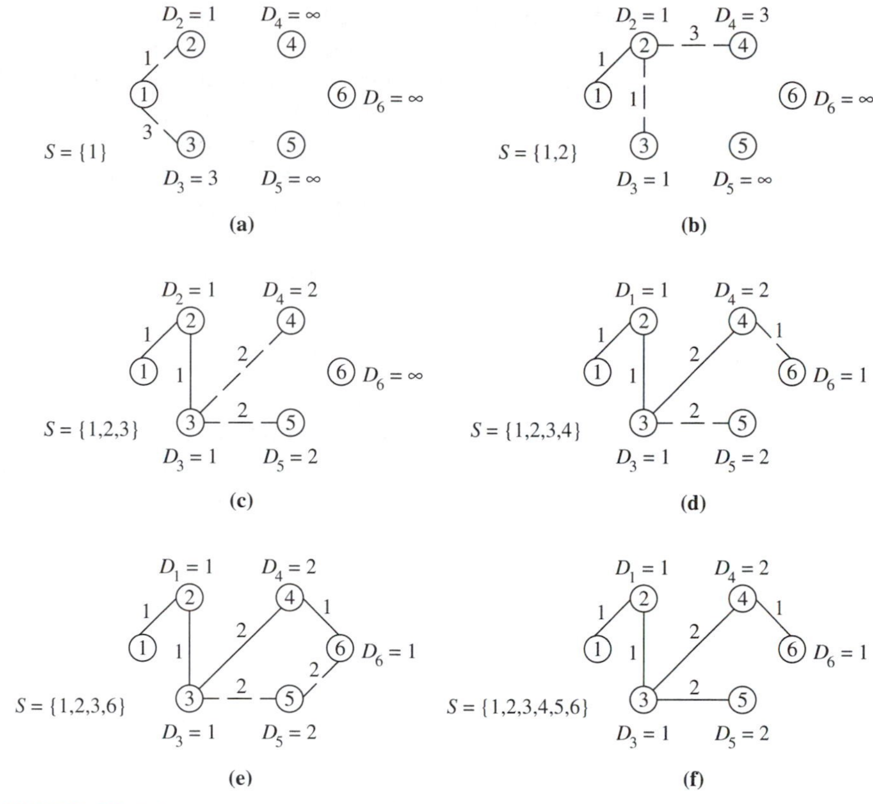

FIGURE 15.5.4 (a) Initial step in Dijkstra's algorithm; (b) step 1 in Dijkstra's algorithm; (c) step 2 in Dijkstra's algorithm; (d) step 3 in Dijkstra's algorithm; (e) step 4 in Dijkstra's algorithm; (f) step 5 in Dijkstra's algorithm.

paths to node 4 with minimum cost of 4, while Dijkstra's algorithm finds only one of the paths. Since the minimum cost is the same, performance is not affected. However, there is an extra hop in the 1–2–3–4 route.

There are advantages and disadvantages to each of the two algorithms discussed thus far. Computationally, for a network of N nodes, the Bellman–Ford algorithm can require on the order of N^3 operations. To see this, note that the minimum on the right of Eq. (15.5.1) must be performed for $N - 1$ values of j, for each of $N - 1$ nodes (i), and for up to $N - 1$ hops (h). However, if the network is not fully connected, the number of paths P to be investigated is less than N^2. Further, the maximum number of hops, say h_m, required to find the optimum routes may be much less than N. Thus we need only about Ph_m operations to find the best routes and we might have $Ph_m \ll N^3$. For Dijkstra's algorithm, Eq. (15.5.3) requires on the order of N^2 operations. Thus, if a network is small and highly interconnected, Dijkstra's algorithm may have a computational advantage.

Another significant difference between the Bellman–Ford and the Dijkstra algorithms is that the former is better suited to distributed processing. To see this, consider Eq. (15.5.1) again, and note that other than the total number of nodes, only local information is needed, such as the c_{ji} and $D_j^{(h)}$, where the $D_j^{(h)}$ can be updated by information exchange with adjacent nodes. Alternatively, Dijkstra's algorithm requires that we know which nodes are connected to which other nodes (called the network topology) for the entire network. Both algorithms have had important applications.

The stability of minimum-cost or shortest-path routing algorithms can be a problem if the cost being minimized is affected by the traffic arriving at individual nodes. For example, if the cost being minimized is average packet delay, the routing algorithm may reroute traffic away from a node if the queue of packets becomes long at the node. This action may cause a long delay to accrue at a node in the alternative route. In response to this situation, the routing algorithm may reroute traffic back through the original node, thus producing a long queue again. Operation may then continue in this oscillatory fashion. It is evident that slower updates of routing tables can reduce the likelihood of this type of instability, but if routing information is updated too slowly, throughput can drop significantly.

Flow control attempts to adjust the offered traffic at network nodes in such a way as to avoid congestion. Figure 15.5.5 shows typical plots of throughput versus offered traffic for no flow control, ideal flow control, and a more reasonable flow-controlled situation. For low values of offered traffic, the throughput is able to keep up. However, for higher values of arriving packets without flow control, the throughput drops, since excessive delays accumulate, which then cause packet retransmission because the sender does not receive an acknowledgment within the allotted time period. In these circumstances,

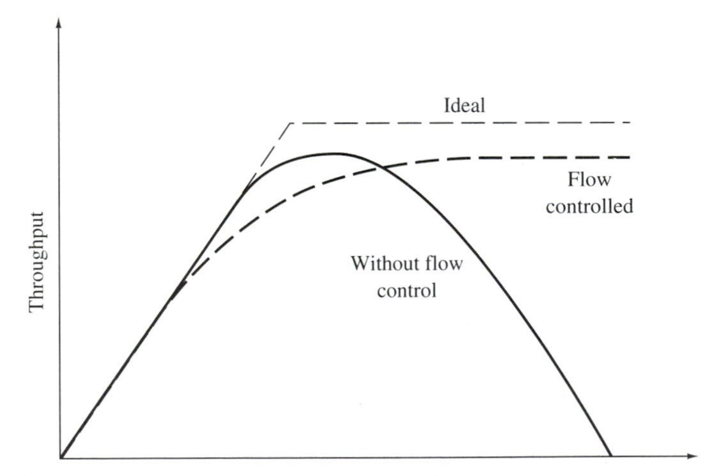

FIGURE 15.5.5 Throughput versus offered traffic with and without flow control. From J. D. Spragins, J. L. Hammond, and K. Paulikowski, *Telecommunications, Protocols and Design*, © 1991 by Addison-Wesley Publishing Company, Inc., p. 395. Reprinted with permission of the publisher.

throughput can eventually drop to zero if offered traffic continues to increase. With flow control, packets are not allowed to enter the network, or some portion of the network, as throughput begins to drop below the offered load. In this way, the throughput for packets in the network can be made to follow the flow-controlled curve in Fig. 15.5.5.

The most commonly used method of flow control is what is called sliding window flow control. This technique allows the receiver to control the rate of arrival of packets by issuing permits. Each permit allows a transmitter to send a window of data that is some fixed number of packets long. After these packets have been sent, the transmitter must wait for another permit before sending an additional window of packets. When the receiver issues new permits, such as at the end of a window or after receiving the first packet in a window, can greatly affect network performance [Spragins, Hammond, and Pawlikowski, 1991].

The ARPANET, developed by the Advanced Research Projects Agency of the U.S. Department of Defense, began service in about 1970 and was one of the first large-scale networks to connect geographically distributed computer systems throughout the United States, and later, the world. ARPANET has had a tremendous impact on research and development of communications networks. We briefly discuss here routing in the ARPANET, since it has been copied in some form or fashion by many subsequent networks.

From 1970 to 1979, ARPANET used an asynchronous, distributed version of the Bellman–Ford algorithm discussed earlier in this section. Distributed means that each node computes its own routing table based on shared information, and asynchronous means that the routing table updates at each node are not computed in synchronism with other nodes. Every 625 msec the nodes transmitted information to their neighbors. The delays used in the routing algorithm were computed approximations to the true packet delays, and this fact coupled with the relatively rapid update of delay information caused routes to change often, thus leading to oscillations and other undesirable behavior. To compensate for this, a large fixed delay was added to computed delays, but this reduced responsiveness to traffic conditions.

The newer ARPANET routing uses a version of Dijkstra's algorithm and time stamping of packets to produce accurate delay measurements. The delay on a link is averaged over a 10-sec interval, and this information is spread throughout the network (called flooding) every 10 sec if the delay estimate changes sufficiently, and at least every 60 sec even if the delay estimate is unchanged. The routing algorithm is asynchronous, since delay estimate updates between nodes are not coordinated.

SUMMARY

The concept of a packet-switched network has been introduced and the situations where packet switching can provide an advantage over TDM and FDM have been established. The OSI model and layered network architectures have been presented and the functions of the several layers have been discussed. The differences between virtual circuit and datagram routing have been described. Little's formula and the fundamentals of an $M/M/1$ queue have been developed and used for the analysis of multiple access methods. In particular, throughput versus offered load for ALOHA, slotted ALOHA, CSMA, CSMA/CD, and token ring access methods were derived or presented and compared. Common algorithms for routing were given, including the Bellman–Ford algorithm and Dijkstra's algorithm, and their computational and information requirements examined. Flow control was defined and how it differs from routing discussed. The characteristics of some common networks such as ARPANET were outlined.

Further details concerning communication networks are pursued in the problems.

PROBLEMS

15.1 Using an argument similar to that used for TDM in Section 15.1, show that the packet delay in FDM is n times longer than in statistical multiplexing.

15.2 Manchester encoding is often used in local area networks. Give the reasons why.

15.3 Virtual circuit and datagram networks each have advantages and disadvantages for different network conditions and different types of traffic. Contrast these two types of networks for
(a) Carrying digitized voice versus a continuous data stream;
(b) Adapting to failures of an intermediate node;
(c) Adapting to network congestion.

15.4 Tabulate the error-detecting capabilities of cyclic codes from Chapter 12.

15.5 Compare the capabilities of the CRC-12 and CRC-16 codes in Table 12.3.3 for detecting single-bit errors, double-bit errors, and error bursts.

15.6 When the throughput expressions in Section 12.5 are used with cyclic codes for error detection as in networking applications, the ratio k/n must be interpreted carefully. In this case, n = frame length and $k = n -$ (number of overhead bits for the frame). Since all bits in the frame are protected by the CRC, for coding studies, the number of overhead bits is just the CRC length. However, different protocols may have different length headers and other overhead bits, so to compare protocols, k must be found by subtracting all of these overhead bits from the frame length.

Letting the number of overhead bits be 48 (the header length in HDLC), plot the throughputs in Eqs. (12.5.1)–(12.5.3) versus frame length for $1 - P = 10^{-4}$ and 10^{-5}. Let $N = 5$ and assume that $r\tau_d = N$.

15.7 For $n = 512$ bits, 48 overhead bits, and $N = 5 = r\tau_d$, plot Eqs. (12.5.1)–(12.5.3) versus error probability, $1 - P$.

15.8 The probability of exactly n users in an $M/M/1$ queueing system is given by p_n in Eq. (15.3.19). Plot p_n versus n for $\rho = 0.1$, 0.5, and 0.9.

15.9 Find an expression for the probability that the number of messages in an $M/M/1$ queueing system is greater than K. Plot this probability versus K for $\rho = 0.1$, 0.5, and 0.9.

15.10 For a finite-length $M/M/1$ queue, the upper limit on the summation in Eq. (15.3.17) is changed. Letting the maximum length of a finite $M/M/1$ queue be K, show that $p_0 = (1 - \rho)/(1 - \rho)^{K+1}$ and $p_n = (1 - \rho)\rho^n/(1 - \rho^{K+1})$.

15.11 In a finite $M/M/1$ queue of length K, the probability of a customer being blocked is the probability that the queue is full. Plot this blocking probability versus ρ in terms of K.

15.12 In several applications, the arriving messages do not have exponentially distributed message lengths as in $M/M/1$ queueing systems. If the message lengths, or equivalently, customer service times, have a general distribution (rather than exponential), but all other assumptions are the same as for $M/M/1$, we have an $M/G/1$ system. For an $M/G/1$ queue, the average number of messages in the queue is

$$E(N) = \frac{\rho}{1 - \rho}\left[1 - \frac{\rho}{2}(1 - \mu^2\sigma^2)\right]$$

and the average time delay is

$$\bar{T} = \frac{1/\mu}{1 - \rho}\left[1 - \frac{\rho}{2}(1 - \mu^2\sigma^2)\right],$$

where σ^2 = variance of the service time [Schwartz, 1987].

(a) Show that if σ^2 = variance of the exponential distribution, we get the $M/M/1$ results in Eqs. (15.3.20) and (15.3.21).

(b) Assume that all packets have the same service time and find $E(N)$ and $\bar{T}$ by letting $\sigma^2 = 0$. This is called an $M/D/1$ queue.

(c) Plot $\bar{T}$ versus ρ for the three cases $\sigma^2 = 1/\mu^2$, $\sigma^2 > 1/\mu^2$, and $\sigma^2 < 1/\mu^2$. Interpret these results.

15.13 As mentioned in Section 15.3, the average time a packet spends in the system is the sum of the average waiting time spent in the queue and the average service time. Letting the average waiting time be denoted by $\bar{W}$, find:

(a) $\bar{W}$ for the $M/M/1$ queue.

(b) $\bar{W}$ for the $M/G/1$ queue (see Problem 15.12).

15.14 By using the $M/G/1$ queue expressions in Problems 15.12 and 15.13, the result in Eq. (15.3.22) for a time-division multiple access (TDMA) system can be made more precise. In addition to the average time spent in the queue and the average service time, another term due to when a packet arrives with respect to the beginning of a time slot must be included.

Assume that the packet arrival time is uniformly distributed over a TDM frame time.

(a) Find an expression for the average packet delay in TDMA for K users.

(b) Plot the results in part (a) versus ρ for the special case of constant packet length.

15.15 For a packet length of τ_p seconds, draw a diagram illustrating the vulnerable period for collisions in pure ALOHA.

15.16 Let the arrival rate of new packets due to each user in an ALOHA system be λ_{new} packets/sec and assume that there are K users of the system.

(a) If the throughput of the system exactly equals the total number of newly arriving packets, show that the maximum number of users that can be accommodated is given by $K_{\text{max}} = 1/(2e\tau_p\lambda_{\text{new}})$.

(b) Find K_{max} if $\tau_p = 34$ msec and $\lambda = \frac{1}{60}$ sec [Abramson, 1973].

15.17 To analyze a slotted ALOHA system with a finite number of users K, let G_i be the probability of an attempted transmission by user i, including both old and new packets, and let S_i be the probability of a successful transmission by user i.

(a) Show that $S_i = G_i \prod^{j \neq i} (1 - G_j)$.

(b) Defining $G = \sum_{i=1}^{K} G_i$, let each of the users be identical so that $S_i = S/K$ and $G_i = G/K$. Show that

$$S = G\left[1 - \frac{G}{K}\right]^{K-1}.$$

Note that as $K \to \infty$, we obtain Eq. (15.4.3) [Tanenbaum, 1988].

15.18 The condition for maximum throughput for a finite number of users in slotted ALOHA is $\sum_{i=1}^{K} G_i = 1$ (see Problem 15.17). If there are K_1 identical interactive users and K_2 identical continuous data transfer users,

(a) Find expressions for S_1, S_2, and the maximum throughput condition.

(b) Consider the two-user situation, $K_1 = K_2 = 1$, and plot S_1, S_2, and $S = S_1 + S_2$ versus G_1 [Tanenbaum, 1988].

15.19 Show that the average number of transmissions for a successful packet transmission is e^{2G} for pure ALOHA.

15.20 Let $\bar{\tau}_b$ be the average amount of time waited before a retransmission after a collision, called the backoff time, in pure ALOHA. Ignoring the propagation delays in the channel, show that the average transmission delay in pure ALOHA is $\bar{T} = \tau_p + (e^{2G} - 1)[\tau_p + \bar{\tau}_b]$.

15.21 Let the backoff delay in pure ALOHA be $\tau_b = k\tau_p$ for $k = 0, 1, \ldots, N - 1$. If the integers k are equally likely, find $\bar{\tau}_b$. Substitute this into $\bar{T}$ in Problem 15.20 and plot $\bar{T}/\tau_p$ versus throughput S for $K = 5$, 10, and 100.

15.22 Sketch curves similar to Fig. 15.4.2 for the stable but lightly loaded case, the stable but overloaded case, and the unstable with an infinite number of users case.

15.23 Plot throughput versus offered load for 1-persistent CSMA with various values of normalized delay.

15.24 Find and compare the relationship between S and G for nonpersistent and 1-persistent CSMA as $a = t_d/\tau_p \to 0$.

15.25 The average transmission delay for a packet in a token ring network with K users is

$$\bar{T} = \frac{\tau}{2} + E[S(t)] + \frac{\tau(1 - \rho/N)}{2(1 - \rho)} + \frac{\rho E[S^2(t)]}{2(1 - \rho)E[S(t)]},$$

where τ is the transmission time between stations, $\rho = \lambda E[S(t)]$, and $S(t)$ represents the service time. Evaluate $\bar{T}$ for $M/M/1$, $M/D/1$, and $M/G/1$ systems.

15.26 Use the Bellman–Ford algorithm to find the routing table for the network in Fig. P15.26.

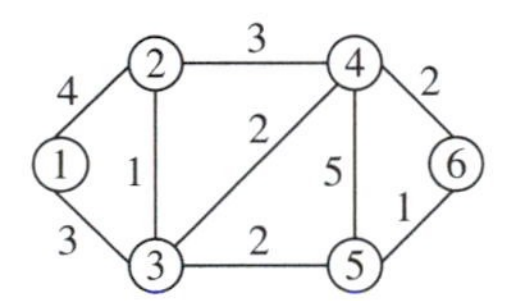

FIGURE P15.26

15.27 Use Dijkstra's algorithm to find the routing table for the network in Fig. P15.26.

15.28 The Bellman–Ford algorithm relies on Bellman's principle of optimality mentioned in Section 10.7. Demonstrate the optimality of the Bellman–Ford algorithm.

Random Variables and Stochastic Processes

A.1 Introduction

Randomness enters into communications problems through the classes of sources, such as speech, images, and music, that we wish to send from one location to another, as well as through the types of distortions, such as additive noise, phase jitter, and fading, that are inserted by real channels. Randomness is the essence of communications; it is that which makes communications both difficult and interesting. Everyone has an intuitive idea as to what "randomness" means, and in this chapter we develop different ways to characterize random phenomena that will allow us to account for randomness in both sources and channels in our communication system analyses and designs.

The treatment in this appendix is necessarily terse and incomplete. Every undergraduate engineer should have at least one course on the topics of probability and random variables, and this appendix does not purport to replace such a course. We simply survey the salient ideas and concepts of probability, random variables, and stochastic processes and define the requisite notation. The minimum mathematical mechanics needed to manipulate random variables and stochastic processes are also included, and after reading this material, the student will be prepared for probability and random processes wherever they are required in the book.

As is typical of undergraduate communications books, theoretical details are suppressed to allow applications to be treated almost immediately. This is certainly justifiable, but the reader should be aware that rigorous proofs of many of the results in this appendix, as well as extensions, may require considerable mathematical maturity beyond our cursory treatment.

A.2 Probability

A *random experiment* is an experiment whose outcome cannot be specified accurately in advance. Typical examples of a random experiment are the toss of a coin, the roll of a die, or the number of telephones in use at a particular time in a given telephone exchange. We denote the possible *outcomes* of a random experiment by ζ_i, $i = 1, 2, \ldots, N$, and the set of all possible outcomes or the *sample space* by $S = \{\zeta_1, \zeta_2, \ldots, \zeta_N\}$. The number of outcomes (N) in the sample space may be finite, countably infinite, or uncountable. Finite sample spaces are the simplest to handle mathematically, and many of the examples familiar to the reader have a finite sample space. As a result, our initial discussion assumes that N is finite.

For a given random experiment, we may only be interested in certain subsets of the sample space or collections of outcomes. We define these subsets of the sample space to be *events*, and we call attention to two special events, the certain event S and the impossible event, which is the empty set $\varnothing$. That is, since S is the set of all possible outcomes of the random experiment, it occurs at every replication of the experiment, while the empty set or null set $\varnothing$ contains no elements, and hence can never occur.

We assign to each event $A \in S$ a nonnegative real number $P(A)$ called the *probability* of the event A. Every probability must satisfy three axioms:

Axiom 1. $P(A) \geq 0$

Axiom 2. $P(S) = 1$

Axiom 3. If $A \cap B = \varnothing$, then $P(A \cup B) = P(A) + P(B)$, where A and B are events.[1]

For simplicity of notation, it is common to write $A \cap B$ as AB and $A \cup B$ as $A + B$. We will use both notations as is convenient. If A^c denotes the complement of set A, it follows from Axioms 1 to 3 that

$$P(A^c) = 1 - P(A). \tag{A.2.1}$$

Further, using set operations and the axioms, it is possible to show that if $A \cap B \neq \varnothing$, then

$$P(A \cup B) = P(A) + P(B) - P(A \cap B). \tag{A.2.2}$$

To define probabilities for a countably infinite number of outcomes, we need the concept of a *field*. A class of sets that is closed under the set operations of complementation and finite unions and intersections is called a *field* $\mathscr{F}$. Thus a field $\mathscr{F}$ is a nonempty class of sets such that:

1. If $A \in \mathscr{F}$, then $A^c \in \mathscr{F}$.
2. If $A \in \mathscr{F}$, and $B \in \mathscr{F}$, then $A + B \in \mathscr{F}$.

From (1) and (2) we can also show that if $A \in \mathscr{F}$ and $B \in \mathscr{F}$, then $A \cap B \in \mathscr{F}$ and $A - B \in \mathscr{F}$. Further, we have that $S \in \mathscr{F}$ and $\varnothing \in \mathscr{F}$. Thus it is evident that any finite number of set operations can be performed on the elements of $\mathscr{F}$ and still produce an element in $\mathscr{F}$.

[1] The symbols "$\cap$" and "$\cup$" represent the set operations intersection and union, respectively. We shall not review set theory here.

A field is called a *Borel field* on S if it is closed under a countably infinite number of unions and intersections. Symbolically, a *Borel field*, denoted by $\mathscr{B}$, is a field defined on S such that if $A_1, A_2, \ldots, A_n, \ldots, \in \mathscr{B}$, then $\bigcup_{i=1}^{\infty} A_i \in \mathscr{B}$. Using this definition, we can also show that if $A_1, A_2, \ldots, A_n, \ldots, \in \mathscr{B}$, then $\bigcap_{i=1}^{\infty} A_i \in \mathscr{B}$. Axioms 1 and 2 remain valid for any event $A \in \mathscr{B}$ and we can extend Axiom 3 to a countably infinite sequence of mutually exclusive events (events that correspond to disjoint sets) in $\mathscr{B}$ as

Axiom 3′. For the mutually exclusive events $A_1, A_2, \ldots, A_n, \ldots, \in \mathscr{B}$,

$$P\left[\bigcup_{i=1}^{\infty} A_i\right] = \sum_{i=1}^{\infty} P(A_i).$$

We cannot proceed to define probabilities for an uncountably infinite number of outcomes without the mathematics of measure theory, which is well beyond the scope of the book. However, Axioms 1, 2, and 3′ give us sufficient tools upon which to base our development.

An important concept in probability and in communications is the concept of independence. Two events A and B are said to be *independent* if and only if

$$P(A \cap B) = P(A)P(B). \tag{A.2.3}$$

Further, it often happens that we wish to calculate the probability of a particular event given the information that another event has occurred. Therefore, we define the conditional probability of B given A as $[P(A) \neq 0]$

$$P(B|A) \triangleq \frac{P(A \cap B)}{P(A)}. \tag{A.2.4}$$

Note that if A and B are independent, then $P(B|A) = P(B)$; that is, knowing that A occurred does not change the probability of B. In calculating the probability of an event, it is often useful to employ what is called the *theorem of total probability*. Consider a finite or countably infinite collection of mutually exclusive ($A_i \cap A_j = \varnothing$ for all $i \neq j$) and exhaustive ($\bigcup_i A_i = S$) events denoted by $A_1, A_2, \ldots, A_n, \ldots$. Then the probability of an arbitrary event B is given by

$$P(B) = \sum_i P(A_i \cap B) = \sum_i P(A_i)P(B|A_i). \tag{A.2.5}$$

EXAMPLE A.2.1

To illustrate the preceding concepts, we consider the random experiment of observing whether two telephones in a particular office are busy (in use) or not busy. The outcomes of the random experiment are thus:

ζ_1 = neither telephone is busy

ζ_2 = only telephone number 1 is busy

ζ_3 = only telephone number 2 is busy

ζ_4 = both telephones are busy.

We assign probabilities to these outcomes according to $P(\zeta_1) = 0.2$, $P(\zeta_2) = 0.4$, $P(\zeta_3) = 0.3$, and $P(\zeta_4) = 0.1$. We can also define the sets (events) $A = \{\zeta_1\}$, $B = \{\zeta_2\}$, $C = \{\zeta_3\}$, and $D = \{\zeta_4\}$, so $P(A) = 0.2$, $P(B) = 0.4$, $P(C) = 0.3$, and $P(D) = 0.1$.

(a) What is the probability of the event that one or more telephones are busy?

$$P(\text{one or more busy}) = P(B \cup C \cup D),$$

which we could evaluate by repeated application of Eq. (A.2.2). However, we note that $A = (B \cup C \cup D)^c$ so we can use Eq. (A.2.1) to obtain

$$P(\text{one or more busy}) = 1 - P(A) = 0.8.$$

(b) What is the probability that telephone number 1 is in use?

$$P(\text{telephone 1 busy}) = P(B \cup D) = P(B) + P(D) = 0.5,$$

since $B \cap D = \varnothing$.

(c) Let E be the event that telephone number 1 is busy and F be the event that telephone number 2 is busy. Are E and F independent?

We need to check if Eq. (A.2.3) is satisfied. From (b) we have $P(E) = 0.5$, and in a similar fashion we obtain $P(F) = 0.4$. Hence

$$P(E)P(F) = 0.2.$$

Since $D = E \cap F$,

$$P(E \cap F) = P(D) = 0.1.$$

Thus $P(EF) \neq P(E)P(F)$ and E and F are not independent.

(d) Referring to event E in part (c), find $P(D|E)$.

From Eq. (A.2.4),

$$P(D|E) = \frac{P(DE)}{P(E)} = \frac{P(E|D)P(D)}{P(E)}.$$

But $P(E|D) = 1$, so

$$P(D|E) = \frac{P(D)}{P(E)} = 0.2.$$

Note that $P(D|E) \neq P(D)$; hence knowing that E has occurred changes the probability that D will occur.

A.3 Probability Density and Distribution Functions

The outcomes of a random experiment need not be real numbers. However, for purposes of manipulation, it would be much more convenient if they were. Hence we are led to define the concept of a random variable.

Definition A.3.1 A real *random variable* $X(\cdot)$ is a function that assigns a real number, called the *value* of the random variable, to each possible outcome of a random experiment.

Therefore, for Example A.2.1 we could define the random variable $X(\zeta_1) = 0$, $X(\zeta_2) = 1$, $X(\zeta_3) = 2$, and $X(\zeta_4) = 3$. Note that for this random variable, values are assigned only for discrete points and the values in between are not allowable. Such a random variable is said to be *discrete*. If a random variable can take on a continuum of values in some range, the random variable is said to be *continuous*. It is also possible to define *mixed* random variables that have both discrete and continuous components.

Of course, to work with random variables, we need a probabilistic description of their behavior. This is available from the probabilities assigned to the individual outcomes of the random experiment. That is, for discrete random variables, we can write

$$P[X(\zeta_i) = x_i] = P[\zeta_i \in S : X(\zeta_i) = x_i], \tag{A.3.1}$$

which simply means that the probability of the value of the random variable x_i is just the probability of the outcome ζ_i associated with x_i through $X(\cdot)$. The right-hand side of Eq. (A.3.1) is read as "the probability of $\zeta_i \in S$ such that $X(\zeta_i) = x_i$." For continuous random variables, the probability of $X(\zeta) = x$ is zero, but the probability of $X(\zeta)$ taking on a range of values is nonzero. Hence we have

$$P[X(\zeta_i) \le x_i] = P[\zeta_i \in S : X(\zeta_i) \le x_i] \tag{A.3.2}$$

for continuous random variables. It is common to drop the explicit dependence on the experimental outcome from the left side of Eq. (A.3.2) and write for some outcome ζ,

$$P[X \le x] = P[X(\zeta) \le x]. \tag{A.3.3}$$

We can just as well talk about other forms of intervals than $X \le x$, but this form has a special meaning.

Definition A.3.2 The *cumulative distribution function* (CDF), or simply *distribution function* of a random variable X, is given by

$$F_X(x) = P[X(\zeta) \le x] = P[X \le x], \tag{A.3.4}$$

for $-\infty < x < \infty$.

A distribution function has the properties

$$0 \le F_X(x) \le 1 \tag{A.3.5a}$$

$$F_X(x_1) \le F_X(x_2) \qquad \text{for } x_1 < x_2 \tag{A.3.5b}$$

$$F_X(-\infty) \triangleq \lim_{x \to -\infty} F_X(x) = 0 \tag{A.3.5c}$$

$$F_X(\infty) \triangleq \lim_{x \to \infty} F_X(x) = 1 \tag{A.3.5d}$$

$$\lim_{\substack{\varepsilon \to 0 \\ \varepsilon > 0}} F_X(x + \varepsilon) \triangleq F_X(x^+) = F_X(x). \tag{A.3.5e}$$

The reader should note that Eq. (A.3.5e) indicates that the distribution function is continuous from the right. Some authors prefer to define the cumulative distribution function in terms of $P[X < x]$, that is, strict inequality, in which case the distribution function is continuous from the left. As long as the specific definition being used is clear, the subsequent details will be consistent. We use Eq. (A.3.4) as our definition throughout the book.

Since $P[X \leq x] + P[X > x] = 1$, then

$$P[X > x] = 1 - F_X(x). \tag{A.3.6}$$

Further, we can write

$$P[x_1 < X \leq x_2] = P[X \leq x_2] - P[X \leq x_1] = F_X(x_2) - F_X(x_1). \tag{A.3.7}$$

If we let

$$F_X(x^-) = \lim_{\substack{\varepsilon \to 0 \\ \varepsilon > 0}} F_X(x - \varepsilon),$$

then

$$P[X = x] = P[X \leq x] - P[X < x] = F_X(x) - F_X(x^-). \tag{A.3.8}$$

For a continuous random variable, $F_X(x^-) = F_X(x)$, so $P[X = x] = 0$. However, for a discrete distribution function, there will be several points where $F_X(x)$ and $F_X(x^-)$ differ, and hence $F_X(x)$ will contain "jumps" or discontinuities at these points.

It is convenient to define another function, called the *probability density function* (pdf), denoted here by $f_X(x)$. For a discrete random variable, the probability density function is given by

$$f_X(x) = \begin{cases} P[X = x_i], & \text{for } x = x_i,\ i = 1, 2, \ldots \\ 0, & \text{for all other } x, \end{cases} \tag{A.3.9}$$

where $P[X = x_i]$ is computed from Eq. (A.3.8). Thus, for a discrete random variable, $f_X(x)$ is nonzero only at the points of discontinuity of $F_X(x)$. Now, if we write $F_X(x)$ in terms of the unit step function, that is,

$$F_X(x) = P[X \leq x] = \sum_{x_i} P[X = x_i]u(x - x_i), \tag{A.3.10}$$

we can obtain an expression for $f_X(x)$ by formally taking the derivative of Eq. (A.3.10) as

$$f_X(x) \triangleq \frac{d}{dx} F_X(x) = \sum_{x_i} P[X = x_i]\delta(x - x_i) \tag{A.3.11}$$

for $-\infty < x < \infty$. We see by inspection that Eqs. (A.3.9) and (A.3.11) agree. Note that it is not strictly mathematically correct to use impulses in Eq. (A.3.11), since the impulses actually have infinite amplitude. However, this is a common methodology and it allows us to treat continuous and discrete random variables in a unified fashion. A more careful approach would be differentiation of the cumulative distribution function to obtain the pdf in the continuous case and differencing as in Eq. (A.3.8) in the discrete case.

For a continuous random variable X with distribution function $F_X(x)$, we have that

$$f_X(x) = \frac{d}{dx} F_X(x) \geq 0 \tag{A.3.12}$$

for all x, and further,

$$F_X(x) = \int_{-\infty}^{x} f_X(\lambda)\, d\lambda. \tag{A.3.13}$$

Using Eq. (A.3.13) with Eqs. (A.3.5c) and (A.3.5d), we find that

$$\int_{-\infty}^{\infty} f_X(\lambda)\, d\lambda = 1. \tag{A.3.14}$$

Similarly, we have for a discrete random variable that

$$\int_{-\infty}^{\infty} f_X(x)\, dx = \sum_{\text{all } i} P[X = x_i] = 1. \tag{A.3.15}$$

EXAMPLE A.3.1

We are given the distribution function of a discrete random variable Y,

$$F_Y(y) = \begin{cases} 0, & y < -1 \\ \frac{1}{8}, & -1 \leq y < 0 \\ \frac{3}{8}, & 0 \leq y < 2 \\ \frac{3}{4}, & 2 \leq y < 3 \\ 1, & y \geq 3. \end{cases}$$

(a) Sketch $F_Y(y)$.

A sketch of $F_Y(y)$ is shown in Fig. A.3.1. The heavy dots at the step points indicate that $F_Y(y)$ is continuous from the right.

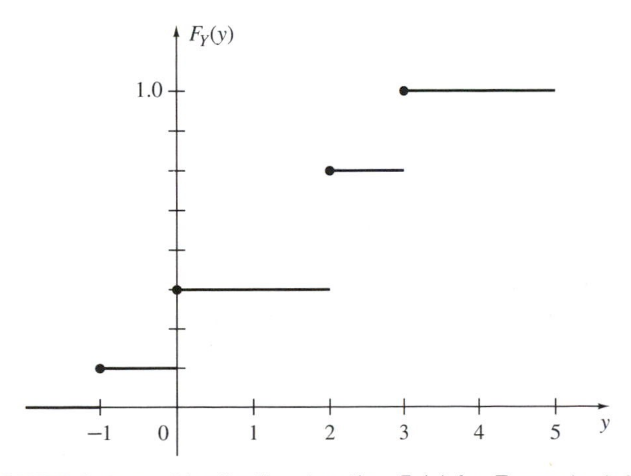

FIGURE A.3.1 Distribution function $F_Y(y)$ for Example A.3.1.

(b) Write an expression for $F_Y(y)$.

In terms of unit step functions, we have

$$F_Y(y) = \tfrac{1}{8}u(y+1) + \tfrac{1}{4}u(y) + \tfrac{3}{8}u(y-2) + \tfrac{1}{4}u(y-3).$$

(c) Write an expression for and sketch the probability density function of Y.

Using Eq. (A.3.11), we can write

$$\begin{aligned} f_Y(y) &= \sum_{y_i} P[Y = y_i]\delta(y - y_i) \\ &= \tfrac{1}{8}\delta(y+1) + \tfrac{1}{4}\delta(y) + \tfrac{3}{8}\delta(y-2) + \tfrac{1}{4}\delta(y-3). \end{aligned}$$

A sketch of this pdf is shown in Fig. A.3.2.

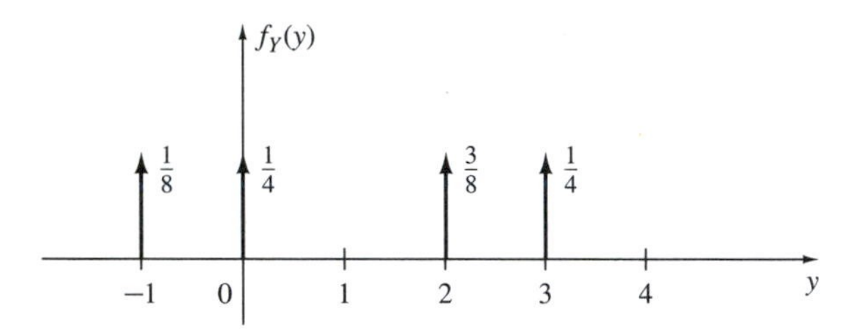

FIGURE A.3.2 Probability density function $f_Y(y)$ for Example A.3.1.

(d) Find $P[Y \leq 0]$, $P[Y \leq 1]$, and $P[Y > 2]$.

$$\begin{aligned} P[Y \leq 0] &= F_Y(0) = \tfrac{1}{8} + \tfrac{1}{4} = \tfrac{3}{8} \\ P[Y \leq 1] &= F_Y(1) = \tfrac{1}{8} + \tfrac{1}{4} = \tfrac{3}{8} \\ P[Y > 2] &= 1 - P[Y \leq 2] = 1 - F_Y(2) \\ &= 1 - [\tfrac{1}{8} + \tfrac{1}{4} + \tfrac{3}{8}] = \tfrac{1}{4}. \end{aligned}$$

EXAMPLE A.3.2

The distribution function of a continuous random variable Z is given by

$$F_Z(z) = \begin{cases} 0, & z < 2 \\ \dfrac{z-2}{4}, & 2 \leq z < 6 \\ 1, & z \geq 6. \end{cases}$$

(a) Sketch $F_Z(z)$. See Fig. A.3.3.

(b) Find an expression for and sketch the pdf of Z.

$$f_Z(z) = \begin{cases} 0, & z < 2 \\ \tfrac{1}{4}, & 2 \leq z < 6 \\ 0, & z \geq 6. \end{cases}$$

See the sketch in Fig. A.3.4.

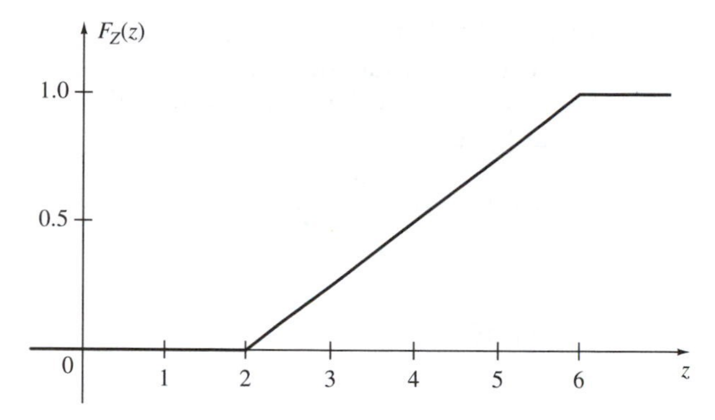

FIGURE A.3.3 Sketch of $F_Z(z)$ in Example A.3.2.

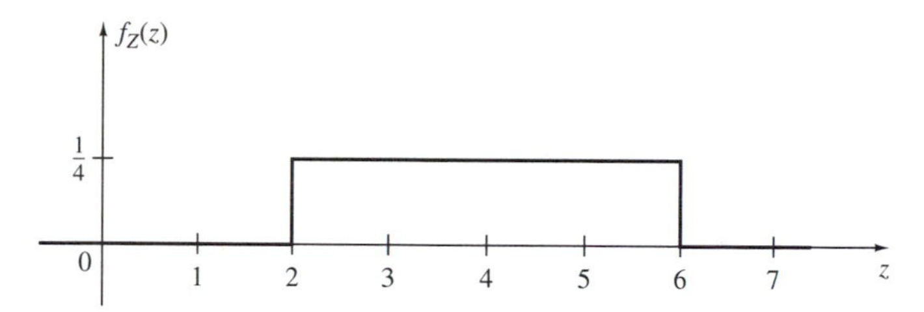

FIGURE A.3.4 Sketch of the pdf of Z for Example A.3.2.

(c) Find $P[Z \leq 3]$ and $P[4 < Z \leq 6]$.

$$P[Z \leq 3] = F_Z(3) = \tfrac{1}{4}$$

$$P[4 < Z \leq 6] = F_Z(6) - F_Z(4) = 1 - \tfrac{1}{2} = \tfrac{1}{2}.$$

It is possible to extend the concepts of distribution functions and density functions to more than one random variable in a straightforward manner. For example, for the case of two random variables X and Y, we can define their *joint distribution*, also called the bivariate distribution of X and Y, as

$$F_{XY}(x, y) = P[X \leq x, Y \leq y], \tag{A.3.16}$$

where properties analogous to the scalar case hold. Assuming that the joint probability density function of X and Y exists, we can also write Eq. (A.3.16) as

$$F_{XY}(x, y) = \int_{-\infty}^{y} \int_{-\infty}^{x} f_{XY}(\lambda, \gamma)\, d\lambda\, d\gamma. \tag{A.3.17}$$

Furthermore, given the joint distribution function $F_{XY}(x, y)$, we can obtain the probability density function from

$$f_{XY}(x, y) = \frac{\partial^2}{\partial x\, \partial y} F_{XY}(x, y). \tag{A.3.18}$$

Of course, it is often the case that we are given a bivariate distribution and we need the univariate distribution function of one of the random variables. The univariate distributions are also called the *marginal* distributions and can be obtained straightforwardly as

$$F_X(x) = F_{XY}(x, \infty) = \int_{-\infty}^{x} \int_{-\infty}^{\infty} f_{XY}(\lambda, \gamma)\, d\gamma\, d\lambda$$
$$= \int_{-\infty}^{x} f_X(\lambda)\, d\lambda. \tag{A.3.19}$$

Similarly, the marginal density is given by

$$f_X(x) = \int_{-\infty}^{\infty} f_{XY}(x, \gamma)\, d\gamma. \tag{A.3.20}$$

For purely discrete random variables, the integrals in Eqs. (A.3.17), (A.3.19), and (A.3.20) are replaced by summations.

There is no difficulty in extending joint distribution and density functions to the multivariate (more than two variables) case.

EXAMPLE A.3.3

Given the joint probability density function

$$f_{XY}(x, y) = \begin{cases} Ke^{-(x+y)}, & \text{for } x \geq 0, \quad y \geq 0 \\ 0, & \text{elsewhere.} \end{cases}$$

(a) Find K such that $f_{XY}(x, y)$ is a valid pdf.

We need

$$\int_{-\infty}^{\infty} \int_{-\infty}^{\infty} f_{XY}(x, y)\, dx\, dy = 1,$$

so we find that

$$K \int_{0}^{\infty} \int_{0}^{\infty} e^{-(x+y)}\, dx\, dy = K = 1.$$

(b) Calculate $F_{XY}(x, y)$.

$$F_{XY}(x, y) = \int_{-\infty}^{y} \int_{-\infty}^{x} f_{XY}(\lambda, \gamma)\, d\lambda\, d\gamma$$
$$= \int_{0}^{y} \int_{0}^{x} e^{-(\lambda+\gamma)}\, d\lambda\, d\gamma = (1 - e^{-x})(1 - e^{-y})$$

for $x \geq 0$, $y \geq 0$, and $F_{XY}(x, y) = 0$ otherwise.

(c) Find the marginal distribution and density functions of Y.

Directly, the distribution function of Y is

$$F_Y(y) = F_{XY}(\infty, y) = (1 - e^{-x})\Big|_{x=\infty} (1 - e^{-y})$$
$$= 1 - e^{-y}, \qquad \text{for } y \geq 0$$

and

$$F_Y(y) = 0, \qquad \text{for } y < 0.$$

We can find the pdf of Y by differentiating $F_Y(y)$ or by integrating over all x in $f_{XY}(x, y)$. Thus

$$f_Y(y) = \frac{d}{dy} F_Y(y) = \begin{cases} e^{-y}, & y \geq 0 \\ 0, & y < 0. \end{cases}$$

(d) Find $P[X \leq 2, Y \leq 5]$.
By substitution in $F_{XY}(x, y)$,

$$P[X \leq 2, Y \leq 5] = F_{XY}(2, 5) = (1 - e^{-2})(1 - e^{-5}).$$

(e) Find $P[X \leq 2, Y > 1]$.
The simplest way to evaluate this probability is to return to the joint density and write

$$P[X \leq 2, Y > 1] = \int_1^\infty \int_0^2 e^{-(x+y)}\, dx\, dy = (1 - e^{-2})e^{-1}.$$

Analogous to independent events in a random experiment, we can have independent random variables.

Definition A.3.3 Two random variables X and Y are said to be *independent* if and only if

$$F_{XY}(x, y) = F_X(x)F_Y(y). \tag{A.3.21}$$

Of course, this condition is equivalent to

$$f_{XY}(x, y) = f_X(x)f_Y(y). \tag{A.3.22}$$

Therefore, two random variables are independent if and only if their joint distribution (or density) factors into the product of their marginal distributions (densities).

EXAMPLE A.3.4

Are the random variables X and Y in Example A.3.3 independent? Examining the pdfs, we know from Example A.3.3 that

$$f_{XY}(x, y) = \begin{cases} e^{-(x+y)}, & x \geq 0, \quad y \geq 0 \\ 0, \cdot & \text{otherwise} \end{cases}$$

and

$$f_Y(y) = \begin{cases} e^{-y}, & y \geq 0 \\ 0, & y < 0. \end{cases}$$

Straightforwardly, we find that

$$f_X(x) = \begin{cases} e^{-x}, & x \geq 0 \\ 0, & x < 0. \end{cases}$$

Hence

$$f_{XY}(x, y) = f_X(x) f_Y(y)$$

and X and Y are independent. It is simple to demonstrate the same result using distribution functions.

Just as we defined the conditional probability of event B occurring given that event A has occurred in Eq. (A.2.4), we can define conditional distribution and density functions. Given two random variables X and Y, we can define the conditional distribution function of X given that $Y \leq y$ as

$$F_{X|Y}(x|y) = \frac{F_{XY}(x, y)}{F_Y(y)}, \tag{A.3.23}$$

where $F_Y(y) \neq 0$. If the conditioning is that $Y = y$ (rather than $Y \leq y$), we have

$$F_{X|Y}(x|Y = y) = \frac{\int_{-\infty}^{x} f_{XY}(\lambda, y)\, d\lambda}{f_Y(y)}. \tag{A.3.24}$$

The distribution functions in Eqs. (A.3.23) and (A.3.24) are both valid univariate distribution functions, and hence satisfy the properties in Eqs. (A.3.5).

From Eq. (A.3.24) we can also find the conditional probability density of X given $Y = y$ by differentiation as

$$f_{X|Y}(x|y) = \frac{f_{XY}(x, y)}{f_Y(y)}. \tag{A.3.25}$$

Similarly, we can obtain

$$f_{Y|X}(y|x) = \frac{f_{XY}(x, y)}{f_X(x)}, \tag{A.3.26}$$

so we can combine Eqs. (A.3.25) and (A.3.26) to get a form of Bayes' rule,

$$f_{Y|X}(y|x) = \frac{f_{X|Y}(x|y) f_Y(y)}{f_X(x)}. \tag{A.3.27}$$

EXAMPLE A.3.5

We are given the joint pdf of X and Y,

$$f_{XY}(x, y) = \begin{cases} x + y, & 0 \leq x \leq 1, \quad 0 \leq y \leq 1 \\ 0, & \text{otherwise.} \end{cases}$$

(a) Find the marginal densities of X and Y.

$$f_X(x) = \begin{cases} \int_0^1 (x + y)\, dy = \left[xy + \frac{y^2}{2} \right]\Bigg|_0^1 = x + \frac{1}{2}, & \text{for } 0 \leq x \leq 1 \\ 0, & \text{otherwise} \end{cases}$$

Similarly, we can show that

$$f_Y(y) = \begin{cases} y + \frac{1}{2}, & 0 \le y \le 1 \\ 0, & \text{otherwise.} \end{cases}$$

(b) Find $f_{X|Y}(x|y)$.
From Eq. (A.3.25),

$$f_{X|Y}(x|y) = \frac{f_{XY}(x, y)}{f_Y(y)},$$

so

$$f_{X|Y}(x|y) = \begin{cases} \dfrac{x + y}{y + \frac{1}{2}}, & 0 \le x \le 1 \\ 0, & \text{otherwise.} \end{cases}$$

(c) Calculate $F_{X|Y}(x|Y = y)$.
From part (b),

$$F_{X|Y}(x|Y = y) = \int_0^x \frac{\lambda + y}{y + \frac{1}{2}} d\lambda = \frac{1}{y + \frac{1}{2}} \left[\frac{\lambda^2}{2} + \lambda y\right]\Bigg|_0^x = \frac{x^2/2 + xy}{y + \frac{1}{2}}$$

for $0 \le x \le 1$, and

$$F_{X|Y}(x|Y = y) = \begin{cases} 0, & x < 0 \\ 1, & x \ge 1. \end{cases}$$

(d) Calculate $F_{X|Y}(x|y)$.
From Eq. (A.3.23), we see that we need $F_Y(y)$, so

$$\begin{aligned} F_Y(y) &= \int_0^y \left(\gamma + \frac{1}{2}\right) d\gamma = \left[\frac{\gamma^2}{2} + \frac{\gamma}{2}\right]\Bigg|_0^y \\ &= \frac{y^2}{2} + \frac{y}{2} = \frac{y}{2}(y + 1), \qquad 0 \le y \le 1 \\ &= 0, \qquad y < 0 \\ &= 1, \qquad y \ge 1. \end{aligned}$$

Also,

$$\begin{aligned} F_{XY}(x, y) &= \int_0^y \int_0^x (\lambda + \gamma)\, d\gamma\, d\lambda \\ &= \frac{xy}{2}(x + y), \qquad 0 \le x \le 1, \quad 0 \le y \le 1 \end{aligned}$$

$$\begin{aligned} F_{XY}(x, y) &= 0, \qquad x < 0, \quad y < 0 \\ &= 1, \qquad x \ge 1, \quad y \ge 1. \end{aligned}$$

$$F_{X|Y}(x|y) = \frac{F_{XY}(x, y)}{F_Y(y)}$$

$$= \frac{x(x+y)}{y+1}, \qquad 0 \leq x \leq 1$$

$$F_{X|Y}(x|y) = 0, \qquad x < 0$$

$$= 1, \qquad x \geq 1.$$

Note that $F_{X|Y}(x|y) \neq F_{X|Y}(x|Y = y)$.

A.4 Mean, Variance, and Correlation

The probability density function or the cumulative distribution function describes in detail the behavior of the particular random variable being considered. However, there are many applications where we need to work with numbers that are representative of the particular distribution function of interest without carrying along the entire pdf or CDF. Perhaps the single most-used parameter of the distribution of a random variable is the expected value, also called the expectation or mean value.

Definition A.4.1 The *expected value* of a random variable X is given by

$$E\{X\} \triangleq \int_{-\infty}^{\infty} x f_X(x)\, dx, \tag{A.4.1}$$

where $f_X(x)$ is the pdf of X. For a discrete random variable X, Eq. (A.4.1) becomes

$$E\{X\} = \sum_i x_i P[X = x_i]. \tag{A.4.2}$$

It is common to represent $E\{X\}$ by the symbol μ_X.

It is also possible to write the expectation of a function of the random variable X, say $g(X)$, in terms of an integral over the pdf of X. This result, sometimes called the *fundamental theorem of expectation*, is given by

$$E\{g(X)\} = \int_{-\infty}^{\infty} g(x) f_X(x)\, dx. \tag{A.4.3}$$

We do not offer a proof of Eq. (A.4.3), and indeed, it is not trivial to prove in all generality. Of course, in Eq. (A.4.3) we could let $g(x) = x$, which would yield Eq. (A.4.1), or $g(x)$ can be any function of x. Of particular importance is when $g(x) = x^2$, so Eq. (A.4.3) becomes

$$E\{X^2\} = \int_{-\infty}^{\infty} x^2 f_X(x)\, dx, \tag{A.4.4}$$

which is called the second moment of X, or the second moment about zero.

Next to the expectation, perhaps the second most widely used parameter of the probability distribution of a random variable is the variance. The *variance* of a random variable X is defined as

$$\text{var}\{X\} \triangleq E\{[X - \mu_X]^2\} = E\{X^2\} - \mu_X^2. \tag{A.4.5}$$

The variance of X is also the second moment about the mean of X. Thus, while μ_X locates the mean or average value of the random variable, the variance is an indicator of the dispersion about the mean. The variance is often denoted by σ_X^2, and the positive square root of the variance is called the *standard deviation* and denoted by σ_X.

EXAMPLE A.4.1

Given a random variable X with pdf

$$f_X(x) = \begin{cases} e^{-x}, & x \geq 0 \\ 0, & x < 0. \end{cases}$$

(a) Find $E\{X\}$.
By Eq. (A.4.1),

$$E\{X\} = \int_0^\infty xe^{-x}\,dx = [xe^{-x} - e^{-x}]\Big|_0^\infty = 1.$$

(b) Find $E\{X^2\}$.
From Eq. (A.4.4),

$$E\{X^2\} = \int_0^\infty x^2e^{-x}\,dx = 2.$$

(c) What is the variance of X?
By Eq. (A.4.5),

$$\text{var}(X) = E\{X^2\} - \mu_X^2 = 2 - 1 = 1.$$

(d) Given $g(X) = aX + b$, what is $E\{g(X)\}$?
Using Eq. (A.4.3), we obtain

$$E\{g(X)\} = E\{aX + b\} = \int_0^\infty [ax + b]e^{-x}\,dx$$

$$= a\int_0^\infty xe^{-x}\,dx + b\int_0^\infty e^{-x}\,dx = a\mu_X + b = a + b.$$

It is important to note that the integral defining the expected value in Eq. (A.4.1) may not exist, in which case the mean does not exist. The most common example of this phenomenon is the Cauchy distribution, defined as

$$f_X(x) = \frac{1}{\pi(1 + x^2)}, \qquad -\infty < x < \infty. \tag{A.4.6}$$

The demonstration that $E\{X\}$ does not exist for Eq. (A.4.6) is left as a problem.

An extremely convenient function for solving various problems involving random variables is obtained by letting $g(X) = e^{j\omega X}$ in Eq. (A.4.3).

Definition A.4.2 The *characteristic function* of a random variable X is defined by

$$\Phi_X(\omega) = E\{e^{j\omega X}\}. \tag{A.4.7}$$

Note that the characteristic function of X is simply the Fourier transform of the pdf of X with ω replaced by $-\omega$. Since the Fourier transform is unique, the characteristic function is also unique, and hence conclusions reached using characteristic functions can be uniquely translated into conclusions concerning pdfs. For example, given two random variables X and Y, if $\Phi_X(\omega) \equiv \Phi_Y(\omega)$, then $F_X(x) \equiv F_Y(y)$. [Since we are assuming that pdfs exist for our work, we can also conclude that $f_X(x) \equiv f_Y(y)$.] Also of great importance is the fact that if we let $E\{X^n\} = m_n$, then

$$\left.\frac{d^n}{d\omega^n}\Phi_X(\omega)\right|_{\omega=0} = j^n m_n. \tag{A.4.8}$$

Therefore, moments of a random variable X are easily obtainable from its characteristic function.

EXAMPLE A.4.2

Given the pdf

$$f_X(x) = \begin{cases} e^{-x}, & x \geq 0 \\ 0, & x < 0. \end{cases}$$

(a) Find $\Phi_X(\omega)$.

$$\Phi_X(\omega) = \int_0^\infty e^{j\omega x} e^{-x}\, dx = \int_0^\infty e^{-(1-j\omega)x}\, dx$$

$$= \left.\frac{-1}{1-j\omega} e^{-(1-j\omega)x}\right|_0^\infty = \frac{1}{1-j\omega}.$$

(b) Calculate $E\{X\}$ and $E\{X^2\}$.
By Eq. (A.4.8),

$$E\{X\} = m_1 = \left.\frac{1}{j}\frac{d}{d\omega}\Phi_X(\omega)\right|_{\omega=0} = \left.\frac{1}{j}\left\{\frac{j}{(1-j\omega)^2}\right\}\right|_{\omega=0} = 1.$$

Also by Eq. (A.4.8),

$$E\{X^2\} = m_2 = \left.\frac{1}{j^2}\frac{d^2}{d\omega^2}\Phi_X(\omega)\right|_{\omega=0} = \left.(-1)\left\{\frac{-j2(1-j\omega)(-j)}{(1-j\omega)^4}\right\}\right|_{\omega=0} = 2.$$

Both results check with Example A.4.1.

We can also consider moments of joint distributions. Given two random variables X and Y with joint pdf $f_{XY}(x, y)$, we can define

$$E\{X^n Y^k\} = \int_{-\infty}^\infty \int_{-\infty}^\infty x^n y^k f_{XY}(x, y)\, dx\, dy. \tag{A.4.9}$$

In fact, we have the more general result, analogous to the univariate case in Eq. (A.4.3), that

$$E\{g(X, Y)\} = \int_{-\infty}^{\infty}\int_{-\infty}^{\infty} g(x, y)f_{XY}(x, y)\,dx\,dy. \tag{A.4.10}$$

We can also define bivariate characteristic functions as

$$\Phi_{XY}(\omega_1, \omega_2) = \int_{-\infty}^{\infty}\int_{-\infty}^{\infty} e^{j\omega_1 x + j\omega_2 y} f_{XY}(x, y)\,dx\,dy. \tag{A.4.11}$$

Letting $E\{X^n Y^k\} = m_{nk}$, we also have that

$$\left.\frac{\partial^k \partial^n \Phi(\omega_1, \omega_2)}{\partial\omega_2^k\,\partial\omega_1^n}\right|_{\omega_1=\omega_2=0} = j^{(n+k)} m_{nk}. \tag{A.4.12}$$

Equations (A.4.9)–(A.4.11) are extended to more than two random variables in the obvious way.

Two random variables X and Y are said to be *uncorrelated* if

$$E\{XY\} = E\{X\}E\{Y\} \tag{A.4.13}$$

and *orthogonal* if

$$E\{XY\} = 0. \tag{A.4.14}$$

The *covariance* of X and Y is given by

$$\text{cov}(X, Y) = E\{[X - \mu_X][Y - \mu_Y]\} = E\{XY\} - \mu_X\mu_Y, \tag{A.4.15}$$

from which we define the *correlation coefficient*

$$\rho_{XY} = \frac{\text{cov}(X, Y)}{\sqrt{\text{var}(X)\,\text{var}(Y)}} = \frac{E\{XY\} - \mu_X\mu_Y}{[(E\{X^2\} - \mu_X^2)(E\{Y^2\} - \mu_Y^2)]^{1/2}}. \tag{A.4.16}$$

We note that

$$-1 \le \rho_{XY} \le 1. \tag{A.4.17}$$

EXAMPLE A.4.3

Let us reconsider the joint pdf of X and Y from Example A.3.5,

$$f_{XY}(x, y) = \begin{cases} x + y, & 0 \le x \le 1, \quad 0 \le y \le 1 \\ 0, & \text{otherwise.} \end{cases}$$

(a) Find $E\{XY\}$.

From Eq. (A.4.9),

$$E\{XY\} = \int_0^1\int_0^1 xy(x + y)\,dx\,dy = \int_0^1\int_0^1 (x^2 y + xy^2)\,dx\,dy$$

$$= \int_0^1 \left[\frac{x^3 y}{3} + \frac{x^2 y^2}{2}\right]\Bigg|_0^1 dy = \int_0^1 \left[\frac{y}{3} + \frac{y^2}{2}\right] dy = \left[\frac{y^2}{6} + \frac{y^3}{6}\right]\Bigg|_0^1 = \frac{1}{3}.$$

(b) Calculate cov(X, Y) and ρ_{XY}.

By Eq. (A.4.15), we need μ_X and μ_Y, so using the marginal pdfs from Example A.3.5,

$$\mu_X = \int_0^1 x\left(x + \frac{1}{2}\right) dx = \left[\frac{x^3}{3} + \frac{x^2}{4}\right]\Bigg|_0^1 = \frac{7}{12}$$

and $\mu_Y = \frac{7}{12}$. Thus

$$\text{cov}(X, Y) = E\{XY\} - \mu_X\mu_Y = \frac{1}{3} - \left(\frac{7}{12}\right)^2 = \frac{-1}{144}.$$

Now

$$E\{X^2\} = \int_0^1 x^2\left(x + \frac{1}{2}\right) dx = \left[\frac{x^4}{4} + \frac{x^3}{6}\right]\Bigg|_0^1 = \frac{5}{12} .$$

and

$$E\{Y^2\} = \frac{5}{12},$$

so

$$\rho_{XY} = \frac{-\frac{1}{144}}{[(\frac{5}{12} - \frac{49}{144})^2]^{1/2}} = -\frac{1}{11}.$$

(c) Are X and Y uncorrelated? Orthogonal?

Since $E\{XY\} \neq 0$, they are not orthogonal, and further, $E\{XY\} = \frac{1}{3} \neq \mu_X\mu_Y = \frac{49}{144}$, so X and Y are not uncorrelated.

(d) Find the joint characteristic function of X and Y.

From Eq. (A.4.11),

$$\Phi_{XY}(\omega_1, \omega_2) = \int_0^1 \int_0^1 e^{j\omega_1 x + j\omega_2 y}(x + y)\, dx\, dy$$

$$= \frac{j}{\omega_2\omega_1^2}[1 - e^{j\omega_2}]\{e^{j\omega_1} - 1 - j\omega_1 e^{j\omega_1}\}$$

$$+ \frac{j}{\omega_1\omega_2^2}[1 - e^{j\omega_1}]\{e^{j\omega_2} - 1 - j\omega_2 e^{j\omega_2}\}.$$

Note that Eq. (A.4.12) is not a good way to obtain joint moments of X and Y for this example.

Finally, we simply note that we can define the mean and variance for conditional distributions straightforwardly as

$$E\{X \mid Y = y\} = \int_{-\infty}^{\infty} x f_{X|Y}(x \mid y)\, dx \triangleq \mu_{X|Y} \qquad \text{(A.4.18)}$$

and

$$\text{var}\{X \mid Y = y\} = E\{X^2 \mid Y = y\} - \mu_{X|Y}^2. \qquad \text{(A.4.19)}$$

The conditional mean plays a major role in the theory of minimum mean-squared error estimation. We will not be able to pursue this connection here.

A.5 Transformations of Random Variables

In electrical engineering and communications, it is often the case that we are given some random variable X with a known pdf $f_X(x)$, and we wish to obtain the pdf of a random variable Y related to X by some function $y = g(x)$. For discrete random variables, the determination of $f_Y(y)$ is quite straightforward. Given a discrete random variable X with a pdf specified by $f_X(x_i) = P[X = x_i]$ and a one-to-one transformation $y = g(x)$, we have

$$f_Y(y_i) = P[Y = y_i = g(x_j)] = P[X = x_j]. \tag{A.5.1}$$

For continuous random variables, we can calculate $f_Y(y)$ by first solving $y = g(x)$ for all of its real roots, denoted $x_i = g^{-1}(y_i)$, $i = 1, 2, \ldots$. Then we can express $f_Y(y)$ in terms of $f_X(x)$ as

$$f_Y(y) = \sum_i \frac{f_X(x)}{|(d/dx)g(x)|}\bigg|_{x=x_i} \tag{A.5.2}$$

$$= \sum_i \left[f_X(x) \left| \frac{dx}{dy} \right| \right] \bigg|_{x=x_i}. \tag{A.5.3}$$

EXAMPLE A.5.1

We are given a random variable X with pdf

$$f_X(x) = \begin{cases} e^{-x}, & x \geq 0 \\ 0, & x < 0 \end{cases}$$

and we wish to find the pdf of a random variable Y related to X by $Y = aX + b$.

Solving $y = g(x)$ for x, we find $x = (y - b)/a$, so from Eq. (A.5.2)

$$f_Y(y) = \frac{e^{-x}u(x)}{|(d/dx)(ax + b)|}\bigg|_{x=(y-b)/a} = \frac{1}{|a|} e^{-((y-b)/a)} u\left(\frac{y-b}{a}\right).$$

We can also use Eq. (A.5.3) to produce the same result, as follows:

$$f_Y(y) = \left[\{e^{-x}u(x)\} \left\{ \left| \frac{d}{dy} \frac{y-b}{a} \right| \right\} \right] \bigg|_{x=(y-b)/a} = \frac{1}{|a|} e^{-((y-b)/a)} u\left(\frac{y-b}{a}\right).$$

One check we have on this result is to use Eq. (A.4.3) and compare the result with $E\{Y\}$ using $f_Y(y)$. From Eq. (A.4.3),

$$E\{aX + b\} = aE\{X\} + b = a + b$$

from Example A.4.1(d). Now, directly,

$$E\{Y\} = \int_{-\infty}^{\infty} y f_Y(y)\, dy = \int_b^{\infty} y \left[\frac{1}{|a|} e^{-((y-b)/a)} \right] dy$$

$$= \frac{1}{|a|} \int_b^{\infty} y e^{-((y-b)/a)}\, dy = a + b.$$

In many applications it is also necessary to find the joint pdf corresponding to transformations of jointly distributed random variables. For instance, consider the bivariate case where we are given two random variables X and Y with joint pdf $f_{XY}(x, y)$ and we wish to find the joint pdf of W and Z that are related to X and Y by $W = g(X, Y)$ and $Z = h(X, Y)$. Similar to the univariate case, we solve the given transformations for x and y in terms of z and w and use the formula

$$f_{WZ}(w, z) = \sum_i \left. \frac{f_{XY}(x, y)}{|J(x, y)|} \right|_{x = x_i,\ y = y_i}, \tag{A.5.4}$$

where x_i and y_i represent the simultaneous solutions of the given transformations and $J(x, y)$ is the Jacobian defined by the determinant

$$J(x, y) = \begin{vmatrix} \frac{\partial}{\partial x} g(x, y) & \frac{\partial}{\partial y} g(x, y) \\ \frac{\partial}{\partial x} h(x, y) & \frac{\partial}{\partial y} h(x, y) \end{vmatrix}. \tag{A.5.5}$$

The extension of this approach to more than two functions of more than two random variables is straightforward.

EXAMPLE A.5.2

Given the two random variables X and Y with joint pdf

$$f_{XY}(x, y) = \begin{cases} e^{-(x+y)}, & \text{for } x \geq 0, \quad y \geq 0 \\ 0, & \text{elsewhere,} \end{cases}$$

we wish to find the joint pdf of the random variables W and Z related to X and Y by the expressions

$$W = X + 2Y \qquad \text{and} \qquad Z = 2X + Y.$$

The Jacobian is

$$J = \begin{vmatrix} 1 & 2 \\ 2 & 1 \end{vmatrix} = [1 - 4] = -3$$

and solving for x and y yields

$$x = \frac{2z - w}{3} \qquad \text{and} \qquad y = \frac{2w - z}{3}.$$

Using Eq. (A.5.4), we obtain

$$f_{WZ}(w, z) = \frac{f_{XY}(x, y)}{|-3|}\bigg|_{x=(2z-w)/3,\ y=(2w-z)/3}$$

$$= \frac{1}{3}\exp\left\{-\left[\frac{2z-w}{3} + \frac{2w-z}{3}\right]\right\}$$

$$= \frac{1}{3}e^{-[(z+w)/3]}, \qquad w \geq 0, \ \frac{w}{2} \leq z \leq 2w$$

$$= 0, \qquad \text{otherwise.}$$

It is sometimes necessary to determine the pdf of one function of two random variables. In this case it is sometimes useful to define an auxiliary variable. For instance, given two random variables X and Y with joint pdf $f_{XY}(x, y)$ and the function $W = g(X, Y)$, suppose that we desire the pdf of W. In this situation, we can define the auxiliary variable Z by $Z = X$ or $Z = Y$ and use Eq. (A.5.4) to obtain the joint pdf of W and Z. We then integrate out the variable Z.

EXAMPLE A.5.3

Given the joint pdf of the random variables X and Y, we wish to find the pdf of $W = X + Y$. To facilitate the solution, we define $Z = Y$, so Eq. (A.5.5) yields

$$J(x, y) = \begin{vmatrix} 1 & 1 \\ 0 & 1 \end{vmatrix} = 1.$$

Solving for x and y, we have $x = w - z$ and $y = z$, so Eq. (A.5.4) gives

$$f_{WZ}(w, z) = \frac{f_{XY}(x, y)}{|1|}\bigg|_{x=w-z,\ y=z} = f_{XY}(w - z, z).$$

Thus

$$f_W(w) = \int_{-\infty}^{\infty} f_{XY}(w - z, z)\, dz. \tag{A.5.6}$$

Note that a special case occurs when X and Y are independent. Then

$$f_W(w) = \int_{-\infty}^{\infty} f_X(w - z) f_Y(z)\, dz. \tag{A.5.7}$$

That is, the pdf of the sum of two independent random variables is the convolution of their marginal densities.

A.6 Special Distributions

There are several distribution functions that arise so often in communications problems that they deserve special attention. In this section we examine five such distributions: the uniform, Gaussian, and Rayleigh distributions for continuous random variables, and the binomial and Poisson distributions for discrete random variables.

The Uniform Distribution

A random variable X is said to have a *uniform* distribution if the pdf of X is given by

$$f_X(x) = \begin{cases} \dfrac{1}{b-a}, & a \le x \le b \\ 0, & \text{otherwise.} \end{cases} \tag{A.6.1}$$

The mean and variance of X are directly

$$E(X) = \int_a^b x \frac{1}{b-a}\, dx = \frac{b+a}{2} \tag{A.6.2}$$

and

$$\operatorname{var}(X) = E(X^2) - E^2(X) = \frac{(b-a)^2}{12}. \tag{A.6.3}$$

The CDF of X is straightforwardly shown to be

$$\begin{aligned} F_X(x) &= \int_a^x \frac{1}{b-a}\, d\lambda = \frac{1}{b-a}\int_a^x d\lambda \\ &= \frac{x-a}{b-a}, \qquad a \le x \le b \\ &= 0, \qquad x < a \\ &= 1, \qquad x > b. \end{aligned} \tag{A.6.4}$$

The characteristic function of a uniform random variable is given by

$$\Phi_X(\omega) = \int_a^b \frac{1}{b-a} e^{j\omega x}\, dx = e^{j\omega[(a+b)/2]} \frac{\sin[(b-a)\omega/2]}{(b-a)\omega/2}. \tag{A.6.5}$$

The Gaussian Distribution

A random variable X is said to have a *Gaussian* or *normal* distribution if its pdf is of the form

$$f_X(x) = \frac{1}{\sqrt{2\pi}\,\sigma} e^{-(x-\mu)^2/2\sigma^2}, \qquad -\infty < x < \infty. \tag{A.6.6}$$

We find that

$$E(X) = \frac{1}{\sqrt{2\pi}\,\sigma}\int_{-\infty}^{\infty} x e^{-(x-\mu)^2/2\sigma^2}\, dx.$$

Letting $y = (x - \mu)/\sigma$, we obtain

$$E(X) = \frac{1}{\sqrt{2\pi}} \int_{-\infty}^{\infty} (\sigma y + \mu) e^{-y^2/2}\, dy$$

$$= \frac{\sigma}{\sqrt{2\pi}} \int_{-\infty}^{\infty} y e^{-y^2/2}\, dy + \frac{\mu}{\sqrt{2\pi}} \int_{-\infty}^{\infty} e^{-y^2/2}\, dy = 0 + \mu = \mu. \tag{A.6.7}$$

Furthermore,

$$E(X^2) = \frac{1}{\sqrt{2\pi}\,\sigma} \int_{-\infty}^{\infty} x^2 e^{-(x-\mu)^2/2\sigma^2}\, dx.$$

Again letting $y = (x - \mu)/\sigma$, we have that

$$E(X^2) = \frac{1}{\sqrt{2\pi}} \int_{-\infty}^{\infty} (\sigma y + \mu)^2 e^{-y^2/2}\, dy$$

$$= \frac{1}{\sqrt{2\pi}} \left\{ \sigma^2 \left[y e^{-y^2/2} \Big|_{-\infty}^{\infty} + \int_{-\infty}^{\infty} e^{-y^2/2}\, dy \right] \right.$$

$$\left. + 2\mu\sigma \int_{-\infty}^{\infty} y e^{-y^2/2}\, dy + \mu^2 \int_{-\infty}^{\infty} e^{-y^2/2}\, dy \right\}$$

$$= \sigma^2 + \mu^2. \tag{A.6.8}$$

Therefore,

$$\operatorname{var}(X) = \sigma^2 + \mu^2 - \mu^2 = \sigma^2. \tag{A.6.9}$$

The CDF of X is written as

$$F_X(x) = \frac{1}{\sqrt{2\pi}\,\sigma} \int_{-\infty}^{x} e^{-(y-\mu)^2/2\sigma^2}\, dy$$

$$= \frac{1}{\sqrt{2\pi}} \int_{-\infty}^{(x-\mu)/\sigma} e^{-z^2/2}\, dz = \frac{1}{2} + \operatorname{erf}\left[\frac{x-\mu}{\sigma}\right], \tag{A.6.10}$$

where $\operatorname{erf}(\cdot)$ is the *error function*, defined by

$$\operatorname{erf}(x) = \frac{1}{\sqrt{2\pi}} \int_{0}^{x} e^{-y^2/2}\, dy. \tag{A.6.11}$$

The reader is cautioned that several different definitions of the error function exist, and hence the exact definition of $\operatorname{erf}(x)$ should be checked before tables or other results are used (see Appendix I).

The characteristic function of a Gaussian pdf is

$$\Phi_X(\omega) = \int_{-\infty}^{\infty} \frac{1}{\sqrt{2\pi}\,\sigma} e^{j\omega x} e^{-(x-\mu)^2/2\sigma^2}\, dx$$

$$= \frac{1}{\sqrt{2\pi}\,\sigma} \int_{-\infty}^{\infty} e^{-(x^2 - 2[j\omega\sigma^2 + \mu]x + \mu^2)/2\sigma^2}\, dx.$$

We can complete the square in the exponent by adding and subtracting $[-\omega^2\sigma^4 + 2j\omega\sigma^2\mu]/2\sigma^2$ to obtain

$$\Phi_X(\omega) = [e^{j\omega\mu - \omega^2\sigma^2/2}] \frac{1}{\sqrt{2\pi}\,\sigma} \int_{-\infty}^{\infty} e^{-[x + (j\omega\sigma^2 + \mu)]^2/2\sigma^2}\, dx$$

$$= e^{j\omega\mu - \omega^2\sigma^2/2}, \tag{A.6.12}$$

since the integral is $\sqrt{2\pi}\,\sigma$. It is important to note that the Gaussian pdf is completely determined by specifying the mean and variance.

Multivariate Gaussian distributions also occur in many applications. We briefly consider here the bivariate Gaussian distribution. Two random variables X and Y are said to be jointly Gaussian distributed or jointly normal if their joint pdf is given by

$$f_{XY}(x, y) = \frac{1}{2\pi\sigma_X\sigma_Y\sqrt{1 - \rho_{XY}^2}} \exp\left\{\frac{-1}{2(1 - \rho_{XY}^2)}\left[\left(\frac{x - \mu_X}{\sigma_X}\right)^2 - 2\rho_{XY}\frac{x - \mu_X}{\sigma_X}\frac{y - \mu_Y}{\sigma_Y} + \left(\frac{y - \mu_Y}{\sigma_Y}\right)^2\right]\right\}, \tag{A.6.13}$$

where ρ_{XY} is the correlation coefficient defined in Eq. (A.4.16). Note that if $\rho_{XY} = 0$ in Eq. (A.6.13), then $f_{XY}(x, y) = f_X(x)f_Y(y)$, and hence X and Y are independent. This result is an important property of Gaussian random variables; namely, that if two Gaussian random variables are uncorrelated, they are also independent. This property does not hold in general for non-Gaussian random variables. The characteristic function corresponding to a bivariate Gaussian distribution is

$$\Phi_{XY}(\omega_1, \omega_2) = \exp\{j[\omega_1\mu_X + \omega_2\mu_Y] - \tfrac{1}{2}[\sigma_X^2\omega_1^2 + 2\omega_1\omega_2\rho_{XY}\sigma_X\sigma_Y + \sigma_Y^2\omega_2^2]\}. \tag{A.6.14}$$

Note that if we are given n independent Gaussian random variables X_i, $i = 1, 2, \ldots, n$, and we form $Z = \sum_{i=1}^n X_i$, then

$$\Phi_Z(\omega) = \prod_{i=1}^{n} \Phi_{X_i}(\omega) = e^{j\omega \sum_{i=1}^n \mu_{X_i} - (\omega^2/2) \sum_{i=1}^n \sigma_{X_i}^2}$$

$$= e^{j\omega\mu_Z - \omega^2\sigma_Z^2/2}, \tag{A.6.15}$$

so that Z is also Gaussian. A form of this result can be extended to dependent Gaussian random variables to yield the conclusion that *any* linear combination of Gaussian random variables is also Gaussian.

The Rayleigh Distribution

A random variable X is said to have a Rayleigh distribution if its pdf is of the form

$$f_X(x) = \frac{x}{\alpha^2} e^{-x^2/2\alpha^2} u(x). \tag{A.6.16}$$

The mean and variance of the Rayleigh distribution are given by

$$E(X) = \alpha \sqrt{\frac{\pi}{2}} \tag{A.6.17}$$

and

$$\operatorname{var}(X) = \left(2 - \frac{\pi}{2}\right)\alpha^2, \tag{A.6.18}$$

respectively.

The Binomial Distribution

The binomial distribution arises when we are considering n independent trials of an experiment admitting only two possible outcomes with constant probabilities p and $1 - p$, respectively. Thus, if the probability of "success" on any one trial is p, the probability of exactly x successes is

$$P[X = x] = b(x; n, p) = \binom{n}{x} p^x(1 - p)^{n-x} \tag{A.6.19}$$

for $x = 0, 1, 2, \ldots, n$. The notation $b(x; n, p)$ is fairly standard in the literature. From the binomial expansion we see that this density sums to 1, since

$$\sum_{x=0}^{n} \binom{n}{x} p^x(1 - p)^{n-x} = (p + 1 - p)^n = 1. \tag{A.6.20}$$

The expected value of X follows as

$$\begin{aligned}
E(X) &= \sum_{x=0}^{n} x\binom{n}{x} p^x(1 - p)^{n-x} \\
&= \sum_{x=1}^{n} np\binom{n-1}{x-1} p^{x-1}(1 - p)^{n-1-(x-1)} \\
&= np \sum_{y=0}^{n-1} \binom{n-1}{y} p^y(1 - p)^{n-1-y} \\
&= np(p + 1 - p)^{n-1} = np.
\end{aligned} \tag{A.6.21}$$

To determine the variance, we find $E\{X(X - 1)\}$ using manipulations similar to those in Eq. (A.6.21), that is,

$$\begin{aligned}
E\{X(X - 1)\} &= \sum_{x=0}^{n} x(x - 1)\binom{n}{x} p^x(1 - p)^{n-x} \\
&= 0 + 0 + \sum_{x=2}^{n} n(n - 1)p^2\binom{n-2}{x-2} p^{x-2}(1 - p)^{n-2-(x-2)} \\
&= n(n - 1)p^2 \sum_{y=0}^{n-2} \binom{n-2}{y} p^y(1 - p)^{n-2-y} \\
&= n(n - 1)p^2 = E(X^2) - E(X).
\end{aligned} \tag{A.6.22}$$

Now since $E(X) = np$,

$$\operatorname{var}(X) = E(X^2) - E^2(X) = n(n-1)p^2 + np - n^2p^2 = np(1-p). \quad \text{(A.6.23)}$$

The characteristic function of X can be evaluated straightforwardly as

$$\begin{aligned}\Phi_X(\omega) = E\{e^{j\omega X}\} &= \sum_{x=0}^{n} e^{j\omega x}\binom{n}{x}p^x(1-p)^{n-x} \\ &= \sum_{x=0}^{n}\binom{n}{x}(pe^{j\omega})^x(1-p)^{n-x} = (pe^{j\omega} + 1 - p)^n. \qquad \text{(A.6.24)}\end{aligned}$$

The moments of a binomial random variable X are usually evaluated from its characteristic function.

Note that unlike a continuous random variable, the equality sign in the CDF is of the utmost importance for discrete random variables. For example, for a binomial random variable X,

$$F_X(\alpha) = P[X \le \alpha] = \sum_{x=0}^{\alpha} b(x; n, p) = \sum_{x=0}^{\alpha}\binom{n}{x}p^x(1-p)^{n-x} \quad \text{(A.6.25)}$$

while

$$P[X < \alpha] = \sum_{x=0}^{\alpha-1}\binom{n}{x}p^x(1-p)^{n-x}. \quad \text{(A.6.26)}$$

The $x = \alpha$ term is missing in Eq. (A.6.26). There are extensive tables of the binomial pdf and CDF.

The Poisson Distribution

The Poisson distribution can be derived as a limiting case of the binomial distribution when n is large and p is small, and Poisson random variables play an important role in queueing problems. A random variable X is said to have a Poisson distribution if its pdf is given by

$$P[X = x] = f(x; \lambda) = \frac{\lambda^x}{x!}e^{-\lambda}, \qquad x = 0, 1, \ldots. \quad \text{(A.6.27)}$$

It is easy to show that $f(x; \lambda)$ sums to 1 as

$$\sum_{x=0}^{\infty} f(x; \lambda) = \sum_{x=0}^{\infty}\frac{\lambda^x}{x!}e^{-\lambda} = e^{-\lambda}\sum_{x=0}^{\infty}\frac{\lambda^x}{x!} = e^{-\lambda}(e^{\lambda}) = 1. \quad \text{(A.6.28)}$$

The mean of X is

$$\begin{aligned}E(X) &= \sum_{x=0}^{\infty} x\frac{\lambda^x}{x!}e^{-\lambda} = 0 + \sum_{x=1}^{\infty}\lambda e^{-\lambda}\frac{\lambda^{x-1}}{(x-1)!} \\ &= \lambda e^{-\lambda}\sum_{y=0}^{\infty}\frac{\lambda^y}{y!} = \lambda. \qquad \text{(A.6.29)}\end{aligned}$$

Similarly, we can show that

$$E\{X(X-1)\} = \lambda^2,$$

so

$$\text{var}(X) = E\{X(X-1)\} + E(X) - E^2(X) = \lambda^2 + \lambda - \lambda^2 = \lambda. \quad \text{(A.6.30)}$$

The characteristic function of a Poisson random variable is

$$\begin{aligned}\Phi_X(\omega) &= \sum_{x=0}^{\infty} e^{j\omega x}\frac{\lambda^x}{x!}e^{-\lambda} = \sum_{x=0}^{\infty}\frac{(e^{j\omega}\lambda)^x}{x!}e^{-\lambda} \\ &= e^{-\lambda}e^{\lambda e^{j\omega}} = e^{\lambda(e^{j\omega}-1)}. \end{aligned} \quad \text{(A.6.31)}$$

A.7 Stochastic Processes and Correlation

Random variables are defined on the outcomes of a random experiment. If we perform the random experiment, we obtain a value or range of values of the random variable. For random processes or stochastic processes, however, the situation is quite different. In the case of a random process, the outcome of the random experiment selects a particular function of time, called a *realization* of the stochastic process. Thus a random or stochastic process depends both on the outcome of a random experiment and on time.

Definition A.7.1 A *random process* or *stochastic process* $X(\xi, t)$ is a family of random variables indexed by the parameter $t \in T$, where T is called the *index set*.

A simple example of a random process is a single frequency sinusoid with random amplitude $A(\xi)$, so that

$$X(\xi, t) = A(\xi)\cos\omega_c t \quad \text{(A.7.1)}$$

for $-\infty < t < \infty$. A few realizations of this random process corresponding to outcomes of the random experiment are sketched in Fig. A.7.1. Note that once we specify $\xi = \xi_1$ (say), we have a deterministic time function $X(\xi_1, t) = A(\xi_1)\cos\omega_c t$, since $A(\xi_1)$ is just a number. Further, if we evaluate $X(\xi, t)$ at some specific value of t, say $t = t_1$, we are left with a random variable

$$X(\xi, t_1) = A(\xi)\cos\omega_c t_1, \quad \text{(A.7.2)}$$

which is no longer a function of time. Although we have only argued these points for the particular random process in Eq. (A.7.1), the conclusions are true for more general random processes.

Since a stochastic process is a function of t, the probability distributions and densities of the stochastic process also depend on t. For example, the first-order pdf of a random process $X(\xi, t)$ may be written as $f_X(x; t)$, where the explicit dependence on ξ is dropped. Often, the dependence on ξ is not expressed in

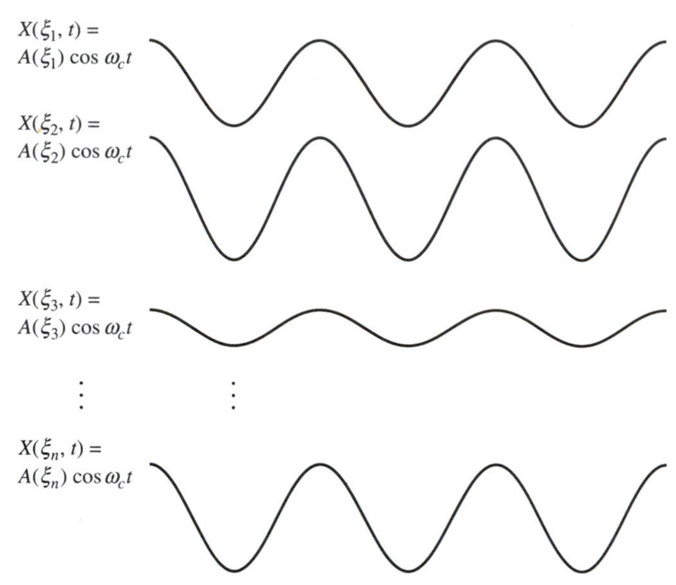

FIGURE A.7.1 Sample functions of the random process $X(\xi, t) = A(\xi) \cos \omega_c t$.

$X(\xi, t)$, so that it is common to write $X(\xi, t) = X(t)$ for a stochastic process. We can write an expression for the CDF of $X(t)$ as

$$F_X(x; t) = P[X(t) \leq x] = \int_{-\infty}^{x} f_X(\alpha; t)\, d\alpha. \tag{A.7.3}$$

Joint distributions of the random variables obtained by evaluating $X(t)$ at t_1, $t_2, \ldots,$ and t_n, are also quite important in characterizing the random process. The nth-order distribution of a random process $X(t)$ is given by

$$\begin{aligned} &F_{X_1 \cdots X_n}(x_1, x_2, \ldots, x_n; t_1, t_2, \ldots, t_n) \\ &\quad = P[X(t_1) \leq x_1, X(t_2) \leq x_2, \ldots, X(t_n) \leq x_n] \\ &\quad = \int_{-\infty}^{x_n} \cdots \int_{-\infty}^{x_2} \int_{-\infty}^{x_1} f_{X_1 \cdots X_n}(\alpha_1, \alpha_2, \ldots, \alpha_n; t_1, t_2, \ldots, t_n)\, d\alpha_1\, d\alpha_2 \cdots d\alpha_n. \end{aligned} \tag{A.7.4}$$

A random process is said to be *strictly stationary of order n* if its nth-order CDF and pdf are invariant to shifts in the time axis. Therefore, a stochastic process $X(t)$ is said to be strictly stationary of order n if

$$\begin{aligned} &F_{X_1 \cdots X_n}(x_1, x_2, \ldots, x_n; t_1, t_2, \ldots, t_n) \\ &\quad = F_{X_1 \cdots X_n}(x_1, x_2, \ldots, x_n; t_1 + \tau, t_2 + \tau, \ldots, t_n + \tau). \end{aligned} \tag{A.7.5}$$

As one might expect, we can also define the mean of a random process $X(t)$ by

$$E\{X(t)\} = \int_{-\infty}^{\infty} x f_X(x; t)\, dx, \tag{A.7.6}$$

where the mean is now in general a function of time. Furthermore, we can define the autocorrelation of a real process $X(t)$ by

$$R_X(t_1, t_2) = E\{X(t_1)X(t_2)\} = \int_{-\infty}^{\infty}\int_{-\infty}^{\infty} x_1 x_2 f_{X_1X_2}(x_1, x_2; t_1, t_2)\, dx_1\, dx_2 \quad \text{(A.7.7)}$$

and the covariance as

$$\text{cov}[X(t_1), X(t_2)] = R_X(t_1, t_2) - E[X(t_1)]E[X(t_2)]. \quad \text{(A.7.8)}$$

As with random variables, these quantities are very good "summary indicators" of random process behavior.

Another important form of stationarity does not require that we examine the nth-order CDF or pdf of the process, but we compute only its mean and autocorrelation. A stochastic process is said to be *wide-sense stationary* (WSS) or *weakly stationary* if

$$E\{X(t_1)\} = E\{X(t_2)\} = \text{constant}, \quad \text{(A.7.9)}$$

and

$$R_X(t_1, t_2) = E\{X(t_1)X(t_2)\} = E\{X(t)X(t + |t_2 - t_1|)\} = E\{X(t)X(t + \tau)\} = R_X(\tau). \quad \text{(A.7.10)}$$

Thus a random process is WSS if its mean is a constant and its autocorrelation depends only on the absolute time difference $|t_2 - t_1|$.

EXAMPLE A.7.1

Consider the random process

$$X(t) = A \cos \omega_c t, \qquad -\infty < t < \infty, \quad \text{(A.7.11)}$$

where ω_c is a constant and A is a zero-mean Gaussian random variable with variance σ_A^2. We would like to find the mean, autocorrelation, and the first-order pdf of $X(t)$.

For the mean we find that

$$E\{X(t)\} = E\{A\} \cos \omega_c t = 0, \quad \text{(A.7.12)}$$

and the autocorrelation is

$$E\{X(t_1)X(t_2)\} = E\{A^2\} \cos \omega_c t_1 \cos \omega_c t_2 = \frac{\sigma_A^2}{2}[\cos \omega_c(t_1 + t_2) + \cos \omega_c(t_1 - t_2)]. \quad \text{(A.7.13)}$$

Letting $t_1 = t_2 = t$, we find that

$$E\{X^2(t)\} = \sigma_A^2[\tfrac{1}{2} + \tfrac{1}{2}\cos 2\omega_c t] = \text{var}\{X(t)\}, \quad \text{(A.7.14)}$$

since $X(t)$ is zero mean from Eq. (A.7.12). Therefore, the first-order pdf of $X(t)$ is Gaussian with zero mean and variance given by Eq. (A.7.14).

We can also ask if this process is wide-sense stationary. Certainly, the mean of $X(t)$ is a constant (0), but from Eq. (A.7.13), it is clear that $R_X(t_1, t_2)$ is not a function of $|t_2 - t_1|$ only. Hence the random process is not WSS.

There are several properties of a WSS stochastic process $X(t)$ that are useful in applications. In stating these properties, we use the notation in Eq. (A.7.10) that $R_X(\tau) = E\{X(t)X(t+\tau)\}$.

Property 1. The mean-squared value of $X(t)$ is

$$R_X(0) = E\{X(t)X(t+\tau)\}\Big|_{\tau=0} = E\{X^2(t)\}. \tag{A.7.15}$$

Property 2. For a WSS process, the autocorrelation is an even function of τ, since

$$R_X(\tau) = E\{X(t)X(t+\tau)\} = E\{X(\alpha-\tau)X(\alpha)\} = R_X(-\tau), \tag{A.7.16}$$

where we let $t = \alpha - \tau$.

Property 3. For a WSS process, the autocorrelation is a maximum at the origin. This can be shown by considering the quantity

$$\begin{aligned} E\{[X(t+\tau) \pm X(t)]^2\} &= E\{X^2(t+\tau)\} + E\{X^2(t)\} \pm 2E\{X(t)X(t+\tau)\} \\ &= 2R_X(0) \pm 2R_X(\tau). \end{aligned} \tag{A.7.17}$$

However, we know that $E\{[X(t+\tau) \pm X(t)]^2\} \geq 0$, so

$$R_X(0) \geq |R_X(\tau)|. \tag{A.7.18}$$

Given two real random processes $X(t)$ and $Y(t)$, we can define two cross-correlation functions as

$$R_{XY}(t_1, t_2) = E\{X(t_1)Y(t_2)\} \tag{A.7.19}$$

and

$$R_{YX}(t_1, t_2) = E\{Y(t_1)X(t_2)\}. \tag{A.7.20}$$

The cross-correlation function does not have the same properties as the autocorrelation, but the following properties of the cross-correlation for WSS processes can prove useful.

Property 4. For two WSS processes with

$$R_{XY}(\tau) = E\{X(t)Y(t+\tau)\} \tag{A.7.21}$$

and

$$R_{YX}(\tau) = E\{Y(t)X(t+\tau)\}, \tag{A.7.22}$$

we have that

$$R_{XY}(-\tau) = R_{YX}(+\tau). \tag{A.7.23}$$

To show this, we begin with Eq. (A.7.21) and consider $R_{XY}(-\tau)$,

$$R_{XY}(-\tau) = E\{X(t)Y(t-\tau)\} = E\{X(\alpha+\tau)Y(\alpha)\} = R_{YX}(\tau), \qquad \text{(A.7.24)}$$

where we have defined $\alpha = t - \tau$.

Property 5. For two WSS processes $X(t)$ and $Y(t)$, the cross-correlation is bounded as

$$|R_{XY}(\tau)| \le \tfrac{1}{2}[R_X(0) + R_Y(0)]. \qquad \text{(A.7.25)}$$

To demonstrate this, we expand the nonnegative quantity

$$E\{[X(t) \pm Y(t+\tau)]^2\} \ge 0 \qquad \text{(A.7.26)}$$

to obtain

$$R_X(0) \pm 2R_{XY}(\tau) + R_Y(0) \ge 0, \qquad \text{(A.7.27)}$$

thus directly yielding Eq. (A.7.25).

Property 6. For two WSS random processes $X(t)$ and $Y(t)$, we can write

$$E\left\{\left[\frac{X(t+\tau)}{\sqrt{R_X(0)}} - \frac{Y(t)}{\sqrt{R_Y(0)}}\right]^2\right\} = \frac{E\{X^2(t+\tau)\}}{R_X(0)} + \frac{E\{Y^2(t)\}}{R_Y(0)} - \frac{2E\{X(t+\tau)Y(t)\}}{\sqrt{R_X(0)R_Y(0)}} \ge 0. \qquad \text{(A.7.28)}$$

Now, the first two terms in Eq. (A.7.28) each have the value 1, so

$$\frac{2R_{XY}(\tau)}{\sqrt{R_X(0)R_Y(0)}} \le 2$$

or

$$R_{XY}^2(\tau) \le R_X(0)R_Y(0). \qquad \text{(A.7.29)}$$

Property 7. If $Z(t) = X(t) + Y(t)$, where $X(t)$ and $Y(t)$ are WSS, then

$$R_Z(\tau) = R_X(\tau) + R_Y(\tau) + R_{XY}(\tau) + R_{YX}(\tau). \qquad \text{(A.7.30)}$$

The proof is direct.

Property 8. If a WSS random process $X(t)$ is periodic of period T, its autocorrelation is also periodic with period T.

To demonstrate this, let $X(t) = X(t+T) = X(t+nT)$ for integer n. Then

$$\begin{aligned} R_X(\tau) &= E\{X(t)X(t+\tau)\} = E\{X(t)X(t+\tau+nT)\} \\ &= R_X(\tau+nT). \end{aligned} \qquad \text{(A.7.31)}$$

A.8 Ergodicity

The concept of stationarity, particularly in the wide sense, is extremely important to the analysis of random processes that occur in communication systems. However, for a specific application, it often occurs that the (ensemble) mean or autocorrelation function is unknown and must be estimated. The most common way of estimating these quantities in engineering applications is to observe a particular sample function of the random process and then compute the mean, autocorrelation, or other required average for this specific sample function. In such a situation, we are actually computing time averages over the (single) sample function.

For example, given a random process $X(\xi, t)$, if we observe a particular sample function of this process, we thus have $X(\xi_1, t) = X(t)$, which is now only a function of t. For this sample function, we can compute its average value as

$$\langle X(\xi_1, t)\rangle = \langle x(t)\rangle = \lim_{T\to\infty} \frac{1}{2T}\int_{-T}^{T} x(t)\,dt, \tag{A.8.1}$$

where $\langle\cdot\rangle$ denotes averaging over time. We can also define a (time) autocorrelation function for this sample function as

$$\mathcal{R}_X(\tau) = \langle x(t)x(t+\tau)\rangle = \lim_{T\to\infty} \frac{1}{2T}\int_{-T}^{T} x(t)x(t+\tau)\,dt. \tag{A.8.2}$$

Now, if we use these time averages as estimates of the ensemble mean and autocorrelation, which we discussed previously, we are claiming that

$$\begin{aligned}\langle x(t)\rangle &= \lim_{T\to\infty} \frac{1}{2T}\int_{-T}^{T} x(t)\,dt = E\{X(\xi; t)\}\\ &= \int_{-\infty}^{\infty} x(\xi; t) f_X(x, \xi; t)\,dx(\xi; t)\end{aligned} \tag{A.8.3}$$

and

$$\begin{aligned}\mathcal{R}_X(\tau) &= \langle x(t)x(t+\tau)\rangle = E\{X(\xi; t)X(\xi; t+\tau)\}\\ &= \int_{-\infty}^{\infty}\int_{-\infty}^{\infty} x_1(\xi; t)x_2(\xi; t+\tau) f_{X_1X_2}(x_1, x_2; t, t+\tau)\,dx_1(\xi; t)\,dx_2(\xi; t+\tau).\end{aligned} \tag{A.8.4}$$

Equation (A.8.3) asserts that the time average of $x(t)$ equals the ensemble average of $X(\xi; t)$. Equation (A.8.4) makes a similar assertion for time and ensemble autocorrelations. Now, time averages and ensemble averages are fundamentally quite different quantities, and when they are the same, we have a very special situation. In fact, we give processes with this property a special name.

Definition A.8.1 We call random processes with the property that time averages and ensemble averages are equal *ergodic*.

Ergodicity does not just apply to the mean and autocorrelation—it applies to all time and ensemble averages. Thus, if we have an ergodic random process, we can determine the ensemble averages by computing the corresponding time average over a particular realization of the random process. Clearly, this is a powerful tool, and ergodicity is often assumed in engineering applications. For a much more careful discussion and development of the ergodic property of random processes, see Gray and Davisson [1986].

We note that for a random process to be ergodic, it must first be stationary. This result is intuitive, since if a time average is to equal an ensemble average, the ensemble average cannot be changing with time.

A.9 Spectral Densities

Electrical engineers are accustomed to thinking about systems in terms of their frequency response and about signals in terms of their frequency content. We would like to be able to extend this "frequency-domain" thinking to incorporate stochastic processes. Unfortunately, the standard Fourier integral of a sample function of a stochastic process may not exist, and hence we must look elsewhere for a frequency-domain representation of random processes. For the restricted class of wide-sense-stationary random processes, it is possible to define a quantity called the *power spectral density*, or *spectral density function*, as the Fourier transform of the autocorrelation, so

$$S_X(\omega) = \mathscr{F}\{R_X(\tau)\} = \int_{-\infty}^{\infty} R_X(\tau)e^{-j\omega\tau}\,d\tau. \tag{A.9.1}$$

The power spectral density is an indicator of the distribution of signal power as a function of frequency. Since the Fourier transform is unique, we can also obtain the autocorrelation function of the WSS process $\{X(t), -\infty < t < \infty\}$ as

$$R_X(\tau) = \mathscr{F}^{-1}\{S_X(\omega)\} = \frac{1}{2\pi}\int_{-\infty}^{\infty} S_X(\omega)e^{j\omega\tau}\,d\omega. \tag{A.9.2}$$

Equations (A.9.1) and (A.9.2) are sometimes called the Wiener–Khintchine relations. We now consider two simple examples.

EXAMPLE A.9.1

We are given a WSS stochastic process $\{X(t), -\infty < t < \infty\}$ with zero mean and $R_X(\tau) = A\cos\omega_c\tau$ for $-\infty < \tau < \infty$. We desire the power spectral density of $X(t)$. Directly from Eq. (A.9.1) we find that

$$\begin{aligned} S_X(\omega) &= \mathscr{F}\{R_X(\tau)\} = \mathscr{F}\{A\cos\omega_c\tau\} \\ &= \pi A[\delta(\omega + \omega_c) + \delta(\omega - \omega_c)]. \end{aligned} \tag{A.9.3}$$

Thus a WSS stochastic process with a purely sinusoidal autocorrelation function with period $T_c = 2\pi/\omega_c$ has a spectral density function with impulses at $\pm\omega_c$.

EXAMPLE A.9.2

A WSS stochastic process $Y(t)$ has the spectral density function

$$S_Y(\omega) = \tau_0\frac{\sin^2(\omega\tau_0/2)}{(\omega\tau_0/2)^2}. \tag{A.9.4}$$

We wish to find the autocorrelation function of the process. Since by Eq. (A.9.2), $R_Y(\tau) = \mathscr{F}^{-1}\{S_Y(\omega)\}$, we have from tables that

$$R_Y(\tau) = \begin{cases} 1 - \dfrac{|\tau|}{\tau_0}, & |\tau| < \tau_0 \\ 0, & |\tau| > \tau_0. \end{cases} \tag{A.9.5}$$

There are several simple properties of the power spectral density that we state here.

Property 1. The mean-square value of a WSS process is given by

$$E\{X^2(t)\} = R_X(0) = \frac{1}{2\pi}\int_{-\infty}^{\infty} S_X(\omega)\, d\omega. \tag{A.9.6}$$

Property 2. The spectral density function of a WSS process is always nonnegative,

$$S_X(\omega) \geq 0 \qquad \text{for all } \omega. \tag{A.9.7}$$

Property 3. The power spectral density of a real WSS random process is an even function of ω:

$$S_X(\omega) = S_X(-\omega). \tag{A.9.8}$$

Property 4. The value of the spectral density function at $\omega = 0$ is [from Eq. (A.9.1)]

$$S_X(0) = \int_{-\infty}^{\infty} R_X(\tau)\, d\tau. \tag{A.9.9}$$

Of course, in communication systems it is often necessary to determine the response of a given linear filter to a random process input. It is problems of this type for which the utility of the power spectral density is most apparent. Specifically, given a linear system with impulse response $h(t)$ and transfer function $H(\omega)$, suppose that we wish to find the power spectral density at its output when its input is a WSS random process $X(t)$ with zero mean and autocorrelation function $R_X(\tau)$. The output stochastic process can be expressed as $Y(t) = \int_{-\infty}^{\infty} h(\alpha)X(t-\alpha)\, d\alpha$, so

$$\begin{aligned} R_Y(\tau) &= E\{Y(t)Y(t+\tau)\} \\ &= \int_{-\infty}^{\infty}\int_{-\infty}^{\infty} h(\alpha)h(\beta)E\{X(t-\alpha)X(t+\tau-\beta)\}\, d\alpha\, d\beta \\ &= \int_{-\infty}^{\infty}\int_{-\infty}^{\infty} h(\alpha)h(\beta)R_X(\tau+\alpha-\beta)\, d\alpha\, d\beta. \end{aligned} \tag{A.9.10}$$

Working first with the integral on β, we note that

$$\int_{-\infty}^{\infty} h(\beta)R_X(\tau+\alpha-\beta)\, d\beta = h(\tau+\alpha) * R_X(\tau+\alpha) \triangleq g(\tau+\alpha). \tag{A.9.11}$$

Using Eq. (A.9.11), we rewrite Eq. (A.9.10) as

$$R_Y(\tau) = \int_{-\infty}^{\infty} h(\alpha)g(\tau + \alpha)\, d\alpha$$

$$= \int_{-\infty}^{\infty} h(-\lambda)g(\tau - \lambda)\, d\lambda = h(-\tau) * g(\tau), \tag{A.9.12}$$

where we have made the change of variables that $\alpha = -\lambda$. Finally, substituting for $g(\tau)$, we obtain

$$R_Y(\tau) = h(-\tau) * h(\tau) * R_X(\tau). \tag{A.9.13}$$

Taking the Fourier transform of both sides of Eq. (A.9.13), we get the useful and simple result

$$S_Y(\omega) = H^*(\omega)H(\omega)S_X(\omega) = |H(\omega)|^2 S_X(\omega). \tag{A.9.14}$$

Thus, given the system transfer function and the power spectral density of the input process, it is straightforward to obtain the power spectral density of the output process via Eq. (A.9.14).

EXAMPLE A.9.3

A WSS stochastic process $X(t)$ has zero mean and autocorrelation function $R_X(\tau) = V\delta(\tau)$. If $X(t)$ is applied to the input of a linear filter with transfer function $H(\omega) = 1/(1 + j\omega RC)$, we wish to find the power spectral density of the output process $Y(t)$. Directly, we have (a process with a flat power spectral density is said to be *white*)

$$S_X(\omega) = V \qquad \text{for } -\infty < \omega < \infty$$

and

$$|H(\omega)|^2 = \frac{1}{1 + \omega^2 R^2 C^2}, \tag{A.9.15}$$

so

$$S_Y(\omega) = \frac{V}{1 + \omega^2 R^2 C^2}, \qquad -\infty < \omega < \infty. \tag{A.9.16}$$

We can also find the autocorrelation function of the output by taking the inverse transform of Eq. (A.9.16) to yield

$$R_Y(\tau) = \frac{V}{2RC} e^{-|\tau|/RC} \tag{A.9.17}$$

for $-\infty < \tau < \infty$.

We can define cross spectral densities of two WSS random processes $X(t)$ and $Y(t)$ as the Fourier transform of their cross-correlation function,

$$S_{XY}(\omega) = \mathscr{F}\{R_{XY}(\tau)\} = \int_{-\infty}^{\infty} R_{XY}(\tau)e^{-j\omega\tau}\, d\tau. \tag{A.9.18}$$

Of course, given $S_{XY}(\omega)$, we have

$$R_{XY}(\tau) = \mathscr{F}^{-1}\{S_{XY}(\omega)\} = \frac{1}{2\pi}\int_{-\infty}^{\infty} S_{XY}(\omega)e^{j\omega\tau}\, d\omega. \tag{A.9.19}$$

Given a linear system with impulse response $h(t)$ and transfer function $H(\omega)$, we find that it is of considerable practical interest to calculate the cross spectral density between the input process $X(t)$ and the output process $Y(t)$. Forming the cross correlation between the WSS process $X(t)$ and $Y(t)$, we have

$$\begin{aligned} R_{XY}(\tau) = E\{X(t)Y(t+\tau)\} &= \int_{-\infty}^{\infty} h(\alpha)E\{X(t)X(t+\tau-\alpha)\}\, d\alpha \\ &= \int_{-\infty}^{\infty} h(\alpha)R_X(\tau-\alpha)\, d\alpha = h(\tau) * R_X(\tau). \end{aligned} \tag{A.9.20}$$

Using Fourier transform properties, we find that the desired cross spectral density is

$$S_{XY}(\omega) = H(\omega)S_X(\omega). \tag{A.9.21}$$

Equation (A.9.21) and its many variants are often valuable when it is necessary to identify an unknown system transfer function.

The Wiener–Khintchine relations in Eqs. (A.9.1) and (A.9.2) can also be developed by starting with a finite sample of a real random process, say $\{X(t), 0 \le t \le T\}$. We start by defining the sample spectral density function

$$S_{X_T}(\omega) = \frac{1}{T}\left|\int_0^T X(t)e^{-j\omega t}\, dt\right|^2 \tag{A.9.22}$$

and the sample autocorrelation function

$$R_{X_T}(\tau) = \begin{cases} \dfrac{1}{2\pi T}\displaystyle\int_0^{T-\tau} X(t)X(t+\tau)\, dt, & 0 \le \tau < T \\ 0, & \tau > T \\ R_{X_T}(-\tau), & \tau < 0, \end{cases} \tag{A.9.23}$$

which can be shown to be a Fourier transform pair,

$$S_{X_T}(\omega) = \int_{-T}^{T} R_{X_T}(\tau)e^{-j\omega\tau}\, d\tau \tag{A.9.24}$$

and

$$R_{X_T}(\tau) = \frac{1}{2\pi}\int_{-\infty}^{\infty} S_{X_T}(\omega)e^{j\omega\tau}\, d\omega \tag{A.9.25}$$

for $-\infty < \tau < \infty$. Taking the expectation of both sides of Eq. (A.9.25) yields

$$E\{R_{X_T}(\tau)\} = \frac{1}{2\pi}\int_{-\infty}^{\infty} E\{S_{X_T}(\omega)\}e^{j\omega\tau}\, d\omega, \tag{A.9.26}$$

$-\infty < \tau < \infty$. Now, for a zero mean, WSS stochastic process, it can be shown that

$$\lim_{T\to\infty} E\{R_{X_T}(\tau)\} = R_X(\tau), \tag{A.9.27}$$

where $R_X(\tau)$ is given by Eq. (A.9.2), assuming that the spectral density exists. Furthermore, it can also be demonstrated that

$$S_X(\omega) = \lim_{T\to\infty} \frac{1}{T} E\left\{\left|\int_0^T X(t)e^{-j\omega t}\,dt\right|^2\right\}. \tag{A.9.28}$$

Equations (A.9.27) and (A.9.28) can sometimes be used to advantage in practical problems.

A.10 Cyclostationary Processes

A common model of transmitted sequences in digital communications systems is given by

$$X(t) = \sum_{n=-\infty}^{\infty} a_n p(t - nT_s), \tag{A.10.1}$$

where $p(t)$ is the pulse shape, T_s is the symbol duration, and $\{a_n\}$ is a WSS sequence with $E\{a_n\} = \mu_a$ and $E\{a_n a_m\} = E\{a_l a_{l+k}\} = R_a(k)$, $k = |n - m|$. We would like to find the power spectral density of $X(t)$. The mean of $X(t)$ is immediately available as

$$E[X(t)] = \mu_a \sum_{n=-\infty}^{\infty} p(t - nT_s), \tag{A.10.2}$$

and the autocorrelation is given by

$$\begin{aligned} R_X(t_1, t_2) &= E[X(t_1)X(t_2)] \\ &= \sum_{n=-\infty}^{\infty} \sum_{m=-\infty}^{\infty} E[a_n a_m] p(t_1 - nT_s)p(t_2 - mT_s) \\ &= \sum_{k=-\infty}^{\infty} R_a(k) \sum_{n=-\infty}^{\infty} p(t_1 - nT_s)P(t_2 - (k + n)T_s). \end{aligned} \tag{A.10.3}$$

From Eqs. (A.10.2) and (A.10.3) it is clear that the sequence $X(t)$ is not WSS. As a result, the power spectral density cannot be defined using Eq. (A.9.1).

Random processes that satisfy the relations

$$E[Y(t_1 + T)] = E[Y(t_1)] \tag{A.10.4}$$

and

$$R_Y(t_1 + T, t_2 + T) = R_Y(t_1, t_2) \tag{A.10.5}$$

are called *cyclostationary* because they are periodic in their time arguments [Franks, 1969]. We see from Eqs. (A.10.2) and (A.10.3) that the sequence $X(t)$ is a cyclostationary process. Fortunately, $X(t)$ can be modified to obtain a WSS process by allowing a random time delay.

Consider a new sequence

$$X(t) = \sum_{n=-\infty}^{\infty} a_n p(t - nT_s - \lambda), \tag{A.10.6}$$

where λ is a uniformly distributed random variable over $0 \leq t < T_s$ independent of a_n. Then

$$E[X(t)] = \sum_{n=-\infty}^{\infty} \mu_a E[p(t - nT_s - \lambda)]$$

$$= \mu_a \sum_{n=-\infty}^{\infty} \frac{1}{T_s} \int_0^{T_s} p(t - nT_s - \lambda)\, d\lambda$$

$$= \frac{\mu_a}{T_s} \sum_{n=-\infty}^{\infty} \int_{t-(n+1)T_s}^{t-nT_s} p(\alpha)\, d\alpha = \frac{\mu_a}{T_s} \int_{-\infty}^{\infty} p(t)\, dt, \tag{A.10.7}$$

which is a constant. Further,

$$R_X(t_1, t_2) = E[X(t_1)X(t_2)]$$

$$= \sum_{n=-\infty}^{\infty} \sum_{m=-\infty}^{\infty} E[a_n a_m] \int_0^{T_s} \frac{1}{T_s} p(t_1 - nT_s - \lambda)p(t_2 - mT_s - \lambda)\, d\lambda$$

$$= \sum_{k=-\infty}^{\infty} R_a(k) \frac{1}{T_s} \sum_{n=-\infty}^{\infty} \int_0^{T_s} p(t_1 - nT_s - \lambda)p(t_2 - (n + k)T_s - \lambda)\, d\lambda$$

$$= \frac{1}{T_s} \sum_{k=-\infty}^{\infty} R_a(k) \sum_{n=-\infty}^{\infty} \int_{t_1-(n+1)T_s}^{t_1-nT_s} p(\alpha)p(\alpha + \tau - kT_s)\, d\alpha$$

$$= \frac{1}{T_s} \sum_{k=-\infty}^{\infty} R_a(k) \int_{-\infty}^{\infty} p(t)p(t + \tau - kT_s)\, dt$$

$$= \frac{1}{T_s} \sum_{k=-\infty}^{\infty} R_a(k)\mathscr{R}_p(\tau - kT_s), \tag{A.10.8}$$

where $\tau = |t_2 - t_1|$ and

$$\mathscr{R}_p(\tau) = \int_{-\infty}^{\infty} p(t)p(t + \tau)\, dt. \tag{A.10.9}$$

Since $R_X(t_1, t_2) = R_X(|t_2 - t_1|)$ and $E[X(t)] =$ constant, $X(t)$ in Eq. (A.10.6) is WSS.

To simplify Eq. (A.10.8) further, assume that the a_n sequence is statistically independent (but not zero mean),

$$R_a(k) = E[a_n a_{n+k}] = \begin{cases} \mu_a^2, & k \neq 0 \\ \sigma_a^2 + \mu_a^2, & k = 0, \end{cases} \tag{A.10.10}$$

where $\sigma_a^2 = E[a_n^2] - \mu_a^2$. Then, Eq. (A.10.8) yields

$$R_X(\tau) = \frac{\sigma_a^2}{T_s} \mathscr{R}_p(\tau) + \frac{\mu_a^2}{T_s} \sum_{k=-\infty}^{\infty} \mathscr{R}_p(\tau - kT_s). \tag{A.10.11}$$

Using Eq. (A.10.9),

$$S_p(\omega) = \mathscr{F}\{\mathscr{R}_p(\tau)\} = |P(\omega)|^2, \tag{A.10.12}$$

where $P(\omega) = \mathscr{F}\{p(t)\}$, we can write several different useful expressions for the power spectral density. Taking the Fourier transform of Eq. (A.10.8), we get the general relationship

$$S_X(\omega) = \frac{1}{T_s}|P(\omega)|^2 R_a(0) + \frac{1}{T_s}\sum_{k=-\infty}^{\infty} R_a(k)|P(\omega)|^2 e^{-j\omega k T_s}$$

$$= \frac{|P(\omega)|^2}{T_s}\left\{R_a(0) + 2\sum_{k=1}^{\infty} R_a(k)\cos k\omega T_s\right\}. \qquad \text{(A.10.13)}$$

Based on the assumptions in Eq. (A.10.10), we can start with Eq. (A.10.11) rewritten as

$$R_X(\tau) = \frac{\sigma_a^2}{T_s}\mathscr{R}_p(\tau) + \frac{\mu_a^2}{T_s^2}\sum_{k=-\infty}^{\infty}\mathscr{R}_p(\tau) * \delta(\tau - kT_s) \qquad \text{(A.10.14)}$$

and take the Fourier transform to get

$$S_X(\omega) = \frac{\sigma_a^2}{T_s}|P(\omega)|^2 + \frac{2\pi\mu_a^2}{T_s^2}\sum_{k=-\infty}^{\infty}\left|P\left(\frac{2k\pi}{T_s}\right)\right|^2 : \delta\left(\omega - \frac{2k\pi}{T_s}\right). \qquad \text{(A.10.15)}$$

Equations (A.10.13) and (A.10.15) find application in several chapters of the book.

SUMMARY

In this appendix we have briefly surveyed the topics of probability, random variables, and stochastic processes. This material is the absolute minimum knowledge required of these fields to continue our study of communication systems.

PROBLEMS

A.1 A binary data source S produces 0's and 1's independently with probabilities $P(0) = 0.2$ and $P(1) = 0.8$. These binary data are then transmitted over a noisy channel that reproduces a 0 at the output for a 0 in with probability 0.9, that is, $P(0|0) = 0.9$. The channel erroneously produces a 0 at its output for a 1 in with probability 0.2, that is, $P(0|1) = 0.2$.

(a) Find $P(1|0)$ and $P(1|1)$.

(b) Find the probability of a 0 being produced at the channel output.

(c) Repeat part (b) for a 1.

(d) If a 1 is produced at the channel output, what is the probability that a 0 was sent?

A.2 Rework Example A.2.1 if $P(\zeta_1) = 0.1$, $P(\zeta_2) = 0.3$, $P(\zeta_3) = 0.5$, and $P(\zeta_4) = 0.1$.

A.3 For Example A.2.1, define the random variables $X(\zeta_1) = 0$, $X(\zeta_2) = 1$, $X(\zeta_3) = 2$, and $X(\zeta_4) = 3$. Find and sketch the cumulative distribution function and the probability density function for this random variable.

A.4 Suppose that for Example A.2.1, we are only interested in the number of telephones in use at any time. Hence we are led to define the random variable $X(\zeta_1) = 0$, $X(\zeta_2) = 1$, $X(\zeta_3) = 1$, and $X(\zeta_4) = 2$. Write and sketch the cumulative distribution function and the probability density function for this random variable.

A.5 We consider the random experiment of observing whether three telephones in an office are busy or not busy. The outcomes of this random experiment are thus:

ζ_1 = no telephones are busy

ζ_2 = only telephone 1 is busy

ζ_3 = only telephone 2 is busy

ζ_4 = only telephone 3 is busy

ζ_5 = telephones 1 and 2 are busy but 3 is not busy

ζ_6 = telephones 1 and 3 are busy but 2 is not busy

ζ_7 = telephones 2 and 3 are busy but 1 is not busy

ζ_8 = all three telephones are busy.

The probabilities of these outcomes are given to be $P(\zeta_1) = 0.3$, $P(\zeta_2) = P(\zeta_3) = P(\zeta_4) = 0.1$, $P(\zeta_5) = P(\zeta_6) = P(\zeta_7) = 0.02$, and $P(\zeta_8) = 0.34$.

(a) What is the probability of the event that one or more telephones are busy?

(b) What is the probability that telephone 3 is in use?

(c) Consider the event that telephone 3 is busy and the event that only telephones 1 and 2 are busy. Are these two events statistically independent?

A.6 Suppose that for the random experiment in Problem A.5, we are only interested in the number of telephones in use at any time. Define an appropriate random variable, find its cumulative distribution function and probability density function, and sketch both functions.

A.7 Given the distribution function of the discrete random variable Z,

$$F_Z(z) = \begin{cases} 0, & z < 0 \\ \frac{1}{4}, & 0 \le z < 2 \\ \frac{3}{8}, & 2 \le z < 3 \\ \frac{1}{2}, & 3 \le z < 5 \\ \frac{7}{8}, & 5 \le z < 8 \\ 1, & z \ge 8. \end{cases}$$

(a) Sketch $F_Z(z)$.

(b) Write an expression for $F_Z(z)$.

(c) Write an expression for and sketch the pdf of Z.

(d) Find $P[Z < 2]$, $P[Z \le 2]$, and $P[2 < Z \le 8]$.

A.8 The cumulative distribution function of a continuous random variable Y is sketched in Fig. PA.8.

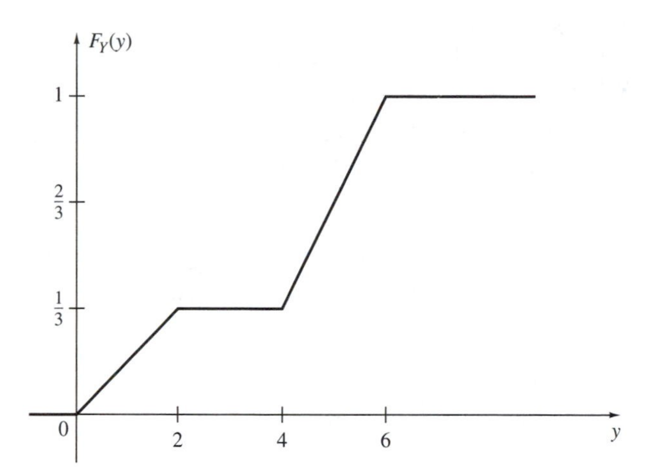

FIGURE PA.8 Cumulative distribution function.

(a) Write an expression for $F_Y(y)$.
(b) Find an expression for and sketch the pdf of Y.
(c) Find $P[Y \leq 1]$, $P[Y \leq 3]$, and $P[2 < Y \leq 5]$.

A.9 Given the joint probability density function

$$f_{XY}(x, y) = \begin{cases} Ke^{-(3x+2y)/6}, & x \geq 0, \quad y \geq 0 \\ 0, & x < 0, \quad y < 0. \end{cases}$$

(a) Find K such that $f_{XY}(x, y)$ is a valid pdf.
(b) Find an expression for $F_{XY}(x, y)$.
(c) Find the marginal distribution and density functions of X.
(d) Calculate $P[X \leq 1, Y \leq 1]$.
(e) Calculate $P[X \leq 2, Y > 1]$.

A.10 Are the random variables X and Y in Problem A.9 statistically independent? Use both distribution and density functions to substantiate your answer.

A.11 Given the joint pdf of Y and Z,

$$f_{YZ}(y, z) = \begin{cases} \dfrac{3y}{2} + \dfrac{z}{2}, & 0 \leq y \leq 1, \quad 0 \leq z \leq 1 \\ 0, & \text{otherwise.} \end{cases}$$

(a) Find the marginal densities of Y and Z.
(b) Find $f_{Y|Z}(y|z)$.
(c) Calculate $F_{Y|Z}(y|Z = z)$.
(d) Calculate $F_{Y|Z}(y|z)$.
(e) Compare the results of parts (c) and (d).

A.12 Given the random variable X with pdf (α a constant)

$$f_X(x) = Ke^{-\alpha x}u(x).$$

(a) Find K in terms of α.
(b) Calculate $E[X]$, $E[X^2]$, and $\text{var}(X)$.
(c) What is $E\{2X^2 + 3X + 1\}$?

A.13 Compute the characteristic function of the pdf in Problem A.12 and calculate the moments using Eq. (A.4.8).

A.14 Given the joint pdf

$$f_{XY}(x, y) = \begin{cases} \frac{1}{12}(2 + xy), & 0 \le x \le 2, \quad 0 \le y \le 2 \\ 0, & \text{otherwise.} \end{cases}$$

(a) Calculate $E\{XY\}$.
(b) Find $\text{cov}(X, Y)$ and ρ_{XY}.
(c) Are X and Y uncorrelated? Orthogonal?

A.15 For the joint pdf in Problem A.11:
(a) Find $E[YZ]$.
(b) Find $\text{cov}(Y, Z)$ and ρ_{YZ}.
(c) Are Y and Z uncorrelated? Orthogonal?

A.16 Given the discrete random variable Z in Problem A.7, find the cumulative distribution function and the probability density function of the random variable $Y = Z + 2$.

A.17 The continuous random variable Y has the pdf

$$f_Y(y) = \frac{1}{2\sqrt{e}} e^{-(y-1)/2}u(y).$$

Find the pdf of the random variable $Z = 2Y + 2$.

A.18 Given the joint pdf of X and Y,

$$f_{XY}(x, y) = \begin{cases} 1, & 0 < x < 1, \quad 0 < y < 1 \\ 0, & \text{otherwise,} \end{cases}$$

find the joint pdf of the random variables $W = X + Y$ and $Z = X - Y$.

A.19 Two random variables X and Y have the joint pdf

$$f_{XY}(x, y) = \begin{cases} \frac{1}{4}e^{-(x+y)/2}, & 0 \le x < \infty, \quad 0 \le y < \infty \\ 0, & \text{otherwise.} \end{cases}$$

We wish to find the pdf of $W = X + Y$.

Hint: Use an auxiliary variable.

A.20 A double-sided exponential or Laplacian random variable has the pdf

$$f_X(x) = K_1 e^{-K_2|x|}, \qquad -\infty < x < \infty.$$

(a) For this to be a pdf, how are K_1 and K_2 related? Let $K_2 = 1$ and find K_1.
(b) Is X zero mean?
(c) Compute the variance of X.

A.21 Find the characteristic function of the Laplacian random variable

$$f_Y(y) = \frac{1}{2\sigma} e^{-|y-\mu|/\sigma}, \qquad -\infty < y < \infty,$$

where $\sigma > 0$. Calculate the mean and variance.

A.22 A Gaussian random variable X has the pdf

$$f_X(x) = \frac{1}{\sqrt{2\pi}\,\sigma} e^{-(x-\mu)^2/2\sigma^2}, \qquad -\infty < x < \infty.$$

Find the pdf of the random variable $Y = aX + b$.

A.23 A random variable X has the uniform pdf given by

$$f_X(x) = \begin{cases} \dfrac{1}{b-a}, & a \le x \le b \\ 0, & \text{otherwise.} \end{cases}$$

(a) Find the pdf of $X_1 + X_2$ if X_1 and X_2 are independent and have the same pdf as X.
(b) Find the pdf of $X_1 + X_2 + X_3$ if the X_i, $i = 1, 2, 3$ are independent and have the same pdf as X.
(c) Plot the resulting pdfs from parts (a) and (b). Are they uniform?

A.24 Let $Y = \sum_{i=1}^{P} X_i$, where the X_i are independent, identically distributed Gaussian random variables with mean 1 and variance 2. Write the pdf of Y.

A.25 Derive the mean and variance of the Rayleigh distribution in Eqs. (A.6.17) and (A.6.18).

A.26 Find the mean and variance of the binomial distribution using the characteristic function in Eq. (A.6.24).

A.27 For a binomial distribution with $n = 10$ and $p = 0.2$, calculate, for the binomial random variable X, $P[X \le 5]$ and $P[X < 5]$. Compare the results.

A.28 Use the characteristic function of a Poisson random variable in Eq. (A.6.31) to derive its mean and variance.

A.29 A random process is given by $X(t) = A_c \cos[\omega_c t + \Theta]$, where A_c and ω_c are known constants and Θ is a random variable that is uniformly distributed over $[0, 2\pi]$.
(a) Calculate $E[X(t)]$ and $R_X(t_1, t_2)$.
(b) Is $X(t)$ WSS?

A.30 For the random process $Y(t) = Ae^{-t}u(t)$, where A is a Gaussian random variable with mean 2 and variance 2,
(a) Calculate $E[Y(t)]$, $R_Y(t_1, t_2)$, and $\text{var}[Y(t)]$.
(b) Is $Y(t)$ stationary? Is $Y(t)$ Gaussian?

A.31 Let X and Y be independent, Gaussian random variables with means μ_x and μ_y and variances σ_x^2 and σ_y^2, respectively. Define the random process $Z(t) = X \cos \omega_c t + Y \sin \omega_c t$, where ω_c is a constant.
(a) Under what conditions is $Z(t)$ WSS?
(b) Find $f_Z(z; t)$
(c) Is the Gaussian assumption required for part (a)?

A.32 A weighted difference process is defined as $Y(t) = X(t) - \alpha X(t - \Delta)$, where α and Δ are known constants and $X(t)$ is a WSS process. Find $R_Y(\tau)$ in terms of $R_X(\tau)$.

A.33 Use the properties stated in Section A.7 to determine which of the following are possible autocorrelation functions for a WSS process ($A > 0$).
(a) $R_X(\tau) = Ae^{-\tau}u(\tau)$

(b) $R_X(\tau) = \begin{cases} A[1 - |\tau|/T], & |\tau| \leq T \\ 0, & \text{otherwise} \end{cases}$

(c) $R_X(\tau) = Ae^{-|\tau|}, \qquad -\infty < \tau < \infty$
(d) $R_X(\tau) = Ae^{-|\tau|} \cos \omega_c \tau, \qquad -\infty < \tau < \infty$

A.34 It is often stated as a property of a WSS, nonperiodic process $X(t)$ that

$$\lim_{\tau \to \infty} R_X(\tau) = (E[X(t)])^2.$$

This follows, since as τ gets large, $X(t)$, $X(t + \tau)$ become uncorrelated. Use this property and property 1 in Section A.7 to find the mean and variance of those processes corresponding to valid autocorrelations in Problem A.33.

A.35 Calculate the power spectral densities for the admissible autocorrelation functions in Problem A.33.

A.36 Calculate the power spectral density for Problem A.31(a).

A.37 If $\mathscr{F}\{R_X(\tau)\} = S_X(\omega)$, calculate the power spectral density of $Y(t)$ in Problem A.32 in terms of $S_X(\omega)$.

A.38 A WSS process $X(t)$ with $R_X(\tau) = V\delta(\tau)$ is applied to the input of a 100% roll-off raised cosine filter with cutoff frequency ω_{co}. Find the power spectral density and autocorrelation function of the output process $Y(t)$.

A.39 A WSS signal $S(t)$ is contaminated by an additive, zero-mean, independent random process $N(t)$. It is desired to pass this noisy process $S(t) + N(t)$ through a filter with transfer function $H(\omega)$ to obtain $S(t)$ at the output. Find an expression for $H(\omega)$ in terms of the spectral densities involving the input and output processes.

Source/User Characteristics and Models

Every communication system takes into account, to a greater or lesser extent, the characteristics of the source or message that it is trying to relay to the user. Such characteristics include the spectral content of the message, the first-order or higher-order probability density functions, and perhaps even a time-domain model of the message. Furthermore, communication systems may also take advantage of the characteristics of the user, such as the capabilities of human hearing and vision. The better the source and user behavior are known, and the more this knowledge is incorporated into the communication system design, the better the performance of the communication system. Of course, if the source and user characteristics are for some reason incorrectly specified but yet are used as a basis for designing a communication system, performance will be poor. In this appendix we briefly describe some commonly employed models of sources and users. Speech, images, facsimile, and television sources as well as human audio and visual system behavior are discussed.

B.1 Speech Models

Telephone speech is usually quoted as occupying one of the frequency ranges 200 to 3200 Hz, 300 to 3300 Hz, or 400 to 3400 Hz, or some slight variation thereof. The peaks in the speech spectrum, called *formants*, are important perceptually, but often, one or two out of about five formants are outside (above 3200 to 3400 Hz) this telephone passband. Despite this fact, users find that most telephone transmissions are highly intelligible and even convey the identity of the talker. AM radio has a passband of about 0 to 5 kHz, so there is little difficulty with speech. However, music can have perceptually significant frequency content up to 20 kHz, which is unavailable on standard AM radio today. The wider message bandwidth of 15 kHz for commercial FM radio tries to take advantage of this situation by providing higher-quality music than AM.

The first-order probability density function of speech is often said to be Laplacian (see Problem A.21) or to have a special form of the gamma density given by [Paez and Glisson, 1972; Jayant and Noll, 1984]

$$f_X(x) = \frac{1}{2}\sqrt{\frac{k}{\pi|x|}}\,e^{-k|x|}, \qquad -\infty < x < \infty, \tag{B.1.1}$$

where k is a constant parameter. Most communication systems today do not utilize these pdf models, although some quantizer designs may rely heavily on such descriptions (see Chapter 9).

An analog time-domain plot of a speech waveform is shown in Fig. B.1.1 [Rabiner and Schafer, 1978]. Although it is misleading to call any short speech segment "typical," this waveform illustrates several terms and properties of speech signals. The (damped) oscillatory portions of the waveform are called *voiced* segments, the lower-amplitude, noisy-looking parts of the plot are called *unvoiced* segments, and the intervals where there are no discernible signals is called *silence*. The time between the largest magnitude positive or negative peaks during voiced segments is often called the *pitch period*, and it is related to the fundamental frequency of oscillation of the vocal cords. For the first voiced segment in Fig. B.1.1, the pitch period is slightly less than 8 msec, while for the

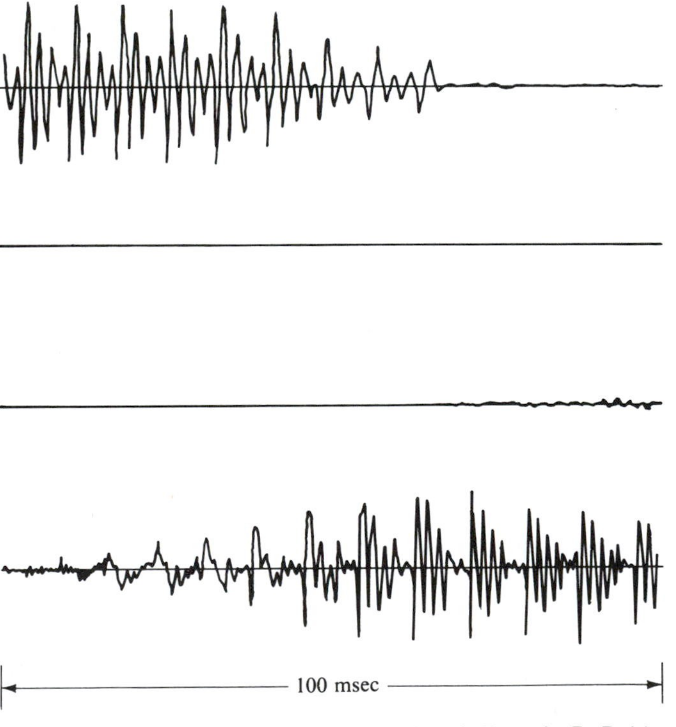

FIGURE B.1.1 Time-domain plot of a speech signal. From L. R. Rabiner and R. W. Schafer, *Digital Processing of Speech Signals*, © 1978, p. 60. Reprinted by permission of Prentice Hall, Inc., Englewood Cliffs, N.J.

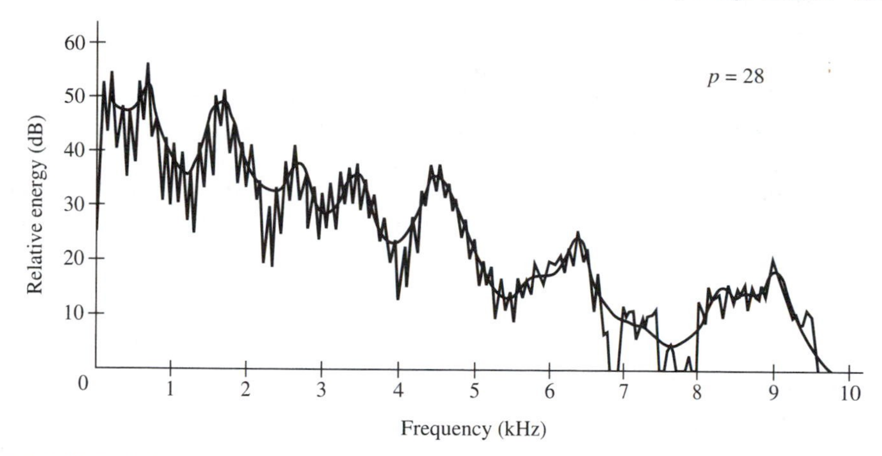

FIGURE B.1.2 Fourier transform and LPC spectrum for a speech segment. From J. Makhoul, "Linear Prediction: A Tutorial Review," *Proc. IEEE*, © 1975 IEEE.

last voiced segment (at the end), the pitch period is about 9 msec. Thus, although the pitch period of a male speaker is longer than that of a female speaker, a particular speaker's pitch changes depending on what is being said.

The Fourier transform of a 20-msec-long speech segment is shown in Fig. B.1.2. The very "squiggly" line is the actual Fourier transform, while the smoother curve is the spectrum of a 28th-order linear predictive coding (LPC) system [Makhoul, 1975]. The peaks in the Fourier transform, which can be considered to be the peaks in the smoother curve, are the formants, and as can be seen from the figure, there are about five below 4 kHz. The deep nulls in the (unsmoothed) Fourier transform occur at integer multiples of the pitch frequency (= reciprocal of the pitch period length). For example, eight nulls occur from 0 to 1 kHz, so the fundamental pitch frequency is 125 Hz, which corresponds to a pitch period length of 8 msec.

Additional details are left to the literature [Rabiner and Schafer, 1978].

B.2 Image Models

The obvious difference between a speech signal and an image is that the image is a two-dimensional source. We can represent the stationary autocorrelation function of an image as $R_X(\tau_h, \tau_v)$, where τ_h and τ_v denote the lags in the horizontal and vertical directions, respectively. It has been found that two particular autocorrelation function models are useful in many image processing tasks. One of these models is the *separable* autocorrelation function model given by

$$R_X(\tau_h, \tau_v) = \sigma_x^2 e^{-\alpha_h|\tau_h| - \alpha_v|\tau_v|}, \tag{B.2.1}$$

where σ_x^2 is the variance of the picture elements and α_h and α_v are parameters. The model in Eq. (B.2.1) is called separable, since the autocorrelation function factors as $R_X(\tau_h, \tau_v) = R_X(\tau_h)R_X(\tau_v)$, which provides clear advantages for many analytical manipulations. Separable autocorrelations are claimed to be useful in modeling images consisting of man-made objects that often have high correlations in the horizontal and vertical directions [Pratt, 1978]. A second common correlation model for images is the *isotropic* autocorrelation function, given by

$$R_X(\tau_h, \tau_v) = \sigma_x^2 e^{-\sqrt{\alpha_h^2\tau_h^2 + \alpha_v^2\tau_v^2}}, \tag{B.2.2}$$

which has greater utility in representing natural scenes that have no preferred directions [Pratt, 1978]. Notice that the autocorrelation function in Eq. (B.2.2) is not separable. These models have not been widely used in commercial image processor and communication system designs, but they have had an impact on several special-purpose systems.

The first-order probability density function of an image is extraordinarily image dependent, varying from unimodal densities such as Gaussian and Rayleigh to uniform and even bimodal densities. However, it is often possible to limit consideration to certain subclasses of images for which a specific pdf is a valid model. Again, commercial image communication systems have not in the past relied on assumptions concerning pdfs, but certain developing systems are moving in that direction.

B.3 Facsimile

The term *facsimile* refers to two-level graphics or black-and-white images. This is an important class of images, which includes business documents, such as letters, line drawings, schematics, and newspapers. The gray-level photographs in newspapers can be easily converted into sequences of black-and-white pixels. Appropriate models for facsimile vary depending on whether the source is a schematic, text, or say, a weather map; however, the common thread is the need to encode sequences of two levels only. If there are long sequences of black or white pixels, or both, a technique called *run-length coding* can be valuable (see Jayant and Noll [1984] for a discussion of run-length coding).

A particular two-state Markov model developed by Capon [1959], which has been useful for modeling weather maps, has the probability of a run of N black or white pixels given by

$$P[NS_S|S] = \frac{1}{E[N|S] - 1}[1 - (E[N|S])^{-1}]^N, \tag{B.3.1}$$

where S = black or white, $N = 1, 2, \ldots, \infty$, and $E[N|S]$ is the average length of a run in state S. Equation (B.3.1) is a familiar discrete pdf called the geometric distribution with parameter $E[N|S]$. Although this model has had some success modeling weather maps, it is not adequate for printed text. For more details on facsimile, see Jayant and Noll [1984] and their references.

B.4 Video

Just as for speech signals, we would like to know something about the spectral shape and bandwidth requirements of a video signal, and perhaps even sketch a "typical" time-domain waveform. The first step in this process is understanding how the sequence of two-dimensional images that we can see are converted to a scalar time-domain waveform. This conversion is effected by raster scanning the video signal as illustrated for a single image frame in Fig. B.4.1. The scanner moves along the lines shown in the figure from left to right, producing an output proportional to the intensity of the image at each location. When the scan reaches the right-hand side of the image, it quickly jumps back to the left-hand side to begin the next line shown. In Fig. B.4.1, both solid and dashed scan lines are shown. For a video signal that is changing continuously with time, we scan the image in one time interval, called a *frame*, along the solid lines shown until the bottom of the frame is reached; then the scanner jumps vertically from the end of the solid line at the bottom to the beginning of the dashed line at the top and continues to scan along the dashed lines. When the dashed line reaches the bottom of the image, a new frame is scanned by starting at the top with the solid lines, and the process is repeated. The set of solid lines and the set of dashed lines is each called a *field*, and therefore, there are two fields per frame.

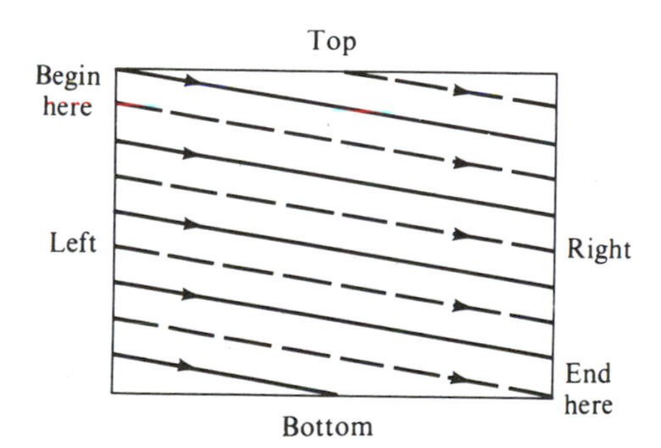

FIGURE B.4.1 Raster scanning of an image frame.

The physical motivation behind this method for scanning an image is the desire to give the impression of continuous motion at the minimum frame rate possible. Experiments with the human eye have shown that by scanning an image twice per frame with interlaced fields as in Fig. B.4.1, the frame rate can be reduced without any perceptible jumping or discontinuities.

Of course, depending on where we are in the world, there are different standards for broadcast television. Our discussion emphasizes the National Television System Committee (NTSC) standard adopted in North America, South America, and Japan. The frame rate for the NTSC standard is 30 frames/sec. Now, in addition to the frame rate, we need to determine the required (or available) *resolution*, which is the smallest element that is just visible in the scanned video signal. Both horizontal and vertical resolution are determined by the scan rate, or time per scan line, since for a given frame rate, the faster the scan, the more lines per frame (vertical resolution) and the more movement can be discerned per line (horizontal resolution). Thus faster scanning is better; unfortunately, faster scanning requires greater bandwidth, so there is a trade-off involved. Therefore, for a given frame rate of 30 frames/sec, if we fix the number of lines/frame to be $N = 525$, the time allocated per horizontal scan line is 63.5 μsec. Recalling from our earlier discussion that when the raster scan reaches the right side of the image, it must move back to the left side to continue the scan, we see that the scan time per horizontal line is reduced from 63.5 μsec by the amount of time required to jump from right to left. This time is 10 μsec and is called the *horizontal retrace time*. The time actually available for scanning a line is thus only 53.5 $\mu\text{sec} \triangleq \tau_l$.

From Chapter 8 we know that in a bandwidth of B hertz, we can have a maximum of $2B$ independent values per second. It follows that the number of elements that can be resolved per line, called the *horizontal resolution*, is

$$n_h = 2B \times 53.5 \times 10^{-6} \text{ elements.} \tag{B.4.1}$$

The vertical resolution would, at first glance, seem to be 525. However, this value is reduced by the number of lines lost during *vertical retrace time*, which occurs twice per frame (once per field) when the raster scan must move from the bottom to the top of an image to continue the scan. The total loss in lines/frame is 21 lines/field $\times$ 2 fields/frame = 42 lines/frame, so that the maximum vertical resolution is 483 lines/frame. This vertical resolution is reduced further by a factor of 0.7, since the raster scan has an arbitrary alignment with respect to the image. Thus the effective vertical resolution is

$$n_v = 0.7 \times 483 \cong 338 \text{ lines.} \tag{B.4.2}$$

As our final step toward obtaining an expression for bandwidth, we note that it is reasonable to require that the horizontal resolution in elements per inch be the same as the vertical resolution in lines/inch. Thus, letting d_h and d_v denote the horizontal and vertical screen size in inches (say), we require that

$$\frac{n_h}{d_h} = \frac{n_v}{d_v} \tag{B.4.3}$$

or

$$n_h = n_v \frac{d_h}{d_v}. \tag{B.4.4}$$

The ratio d_h/d_v is called the *aspect ratio* and is set at $\frac{4}{3}$ for NTSC video. Therefore, from Eqs. (B.4.1), (B.4.2), and (B.4.4), we find that the bandwidth requirement for NTSC video is

$$B \cong \frac{388 \times \frac{4}{3}}{2 \times 53.5 \times 10^{-6}} \cong 4.2 \text{ MHz.} \tag{B.4.5}$$

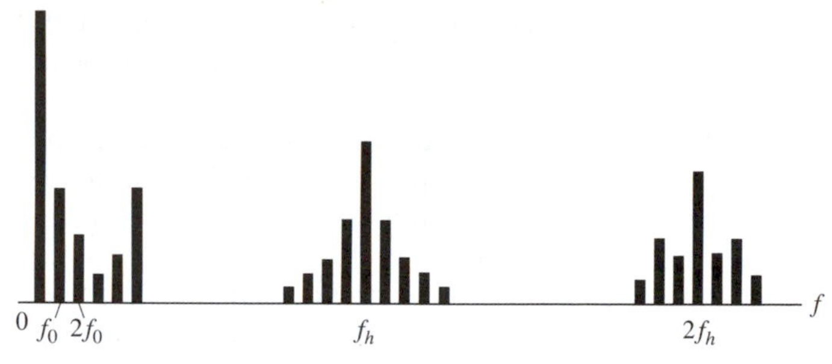

FIGURE B.4.2 Approximate frequency content for an image.

To get some idea of the spectral distribution within this 4.2-MHz bandwidth, we note that the raster scanned image is approximately periodic in the horizontal direction at the line rate, call this period T_h, and it is approximately periodic in the vertical direction at the field rate, call this period T_v. We then approximate the scalar scanned version of the image by the product of two periodic functions, so

$$s_I(t) = \underbrace{\left[\sum_{k=-\infty}^{\infty} c_k e^{jk(2\pi/T_h)t}\right]}_{s_h(t)} \cdot \underbrace{\left[\sum_{m=-\infty}^{\infty} c_m e^{jm(2\pi/T_v)t}\right]}_{s_v(t)}. \tag{B.4.6}$$

Since $T_v \gg T_h$, $s_v(t)$ is slowly varying with respect to $s_h(t)$, so we can write

$$s_I(t) = \sum_{k=-\infty}^{\infty} c_k s_v(t) e^{jk(2\pi/T_h)t}, \tag{B.4.7}$$

which implies that $s_v(t)$ modulates an infinite number of carriers at the frequencies $\pm k(2\pi/T_h)$. The Fourier transform of $s_I(t)$ thus takes the form shown in Fig. B.4.2. Note that $f_v = 1/T_v = 60$ Hz and $f_h = 1/T_h = 15{,}750$ Hz, and that since $c_k c_m$ usually decreases with increasing km product, there is very little spectral content between the concentrations at $\pm kf_h$. Of course, for a true video signal the periodicity is not exact and the spectral lines are broadened slightly; however, the frequency bands occupied are still well represented by Fig. B.4.2. Discussion of color video signals and the development of television transmission methods is left to Appendix F.

B.5 Hearing Models and Perception

Models of the human auditory process are beginning to find a broader application in data compression system design than previously. However, one perceptual property has already had a substantial impact on communication system design for voice transmission. This property is that low-amplitude sound can be extremely critical to the intelligibility and quality of an utterance. This property is the principal motivation for the use of logarithmic companding and the μ-law quantizing characteristic described in Chapter 9.

Another principle that is receiving attention in conjunction with the speech data compression systems is that of auditory masking. The idea behind auditory masking is that any distortion present in a reconstructed or synthesized speech signal should be redistributed in the frequency domain to have approximately the same spectral shape as the speech spectrum itself. Thus, in frequency bands where the undistorted speech has a large amplitude, the distortion will also have a large amplitude, and when the speech spectrum has a lower amplitude, the distortion will have a relatively lower amplitude, the concept being that the distortion will be covered up or masked by the desired speech spectrum. At this time auditory masking is finding wider applications.

B.6 Human Visual System Models

The main idea behind much of the work on human vision system models has been that the eye is not equally sensitive to all frequencies. Hence effort has been expended to devise representative transfer functions for the behavior of the human eye. Mostly, these transfer functions have taken the form of low-pass filters with some amplitude undulations. No model has emerged as being universally preferred over others.

Of course, TV takes advantage of the response time of the eye in its selection of a frame rate and the use of interlaced scan lines. There is also the possibility of "hiding" distortion in images similar to the practice of auditory masking in speech. For example, granular distortion should be suppressed (if possible) in constant areas of an image, whereas in high-activity regions, granular distortion may well go unnoticed.

Channel Models

We briefly present models, or mathematical representations, for some commonly occurring communication system channels. As always, it is important that mathematical models be simple enough to be tractable to analysis, yet at the same time, accurately reflect the physical situation. Although the designation *channel* sometimes includes everything between the source and the user (i.e., modulators, antennas, transmission medium, demodulator, etc.), here we use channel to represent the transmission media plus the receiver "front end," which usually means down to the first mixer. Channel effects can include a variable gain with frequency, limited bandwidth, additive noise, multiplicative noise, crosstalk between channels, and fading (time-varying gain). In this appendix and the rest of the book, our attention is focused primarily on channels with a time-invariant frequency response, bandwidth limitations, and additive noise.

C.1 Additive Noise Channels

Many times in the book we assume that the additive noise in a communication system is white with a two-sided power spectral density

$$S_n(\omega) = 2\pi \frac{\mathcal{N}_0}{2} \text{ watts/rad/sec}, \qquad -\infty < \omega < \infty. \tag{C.1.1}$$

This assumption is valid for frequencies up to about 10^{13} Hz when thermal noise that is presented to the receiver antenna input and thermal noise generated within the receiver are the dominant noise sources. This is often the case, and in this situation the power spectral density can be expressed as

$$S_n(\omega) = 2\pi \frac{kT}{2}, \qquad -\infty < \omega < \infty, \tag{C.1.2}$$

where $k = 1.38 \times 10^{-23}$ J/kelvin (K) and T is the overall *system noise temperature* (K). The *thermal noise* is a result of the random fluctuations of molecules in the "air" and of fluctuating electrons within conductors, resistors, and so on, in the receiver. Thus the overall system noise temperature is the sum of two components. One part is due to blackbody radiation, which produces thermal noise at the antenna input, and this thermal noise is represented by an *equivalent source noise temperature* (T_s). The other component is the thermal noise generated within the receiver, which is denoted by the *effective noise temperature* (T_e). Therefore, $T = T_s + T_e$, but for a very low noise receiver, $T_e \ll T_s$. If the channel frequency response is not flat but has a transfer function $H_c(\omega)$, the power spectral density expression in Eq. (C.1.2) becomes

$$S_n(\omega) = \pi k T |H_c(\omega)|^2, \qquad -\infty < \omega < \infty. \tag{C.1.3}$$

Another quantity, called the *noise figure* (NF), is the ratio of the receiver input signal-to-noise ratio to the output signal-to-noise ratio. This is equivalent to defining the noise figure as the ratio of the output noise power to the input noise power. The noise figure is thus a measure of the noise contributed by the receiver, and it can be expressed in terms of noise temperatures as

$$\text{NF} = 1 + \frac{T_e}{T_s}. \tag{C.1.4}$$

Often, for standardization purposes, T_s is taken to be "room temperature" or 290 K. In this case,

$$\text{NF} = 1 + \frac{T_e}{290}. \tag{C.1.5}$$

From Eqs. (C.1.4) and (C.1.5), we see that NF ≥ 1 and that the noise figure of a very low noise receiver is only slightly greater than 1.

For communications channels that do not have a flat frequency response, it is sometimes convenient for noise analyses to define what is called the *equivalent noise bandwidth.* The equivalent noise bandwidth is the bandwidth of an ideal filter, denoted here by Ω_n rad/sec, which has the same total noise power as the original system. Thus, for some channel with transfer function $H_c(\omega)$, the total noise power is

$$\frac{1}{2\pi}\int_{-\infty}^{\infty} S_n(\omega)\, d\omega = \frac{kT}{2}\int_{-\infty}^{\infty} |H_c(\omega)|^2\, d\omega, \tag{C.1.6}$$

which we must equate to the noise power of an ideal filter of bandwidth Ω_n that is given by $(kT/2)|H_c(\omega_0)|^2\Omega_n$. The scaling constant $|H_c(\omega_0)|^2$ depends on the channel gain at midband (ω_0) or the maximum channel gain. Therefore,

$$\Omega_n = \frac{(1/2\pi)\int_{-\infty}^{\infty} S_n(\omega)\, d\omega}{(kT/2)|H_c(\omega_0)|^2} = \frac{\int_{-\infty}^{\infty} |H_c(\omega)|^2\, d\omega}{|H_c(\omega_0)|^2}. \tag{C.1.7}$$

The thermal noise in a communication system thus has a flat power spectral density given by Eq. (C.1.2) and is usually considered to be additive. In addition, this noise is often assumed to have a Gaussian probability density function, and indeed, in many physical situations, this can be shown to be true. Therefore, we are led to the familiar and important additive white Gaussian noise (AWGN) channel model, which appears often in this book and is prevalent in communication system analyses today.

C.2 Deep Space, Satellite, and Radio Channels

The additive white Gaussian noise channel is generally a good model for deep space, satellite, and radio channels. Furthermore, all these channels can have bandwidths in the several megahertz range. Radio channels can encounter fading, which causes the received power to vary with time, and such phenomena have to be accounted for in the design process. We do not discuss fading channel models in this book primarily because of their complexity, but also because they must be carefully matched to the particular physical situation of interest.

C.3 Telephone Channels

Two long-distance telephone calls placed from the same telephone to the same number at different times may take different routes, and hence have drastically different transmission characteristics. Standard dial-up connections over the switched telephone network are classified as 3002 unconditioned voice-grade private lines, and according to *Telecommunications Transmission Engineering*, Vol. 3 [AT&T, 1977], they have the attenuation distortion characteristics shown in Table C.3.1. Notice that the attenuation is with respect to a 1000-Hz reference, and all that is guaranteed is that the attenuation will not vary more than the given amounts for the stated frequency ranges. For the frequency ranges not listed, no promises are made. The variation in envelope delay distortion (EDD) for a 3002 line is also shown in the table.

TABLE C.3.1 AT&T Grades of Conditioning

Type	Frequency Range (Hz)	Attenuation[a] (dB)	Frequency Range (Hz)	EDD Variation (μsec)
3002	500–2500	−2 to +8	800–2600	1750
	300–3000	−3 to +12		
C1	1000–2400	−1 to +3	800–2600	1750
	300–2700	−2 to +6	1000–2400	1000
	2700–3000	−3 to +12		
C2	500–2800	−1 to +3	1000–2600	500
	300–3000	−2 to +6	600–2600	1500
			500–2800	3000
C4	500–3000	−2 to +3	1000–2600	300
	300–3200	−2 to +6	800–2800	500
			600–3000	1500
			500–3000	3000
C5	500–2800	−0.5 to +1.5	1000–2600	100
	300–3000	−3 to +3	600–2600	300
			500–2800	600

[a] Relative to 1000 Hz.

If the 3002 line does not provide sufficient transmission quality for a given application, it may be possible to lease a private conditioned line that guarantees tighter constraints on attenuation and EDD variation. The several grades of conditioning available are listed in Table C.3.1 with their associated distortion characteristics. When a customer leases a conditioned private line, the line is dedicated to that user and if necessary, fixed equalization is added to achieve the desired grade of conditioning. Calls placed at different times take the same route and thus have the same specifications on the channel transfer function.

A more recent type of conditioning, not given here, is called *D-conditioning*. This type of conditioning guarantees a minimum signal-to-noise ratio and minimum values for signal-to-distortion ratio for second and third harmonic distortion.

The predominant causes of errors in the telephone network are impulsive noise and bursty transients. However, it has been discovered through years of experience that the AWGN channel model can be employed successfully for analysis and design on the telephone channel. This is primarily because at higher data rates, the AWGN model is fairly accurate, while at lower data rates, a transmitter and receiver designed to work well against Gaussian noise will also perform acceptably in the telephone network. Further, impulsive and burst noise may be difficult to handle analytically and may lead to a very conservative design. In any event, the AWGN channel model has proven quite adequate for many telephone network applications.

Communication Link Design

As is evident from Chapters 7 and 10, analog and digital communication system performance depends quite heavily on the received signal-to-noise ratio (SNR). For line-of-sight radio links, the received SNR can be ascertained by performing what is called a *link budget calculation.* This calculation allows us to see how the transmitted power is "budgeted" over the various portions of the communication link. Although they are quite similar, we consider analog and digital link budgets separately.

We begin the appendix, however, with a section showing how we can analyze thermal noise effects (see Section C.1) in a cascade connection of amplifiers. This is particularly important since the receiver front ends for many communication systems often have several stages of amplification.

D.1 A Cascade of Amplifiers

We consider a series connection of three amplifiers, denoted A_1, A_2, A_3, shown in Fig. D.1.1, each with respective gains, noise temperatures, and noise figures, G_i, T_i, and NF_i, $i = 1, 2, 3$. We assume that the input signal power is P_{in}, the input noise power in bandwidth Ω is $kT_0\Omega/2$, and that all amplifiers have the same bandwidth Ω. We wish to find expressions for the overall effective noise temperature (T_e), the overall noise figure (NF), and the total output noise power (N_{out}).

Since the noise at each stage is additive and independent of the other stages,

$$\begin{aligned} N_{\text{out}} &= \frac{kT_0\Omega G_1 G_2 G_3}{2} + \frac{kT_1\Omega G_1 G_2 G_3}{2} + \frac{kT_2\Omega G_2 G_3}{2} + \frac{kT_3\Omega G_3}{2} \\ &= \frac{k\Omega G_1 G_2 G_3}{2}\left\{T_0 + T_1 + \frac{T_2}{G_1} + \frac{T_3}{G_1 G_2}\right\}. \end{aligned} \tag{D.1.1}$$

By inspection of Eq. (D.1.1), the effective noise temperature of the cascade connection is

$$T_e = T_0 + T_1 + \frac{T_2}{G_1} + \frac{T_3}{G_1 G_2}. \tag{D.1.2}$$

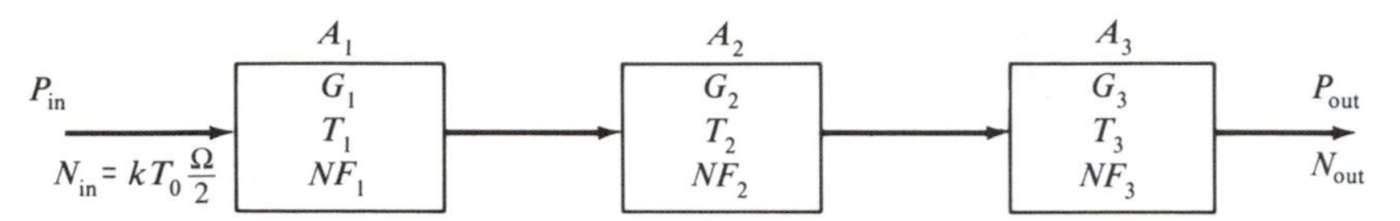

FIGURE D.1.1 Cascade of three amplifiers.

Dividing both sides of Eq. (D.1.2) by 290 K and using Eq. (C.1.5), the overall noise figure is found as

$$\mathrm{NF} = \mathrm{NF}_1 + \frac{\mathrm{NF}_2 - 1}{G_1} + \frac{\mathrm{NF}_3 - 1}{G_1 G_2}. \tag{D.1.3}$$

Equations (D.1.2) and (D.1.3) carry the same message, namely that if G_1 is chosen to be large, the first stage amplifier essentially determines the overall system noise content. In fact, if $G_1 \gg 1$ and T_1 is small, there is little noise contributed by the three stages of amplification. This result indicates why much emphasis is placed on preamplifier (first-stage amplifier) design in communications receivers and why these preamplifiers are placed close to the receiving antenna.

D.2 Analog Link Budget

To determine the received SNR in dB for an analog radio communication system, we simply start with the transmitted power and add in gains and losses in dB until we reach the receiver. We then subtract the noise power in dB to obtain the SNR. To specify the gain of the transmitting antenna, we need the concept of *effective isotropic radiated power* (EIRP). An isotropic antenna is simply an ideal reference antenna that radiates uniformly in all directions. The uniform gain of the isotropic antenna is 0 dB or 1, while a real antenna will have a gain that varies as a function of direction. To determine the EIRP in a particular direction, we need to know the transmitted power in dBW (or dBm) and the gain (dB) of the antenna in the desired direction.

Letting S_R denote the total received power in dBW, we have that

$$(S_R)_{\mathrm{dBW}} = (\mathrm{EIRP})_{\mathrm{dBW}} - (PL)_{\mathrm{dB}} + (G_R)_{\mathrm{dB}}, \tag{D.2.1}$$

where EIRP was defined previously, $(PL)_{\mathrm{dB}}$ is the path loss in dB, usually for free space propagation, and $(G_R)_{\mathrm{dB}}$ is the gain of the receiving antenna. The thermal noise power at the receiver is given by

$$N = \frac{k(T_s + T_e)_{\mathrm{DF}}}{2}, \tag{D.2.2}$$

where k is Boltzmann's constant, $T = T_s + T_e$ is the overall system noise temperature, and $_{\mathrm{DF}}$ is the IF bandwidth in rad/sec. Writing the noise power in dB, we have

$$\begin{aligned} N_{\text{dB}} &= k_{\text{dB}} + T_{\text{dB}} + (_{\text{DF}})_{\text{dB}} \\ &= 10 \log_{10} (1.38 \times 10^{-23}) + 10 \log_{10} T + 10 \log_{10\,\text{DF}} + 3 \text{ dB}. \end{aligned} \quad (\text{D.2.3})$$

The received SNR in dBW is thus

$$\begin{aligned} \text{SNR}_{\text{dB}} &= (S_R)_{\text{dBW}} - N_{\text{dB}} \\ &= (\text{EIRP})_{\text{dBW}} - (PL)_{\text{dB}} + (G_R)_{\text{dB}} - k_{\text{dB}} - T_{\text{dB}} - (\Omega_{\text{IF}})_{\text{dB}} + 3 \text{ dB}. \end{aligned} \quad (\text{D.2.4})$$

Equation (D.2.4) is the desired link budget calculation, and it represents the signal-to-noise ratio presented to the demodulator in the receiver.

Our calculation is greatly simplified, and we have neglected connector losses, coupler losses, fading losses, and other gains and losses within the transmitter and receiver. The terms that are included in Eq. (D.2.4) are the major components in most link budget analyses. More details are provided in Freeman [1981].

D.3 Digital Link Budget

For a digital communication system, the quantity of interest at the receiver is $E_b/\mathcal{N}_0$ rather than the SNR as in analog systems (see Chapter 10). We can relate $E_b/\mathcal{N}_0$ to the SNR in Section D.2 by noting that $E_b = ST_b$, where S is the received signal power and T_b is the time interval allocated to one bit, and by recalling from Appendix C that $\mathcal{N}_0 = kT/2$. The total noise power is thus $N = kT_{\text{DF}}/2$. Therefore, we have that

$$\text{SNR}_{\text{dBW}} = S_{\text{dBW}} - N_{\text{dB}} = (E_b)_{\text{dBW}} - (\mathcal{N}_0)_{\text{dBW}} - (T_b)_{\text{dB}} - (_{\text{DF}})_{\text{dB}} + 3 \text{ dB}, \quad (\text{D.3.1})$$

so

$$\begin{aligned} \left(\frac{E_b}{\mathcal{N}_0}\right)_{\text{dB}} &= \text{SNR}_{\text{dBW}} + (T_b)_{\text{dB}} + (_{\text{DF}})_{\text{dB}} \\ &= (\text{EIRP})_{\text{dBW}} - (PL)_{\text{dB}} + (G_R)_{\text{dB}} - k_{\text{dB}} - T_{\text{dB}} - (T_b)_{\text{dB}} + 3 \text{ dB}, \end{aligned} \quad (\text{D.3.2})$$

where we have used Eq. (D.2.4) for SNR_{dB}.

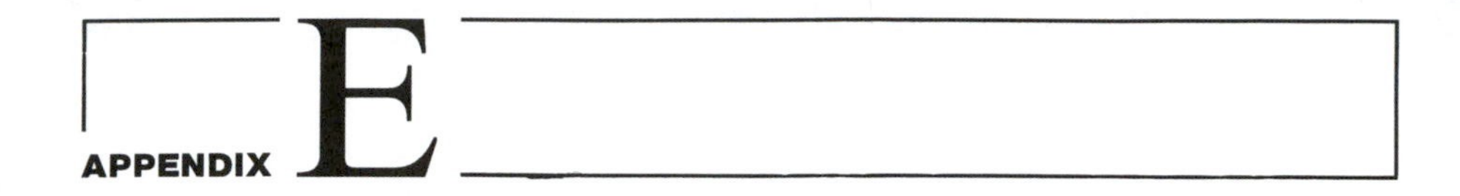

APPENDIX E

Switching Systems

The purpose of switching is to provide connections for all possible telephone calls with a minimum probability that the call will not go through, called *blocking*. Obviously, this problem cannot be solved simply by providing a pair of wires for all possible connections between telephones. Switching allows the load to be distributed somewhat. Calling data indicate that very few people need to be connected with overseas phones and that the highest density of calls go to our present geographical area. Thus switching centers are distributed geographically, and we are able to obtain some estimate of the number of connections that must be provided between the various centers. The problem of switching thus reduces to providing the necessary connections between these centers.

Switching constitutes a substantial fraction of the cost of a telephone call and is absolutely essential to the successful operation of the telephone network as we know it today. Yet, few undergraduate communications texts discuss switching. This is primarily because telecommunications switching is almost a subject unto itself, topically quite separate from other communications disciplines. In this appendix we briefly develop some of the salient features of telephone switching systems, with the details left primarily to the references.

E.1 Typical Telephone Call

Insight into the elementary functions required of a switching system is provided by considering the steps required to complete a typical telephone call.

STEP 1. The customer picks up the telephone, thus originating the call and sending an "off-hook" signal. The switching system must detect this off-hook signal as a request for service.

STEP 2. The switching system then prepares to receive information from the customer and notifies the customer that it is ready by returning a dial tone.

STEP 3. The number "dialed" by the customer is received and the address of the called telephone is determined.

STEP 4. An unused or idle path between the two telephones is found, or if an idle path does not presently exist, a "fast busy" signal is returned to the calling customer.

STEP 5. If an idle path is found in step 4, the system checks to see if the called telephone is idle.

STEP 6. If the called telephone is idle, ringing takes place. If the called telephone is busy, a busy signal is returned to the caller.

STEP 7. If the called telephone is idle, the line is monitored for an off-hook condition, indicating that the telephone has been answered. The connection is then completed.

STEP 8. Release all equipment and return it to the idle state when the caller telephone on-hook signal is detected.

This sequence indicates that the switching system must not only determine the route of the call and make the connection, but must also supervise the call from beginning to end in order to make equipment available when needed and to release equipment when the call is over. The on-hook, off-hook, and dial tone are *supervisory* signals, while the dial pulses from a dialed number or the tones from a push-button telephone are *control* signals. The methods used to accomplish the various steps in completing a call differ substantially depending on the type of switching system. We shall only be able to hint at these differences here.

E.2 Types of Switches

There are several ways that switches can be classified. One way is in terms of the technologies used to implement the switch. Early switches were electromechanical and of the step-by-step type in that they responded to the individual dial pulses themselves. It is noted here that dial pulses occur at a rate of 10 pulses/sec and the number of pulses corresponds to the actual number dialed. For example, if a 6 is dialed, six pulses are produced. The next broad class of switches combined electromechanical switching technology with electronics via stored program control. That is, electronics was used to implement the control features required for switching. The third class of switching systems are all electronic in that both the switching and control functions are implemented electronically. The touch-tone or dual-tone multifrequency (DTMF) control signaling is easily used with the electronically controlled switches and has often been used with step-by-step switches with the aid of a translator that changes the tones to dial pulses. For push-button DTMF signaling, pairs of tones are used to designate the desired digit as illustrated in Fig. E.2.1.

Another classification of switching systems is in terms of whether the same physical connection is maintained throughout the duration of the call. When the same physical connection is maintained, the call is said to be *circuit switched*, and circuit switching is the type of switching used when analog signals are switched. Hence circuit switching is sometimes called *analog switching*, and the

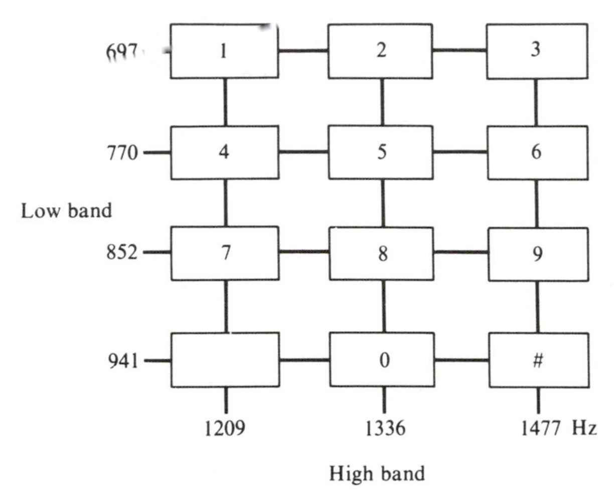

FIGURE E.2.1 Push-button DTMF digit and tone assignments.

electromechanical switches are basically analog switches. When the signal to be switched is digitized, for example, using PCM, the time-division-multiplexed sequence of binary words can be switched one word at a time by placing the word into a memory and then reading it out of memory at the correct time for transmission in another TDM sequence. In this type of switching no one physical connection is dedicated to a particular voice channel as in circuit switching.

E.3 Switch Design and Analysis

To design a switch, we readdress the basic problem of switching as stated at the beginning of this appendix: namely, how do we provide the required connections between switching centers with a minimum blocking probability? The simplest structure that might be devised to provide the necessary paths is the square array shown in Fig. E.3.1, where the $\times$'s denote possible connections. There are only N customers L_i, $i = 1, 2, \ldots, N$, for this switch, but they are shown at both the input and output since the communication is bidirectional but the switch is shown as unidirectional. This square switching array requires $N^2 - N$ switching elements, called *crosspoints* (since all inputs are also outputs), and is called *strictly nonblocking* because as long as a pair of lines remains unconnected, a free path remains to form the required connection. Further, distinct paths are available for each direction of transmission, and hence this kind of switch is called *four-wire* because a separate pair of wires and switch contacts is used for each direction of transmission.

It is common practice to use a single pair of wires for subscriber lines in a local exchange area, and therefore local offices and other local switches, called tandem switches, are mostly *two-wire*. The square array can thus be "folded" to obtain the triangular configuration in Fig. E.3.2, which has $N(N - 1)/2$ crosspoints and strictly nonblocking operation. *Strictly nonblocking* means that there is always a path through the switch for each available input and output.

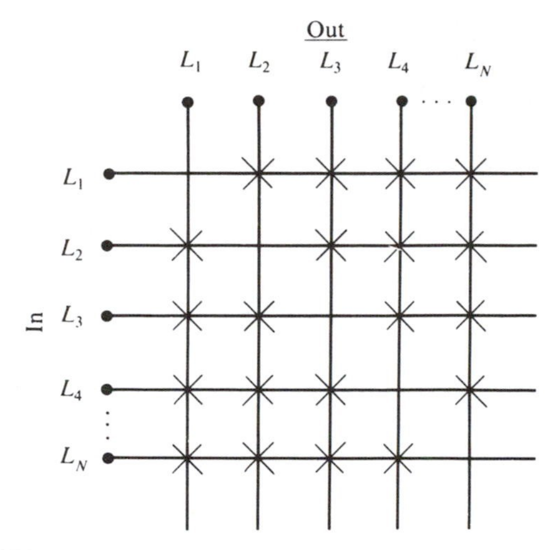

FIGURE E.3.1 Square nonblocking switch array.

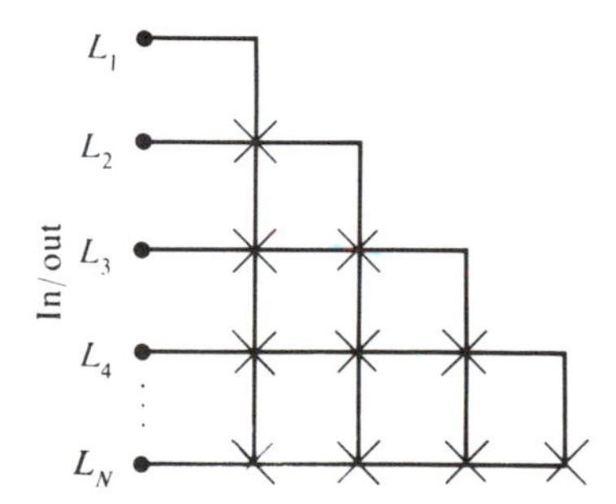

FIGURE E.3.2 Triangular switch configuration.

The square and triangular switch structures have the problem that they do not allow crosspoints to be shared. That is, each crosspoint can only be used by one path through the switch, thus resulting in a low utilization of crosspoints. To make more efficient use of crosspoints, it has become standard practice to construct switches in several stages. A typical three-stage switch is shown in Fig. E.3.3, where square blocks with n "inputs" and k "outputs" represent $n \times k$ arrays of crosspoints. The N inlets and outlets are partitioned into groups of size n. Thus there are N/n arrays in the first and third stages. Each array in the first and third stages has one connection to each of k arrays in the second stage. The arrays in the first and third stages are $n \times k$, while the second-stage arrays are $N/n \times N/n$. The ratio k/n is called the *space expansion ratio.* Suppose that a connection is to be made from a given input array to a given output array. This will be possible only if an array in the middle stage can be found that has idle connections to both arrays of interest. If no such path exists, the call is *blocked.* If $k = 1$, only one call can be completed from any inlet array or to any outlet array at any instant of time.

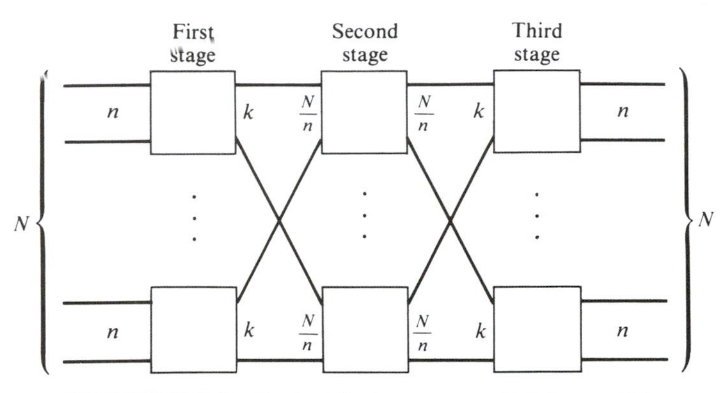

FIGURE E.3.3 Typical 3-stage space-division switch.

The condition for this switch to be strictly nonblocking is as follows: In the worst case, $n-1$ inlets sharing the first-stage array of interest would be busy, connected to $n-1$ second-stage arrays. Furthermore, the other $n-1$ outlets sharing the third-stage array of interest would be busy because they are connected to $n-1$ second-stage arrays. If these two sets of second-stage arrays are completely disjoint, there are $2n-2$ arrays that cannot be used to make the connection, since each second stage has only one connection per outer stage. Thus, at least one more second-stage array is required, giving a total for k of $2n-1$. Since there are N/n inlets and outlets on each second-stage array, the selection for k above guarantees strictly nonblocking operation.

The number of crosspoints in the three-stage switch in Fig. E.3.3 is

$$C = 2\left(\frac{N}{n}\right)nk + k\left(\frac{N}{n}\right)^2 = 2Nk + k\left(\frac{N}{n}\right)^2. \tag{E.3.1}$$

For nonblocking operation, $k = 2n - 1$. For a given N, the value n must be chosen. A common choice is $n = \sqrt{N}$, in which case

$$C = 6N^{3/2} - 3N. \tag{E.3.2}$$

Table E.3.1 compares the number of square array switch crosspoints with Eq. (E.3.2). It is evident that substantial savings in crosspoints are possible for large N (large switches). The value of n that minimizes C for a given N is approximately $n \cong \sqrt{N/2}$. There is much more that can be said concerning the optimization of the three-stage switch and designing multiple-stage switches; however, we must leave further discussion to the references [Clos, 1953; Collins and Pedersen, 1973].

The switches discussed thus far are called *space-division switches*, since they allocate space within a switching array to a particular connection. Another class of switches are *time-division switches*, which basically allocate time slots to the various required connections. There are two types of time-division switches, those that switch unquantized and uncoded samples of the voice conversations and those that switch PCM words. The first type of digital switching is sometimes called analog time-division switching or time-multiplexed space-division switching, while the latter is called digital time-division switching or

TABLE E.3.1 Comparison of the Number of Crosspoints Between a Square Array and a Three-Stage Nonblocking Array

N	Square Array (N^2)	Three-Stage Array ($6N^{3/2} - 3N$)
4	16	36
9	81	135
16	256	336
25	625	675
36	1,296	1,188
49	2,401	1,911
64	4,096	2,880
81	6,561	4,131
100	10,000	5,700
...	...	...
1,000	1,000,000	186,737
10,000	100,000,000	5,970,000

Source: C. Clos, "A Study of Non-blocking Switching Networks," *Bell System Technical Journal,* © 1953 AT&T Bell Laboratories.

simply digital switching. Here we briefly discuss time-division switching, hereafter designated *digital switching.* In digital switching, an input stream of PCM words is read sequentially into memory; then the words are read out of memory in the order needed to fill its designated slot on an output stream of PCM words. This is often referred to as *time slot interchange.*

It is common to combine time-division and space-division stages to obtain a more efficient switch. One such important switching structure is the time–space–time (TST) switch represented by Fig. E.3.4. The details of time slot interchange for a TST switch are illustrated by Fig. E.3.5. Each input line carries 120 time-division-multiplexed (TDMed) channels, which are then sequentially

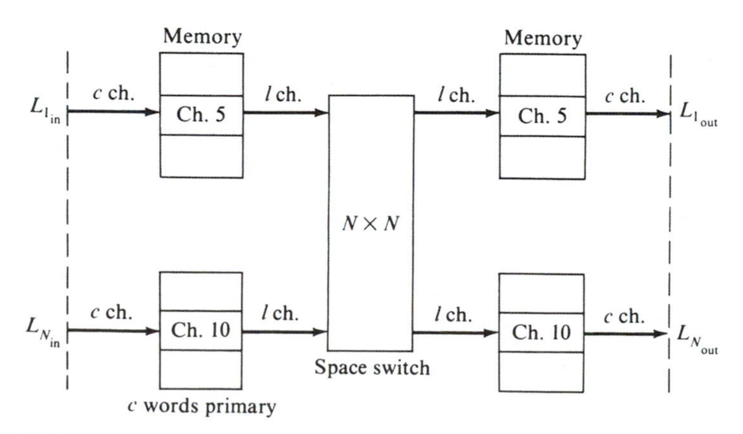

FIGURE E.3.4 General time-space-time switch structure. From A. A. Collins and R. D. Pederson, *Telecommunications: A Time for Innovation,* 1973. Dallas, Tex.: Merle Collins Foundation.

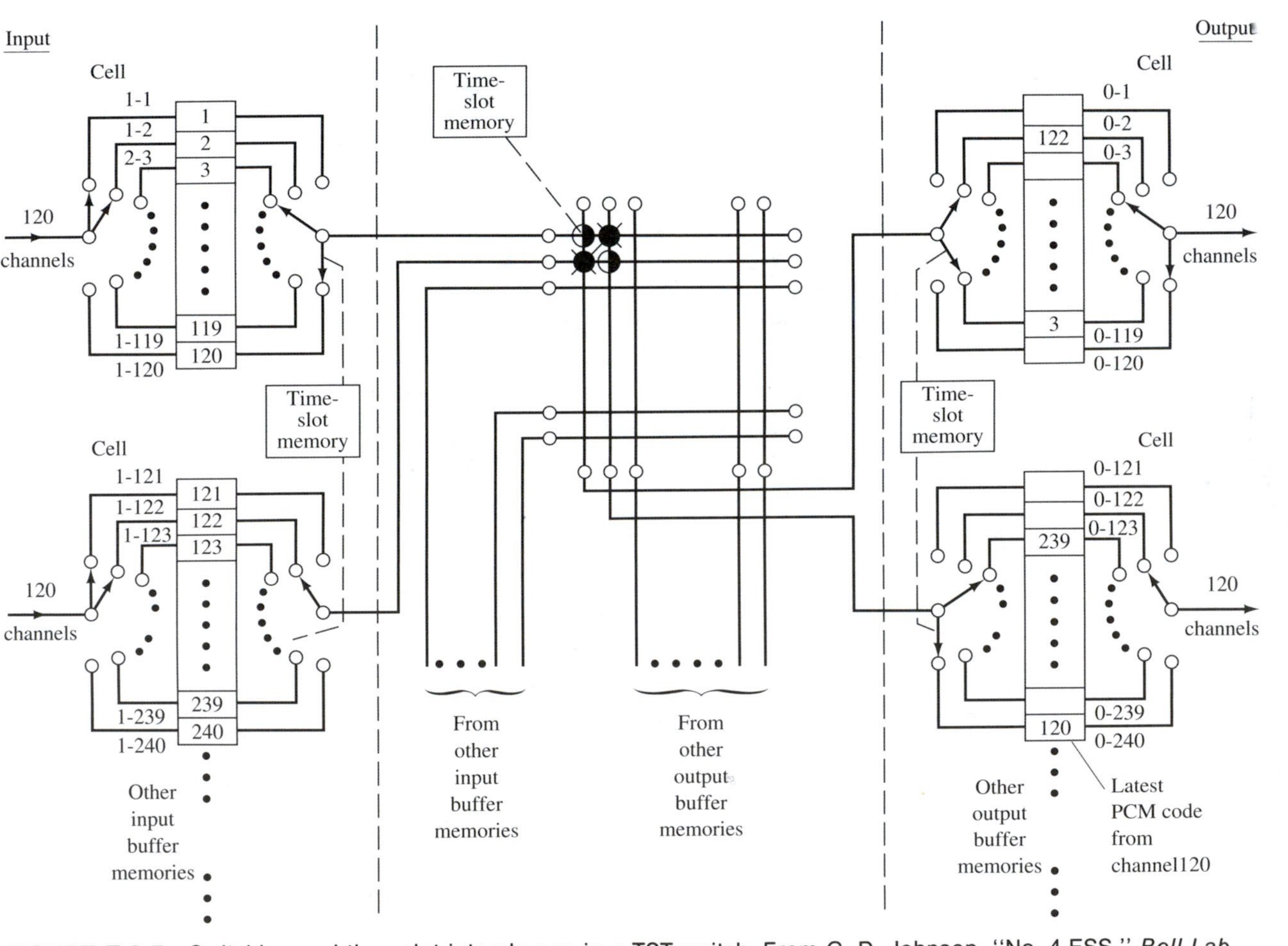

FIGURE E.3.5 Switching and time slot interchange in a TST switch. From G. D. Johnson, "No. 4 ESS," *Bell Lab. Rec.*, © 1973 AT&T Bell Laboratories.

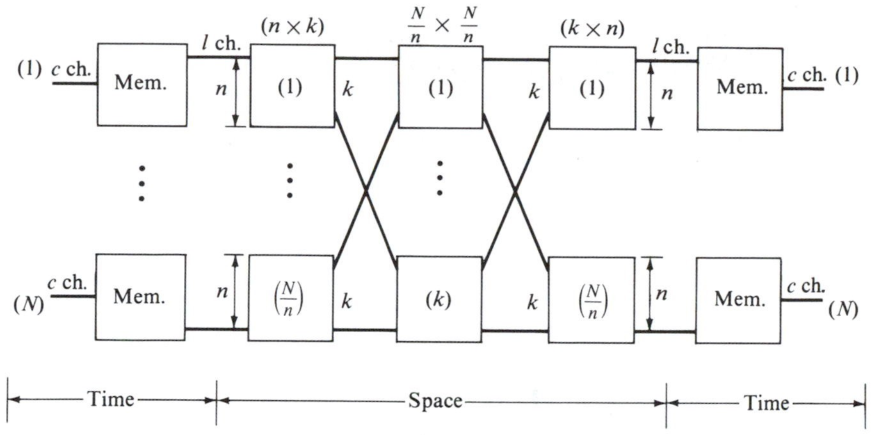

FIGURE E.3.6 TST structure with a 3-stage space-division switch. From A. A. Collins and R. D. Pederson, *Telecommunications: A Time for Innovation*, 1973. Dallas, Tex.: Merle Collins Foundation.

read into memory. On the output side, there are lines with 120 TDMed channels that are sequentially reading the words from the output buffer memories. These output memory locations are filled by reading the words from the input memories in a nonsequential manner such that the words are routed to the desired destinations. Figure E.3.6 shows a TST structure where the space-division switch has the form of the three-stage array in Fig. E.3.3. Note that Figs. E.3.4 and E.3.6 show c channels per unit time entering the input memories and l channels per unit time leaving the memories. The number of channels l plays a role analogous to k for a three-stage space-division switch, and therefore for strictly nonblocking operation, $l = 2c - 1$. This indicates that the bit rate on the lines labeled by l is l/c times the rate of those lines labeled by c channels. The quantity l/c is sometimes called the *time expansion ratio* of the switch.

Many commercial switches are designed today to have multiple time and space stages, such as TST, STS, TSSST, and others. Time-division switching is generally less costly than space-division switching, but of course, there are practical limits as to how many channels can be TDMed on a single line. Hence, once this limit is approached, it is advantageous to interleave time- and space-division switching stages. We are thus lead to the configurations mentioned previously. There are many interesting details omitted here, including example switch designs; however, length limitations preclude their examination. The interested reader is encouraged to consult the literature [Collins and Pederson, 1973; Briley, 1983].

E.4 Blocking Probability and Switch Designs

Both space-division and time-division switch complexity can be greatly reduced by allowing a small positive probability that a call will be *blocked*; that is, that there will not be a switching connection available to complete the call.

TABLE E.4.1 Comparison of the Number of Crosspoints Required by a Three-Stage Space-Division Switch Under Nonblocking and Blocking Operation

Switch Size N	n	k	Number of Crosspoints with Blocking	Number of Crosspoints in Nonblocking Design
128	8	5	2,560	7,680 ($k = 15$)
512	16	7	14,336	63,488 ($k = 31$)
2,048	32	10	81,920	516,096 ($k = 63$)
8,192	64	15	491,520	4.2 million ($k = 127$)
32,768	128	24	3.1 million	33 million ($k = 255$)
131,072	256	41	21.5 million	268 million ($k = 511$)

Source: J. Bellamy, *Digital Telephony*, copyright © 1982 John Wiley & Sons, Inc., New York. Reprinted by permission of John Wiley & Sons, Inc.

Commercial telephone networks are usually designed such that there is a certain probability of blocking during the peak load time of the day, and everyone has experienced blocked telephone calls on special days such as Mother's Day and Valentine's Day. The striking reduction in switch size possible by allowing a certain blocking probability is well illustrated by Table E.4.1, which compares the number of crosspoints required by a three-stage space-division switch, as in Fig. E.3.3, with no blocking and a probability of blocking of 0.002. The probability of a switch input line being busy is assumed to be 0.1.

The calculation of blocking probabilities for multistage switch designs is difficult indeed. Lee [1955] and Jacobaeus [1950] presented methods that, although requiring several simplifying assumptions, have been applied rather successfully. However, these technologies are basically static methods and cannot take into account the dynamics of a switching system, and hence simulations are also very valuable in evaluating switch designs. Both static techniques and simulations become complicated for large multistage switch designs. Although we have the necessary probability background to develop the Jacobaeus and Lee methods here, we must, because of length constraints, leave this material to the references (see Jacobaeus [1950]; Lee [1955]; Collins and Pederson [1973]; and Bellamy [1982]).

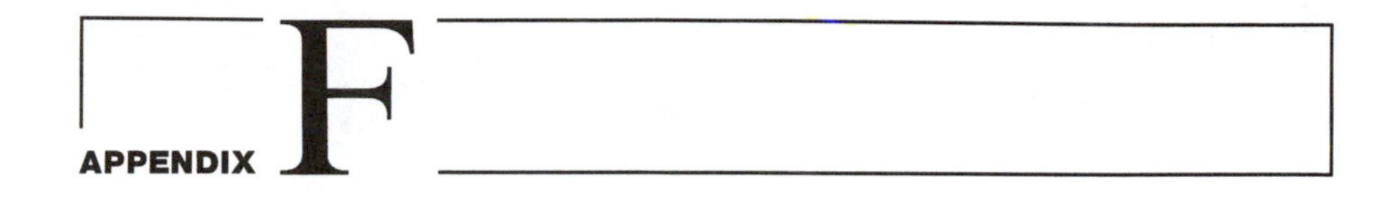

Television

In this appendix we describe briefly some of the salient features of monochrome and color television (TV) transmitters and receivers. As will be seen, TV relies heavily on analog modulation techniques discussed in Chapters 5 and 6 and on properties of the human eye. Appendix B should be read in conjunction with this appendix.

F.1 Monochrome or Black-and-White TV

The baseband bandwidth required for a monochrome or black-and-white TV signal in the United States is about 4.2 MHz, and can be computed using, among other quantities, the number of horizontal and vertical resolution cells, the scan time per line, and the number of lines. An approximate calculation of this bandwidth is presented in Section B.4. The actual video signal that is transmitted is called *composite video*, since it consists of the raw video obtained by raster scanning the image and blanking pulses and sync pulses that occur at the end of each horizontal scan line. A "typical" baseband composite video signal is shown in Fig. F.1.1. This is actually the inverted video signal, but this is a minor detail, since $m_v(t)$ in Fig. F.1.1 is the message signal applied to the modulator. The blanking level and the sync level occur during the horizontal retrace time. The relative amplitude levels shown in the figure are not to scale, so that important concepts can be clearly accentuated. The blanking level is actually "blacker than black" and the sync pulse is the maximum amplitude value. There is a specification on the minimum amplitude level (which represents white), which plays a critical role in demodulation. At the end of each field, the sync pulse is widened to blank video during the 21-line vertical retrace time. This is the origin of the black bar at the bottom (or top) of a TV picture when the vertical hold is misadjusted.

The 4.2-MHz bandwidth composite video information is transmitted using vestigial sideband AM plus a carrier signal to allow envelope detection. The audio is transmitted using frequency modulation of a separate carrier located 4.5 MHz above the video carrier. The transmitted spectrum is represented by

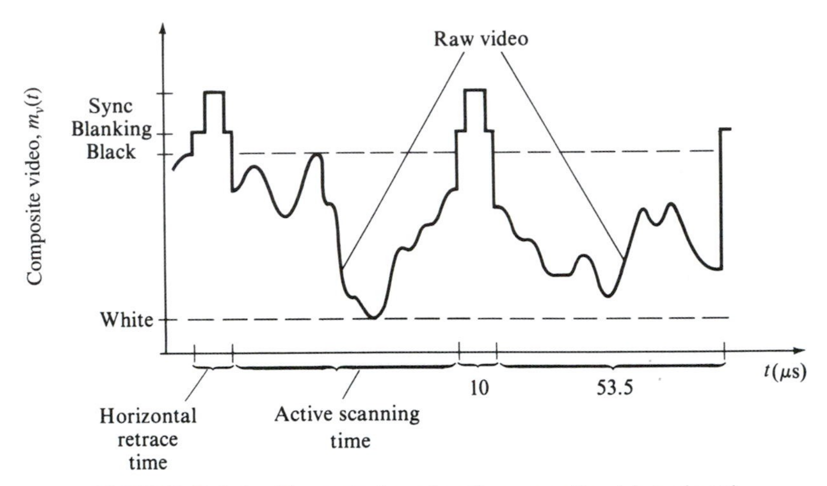

FIGURE F.1.1 Example baseband composite video signal.

the diagram in Fig. F.1.2. The VSB shaping of the lower sideband is as shown in Fig. F.1.3 and is accomplished at the receiver. The FM signal has a frequency deviation of $\Delta f = 25$ kHz and a baseband audio bandwidth of about 15 kHz. Hence the required FM waveform bandwidth from Carson's rule is about 80 kHz. Thus, almost all of the 250 kHz above 4.5 MHz can be considered to be a guardband. The total bandwidth allocated to a TV channel is 6 MHz, as is evident from Fig. F.1.2.

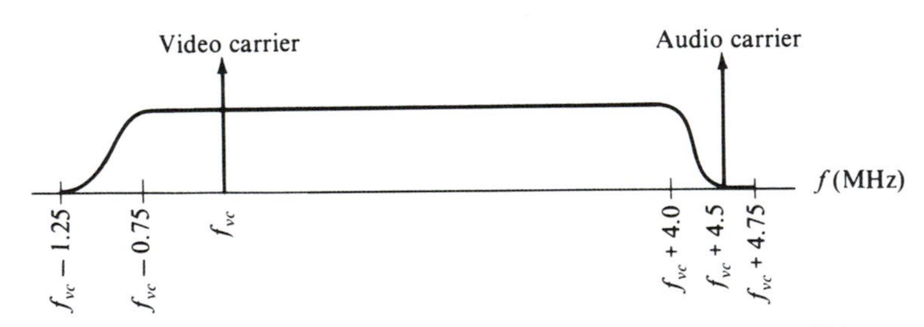

FIGURE F.1.2 Transmitted spectrum for monochrome TV.

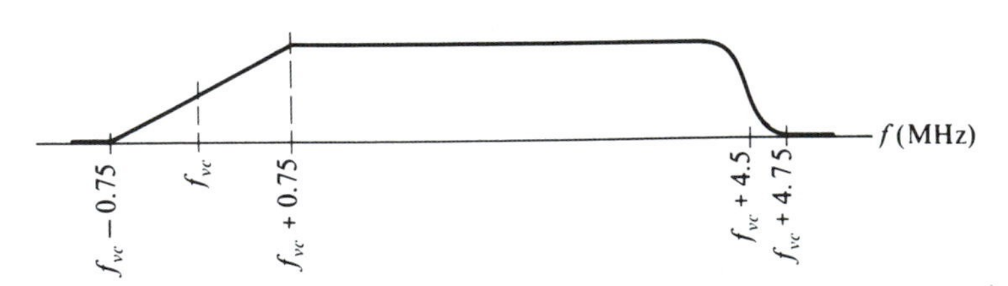

FIGURE F.1.3 VSB shaping of monochrome TV signal at the receiver.

A monochrome TV receiver has a superheterodyne receiver with an IF center frequency in the range 41 to 46 MHz. The output of the IF amplifier, which includes the VSB shaping in Fig. F.1.3, is given by

$$s(t) = \{A_{vc}[1 + am_v(t)] + A_{ac} \cos [\omega_a t + \phi(t)]\} \cos \omega_{vc} t \\ - \{aA_{vc}[m_{hv}(t) + m_{\beta v}(t)] + A_{ac} \sin [\omega_a t + \phi(t)]\} \sin \omega_{vc} t, \quad \text{(F.1.1)}$$

where A_{vc} is the video carrier amplitude, a the modulation index $(= 0.875)$, $m_v(t)$ the baseband video information, A_{ac} the audio carrier amplitude, ω_a the audio carrier frequency, ω_{vc} the video carrier frequency, $m_{hv}(t)$ the Hilbert transform of $m_v(t)$, $m_{\beta v}(t)$ the vestige component, and $\phi(t)$ contains the audio information. Equation (F.1.1) is also the input to the envelope detector. The envelope detector output can be expressed straightforwardly as

$$\rho(t) = [\{A_{vc}[1 + am_v(t)] + A_{ac} \cos [\omega_a t + \phi(t)]\}^2 \\ + \{aA_{vc}[m_{hv}(t) + m_{\beta v}(t)] + A_{ac} \sin [\omega_a t + \phi(t)]\}^2]^{1/2}. \quad \text{(F.1.2)}$$

Now, $|a[m_{hv}(t) + m_{\beta v}(t)]|^2 \ll 1$ and $A_{ac} \ll A_{vc}$, so

$$\begin{aligned} \rho(t) &\cong A_{vc}^2[1 + am_v(t)]^2 + 2A_{vc}A_{ac}[1 + am_v(t)] \cos [\omega_a t + \phi(t)] \\ &\quad + 2aA_{vc}\underbrace{A_{ac}[m_{hv}(t) + m_{\beta v}(t)]}_{\text{small}} \sin [\omega_a t + \phi(t)] \\ &\quad + A_{vc}^2\underbrace{a^2[m_{hv}(t) + m_{\beta v}(t)]^2}_{\text{small}}]^{1/2} \\ &\cong [A_{vc}^2[1 + am_v(t)]^2 + 2A_{vc}A_{ac}[1 + am_v(t)] \cos [\omega_a t + \phi(t)]]^{1/2} \\ &= A_{vc}[1 + am_v(t)]\left\{1 + \frac{2A_{ac} \cos [\omega_a t + \phi(t)]}{A_{vc}[1 + am_v(t)]}\right\}^{1/2}. \end{aligned} \quad \text{(F.1.3)}$$

Since the second term under the square root is small, we use $[1 + x]^{1/2} \cong 1 + x/2$ to find

$$\rho(t) \cong A_{vc}[1 + am_v(t)] + A_{ac} \cos [\omega_a t + \phi(t)]. \quad \text{(F.1.4)}$$

The assumption that $A_{ac} \ll A_{vc}$ is guaranteed by the minimum value set for "white" in Fig. F.1.1 and the gain of the receiver filter at $f_a = 4.5$ MHz in Fig. F.1.3.

The video signal is obtained from $\rho(t)$ by low-pass filtering. An IF amplifier tuned to 4.5 MHz picks the audio component out of $\rho(t)$, which is then passed to a standard FM detector.

F.2 Color TV

The three primary colors are red, green, and blue, and hence any color image can be broken down into these components, denoted here by $m_r(t)$, $m_g(t)$, and $m_b(t)$, respectively. However, to be compatible with monochrome TV, these signals are not transmitted directly. Rather, a linear combination of $m_r(t)$, $m_g(t)$, and $m_b(t)$ is constructed that is virtually the same as the video signal in black-and-white TV, called the *luminance* signal, and that is given by

$$m_v(t) = 0.3m_r(t) + 0.59m_g(t) + 0.11m_b(t). \tag{F.2.1}$$

Of course, two more waveforms are needed at the receiver to obtain the red, blue, and green components, and hence *chrominance signals* are also transmitted, which have the form

$$m_i(t) = 0.60m_r(t) - 0.28m_g(t) - 0.32m_b(t) \tag{F.2.2}$$

and

$$m_q(t) = 0.21m_r(t) - 0.52m_g(t) + 0.31m_b(t). \tag{F.2.3}$$

Since the luminance signal is the entire video signal in monochrome TV, it is allocated a bandwidth of 4.2 MHz. The spectrum of the luminance signal resembles that in Fig. B.4.2. The chrominance signals $m_i(t)$ and $m_q(t)$ can be low-pass filtered to 1.5 MHz and 0.5 MHz, respectively, without a noticeable loss in color image quality. Of course, to be compatible with monochrome TV, we must not use additional bandwidth, and the luminance signal already occupies the total allocated bandwidth of 4.2 MHz. How do we transmit the chrominance components, then? The answer lies in exploiting the form of the spectrum for the luminance signal as given in Fig. B.4.2. More specifically, the chrominance signals have periodic components with periods $T_h = 1/f_h$ and $T_v = 1/f_v$, just as $m_v(t)$ does. Hence, if a subcarrier is chosen for the chrominance components that falls halfway between two spectral lines in the luminance, the chrominance signal spectral lines with respect to the subcarrier will fall in the gaps between the luminance component spectral lines. Consequently, the chrominance subcarrier frequency (f_{cc}) is selected to be halfway between the 227th and 228th harmonics of $f_h = 15{,}750$ Hz, or $f_{cc} = 227.5 \times 15{,}750 \text{ Hz} \cong 3.58$ MHz.

The modulation methods used for $m_i(t)$ and $m_q(t)$ are somewhat different. The chrominance signals are transmitted in phase quadrature, with AMDSB-SC being used for all frequencies below 0.5 MHz. Those frequencies in $m_i(t)$ above 0.5 MHz are transmitted via lower-sideband SSB-SC. The chrominance-modulated signals are then added to the luminance to obtain the baseband signal

$$m(t) = m_v(t) + m_i(t) \cos \omega_{cc}t + m_q(t) \sin \omega_{cc}t + m_{ihh}(t) \sin \omega_{cc}t, \tag{F.2.4}$$

where $m_{ihh}(t)$ is the Hilbert transform of that portion of $m_i(t)$ in the frequency range 0.5 to 1.5 MHz. The baseband signal $m(t)$ in Eq. (F.2.4) is then transmitted with AMVSB-TC as in monochrome TV.

When the color signal is applied to a monochrome receiver, the envelope detector output is $m(t)$ in Eq. (F.2.4), which we would like to be just $m_v(t)$. Fortunately, since f_{cc} is an odd multiple of half the line rate (f_h), that is, $f_{cc} = 227.5f_h$, the extra components in Eq. (F.2.4) have a phase reversal from line to line that is averaged out by the eye, and hence goes unnoticed. More precisely, the red, blue, and green components are normalized such that all three lie between 0 and +1, so $m_v(t)$ is always positive, but $m_i(t)$ and $m_q(t)$ may be positive or negative. The signal $m_i(t) \cos \omega_{cc}t + m_q(t) \sin \omega_{cc}t + m_{ihh}(t) \sin \omega_{cc}t$ thus appears as a modulated sinusoid with frequency f_{cc} about an offset $m_v(t)$. The line-to-line phase reversal causes this variation about $m_v(t)$ to be virtually invisible to the human eye.

A color TV receiver obtains the luminance component by envelope detection but follows this with a notch filter at f_{cc} to reject the principal subcarrier component. Since the chrominance components are in quadrature, an accurate local oscillator is needed for their demodulation. This is made available by transmitting at least eight cycles of a sinusoid with frequency $f_{cc} \cong 3.58$ MHz right after the sync pulse on "top" of the blanking pulse (see Fig. F.1.1). This sinusoidal burst is tracked by a PLL with a phase reference that can be manually adjusted by the hue or tint controls on the TV. Thus $m_i(t)$ and $m_q(t)$ are obtained by bandpass filtering $m(t)$ between about $3.58 - 1.5 = 2.08$ to $3.58 + 0.5 \cong 4.1$ MHz and then using coherent demodulation. Note that there are some luminance terms in the frequency range 2.08 to 4.1 MHz, but they are not visible to the human eye because of the line-to-line phase reversal property discussed previously.

The red, blue, and green components can then be obtained as

$$m_r(t) = m_v(t) - 0.96m_i(t) + 0.62m_q(t) \tag{F.2.5}$$

$$m_g(t) = m_v(t) - 0.28m_i(t) - 0.64m_q(t) \tag{F.2.6}$$

$$m_b(t) = m_v(t) - 1.10m_i(t) + 1.70m_q(t). \tag{F.2.7}$$

Note that if a color TV receives a monochrome signal, then $m_r(t) = m_g(t) = m_b(t) = m_v(t)$, the luminance only.

Finally, we note that it is desirable to have the audio carrier signal (4.5 MHz) fall between chrominance carrier spectral lines, and to achieve this, f_h is adjusted to 15,734.266 Hz, so $f_{cc} = 3.579545$ MHz. This slight change does not materially affect our discussion, so we do not pursue it further here.

Pseudorandom Sequences

Pseudorandom sequences, or pseudonoise (PN) sequences as they are sometimes called, have applications to data scrambling, spread spectrum, and encryption, just to name a few. Desirable characteristics of pseudorandom sequences are that they must (1) be easy to implement, (2) have certain randomness properties, (3) have long periods, and (4) be difficult to reconstruct from a short subsequence [Pickholtz, Schilling, and Milstein, 1982]. Sequences that have one or more of these characteristics are maximal-length linear shift register sequences, Kasami sequences [see Sarwate and Pursley, 1980], and Gold sequences [Gold, 1967]. We confine the discussion here to binary maximal-length linear shift register sequences, also called *m*-sequences. A full exposition on *m*-sequences requires a working knowledge of finite fields, often called Galois fields [Ziemer and Peterson, 1985]. However, for conciseness and simplicity, the development in this appendix does not require such a background, and most properties are therefore given without proofs.

G.1 *m*-Sequences

A generator for a binary linear shift register sequence is shown in Fig. G.1.1. The sequence $\{b_i\}$ and the coefficients take on the two values 0 or 1 and the summations shown are modulo-2. From the figure we have

$$b_k = \sum_{i=1}^{n} a_i b_{k-i} (\text{mod } 2), \tag{G.1.1}$$

where $a_n = 1$ for an n-stage shift register. After each clock cycle, the contents of the nth stage are discarded and the contents of all other stages are shifted one stage to the right. The last b_k is placed in stage 1 and a new b_k is formed according to Eq. (G.1.1). The output sequence depends on the initial contents of the shift register and the choice of the feedback coefficients $\{a_i, i = 1, 2, \ldots, n\}$.

The output (b_k) of the generator is periodic with maximum period $N = 2^n - 1$. The all-zeros sequence can only occur if the initial shift register contents are all zero or the coefficients are all zero. A feedback connection that gives a period of $2^n - 1$ can always be found, and the resulting sequences are

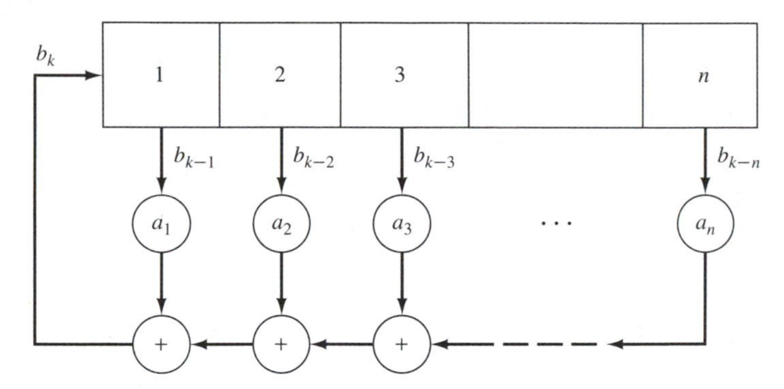

FIGURE G.1.1 Binary linear shift register sequence generator.

called *maximal-length sequences* or *m-sequences.* Since no linear feedback connection can have a longer period for an n-stage shift register, this constitutes a very efficient implementation.

G.2 Properties of m-Sequences

Maximal-length sequences have several randomness properties that make them attractive in applications. These properties are [Pickholtz, Schilling, and Milstein, 1982; Ziemer and Peterson, 1985]:

Property 1. A maximal-length sequence has $\frac{1}{2}(N+1) = 2^{n-1}$ ones and $2^{n-1} - 1$ zeros.

Property 2. If we slide a window of length n along the output sequence for N shifts, each n-tuple, except the all-zeros sequence, will appear exactly once.

Property 3. If we define a run as a sequence of identical symbols, then for any maximal length sequence, there is
(1) One run of 1's of length n,
(2) One run of 0's of length $n - 1$,
(3) One run of 1's and one run of 0's of length $n - 2$,
(4) Two runs of 1's and two runs of 0's of length $n - 3$,
⋮
(n) 2^{n-3} runs of 1's and 2^{n-3} runs of 0's of length 1.

For many communications applications, the 0,1 sequence is changed to $a \pm 1$ sequence by defining $b'_k = 1 - 2b_k = (-1)^{b_k}$ for $b_k = 0$ or 1. Defining the periodic autocorrelation of this new primed sequence as

$$\theta_{b'}(l) = \frac{1}{N} \sum_{k=1}^{N} b'_k b'_{k+l}, \tag{G.2.1}$$

we have

Property 4. The periodic autocorrelation of a maximal length ± 1 sequence is

$$\theta_{b'}(l) = \begin{cases} 1, & l = 0, N, 2N, \ldots \\ -\dfrac{1}{N}, & \text{otherwise.} \end{cases} \tag{G.2.2}$$

A property of m-sequences that is useful in obtaining various results is the shift-and-add property, which says

Property 5. The modulo-2 sum of an m-sequence and any shifted version of the same sequence is itself a shifted version of the original sequence.

Note that of the four desirable characteristics mentioned in the introduction to this appendix, maximal-length sequences are easy to implement, they have certain randomness properties, and for a given shift register length n, they have long periods. Unfortunately, however, m-sequences do not possess the fourth characteristic, since it is relatively easy to reconstruct the entire sequence by observing only $2n - 2$ consecutive bits in the sequence [Pickholtz, Schilling, and Milstein, 1982; Ziemer and Peterson, 1985].

G.3 Partial Correlations

Because of delay and complexity constraints, it is not always possible to perform the discrete autocorrelation in Eq. (G.2.1) over an entire period N. In these situations, the autocorrelation may be calculated over only N_w bits, where $N_w < N$, so we have instead of Eq. (G.2.1) the partial autocorrelations

$$\theta_{b'}(l, l', N_w) = \frac{1}{N_w} \sum_{k=l'}^{l'+N_w} b'_k b'_{k+l}. \tag{G.3.1}$$

In Eq. (G.3.1), l' is included to allow for an arbitrary alignment of the window N_w bits long. Thus the partial autocorrelations are a function of both window location and window length.

It is useful for the system designer to know the mean and variance of $\theta_{b'}(l, l', N_w)$ averaged over all l'. These quantities are [Ziemer and Peterson, 1985]

$$\frac{1}{N} \sum_{l'=0}^{N-1} \theta_{b'}(l, l', N_w) = -\frac{1}{N} \tag{G.3.2}$$

and

$$\text{var}[\theta_{b'}(l, l', N_w)] = \frac{1}{N_w}\left[1 - \frac{N_w - 1}{N}\right] - \frac{1}{N^2}. \tag{G.3.3}$$

If $N_w = N$, Eq. (G.3.3) is zero.

The Union Bound

It is often the case that we are given a set of transmitted signal vectors, $\mathbf{s}_i$, $i = 1, 2, \ldots, N$, and we need to calculate $P[\mathscr{E} | m_k]$, the probability of error given that message m_k (and hence, $\mathbf{s}_k$) was transmitted. If the messages are equally likely and the effect of the channel is to add white Gaussian noise, then what we need to find is the probability that the received vector $\mathbf{r}$ is closer to one or more of the $\mathbf{s}_i$, $i \neq k$, than it is to $\mathbf{s}_k$. Adopting the notation from Wozencraft and Jacobs [1965] that $\mathscr{E}_{ki}$ denotes the event that $\mathbf{r}$ is closer to $\mathbf{s}_i$ than to $\mathbf{s}_k$ when $\mathbf{s}_k$ is transmitted, we can write

$$P[\mathscr{E} | m_k] = P[\mathscr{E}_{k1} \cup \mathscr{E}_{k2} \cup \cdots \cup \mathscr{E}_{k,k-1} \cup \mathscr{E}_{k,k+1} \cup \cdots \cup \mathscr{E}_{k,N}]. \quad \text{(H.1.1)}$$

Using Eq. (A.2.2) and axiom 3′ in Appendix A, we find that

$$P[\mathscr{E} | m_k] \leq \sum_{\substack{i=1 \\ i \neq k}}^{N} P[\mathscr{E}_{ki}], \quad \text{(H.1.2)}$$

which is called the *union bound*. Note that the quantities $P[\mathscr{E}_{ki}]$ on the right side of Eq. (H.1.2) depend on only two signals, and hence are very easy to calculate.

For the special case when the signal set is symmetric, $P[\mathscr{E}] = P[\mathscr{E} | m_k]$, and many of the $P[\mathscr{E}_{ki}]$ are the same, so the union bound greatly simplifies system analyses in these situations.

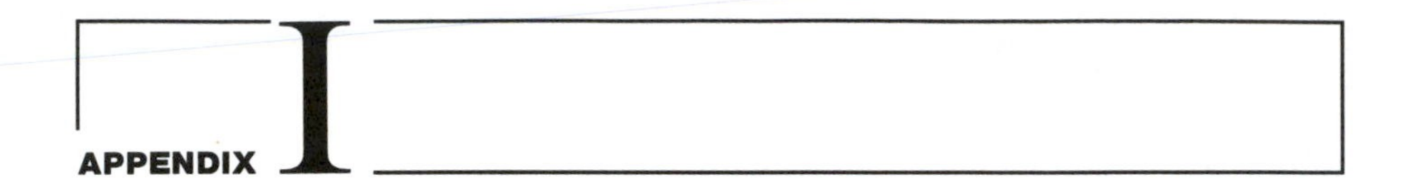

Bessel Functions and the Error Function

TABLE I.1 Bessel Functions of the First Kind

β	J_0	J_1	J_2	J_3	J_4	J_5	J_6	J_7	J_8	J_9	J_{10}
0.0	1.00										
.2	.99	.10									
.4	.96	.20	.02								
.6	.91	.29	.04								
.8	.85	.37	.08	.01							
1.0	.77	.44	.11	.02							
.2	.67	.50	.16	.03	$.01^-$						
.4	.57	.54	.21	.05	$.01^-$						
.6	.46	.57	.26	.07	.01						
.8	.34	.58	.31	.10	.02						
2.0	.22	.58	.35	.13	.03	$.01^-$					
.2	.11	.56	.40	.16	.05	.01					
.4	.00	.52	.43	.20	.06	.02					
.6	−.10	.47	.46	.24	.08	.02	$.01^-$				
.8	−.19	.41	.48	.27	.11	.03	$.01^-$				
3.0	−.26	.34	.49	.31	.13	.04	.01				
.2	−.32	.26	.48	.34	.16	.06	.02				
.4	−.36	.18	.47	.37	.19	.07	.02	$.01^-$			
.6	−.39	.10	.44	.40	.22	.09	.03	$.01^-$			
.8	−.40	.01	.41	.42	.25	.11	.04	.01			
4.0	−.40	−.07	.36	.43	.28	.13	.05	.02			
.2	−.38	−.14	.31	.43	.31	.16	.06	.02	$.01^-$		
.4	−.34	−.20	.25	.43	.34	.18	.08	.03	$.01^-$		
.6	−.30	−.26	.18	.42	.36	.21	.09	.03	.01		
.8	−.24	−.30	.12	.40	.38	.23	.11	.04	.01		

TABLE I.1 (Continued)

β	J_0	J_1	J_2	J_3	J_4	J_5	J_6	J_7	J_8	J_9	J_{10}
5.0	−.18	−.33	.05	.36	.39	.26	.13	.05	.02	$.01^-$	
.2	−.11	−.34	−.02	.33	.40	.29	.15	.07	.02	$.01^-$	
.4	−.04	−.35	−.09	.28	.40	.31	.18	.08	.03	$.01^-$	
.6	.03	−.33	−.15	.23	.39	.33	.20	.09	.04	.01	
.8	.09	−.31	−.20	.17	.38	.35	.22	.11	.05	.02	$.01^-$
6.0	.15	−.28	−.24	.11	.36	.36	.25	.13	.06	.02	$.01^-$
.2	.20	−.23	−.28	.05	.33	.37	.27	.15	.07	.03	$.01^-$
.4	.24	−.18	−.30	−.01	.29	.37	.29	.17	.08	.03	.01
.6	.27	−.12	−.31	−.06	.25	.37	.31	.19	.10	.04	.01
.8	.29	−.07	−.31	−.12	.21	.36	.33	.21	.11	.05	.02
7.0	.30	−.00	−.30	−.17	.16	.35	.34	.23	.13	.06	.02
.2	.30	.05	−.28	−.21	.11	.33	.35	.25	.15	.07	.03
.4	.28	.11	−.25	−.24	.05	.30	.35	.27	.16	.08	.04
.6	.25	.16	−.21	−.27	−.00	.27	.35	.29	.18	.10	.04
.8	.22	.20	−.16	−.29	−.06	.23	.35	.31	.20	.11	.05
8.0	.17	.23	−.11	−.29	−.11	.19	.34	.32	.22	.13	.06
.2	.12	.26	−.06	−.29	−.15	.14	.32	.33	.24	.14	.07
.4	.07	.27	−.00	−.27	−.19	.09	.30	.34	.26	.16	.08
.6	.01	.27	.05	−.25	−.22	.04	.27	.34	.28	.18	.10
.8	−.04	.26	.10	−.22	−.25	−.01	.24	.34	.29	.20	.11
9.0	−.09	.25	.14	−.18	−.27	−.06	.20	.33	.31	.21	.12
.2	−.14	.22	.18	−.14	−.27	−.10	.16	.31	.31	.23	.14
.4	−.18	.18	.22	−.09	−.27	−.14	.12	.30	.32	.25	.16
.6	−.21	.14	.24	−.04	−.26	−.18	.08	.27	.32	.27	.17
.8	−.23	.09	.25	.01	−.25	−.21	.03	.25	.32	.28	.19
10.0	−.25	.04	.25	.06	−.22	−.23	−.01	.22	.32	.29	.21

TABLE I.2 Error Function

$$\operatorname{erf}(x) = \frac{1}{\sqrt{2\pi}} \int_0^x e^{-y^2/2}\, dy$$

x	erf x	x	erf x
0.05	0.01994	1.55	0.43943
0.10	0.03983	1.60	0.44520
0.15	0.05962	1.65	0.45053
0.20	0.07926	1.70	0.45543
0.25	0.08971	1.75	0.45994
0.30	0.11791	1.80	0.46407
0.35	0.13683	1.85	0.46784
0.40	0.15542	1.90	0.47128
0.45	0.17364	1.95	0.47441
0.50	0.19146	2.00	0.47725
0.55	0.20884	2.05	0.47982
0.60	0.22575	2.10	0.48214
0.65	0.24215	2.15	0.48422
0.70	0.25804	2.20	0.48610
0.75	0.27337	2.25	0.48778
0.80	0.28814	2.30	0.48928
0.85	0.30234	2.35	0.49061
0.90	0.31594	2.40	0.49180
0.95	0.32894	2.45	0.49286
1.00	0.34134	2.50	0.49379
1.05	0.35314	2.55	0.49461
1.10	0.36433	2.60	0.49534
1.15	0.37493	2.65	0.49597
1.20	0.38493	2.70	0.49653
1.25	0.39435	2.75	0.49702
1.30	0.40320	2.80	0.49744
1.35	0.41149	2.85	0.49781
1.40	0.41924	2.90	0.49813
1.45	0.42647	2.95	0.49841
1.50	0.43319	3.00	0.49865

APPENDIX J

Mathematical Formulas

Trigonometric Identities

$$e^{\pm j\theta} = \cos\theta \pm j\sin\theta$$

$$\cos\theta = \frac{1}{2}(e^{j\theta} + e^{-j\theta}) = \sin(\theta + 90^\circ)$$

$$\sin\theta = \frac{1}{2j}(e^{j\theta} - e^{-j\theta}) = \cos(\theta - 90^\circ)$$

$$\sin^2\theta + \cos^2\theta = 1$$

$$\cos^2\theta - \sin^2\theta = \cos 2\theta$$

$$\cos^2\theta = \frac{1}{2}(1 + \cos 2\theta)$$

$$\cos^3\theta = \frac{1}{4}(3\cos\theta + \cos 3\theta)$$

$$\sin^2\theta = \frac{1}{2}(1 - \cos 2\theta)$$

$$\sin^3\theta = \frac{1}{4}(3\sin\theta - \sin 3\theta)$$

$$\sin(\alpha \pm \beta) = \sin\alpha\cos\beta \pm \cos\alpha\sin\beta$$

$$\cos(\alpha \pm \beta) = \cos\alpha\cos\beta \mp \sin\alpha\sin\beta$$

$$\tan(\alpha \pm \beta) = \frac{\tan\alpha \pm \tan\beta}{1 \mp \tan\alpha\tan\beta}$$

$$\sin\alpha\sin\beta = \frac{1}{2}\cos(\alpha - \beta) - \frac{1}{2}\cos(\alpha + \beta)$$

$$\cos\alpha\cos\beta = \frac{1}{2}\cos(\alpha - \beta) + \frac{1}{2}\cos(\alpha + \beta)$$

$$\sin\alpha\cos\beta = \frac{1}{2}\sin(\alpha - \beta) + \frac{1}{2}\sin(\alpha + \beta)$$

Series Expansions and Approximations

$$(1 + x)^n = 1 + nx + \frac{n(n-1)}{2!}x^2 + \cdots, \qquad |nx| < 1$$

$$e^x = 1 + x + \frac{1}{2!}x^2 + \cdots$$

$$a^x = 1 + x \ln a + \frac{1}{2!}(x \ln a)^2 + \cdots$$

$$\ln(1 + x) = x - \frac{1}{2}x^2 + \frac{1}{3}x^3 + \cdots$$

$$\sin x = x - \frac{1}{3!}x^3 + \frac{1}{5!}x^5 - \cdots$$

$$\cos x = 1 - \frac{1}{2!}x^2 + \frac{1}{4!}x^4 - \cdots$$

$$\tan x = x + \frac{1}{3}x^3 + \frac{2}{15}x^5 + \cdots$$

$$\arcsin x = x + \frac{1}{6}x^3 + \frac{3}{40}x^5 + \cdots$$

$$\arctan x = \begin{cases} x - \frac{1}{3}x^3 + \frac{1}{5}x^5 - \cdots, & |x| < 1 \\ \frac{\pi}{2} - \frac{1}{x} + \frac{1}{3x^3} - \cdots, & x > 1 \end{cases}$$

$$J_n(x) = \frac{1}{n!}\left(\frac{x}{2}\right)^n - \frac{1}{(n+1)!}\left(\frac{x}{2}\right)^{n+2} + \frac{1}{2!\,(n+2)!}\left(\frac{x}{2}\right)^{n+4} - \cdots$$

$$J_n(x) \approx \sqrt{\frac{2}{\pi x}} \cos\left(x - \frac{\pi}{4} - \frac{n\pi}{2}\right), \qquad x \gg 1$$

$$I_0(x) \approx \begin{cases} e^{x^2/4}, & x^2 \ll 1 \\ \frac{e^x}{\sqrt{2\pi x}}, & x \gg 1 \end{cases}$$

Fourier Transform Table and Properties

TABLE K.1 Fourier Transform Pairs

$f(t)$	$F(\omega) = \mathscr{F}\{f(t)\}$
$e^{-at}u(t)$	$\dfrac{1}{a + j\omega}$
$te^{-at}u(t)$	$\dfrac{1}{(a + j\omega)^2}$
$e^{-a\|t\|}$	$\dfrac{2a}{a^2 + \omega^2}$
$e^{-t^2/2\sigma^2}$	$\sigma\sqrt{2\pi}\, e^{-\sigma^2\omega^2/2}$
$\operatorname{sgn}(t)$	$\dfrac{2}{j\omega}$
$u(t)$	$\pi\delta(\omega) + \dfrac{1}{j\omega}$
$\delta(t)$	1
1	$2\pi\delta(\omega)$
$e^{\pm j\omega_c t}$	$2\pi\delta(\omega \mp \omega_c)$
$\cos \omega_c t$	$\pi[\delta(\omega - \omega_c) + \delta(\omega + \omega_c)]$
$\sin \omega_c t$	$\dfrac{\pi}{j}[\delta(\omega - \omega_c) - \delta(\omega + \omega_c)]$
$f(t) = \begin{cases} 1, & \|t\| < \dfrac{\tau}{2} \\ 0, & \|t\| > \dfrac{\tau}{2} \end{cases}$	$\dfrac{\tau \sin(\omega\tau/2)}{(\omega\tau/2)}$
$\dfrac{\Omega}{2\pi} \dfrac{\sin(\Omega t/2)}{(\Omega t/2)}$	$F(\omega) = \begin{cases} 1, & \|\omega\| < \dfrac{\Omega}{2} \\ 0, & \|\omega\| > \dfrac{\Omega}{2} \end{cases}$
$f(t) = \begin{cases} 1 - \dfrac{\|t\|}{\tau}, & \|t\| < \tau \\ 0, & \|t\| > \tau \end{cases}$	$\tau\left[\dfrac{\sin(\omega\tau/2)}{(\omega\tau/2)}\right]^2$

TABLE K.2 Fourier Transform Properties

If → then:	$f(t)$ ↔	$F(\omega)$
Linearity	$af(t) + bg(t)$	$aF(\omega) + bG(\omega)$
Scaling	$f(at)$	$\frac{1}{\lvert a \rvert} F\left(\frac{\omega}{a}\right)$
Time delay	$f(t - t_d)$	$e^{-j\omega t_d} F(\omega)$
Frequency shifting	$e^{j\omega_c t} f(t)$	$F(\omega - \omega_c)$
Modulation	$f(t) \cos \omega_c t$	$\frac{1}{2}[F(\omega + \omega_c) + F(\omega - \omega_c)]$
Time convolution	$f(t) * g(t)$	$F(\omega)G(\omega)$
Frequency convolution	$f(t)g(t)$	$\frac{1}{2\pi} F(\omega) * G(\omega)$
Symmetry (duality)	$F(t)$	$2\pi f(-\omega)$
Time differentiation	$\frac{d}{dt} f(t)$	$j\omega F(\omega)$
Time integration	$\int_{-\infty}^{t} f(\tau)\, d\tau$	$\frac{1}{j\omega} F(\omega) + \pi F(0)\delta(\omega)$

References

Abramson, N. 1973. "The Aloha System." Chapter 14 in *Computer Communication Networks*. N. Abramson and F. F. Kuo, eds. Englewood Cliffs, N.J.: Prentice Hall, pp. 501–517.

Amoroso, F. 1980. "The Bandwidth of Digital Data Signals." *IEEE Commun. Mag.*, Vol. 18, Nov.

Anderson, J. B., and J. R. Lesh, eds. 1981. "Special Section on Combined Coding and Modulation." *IEEE Trans. Commun.*, Vol. COM-29, Mar.

Anderson, J. B., C.-E. W. Sundberg, T. Aulin, and N. Rydbeck. 1981. "Power-Bandwidth Performance of Smoothed Phase Modulation Codes," *IEEE Trans. Commun.*, Vol. COM-29, Mar., pp. 187–195.

Anderson, R. R., and J. Salz. 1965. "Spectra of Digital FM." *Bell Syst. Tech. J.*, Vol. 44, July–Aug., pp. 1165–1189.

AT&T, *Telecommunications Transmission Engineering*, Vols. 1–3. New York: AT&T, 1977.

Bellamy, J. 1982. *Digital Telephony*. New York: Wiley.

Berger, T. 1971. *Rate Distortion Theory: A Mathematical Basis for Data Compression*. Englewood Cliffs, N.J.: Prentice Hall.

Bertsekas, D., and R. Gallager. 1987. *Data Networks*. Englewood Cliffs, N.J.: Prentice Hall.

Bhargava, V. K., D. Haccoun, R. Matyas, and P. Nuspl. 1981. *Digital Communications by Satellite*. New York: Wiley.

Blahut, R. E. 1983. *Theory and Practice of Error Control Codes*. Reading, Mass.: Addison-Wesley.

Blahut, R. E. 1990. *Digital Transmission of Information*. Reading, Mass.: Addison-Wesley.

Briley, B. E. 1983. *Introduction to Telephone Switching*. Reading, Mass.: Addison-Wesley.

Capon, J. 1959. "A Probabilistic Model for Run-Length Coding of Pictures." *IRE Trans. Inf. Theory*, Vol. IT-5, pp. 157–163.

Carlson, A. B. 1975. *Communication Systems*. New York: McGraw-Hill.

Chen, W. H. 1963. *The Analysis of Linear Systems*. New York: McGraw-Hill

Clark, G. C., Jr., and J. B. Cain. 1981. *Error-Correction Coding for Digital Communications*. New York: Plenum Press.

Clos, C. 1953. "A Study of Non-blocking Switching Networks," *Bell Syst. Tech. J.*, Vol. 32, Mar., pp. 406–424.

Collins, A. A., and R. D. Pedersen. 1973. *Telecommunications: A Time for Innovation*. Dallas, Tex.: Merle Collins Foundation.

Cooper, G. R., and C. D. McGillem. 1986. *Modern Communications and Spread Spectrum*. New York: McGraw-Hill.

Davies, D. W., and D. L. A. Barber. 1973. *Communication Networks for Computers*. New York: Wiley.

de Buda, R. 1972. "Coherent Demodulation of Frequency Shift Keying with Low Deviation Ratio." *IEEE Trans. Commun.*, Vol. COM-20, June, pp. 429–436.

Dixon, R. C. 1976. *Spread Spectrum Techniques.* New York: IEEE Press.

Dixon, R. C. 1984. *Spread Spectrum Systems.* New York: Wiley.

Farvardin, N., and J. W. Modestino. 1984. "Optimum Quantizer Performance for a Class of Non-Gaussian Memoryless Sources." *IEEE Trans. Inf. Theory*, Vol. IT-30, May, pp. 485–497.

Forney, G. David, Jr. 1989. "Introduction to Modem Technology: Theory and Practice of Bandwidth Efficient Modulation from Shannon and Nyquist to Date." University Video Communications and the IEEE; Copyright: Motorola.

Foschini, G. J., R. D. Gitlin, and S. B. Weinstein. 1974. "Optimization of Two-Dimensional Signal Constellations in the Presence of Gaussian Noise." *IEEE Trans. Commun.*, Vol. COM-22, Jan., pp. 28–38.

Franks, L. E. 1969. *Signal Theory.* Englewood Cliffs, N. J.: Prentice Hall.

Franks, L. E. 1980. "Carrier and Bit Synchronization in Data Communication: A Tutorial Review." *IEEE Trans. Commun.*, Vol. COM-28, Aug., pp. 1107–1121.

Freeman, R. L. 1981. *Telecommunications Transmission Handbook*, 2nd ed. New York: Wiley.

Gallager, R. G. 1968. *Information Theory and Reliable Communication.* New York: Wiley.

Gersho, A., and R. M. Gray. 1991. *Vector Quantization and Signal Compression.* Hingham, Mass.: Kluwer.

Gilhousen, K. S., I. M. Jacobs, R. Padovani, and L. A. Weaver, Jr. 1990. "Increased Capacity Using CDMA for Mobile Satellite Communication." *IEEE J. Sel. Areas Commun.*, Vol. SAC-8, May, pp. 503–514.

Gold, R. 1967. "Optimal Binary Sequences for Spread-Spectrum Multiplexing." *IEEE Trans. Inf. Theory*, Vol. IT-13, pp. 619–621.

Gold, R. 1968. "Maximal Recursive Sequences with 3-Valued Recursive Cross Correlation Functions." *IEEE Trans. Inf. Theory*, Vol. IT-14, Jan., pp. 154–156.

Golomb, S. W., ed. 1964. *Digital Communications with Space Applications.* Englewood Cliffs, N.J.: Prentice Hall.

Gray, R. M., and L. D. Davisson. 1986. *Random Processes: A Mathematical Approach for Engineers.* Englewood Cliffs, N.J.: Prentice Hall.

Hammond, J. L., and P. J. P. O'Reilly. 1986. *Performance Analysis of Local Computer Networks.* Reading, Mass.: Addison-Wesley.

Haykin, S. 1983. *Communication Systems.* New York: Wiley.

Heller, J. A., and I. M. Jacobs. 1971. "Viterbi Decoding for Satellite and Space Communications." *IEEE Trans. Commun. Technol.*, Vol. COM-19, Oct., pp. 835–848.

Hirsch, D., and W. J. Wolf. 1970. "A Simple Adaptive Equalizer for Efficient Data Transmission." *IEEE Trans. Commun. Technol.*, Vol. COM-18, Feb., pp. 5–12.

Holmes, J. K. 1982. *Coherent Spread Spectrum Systems.* New York: Wiley.

Holzman, L. N., and W. J. Lawless. 1970. "Data Set 203: A New High-Speed Voiceband Modem." *Computer*, Sept.–Oct., pp. 24–30.

Houston, S. W. 1975. "Modulation Techniques for Communication, Part I: Tone and Noise Jamming Performance for Spread Spectrum *M*-ary FSK and 2, 4-ary DPSK

Waveforms." *Proceedings of the IEEE National Aerospace and Electronics Conference* (NAECON '75), Dayton, Ohio, June 10–12, pp. 51–58.

Huffman, D. A. 1952. "A Method for the Construction of Minimum Redundancy Codes." *Proc. IRE*, Vol. 40, Sept., pp. 1098–1101.

Jackson, D. 1941. *Fourier Series and Orthogonal Polynomials.* Washington, D.C.: The Mathematical Association of America.

Jacobaeus, C. 1950. "A Study of Congestion in Link Systems." *Ericsson Tech.*, No. 48, Stockholm.

Jahnke, E., and F. Emde. 1945. *Tables of Functions.* New York: Dover.

Jayant, N. S., and P. Noll. 1984. *Digital Coding of Waveforms.* Englewood Cliffs, N.J.: Prentice Hall.

Johnson, G. D. 1973. "No. 4 ESS." *Bell Lab. Rec.*, Sept., pp. 226–232.

Kaplan, W. 1959. *Advanced Calculus.* Reading, Mass.: Addison-Wesley.

Keiser, G. E. 1989. *Local Area Networks.* New York: McGraw-Hill.

Kernighan, B. W., and S. Lin. 1973. "Heuristic Solution of a Signal Design Optimization Problem." *Proc. 7th Annual Princeton Conference on Information Science and Systems*, Mar.

Kleinrock, L., and S. S. Lam. 1975. "Packet Switching in a Multiaccess Broadcast Channel: Performance Evaluation." *IEEE Trans. Commun.*, Vol. COM-23, Apr., pp. 410–423.

Kotel'nikov, V. A. 1947. *The Theory of Optimum Noise Immunity.* Doctoral dissertation, Molotov Energy Institute, Moscow. Also published by McGraw-Hill, New York, 1959.

Kretzmer, E. R. 1965. "Binary Data Communication by Partial Response Transmission." *Conf. Rec.*, 1965 IEEE Annual Communications Conference, pp. 451–455.

Kretzmer, E. R. 1966. "Generalization of a Technique for Binary Data Communication." *IEEE Trans. Commun. Technol.*, Feb., pp. 67–68.

Kuo, F. F. 1962. *Network Analysis and Synthesis.* New York: Wiley.

Lathi, B. P. 1968. *Communication Systems.* New York: Wiley.

Lee, C. Y. 1955. "Analysis of Switching Networks." *Bell Syst. Tech. J.*, Vol. 34, Nov., pp. 1287–1315.

Lender, A. 1963. "The Duobinary Technique for High Speed Data Transmission." *IEEE Trans. Commun. Electron.*, Vol. 82, May, pp. 214–218.

Lender, A. 1964. "Correlative Digital Communication Techniques." *IEEE Trans. Commun. Technol.*, Dec., pp. 128–135.

Lender, A. 1966. "Correlative Level Coding for Binary Data Transmission." *IEEE Spectrum*, Vol. 3, Feb., pp. 104–115.

Lender, A. 1981. Chapter 7 in *Digital Communications: Microwave Applications.* K. Feher, ed. Englewood Cliffs, N.J.: Prentice Hall, pp. 144–182.

Levitt, B. K. 1985. "Strategies for FH/MFSK Signaling with Diversity in Worst-Case Partial Band Noise." *IEEE J. Sel. Areas Commun.*, Vol. SAC-3, Sept., pp. 622–626.

Lin, S., and D. J. Costello, Jr. 1983. *Error Control Coding: Fundamentals and Applications.* Englewood Cliffs, N.J.: Prentice Hall.

Lloyd, S. P. 1982. "Least Squares Quantization in PCM." *IEEE Trans. Inf. Theory*, Vol. IT-28, Mar., pp. 129–137 (unpublished memorandum, Bell Laboratories, 1957).

Lucky, R. W. 1965. "Automatic Equalization for Digital Communications." *Bell Syst. Tech. J.*, Vol. 44, Apr., pp. 547–588.

Lucky, R. W. 1966. "Techniques for Adaptive Equalization of Digital Communication." *Bell Syst. Tech. J.*, Vol. 45, Feb., pp. 255–286.

Lucky, R. W., and H. Rudin. 1967. "An Automatic Equalizer for General-Purpose Communication Channels." *Bell Syst. Tech. J.*, Vol. 46, Nov., pp. 2179–2207.

Lucky, R. W., J. Salz, and E. J. Weldon, Jr. 1968. *Principles of Data Communication.* New York: McGraw-Hill.

Makhoul, J. 1975. "Linear Prediction: A Tutorial Review." *Proc. IEEE*, Vol. 63, Apr., pp. 561–580.

Martin, D. R., and P. L. McAdam. 1980. "Convolutional Code Performance with Optimal Jamming." *Conf. Rec.*, 1980 IEEE International Conference on Communications, Seattle, Wash., June 8–12, pp. 4.3.1–4.3.7.

Max, J. 1960. "Quantizing for Minimum Distortion." *IRE Trans. Inf. Theory*, Vol. IT-6, Mar., pp. 7–12.

Mazur, B. A., and D. P. Taylor. 1981. "Demodulation and Carrier Synchronization of Multi-*h* Phase Codes." *IEEE Trans. Commun.*, Vol. COM-29, Mar., pp. 257–266.

McEliece, R. J. 1977. *The Theory of Information and Coding.* Reading, Mass.: Addison-Wesley.

Newman, D. B., Jr., and R. L. Pickholtz, eds. 1987. Special Issue on "Network Security." *IEEE Network*, Vol. 1, Apr.

Noll, P., and R. Zelinski. 1978. "Bounds on Quantizer Performance in the Low Bit-Rate Region." *IEEE Trans. Commun.*, Vol. COM-26, Feb., pp. 300–304.

Nyquist, H. 1924. "Certain Factors Affecting Telegraph Speed." *Bell Syst. Tech. J.*, Vol. 3, Apr., pp. 324–346.

Nyquist, H. 1928. "Certain Topics in Telegraph Transmission Theory." *Trans. AIEE*, Vol. 47, Apr., pp. 617–644.

Owen, F. F. E. 1982. *PCM and Digital Transmission Systems.* New York: McGraw-Hill.

Paez, M. D., and T. H. Glisson. 1972. "Minimum Mean-Squared-Error Quantization in Speech PCM and DPCM Systems." *IEEE Trans. Commun.*, Vol. COM-20, Apr., pp. 225–230.

Pahlavan, K., and J. L. Holsinger. 1988. "Voice-Band Data Communication Modems: A Historical Review: 1919–1988." *IEEE Commun. Mag.*, Vol. 26, Jan., pp. 16–27.

Pasupathy, S. 1977. "Correlative Coding: A Bandwidth-Efficient Signaling Scheme." *IEEE Commun. Mag.*, Vol. 15, July, pp. 4–11.

Pasupathy, S. 1979. "Minimum Shift Keying: A Spectrally Efficient Modulation." *IEEE Commun. Mag.*, Vol. 17, July, pp. 14–22.

Pickholtz, R. L., D. L. Schilling, and L. B. Milstein. 1982. "Theory of Spread-Spectrum Communications: A Tutorial." *IEEE Trans. Commun.*, Vol. COM-30, May, pp. 855–884.

Pickholtz, R. L., D. L. Schilling, and L. B. Milstein. 1984. "Revisions to 'Theory of Spread Spectrum Communications: A Tutorial'," *IEEE Trans. Commun.*, Vol. COM-32, Feb., pp. 211–212.

Pratt, W. K. 1978. *Digital Image Processing.* New York: Wiley.

Proakis, J. G. 1989. *Digital Communications.* New York: McGraw-Hill.

Qureshi, S. U. H. 1985. "Adaptive Equalization." *Proc. IEEE,* Vol. 73, Sept., pp. 1349–1387.

Rabiner, L. R., and R. W. Schafer. 1978. *Digital Processing of Speech Signals.* Englewood Cliffs, N.J.: Prentice Hall.

Rice, S. O. 1963. "Noise in FM Receivers." Chapter 25 in *Proc., Symposium on Time Series Analysis.* M. Rosenblatt, ed. New York: Wiley, pp. 395–424.

Rice, S. O. 1982. "Envelopes of Narrow-Band Signals." *Proc. IEEE,* Vol. 70, July, pp. 692–699.

Sarwate, D. V., and M. B. Pursley. 1980. "Crosscorrelation Properties of Pseudorandom and Related Sequences." *Proc. IEEE,* Vol. 68, May, pp. 593–619.

Schilling, D. L., L. B. Milstein, R. L. Pickholtz, and R. W. Brown. 1980. "Optimization of the Processing Gain of an *M*-ary Direct Sequence Spread Spectrum Communication System." *IEEE Trans. Commun.,* Vol. COM-28, Aug., pp. 1389–1398.

Schilling, D. L., R. L. Pickholtz, and L. B. Milstein, guest eds. 1990. "Spread Spectrum Communications I." Special issue, *IEEE J. Sel. Areas Commun.,* Vol. SAC-8, May.

Scholtz, R. A. 1982. "The Origins of Spread-Spectrum Communications." *IEEE Trans. Commun.,* Vol. COM-30, May, pp. 822–854.

Schwartz, L. 1950. *Théorie des Distributions,* Vol. 1. Paris: Hermann.

Schwartz, M. 1987. *Telecommunications Networks: Protocols, Modeling, and Analysis.* Reading, Mass.: Addison-Wesley.

Schwartz, M. 1990. *Information Transmission, Modulation, and Noise,* 4th ed. New York: McGraw-Hill.

Shannon, C. E. 1948. "A Mathematical Theory of Communication." *Bell Syst. Tech. J.,* Vol. 27, July, pp. 379–423; Oct., pp. 623–656.

Shannon, C. E. 1959. "Coding Theorems for a Discrete Source with a Fidelity Criterion." *IRE Natl. Conv. Rec.,* Pt. 4, Mar., pp. 142–163.

Shannon, C. E., and W. Weaver. 1949. *The Mathematical Theory of Communication.* Urbana, Ill.: University of Illinois Press.

Simon, M. K., J. K. Omura, R. A. Scholtz, and B. K. Levitt. 1985. *Spread Spectrum Communications,* Vols. I and II, Rockville, Md.: Computer Science Press.

Sklar, B. 1988. *Digital Communications: Fundamentals and Applications.* Englewood Cliffs, N.J.: Prentice Hall.

Spragins, J. D., J. L. Hammond, and K. Pawlikowski. 1991. *Telecommunications: Protocols and Design.* Reading, Mass.: Addison-Wesley.

Stallings, W. 1984. "Local Network Performance." *IEEE Commun. Mag.,* Vol. 22, Feb., pp. 27–36.

Sunde, E. D. 1961. "Pulse Transmission by AM, FM, and PM in the Presence of Phase Distortion." *Bell Syst. Tech. J.,* Vol. 40, Mar., pp. 353–422.

Tanenbaum, A. S. 1988. *Computer Networks.* Englewood Cliffs, N.J.: Prentice Hall.

Taub, H., and D. L. Schilling. 1971. *Principles of Communication Systems.* New York: McGraw-Hill.

Temple, G. 1953. "Theories and Applications of Generalized Functions." *J. London Math. Soc.*, Vol. 28, pp. 134–148.

Thapar, H. K. 1984. "Real-Time Application of Trellis Coding to High-Speed Voiceband Data Transmission." *IEEE J. Sel. Areas Commun.*, Vol. SAC-2, Sept., pp. 648–658.

Thomas, G. B., Jr. 1968. *Calculus and Analytic Geometry.* Reading, Mass.: Addison-Wesley.

Ungerboeck, G. 1982. "Channel Coding with Multilevel/Phase Signals." *IEEE Trans. Inf. Theory*, Vol. IT-28, Jan., pp. 55–67.

Viswanathan, R., and K. Taghizadeh. 1988. "Diversity Combining in FH/BFSK Systems to Combat Partial Band Jamming." *IEEE Trans. Commun.*, Vol. COM-36, Sept., pp. 1062–1069.

Viterbi, A. J., and I. M. Jacobs. 1975. "Advances in Coding and Modulation for Noncoherent Channels Affected by Fading, Partial Band, and Multiple Access Interference." In *Advances in Communication Systems*, Vol. 4. A. V. Balakrishnan, ed. New York: Academic Press.

Wozencraft, J. M., and I. M. Jacobs. 1965. *Principles of Communication Engineering.* New York: Wiley.

Wyner, A. D. 1981. "Fundamental Limits in Information Theory." *Proc. IEEE*, Vol. 69, Feb., pp. 239–251.

Ziemer, R. E., and R. L. Peterson. 1985. *Digital Communications and Spread Spectrum Systems.* New York: Macmillan.

Index

A

B

C

D

G

H

I

J

L

M

N

O

P

Q

R

S